Lecture Notes in Artificial Intelligence   12144

Subseries of Lecture Notes in Computer Science

More information about this series at http://www.springer.com/series/1244

Hamido Fujita · Philippe Fournier-Viger ·
Moonis Ali · Jun Sasaki (Eds.)

Trends in Artificial Intelligence Theory and Applications

Artificial Intelligence Practices

33rd International Conference
on Industrial, Engineering and Other Applications
of Applied Intelligent Systems, IEA/AIE 2020
Kitakyushu, Japan, September 22–25, 2020
Proceedings

 Springer

Editors
Hamido Fujita ⓘ
Iwate Prefectural University
Takizawa, Japan

Philippe Fournier-Viger ⓘ
Harbin Institute of Technology (Shenzhen)
Shenzhen, China

Moonis Ali
Texas State University
San Marcos, TX, USA

Jun Sasaki
Iwate Prefectural University
Takizawa, Japan

ISSN 0302-9743   ISSN 1611-3349 (electronic)
Lecture Notes in Artificial Intelligence
ISBN 978-3-030-55788-1   ISBN 978-3-030-55789-8 (eBook)
https://doi.org/10.1007/978-3-030-55789-8

LNCS Sublibrary: SL7 – Artificial Intelligence

This Springer imprint is published by the registered company Springer Nature Switzerland AG
The registered company address is: Gewerbestrasse 11, 6330 Cham, Switzerland

Preface

In recent decades, society has entered a digital era where computers have become ubiquitous in all aspects of life, including education, governance, science, healthcare, and industry. Computers have become smaller, faster and the cost of data storage and communication have greatly decreased. As a result, more and more data is being collected and stored in databases. Besides, novel and improved computing architectures have been designed for efficient large-scale data processing such as big data frameworks, FPGAs, and GPUs. Thanks to these advancements and recent breakthroughs in artificial intelligence, researchers and practitioners have developed more complex and effective artificial intelligence-based systems. This has led to a greater interest in artificial intelligence to solve real-world complex problems, and the proposal of many innovative applications.

This volume contains the proceedings of the 33rd edition of the International Conference on Industrial, Engineering, and other Applications of Applied Intelligent Systems (IEA AIE 2020), which was held during September 22–25, 2020, in Kitakyushu, Japan. IEA AIE is an annual event that emphasizes applications of applied intelligent systems to solve real-life problems in all areas including engineering, science, industry, automation and robotics, business and finance, medicine and biomedicine, bioinformatics, cyberspace, and human-machine interactions. This year, 119 submissions were received. Each paper was evaluated using a double-blind peer review by at least three reviewers from an international Program Committee consisting of 82 members from 36 countries. Based on the evaluation, a total of 62 papers were selected as full papers and 17 as short papers, which are presented in this book. We would like to thank all the reviewers for the time spent to write detailed and constructive comments to authors, and to these latter for the proposal of many high-quality papers.

In the program of IEA AIE 2020, two special sessions were organized named Collective Intelligence in Social Media (CISM 2020) and Intelligent Knowledge Engineering in Decision Making Systems (IKEDS 2020). Moreover, three keynote talks were given by distinguished researchers, one by Prof. Tao Wu from Shanghai Jiao Tong University School of Medicine (China), one by Enrique Herrera Viedma from the University of Granada (Spain), and another by Ee-Peng Lim from Singapore Management University (Singapore). Lastly, we would like to thank everyone who contributed to the success of this year's edition of IEA AIE that is authors, Program Committee members, reviewers, keynote speakers, and organizers.

September 2020

Hamido Fujita
Philippe Fournier-Viger
Moonis Ali
Jun Sasaki

Organization

General Chair

Hamido Fujita   Iwate Prefectural University, Japan

General Co-chairs

Moonis Ali   Texas State University, USA
Franz Wotawa   TU Graz, Austria

Organizing Chair

Jun Sasaki   Iwate Prefectural University, Japan

Program Chairs

Philippe Fournier-Viger   Harbin Institute of Technology (Shenzhen), China
Hideyuki Takagi   Kyushu University, Japan

Special Session Chairs

Yinglin Wang   Shanghai University of Finance and Economic, China
Ali Selamat   Universiti Teknologi Malaysia, Malaysia
Prima O.D.A.   Iwate Prefectural University, Japan

Special Session Organizers

Jerry Chun-Wei Lin   Western Norway University of Applied Sciences, Norway
Philippe Fournier-Viger   Harbin Institute of Technology (Shenzhen), China
Rage Uday Kiran   University of Aizu, Japan
Ngoc-Thanh Nguyen   Wroclaw University of Science and Technology, Poland
Van Du Nguyen   Nong Lam University, Vietnam

Publicity Chair

Toshitaka Hayashi   Iwate Prefectural University, Japan

Program Committee

Rui Abreu	University of Lisbon, Portugal
Otmane Ait Mohamed	Corcordia University, Canada
Hadjali Allel	ENSMA, France
Xiangdong An	The University of Tennessee, USA
Artur Andrzejak	Heidelberg University, Germany
Farshad Badie	Aalborg University, Denmark
Ladjel Bellatreche	ENSMA, France
Fevzi Belli	Paderborn University, Germany
Adel Bouhoula	University of Carthage, Tunisia
Ivan Bratko	University of Ljubljana, Slovenia
João Paulo Carvalho	University of Lisbon, Portugal
Chun-Hao Chen	National Taipei University of Technology, Taiwan
Shyi-Ming Chen	National Taiwan University of Science and Technology, Taiwan
Flávio Soares Corrêa da Silva	University of São Paulo, Brazil
Giorgos Dounias	University of the Aegean, Greece
Alexander Ferrein	Aachen University of Applied Science, Germany
Philippe Fournier-Viger	Harbin Institute of Technology (Shenzhen), China
Hamido Fujita	Iwate Prefectural University, Japan
Vicente García Díaz	University of Oviedo, Spain
Alban Grastien	The Australian National University, Australia
Maciej Grzenda	Warsaw University of Technology, Poland
Jun Hakura	Iwate Prefectural University, Japan
Tim Hendtlass	School of Biophysical Sciences and Electrical Engineering, Australia
Dinh Tuyen Hoang	Yeungnam University, South Korea
Tzung-Pei Hong	National University of Kaohsiung, Taiwan
Wen-Juan Hou	National Central University, Taiwan
Ko-Wei Huang	National Kaohsiung University of Science and Technology, Taiwan
Quoc Bao Huynh	Ho Chi Minh City University of Technology, Vietnam
Said Jabbour	University of Artois, France
He Jiang	Dalian University of Technology, China
Rage Uday Kiran	University of Aizu, Japan
Yun Sing Koh	The University of Auckland, New Zealand
Adrianna Kozierkiewicz	Wroclaw University of Science and Technology, Poland
Dariusz Krol	Wroclaw University of Science and Technology, Poland
Philippe Leray	University of Nantes, France
Mark Levin	Russian Academy of Sciences, Russia
Jerry Chun-Wei Lin	Western Norway University of Applied Sciences, Norway

Yu-Chen Lin	Feng Chia University, Taiwan
Jose Maria-Luna	University of Cordoba, Spain
Wolfgang Mayer	University of South Australia, Australia
Joao Mendes-Moreira	University of Porto, Portugal
Engelbert Mephu Nguifo	Université Clermont Auvergne, France
Mercedes Merayo	Universidad Complutense de Madrid, Spain
Abidalrahman Moh'D	Eastern Illinois University, USA
Anirban Mondal	Ashoka University, India
Saqib Nawaz	Harbin Institute of Technology (Shenzhen), China
Roger Nkambou	Université du Québec à Montréal, Canada
Ngoc-Thanh Nguyen	Wroclaw University of Science and Technology, Poland
Quang Vu Nguyen	Vietnam-Korea Friendship Information Technology College, Vietnam
Van Du Nguyen	Nong Lam University, Vietnam
Ayahiko Niimi	Future University Hakodate, Japan
Xinzheng Niu	University of Electronic Science and Technology of China, China
Farid Nouioua	Aix-Marseille Université, France
Mourad Nouioua	Harbin Institute of Technology (Shenzhen), China
Barbara Pes	University of Cagliari, Italy
Marcin Pietranik	Wroclaw University of Science and Technology, Poland
Ingo Pill	TU Graz, Austria
Matin Pirouz	California State University, USA
Krishna P. Reddy	International Institute of Information Technology, Hyderabad, India
Gregorio Sainz-Palmero	University of Valladolid, Spain
Eugene Santos Jr.	Dartmouth College, USA
Jun Sasaki	Iwate Prefectural University, Japan
Ali Selamat	Universiti Teknologi Malaysia, Malaysia
Nazha Selmaoui-Folcher	University of New Caledonia, New Caledonia
Sabrina Senatore	University of Salerno, Italy
Neal Snooke	Aberystwyth University, UK
Gerald Steinbauer	TU Graz, Austria
Ahmed Tawfik	Microsoft Research, USA
Trong Hieu Tran	Hanoi University of Engineering and Technology, Vietnam
Van Cuong Tran	Quang Binh University, Vietnam
Chun-Wei Tsai	National Sun Yat-sen University, Taiwan
Alexander Vazhenin	University of Aizu, Japan
Bay Vo	HCM City University of Technology, Vietnam
Toby Walsh	NICTA, Australia
Yutaka Watanobe	University of Aizu, Japan
Tomasz Wiktorski	University of Stavanger, Norway
Cheng Wei Wu	National Ilan University, Taiwan

Franz Wotawa	TU Graz, Austria
Jimmy Ming-Tai Wu	Shandong University of Science and Technology, China
Mu-En Wu	National Taipei University of Technology, Taiwan
Unil Yun	Sejong University, South Korea
Wei Zhang	Adobe Systems, USA

Contents

Knowledge Based Systems

Innovative Applications of Intelligent Systems

Industrial Applications

Financial Applications and Blockchain

Medical and Health-Related Applications

Machine Learning

Data Management and Data Clustering

Pattern Mining

System Control, Classification, and Fault Diagnosis

Natural Language Processing

Question Generation Through Transfer Learning

Yin-Hsiang Liao and Jia-Ling Koh$^{(\boxtimes)}$ iD

Department of Computer Science and Information Engineering,
National Taiwan Normal University,
Taipei, Taiwan, R.O.C.
zenonliao@gmail.com, jlkoh@csie.ntnu.edu.tw

Abstract. An automatic question generation (QG) system aims to produce questions from a text, such as a sentence or a paragraph. Traditional
approaches are mainly based on heuristic, hand-crafted rules to trans-
duce a declarative sentence to a related interrogative sentence. However,
creating such a set of rules requires deep linguistic knowledge and most
of these rules are language-specific. Although a data-driven approach
reduces the participation of linguistic experts, to get sufficient labeled
data for QG model training is still a difficult task. In this paper, we
applied a neural sequence-to-sequence pointer-generator network with
various transfer learning strategies to capture the underlying information of making a question, on a target domain with rare training pairs.
Our experiment demonstrates the viability of domain adaptation in QG
task. We also show the possibility that transfer learning is helpful in
a semi-supervised approach when the amount of training pairs in the
target QG dataset is not large enough.

Keywords: Question generation · Sequence-to-sequence model ·
Transfer learning

1 Introduction

Nowadays, teachers in school are far from the only way in which students can
get knowledge. Besides from the traditional education such as classrooms, there
are plenty of sources to be chosen, like massive open online courses (MOOCs)
or open educational materials. Sufficient and frequent quizzes help students get
better learning outcomes than just studying textbooks or notes [3,8]. However,
creating reasonable and meaningful questions is a costly task in both time and
money. The amount of related quizzes is not comparable with the amount of
growing online educational materials. Accordingly, it is worthwhile to build a

This research is partially supported by the "Aim for the Top University Project" of
National Taiwan Normal University (NTNU), sponsored by the Ministry of Education
and Ministry of Science and Technology, Taiwan, R.O.C. under Grant no. MOST 108-
2221-E-003-010.

H. Fujita et al. (Eds.): IEA/AIE 2020, LNAI 12144, pp. 3–17, 2020.
https://doi.org/10.1007/978-3-030-55789-8_1

reliable automatic question generation (QG) system for educational purpose. Meanwhile, the previous works are generally based on English [6, 7, 17–20]. In this paper, we aim to create a system to utilize educational resources in traditional Chinese for middle-school students.

In computational linguistics, there are two mainstream approaches for QG – rule-based approaches and data-driven approaches. Rule-based approaches basically utilize grammars to generate texts. Ideally, the rule-based approaches can generate good questions if researchers design rules well. However, creating such a set of rules requires deep linguistic knowledge and most of these rules are language-specific. In order to reduce the participation of experts, a data-driven approach is required to alleviate the need of strong linguistic background for a certain language. In addition, the proposed methods should fit our application scenario in traditional Chinese.

A data-driven question generation system for educational purpose has difficulties as following. 1) Proper labeled data is insufficient. Training question generation models need input texts, paragraphs or sentences, and corresponding questions in supervised approaches. Quiz questions are relatively few with respect to the educational content. Even if quizzes are abundant because of an existing question bank, the annotated sentence-question pairs remain rare. 2) The absence of an effective evaluation metrics. In previous works, the most popular evaluation metrics include BLEU [11] and ROUGE [9]. So far, there has been no explicit and correct way to objectively evaluate whether the generated questions of a model are good for the given content.

To materialize the aforementioned system, we mainly study how to apply a supervised approach of transfer learning on a Seq2seq pointer network (PTN) model for solving the QG task with rare sentence-question training pairs. We also consider a semi-supervised approach combined with domain adoption. Our main contributions are as follows:

1) We demonstrate the viability of domain adaptation in QG task.
2) We perform several fine-tuning strategies to construct a Seq2seq PTN model for QG with a better performance.
3) We investigate the feasibility of semi-supervised fine-tuning for the Seq2seq PTN model.
4) In addition to BLEU and ROUGE, we use an additional simple evaluation, the proportion of generated questions with interrogatives, to evaluate the generate questions.
5) The proposed framework demonstrates the possibility of applying the QG system to school teaching materials.

This paper is organized as follows. In Sect. 2, we introduce the related works. In Sect. 3, we describe the proposed approaches in detail. Section 4 shows performance evaluation of the proposed methods and discusses the results of experiments. In Sect. 5, we conclude this research and point out the future works.

2   Related Works

Researchers have dealt with question generation by rule-based approaches in the past [12]. These solutions depended on well-designed rules, based on profound linguistic knowledge, to transform declarative sentences into their syntactic representations and then generate interrogative sentences. [7] took an "overgenerate-and-rank" strategy, which used a set of rules to generate more-than-enough questions and leveraged a supervised learning method to rank the produced questions. The rule-based approaches performed well on well-structured input text. However, because of the limitation of hand-crafted rules, the systems failed to deal with subtle or complicated text. In addition, these heuristic rule-based approaches focused on the syntactic information of input words, most of which ignored the semantic information.

Du et al. [6] first proposed a Seq2seq framework with attention mechanism to model the question generation task for reading comprehension. Their work considered context information both from a sentence and a paragraph. [18] presented another QG model with two kinds of decoders. By considering the types of words including interrogatives, topic words, and ordinary words, the model aimed to generate questions for an open-domain conversational system. To leverage more data with potentially useful information, the answer of a question was considered in [15] and [17]. The work of Tang et al. showed that the QA and QG tasks enhanced each other in their training framework.

The aforementioned QG works were remarkable; nevertheless, their successes were inseparable with SQuAD, a relatively large, publicly available, general purpose dataset. If there is a shortage of training data in a domain where we are interested, e.g., teaching materials in middle school, it is difficult to train the models to an acceptable level. For educational purpose, Z. Wang et al. [19] proposed a Seq2seq-based model, QG-net, that captures the way how humans ask questions from a general purpose dataset, SQuAD, and directly applied the built model to the learning material, the OpenStax textbooks. Similarly, our study aims to build a system for middle school education but there does not exist a large-scale dataset with manually labeled sentence-question pairs. However, our work mainly differs from QG-net in the following aspects. First, we focus on the effectiveness of domain adaptation: we tune the proposed model by hundreds of labeled pairs in our target domain, middle school textbooks. Second, since we have labeled data in target domain, we therefore do quantitative evaluations, which were unseen for the generated questions in the target domain of QG-net. Furthermore, our study proposes a semi-supervised approach to leverage more generated questions as training pairs to fine-tune the baseline model.

Recently, in natural language processing (NLP) community, there were various applications of the Seq2Seq model. A Seq2seq model typically consists an encoder and a decoder. Most of the frameworks are implemented by RNNs. The encoder looks through the input text as a context reader, and converts it to a context vector with textual information. The vector is then decrypted by the decoder as a question generator. The decoding procedure often take the attention mechanism [1,10] to generate a meaningful question corresponding to the input text. [16] proposed the pointer network, a modification of Seq2seq, to deal with the words absent in the training set. Their work was later used by Z. Wang et al. [19] to point out which part of an input content is more possible to appear in the output question. We will describe in detail how we apply the Seq2seq model and its variations in Sect. 3.

Deep neural networks (DNNs) often benefit from transfer learning. In NLP, transfer learning has also been successfully applied in tasks like QA [5], among other things. [5] demonstrated that a simple transfer learning technique can be very useful for the task of multi-choice question answering. Besides, the paper showed that by an iterative self-labeling technique, unsupervised transfer learning is still useful. Inspired by [5], we performed experiments to investigate the transferability of encoder and decoder learned from a source QG dataset to a target dataset using a sequence-to-sequence pointer-generator network. The size of the target dataset considered in our study is even smaller than that used in [5]. Although unsupervised transfer learning for QG is still a challenge, we showed that transfer learning is helpful in a semi-supervised approach.

3   Methods

The proposed method consists of three part of processing: data preparation, training baseline model, and domain adaptation.

3.1   Task Definition

Given an input sentence $S = \{w_k\}_{k=1}^{L^S}$, where w_k is a word in the input sentence S with length L^S. The goal of question generation is to generate a natural question Q, to maximize $P(Q|S, \theta)$ [6].

$$P(Q|S, \theta) = \prod_{t=1}^{L^Q} P(q_t|S, \{q_\tau\}_{\tau=1}^{t-1}\}, \theta), \tag{1}$$

where L^Q denotes the length of output question Q, and q_t denotes each word within Q, respectively. Besides, θ denotes the set of parameters of a prediction model to get $P(Q|S, \theta)$, the conditional probability of the predicted question sequence Q, given the input S. A basic assumption is that the answer of the generated question should be a consecutive segment in S. Accordingly, we mainly consider how to generate the factual questions.

3.2   Data Preparation

For preparing the source domain data and target domain data, the data preprocessing steps are required as follows.

Source Domain Data. In the DRCD dataset, each given data consists of a triple (P, Q, A), where P is a paragraph, Q is a question from the given paragraph, and A is the answer in the paragraph. The following processing is performed to get a pair of input sentence S and the corresponding question Q.

1. Extract the sentence S that contains the answer A from the paragraph P.
2. Generate the sentence-question pairs (S, Q).
3. Segment the texts in each sentence-question pair via Jieba [14] .

The constructed dataset is denoted by DB_{DRCD}.

Target Domain Data. We collected a dataset of sentences in textbooks of junior high school in science subject and a dataset of multi-selection questions from a question bank. Then the proper questions are selected and matched with the sentences in the textbooks semi-automatically as the following:

1. Exclude the questions with a figure(s).
2. Exclude the questions with the form "which of the following is correct/wrong" and keep the factual questions.
3. Match each remaining question Q and its answer A with each sentence S in the textbook as follows:
   - If A appears in the sentence S, generate a candidate sentence-question pair (S, Q) and compute BLEU-4 scores [11] between S and Q.
4. Select the sentence-question pairs with BLEU-4 scores higher than a threshold value and manually make the proper pairs remaining.
5. Exclude duplicating sentence-question pairs.
6. Perform word segmentation on the texts of the pairs via Jieba.

3.3 Baseline Model

The baseline neural model used in this study is the pointer network model. We describe details of the model as follows.

Our encoder and decoder are implemented by a bidirectional Gated Recurrent Unit (bi-GRU) [1] and a GRU, respectively [4]. The words in S, i.e. w_i, are inputted to an embedding layer, and then inputted into the encoder one-by-one. Then a sequence of hidden states h_i of the encoder is produced.

In the Seq2seq model, the attention distribution indicates the weights of hidden states in GRU encoder cells for predicting the result of decoder. On each step, the decoder receives the word embedding of the previously generated word q_{t-1} and produces the hidden state of decoder at step t, h_t^d. The attention distribution a_t^e is calculated as:

$$e_{ti} = h_i^T W_{attn} h_t^d + b_{attn}, \tag{2}$$

$$a_t^e = softmax(e_t). \tag{3}$$

Then the hidden states of encoder, h_i, are weighted summed to get the context vector: $c_t^e = \sum_i a_{ti}^e h_i$.

Moreover, we apply the intra decoder attention mechanism [20], where the attention distribution a_t^d for each decoding step t is calculated as:

$$e_{tt'}^d = h_{t'}^{d^T} W_{attn}^d h_t^d + b_{attn}^d, \tag{4}$$

$$a_t^d = softmax(e_t^d), \tag{5}$$

where h_t^d and $h_{t'}^d$ denote the hidden states of decoder at step t and the previous step t', respectively. Accordingly, another context vector c_t^d is generated by the intra decoder mechanism as follows: $c_t^d = \sum_{t'} a_{tt'}^d h_{t'}^d$. After concatenating h_t^d with c_t^e and c_t^d, denoted by $\parallel$, an output layer generates the token by computing the following probability distribution:

$$P_{voc}(q_t) = softmax(W_{out}[h_t^d \parallel c_t^e \parallel c_t^d] + b_{out}). \tag{6}$$

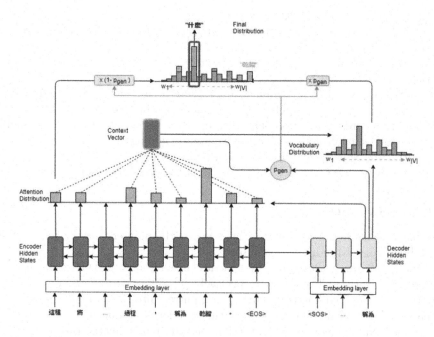

Fig. 1. The framework of pointer-generator network.

Pointer Network (PTN) is a variation of the Seq2seq model. It calculates the output word probabilities as a weighted sum of two probabilities, one comes from the output of Seq2seq model and another comes from the attention weights of input sentence. As shown in Fig. 1, the proportion of the two probabilities are controlled by the tunable parameter P_{gen}.

$$P_{gen} = \sigma(W_u[h_t^d \parallel c_t^e \parallel c_t^d] + b_u). \tag{7}$$

The PTN will create an extra dictionary for the unseen words occurring in the input sentence. Those words will be probably chosen to be the output when their probabilities from the attentions weights of input are large. $P_{PTN}(q_t = x_i) = a_{ti}^e$, where x_i is in S. Finally, $P(q_t) = P_{gen} \times P_{voc}(q_t) + (1 - P_{gen}) \times P_{PTN}(q_t)$. By the usage of point network, this model can effectively deal with out-of-vocabulary problem. In order to prevent repetition in the generation model, coverage mechanism [13] is applied. A coverage value cov_t is computed, which is the sum of attention distributions over all previous decoder time steps and contributes to the loss:

$$cov_t = \sum_{i<t} a_{ti}^e, \tag{8}$$

$$covloss = \sum_{i<t} min(a_{ti}^e, cov_t). \tag{9}$$

The total loss is $loss_t = logP(w_t^*) + \lambda \times covloss_t$, where $P(w_t^*)$ is the negative log likelihood (NLL) of the target word q_t for the time step t. λ is a given hyperparameter to control the weight of coverage loss. The overall loss for the whole sequence is: $loss = 1/L^S \times \sum_t loss_t$. Note that we did not put Chinese characters into GRU directly. Instead, we first built a dictionary of frequent words from the word segmentation results of the training set. For each word in the dictionary, if it is in the vocabulary of fastText [2], its corresponding pre-trained word embedding is loaded.

In the training process of a Seq2seq model, the inference of a new token is based on the current hidden state and the previous predicted token. A bad inference will then make the next inference worse. This phenomenon is a kind of error propagation. D. Bahdanau et al. [1] thus proposed a learning strategy, named teacher forcing, to ease the problem. Instead of always using the generated tokens, the strategy gently changed the training process from fully using the true tokens, toward mostly using the generated tokens. This method can yield performance improvement for sequence prediction tasks such as QG. In the proposed model, we guide the training by 0.75 at beginning, and decay the ratio by multiplying 0.9999 after each epoch.

3.4   Domain Adaptation

Supervised Domain Adaptation. We take the supervised transfer learning approach as the following three steps:

1. Given an epoch number *epon* to train the PTN on the source domain, which contains abundant training data, and save the model of each epoch until reach the number *epon*.
2. Select a best model according to a selection strategy as the base model M_b. In the experiment, we choose the model with the highest average BLEU-4 on the validation set of target domain.
3. Fine-tune the model M_b. That is, we initialize another training process with learnt parameters of the model M_b on the dataset set of target domain.

The PTN consists of the following layers: the embedding layer for encoder/decoder, the bi-GRU layer and attention network for encoder, the GRU layer and attention network for decoder, the output layer for question generation, and the parameter for computing P_{gen}. We try various strategies of domain adoptions by freezing or retraining some layers of the model. The details of various fine-tuning strategies are described in the experiments.

Semi-supervised Domain Adaptation. Algorithm 1 shows our basic semi-supervised approach.

Algorithm 1: Semi-supervised QG Domain Adaptation

Input: Source dataset DB_{source}: (sentence, question) pairs;
Target datasets DB_{target}: DB_{target}^{l} :(sentence, question) pairs;
$\quad\quad\quad DB_{target}^{ul}$: sentences; Number of training iterations n.
Output: QG model M*

---
1. Pre-train QG models on the source dataset.
2. Select a best Pre-trained model M_b.
3. Repeat
4.   For each sentence S in the target dataset DB_{target}^{ul}, use M_b to generate its question Q.
5.   Combine the predicted question Q with S to generate a training pair of DB_{target}^{ul}.
6.   Fine-tune M_b according to $DB_{target}^{l} \cup$ the training pairs of DB_{target}^{ul}.
7. Until reach the nth iteration

We also proposed an enhanced version of semi-supervised QG domain adaptation. Instead of applying M_b to predict the questions of DB_{target}^{ul} directly, M_b is fine-tuned by DB_{target}^{l} after step 2. The model with the highest average BLEU-4 on the validation set of target domain is then selected to perform the iteration from step 3 to 7.

4   Performance Evaluation

4.1   Experiments Setup

For training the base model, we used the DB_{DRCD}, which is an open domain traditional Chinese machine reading comprehension (MRC) dataset. The dataset contains 10,014 paragraphs from 2,108 Wikipedia articles and 30,000+ questions generated by annotators. We excluded those sentence-question pairs over 80 and 50 words, respectively, so 26,175 pairs are extracted. Moreover, the dictionary of vocabulary contains the ones with frequency no less than 3. Therefore, the vocabulary size of DB_{DRCD} is 28,981. In the target domain, the dataset $DB_{textbook}$ contains 480 labeled pairs. We applied random sampling to separate the data into train/test/validation sets in the proportion 7/2/1. Accordingly, there are 336 pairs with vocabulary size 787 in the training set, 48 pairs for validation, and 96 pairs for testing. The effectiveness of transfer learning is evaluated by the model's performance on the test sets of target domain. For the training set consisting of both DB_{DRCD} and $DB_{textbook}$, the pairs/vocabulary size is 26,511/29,201. It implies that there are 220 words in the vocabulary of $DB_{textbook}$ but not in the vocabulary of DB_{DRCD}. For all the experiments, our models have 300-dimensional hidden states and 300-dimensional pre-trained word embedding. We set learning rate to be 0.0001, teacher forcing rate 0.75, batch size 32, and use Adam as the optimizer.

We report the evaluation results with the following metrics.

- BLEU-4 [11]. BLEU measures precision by how much the words in prediction sentences appear in reference sentences. BLEU-4 measures the average for calculation on 1-gram to 4-gram.
- ROUGE- L [9]. ROUGE-L measures recall by how much the words in reference sentences appear in prediction sentences using Longest Common Subsequence (LCS) based statistics.
- Ratio of interrogatives, denoted by '?%'. In Chinese, there are various interrogatives corresponding to the 5W1H, i.e. 'Who, Why, When, Where, What, and How', questions. A question with an interrogative usually shows the explicit target of a question instead of a yes/no question, which is a question with higher quality. Accordingly, we measure how many percentage of sentences in the generated questions having one of the interrogatives.

4.2 Experiment Results

Experiment 1. In this experiment, three QG models, denoted as M_1, M_2, and M_3, are constructed by varying the training dataset. Moreover, for each model, two versions are trained by fixing or not fixing the pre-trained embedding layer as shown in Table 1. The I/O Voc. denote the input and output vocabulary of the model. The evaluation results of Experiment 1 are shown in Table 2. The results show that, for the models constructed by training DB_{DRCD} or $DB_{Textbook}$ only, fixing the pre-trained embedding layer gets the better result on the test data of $DB_{Textbook}$ than fine-tuning the embedding layer. Besides, the model trained by DB_{DRCD} (Model M_1) outperforms the model trained directly on the target dataset $DB_{textbook}$ (Model M_2). It indicates that the limited size of the target dataset could not provide enough data for the complex structure of the PTN generation model. Moreover, by using the training data of $DB_{DRCD} \cup DB_{Textbook}$ could not solve the problem properly, because the ratio of $DB_{Textbook}$ to DB_{DRCD} is tiny. It implies the necessarily of transfer learning.

Table 1. The descriptions of the constructing models of Experiment 1.

Model	I/O Voc.	Training dataset	Fixed_emb
Model M_1/M_1'	DB_{DRCD}	DB_{DRCD}	Yes/No
Model M_2/M_2'	$DB_{Textbook}$	$DB_{Textbook}$	Yes/No
Model M_3/M_3'	$DB_{DRCD} \cup DB_{Textbook}$	$DB_{DRCD} \cup DB_{Textbook}$	Yes/No

Table 2. The evaluation results of Experiment 1.

Fix embedding layer				Fine-tune embedding layer			
Model	BLEU-4	ROUGE_L	?%	Model	BLEU4	ROUGE_L	?%
Model M_1	**0.344**	**0.430**	**0.0625**	Model M_1'	0.295	0.401	0.115
Model M_2	0.259	0.388	0.208	Model M_2'	0.193	0.336	0.083
Model M_3	0.299	0.414	0.052	Model M_3'	0.314	0.418	0.0625

Experiment 2. The purpose of Experiment 2 is to demonstrate the effectiveness of domain adaptation. In the following experiments, the best model in M_1's training process is selected as M_b. Various versions of domain adaptation are designed by deciding whether transferring the learnt parameters of layers in M_b related to the input/output vocabulary.

Table 3. The descriptions of the constructing models of Experiment 2.

Model	I/O Voc	Retraining layers
Model M_4	DB_{DRCD}	Embedding for encoder & decoder, Out
Model M_5	DB_{DRCD}	Out
Model M_6	DB_{DRCD}	None
Model M_7	$DB_{DRCD}/DB_{Textbook}$	Embedding for decoder, Out
Model M_{7*}	$DB_{DRCD}/DB_{Textbook}$	Out
Model M_{7a}	$DB_{DRCD}/DB_{Textbook}$ + '?s'	Embedding for decoder, Out
Model M_8	$DB_{Textbook}$	Embedding for encoder & decoder, Out
Model M_{8a}	$DB_{Textbook}$ + '?s'	Embedding for encoder & decoder

Table 4. The evaluation results of Experiment 2.

Model	BLEU-4	ROUGE_L	?%	Model	BLEU4	ROUGE_L	?%
Model M_1	0.344	0.430	0.0625	Model M_7	0.340	0.456	0.365
Model M_4	0.352	0.438	0.010	Model M_7^*	0.305	0.426	0.0625
Model M_5	0.366	0.443	0.010	Model M_{7a}	0.355	0.470	0.177
Model M_6	**0.426**	**0.526**	**0.323**	Model M_8	0.438	0.535	0.302
				Model M_{8a}	**0.447**	**0.545**	**0.354**

Five strategies for constructing the models by fine-tuning M_1, to get models M_4 to M_8, are described as Table 3. The retraining layers means the learnt parameters of the layers in M_b were not loaded, which are retrained by $DB_{Textbook}$. For models M_7 and M_8, in order to provide more vocabulary of interrogatives, the variations of the models, denoted as M_{7a} and M_{8a}, are trained by adding a set of interrogatives selected from DB_{DRCD} to the output vocabulary of $DB_{textbook}$. Besides, in order to compare with M_5, the model version M_7* only change the output layer but keep the embedding layer of the decoder mapping to the vocabulary of DB_{DRCD}.

The evaluation results of Experiment 2 are shown in Table 4. To compare with the result of model M_1, for the models with I/O Voc $= DB_{DRCD}$ or $DB_{textbook}$, domain adoption has improvement on all the evaluation metrics. Regarding the models M_4, M_5, and M_6, the results show that to load the pre-trained parameters by DB_{DRCD} of both the embedding layer and output layer (M_6) is helpful significantly for fine-tuning by $DB_{textbook}$. Besides, by changing the I/O vocabulary to $DB_{textbook}$ and retrain both the embedding and output layers (M_8) also achieves significant improvement. Furthermore, adding the set of interrogatives to the output vocabulary (M_{7a} and M_{8a}) is useful to improve the evaluation result on BLEU4 and ROUGE_L scores for both models M_7 and M_8. To compare the results of models M_7 and M_5, to retrain the output layer mapping to the vocabulary of DB_{DRCD} outperform to retrain the output layer mapping to the vocabulary of $DB_{textbook}$. We attribute this to the layers of the encoder/decoder in the pre-trained model have to be modified significantly when the vocabulary of input and output layer is not consistent. By changing the embedding layer of decoder (model M_7) and adding interrogatives to the output vocabulary (model M_{7a}) is helpful.

Experiment 3. The purpose of Experiment 3 is to control the tunable layers when performing domain adoption as shown in Table 5. For the same I/O Vocabulary setting, two outperforming strategies in the previous experiment are selected, i.e. M_6 and M_{8a}. For the model M_{8a}, the output vocabulary is different from M_b. Therefore, the output layer of M_{8a} should be tunable.

Table 5. The descriptions of the constructing various models of Model M_6/M_{8a} in Experiment 3.

Model	I/O Voc	Fine_tuned layers
Model M_6/M_{8a}	Source Voc/$DB_{Textbook}$ + '?s'	All
Model M_{6_2}/M_{8a_2}	Source Voc/$DB_{Textbook}$ + '?s'	Out
Model M_{6_3}	Source Voc/$DB_{Textbook}$ + '?s'	Pgen
Model M_{6_4}/M_{8a_4}	Source Voc/$DB_{Textbook}$ + '?s'	Out, Pgen
Model M_{6_5}/M_{8a_5}	Source Voc/$DB_{Textbook}$ + '?s'	Encoder, Decoder, Out, Pgen

The evaluation results of Experiment 3 are shown in Table 6. Fine-tuning the last two layers usually gets a better result than only fine-tuning the last layer (version 4 v.s. version 2). Overall, for domain adoption on models M_6 and M_{8a}, only fine-tuning the last two layers is not enough. However, fine-tuning the entire model is not always best. To fix the pre-trained embedding layer gets the same or even a better result (version 5).

Table 6. The evaluation results of Experiment 3.

Model	BLEU-4	ROUGE_L	?%	Model	BLEU4	ROUGE_L	?%
Model M_6	0.426	0.526	0.323	Model M_{8a}	0.455	0.553	0.167
Model M_{6_2}	0.382	0.488	0.323	Model M_{8a_2}	0.090	0.180	0
Model M_{6_3}	0.365	0.447	0.052				
Model M_{6_4}	0.387	0.487	0.396	Model M_{8a_4}	0.103	0.195	0.073
Model M_{6_5}	**0.421**	**0.528**	**0.406**	Model M_{8a_5}	**0.470**	**0.565**	**0.188**

Experiment 4. For Models M_{6_5} and M_{8a_5}, we manually gave survey scores for evaluating the quality of the generated questions according to the following aspects: 1) the fluency, 2) the appropriate interrogatives, 3) the clarity and correctness. The given score for each aspect is 0, 0.5, or 1.

The average scores for the generated questions by the two models are shown in Table 7. Overall, the quality of questions generated by M_{6_5} is better than M_{8a_5}, which have higher clarity/correctness and more have appropriate interrogatives. The results show that the BLEU-4 and ROUGE_L scores as shown in Table 6 are consistent with the quality of fluency. However, these popular metrics can not objectively evaluate clarity/correctness and appropriate usage of interrogatives in the generated questions. The ratio of interrogatives in the generated questions appears to provide a good measuring to show the performance of QG on the other two aspects.

Table 7. The human evaluation results for M_{6_5} and M_{8a_5}.

Model	Fluency	Proper '?'	Clarity correctness
Model M_{6_5}	0.86	0.29	0.52
Model M_{8a_5}	0.87	0.19	0.47

Experiment 5. In this experiment, we study the relationship between the amount of training data from the target dataset for fine-tuning the model and the performance.

The results of experiments are shown in Fig. 2(a), where the solid lines and dashed lines represent the results of M_6 and M_{8a}, respectively. As expected, the

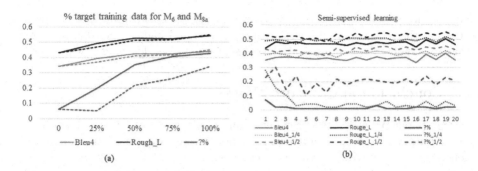

Fig. 2. The evaluation results of Experiment 4.

more training data is used for fine-tuning, the better the model's performance is. For model M_{6_5}, the improvement from using 0 to using 25% of target training data is larger than that from 25 to 50, 50 to 75, and 75 to 100. However, for model M_{8a_5}, the improvement from 25 to 50% of target training data is larger than the others.

The solid lines shown in Fig. 2(b) denote the evaluation results of the semi-supervised approach by applying Algorithm 1 according to the domain adoption strategy of model M_{6_5}. Besides, the ratio of data in DB^l_{target} and DB^{ul}_{target} is 1:3. The results show the semi-supervised approach helps the PTN model on measurements such as BLEU-4, but decrease the ratio of interrogatives.

Since the central dogma of BLEU is to measure the similarity between references and the prediction, when the reference questions are similar to the input sentences, the loss function, NLL, may decrease due to the copy mechanism rather than the implicit information of asking questions. To copy almost the same content with the given sentence as the predicted query will get a high BLEU score. However, it is not necessarily that the model generates a complete question. According to the results of semi-supervised approach starting from the pre-trained model M_b of source domain, although the constructed model can get a higher BLEU-4 score incrementally, the ratio of interrogatives in the self-generated questions is decreasing. The reason is that the generated questions are not representative for the target domain. It will gradually learn a model tending to copy the content for reducing the NLL loss function. Therefore, it is important to start tuning from a suitable model for the target domain.

The dot dashed lines and period dashed lines in Fig. 2(b) show the enhanced versions of semi-supervised transfer learning, where the ratio of data in DB^l_{target} and DB^{ul}_{target} is 1:3 and 1:1, denoted by Ex1/4 and Ex1/2, respectively. Because the enhanced versions of semi-supervised transfer learning use DB^l_{target} to fine-tune the pre-trained model M_b of source domain before using the self-generated questions as training pairs, the model is better fitting the target domain. From the results of Ex1/4, it achieves higher BLEU-4 and Rouge_L scores, but the ratio of interrogatives in the self-generated questions still decrease dramatically from iteration 1 to 4. However, when the ratio of data in and is 1:1, not only the measurements such as BLEU4 get improvement, but also the ratio of interrogatives in the self-generated questions keep stable and higher than 0.2. It shows that to combine domain adaptation and semi-supervised learning is workable.

5   Conclusion and Future Works

In this paper, in order to construct a question generation system for educational purpose, we propose a neural sequence-to-sequence framework of pointer-generator network with various transfer learning strategies to construct the suitable model on a target domain with rare training pairs. The results of experiments show, two of the transfer learning strategies work well to generate satisfactory results to some extent. We also show the possibility that transfer learning is helpful in a semi-supervised approach when the amount of questions in the target QG dataset is not large enough. Furthermore, though the performance measurements such as BLEU or ROUGE are widely used in previous researches of questions generation, we find their deficits for the PTN model in a task with high intra-similarity pairs, i.e., the input sentence and question. Accordingly, we raise another way for measuring the performance of QG task and hope it could inspire more researches of QG task. We will study the QG problem further in three possible approaches. First, the answer will provide more information for generating the corresponding question, which thus could be a part of input feature. Second, the potential of our proposed measurement does not end. We will study how to use the ratio of interrogatives as a reward for reinforcement learning. Finally, for the semi-supervised approach, we will study how to design an evaluation metric to select training pairs from the generated questions for better fine-tuning the base model according to a given order.

References

1. Bahdanau, D., Cho, K., Bengio, Y.: Neural machine translation by jointly learning to align and translate. In: ICLR 2015 Oral Presentation (2015). https://arxiv.org/abs/1409.0473v7
2. Bojanowski, P., Grave, E., Joulin, A., Mikolov, T.: Enriching word vectors with subword information. Trans. Assoc. Comput. Linguist. 5, 135–146 (2017)
3. Chen, G., Yang, J., Hauff, C., Houben, G.: LearningQ: a large-scale dataset for educational question generation. In: Proceedings of ICWSM-2018 (2018)
4. Cho, K., Merrienboer, B., Bahdanau, D., Bengio, Y.: On the properties of neural machine translation: encoder-decoder approaches. In: Proceedings of the Eighth Workshop on Syntax, Semantics and Structure in Statistical Translation SSST-8, pp. 103–111 (2014)
5. Chung, Y., Lee, H., Glass, J.: Supervised and unsupervised transfer learning for question answering. In: Proceedings of NAACL-HLT, pp. 1585–1594 (2018)
6. Du, X., Shao, J., Cardie, C.: Learning to ask: neural question generation for reading comprehension. In: Proceedings of the 55th Annual Meeting of the Association for Computational Linguistics, pp. 1342–1352 (2018)

7. Heilman, M., Smith, N. A.: Good question! statistical ranking for question generation. In: Proceedings of the Human Language Technologies: The 2010 Annual Conference of the North American Chapter of the Association for Computational Linguistics, pp. 609–617. Association for Computational Linguistics, Los Angeles (2010)
8. Karpicke, J.: Retrieval-based learning: active retrieval promotes meaningful learning. Curr. Dir. Psychol. Sci. **21**(3), 157–163 (2012)
9. Lin, C.-Y.: ROUGE: a package for automatic evaluation of summaries. In: Proceedings of the Workshop on Text Summarization Branches Out (WAS 2004), pp. 74–81 (2004)
10. Luong, M., Pham, H., Manning, C.D.: Effective approaches to attention-based neural machine translation. In: Proceedings of the EMNLP, vol. 2015, pp. 1412–1421 (2015)
11. Papineni, K., Roukos, S., Ward, T., Zhu, W.J.: BLEU: A method for automatic evaluation of machine translation. In: Proceedings of the 40th Annual Meeting of the Association for Computational Linguistics (2002)
12. Rus, V., Wyse, B., Piwek, P., Lintean, M., Stoyanchev, S., Moldovan, C.: The first question generation shared task evaluation challenge. In: Proc. the 6th International Natural Language Generation Conference. Association for Computational Linguistics, Stroudsburg, PA, USA, pp. 251–257, (2010)
13. See, A., Liu, P. J., Manning, C.D.: Get to the point: summarization with pointer-generator networks. In: Proceedings of the 55th Annual Meeting of the Association for Computational Linguistics, pp. 1073–1083 (2017)
14. Sun, J.: "Jieba" (Chinese for "to stutter") Chinese text segmentation: built to be the best Python Chinese word segmentation module (2012)
15. Tang, D., Duan, N., Qin, T., Yan, Z., Zhou, M.: Question answering and question generation as dual tasks (2017). ArXiv e-prints. https://arxiv.org/abs/1706.02027
16. Vinyals, O., Fortunato, M., Jaitly, N.: Pointer networks. In: Proceedings of the Advances in Neural Information Processing Systems, pp. 2674–2682 (2015)
17. Wang, T., Yuan, X., Trischler, A.: A joint model for question answering and question generation. In: Proceedings the 1st Workshop on Learning to Generate Natural Language (2017)
18. Wang, Y., Liu, C., Huang, M., Nie, L.: Learning to ask questions in open-domain conversational systems with typed decoders. In: Proceedings of the 56th Annual Meeting of the Association for Computational Linguistics, pp. 2193–2203 (2018)
19. Wang, Z., Lan, A.S., Nie, W., Waters, A.E., Grimaldi, P.J., Baraniuk, R.G.: QG-net: a data-driven question generation model for educational content. In: Proceedings of the Fifth Annual ACM Conference on Learning at Scale, pp. 7:1–7:10 (2018)
20. Zhou, Q., Yang, N., Wei, F., Tan, C., Bao, H., Zhou, M.: Neural question generation from text: a preliminary study (2017). ArXiv e-prints. https://arxiv.org/abs/1704.01792

KIDER: Knowledge-Infused Document Embedding Representation for Text Categorization

Yu-Ting Chen[1], Zheng-Wen Lin[2], Yung-Chun Chang[3(✉)], and Wen-Lian Hsu[4]

[1] Department of Statistics, National Taiwan University, Taipei 10617, Taiwan
[2] Department of Information Science and Applications, National Tsing Hua University, Hsinchu 300, Taiwan
[3] Department of Data Science, Taipei Medical University, Taipei 10617, Taiwan
changyc@tmu.edu.tw
[4] Institute of Information Science, Academia Sinica, Taipei 10617, Taiwan

Abstract. Advancement of deep learning has improved performances on a wide variety of tasks. However, language reasoning and understanding remain difficult tasks in Natural Language Processing (NLP). In this work, we consider this problem and propose a novel Knowledge-Infused Document Embedding Representation (KIDER) for text categorization. We use knowledge patterns to generate high quality document representation. These patterns preserve categorical-distinctive semantic information, provide interpretability, and achieve superior performances at the same time. Experiments show that the KIDER model outperforms state-of-the-art methods on two important NLP tasks, i.e., emotion analysis and news topic detection, by 7% and 20%. In addition, we also demonstrate the potential of highlighting important information for each category and news using these patterns. These results show the value of knowledge-infused patterns in terms of interpretability and performance enhancement.

Keywords: Text categorization · Natural Language Processing · Knowledge representation

1 Introduction

Text categorization possesses a lot of applications, including fake news detection, sentiment analysis, and news topic detection. Many tasks can be solved if we can conquer this specific area in natural language processing (NLP). Various methods have been proven effective [1]. However, by regarding documents as bag of words (BoW), features are limited to independent word level. Classical features are word counts and term frequency-inverse document frequency (TF-IDF) [2]. These features hardly contain semantic level information. Also, the knowledge can only be extracted from available training data. Therefore, the performance of text categorization tasks cannot achieve industry acceptable level due to limitation of the text representation.

© Springer Nature Switzerland AG 2020
H. Fujita et al. (Eds.): IEA/AIE 2020, LNAI 12144, pp. 18–29, 2020.
https://doi.org/10.1007/978-3-030-55789-8_2

The problem has been addressed in two ways - incorporating external knowledge and changing representation mechanism. In the case of using external knowledge, we can either enrich a document using existing knowledge repository like Wikipedia or we can conceptualize a document. However, few papers have discussed the usage of external knowledge beyond their capability of performance enhancement, which is what we examine exclusively in this paper. In the case of changing representation mechanism, Mikolov et al. [3] proposed a new vector representation method - word embedding. Word embedding serves as a more sophisticated semantic-learning representation for texts. By combining it with dependency-catching neural networks such as convolutional neural network (CNN) [4], long short-term memory (LSTM) networks [5], and Bidirectional Encoder Representations from Transformers (Bert) [6], performance for text categorization has improved remarkably. Further studies came up with variant like combining CNN and LSTM [7] to raise performance, or introduce attention mechanism [8] to highlight attended unit, usually words. Nevertheless, even after adding attention mechanism, the model still fail to provide crucial patterns whose complete meaning cannot be derived directly from its components, such as events, collocations, etc. Therefore, there are still difficulties in deeper semantic interpretation, which is a serious problem in many situations [9].

In light of this, our paper focuses on combining both approaches mentioned above to sophisticate feature representation. In the representation stage, We utilize external knowledge and design interpretable domain linguistic knowledge patterns to represent semantic categorical-distinctive information. In the classification stage, we transfer a document's patterns into embedded vector to preserve their semantic knowledge. Linguistic patterns, compared to plain words or vectors, fit how humans determine the category of a document. We consider the phenomenon where one does not categorize a document after human reads through it; instead, human captures some key concepts during the process of reading and uses them to identify the category efficiently [10]. These key concepts can be thought of as linguistic patterns. We believe these patterns can help us better understand essence of documents and increase performance. This paper makes following contributions:

- Knowledge Integration and Inference: KIDER incorporates external knowledge to gain interpretable semantic knowledge per category and per document.
- Model: KIDER reaches state-of-the-art performances by designing a whole new document representation method.
- Extensive Usage: The domain knowledge linguistic patterns generated can be further exploited in other NLP tasks such as text summarization.

2 KIDER Model

Figure 1 gives an overview of the KIDER model. First, we label sentences through three layers, represent document by tag diagrams, and select informative paths as linguistic patterns. Second, we merge and sort patterns to construct the KP list. Then we match a document with the list, and feed the tagged document and matched-patterns called KP Attention into a Doc2Vec model. Finally, a feed-forward neural network serves as the classifier. We introduce each component in detail in the following sections.

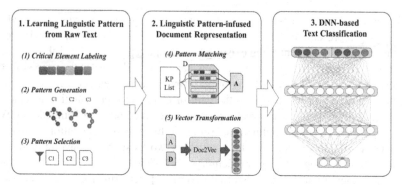

Fig. 1. The architecture of KIDER

2.1 Learning Linguistic Patterns from Raw Text

We design our sentence labeling scheme by mimicking human inferencing process. Our brain tends to group words based on same concepts they share [11]. For instance, " 棒球(baseball)" and " 籃球(basketball)" both belong to the concept " 運動(sport)". In light of this, we abstract words not in the keyword dictionary by a NEOntology layer collected from the open source of Stanford NLP Group [12] and an EHowNet [13] layer provided from CKIP Group. We use training corpus to generate the keyword dictionary. The dictionary consists of 0.01% words in each category with highest Log-Likelihood Ratio (LLR) [14]. Keywords are preserved since their categorical-distinctive property can help machine quickly identify the category.

To illustrate whole labeling process, consider the sentence "陳偉殷成為大聯盟單場最多三振的台灣投手(Chen Wei Ying became the Taiwanese pitcher in MLB with most strike outs)" in Fig. 2. Firstly, " 陳偉殷(Chen Wei Ying)", " 大聯盟(MLB)", " 三振(strike out)", and " 投手(pitcher)" are found in the keyword dictionary and tagged. Subsequently, name entities like " 台灣(Taiwanese)" is tagged as "{國家(country)}"by NEOntology. Afterwards, remaining terms like " 成為(became)" and " 最多(most)" are abstracted with their corresponding EHowNet senses if they exist. Finally, we can obtain a tagged sentence enabling us to generate prominent patterns in the next stage.

To generate linguistic patterns, we construct a semantic graph [15] for each category. Each sentence is a directed path with tags as nodes. Directed edges point from tag i to tag j if tag j immediately follows tag i in a tagged sentence. Weight of each edge composes two values: Critical Degree Sum (CDS) and Walk Frequency (WF). CDS is the sum of LLR of tags on both ends of an edge. It represents sum of importance of two tags. We still lack co-occurrence information. Therefore, we combine WF - number of times two adjacent and ordered tags occurs in a sentence. Altogether, edge weight is obtained by dividing CDS by WF and is called Critical Collated Degree (CCD). The reason we put WF in the denominator is because we observe that tags with high WF often contain less information compared to moderate or few occurrences. For instance, " 由於(Due to)" are less representative than " 棒球員(Baseball player)" in the sport category. Edge weight CCD's formulas are listed in Eq. (1) and (2):

$$CDS_{i,j} = LLR_i + LLR_j \tag{1}$$

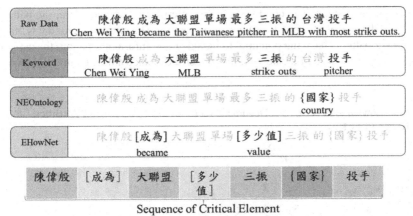

Fig. 2. Critical element labeling process

$$CCD_{i,j} = \frac{CDS_{i,j}}{WF_{i,j}}, i : tag\ i, \quad j : tag\ j \qquad (2)$$

After giving each edge its weight, we start to extract a candidate knowledge pattern for category c, which represent as CKP_c. First, we reverse signs of edge weights. Next, we start from an arbitrary tag i, find the shortest path from tag i to tag j which $j \neq i$. Again, we select another tag k where $k \neq i, j$, and find the shortest path between tag i and tag k. This iterative process will stop when all tags have been selected as the starting node. Eventually, we will get $n * (n - 1)$ shortest paths, n is number of tags in a tag diagram. This process makes sure we find most informative linguistic patterns for arbitrary two different tags.

Linguistic patterns selected may still not be representative enough. For instance, they might be too long or too short to be representative, might not be coherent enough to match any tag sentence, and might occur in many categories simultaneously. All these factors make patterns less prominent. Consequently, we address these problems by filtering our final knowledge pattern (KP_c) by three criterions: 1) the length of a pattern must be between two to five; 2) a pattern must exist in at least one of the tag sentences in its own category; 3) a pattern should not exist in tag sentences in other categories. The definition of 'exist' is defined in the following Eq. (3).

$$Set(CKPc) \cap any\ Set(Sci) = Set(CKPc),\ Sci : tag\ sentence\ i\ in\ category\ c \quad (3)$$

2.2 Knowledge-Infused Document Embedding Representation (KIDER)

The pattern matching process is shown in Fig. 3. First, we discuss the construction of KP list. Since the category of a new document is unknown, we need to merge and sort all categories' selected patterns and give each pattern an importance score, we adopt the normalized LLR sum of tags as a pattern's score. Then, we sort all patterns according to three ordered conditions. Only patterns with same value in preceding conditions will be passed to next stage. The conditions are a pattern's normalized score, the pattern's length, and the size of training documents in a category. The intuition behind it is that we think longer patterns have the potential to include more information about the category. Also, the larger the size of a category, the higher the chance a document will belong to it.

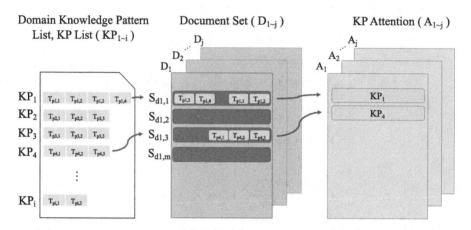

KP List: sorted in order by "Normalize CCD", "KP length", "Cluster size"

Match: $set(KP_i) \cap set(S_{dj, m}) = set(KP_i)$

Fig. 3. Pattern matching process

Second, we match each tag sentence in the tagged document with the KP list. The definition of match is the same as Eq. (3). If a sentence does not match any pattern, it's deleted; if a sentence is matched with a pattern with some additional tagged words, they are discarded. We repeat this process until all sentences are matched. Finally, we obtain a new document representation - KP Attention by concatenating all matched tags within tagged sentences. The difference between a tagged document and a KP Attention is that the latter removes unmatched tagged sentences and tags.

To feed a document representation into the classifier, we convert both tagged documents and KP Attention into vectors. We illustrate the document embedding generation process in Fig. 4. Document embeddings generation process. We adopt Doc2Vec method proposed by [16] as our representation model. Two different network designs are used to enrich semantic information, distributed memory (DM) and distributed bag of words (DBOW). DM considers more of word's neighbors' information by using the document id and randomly sample consecutive words to predict next word. DBOW considers more of BoW information - using the document id to predict words randomly sampled from the document.

For a document representation, we use DM and DBOW to each generate a 50-dimensional vector[1] and concatenate them together, getting a 100-dimensional vector.

---

[1] We empirically set the window size to 3, dimension of vector to 50, and epochs to 100.

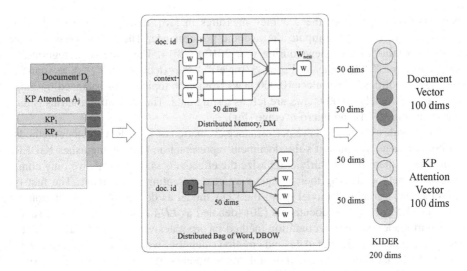

Fig. 4. Document embeddings generation process

We repeat this process twice, using a document's two different document representations – tagged document D_j and KP attention A_j, and concatenate their own 100-dimensional vector together. Therefore, each document is now represented by a 200-dimensional vector.

2.3 The Architecture of DNN for Text Classification

The main contribution of the KIDER is the document representation mechanism described above. Therefore, we except the classifier we choose doesn't matter much. This expectation is proved when we tested results on different classifiers. Performances vary little. Thus, we pick the classifier with best performance – DNN. In the design of neural network, we apply a feed forward network with 2 hidden layers, each with 512 nodes. We chose nonlinear leaky RELU [17] and the softmax activation function for hidden and output layers, respectively. Our learning rate is set to 0.01, batch size 128, and epochs 100. These settings are determined empirically. Model inputs are document vectors obtained in Sect. 2.2, and the outputs are probability distribution of the document's category. We select the category with highest probability as our prediction.

3 Experiment

3.1 Dataset and Setting

Experiments are conducted on two different Chinese datasets, one for emotion analysis and the other for news topic detection. They are collected from Yahoo News by [18] and [19]. The Yahoo News Reader Emotion Corpus contains approximately 47,000 news from eight categories, including " 擔心(worried)", " 感動(warm)", " 無聊(boring)", " 怪異(odd)", " 沮喪(depressed)", " 生氣(angry)", " 開心(happy)", and " 實用(useful)".

For evaluation, we adopt the accuracy measures in [18], macro-average (AM) and micro-average (Aμ) to compute the average performance. The Yahoo News Topic Corpus contains 140,000 news during year 2010 to 2014. There are six categories, " 教育(education)", " 健康(health)", " 旅遊(travel)", " 科技(technology)", " 運動(sport)", and " 政治(politics)". We select 10,000 news from each topic as our balanced training data, approximately 80,000 news are left as testing data. The evaluation metrics is the macro-average (FM) and micro-average (Fμ).

We provide a comprehensive performance evaluation of KIDER with other methods on both datasets. Because it's the document representation that distinguishes KIDER from others, therefore, to fairly emphasize the effectiveness of KIDER, we only compare with methods having their unique way of representing a document. The first is a probabilistic graphical model using the LDA model as document representation to train an SVM to classify documents [20] (denoted as *LDA-SVM*). Another is a state-of-the-art reader-emotion recognition method incorporates various knowledge sources such as bigrams, words, metadata, and emotion categories words [18] to learn semantic frames that characterize an emotion and are comprehensible for humans (denoted as FBA). Generated frames are adopted to predict emotion through an alignment-based matching algorithm. Moreover, two well-known deep learning-based approaches were also included: convolutional neural network [21] (denoted as *TCNN*), which learns a neighbor-catching document representation, and recurrent neural network with long short-term memory [22] (denoted as *LSTM*), which incorporate time-dependency into document representation. We also include results of Naive Bayes (denoted as *NB*) [23] as a baseline, which represent documents by BoW features.

3.2 Results and Discussion

Table 1 comparison of the accuracies of six reader emotion classification systems displays the performance of reader emotion recognition and topic detection. We observe that LDA-SVM, TextRNN, and TextCNN perform relatively poor in terms of micro accuracy, which are 58.68%, 61.05%, 64.31% respectively, meaning the performance is low in major categories but reach better performance in minor ones. On the other hand, methods which combine more complex and richer features like FBA and our method - KIDER perform better in terms of macro- and micro-accuracy, achieving higher than 85% overall performance. Our KIDER method reaches best performance, outperforming state-of-the-art method by almost 7%.

For the results of Yahoo News topic classification, NB performs relatively low with approximating 45% overall performance. FBA reaches promising performances with about 73% micro F1-score. These comparisons show we can enhance performance by incorporating various groups of features. Moreover, the well-known deep learning methods further increase performance by 4% to 5%. LSTM ranked third, reaching 77.95% and 74.91% in micro and macro F1-score. TCNN improved a little with micro and macro F1-score being 78.65% and 75.81%. Finally, KIDER achieves amazing results. For each category, F1-score is above 96%, which is 6.55% to 34.91% higher than second-best method. In terms of overall performance, KIDER approach still dominates the result. Our macro F1-score is 97.75%; micro F1-score is 97.97%, only 2–3% away from perfect classification.

Table 1. Comparison of the accuracies of six reader emotion classification systems

	NB	SVM	FBA	LSTM	TCNN	KIDER	Supp.
Emotion	*Accuracy (%)*						
Worried	69.56	92.83	75.80	55.94	**97.32**	86.21	261
Warm	15.09	87.09	91.91	89.58	**98.80**	94.24	835
Boring	75.67	76.21	90.52	52.07	**98.30**	78.87	1473
Odd	73.90	85.40	83.34	72.35	**95.54**	93.11	1526
Depressed	73.76	81.43	92.15	87.98	**95.49**	91.16	1573
Angry	47.00	74.21	87.83	54.46	61.90	**92.97**	4326
Happy	37.90	37.59	**88.94**	45.29	45.93	85.03	7343
Informative	20.60	44.02	80.92	65.17	62.19	**93.27**	18265
A^M	51.69	76.10	86.43	65.36	81.93	**89.07**	35602
A^μ	34.52	58.68	84.63	61.05	64.31	**91.37**	35602
Topic	*F_1-score (%)*						
Education	46.69	51.17	62.33	52.08	50.48	**97.24**	5023
Health	49.53	54.41	80.88	86.29	87.37	**97.72**	5844
Travel	20.97	70.16	59.21	61.91	65.96	**96.35**	12256
Technology	40.02	56.69	56.34	72.98	73.88	**97.93**	17031
Sport	56.45	89.92	71.38	90.48	91.94	**98.49**	18919
Politics	59.31	81.62	84.69	85.09	85.23	**98.79**	19023
F^M	45.50	67.35	69.14	74.81	75.81	**97.75**	78096
F^μ	46.85	72.41	74.51	77.95	78.65	**97.97**	78096

According to the experimental results, we demonstrate that KIDER method is able to provide stably high performances under different categories and categorization tasks.

We further examine the value of several document representations approach on the reader emotion dataset. As shown in Table 2. Different Document Representation Combinations, we compare different type of text representation methods. *Raw Document*'s sentences are all transformed into tag sentences through critical element labeling described in Sect. 2.1. *Filtering Document* and *KP Attention* further removed sentences with unmatched patterns in the KP list. The difference between them is that tags in matched sentences but not in matched patterns are preserved in *Filtering Document*. We conducted right-tailed t-tests between different settings of our KIDER method with 90%(*), 95%(**), and 99.9%(***) confidence level.

We do this evaluation study for two purposes - to know where main performance comes from and to determine the best combination of document representation. After using external knowledge to label sentences, F_1-score already beat other methods with no use of external source by a large margin. This result shows that external knowledge plays an important role in raising performance. After removing sentences without a matched

Table 2. Different document representation combinations

Combination of document representation	F_1-score
Raw document	87.84% ***
Filtering document	88.45% ***
Filtering document + KP attention	89.67% ***
Raw document + Filtering document + KP attention	90.82% ***
Raw document + KP attention, KIDER	**91.10%**

pattern, performance increases. This shows that the KP List is able to preserve categorical - distinctive sentences while reduce noises. F_1-score increases again when using all three document representations together. However; by ditching Filtering Document, F_1-score increases furthermore. Therefore, KP Attention itself paired with Raw Document is the setting in KIDER.

Table 3. Domain knowledge linguistic pattern examples of eight emotions

Emotion	*Linguistic Patterns*
angry	{政客} 政府 表示 [阻動] ({politicians} government indicate [resist]) [嚴重性值] 表示 漲價 [事務] ([severity] indicate price rise [affair])
worry	颱風 氣象局 [方位值] 超大 (typhoon Weather Bureau [azimuth] big) 氣象局 增強 海面 [速度] (Weather Bureau strengthen sea level [speed])
bored	女星 兩人 日前 爆料 (female celebrity two people previously spill it) 朋友 兩人 笑說 [煩惱] (friends two people laugh and tell [worries])
happy	[表示] 安打 比賽 [良好] ([indicate] homerun game [well]) [非常好] 比賽 冠軍 [數量值] ([very well] game champion [quantity])
odd	[使動] 女子 發現 店員 ([move] lady find clerk) {賽事} 因為 [狀況] ({game} because [situation])
depressed	[劫難] 附近 造成 [害怕] ([calamity] nearby cause [fear]) [數量值] 地區 造成 土石流 ([quantity] region cause mudslide)
warm	[繼續] 自己 社會 [功用值] ([continue] myself society [function]) [時間值] 以及 經濟 預期 ([time value] and economic prospect)
informative	[時間值] 以及 經濟 預期 ([time value] and economic prospect) [連續] 投資 產業 [先進性值]([continuous] invest industry [advantage])

Finally, we note that it takes around 3 s to classify a new 300-word document. The main cost comes from the pattern matching process. The time is short despite our intensive work. That is because the most computational expensive process is to prepare the KP list, which only needs to be done once during training period.

4 Knowledge Representation

In this section, we investigate the pattern inference procedure for each category. We randomly select two patterns with high scores from all eight emotions, translate them into English and display in Table 3. Words without brackets are keywords; with curly brackets ({}) are named entities; with square brackets ([]) are general concepts. We take patterns in "開心(happy)" to demonstrate their interpretability. "［表示］安打 比賽 ［良好］([indicate] homerun game [well])" and "［非常好］比賽冠軍 ［數量值］([very well] game champion [quantity])" all imply winning events, which is a reasonable cause to feel "開心(happy)". These patterns are interpretable and conform to human's understanding. They provide crucial semantic information for each category.

Table 4. The matched linguistic patterns of our example news article

Original Sentence	Matched Pattern	Score
西班牙財長表示，信用市場正逐漸將西班牙拒絕於門外。(Spain's financial minister indicated that credit market is gradually shutting out Spain.)	［程度］市場 ［對待］([degree] market treat)	108
因此，西班牙銷售公債的舉動正是市場對西班牙償債能力的信心測試。(Therefore, the selling-bond movement of Spain is the test of market's confidence in Spain's solvency.)	［人格特質］市場 能力 ([characteristic] market ability)	137
各方抱持懷疑態度的同時，西班牙經濟部長出面澄清，馬德里當局目前並未決定要向歐盟尋求紓困。(While people around the world remain skeptical, Spain's economic minister stepped out and clarified that Madrid authorities have not decided to ask for EU's bailout.)	［控制］西班牙 ([control] Spanish)	41
此外，歐洲央行也表明將努力紓解眼前的歐債危機。(In addition, The European Central Bank said they will work hard to solve the European debt crisis.)	主觀描述 歐洲(subjective description Europe)	72
金融市場投資人深受鼓舞，使歐美股市周三收紅。(Financial market investors are encouraged. As a result, Europe and U.S. stock market closed in the red on Wednesday.)	［增多］市場 股市 ([rise] market stock market)	145

We further examine the above linguistic patterns in units of news. Sentences with high-scored matching patterns can be viewed as a piece of news' highlight. We select single news to display our findings in Table 4. The last sentence illustrates the phenomenon that stock market was getting better, which is the crucial useful part of this news for readers. Its corresponding pattern "[增多] 市場 股市([rise] market stock market)" reasonably receives highest score 145. This demonstration shows that our scores have the potential to imply highlights of an article. In addition, knowledge patterns themselves are capable of simplifying complex sentences and capturing essential meaning of sentences. Therefore, the generated linguistic patterns and their scores can further be used in other applications such as text summarization.

5 Conclusion

We present KIDER, a document representation method that utilizes linguistic knowledge and patterns to enhance performance and increase interpretability. We conduct extensive experiments to show that, compared to a rich set of existing methods; KIDER is capable of reaching superior classification results. Micro accuracy for emotion analysis is 91.37%; micro F1-score for topic detection is 97.97%, which are all remarkably high. Moreover; we examine reasons behind high performance. We also demonstrate the interpretability by showing additional usage of patterns in finding highlights for each category and news. We anticipate the proposed method can inspire more work to focus more on using and analyzing linguistic knowledge. Ultimately, we believe that together we can pave the road to interpretable and high-performance artificial intelligence.

In the future, we plan to refine KIDER and employ it to other NLP applications. Additionally, we would like to utilize sense-clustering techniques to label crucial elements in text, with the goal of contributing a fully automatic KIDER to the whole NLP society.

References

1. Aggarwal, C.C., Zhai, C.: Mining Text Data. Springer Science & Business Media, Heidelberg (2012)
2. Wu, H., Salton, G.: A comparison of search term weighting: term relevance vs. inverse document frequency. In: ACM SIGIR Forum, vol. 16, pp. 30–39. ACM (1981)
3. Mikolov, T., Sutskever, I., Chen, K., Corrado, G.S., Dean, J.: Distributed representations of words and phrases and their compositionality. In: Advances in Neural Information Processing Systems, pp. 3111–3119 (2013)
4. LeCun, Y., et al.: Backpropagation applied to handwritten zip code recognition. Neural Comput. 1(4), 541–551 (1989)
5. Hochreiter, S., Schmidhuber, J.: Long short-term memory. Neural Comput. 9(8), 1735–1780 (1997)
6. Devlin, J., Chang, M.W., Lee, K., Toutanova, K.: BERT: pre-training of deep bidirectional transformers for language understanding. arXiv preprint arXiv:1810.04805 (2018)
7. Lee, J.Y., Dernoncourt, F.: Sequential short-text classification with recurrent and convolutional neural networks. arXiv preprint arXiv:1603.03827 (2016)

8. Bahdanau, D., Cho, K., Bengio, Y.: Neural machine translation by jointly learning to align and translate. arXiv preprint arXiv:1409.0473 (2014)

9. Alexander, J.A.: Template-based procedures for neural network interpretation. Ph.D. thesis, University of Colorado (1994)

10. Seidenberg, M.: Language at the Speed of Sight How We Read, Why So Many Can't, and What Can Be Done About It, 1st edn. Basic Books, New York (2017)

11. Landauer, T.K., Foltz, P.W., Laham, D.: An introduction to latent semantic analysis. Discourse Process. **25**(2–3), 259–284 (1998)

12. Finkel, J.R., Grenager, T., Manning, C.: Incorporating non-local information into information extraction systems by Gibbs sampling. In: Proceedings of the 43rd Annual Meeting on Association for Computational Linguistics, pp. 363–370. Association for Computational Linguistic (2005)

13. Chen, K.J., Huang, S.L., Shih, Y.Y., Chen, Y.J.: Extended-Hownet: a representational framework for concepts. In: Proceedings of OntoLex 2005-Ontologies and Lexical Resources (2005)

14. Dunning, T.: Accurate methods for the statistics of surprise and coincidence. Comput. Linguist. **19**(1), 61–74 (1993)

15. Chang, Y.C., Chu, C.H., Su, Y.C., Chen, C.C., Hsu, W.L.: PIPE: a BIOC module for protein-protein interaction passage extraction. Database (Oxford) **2016**, baw101 (2016)

16. Le, Q., Mikolov, T.: Distributed representations of sentences and documents. In: International Conference on Machine Learning, pp. 1188–1196 (2014)

17. Maas, A.L., Hannun, A.Y., Ng, A.Y.: Rectifier nonlinearities improve neural network acoustic models. In: Proceedings of ICML, vol. 30, p. 3 (2013)

18. Chang, Y.C., Chen, C.C., Hsieh, Y.L., Chen, C.C., Hsu, W.L.: Linguistic template extraction for recognizing reader-emotion and emotional resonance writing assistance. In: ACL, pp. 775–780 (2015)

19. Chang, Y.-C., Hsieh, Y.-L., Chen, C.-C., Hsu, W.-L.: A semantic frame-based intelligent agent for topic detection. Soft. Comput. **21**(2), 391–401 (2015). https://doi.org/10.1007/s00500-015-1695-4

20. Blei, D.M., Ng, A.Y., Jordan, M.I.: Latent Dirichlet allocation. J. Mach. Learn. Res. **3**, 993–1022 (2003)

21. Kim, Y.: Convolutional neural networks for sentence classification. arXiv preprint arXiv:1408.5882 (2014)

22. Yogatama, D., Dyer, C., Ling, W., Blunsom, P.: Generative and discriminative text classification with recurrent neural networks. arXiv preprint arXiv:1703.01898 (2017)

23. McCallum, A., Nigam, K.: A comparison of event models for Naïve Bayes text classification. In: AAAI/ICML-1998 Workshop on Learning for Text Categorization, pp. 41–48 (1998)

Discriminative Features Fusion with BERT for Social Sentiment Analysis

Duy-Duc Le Nguyen[1], Yen-Chun Huang[1], and Yung-Chun Chang[1,2(✉)] ⓘ

[1] Graduate Institute of Data Science, Taipei Medical University, Taipei, Taiwan
{m946108006,m946108007,changyc}@tmu.edu.tw
[2] Pervasive AI Research Labs, Ministry of Science and Technology, Hsinchu, Taiwan

Abstract. The need for sentiment analysis in social networks is increasing. In recent years, many studies have shifted from author sentiment research to reader sentiment research. However, the use of words that hinders sentiment analysis is very diverse. In this paper, we provide a model that combines the latest and most recent contextual text embedding technology and feature selection to more accurately detect the emotional intent of an article. We named it DF2BERT (Discriminative Features Fusion with Bert), and extensively applied datasets in different languages and different text classification tasks to validate our method, and compared it with several well-known approaches. Experimental results show that our model can effectively predict sentiment behind the text which outperform comparisons.

Keywords: Natural language processing · Sentiment analysis · Text representation · Feature fusion · Deep neural network

1 Introduction

At present, big data analysis technology has rapidly developed. According to previous research, most of the unstructured data are currently buried in text data, and they are mainly distributed on the Internet. Today, the online community and news media have become the main field of information acquisition and transmission for modern people. In addition, sentiment analysis research is getting more and more attention with an attempt to obtain trends in public by mining opinions that are subjective statements that reflect people's sentiments or perceptions about topics [11]. Thanks to the optimization of natural language processing (NLP) technology. It helps us to accurately analyze the sentiment and opinions of massive texts. While previous researches on emotions mainly focused on detecting the emotions that the authors of the documents were expressing. It is worthy of note that the reader-emotions, in some aspects, differ from that of the authors and may be even more complex [6,12]. For instance, a news article with the title "The price of crude oil will rise 0.5% next week" is just objectively reporting an event without any emotion, but it may invoke emotions like

© Springer Nature Switzerland AG 2020
H. Fujita et al. (Eds.): IEA/AIE 2020, LNAI 12144, pp. 30–35, 2020.
https://doi.org/10.1007/978-3-030-55789-8_3

angry or worried in its readers. Furthermore, it is possible to get more sponsorship opportunities from the company or manufacturer if the articles describing a certain product are able to promote greater emotional resonance in the readers.

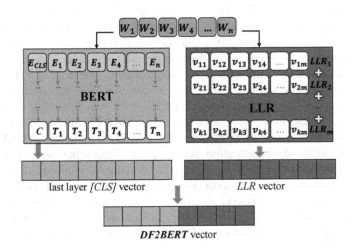

Fig. 1. Architecture of the discriminative feature fusion with BERT model (DF2BERT).

In this paper, we proposed the method which is discriminative features fusion with BERT (DF2BERT) for social sentiment analysis. We first extract the discriminative features from raw text. Then, we integrate the features into BERT for sentiment classification. Our experiments demonstrate that DF2BERT can achieve a higher performance than other well-known methods of text categorization on two different sentiment categorization tasks. Furthermore, the proposed method embraces the advantages of both feature-based and deep learning-based approaches. First of all, as shown in our experiments, it is language-independent. Second, compared to feature-based systems, DF2BERT is automatic and scalable. We believe DF2BERT points to a promising direction for many natural language applications.

2   Discriminative Features Fusion with BERT Model

Figure 1 illustrates an overview of the DF2BERT model. E is the embedding representation and T is the final output of BERT architecture. There are 12 layers in the original BERT model where each token will have 12 intermediate representations. Every sequence takes the special classification token $[CLS]$ as the first token which is followed by the WordPiece tokens. For classification tasks, $[SEP]$, the separator token, can be ignored. The maximum sequence of length of the input are 512 tokens. The final hidden state of the $[CLS]$ token is taken as a fixed pooled representation of the input sequence. This is a vector with 768 size represent the whole input sequence for classification tasks. Since BERT has been pre-trained for Masked Language Modeling and Next Sentence Prediction, it also takes masking and next sentence representation as inputs. But classification tasks do not need to consider about masking and next sentence predicting, masking representation is an array of 1 value and next sentence representation is a 0 one.

BERT for sequence classification takes one segments as input. The segment is re-presented as a single input sequence to BERT with an special tokens: $[CLS], x_1, \ldots, x_N$. N need to be less or equal to L which is a parameter that controls the maximum sequence length during training. At the output, the $[CLS]$ representation is fed into an output layer for the sentiment analysis. Adam [4] is the optimization of BERT with the following parameters: $\beta_1 = 0.9, \beta_2 = 0.999, \epsilon = 1e^{(-5)}$ and $L_2 = 0.01$. BERT trains with 0.1 dropout rate on all layers and attention weights with activation function as GELU. Furthermore, to increase classification accuracy by eliminating noise features. A noise feature is kind of that when added to the document representation, increase the classification error on new data. For instance, the Bernoulli model is sensitive to noise features. Moreover, sentiment lexicons are important information for identifying sentiment behind the text. In light of this, we use the log likelihood ratio (LLR) [10], an effective feature selection method, to learn a set of sentiment-specific lexicons. Given a training dataset, we first obtain four frequencies $k = N(L \wedge S)$, $l = N(L \wedge \neg S)$, $m = N(\neg L \wedge S)$, and $n = N(\neg L \wedge \neg S)$, in which $N(L \wedge S)$ denotes the number of documents that contain l and belong to sentiment S, $N(L \wedge \neg S)$ denotes the number of documents that contain L but does not belong to sentiment S, and so on. LLR employs Equation to calculate the likelihood of the assumption that the occurrence of a lexicon l in sentiment T is not random.

$$LLR(L, S) = 2log \left(\frac{p(L|S)^k (1 - p(L|S))^m p(L|\neg S)^l (1 - p(L|\neg S))^n}{p(L)^{k+l}(1 - p(L))^{m+n}} \right) \quad (1)$$

In (1), probabilities $p(L)$, $p(L|S)$, and $p(L| \wedge S)$ are approximated using maximum likelihood estimation. A lexicon with a large LLR value is closely associated with the sentiment. We rank the lexicons in the training data based on their LLR values and select lexicons with high LLR values to compile a sentiment lexicon list.

3   Experimental Results and Discussions

In order to examine the flexibility and robustness of the proposed method, we use publicly available datasets as well as those collected on our own for evaluation. Table 1 shows the descriptive statistics of three datasets used in our experiments. We tested the proposed method on a sentiment classification of IMDB movie review dataset [9] (denoted as IMDB). This is a balance dataset which containing the same proportion of positive and negative in 25,000 movie reviews for training, and 25,000 for testing. In order to evaluate the reliability, we tested DF2BERT on a small Chinese movie review corpus [1], which collected from an electronic bulletin board system that provides a social platform for academic purposes (denoted as PTT).

Table 1. Descriptive statistics of three datasets used in text categorization experiments. Total amount of documents in each dataset is listed in parentheses.

Dateset	Category	#Train	#Test	Total
IMDB	Positive	12,500	12,500	12,500
(50,000)	Negative	12,500	12,500	12,500
PTT	Positive	1,132	1,132	2,246
(4,492)	Negative	1,132	1,132	2,246

To evaluate the effectiveness of the compared systems, we adopt the convention of classification accuracy for sentiment analysis [5], and use macro-average for the estimation of overall performance. We re-implement BERT with RoBERTa setting [8]. We primarily follow the original BERT optimization hyper-parameters given in Sect. 3, except β_2 is set as 0.98 to improve stability when training. The maximum sequence length is 512 tokens where padding or truncating at the end of segment. As an out, the last hidden layer of $[CLS]$ which represent a whole sentence will concatenate with the output layer of feature extraction step. Here, we extract top 70 features affected to each label. During training time, we have found that 70 extracted features achieved better performance compare with 30, 50, 100, 150 and 200 features.

A comprehensive performance evaluation of the DF2BERT with 10 well-known text classification methods is provided. The first is a keyword-based model which adopts TF-IDF term weighting and is trained by SVM (denoted as SVM) with linear kernel. Another is the random forest approach which consist a large number of relatively uncorrelated decision trees operating as an ensemble (denoted as RF). Next, is a gradient boosting decision tree that integrates multiple learners for classification problems (denoted as XGB) Two well-known deep learning-based text classification approaches were also included: convolutional neural network for text classification (denoted as TCNN) [3] and recurrent neural network with long short-term memory (denoted as LSTM) [7]. Moreover, comparing DF2BERT with BERT enables us to verify the contribution of the proposed discriminative feature fusion techniques. To serve as a standard for comparison, we also included the results of Naive Bayes (denoted as NB), Decision Tree (denoted as DT), Logistic Regression (denoted as LR), and k-Nearest Neighbors (denoted as KNN) as baselines. The performances of sentiment analysis and topic detection systems are listed in Table 2.

As the results, most of machine learning-based approaches can achieve about 80% classification accuracy. For baseline models, regarding the binary-class dataset, we can see that in Logistic Regression and SVM, the performance is better than the tree structure model, and the performance results of Naive Bayes

Table 2. Comparison of the performance of 11 classification systems of IMDB and PTT corpus. Bold numbers indicate the best performance in the category.

	NB	DT	LR	KNN	SVM	RF	XGB	TCNN	LSTM	BERT	DF2BERT
IMDB	83.0	70.4	88.3	66.2	88.3	73.0	81.0	82.6	85.4	93.2	**93.5**
PTT	80.0	70.0	84.4	71.0	87.8	74.3	81.9	53.5	78.1	87.4	**88.2**

and XGBoost are closely behind. Text classification contains many features and is mostly linearly separable [2]. Logistic Regression is better than the tree structure model in dealing with linear separable classification problems. It is more effective in dealing with outliers in the data and avoids over-fitting the training set to the test set. It is worth noting that the SVM can achieve good performances. This can be partially explained by using the statistical learning theory argument of SVM in. If there exists a linear separator w with a small 2-norm that separates in-class data from out-of-class data with a relative large margin, an SVM can perform well.

In general, deep learning-based methods can achieve about 85% classification accuracy on IMDB dataset. However, deep learning-based methods inferior in PTT dataset. This may be due to the insufficient amount of data in the PTT dataset itself, which is difficult to reflect the power of deep learning. Moreover, the limitation of the neural network-based methods is that it cannot take the order of the inputs into account like recurrent neural network. If the first and last token are the same word, it will treat those tokens exactly the same. By contrast, BERT overcomes this problem through learning the context and semantic information. Consequently, it performs better than other machine learning and deep learning methods. Notably, DF2BERT achieves the best performance with approximating 94% accuracy since we further integrate the discriminative features into BERT model. This is because the proposed method can extract useful words and terms in each class so that the vector represents a sequence generated by BERT with the extra LLR vector at the end will orient to the right class.

To summarize, the proposed discriminative feature fusioin with BERT approach successfully integrates the syntactic, semantic, and content information in text to recognize sentiment behind document. Hence, it achieves the best classification accuracy, as shown in Table 2.

4   Conclusion

In this paper, we propose a new model to improve text embedding results, combining feature selection algorithms with the latest natural language processing method. The model was trained to deal with cross-language classification in English and Chinese data. The results show that our method is effective in sentiment analysis classification tasks. In the future, we will refine DF2BERT and employ it to other NLP applications. We will also investigate the syntactic dependency information in social media to incorporate further syntactic and semantic information into the BERT structure.

Acknowledgement. This research was supported by the Ministry of Science and Technology of Taiwan under grant 107-2410-H-038-017-MY3 and 109-2634-F-001-008-.

References

1. Chu, C.H., Wang, C.A., Chang, Y.C., Wu, Y.W., Hsieh, Y.L., Hsu, W.L.: Sentiment analysis on chinese movie review with distributed keyword vector representation. In: 2016 Conference on Technologies and Applications of Artificial Intelligence (TAAI), pp. 84–89. IEEE (2016)
2. Joachims, T.: Text categorization with support vector machines: learning with many relevant features. In: Nédellec, C., Rouveirol, C. (eds.) ECML 1998. LNCS, vol. 1398, pp. 137–142. Springer, Heidelberg (1998). https://doi.org/10.1007/BFb0026683
3. Kim, Y.: Convolutional neural networks for sentence classification (2014). arXiv preprint arXiv:1408.5882
4. Kingma, D.P., Ba, J.: Adam: a method for stochastic optimization (2014). arXiv preprint arXiv:1412.6980
5. Lin, K.H.Y., Yang, C., Chen, H.H.: What emotions do news articles trigger in their readers? In: Proceedings of the 30th Annual International ACM SIGIR Conference on Research and Development in Information Retrieval, pp. 733–734. Citeseer (2007)
6. Lin, K.H.Y., Yang, C., Chen, H.H.: Emotion classification of online news articles from the reader's perspective. In: Proceedings of the 2008 IEEE/WIC/ACM International Conference on Web Intelligence and Intelligent Agent Technology, vol. 01, pp. 220–226. IEEE Computer Society (2008)
7. Liu, P., Qiu, X., Huang, X.: Recurrent neural network for text classification with multi-task learning (2016). arXiv preprint arXiv:1605.05101
8. Liu, Y., et al.: Roberta: a robustly optimized BERT pretraining approach (2019). arXiv preprint arXiv:1907.11692
9. Maas, A.L., Daly, R.E., Pham, P.T., Huang, D., Ng, A.Y., Potts, C.: Learning word vectors for sentiment analysis. In: Proceedings of the 49th Annual Meeting of the Association for Computational Linguistics: Human Language Technologies, vol. 1, pp. 142–150. Association for Computational Linguistics (2011)
10. Manning, C., Schütze, H.: Lexical acquisition. In: Foundations of Statistical Natural Language Processing, vol. 999, pp. 296–305. MIT press, Cambridge (1999)
11. Pang, B., Lee, L., Vaithyanathan, S.: Thumbs up?: sentiment classification using machine learning techniques. In: Proceedings of the ACL-02 Conference on Empirical Methods in Natural Language Processing, vol. 10, pp. 79–86. Association for Computational Linguistics (2002)
12. Tang, Y.J., Chen, H.H.: Mining sentiment words from microblogs for predicting writer-reader emotion transition. In: LREC, pp. 1226–1229 (2012)

Text Sentiment Transfer Methods by Using Sentence Keywords

Shengwei Hu, Bicheng Li$^{(\boxtimes)}$, Kongjie Lin, Rui Wang, and Kai Liu

School of Computer Science and Technology, Huaqiao University,
Xiamen, Fujian, China
lbclm@163.com

Abstract. Text sentiment transfer models modify sentence sentiments while retaining their semantic content. The main challenge is to separate sentiment-independent content information from the semantic information of the sentence. The previous works usually expect to utilize the model encoder to infer the sentence-level representation that removed sentiment information. However, the strength of the models' abilities to reconstruct text is difficult to control which resulting encoder infer sentence-level sentiment-independent content embedding failed. In this paper, we address this challenge by using word-level representation. We first use the POS-Tagging technique to tag the part of speech of word sequence, then extracting content keywords by three schemes and use them as input, rather than the entire sentence, to obtain a purer word-level sentiment-independent content representation. In this way, the model does not require to infer the sentiment-independent representation, which avoids the instability of the adversarial training process. Experiments show that our method achieves the state-of-the-art performance and is also effective in long text sentiment transfer tasks.

Keywords: Text sentiment transfer · Text generation · Natural language processing

1 Introduction

Text sentiment transfer is an important task in designing intelligent and controllable natural language generation (NLG) systems. In the supervised field, recent related work such as machine translation [1] and dialogue systems [2] with a parallel corpus have demonstrated considerable success. However, a parallel corpus may be not available for a sentiment transfer task. Therefore, it is usually an unsupervised learning task, which makes the text sentiment transfer task more challenging.

There is a common strategy for unsupervised text sentiment transfer models, which is to separate the sentiment-independent content from the semantic information of the text. Then, based on the separated content and the target

Supported by China National Social Science Fund (19BXW110).

H. Fujita et al. (Eds.): IEA/AIE 2020, LNAI 12144, pp. 36–48, 2020.
https://doi.org/10.1007/978-3-030-55789-8_4

sentiment a new sentence is generated. Consequently, the key challenge in these approaches is to model the sentiment-independent content representation.

In the conditional generative modeling, most latent representations are either given or automatically inferred from the data distribution. Almost all of the previous work [3–10] on unsupervised sentiment transfer have pursued the latter approach. The previous models usually used an auto-encoder with sentiment discriminator, expect to use the sentiment discriminator loss to guide encoder to infer the sentence-level embedding that removed sentiment information. Then the sentence-level embedding combine with the target sentiment embedding for decoding. And during the training phase, reconstruction loss as one of the objective functions to optimize the model. However, it is difficult to learn a pure sentiment-independent content representation based on this architecture, whether adversarial training is used or not. There are two situations of concern: the ability of the model to reconstruct text is overly strong or it is insufficient. If the reconstruction ability is overly strong, the model transfer sentiment usually forces the replacement of several sentiment words by an optimizing sentiment discriminator loss while keeping the reconstruction loss low. In this case, the representation inferred by the encoder is entangled, which makes the transferred sentence unreasonable. Contrarily, if the reconstruction ability is insufficient, the transferred sentences usually lose or change important content information of the original sentence. For example, after transferring a sentence with a negative meaning, e.g., "the desserts were very bland" into "the environment was very excellent", the object described has been unreasonably changed from food to the environment. Therefore, using model infer the sentence-level embeddings that remove emotional information are often difficult to achieve. Besides, the previous models require a sentiment discriminator to guide the generator optimization. This type of adversarial training leads to the model training phase being unstable because discrete text data is undifferentiable.

In this paper, we propose a novel text sentiment transfer method to address the above problems identified in the previous methods. We propose using word-level representation instead of sentence-level representation as model input, which avoids the instability of the adversarial training process. Firstly, we extract sentence keywords that do not contain sentiment information. Then Based on these keywords to reconstruct sentences with target sentiment. Through this method, word-level representation instead of sentence-level representation as model input, and the model is induced to understand the intrinsic semantic relationship between keywords, the transferred sentences are more in line with human logic. And we built two Amazon corpus with lengths of 15 and 50 words, respectively, to test the effect of our method when transferring long text sentiments. The experiments show that our method achieves the state-of-the-art performance in automatic and human evaluation results.

2   Related Work

In recent research, a common view is that sentiment transfer can be achieved by disentangling sentence representations in a shared latent space. Most of the solutions propose a two-step approach. Firstly, an adversarial approach to learn a latent sentiment-independent content representation of input sentences is used. Secondly, a decoder takes the latent and the target sentiment to generate a new sentence. Unfortunately, the discrete nature of the sentence makes training difficult. To address this issue, most of these models utilize a Gumbel-Softmax layer [4–6] or a reinforced learning method [12,13] to provide a more stable training process. However, these models tend to be slow to train and challenging to tune in practice.

According to the observation that text sentiment is often marked by distinctive phrases (e.g., "so terrible"), Li et al. [14] first deletes phrases associated with the sentences original sentiment using statistical methods, then uses a neural model to either directly generate the deleted phrase or retrieve new phrases associated with the target sentiment from corpus. These phrases are smoothly combined by the model into a final output. However, as the phrases are deleted using statistical methods, they are not necessarily sentiment related. In addition, the searched phrases also do not necessarily match the semantics of the original sentence.

There are also some methods based on back-translation [11,15]. These methods perform two translation processes. Prabhumoye et al. [11] uses the first-translation to rewrite the sentence to remove the original sentiment. In the second-translation, the re-written input sentence is encoded and a new sentence is generated based on its representation using a separate sentiment-specific generator. Lample et al. [15] combine the target sentiment with the input sentence during the first translation process to generate a sentence with the target sentiment, then use the generated sentence in combination with the original sentiment to reconstruct the original sentence. Obviously, back-translation is very time consuming, and it is more difficult to ensure that the sentence content is unchanged after performing two translation processes.

Besides, Liao et al. [16] exploits pseudo-parallel data via heuristic rules, thus turns this task to a supervised setting. However, the quality of the pseudo-parallel data is not quite satisfactory, which seriously affects the performance of the model.

3   Text Sentiment Transfer Method

Recent researches indicate that learn a latent sentiment-independent content representation z is not necessary, nor is it easily achieved Lample et al. [15]. Therefore, we do not seek disentanglement and do not include any adversarial training in our methods.

Different from image data, natural language text is discrete data; the removal of sentiment information can be achieved by extracting or removing some words. In this way, we can achieve the purpose that separate sentiment-independent content information from the semantic information without model inference. To be sure, the semantics of a sentence is usually composed of the described objects, actions, and emotions held by the author. Empirically, these are usually reflected by nouns, verbs, and adjectives in sentences, respectively. So, our method is that using the POS-Tagging technique to tag the word sequence, and based on the part of speech of word, we propose three content keywords extraction schemes to get sentiment-independent content keywords. The first scheme is 'KeepN', in which we retain the nouns, personal pronouns, and possessive pronouns in the sentence; these keywords contain the characters and the main description objects. The second scheme is 'KeepNV'. Based on the first scheme, we additionally retain the verb; these keywords contain the predicate information of the sentence. The third scheme is 'StyleWord', in which we only filter out the stopwords in the sentences and use 'StyleWord' instead of sentiment words matched by the sentiment lexicon proposed by Hu et al. [17]; these keywords contain the most semantic components of the sentence.

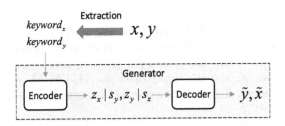

Fig. 1. The model architecture is an auto-encoder, both the encoder and decoder are LSTM. x, y represent sentences from two different sentiment domains, and $\tilde{y}, \tilde{x}$ represent the generated sentences transferred from x, y, respectively, $keyword_x, keyword_y$ represent the keyword extracted from x, y.

After extraction, the sentiment-independency content keywords are obtained from the input sentence. Then, the model generates new sentence with target sentiment based on these keywords. Our model presented in Fig. 1. During the training phase, firstly, we extract the keywords from original sentence and encode it to obtain the latent representation which does not contain any sentiment information. Then, the latent representation is combined with their self-sentiment representation to produce a joint to reconstruct the original sentence. Through this training process, the model learns to reconstruct sentence with self-sentiment according to some keywords. The optimization objective function is:

$$
\begin{aligned}
L_{res}(\theta_E, \theta_D) = \; & \mathbb{E}_{x \sim X}[-log p_D(x|p_E(KeyWord_x), s_x)] \\
& + \mathbb{E}_{y \sim Y}[-log p_D(y|p_E(KeyWord_y), s_y)]
\end{aligned}
\tag{1}
$$

Certainly, parallel sentences with similar sentence content or keyword but different sentiments exist in the corpus. These parallel sentences have similar content keywords or similar sentiment-independent content embedded representations. Therefore, after the training phase, similar keywords can reconstruct sentences with different sentiments. In this way, the model can automatically learn the expression of sentences with different sentiments. During the test phase, sentiment transfer can be achieved by combined target sentiment with to produce a new joint representation. Lastly, according to the new joint representations, the decoder generates a new sentence with the target sentiment.

Our model is similar to Hu et al. [4], the key difference is that there is no CNN-based sentiment discriminator. As the input keywords do not contain emotional information, it is not necessary to use a sentiment discriminator to provide information to optimize the encoder to learn the disentanglement representation. In addition, the model avoids adversarial training. Therefore, training in the proposed model is more stable than in previous models.

4 Experiment and Analysis

4.1 Datasets

Yelp Review Dataset[1]. The Yelp dataset consists of restaurant and business reviews from Yelp Dataset Challenge. Each review has a sentiment rating of 1 to 5 stars, over three stars are considered positive and below three stars are considered negative. We use the sub-dataset provided by Shen et al. [5], which has filtered out reviews that have three-stars or are longer than 15 words. The vocabulary size is 9357.

Amazon Food Review dataset[2]. We use the dataset provided by McAuley et al. [18]. It consists of food reviews from Amazon; as with the Yelp reviews, they have a sentiment rating of 1 to 5. Different from previous work experimenting with sentence-level reviews within 15 words, we experiment with a longer paragraph-level review dataset. We preprocessed the dataset and built two sub-datasets with maximum lengths of 15 and 50 words. Since the paragraph-level reviews contain a large number of neutral-sentiment sentences, we only retained rating scores of 1 and 5 of reviews as extremely negative and positive reviews, respectively. Finally, we filter out sentences that contain words that appear less than five times. The resulting vocabularies of the two sub-datasets contain 12450 and 32510 words, respectively.

The statistical results of the above datasets are shown in Table 1.

Table 1. Datasets statistical results. The left side of '/' indicates the number of positive sentences and the right side indicates the number of negative sentences.

Datasets	Train	Dev	Test
Yelp	270K/180K	38K/25K	76K/50K
Amazon-15	110K/110K	\	6K/6K
Amazon-50	146K/140K	\	7K/7K

---

[1] https://www.yelp.com/dataset.
[2] https://github.com/345074893/TSTMBUSK.

4.2 Evaluation Metrics

The evaluation consists of two parts: automatic evaluation and human evaluation.

Automatic Evaluation

Sentiment control. We measure the sentiment transfer accuracy using a CNN-based classifier Kim et al. [19] that employs convolutional layers with multiple filters of different sizes to capture the sentence features.

Content preservation. A large amount of previous work to measure content preservation usually calculates the self-Bleu score between the original sentence and the transferred sentence and compares which approach achieves a higher score. In fact, a sentence after the sentiment transfer is not exactly the same as the original sentence on the N-gram. Therefore, the transferred sentences that have copied a large number of original sentence words and have achieved a high self-bleu score does not mean that this is a good transferred sentence. A better comparison is to compare the self-Bleu scores in a range. We use a parallel corpus constructed by Li et al. [14] on the yelp dataset to measure the self-bleu scores between the reference sentence and the original sentence as the gold value. This parallel corpus consists of 1000 sentences and its transferred sentences (including 500 positive sentences and 500 negative sentences). We compare the self-Bleu scores to determine which model is closest to the gold value. In order to conduct a comprehensive evaluation, we additionally use the Word Mover's Distance (WMD) based on the Earth Mover's Distance, to measure content preservation as one aspect of semantics; this metric is proposed by Mir et al. [20]. They have released a sentiment lexicon on the yelp dataset and used it to match the sentiment words in the input sentence and the transferred sentence. They have replaced these sentiment words with a 'custom-style' and calculate the minimum distance between the embeddings of the input sentence and the transferred sentence. Same with Self-Bleu, we compare the score of which model is closest to the gold value that is evaluated on the parallel corpus.

Human Evaluation

Similar to Li et al. [14], we have hired three workers to rate the outputs for all of the models. We perform the human evaluation using three aspects: text readability, content preservation, and sentiment matching. In terms of readability, we evaluate whether the text contains syntax or context semantic errors. In order to reduce the manual evaluation error, we set the score range to 0–2 in the three aspects instead of 0–4, but on Amazon-50 we set the score range to 0–4. For example, in readability, score 2 is used when the transferred text does not have any grammatical or context semantic errors, or there is an error that hardly affects reading. Score 1 is used when there is an obvious error and score 0 is used when the generated text is very poor. In addition, we also set an overall rating score. When the transferred sentence is perfect in all three aspects and humans cannot distinguish between input sentences and transferred sentences then the overall rating score is 1, otherwise the score is 0. For each dataset, we evaluated

200 randomly sampled examples (100 negative and 100 positive sentences). At last, we compute the correlation of the results between automatic evaluation and human judgments.

4.3   Experiment Result

In this section, we compare our proposed method with different baseline models recently proposed Hu et al. [4]; Shen et al. [5]. The model of Hu et al. [4] is an improved version of the original paper, which is released in Texar [21]. Compared with the original model, the basic architecture is changed from VAE to AE, it avoids the sampling process before decoding, which makes the model better in terms of content preservation.

Evaluation Results on the Yelp Dataset. Before measuring the transferred text in terms of accuracy we first evaluated the performance of sentiment discrimination. It achieved an accuracy of 97.28%. So, the sentiment transfer accuracy calculated by the discriminator is reliable. The automatic evaluation result are shown in Table 2.

Table 2. Automatic evaluation results on the Yelp dataset.

Yelp	Accuracy	Self-Bleu	Self-Bleu3	Self-Bleu4	WMD
(Hu et al.)	**88.8%**	60.5(35.2)	54.2(29.3)	44.6(26.7)	0.16(−0.39)
(Shen et al.)	83.7%	8.2(−17.1)	5.4(19.5)	1.9(−16.0)	0.72(0.17)
KeepN	83.0%	11.1(−14.2)	8.5(−16.4)	3.6(−14.3)	**0.62(0.07)**
KeepNV	64.1%	23.7(−1.6)	**23.2(−1.7)**	12.6(−5.7)	0.44(−0,11)
Styleword	72.7%	**24.1(−1.2)**	22.8(−2.1)	**13.2(−4,7)**	0.43(−0.12)
Parallel Corpus	\	25.3	24.9	17.9	0.55

In terms of accuracy, as can be seen from the Table 2, Hu et al. [4] achieve the highest accuracy because only they use the discriminator to guide the model optimization. Notice the accuracy of 'KeepNV' is the lowest because some extracted verbs are sentimentally inclined, such as 'enjoy, love, like', resulting in input keywords that are not sentiment-independent. As a result, these keywords, which are usually retained in the transferred sentence, cause the model to make a false judgment when controlling the sentence sentiments only through sentiment labels. Therefore, extracting keywords without any sentiment noise is crucial in our methods.

In terms of content preservation, we have calculated the Bleu, Bleu3 and Bleu4. The results show that our model performs better on both self-Blue and WMD. In the results of Hu et al. [4] we can see that they reach the highest scores in the five Bleu metrics and the lowest in the WMD. The differences with the manually parallel corpus reference values are the largest compared to all models, in the Bleu-4 it reaches a score of 44.60. So, whether it is from the N-Gram or the semantic vector of view, it is obvious that a large number of words in the transferred sentence are copied from the original sentence. Although the model by Hu et al. [4] still achieves a higher value in the accuracy, it is because it simply replaces some sentiment words under the optimization of the discriminator loss. Therefore, the model did not learn the different sentiment expressions of the same content correctly; this is caused by an overly strong text reconstruction ability. Similar to Hu et al. [4], the differences between the test results of Shen et al. [5] and the reference values are also large. This is caused by an insufficient text reconstruction ability; the adversarial losses may dominate the training process, which makes the model prefer to generate sentences that look real, and the model does not take any measures to preserve the content.

However, the results of our three methods are similar to the manually parallel corpus evaluation results because our model generates sentences according to the intrinsic semantic relationships between the keywords. In this way, there is no problem of excessive or insufficient optimization of the reconstruction loss.

Table 3. The human evaluation results on the Yelp dataset. Rea-S is readability score, Con-S is content preservation score, Sen-S is sentiment match score, and Ove-S is the overall score.

Yelp	Rea-S	Con-S	Sen-S	Ove-S
(Hu et al.)	1.000	1.700	1.275	0.150
(Shen et al.)	0.475	0.375	0.925	0.075
KeepN	1.275	1.475	**1.425**	0.375
KeepNV	**1.450**	1.775	1.135	**0.385**
Styleword	1.200	**1.800**	1.175	0.325

The results of the human evaluation are shown in Table 3. It can be seen that the human evaluation results of Shen are the worst. During the evaluation process, we found that many sentences generated by the model are incomplete, the entire object component or other sentence component is lost. For example, the review 'The cake portion was extremely light and bit dry.' changed to 'the portions are' after the transfer. Although the sentence is incomplete, the generated part is fluent. Perhaps this type of generated sentence is true for the discriminator of GAN, which reflects the difficulty of training GAN and usually cannot measure whether the network is trained well. We can see that the sample of Hu et al. [4] is also poor in readability because we found that the transferred sentences do not reasonably convey the sentiment expression. We can see this problem from the two samples from the Hu et al. [4] model included in Table 4.

Comparing the three methods we proposed, 'KeepN' performs the best in sentiment matching because the keywords extracted by this method are almost without sentiment noise. 'KeepNV' performs the best in readability because the keywords include verbs, and the model obtains more predicate information from the keywords. This makes the generated sentences more logical. 'Styleword' performs best in content preservation, because the number of sentences extracted in this way is usually the most. In terms of the overall score, 'KeepNV' reached 0.385 points, which comprehensively indicates that our model achieved better performance.

Table 4. Example transferred sentences on Yelp dataset. Keywords are extracted from sentences using different schemes.

Negative to Positive	
Input	the bad service of the waitresses make me dread going sometimes
(Hu et al.)	the great service of the waitresses make me viva going sometimes
(Shen et al.)	the service is all of the staff are well organized
KeepN	Keywords: service, waitresses, me
	the service is great and waitressesare are always friendly and helpful to me
KeepNV	Keywords: service, waitresses, make, me, dread, going
	Service and waitresses make me to love going back
Styleword	Keywords: 'sentiment_word', service, waitresses, make, "sentiment_word", going, sometimes
	Great service waitresses make great going sometimes
Positive to Negative	
Input	also very knowledgeable on menu and suggestions!
(Hu et al.)	Also very slow on menu and replaced!
(Shen et al.)	Also have a menu and charge up
KeepN	Keywords: menu, suggestions
	The menu was very limited and the suggestions was very lacking
KeepNV	Keywords: menu, suggestions
	terrible menu suggestions
Styleword	Keywords: also 'sentiment_word', menu, suggestions
	also bad menu suggestions!

Table 4 presents samples of results for all of the models. As can be seen in the table, as we just analyzed above, the model of Hu et al. [4] does not really understand the semantics of the sentence, just simply replaces some words to achieve the purpose of the sentiment transfer. The model of Shen et al. [5] has serious attribute entanglement problems, and the readability of the generated samples is also very poor. However, our three methods overcome the above problems, by understanding the semantic relationship between keywords, the model can generate sentences that are more reasonable, and in line with human logic.

Evaluation Results on the Amazon Datasets. Similarly, we tested the performance of the sentiment discriminator on the two datasets Amazon-15 and Amazon-50. The accuracy rates reached 90.40% and 94.39%, respectively.

Different from the Yelp dataset, the Amazon datasets are paragraph-level review, which places higher demands on the ability of the model to understand sentence semantics, including the semantic relationships between contextual sentences.

Table 5. Automatic evaluation results on the Amazon datasets.

Amazon-15	Accuracy	Self-Bleu	Self-Bleu3	Self-Bleu4	WMD
(Hu et al.)	81.46%	62.65	57.00	47.90	0.17
(Shen et al.)	**85.85%**	3.96	2.30	0.70	0.81
KeepN	74.61%	11.98	9.65	4.50	0.63
KeepNV	55.60%	28.91	28.80	17.80	0.40
Styleword	66.12%	31.20	28.30	18.10	0.39
Amazon-50	Accuracy	Self-Bleu	Self-Bleu3	Self-Bleu4	WMD
KeepN	**87.08%**	8.32	6.00	2.20	0.60
KeepNV	56.53%	24.17	25.10	14.50	0.40
Styleword	71.40%	26.45	25.20	14.90	0.39

The results of automatic evaluations are shown in Table 5. We only compared other baseline models on the Amazon-15, because their models exhibited different degrees of model collapse when training on the Amazon-50 dataset. In the Amazon-15 dataset, we can see from the automatic evaluation results that the accuracy of Shen et al. [5] is the highest, and the accuracy of both Hu et al. [4] and our models is lower than the Yelp dataset. With respect to our method, this is because paragraph-level reviews usually contain several sentences to describe different objects, and these sentences may express inconsistent sentiments on different objects. For example, a negative review in the test dataset 'the restaurant was relatively clean. the service was nonexistent.', this review was scored 1 point due to strong dissatisfaction with the service. This kind of sample with sentiment noise has a very serious impact on our model. We can see that Shen et al. [5] performs worse on the Amazon-15 dataset than it performs on the Yelp dataset in terms of content preservation reflected by all of the Self-Bleu scores and the WMD value.

From the human evaluation results presented in Table 6, we can still see the superiority of our three methods. The results are basically consistent with their performance on the Yelp dataset, which reflects the robustness of our methods. Notice that our methods are still the best on the overall score.

In the Amazon-50 dataset, only our three methods work; the other models collapse. The model of Hu et al. [4], in order to optimize the discriminator loss, reconstruction losses rose sharply, which cause the model to collapse (refer to

Table 6. Human evaluation results on the Amazon datasets.

Amazon-15	Rea-S	Con-S	Sen-S	Ove-S
(Hu et al.)	0.600	1.700	0.900	0
(Shen et al.)	0.900	0.350	0.750	0
KeepN	1.100	1.600	**1.250**	0.200
KeepNV	**1.200**	1.850	1.000	**0.250**
Styleword	1.050	**1.950**	1.100	0.150
Amazon-50	Rea-S	Con-S	Sen-S	Ove-S
KeepN	0.800	2.500	**2.250**	**0.050**
KeepNV	**1.350**	**3.550**	1.000	0
Styleword	0.450	1.700	0.650	0

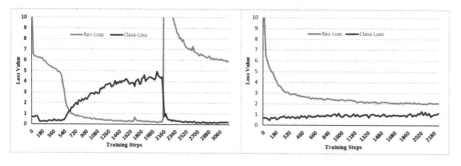

Fig. 2. Loss curve of Hu et al. [4] (left) and our 'KeepN' (right). Rec-loss represents the text reconstruction loss, and Class-loss represents the sentiment loss given by discriminator. The model only optimizes the rec-loss before step 2160. When the optimization of class-loss begins, the rec-loss rises sharply and makes the model collapse. Compared to Hu, our model training is stable and does not collapse.

Fig. 2). In a long review, sentiments are usually expressed through the description of events, and the transfer of the polarity of sentiments usually changes the original expression of the sentence. Therefore, we can conclude that in the transfer of long text sentiment under unsupervised conditions, using complete texts as input, most methods are affected to some extent by the reconstruction loss optimization, resulting in generated text that is poorly readable. So, finding the key content information in a sentence and expressing it with different sentiments through alternative expressions offers a solution to this problem. In addition, since the GAN does not perform well on discrete data, it is more difficult to optimize for longer paragraph-level text.

As can be seen from the results in Table 5, the method 'KeepN' achieves the highest accuracy, the 'KeepNV' and 'Styleword' results are similar to their performance on the Yelp dataset. In the human evaluation results, the overall score of each method has dropped significantly. It shows that long sentence sentiment transfer is still an important research direction in the future.

5   Conclusion

In this paper, we point out that previous work used sentence-level representation are usually entangled, which resulting transferred sentence is usually unreasonable, we propose a novel method to address the problem that using word-level representation. The experiments show that our method achieves the state-of-the-art performance in automatic and human evaluation results. However, there are also shortcomings in this type of method. The sentiment noise in keywords has a significant impact on the model's performance. In future work, we would like to explore other kinds of possibilities ways to separate sentiment-independent content information more accurately from the semantic information of the sentence.

References

1. Bahdanau, D., Cho, K., Bengio, Y.: Neural machine translation by jointly learning to align and translate (2014). arXiv:1409.0473v7
2. Wen, T.H., Vandyke, D., Mrksic, N., et al.: A network-based end-to-end trainable task-oriented dialogue system. In: Proceedings of the 15th Conference of the European Chapter of the Association for Computational Linguistics, pp. 438–449. EACL, Valencia (2017)
3. Fu, Z., Tan, X., Peng, N., et al.: Style transfer in text: exploration and evaluation. In: Thirty-Second AAAI Conference on Artificial Intelligence, pp. 663-670. AAAI, San Francisco (2018)
4. Hu, Z., Yang, Z., Liang, X., et al.: Toward controlled generation of text. In: Proceedings of the 34th International Conference on Machine Learning, pp. 1587–1596. PMLR, Sydney (2017)
5. Shen, T., Lei, T., Barzilay, R., et al.: Style transfer from non-parallel text by cross-alignment. In: Proceedings of 31st Conference on Neural Information Processing Systems, pp. 6830–6841. NeurIPS, Long Beach (2017)
6. Yang, Z., Hu, Z., Dyer, C., Berg-Kirkpatrick, T., et al.: Unsupervised text style transfer using language models as discriminators. In: Proceedings of 32st Conference on Neural Information Processing Systems, pp. 7298–7309. NeurIPS, Montré al (2018)
7. Xu, J., Sun, X., Zeng, Q., et al.: Unpaired sentiment-to-sentiment translation: a cycled reinforcement learning approach. In: Proceedings of the 56th Annual Meeting of the Association for Computational Linguistics, pp. 979–988. ACL, Melbourne (2018)
8. John, V., Mou, L., Bahuleyan, H., et al.: Disentangled representation learning for text style transfer. In: Proceedings of the 57th Conference of the Association for Computational Linguistics, pp. 424–434. ACL, Florence (2019)
9. Zhao, Y., Bi, W., Cai, D., et al.: Language style transfer from sentences with arbitrary unknown styles (2018). arXiv preprint arXiv:1808.04071
10. Gong, H., Bhat, S., Wu, L., et al.: Reinforcement learning based text style transfer without parallel training corpus. In: Proceedings of the 2019 Conference of the North American Chapter of the Association for Computational Linguistics, pp. 3168–3180. NAACL-HLT, Minneapolis (2019)

11. Prabhumoye, S., Tsvetkov, Y., Salakhutdinov, R., et al.: Style transfer through back-translation. In: Proceedings of the 56th Annual Meeting of the Association for Computational Linguistics (Long Papers), pp. 866–876. ACL, Melbourne (2018)

12. Yu, L., Zhang, W., Wang, J., et al.: Seqgan: sequence generative adversarial nets with policy gradient. In: Proceedings of the Thirty-First Conference on Artificial Intelligence, pp. 2852–2858. AAAI, San Francisco (2017)

13. Luo, F., Li, P., Yang, P., et al.: Towards fine-grained text sentiment transfer. In: Proceedings of the 57th Annual Meeting of the Association for Computational Linguistics, pp. 2013–2022 (2019)

14. Li, J., Jia, R., He, H., et al.: Delete, retrieve, generate: a simple approach to sentiment and style transfer. In: Proceedings of the 16th Conference of the North American Chapter of the Association for Computational Linguistics: Human Language Technologies, pp. 1865–1874. NAACL, New Orleans (2018)

15. Lample, G., Subramanian, S., Smith, E., et al.: Multiple-attribute text rewriting. In: ICLR (Poster) (2019)

16. Liao, Y., Bing, L., Li, P., et al.: Quase: sequence editing under quantifiable guidance. In Proceedings of the 2018 Conference on Empirical Methods in Natural Language Processing, pp. 3855–3864 (2018)

17. Hu, M., Liu, B.: Mining and summarizing customer reviews. In: Proceedings of the Tenth International Conference on Knowledge Discovery and Data Mining, pp. 168–177. SIGKDD, Seattle (2004)

18. McAuley, J.J., Leskovec, J.: From amateurs to connoisseurs: modeling the evolution of user expertise through online reviews. In Proceedings of the 22nd International Conference on World Wide Web, pp. 897–908. WWW, Rio de Janeiro (2013)

19. Kim, Y.: Convolutional neural networks for sentence classification. In: Proceedings of the 2014 Conference on Empirical Methods in Natural Language Processing, pp. 1746–1751. EMNLP, Doha (2014)

20. Mir, R., Felbo, B., Obradovich, N., et al.: Evaluating style transfer for text. In: Proceedings of the 2019 Conference of the North American Chapter of the Association for Computational Linguistics: Human Language Technologies, pp. 495–504. NAACL-HLT, Minneapolis (2019)

21. Hu, Z., Shi, H., Yang, Z., et al.: Texar: a modularized, versatile, and extensible toolkit for text generation. In: Proceedings of the 57th Conference of the Association for Computational Linguistics, pp. 159–164. ACL, Florence (2019)

Robotics and Drones

Path Planning of Mobile Robot Group Based on Neural Networks

Mikhail Medvedev$^{(\boxtimes)}$ and Viacheslav Pshikhopov

R&D Institute of Robotics and Control Systems, Taganrog, Russia
medvmihal@sfedu.ru

Abstract. This article is devoted to development of a neural control system for group of drones. The control system estimates the environmental complexity, selects the best path planning algorithm, and plans the path of the copters. Algorithm of the complexity estimation is a logical classification of the environmental state. The complexity depends of obstacle's location in the environment. Selection of the best path planning algorithm is a neural classification of the environment. The neural network selects the best path planning algorithm from two algorithms. The first one is the shortest path planning algorithm, and the second one is the safest path planning algorithm. A hybrid learning algorithm is proposed for the neural classifier of the environment. This learning algorithm consists of a supervised and an unsupervised unit. Evaluation of the neural classifier is performed by the multi criterion including the motion time, the path length, and the minimal distance from the path to the obstacles. The path planning algorithms are implemented by neural networks also. These networks are trained by supervised algorithms. The neural planner output is the next desired location of the drone, but not whole path to the target point. This approach allows the group of drones to operate in a dynamical environment without recalculating the whole path. Also this research proposes algorithms of drone's formation in an uncertain unmapped 3-D environment.

Keywords: Path planning · Group control · Neural network · Machine learning

1 Introduction

A group control is the most urgent problem of robotics. In papers [1–4] effectively solved by group of robots tasks are studied. Planning and synchronization of robot's actions are the most important problems of a group control. The reviews of group and cooperative control approaches are presented in works [1–4].

The problem of planning is solved at a sufficiently high level in preliminary mapped environments [5]. In unmapped and obstructed environments robots are supervised by operator. Therefore the methods of planning for uncertain obstructed environments are to be improved. Researches [1–5] note the main problems of group control, including uncertainty, dynamic of environment, and high computational cost of the optimal path calculation. A significant number of papers are devoted to the problem of group control in uncertain environment. Most of the papers consider 2-D environment. But for 3-D

© Springer Nature Switzerland AG 2020
H. Fujita et al. (Eds.): IEA/AIE 2020, LNAI 12144, pp. 51–62, 2020.
https://doi.org/10.1007/978-3-030-55789-8_5

environment the problem of group of robots motion planning is more difficult. Artificial intelligence and bio-inspired algorithms are the perspective means to solve the problem of group control.

Article [6] is the review of planning algorithms for three-dimensional environments. A taxonomy and description of analytical and bio-inspired algorithms are presented. The problem of path planning for a group of robots is studied also.

In paper [7] the method of path planning for an aircraft - type UAV is proposed. The method divides 3-D space into cells by Voronoi diagrams and Delaunay triangulation. Then genetic algorithm searches the optimal path in the cellular space. The research of the proposed algorithms is performed by numerical simulations. The article considers an algorithm for planning the movement of a single UAV. The cost of passing areas of the environment is minimized. However, the issue of converting the cost into physically interpreted indicators is not considered.

Proposed in research [8] hybrid algorithm includes a deliberative unit and a reactive bio-inspired. The hybrid algorithm is used for the group control of mobile robots. The problem of coordinated movement of the group of drones in an uncertain three-dimensional environment is investigated. Formation of the group is designed by Delaunay triangulation and positions of the robots optimization. Preliminary positions of the robots are not determined. The bio-inspired algorithm based on unstable modes avoids collisions. The proposed algorithms are implemented in the control system of the group of drones. The group operates in an environment with stationary obstacles. In report [9] presented in paper [8] algorithms are enhanced for environments with moving obstacles. In this paper, the group is considered a 2-D environment. In addition, the speed of the group is not depended on the state of the environment.

Different algorithms of a swarm and group control for 3-D environments are developed [10–12]. In report [10] a PSO algorithm is proposed. The algorithm ensures low computational cost. In article [11] a pigeon algorithm is proposed. The algorithm plans path of the group of UAV's. Paper [12] solves the problem of path planning in uncertain dynamical environments by a wolf flock algorithm. Proposed algorithm takes into account a distance to the target point, and distances between the robots and obstacles. Simulation results for the group of drones fly in a three-dimensional environment are presented. The considered methods use search algorithms. They are not always applicable in practice, due to physical limitations.

Neural network-based control systems are a perspective field of robotics [13]. Deep learning networks are promising means of an intelligent control nowadays [14]. Different problems of a group control are solved by neural networks [15–22].

The group of robots motion problem is studied in article [15]. Neural controller solves the problem in an uncertain environment with obstacles. Neural network tunes the coefficients of the adaptive controller. The group of robots achieves a target area, saves a given formation, and avoids collisions. The paper uses a pre-set formation and considers only flat motion.

In article [16] the problem of an uncertain environment survey is considered. Developed neural network works out the decentralized searching algorithm based on pairwise interaction. The simulation results for a disaster area survey by the group of robots are presented. The advantages of proposed algorithms are demonstrated.

Agrawal P. and Agrawal H., [17] solve the problem of the evasive objects group persecution. Hunters have to recognize the friend group robots, and perform the path planning to persecution of the evasive objects. Developed neural network solves the persecution problem taking into account the delays and limitations of communication channels. The network predicts the evasive objects motion. The proposed approach advantages in an obstructed environment are demonstrated by numerical simulations.

In paper [18] the problem of group detection and tracking of people and animals by drones is solved. Deep learning neural network detects the objects on the board of a single drone. The proposed algorithms performance is demonstrated by numerical simulations and experiments.

In article [19] the decentralized robotic system of a store is considered. The control system of a robot consists of two neural networks. The first network classifies the state of environment into five classes. Every class corresponds to different actions of the robot. The second neural network detects errors. Performance of the developed robotic system is estimated by numerical simulations.

Wang et al. [20] solve the problem of robot's formation constructing. The formation has to be optimal for a given shape. The solution is based on a recurrent neural network application. The problem of formation constructing is transformed to the convex optimization problem for a non-smooth criterion. Proposed method is studied by numerical simulations and experiments.

Paper [21] presents the motion control system of a wheeled mobile robot in an uncertain environment. Deep learning networks plan the robot path. The novel cascade topology of the neural network is developed. In different situations the different number of cascades is involved into planning process. Simulation results and experiments are presented for the wheeled robot in uncertain obstructed environment.

In papers [16–21], a 2-D stationary environment is considered. This the main limitation of these papers.

Article [22] presents an onboard machine vision system with an aim of navigating and controlling an unmanned aerial vehicle. New original methods are proposed for multiband images fusion based on diffuse morphology. The original methods are developed for deep machine learning.

This paper presents the development of a neural based group control system. The algorithms for estimation of an environment complexity, searching for the best planning algorithm, and planning the path of drones for 3-D environment are described.

2 Problem Statement

The group consists of N_R drones. The mathematical model of the drone is described as follows [8, 9].

$$\dot{y}_i = R(y_i)x_i, \tag{1}$$

$$M_i\dot{x}_i(t) = B_iu_i + F_{di}, \tag{2}$$

where $y_i = [y_{1i}\ y_{2i}\ y_{3i}\ y_{4i},\ y_{5i},\ y_{6i}]^T$ is a vector of position (y_{1i}, y_{2i}, y_{3i}) and orientation (y_{4i}, y_{5i}, y_{6i}) of i-th drone in the fixed frame $O_gY_{1g}Y_{2g}Y_{3g}$ (Fig. 1), $x_i = [x_{1i}\ x_{2i}\ x_{3i}$

x_{4i}, x_{5i}, $x_{6i}]^T$ is a vector of linear (x_{1i}, x_{2i}, x_{3i}) and angular (x_{4i}, x_{5i}, x_{6i}) velocities of i-th robot in the moving frame $OY_1Y_2Y_3$, $R(y_i)$ is the matrix of kinematics described by Euler's angles, M_i is the matrix of i-th robot inertia parameters, F_{di} is the vector of i-th robot dynamical and external forces and torques, u_i is a vector of i-th robot control inputs, B_i is the input matrix of i-th robot, $i = 1, 2, ..., N_R$; N_R is number of robots in the group.

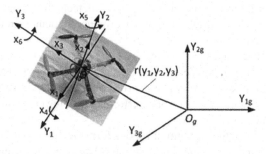

Fig. 1. Frames of a drone.

The navigation system of the drone measures vectors y_i and x_i. The vision system of the drone detects obstacles located at distances from 0 to R_{vs}. Field of view of the vision system is bounded by angles α_{vs} and β_{vs}. Likewise the communication system of the drones ensures communication with each other.

3-D environment consists of three areas Ω_0, Ω_1, and Ω_2 shown in Fig. 2.

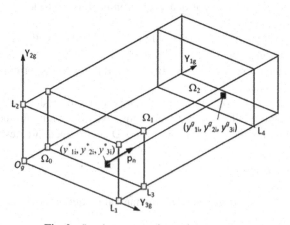

Fig. 2. Starting, uncertain, and target areas.

At start time $t = t_0$ the drones are within area Ω_0. Area Ω_2 is a target area. All drones must be in the target area Ω_2. Area Ω_1 is an uncertain environment with moving and stationary obstacles. At time $t = t_0$ information about the obstacles in area Ω_1 is absent in the control system of the drones. Thus task of the group is to move from the

Ω_0 through area Ω_1 to area Ω_2. Also the drones have to locate in area Ω_0 such way to optimize a given criterion. The drones transmit each other vectors y_i, x_i, distances to obstacles $r_{obs}(i, j)$, and vector E_i, describing area Ω_1.

Block diagram of the group control system is shown in Fig. 3. The inputs of Planning System block are estimation of environmental state C_j (j is number of other group robots), planning algorithms E_j, positions and velocities (y_j, x_j) of the drones, positions and velocities (y_{oj}, x_{oj}) of the detected by drones obstacles, data (y_{iN}, x_{iN}) of the navigation system (Navigation System), and data (y_{oi}, x_{oi}) of the vision system (Vision System). The outputs of Planning System block are desired position and velocity (y^*_i, x^*_i) of i-th drone. The outputs of Planning System block feed to the motion control system block of i-th drone (Motion Control System block). The output of Motion Control System block is a vector of the control actions u_i.

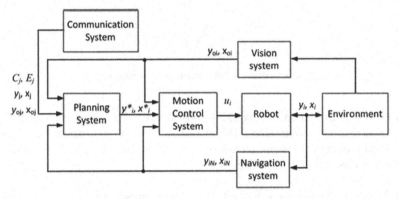

Fig. 3. Block diagram of the planning and control system.

The objectives of the presented planning and control system are described as follows. The first objective is estimation of the environmental state in terms of the presence and location of obstacles in the field of the vision system. The second objective is calculation of the optimal position y^*_i of i-th drone. The optimization criterion is described in Sect. 3. The third objective is the best path planning algorithm searching in the database of available algorithms. The fourth objective is the path planning of the drones from current position to target point y^*_i. The fifth objective is calculation of desired velocity x^*_i of movement to target point y^*_i.

3 Algorithms of the Environmental State Estimation and the Optimal Position Calculation

Estimation of the environmental state is classification based on data of the vision system. The classification result is integer C_i. If obstacles are not detected, then integer C_i is equal to 0. If obstacles are detected by the vision system, sector I is free (Fig. 4), and the distance from the robot to the obstacles are safe, then integer C_i is equal to 1. If obstacles are detected by the vision system and sector I is occupied, then integer C_i is equal to 2.

If all sectors are occupied by obstacles, then integer C_i is equal to 3. Thus integer C_i is estimation of local area Ω_1 complexity in term of geometry [23].

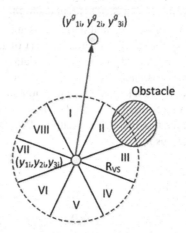

Fig. 4. Estimation of the environmental complexity.

Formation of the group in area Ω_0 is constructed by Delaunay triangulation and optimization of the robots locations. This algorithm for a two-dimensional environment is proposed in paper [8]. In this paper a three-dimensional environment Ω_0 is considered. The algorithm includes the following steps.

Step 1. The aggregated group is formed. This group consists of the drones, detected by the vision system obstacles, and vertices of area Ω_0. In Fig. 2 the vertices are presented as white squares.
Step 2. Neighbors of i-th drone are calculated using Delaunay triangulation [10, 11]. The result of Delaunay triangulation is set M_{Ri} of adjacent objects to which i-th drone is connected by edges.
Step 3. The distances from i-th robot to the adjacent objects are calculated.

$$r_{ij} = k_{ij} \left[\left(y_{1i} - y_{1j}\right)^2 + \left(y_{2i} - y_{2j}\right)^2 + \left(y_{3i} - y_{3j}\right)^2 \right]^{0.5}, \tag{3}$$

where $i = \overline{1, n_i}$, $j = \overline{1, N_G}$, n_i is amount of adjacent to i-th drone objects, r_{ij} is the distance from i-th drone to j-th object.
Weight coefficients k_{ij} depends from integer C_i. Coefficients k_{ij} tends to 1 if $C_i = 0$, and tends to 0 if $C_i = 3$. The values of coefficients k_{ij} for $C_i = 1$ and $C_i = 2$ are determined by numerical or experimental investigations.
Step 4. Distances (3) are optimized. Free variables are y_{i1}, y_{i2}, y_{i3}. The optimization criterion is described as follows.

$$\left(\min_{i,j}\left(r_{ij}\right) \right) \rightarrow \max_{y_{1i}, y_{2i}, y_{3i}} . \tag{4}$$

4 Path Planning Algorithms

The problem of the best path planning algorithm searching is described as follows. There are two algorithms in the database. The first one is the shortest path planning algorithm. The second one is the safest path planning algorithm.

The planned path is evaluated by the following criterion.

$$J_{\Sigma} = k_1 J_1 + k_2 J_2 + k_2 J_3 \rightarrow \max, \tag{5}$$

where k_1, k_2, k_3 are weight coefficients determined by experts, $k_1 + k_2 + k_3 = 1, J_1$ is a normalized time of the movement to target point taken with the opposite sign, J_2 is a minimal normalized distance between the drones and obstacles, J_3 is a normalized path length taken with the opposite sign.

The problem is to search the optimal algorithm in term of criterion (5) maximum. The searching is based on the vision system data. A deep learning network processes the data from vision system of the drone. Block diagram of the network learning is shown in Fig. 5. It could be used for a supervised and unsupervised learning [24].

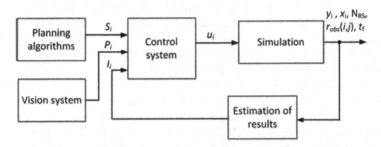

Fig. 5. Block diagram of the neural network learning.

The inputs of the neural network are image P_i from the vision system and planning algorithm S_i. Based on the results of simulations, the evaluation of criterion J_{Σ} (5) is formed. Criterion J_{Σ} (5) also feeds the control system. Planning algorithm S_i is D* or modified D*. Data from the vision system feeds to the control system as a matrix. Elements of the matrix are distances from the robot to the obstacle (Fig. 6). The matrix dimensions are $N\alpha + 2 \times N\beta + 2$. If there is no obstacle in the cell, then in the corresponding element of the matrix is 255. The first column and the last row of the matrix contain the drone position. The second column and the second last row of the matrix contain the drone target position.

The training sample should contain all the situations required for training, and in sufficient quantity. In practice different situations occur with different frequency. Thus, accumulation of rare situations is possible either through their special creation or a significant increase in the training sample. Special creation of the samples for training is a complex, hard, and laborious process. In fact, it is necessary to determine the most important situations manually. Control of accumulation of the training sample is also a manual operation. Random accumulation of the training sample leads to a significant increase of the training sample.

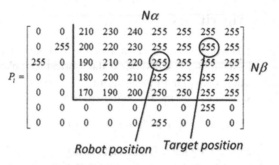

Fig. 6. A matrix representation of the environment.

In this article the following iteration algorithm is proposed.

1) Sample P_k is generated in step k of the iterations.
2) Algorithm D* (Modified D*) calculates the drone path for every element of learning sample P_k.
3) Simulation of the drone's motion is performed for D* (Modified D*). Dependence of the drones's velocity from the distance to obstacles is simulated also.
4) Results of the simulation are values of criterion J_Σ (5) for D* (Modified D*).
5) Generated elements of sample P_k are classified into two classes depending from the values of criterion J_Σ (5) for D* and Modified D*. If the value of criterion J_Σ (5) for D* is more than the value of criterion J_Σ (5) for Modified D*, then the elements of sample P_k belongs to the class 1. In other case one belongs to the class 2.
6) If $k > 1$, then the element of sample P_k feeds to neural network NN_{k-1} trained in the previous step. If network NN_{k-1} classifies the element of sample P_k correctly, then one is not saved to sample P_k. In other case the element is saved to sample P_k.
7) Iteration k of the neural network learning is performed using sample P_k.
8) The iterations are interrupted if the desired accuracy of learning is achieved.

Proposed algorithm selects situations for further learning that the neural network classifies with errors. The algorithm is similar to reinforcement learning [24] because selection of the sample elements P_k is performed by results of the network operating.

Paper [25] also proposes a similar algorithm implemented by two neural agents. Therefore it is necessary to synchronize the learning process of the two neural networks. Unlike the work [25] in our algorithm, the neural network trains itself.

Optimal planning algorithms require significant computational costs because an optimal path has to be calculated as a whole starting from a target point. There are algorithms of the computational costs decreasing. But efficiency of these algorithms is low for dynamical environments or moving target point. In this paper proposed neural planner produces not the entire trajectory. The planner calculates the current direction of movement. In other words the network plans a local (suboptimal) path. Decreasing the computational cost of path calculation algorithm by planning a suboptimal path is a widespread approach [26, 27].

In this article a reference algorithm is D* [28, 29]. The first algorithm is the shortest path finding algorithm D* for the original map. The second algorithm is the path finding algorithm D* for the transformed map of environment. The weights of the cells in the vicinity of obstacles are increased in the transformed map. Algorithm D* (Modified D*) classifies shown in Fig. 6 matrix into 26 categories classes [29]. Each category corresponds to one of the cells through which the planned path passes. Classified matrices or images are used for the neural networks learning. The neural networks are learned by supervisor (D* or Modified D*).

Planned path feeds to the position-path motion control system described in book [30]. The position-path control system includes the block of unstable modes of movement in the vicinity of obstacles [31]. The unstable modes block allows avoid obstacles in uncertain dynamical environments.

5   The Results of Simulations and Experiments

Planning algorithm selection is implemented on the first deep neural network. The first network consists of an input layer, ten convolutional layers, and an output layer. The number of the input layer neurons is $N\alpha + 2 \times N\beta + 2$. Parameters $N\alpha = 180$ and $N\beta = 30$ are horizontal and vertical resolutions of the vision system. The numbers of the convolutional layers filters are 64 for layer 1, 64 for layer 2, 128 for layer 3, 128 for layer 4, 256 for layer 5, 256 for layer 6, 256 for layer 7, 512 for layer 8, 512 for layer 9, and 512 for layer 10. The numbers of the output layer neurons are 2.

The first network is learned by algorithm 1)–8) of Sect. 4 and criterion J_Σ (5). Values of the criterion coefficients are $k_1 = 1, k_2 = k_3 = 0$. Thus the performance of the control system is evaluated by the time of the drone's movement.

Velocity of the drone is calculated in accordance of the following expression.

$$
V_R = \begin{cases} V_{\min} + (V_{\max} - V_{\min})\frac{r}{r_s}, & V_R \leq V_{\max}, \\ V_{\max}, & V_R > V_{\max}, \end{cases} \tag{6}
$$

where r are the distance to obstacle, $r_s = 3$ m is the safety distance, $V_{min} = 0.25$ m/s is the minimal velocity of the group, $V_{max} = 1.5$ m/s is the maximal velocity of the group.

During the learning of the first network 4 iterations were performed. Created sample includes about 50 000 elements. Accuracy of the learning is about 90%. For comparison, learning accuracy on an unfiltered sample of 50,000 elements is about 78%.

Similar neural networks are used as path planners. The second network is supervised by D*. The third network is supervised by Modified D*. A sample of 100,000 elements has been created. The resulting accuracy is about 90%.

Simulation results of the developed neural control system for the group including five drones are shown in Fig. 7a.

Starting locations of the drones are [2; 1; 1], [2; 2; 2], [4; 3; 1], [5; 2; 2], and [6; 1; 1]. Vertices of area Ω_0 are [0; 0; 0], [0; 0; 5], [0; 5; 5], [0; 5; 0], [10; 5; 0], [10; 5; 5], [10; 0; 5] and [10; 0; 0]. Area Ω_0 is moving along axis Oy_2 with velocity 1 m/s. Locations of the obstacles are absent in the control system of drones. The group performs two maneuvers during the movement. The reference positions of the drones are the result of minimax problem (4) solution. The second and the third neural networks calculate paths to the reference positions.

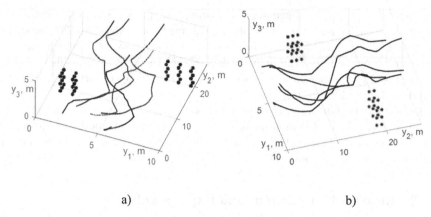

a)   b)

Fig. 7. Simulation results for the group of drones flight.

Influence of coefficients k_{ij} (3) is evaluated for the following values: $k_{ij} = 1.0$ for $C_i = 0$; $k_{ij} = 0.9$ for $C_i = 1$; $k_{ij} = 0.8$ for $C_i = 2$; $k_{ij} = 0.7$ for $C_i = 3$. The simulation results are shown in Fig. 7b. If this case the whole group is concentrated in a central area of the environment where there are no obstacles.

During the simulation, the influence of the planning algorithm selection on the motion time was evaluated. Without the selection averaged time of the group motion is about 24 s. With the first neural network averaged time is about 22.7 s. Thus the time of the group motion is decreased by 5.5%.

6 Conclusions

The novelty of the proposed algorithms is described as follows.

Algorithm of the group formation constructing is proposed. The algorithm includes 3-D Delaunay triangulation and optimization of i-th drone position. This result is development of algorithms presented at [11], but in this article a maximin optimization problem for a 3-D environment is used.

Coefficients k_{ij} as function of the state complexity are introduced. These coefficients shift the drones from obstacles to free areas of a 3-D environment.

The neural system of the planning algorithm selection is developed. Due to the planning algorithm selection it is possible to increase performance of the control system. Numerical simulation shows the time of the group motion is decreased by 5.5%.

Neural path planning algorithm is proposed for 3-D environment. The algorithm is characterized by the use of a neural network that produces the current direction of movement. The algorithm decreases time of the path planning by three times. The effect is achieved for 3-D uncertain dynamical environments.

A new iterative procedure is proposed for training a neural network. It consists of filtering a new training sample at each iteration. This reduces the sample size required by the network for training.

Acknowledgment. The study is supported by Russian Science Foundation, grant 16-19-00001, executed at Southern Federal University.

References

1. Cao, Y.U., Fukunaga, A.S., Kahng, A.B.: Cooperative mobile robotics: antecedents and directions. Auton. Robots **4**, 1–23 (1997)
2. Tan, Y., Zheng, Z.-Y.: Research advance in swarm robotics. Defense Technol. **9**(1), 18–39 (2013)
3. Francesca, G., Birattari, M.: Automatic design of robot swarms: achievements and challenges. Front. Robot. AI. **3**(1) (2016)
4. Rizk, Y., Awad, M., Tunstel, E.W.: Cooperative heterogeneous multi-robot systems: a survey. ACM Comput. Surv. **52**(2) (2019)
5. LaValle, S.: Planning Algorithms, p. 842. University Press, Cambridge (2006)
6. Yang, L., Qi, J., Song, D., Han, J., Xia, Y.: Survey of robot 3D path planning algorithms. J. Control Sci. Eng. **2016**, 22 (2016)
7. Qu, Y., Zhang, Y., Zhang, Y.: A global path planning algorithm for fixed-wing UAVs. J. Intell. Rob. Syst. **91**(3), 691–707 (2017). https://doi.org/10.1007/s10846-017-0729-9
8. Pshikhopov, V., Medvedev, M.: Group control of autonomous robots motion in uncertain environment via unstable modes. SPIIRAS Proc. **60**(5), 39–63 (2018)
9. Pshikhopov, V., Medvedev, M., Kulchenko, A., Lazarev, V.: The decentralized control method for a multicopters group using unstable modes. In: ACM International Conference Proceeding Series. Proceedings of 2019 11th International Conference on Computer and Automation Engineering, Perth, Australia, 23–25 February, pp. 93–96 (2019)
10. Shi, Y.: Particle swarm optimization: developments, applications and resources. In: International Conference on Evolutional Computations, vol. 1, pp. 81–86 (2001)
11. Duan, H., Qiao, P.: Pigeon-inspired optimization: a new swarm intelligence optimizer for air robot path planning. Int. J. Intell. Comput. Cybern. **7**(1), 24–37 (2014)
12. Radmanesha, M., Kumar, M., Sarim, M.: Grey wolf optimization based sense and avoid algorithm in a Bayesian framework for multiple UAV path planning in an uncertain environment. Aerosp. Sci. Technol. **77**, 168–179 (2018)
13. Gupta, M.M., Jin, L., Homma, N.: Static and Dynamic Neural Networks: From Fundamentals to Advanced Theory. Wiley, Hoboken (2003)
14. LeCun, Ya., Yoshua, B., Geoffrey, H.: Deep learning. Nature. **521**, 436–444 (2015)
15. Yu, J., Ji, J., Miao, Z., Zhou, J.: Neural network-based region reaching formation control for multi-robot systems in obstacle environment. Neurocomputing **333**, 11–21 (2019)
16. Geng, M., Xu, K., Zhou, X., Ding, B., Wang, H., Zhang, L.: Learning to cooperate via an attention-based communication neural network in decentralized multi-robot exploration. Entropy **21**(3), 294 (2019)
17. Agrawal, P., Agrawal, H.: Adaptive algorithm design for cooperative hunting in multi-robots. Int. J. Intell. Syst. Appl. **10**(12), 47–55 (2018)
18. Price, E., Lawless, G., Ludwig, R., Martinovic, I., Bulthoff, H.H., Black, M.J., Ahmad, A.: Deep neural network-based cooperative visual tracking through multiple micro aerial vehicles. IEEE Robot. Autom. Lett. **3**(4), 3193–3200 (2018)
19. Martínez-García, E.A., Torres-Córdoba, R., Carrillo-Saucedo, V.M., López-González, E.: Neural control and coordination of decentralized transportation robots. Proc. Inst. Mech. Eng. Part I. J. Syst. Control Eng. **232**(5), 519–540 (2018)

20. Wang, Y., Cheng, L., Hou, Z.-G., Yu, J., Tan, M.: Optimal formation of multirobot systems based on a recurrent neural network. IEEE Trans. Neural Netw. Learn. Syst. **27**(2), 322–333 (2016)
21. Pshikhopov, V., Medvedev, M., Vasileva, M.: Neural network control system of motion of the robot in the environment with obstacles. In: Wotawa, F., Friedrich, G., Pill, I., Koitz-Hristov, R., Ali, M. (eds.) IEA/AIE 2019. LNCS (LNAI), vol. 11606, pp. 173–181. Springer, Cham (2019). https://doi.org/10.1007/978-3-030-22999-3_16
22. Knyaz, V.A., Vishnyakov, B.V., Gorbatsevich, V.S., Vizilter, Y., Vygolov, O.V.: Intelligent data processing technologies for unmanned aerial vehicles navigation and control. SPIIRAS Proc. **45**(2), 26–44 (2016)
23. Karkishchenko, A.N., Pshikhopov, V.: On finding the complexity of an environment for the operation of a mobile object on a plane. Autom. Remote Control **80**(5), 897–912 (2019)
24. Li, Z., Zhao, T., Chen, F., Hu, Y., Su, C.-Y., Fukuda, T.: Reinforcement learning of manipulation and grasping using dynamical movement primitives for a humanoid like mobile manipulator. IEEE/ASME Trans. Mechatron. **23**(1), 121–131 (2018)
25. Zoph, B., Vasudevan, V., Shlens, J., Le, Q.V.: Learning transferable architectures for scalable image recognition. In: Proceeding of 2017 IEEE/CVF Conference on Computer Vision and Pattern Recognition, Honolulu, Hawaii, 22–25 July 2017 (2017)
26. Han, J.: An efficient approach to 3D path planning. Inf. Sci. **478**, 318–330 (2019)
27. Lavrenov, R.O., Magid, E.A., Matsuno, F., Svinin, M.M., Sutakorn, J.: Development and implementation of spline-based path planning algorithm in ROS/gazebo environment. SPIIRAS Proc. **18**(1), 57–84 (2019)
28. Stentz, A.: Optimal and efficient path planning for partially-known environments. In: Proceedings of the International Conference on Robotics and Automation, pp. 3310–3317 (1994)
29. Carsten, J., Ferguson, D., Stentz, A.: 3D Field D*: improved path planning and replanning in three dimensions. In: Proceedings of the 2006 IEEE/RSJ International Conference on Intelligent Robots and Systems, Beijing, China, pp. 3381–3386 (2006)
30. Pshikhopov, V., Medvedev, M.: Position-path control of a vehicle. In: Path Planning for Vehicles Operating in Uncertain 2D Environments, pp. 1–23. Butterworth-Heinemann, Oxford (2017)
31. Pshikhopov, V., Medvedev, M.: Motion planning and control using bionic approaches based on unstable modes. In: Path Planning for Vehicles Operating in Uncertain 2D Environments, pp. 239–280. Butterworth-Heinemann, Oxford (2017)

Push Recovery and Active Balancing for Inexpensive Humanoid Robots Using RL and DRL

Amirhossein Hosseinmemar[1(✉)], John Anderson[1], Jacky Baltes[1,2], Meng Cheng Lau[1], and Ziang Wang[1]

[1] University of Manitoba, Winnipeg, Canada
amirhossein.memar@gmail.com
[2] National Taiwan Normal University, Taipei, Taiwan
http://aalab.cs.umanitoba.ca

Abstract. Push recovery of a humanoid robot is a challenging task because of many different levels of control and behaviour, from walking gait to dynamic balancing. This research focuses on the active balancing and push recovery problems that allow inexpensive humanoid robots to balance while standing and walking, and to compensate for external forces. In this research, we have proposed a push recovery mechanism that employs two machine learning techniques, Reinforcement Learning and Deep Reinforcement Learning, to learn recovery step trajectories during push recovery using a closed-loop feedback control. We have implemented a 3D model using the Robot Operating System and Gazebo. To reduce wear and tear on the real robot, we used this model for learning the recovery steps for different impact strengths and directions. We evaluated our approach in both in the real world and in simulation. All the real world experiments are performed by Polaris, a teen-sized humanoid robot.

Keywords: Inexpensive humanoid robots · Push recovery · Active balancing · Reinforcement learning · Deep reinforcement learning

1 Introduction and Related Work

One of the ultimate goals of the human is to instruct robots to do variety of tasks in human environments, e.g., home or work. One of the main requirement for such a robot is to be able to follow a human everywhere, and can walk on different surfaces. It should also be able to recover from external pushes while it's standing or walking. Over decades researchers have introduced and implemented different state of the art approaches to this specific problem. Most of the previous studies requires very expensive hardware [2,12] to be able to absorb an external push. Moreover, push recovery with inexpensive hardware is more challenging [8–10].

© Springer Nature Switzerland AG 2020
H. Fujita et al. (Eds.): IEA/AIE 2020, LNAI 12144, pp. 63–74, 2020.
https://doi.org/10.1007/978-3-030-55789-8_6

Morimoto et al. [14] employed reinforcement learning (RL) to teach a simulated biped robot how to walk with changing the frequency of a simulated robot's walk (timing of the steps) in a two-dimensional (2D) simulation environment. Gu et al. [5] investigated Deep Reinforcement Learning (DRL) and Q-learning in one particular setup in both simulation and the real world. They use robotic arms with 7 degrees of freedom (DOF) in a way that the origin of the arms were installed to fixed table. Kim et al. [11] used the simulated NASA's Valkyrie humanoid robot and employed RL to obtain recovery steps after being hit by an external force. This robot worth more than USD $2000,000. Kumar et al. [13] investigated a walking mechanism for a 2D simulated bipedal robot without the upper body with 6 DOF. The authors used RL and deep neural network (DNN) to achieve this task. Haarnoja et al. [6] employed DRL to train Minitaur a quadrupedal legged robot for walking. The authors investigated their approach in both simulation and the real world. Peng et al. [15] used DRL in a simulation to train different 3D animation characters such as human, dinosaur, and etc. The authors used imitating reference motion capture to achieve highly dynamic motions. Gil et al. [3] studied RL and Q-learning for teaching a simulated humanoid robot (Nao), how to stand up and walk a short distance in a straight line. The ultimate goal of the authors was the simulated robot traverse from point A to point B in a short period of time without falling.

The motivation of this work is to adopt affordable robotics devices and technologies to build a humanoid robot that is able to recover from external pushes similar to humans. Building a humanoid robot with less expensive equipment means that all elements of software, from vision to control, must be much more robust. Also, employing RL techniques to model the robot in situations that the traditional control methods fail or do not perform as good. It is necessary to read our previous works [9,10] before reading this paper.

2   Machine Learning

For the machine learning part of this research we used the TensorFlow (version 1.9) [4] framework with Python (version 2.7). We used Python as the programming language to implement the learning and testing parts of our research.

RL is learning what to do - how to map situations to appropriate actions - so as to maximize a numerical signal [18]. The learner is not told what set of actions to take in different situations, but must instead discover which actions yield the best reward over time. We used the Q-learning technique as the RL mechanism in our work. Equation 1 shows the components of the Q-learning formula [19] .

$$Q(state, action) = R(state, action) + gamma$$
$$*Max[Q(next\_state, all\_possible\_actions)] \tag{1}$$

In this equation, gamma (γ) ranges between 0 and 1. If γ is closer to 0, we are more interested in immediate rewards and if this number is closer to 1 we are more interested in delayed rewards. We can then rewrite Eq. 1 as:

$$Q(s, a) = R(s, a) + \gamma max_{a'}(Q(s', a')) \text{ [20]} \tag{2}$$

where $R(s,a)$ denotes the reward function, s' denotes the next state after the action a was taken, a denotes the action in state s and a' denotes action in state s'.

Polaris was rewarded $+100$ if it was pushed and took a proper set of actions that did not let it fall. Otherwise, -100 was given, as the punishment. Before choosing these two arbitrary numbers, we tested the learning process with $+100$ and -10000. However, we did not find this set of numbers to be a good combination. Since the robot's learning process was in real-time (no fast forwarding in the simulation), the procedure of every push took 10 s from resetting to the fall or stable state in the simulation environment. Because of this constraint, we set the initial number of random tries to 500 for every direction and distance of a push. For example when Polaris started learning the actions for pushes from front direction and the distance of 30 cm, it was allowed to take only 500 random sets of actions (in the given threshold: Table 1, [10]) for the front direction with the distance of 30 cm. After it explored randomly for 500 times, it started to exploit from its previous exploration. Based on our previous experiences, we chose the learning rate of 0.05 for the next step - that is, taking actions based on the robot's previous experience. All the discretized domains had a chance of having 500 random actions at the very beginning of learning. However, the total number of successful and unsuccessful actions that the robot made for each direction and each distance was not necessarily the same.

For every direction and distance, we created a look-up table to store the robot's step parameters [10]. After an action was taken by the robot, the look-up table was updated to reflect the newly-earned reward. If the action was successful, 100 was added to the previous reward, otherwise 100 was deducted from the previous reward. The goal is to maximize the cumulative rewards in the long run. As the look-up table is improved over time, we must also decide when further training would not be helpful. We used a fixed value of 10,000 iterations. To choose this number, we tested some arbitrary numbers of iterations such as 500, 1000, 5000, and 10000, on one category of push (front 40 cm). The results for 10000 iterations was better than others. We chose the front direction to test, because based on our preliminary tests, recovery from the front side is harder than other sides. Also, we chose the 40 cm distance, as an average external force.

The robot's environment is discretized into 14 discrete states (Fig. 10, [10]). Each of these discrete states was explained in [10]. By employing RL, Polaris learned how to recover from various pushes $\{30, 40, 50 \}_{cm}$ from different directions {left, right, front, back}. The final output of each model is a set of actions (walking engine parameters). Each learned action corresponds to a discrete state, and Polaris selected the corresponding action when it found itself in that discrete state.

In our approach to recover from every push, the robot takes one right and one left step. The first step must be able to absorb the majority of the push, since it is an initial step. If a push comes from the front or back sides, the first step will be the right foot. Also, if the push is applied from the right side, the robot takes the right foot as the first step. The only time that the robot takes the left step as the first step is if it is pushed from the left side. After the first step is learned, the learning process for the second step will begin. The first step is considered as learned step, only if by taking it, the robot will be in a stable state in the simulation environment. However, a learned step might not be as good in the real world environment for various reasons. In the simulation environment, many limitations are not considered, such as over torquing servo motors or voltage oscillation. For example the robot might take one right step after the push is applied and it recovers from the impact. However, there might be lots of weight (pressure) on a single servo (that is, the right ankle servo). Even though this might not affect the robot in the simulation, it will over torque the right ankle of Polaris in the real world, which will cause Polaris to fall. To overcome this possible issue, the second step will be used to reduce the pressure on the servos. Learning of the second step is started upon completion of the learning of the first step. In the learning process, for every attempt (push), only one parameter will be altered at a time. This is similar to the hand-tuned approach [9] to push recovery.

We implemented each model (Fig. 10, [10]) (for example, direction: front, distance: 40 cm) as a look-up table and constructed it as a multi-dimensional array. In the look-up table, all the rows refer to the discrete states and all the columns refer to the discrete actions . After this step, we implemented Q-learning as a function which receives the look-up table as its parameter and updates the table's value(s) in every iteration. In every iteration, the Q-learning function selects an action from the 9 action categories, with a value between the minimum and maximum of that action category (Table 1, [10]). Depending on the learning rate, the strategy (behaviour) of the robot will be different.

Polaris learned different combinations of walking parameters that it used in push recovery to replace the hand-tuned entries (discussed in [10]). Figure 1 shows the system diagram of our adaptive closed-loop control using RL for one step of a stride. In this diagram a disturbance represents an external push that can be added to the closed-loop control at anytime (before or after a set of instructions send to the motors) during a walking step. The IMU measures a potential disturbance signal along with some added ambience noise and send it to the controller module. After that the controller measures the strength and the direction of the disturbance. Then using the look-up table that was generated by RL module it matches a category that is appropriate for the push and send an updated set of walking parameters to the motors.

We discussed the transition of our hand-tuned approach [9] to a fully autonomous approach that used RL to generate walking step trajectories. However, there is room for further improvement. The choice of reward function in our RL implementation was based on the simulations' assessment (Gazebo simulation). The main issue with the reward function for this approach is that Polaris can only distinguish between the best and worst actions. The best actions lead the robot to a stable state. The worst actions are actions that lead the robot to a fall state. However, some action choices for which the robot is penalized do actually have the potential to lead the robot to a stable state, but a fall results

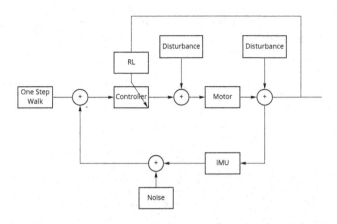

Fig. 1. System diagram of our adaptive closed-loop control using RL.

in negative rewards and the robot then avoids attempting those actions in the future. As a result, many possible step trajectories will not likely be learned. We designed a DNN structure with one input layer, 20 hidden layers, and one output layer. The input layer takes the 9 robot walking engine parameters (Table 1, [10]), and learns the reward based on the actions taken by the robot.

Training and testing the DNN involves using both the RL and DRL components running in tandem. We reserved the final output of the previously-trained RL component (i.e. the finished policy using only RL), and then reset the RL to start fresh. The RL component begins training as previously. However, instead of using the simulation to assess an action, it uses the currently learned reward function from the DNN. We have divided the whole DNN process into three steps.

First, gathering a rich dataset for training the DNN (that is, the resulting look-up table). Second, using this dataset to train the DNN. The network needs to learn a model based on the sample data (our sample data consisted of 24000 data points). Third, the trained DNN then predicts if an action will lead the robot to a fall state or a stable state. we used a sigmoid activation function (Eq. 3) to keep the output of the DNN between 0 and 1. Also, we used the binary-cross-entropy loss function [1] to measure the loss during training. Equation 4 shows the binary-cross-entropy function.

$$S(x) = \frac{1}{1 + e^{-x}} \quad [7] \tag{3}$$

$$- (l_i * log(p) + (1 - l_i) * log(1 - p)) \quad [1] \tag{4}$$

If the output p (the probability of the action being successful) of the DNN (the output of the network) is $0 \geq p < 0.9$, the action will be classified as a fall and a -100 will be generated. However, if the output is $0.9 \geq p \leq 1.0$, the action will be classified as stable and $+100$ will be returned. The generated values will be accumulated and go towards a course of action at the end, and assign a −100 or +100 for the reward.

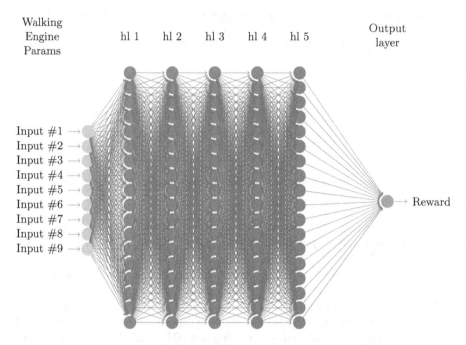

Fig. 2. Partial view (instead of 20 hidden layers, only 5 are shown) of the architecture of the Deep Neural Network

Figure 2 shows the structure of our DNN showing 5 of the 20 hidden layers. In this figure, the green layer is the input layer of the network and the parameters {Input #1,..., Input #9} refer to the robot actions that were listed in Table 1, [10]. We implemented the network with 20 hidden layers with 18 neurons in each layer. The number of the neurons in the hidden layers are usually related to the quantity of inputs to the network and are often double the number of inputs. However, this is not a universal rule, and requires some tests and modifications to parameters such as the number of neurons in the hidden layer. For choosing the number of hidden layers, we examined the performance of the network with varying numbers of hidden layers {2, 9, 15, 20} and neurons on each layer {9, 18}. For example, we checked the performance of the network with 15 hidden layers, once with 9 neurons on each layer and with 18 neurons on each hidden layer. The last layer is the output layer, which produces a floating point number between [0,1] which is a learned reward for an action, representing the probability that the action will result in a stable state.

Selecting a correct set of parameters for building, training and testing a DNN is a time consuming process. The network needs to be tested and evaluated with different sets of parameters to check the accuracy and performance of the network. For example, there are parameters such as *learning rate* (how fast the network should learn), *batch size* (the number of training samples in which will be given to the network during training), *epoch* (one forward pass and one backward pass for all the training sample data), and *number of hidden layers*, which play an important role in every DNN. After testing the performance of the network with many different combinations, we chose the learning rate of 0.1, batch size of 20000, with the epoch of 10000 and 20 hidden layers. The system diagram for the DRL is similar to the Fig. 1, only the RL component is replaced by the DRL component.

3   Evaluation

In order to examine the effectiveness of our approach, we ran various experiments in simulation and in the real world. We compared the performance of our approach against two baseline cases in simulation and one base case in the real world. In the first base case, the robot was not able to use any sensor feedback (IMU) during the experiment (open-loop control feedback [10]) (real world and simulation). In the second base case, the robot took random actions within the given thresholds of walking engine parameters and was able to use the IMU (closed-loop control feedback; simulation). During the experiments, the robot did not know anything about the environment (different surfaces). The robot was examined on concrete, carpet, and artificial turf in the real world. In the simulation, it was examined on simulated concrete and turf. The type of surface, and the directions and strength of the pushes were all chosen randomly.

3.1   Experimental Environment

In this section we describe the experimental environments in two worlds. The first world is in simulation and second is the real world. Every world has its own environments and the robot was examined in those environments.

To evaluate our approach quantitatively, we set up various experimental environments to control the push force applied to the robot, based on the international robotics competition, RoboCup 2018 push recovery technical challenge [16].

In the simulation we generated two environments. The only differences between the two environments are the surface the robot was walking on and the strength and direction of the pushes applied. In the first environment, the surface is simulated concrete, and the second environment is a simulated soccer field with artificial turf. Every experiment consisted of 360 trials. For all 4 sides (back, right, front, left) and 3 distances (30, 40, 50)$_{cm}$, 30 pushes were applied. For every environment, we examined our RL and DRL approaches. In total, 2880 pushes were applied to the robot in the simulation for all the experiments. Figure 5 bottom, shows 1 of the 2880 pushes with the recovery reaction in the simulation on concrete.

In the real world we designed three environments. As in the simulation, the only differences between the environments are the surface on which the robot was walking and the strength and direction of the pushes applied. In the first environment, the surface is concrete, while the second is a low carpet, and the last is an artificial soccer turf. Figure 3 shows these three surfaces upon which the robot walked during the experiments in the real world. Figure 4 shows the experimental design using the two worlds. We generated all the push directions and distances randomly. This is again to avoid the bias that would result if one environment was always chosen first and another chosen last. We examined every direction and distance three times, resulting in a total of 36 pushes for every experiment. For all of the 4 sides (back, right, front, left) and 3 distances $(30, 40, 50)_{cm}$, 3 pushes were applied. For every environment, we examined the RL and DRL approaches. In total, 324 pushes were applied to the robot in the real world for all the experiments. Figure 5 top, shows 1 of the 324 pushes with the recovery reaction in the real world on concrete.

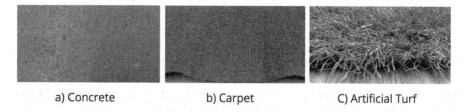

a) Concrete   b) Carpet   C) Artificial Turf

Fig. 3. Three different surfaces for three different real world environments. From left to right: concrete, carpet, and artificial soccer turf.

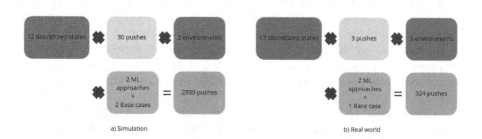

a) Simulation   b) Real world

Fig. 4. Experimental design in the simulation and the real world

4   Experimental Results

Based on the experimental results, our RL and DRL approaches performed better, by a large margin, than the base cases for every surface in both simulation and the real world. For the real world experiments, the results show that the best surface in which our approaches outperformed is the concrete surface. This is due to two main reasons: 1) all of the actions were learned on the concrete in the simulation during the learning process, and 2) Concrete is the stiffest surface and it is easier to recover on such a surface. The second easiest surface among the three is carpet. Finally, the hardest surface for recovery is the artificial turf. Figure 6 top, compares the results of RL, DRL, and the base case on three different surfaces based on four different directions that the forces were applied.

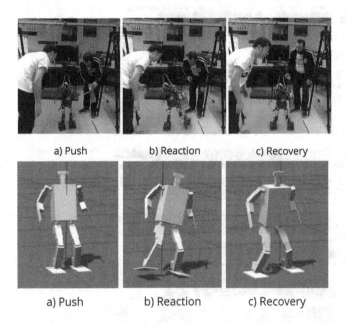

Fig. 5. Push, reaction, and recovery on a concrete surface

For the simulation experiments, the results show that the best surface on which our approaches performed is the concrete surface, which is similar to the real world experiment. Artificial turf is the next more difficult surface in simulation. Figure 6 bottom, compares the results of RL, DRL, and the base case on these two different surfaces.

5   Conclusion and Future Work

This research is concentrated on active balancing and push recovery for an inexpensive humanoid robot. This research makes a number of important contributions to the state of the art in push recovery and active balancing: Our main contribution is a new approach to push recovery based on learning appropriate

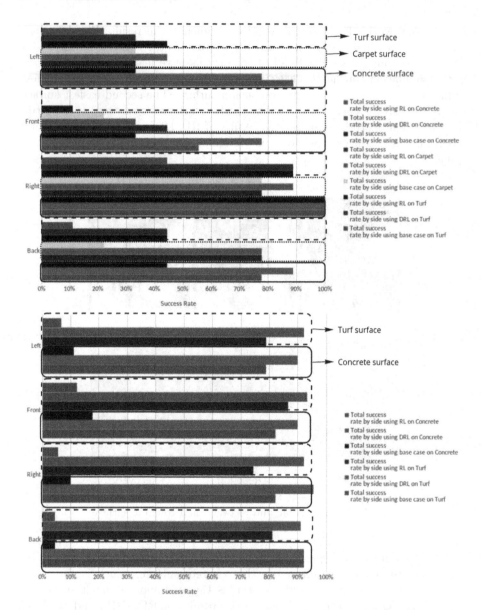

Fig. 6. Performance comparison of all approaches on all surfaces in the real world (top) and simulation (bottom), using the open-loop feedback control.

step responses using RL and DRL. It is important to note that this approach is specifically intended for inexpensive humanoid robots that use more fragile, low-torque, limited-power servos with low computational power. No research in the literature provides a solution that recovers from small, medium, and large disturbances for inexpensive robots. Our methods were also tested in the real world using a physical robot, as opposed to other methods examined only in simulation or in restricted settings. We introduced a novel way of using RL, DRL, and stepping (taking steps) for push recovery. This generates the walking engine parameters and uses these for push recovery of a humanoid robot. These parameters then will be used in the inverse kinematics model to generate a trajectory of a step. We improved our previous work [10] in this research.

Push recovery and active balancing of humanoid robots are important ongoing research topics in robotics. There is no perfect solution to all these problems even on expensive equipment, and the challenges are more significant for inexpensive humanoid robots that use more fragile, low-torque, limited-power servos along with low computational power. In this research, we have described introduced a novel push recovery approach using RL and DRL using stepping on an inexpensive 20-DOF humanoid robot [17].

Although the evaluation of our research has successfully highlighted the benefit of our approach, our research and implementation also raise a number of areas in which improvement could be made. As future work, we would like to have another DNN that outputs walking engine parameters. Another interesting direction of future work would be to replace the look-up tables that we used for RL and DRL with DNNs.

References

1. Buja, A., Stuetzle, W., Shen, Y.: Loss functions for binary class probability estimation and classification: Structure and applications. Working draft. Accessed 3 Nov 2005
2. Feng, S., Whitman, E., Xinjilefu, X., Atkeson, C.G.: Optimization based full body control for the atlas robot. In: 2014 14th IEEE-RAS International Conference on Humanoid Robots (Humanoids), pp. 120–127. IEEE (2014)
3. Gil, C.R., Calvo, H., Sossa, H.: Learning an efficient gait cycle of a biped robot based on reinforcement learning and artificial neural networks. Appl. Sci. **9**(3), 502 (2019)
4. Google. Tensorflow (2017). URL https://www.tensorflow.org
5. Gu, S., Holly, E., Lillicrap, T., Levine, S.: Deep reinforcement learning for robotic manipulation with asynchronous off-policy updates. In: 2017 IEEE International Conference on Robotics and Automation (ICRA), pp. 3389–3396. IEEE (2017)
6. Haarnoja, T., Zhou, A., Ha, S., Tan, J., Tucker, G., Levine, S.: Learning to walk via deep reinforcement learning (2018). arXiv preprint arXiv:1812.11103
7. Han, J., Moraga, C.: The influence of the sigmoid function parameters on the speed of backpropagation learning. In: Mira, J., Sandoval, F. (eds.) IWANN 1995. LNCS, vol. 930, pp. 195–201. Springer, Heidelberg (1995). https://doi.org/10.1007/3-540-59497-3_175
8. Hosseinmemar, A.: Push recovery and active balancing for inexpensive humanoid robots. PhD thesis, Department of Computer Science, University of Manitoba, Winnipeg, Canada (2019)
9. Hosseinmemar, A., Baltes, J., Anderson, J., Lau, M.C., Lun, C.F., Wang, Z.: Closed-loop push recovery for an inexpensive humanoid robot. In: Mouhoub, M., Sadaoui, S., Ait Mohamed, O., Ali, M. (eds.) IEA/AIE 2018. LNCS (LNAI), vol. 10868, pp. 233–244. Springer, Cham (2018). https://doi.org/10.1007/978-3-319-92058-0_22. ISBN 978-3-319-92058-0

10. Hosseinmemar, A., Baltes, J., Anderson, J., Lau, M.C., Lun, C.F., Wang, Z.: Closed-loop push recovery for inexpensive humanoid robots. Appl. Intell. **49**(11), 3801–3814 (2019). https://doi.org/10.1007/s10489-019-01446-z
11. Kim, D., Lee, J., Sentis, L.: Robust dynamic locomotion via reinforcement learning and novel whole body controller (2017). arXiv preprint arXiv:1708.02205
12. Kuindersma, S., et al.: Optimization-based locomotion planning, estimation, and control design for the atlas humanoid robot. Auton. Rob. **40**(3), 429–455 (2015). https://doi.org/10.1007/s10514-015-9479-3
13. Kumar, A., Paul, N., Omkar, S.: Bipedal walking robot using deep deterministic policy gradient (2018). arXiv preprint arXiv:1807.05924
14. Morimoto, J., Cheng, G., Atkeson, C.G., Zeglin, G.: A simple reinforcement learning algorithm for biped walking. In: IEEE International Conference on Robotics and Automation, Proceedings, ICRA 2004, vol. 3, pp. 3030–3035. IEEE (2004)
15. Peng, X.B., Abbeel, P., Levine, S., van de Panne, M.: Deepmimic: example-guided deep reinforcement learning of physics-based character skills. ACM Trans. Graph. (TOG) **37**(4), 143 (2018)
16. Robocup. Robocup technical challenge (2018). http://www.robocuphumanoid.org/wp-content/uploads/RCHL-2018-Rules-Proposal_final.pdf
17. Sadeghnejad, J.S., et al.: Autman humanoid kid size team description paper (2016)
18. Sutton, R.S., Barto, A.G.: Reinforcement Learning: An Introduction, vol. 1. MIT Press, Cambridge (1998)
19. Watkins, C.J., Dayan, P.: Q-learning. Mach. Learn. **8**(3–4), 279–292 (1992)

Optimal Control Problem of a Differential Drive Robot

Luis F. Recalde[1]([✉]), Bryan S. Guevara[1]([✉]), Giovanny Cuzco[2]([✉]),
and Víctor H. Andaluz[1]([✉])

[1] Universidad de las Fuerzas Armadas ESPE, Sangolquí, Ecuador
{lfrecalde1,bsguevara,vhandaluz1}@espe.edu.ec
[2] Universidad Nacional de Chimborazo, Riobamba, Ecuador
gcuzco@unach.edu.ec

Abstract. This paper proposes an optimal control to track trajectories applied in a differential drive robot. For this control, the kinematic model of the system was developed considering the position of interest as the center of gravity of the mobile robot. Additionally, the optimization problem with restrictions developed through the Sequential Least Squares Programming method was used to keep the robot on the trajectory and to find the optimal values. Moreover, the stability of the proposed controller for tracking trajectories applying optimal controllers is demonstrated analytically; this allows to prove that the control problem has a unique solution for the desired trajectory. Finally, the results are compared with a controller based on inverse kinematics.

Keywords: Optimal control · Kinematic · Differential drive robot

1 Introduction

The latest advances in robotics focus on the development of functional robots, which are used in industries in either structured or partially structured environments. In structured environments the applications are welding, cutting, manufacturing among others, and in unstructured environments like space exploration applications where robots are sent to exploration missions; military applications where robots are used for mine detection; exploration of hostile places where robots are used for rescue operations. In the area of agriculture there are robots used for pest control, fumigation, soil preparation, [1–3]. The integration of robotics in the medical field has been a point of interest in recent years. Service, assistance, rehabilitation and surgery are the areas of human health most benefited by advances in medical service robotics [6,7].

Service robotics not only applies to the medical field, in recent years there has been a noticeable increase in the use of service robotics in tasks in the industrial field which has been one of the main precursors of more sophisticated control systems in order to perform autonomous tasks in a robust and efficient way [10,14]. Mobile robots are widely used in service robotics for product transport,

© Springer Nature Switzerland AG 2020
H. Fujita et al. (Eds.): IEA/AIE 2020, LNAI 12144, pp. 75–82, 2020.
https://doi.org/10.1007/978-3-030-55789-8_7

cleaning, inspection, maintenance among other applications, these tasks require high precision controllers, so the implementation of electronic systems, reliable control methods and intelligent, people are required to fulfill the tasks that demand a high degree of precision [11,12].

Path tracking is one of the main problems of mobile robots and focuses on how to generate the appropriate linear and angular speed for the mobile robot to follow a predefined path [9]. In recent years, many investigations have been carried out on the control of trajectories the majority of these investigations include the kinematic model of the robot. The main control algorithms can be classified into 6 categories: (1) Back-stepping; (2) Linearization; (3) Sliding Mode; (4) Fuzzy systems; (5) Neural Networks and (6) Neuro Fuzzy Systems [8, 15,17,18].

In this context, a new method is proposed to solve the trajectory tracking problem for a mobile robot of the differential type. Based on the optimal control problem where the objective is to minimize a cost function based on the errors of control, finding the optimal values of the control signals that allows the mobile robot to follow the predefined path. Therefore, in this sense an analysis of the objective function is performed to define if there is a solution for the desired path.

This article is divided into six sections, including the introduction. In Sect. 2 the kinematic model of a differential type robot is presented. The design, stability analysis and the system optimizer are presented in Sect. 3. The results and discussions are shown in Sect. 4. Finally, the conclusions are presented in Sect. 5.

2   Kinematic Transformation

The mobile platform has two driven wheels which are controlled independently by two DC motors and one castor wheel that contribute with physical stability of the robot. A way to represent the differential drive robot is using external kinematics (see Fig. 1). [1].

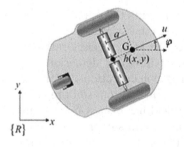

Fig. 1. Differential drive robot with displaced mass center.

The general coordinate system and position of the mobile robot which is defined by point G with location $\mathbf{h} = \begin{bmatrix} x & y \end{bmatrix}^T$ where a is the distance from the center of axes of the wheels in relation to the absolute coordinate system $\{\mathbf{R}\}$. The differential drive robot has a non-holonomic restriction, so it can only move perpendicular to the wheel axes [1].

$$\dot{x}\sin(\varphi) - \dot{y}\cos(\varphi) + a\dot{\varphi} = 0 \tag{1}$$

The kinematic model considering the non-holonomic restrictions is represented as:

$$\begin{bmatrix} \dot{x} \\ \dot{y} \\ \dot{\varphi} \end{bmatrix} = \begin{bmatrix} \cos(\varphi) & -a\sin(\varphi) \\ \sin(\varphi) & a\cos(\varphi) \\ 0 & 1 \end{bmatrix} \begin{bmatrix} \mu \\ \omega \end{bmatrix} \tag{2}$$

where, μ is the linear velocity to the frontal direction and the angular velocity ω that rotates the referential system counterclockwise around the axis Z.

Also, kinematic model can be written in a compact form as $\dot{\mathbf{h}}(t) = \mathbf{f}(\varphi)\nu(t)$

$$\dot{\mathbf{h}}(t) = \mathbf{J}(\varphi)\nu(t) \tag{3}$$

$$\dot{\varphi} = \omega \tag{4}$$

where $\dot{\mathbf{h}}(t) = \begin{bmatrix} \dot{x} & \dot{y} \end{bmatrix}^T$ is the velocity vector of the point of interest, $\mathbf{J}(\varphi)$ represents the Jacobian matrix and $\nu(t) = \begin{bmatrix} \mu & \omega \end{bmatrix}^T$ is the control vector of the point of interest.

3   Controller Design

It is proposed an optimal control system that minimizes a performance index based on system errors. In order to know the errors of the system it is necessary to use numerical methods to be able to know the evolution of the system, it is based on an approximation of the states at the instant $k+1$, if the state and the control action at instant k are known, then the approximation called Euler's method can be used,

$$\mathbf{h}(k+1) = \mathbf{h}(k) + \mathbf{J}(\varphi)\nu(k)T_o \tag{5}$$

where T_o represents the sampling period and $k \in \{1, 2, 3, 4, 5..\}$.

The function of the control algorithm allows the mobile robot to follow the specific path. The control errors can be expressed in their matrix form as shown below,

$$\tilde{\mathbf{h}}(k) = \mathbf{h_d}(k) - \mathbf{h}(k) \tag{6}$$

where $\tilde{\mathbf{h}}(k) = \begin{bmatrix} \tilde{h}_x & \tilde{h}_y & \tilde{h}\varphi \end{bmatrix}^T$ it's vector of control errors, $\mathbf{h_d}(k) = \begin{bmatrix} h_{x_d} & h_{y_d} & h_{\varphi_d} \end{bmatrix}^T$ it is the vector of desired states of the system and $\mathbf{h}(k) = \begin{bmatrix} h_x & h_y & h_\varphi \end{bmatrix}^T$ it is the vector of system's states.

In Fig. 2 shows the schematic of the proposed controller in order to track error tend to zero, it is required to minimize a quadratic cost function that is made up of control errors, under the system's own equality conditions and its actuator limitations, and its representation would be as follows:

$$\min \mathbf{G} = \frac{1}{2} \sum_{n=k}^{k+1} \tilde{\mathbf{h}}(k)^T \mathbf{Q}\tilde{\mathbf{h}}(k) \tag{7}$$

$$\text{subject to: } \mathbf{h}(k+1) = \mathbf{h}(k) + \mathbf{J}(\varphi)\nu(k)T_o \tag{8}$$

with the following inequality restrictions

$$\nu_{min} \leq \nu(t) \leq \nu_{max} \tag{9}$$

where $\mathbf{G}$ is the cost function, $\mathbf{Q}$ is a positive definite diagonal matrix that will weigh the errors of the system, $\nu_{\min} = \begin{bmatrix} \mu_{\min} & \omega_{\min} \end{bmatrix}^T$ is a vector that represents the lower limits of the mobile robot and $\nu_{\max} = \begin{bmatrix} \mu_{\max} & \omega_{\max} \end{bmatrix}^T$ is a vector that represents the upper limits of the mobile robot. At each sampling moment, the model is used to determine the behavior of the system and, depending on this behavior, the objective function is minimized.

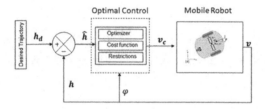

Fig. 2. Proposed control scheme.

3.1 Stability Analysis

Assuming that the functions are continuously differentiate [5,16],

$$\mathbf{G} = \frac{1}{2} \int_{t_0}^{t_f} \tilde{\mathbf{h}}(t)^T \mathbf{Q}\tilde{\mathbf{h}}(t)dt \tag{10}$$

$$\dot{\mathbf{h}}(t) = \mathbf{J}(\varphi)\nu(t) \tag{11}$$

$$\mathbf{H}(\tilde{\mathbf{h}}, \gamma, l) = \frac{l}{2}\tilde{\mathbf{h}}(t)^T \mathbf{Q}\tilde{\mathbf{h}}(t) \tag{12}$$

$$\nabla_{\tilde{\mathbf{h}}}\mathbf{H} = l\mathbf{Q}\tilde{\mathbf{h}} \tag{13}$$

$$\nabla_{\tilde{\mathbf{h}}}^2\mathbf{H} = \mathbf{Q} \in \{\mathbf{R}^{n,n}\} \tag{14}$$

where $\mathbf{H}(\tilde{\mathbf{h}}, \gamma, l)$ is the Hamiltonian of the optimization problem, $\nabla_{\tilde{\mathbf{h}}}\mathbf{H}$ is the gradient of the Hamiltonian and $\nabla_{\tilde{\mathbf{h}}}^2\mathbf{H}$ is the Hamiltonian Hessian. If $\mathbf{Q} > 0$ then $\nabla_{\tilde{\mathbf{h}}}^2\mathbf{H} > 0$ and if Hamiltonian of the optimization problem is a strictly convex function with respect to $\tilde{\mathbf{h}}(t)$, so the optimization problem has a unique solution for the desired trajectory based on Theorem 11.12 [4].

Hence the error approaches asymptotically to zero when the optimization problems have a unique solution. This implies that the equilibrium point of the closed loop is asymptotically stable, so that the position error of the system $\tilde{\mathbf{h}}(k) \to 0$ asymptotically when $t \to 0$.

4   Results and Discussion

This section presents the performance of the proposed controller which is implemented using Python and its scientific libraries, that uses Sequential Least Squares Programming to minimize a function of several variables with any combination of bounds, equality and inequality constraints [13]. The result is compared to a controller using inverse kinematics, a sampling period of 0.1 s which was considered as the maximum application in robotics and a distance to the center of mass of 0.1 m. For the optimal controller is defined $\nu_{\min} = \begin{bmatrix} -2.5 & -2.5 \end{bmatrix}^T$, $\nu_{\max} = \begin{bmatrix} 2.5 & 2.5 \end{bmatrix}^T$ and $\mathbf{Q}$ is a diagonal matrix $\mathbf{diag} \left(0.5\ 0.5\ 0.5 \right)$. Finally for the inverse kinematics controller the control law is defined as $\nu_{ref} = \mathbf{J}^{-1}(\dot{\mathbf{h}}_d + \mathbf{K}\tilde{\mathbf{h}})$, where ν_{ref} is a vector of control signals, $\mathbf{K}$ is a diagonal matrix $\mathbf{diag} \left(0.5\ 0.5 \right)$ and $\dot{\mathbf{h}}_d$ is the desired velocity of the desired trajectory [2,3].

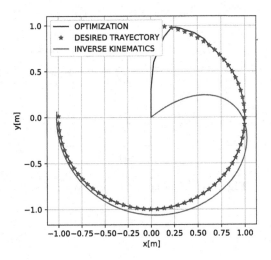

Fig. 3. Circular trajectory.

Circular path: the objective for the mobile robot is to follow a path described by $\mathbf{h_d} = \begin{bmatrix} h_{x_d} & h_{y_d} & h_{\varphi_d} \end{bmatrix}^T$, where $h_{y_d} = \sin(0.5t)$, $h_{x_d} = \cos(0.5t)$, $h_{\varphi_d} = \arctan(\frac{\dot{h}_{y_d}}{\dot{h}_{x_d}})$, the initial conditions of the mobile robot are $\begin{bmatrix} 0 & 0 & \frac{\pi}{2} \end{bmatrix}$. Figure 3 shows the results of the controllers with the circular path. Figure 4 presents the errors with each controller respectively, this graph shows that the errors tends to zero when $t \rightarrow 0$. Finally Fig. 5 show the maneuverability commands to the differential drive robot with each controller respectively.

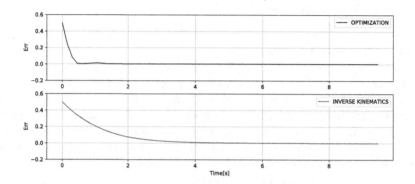

Fig. 4. Errors comparation.

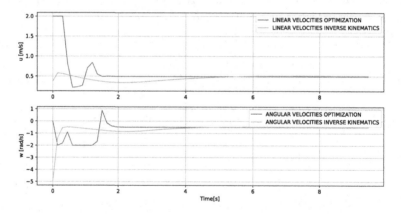

Fig. 5. Differential drive robot velocities.

5   Conclusions

In this work the design of an optimal controller based on the errors of the system for a trajectory tracking was presented. The design of the controller is based on a kinematic model of the system and the resolution of the optimization problem with restrictions to help finding the optimal values for the mobile platform and stability was analytically tested.

Two controllers analyzed are Optimal controller and Inverse Kinematics controller. They were tested by simulations in the tracking control trajectory for differential drive robot. Both techniques have acceptable results, the trajectory tracking control simulations showed that inverse Kinematics controller has a settling time error higher than the Optimal controller. This paper can be used as a starting point in future researches.

Acknowledgements. The authors would like to thank the Coorporación Ecuatoriana para el Desarrollo de la Investigación y Academia CEDIA for their contribution in innovation, through the CEPRA projects, especially the project CEPRA-XIII-2019-08; Sistema Colaborativo de Robots Aéreos para Manipular Cargas con Óptimo Consumo de Recursos; also the Universidad Nacional de Chimborazo, Universidad de las Fuerzas Armadas ESPE and the Research Group ARSI, for the support for the development of this work.

References

1. Andaluz, G.M., et al.: Modeling dynamic of the human-wheelchair system applied to NMPC. In: Kubota, N., Kiguchi, K., Liu, H., Obo, T. (eds.) ICIRA 2016. LNCS (LNAI), vol. 9835, pp. 179–190. Springer, Cham (2016). https://doi.org/10.1007/978-3-319-43518-3_18
2. Andaluz, V.H., et al.: Modeling and control of a wheelchair considering center of mass lateral displacements. In: Liu, H., Kubota, N., Zhu, X., Dillmann, R., Zhou, D. (eds.) ICIRA 2015. LNCS (LNAI), vol. 9246, pp. 254–270. Springer, Cham (2015). https://doi.org/10.1007/978-3-319-22873-0_23
3. Andaluz, V.H., et al.: Robust control with dynamic compensation for human-wheelchair system. In: Zhang, X., Liu, H., Chen, Z., Wang, N. (eds.) ICIRA 2014. LNCS (LNAI), vol. 8917, pp. 376–389. Springer, Cham (2014). https://doi.org/10.1007/978-3-319-13966-1_37
4. Beck, A.: Introduction to Nonlinear Optimization. Society for Industrial and Applied Mathematics, Philadelphia (2014). https://doi.org/10.1137/1.9781611973655
5. Betts, J.T.: Practical Methods for Optimal Control and Estimation Using Nonlinear Programming. Society for Industrial and Applied Mathematics, Philadelphia (2010)
6. Biswas, K., Mazumder, O., Kundu, A.S.: Multichannel fused EMG based biofeedback system with virtual reality for gait rehabilitation. In: 4th International Conference on Intelligent Human Computer Interaction: Advancing Technology for Humanity, IHCI 2012 (2012). https://doi.org/10.1109/IHCI.2012.6481834
7. De La Cruz, C., Bastos, T.F., Carelli, R.: Adaptive motion control law of a robotic wheelchair. Control Eng. Pract. **19**(2), 113–125 (2011). https://doi.org/10.1016/j.conengprac.2010.10.004
8. Dierks, T., Jagannathan, S.: Control of nonholonomic mobile robot formations: backstepping kinematics into dynamics. In: Proceedings of the IEEE International Conference on Control Applications, pp. 94–99 (2007). https://doi.org/10.1109/CCA.2007.4389212
9. Elsayed, M., Hammad, A., Hafez, A., Mansour, H.: Real time trajectory tracking controller based on lyapunov function for mobile robot. Int. J. Comput. Appl. **168**(11), 1–6 (2017). https://doi.org/10.5120/ijca2017914540

10. Ferrara, A., Incremona, G.P.: Design of an integral suboptimal second-order sliding mode controller for the robust motion control of robot manipulators. IEEE Trans. Control Syst. Technol. **23**(6), 2316–2325 (2015). https://doi.org/10.1109/TCST. 2015.2420624

11. Kand, M.S.T., Sadeghian, R., Masouleh, M.T.: Design, analysis and construction of a novel flexible rover robot. In: International Conference on Robotics and Mechatronics, ICROM 2015, pp. 377–382. Institute of Electrical and Electronics Engineers Inc. (2015). https://doi.org/10.1109/ICRoM.2015.7367814

12. Kantharak, K., Somboonchai, C., Tuan, N.T., Thinh, N.T.: Design and development of service robot based human - robot interaction (HRI). In: Proceedings - 2017 International Conference on System Science and Engineering, ICSSE 2017, pp. 293–296. Institute of Electrical and Electronics Engineers Inc. (2017). https://doi.org/10.1109/ICSSE.2017.8030884

13. Kraft, D.: Institut für Dynamik der Flugsysteme, Deutsche Forschungs-und Versuchsanstalt für Luft-und Raumfahrt (DFVLR). A software package for sequential quadratic programming. Forschungsbericht. Deutsche Forschungs- und Versuchsanstalt für Luft- und Raumfahrt, DFVLR, Braunschweig (1988)

14. Osusky, J., Ciganek, J.: Trajectory tracking robust control for two wheels robot. In: Proceedings of the 29th International Conference on Cybernetics and Informatics, K and I 2018, vol. 2018, no. (January), pp. 1–4. Institute of Electrical and Electronics Engineers Inc. (2018). https://doi.org/10.1109/CYBERI.2018.8337559

15. Pandey, A., Parhi, D.R.: Optimum path planning of mobile robot in unknown static and dynamic environments using fuzzy-wind driven optimization algorithm. Defen. Technol. **13**(1), 47–58 (2017). https://doi.org/10.1016/j.dt.2017.01.001

16. Pytlak, R.: Numerical Methods for Optimal Control Problems with State Constraints. Lecture Notes in Mathematics, vol. 1707. Springer, Heidelberg (1999). https://doi.org/10.1007/BFb0097244

17. Singh, M.K., Parhi, D.R.: Intelligent neuro-controller for navigation of mobile robot. In: Proceedings of the International Conference on Advances in Computing, Communication and Control, ICAC3 2009, pp. 123–128 (2009). https://doi.org/10.1145/1523103.1523129

18. Yuan, G., Yang, S.X., Mittal, G.S.: Tracking control of a mobile robot using a neural dynamics based approach. In: Proceedings - IEEE International Conference on Robotics and Automation, vol. 1, pp. 163–168 (2001). https://doi.org/10.1109/robot.2001.932547

Optimal Trajectory Tracking Control for a UAV Based on Linearized Dynamic Error

Christian P. Carvajal[1]([⊠]), Víctor H. Andaluz[2]([⊠]), Flavio Roberti[1]([⊠]), and Ricardo Carelli[1]([⊠])

[1] Instituto de Automática, Universidad Nacional de San Juan, San Juan, Argentina
{cpcarvajal,froberti,rcarelli}@inaut.unsj.edu.ar
[2] Universidad de Las Fuerzas Armadas ESPE, Sangolquí, Ecuador
vhandaluz1@espe.edu.ec

Abstract. This work proposes a solution method for tracking procedures for Unmanned Aerial Vehicles (UAVs). The proposed controller is based on the dynamics of the error obtained from the kinematic model of the UAV, *i.e.*, on linearized error behavior during the tracking task. For the correction of the trajectory tracking error, an optimal controller is used that provides a gain to compensate the errors and disturbances during the task proposed by using LQR algorithm. The experimental results are presented with several weight options in the proposed functional cost for analysis the UAV behavior.

Keywords: Trajectory tracking · Optimal control · LQR · Kinematic model · UAV

1 Introduction

In recent years, robotics research has developed new technologies that improve the intelligence and mobility of aerial robots. In this context, the area of unmanned aerial vehicles (UAVs) has generated incredible technological advances in terms of autonomous navigation, perception, mapping, obstacle avoidance and the auto-localization; topics that have been of great help so that robots do not have workspace limits in the execution of tasks such as: *i)* inspection [1]; *ii)* objects transport to inaccessible areas [2]; *iii)* precision agriculture *iv)* surveillance, mapping and modeling 3D [3]; *iv)* dangerous or inaccessible tasks for humans; *v)* river and lagoon monitoring, among others [4, 5]. In order to execute various tasks with UAVs, multiple control strategies have been developed that solve tasks for positioning, path following and trajectories tracking.

Autonomous controllers for a UAV can be classified into three large groups: *i)* based on learning or that a model is not used, where algorithms have been developed with Fuzzy Logic for trajectory tracking [6], controls with neural networks to avoid obstacles and collisions in navigation [7]. *ii)* Non-linear controllers that are practically based on the non-linear UAV model [8, 9]; In [10] the controller they propose allows to meet the problem of positioning, trajectory tracking and path following, applying modifications in the controller reference parameters. Similarly, there are other sliding mode controllers

© Springer Nature Switzerland AG 2020
H. Fujita et al. (Eds.): IEA/AIE 2020, LNAI 12144, pp. 83–96, 2020.
https://doi.org/10.1007/978-3-030-55789-8_8

[11], predictive [12], adaptive and linearization in feedback [13] which are based on kinematic or dynamic model; *iii)* Linear controllers, their great advantage constitutes the ease of implementation in a real platform and employ a linearized model of the quadrotor, applications were found where PID, LQR/LQG controllers are used [14], that perform feedback controller for UAV attitude behavior and trajectory tracking. Other applications have focused on the continuous trajectory tracking using linear algebra controllers to minimize the computation time of the on-board controller. In [15] linear algebra control is an advanced control technique that does not require complex calculations and the computational requirement is low, which allows the implementation of low-performance processors and maintaining a correct performance for the execution of the different tasks.

Most controllers manage to reduce the tracking error to zero, but do not consider the behavior of control actions, control actions are high when the error is proportionally large and when disturbances occur in the system. In order to solve this efficiency, so optimization techniques are used for that the performance of the UAV is optimal while the task proposed is executed, *i.e.,* complete the task with appropriate control actions and without generating much effort the quadcopter; these techniques are known as optimal control that is used in systems that evolve over time and are susceptible to external forces [16, 17]. The controller proposed in this work is based on the linearized model of the error behavior for the trajectory tracking of a UAV, the proportional controller with an optimal error correction gain, where the gain is obtained by means of LQR [18], this method minimizes a proposed cost functional that weighs the system states.

This article is organized into 5 Sections. Section 2 presents kinematic model for Unmanned Aerial Vehicles (UAVs). The design of the control algorithm in Sect. 3. The discussion of results is shown in Sect. 4, and finally the conclusions of the paper are presents in Sect. 5.

2 Kinematic Model

To obtain the kinematic model of the UAV, the point of interest in the robot center is considered. The kinematic model of the aerial robot contains the four maneuverability velocities to achieve displacement and four velocities that represent the velocities within the frame of reference. Figure 1 represents the kinematic variables of the quadcopter on the fixed reference frame $< \mathcal{R} >$.

In other words, the movement of the quadcopter over the fixed frame of the reference $< \mathcal{R} >$ is defined as:

$$\begin{bmatrix} \dot{x}_a \\ \dot{y}_a \\ \dot{z}_a \\ \dot{\theta}_a \end{bmatrix} = \begin{bmatrix} \cos(\theta_a) & -\sin(\theta_a) & 0 & 0 \\ \sin(\theta_a) & \cos(\theta_a) & 0 & 0 \\ 0 & 0 & 1 & 0 \\ 0 & 0 & 0 & 1 \end{bmatrix} \begin{bmatrix} u_{la} \\ u_{ma} \\ u_{na} \\ \omega_a \end{bmatrix} \tag{1}$$

Thus, the kinematic model can be written as follows:

$$\dot{\chi}_a = \Gamma_a U_a \tag{2}$$

where, $\dot{\chi}_a = \begin{bmatrix} \dot{x}_a & \dot{y}_a & \dot{z}_a & \dot{\theta}_a \end{bmatrix}^T$ is the velocity vector over the fixed frame of reference $< \mathcal{R} >$, Γ_a is the matrix that transforms the control velocities into the reference frame velocities and $\mathbf{U_a} = \begin{bmatrix} u_{la} & u_{ma} & u_{na} & \omega_a \end{bmatrix}^T$ represents the vector of aerial robot control velocities: u_{la} is the linear velocity of front control, u_{ma} linear velocity of lateral control, u_{na} linear velocity of altitude control and ω_a is the angular rotation velocity.

Remark: The model described is a simplified kinematic model because the pitch and roll angles of the AUV is not considered in these work.

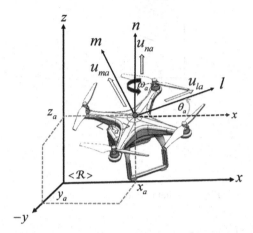

Fig. 1. Kinematic schematic of the autonomous quadcopter.

3 Controller Design

The trajectory tracking problem result when the robot has to follow a virtual point that moves continuously over time along the fixed frame reference $< \mathcal{R} >$, Fig. 2 shows the trajectory tracking behavior of the UAV on the fixed frame.

Therefore, the virtual model of the trajectory is similar to the simplified kinematic model of the UAV as following:

$$\begin{bmatrix} \dot{x}_r \\ \dot{y}_r \\ \dot{z}_r \\ \dot{\theta}_r \end{bmatrix} = \begin{bmatrix} \cos(\theta_r) & -\sin(\theta_r) & 0 & 0 \\ \sin(\theta_r) & \cos(\theta_r) & 0 & 0 \\ 0 & 0 & 1 & 0 \\ 0 & 0 & 0 & 1 \end{bmatrix} \begin{bmatrix} u_{lr} \\ u_{mr} \\ u_{nr} \\ \omega_r \end{bmatrix} \tag{3}$$

So, (3) rewritten in reduced terms can be represented as:

$$\dot{\chi}_r = \mathbf{\Gamma}_r \mathbf{U}_r \tag{4}$$

where, $\dot{\chi}_\mathbf{r} = \begin{bmatrix} \dot{x}_r & \dot{y}_r & \dot{z}_r & \dot{\theta}_r \end{bmatrix}^T$ is the velocity vector over the fixed frame of reference $< \mathcal{R} >$, $\mathbf{\Gamma}_\mathbf{r}$ matrix that transforms the control velocities into the driving velocities with respect to the frame reference of the virtual point and $\mathbf{U}_\mathbf{r} = \begin{bmatrix} u_{lr} & u_{mr} & u_{nr} & \omega_r \end{bmatrix}^T$ represents the velocity vector of the virtual model.

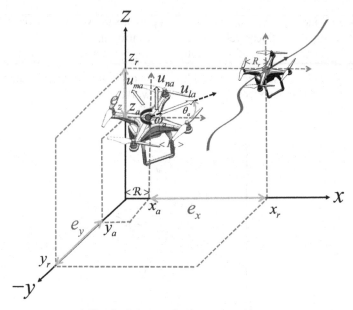

Fig. 2. Scheme of trajectory tracking.

Now, the positional error of aerial robot with respect to the fixed reference frame $< \mathcal{R} >$ is calculated as:

$$\tilde{\chi} = \chi_r - \chi_a = \left[x_r - x_a \; y_r - y_a \; z_r - z_a \; \theta_r - \theta_a \right]^T \tag{5}$$

where, χ_r is the posture of the virtual model, in other words the reference to follow; χ_a is the position of the aerial robot and is determined as follows:

$$\chi_a(t) = \chi_a(0) + \int_{t_0}^{t} \dot{\chi}_a(\gamma) d\gamma \tag{6}$$

The controller proposed in this work is available to work within the mobile reference frame $< \mathcal{R}_a >$, *i.e.*, in the referential frame of the UAV, so the error (5) in the aerial robot is determined by:

$$\tilde{\chi} = \Gamma \tilde{\chi} \tag{7}$$

where, $\tilde{\chi} = \left[e_x \; e_y \; e_z \; e_\theta \right]^T$ is the vector of errors in the referential frame $< \mathcal{R}_a >$ and $\Gamma = \Gamma_a^{-1}$ represents the coordinate transformation matrix of the fixed reference frame to the UAV reference frame, therefore, the robot displacement error is calculated as:

$$\begin{bmatrix} e_x \\ e_y \\ e_z \\ e_\theta \end{bmatrix} = \begin{bmatrix} \cos(\theta_a) & \sin(\theta_a) & 0 & 0 \\ -\sin(\theta_a) & \cos(\theta_a) & 0 & 0 \\ 0 & 0 & 1 & 0 \\ 0 & 0 & 0 & 1 \end{bmatrix} \begin{bmatrix} x_r - x_a \\ y_r - y_a \\ z_r - z_a \\ \theta_r - \theta_a \end{bmatrix} \tag{8}$$

In order to obtain a model that represents the trajectory tracking, the error dynamic behavior is obtained by differentiating the error (7):

$$\dot{\tilde{\chi}} = \dot{\Gamma}\tilde{\chi} + \Gamma\dot{\tilde{\chi}} \tag{9}$$

substituting the corresponding matrices and vectors in (4) you get:

$$\dot{\tilde{\chi}} = -\dot{\theta}_a \begin{bmatrix} \sin(\theta_a) & -\cos(\theta_a) & 0 & 0 \\ \cos(\theta_a) & \sin(\theta_a) & 0 & 0 \\ 0 & 0 & 0 & 0 \\ 0 & 0 & 0 & 0 \end{bmatrix} \tilde{\chi} + \Gamma \begin{bmatrix} \dot{x}_r \\ \dot{y}_r \\ \dot{z}_r \\ \dot{\theta}_r \end{bmatrix} - \Gamma \begin{bmatrix} \dot{x}_a \\ \dot{y}_a \\ \dot{z}_a \\ \dot{\theta}_a \end{bmatrix} \tag{10}$$

Replacing the kinematic model of the UAV in (10) and taking into account that $\Gamma = \Gamma_a^{-1}$ it results:

$$\dot{\tilde{\chi}} = -\dot{\theta}_a \begin{bmatrix} \sin(\theta_a) & -\cos(\theta_a) & 0 & 0 \\ \cos(\theta_a) & \sin(\theta_a) & 0 & 0 \\ 0 & 0 & 0 & 0 \\ 0 & 0 & 0 & 0 \end{bmatrix} \tilde{\chi} + \Gamma \begin{bmatrix} \dot{x}_r \\ \dot{y}_r \\ \dot{z}_r \\ \dot{\theta}_r \end{bmatrix} - \begin{bmatrix} u_{la} \\ u_{ma} \\ u_{na} \\ \omega_a \end{bmatrix} \tag{11}$$

Now, in (11) the kinematic model of the virtual point model is replaced (8) and the errors in the reference system of the aerial robot (3) result:

$$\dot{\tilde{\chi}} = -\dot{\theta}_a \begin{bmatrix} -e_y \\ e_x \\ 0 \\ 0 \end{bmatrix} + \begin{bmatrix} \cos(e_\theta) & -\sin(e_\theta) & 0 & 0 \\ \sin(e_\theta) & \cos(e_\theta) & 0 & 0 \\ 0 & 0 & 1 & 0 \\ 0 & 0 & 0 & 1 \end{bmatrix} \begin{bmatrix} u_{lr} \\ u_{mr} \\ u_{nr} \\ \omega_r \end{bmatrix} - \begin{bmatrix} u_{la} \\ u_{ma} \\ u_{na} \\ \omega_a \end{bmatrix} \tag{12}$$

Then, developing (12) and considering that $\dot{\theta}_a = \omega_a$ the error dynamics are obtained, that is, the error behavior when the quadcopter follows the desired trajectory:

$$\begin{bmatrix} \dot{e}_x \\ \dot{e}_y \\ \dot{e}_z \\ \dot{e}_\theta \end{bmatrix} = \begin{bmatrix} \omega_a e_y - u_{la} + u_{lr}\cos(e_\theta) - u_{mr}\sin(e_\theta) \\ -\omega_a e_x - u_{ma} + u_{lr}\sin(e_\theta) + u_{mr}\cos(e_\theta) \\ u_{nr} - u_{na} \\ \omega_r - \omega_a \end{bmatrix} \tag{13}$$

As you can see the dynamics of the error is nonlinear (13). In order to design a linear control that will bring the error to zero, the dynamic error model is linearized around the operation point of the trajectory tracking in $:e_x = e_y = e_z = e_\theta = 0$. In other words it is considered that: $e_\theta \approx 0$, thus $\cos(e_\theta) \approx 1$ and $\sin(e_\theta) \approx e_\theta$, therefore, the linearized model (13) is expressed as follows:

$$\begin{bmatrix} \dot{e}_x \\ \dot{e}_y \\ \dot{e}_z \\ \dot{e}_\theta \end{bmatrix} = \begin{bmatrix} 0 & \omega_r & 0 & 0 \\ -\omega_r & 0 & 0 & u_{lr} \\ 0 & 0 & 0 & 0 \\ 0 & 0 & 0 & 0 \end{bmatrix} \begin{bmatrix} e_x \\ e_y \\ e_z \\ e_\theta \end{bmatrix} + I \begin{bmatrix} \Delta u_l \\ \Delta u_m \\ \Delta u_n \\ \Delta \omega \end{bmatrix} \tag{14}$$

where, I is an identity matrix of dimension four and $\Delta u_l = u_{lr} - u_{la}$, $\Delta u_m = u_{mr} - u_{ma}$, $\Delta u_l = u_{nr} - u_{na}$, $\Delta \omega = \omega_r - \omega_a$ represent the variation of the control actions. The linearized system can be represented as a model of states:

$$\dot{\tilde{\chi}} = \mathbf{A}\tilde{\chi} + \mathbf{B}\Delta \mathbf{U} \tag{15}$$

3.1 Control Scheme

The proposed control loop is used to find an entry $\Delta \mathbf{U}^*$ such that the position error converges to zero in an optimal way. A Quadratic Linear Regulator (LQR) is proposed [11], based on linearized error dynamic, for which the feedback gain results from the minimization of the cost functional or performance index:

$$J = \frac{1}{2} \int_0^\infty \left(\tilde{\chi}_{(t)}^T \mathbf{Q}\tilde{\chi}_{(t)} + \Delta \mathbf{U}_{(t)}^T \mathbf{R}\Delta \mathbf{U}_{(t)} \right) dt \tag{16}$$

J represents the cost functional to minimize, $\mathbf{Q}$ it is a positive semi-defined matrix of $\mathbb{R}^{4\times4}$ and weigh the system states; $\mathbf{R}$ is a definite positive matrix of $\mathbb{R}^{4\times4}$ which weighs the error correction actions. The trajectory tracking control law that provides UAV driving velocities is defined as [19]:

$$\bar{\mathbf{U}} = -K\chi + \mathbf{U}_{ref} \tag{17}$$

where $\mathbf{K}$ it is the optimal gain that results from the minimization of the functional cost J, the resulting loop feedback:

$$\begin{bmatrix} \bar{u}_{la} \\ \bar{u}_{ma} \\ \bar{u}_{na} \\ \bar{\omega}_a \end{bmatrix} = - \begin{bmatrix} \Delta u_l^* \\ \Delta u_m^* \\ \Delta u_n^* \\ \Delta \omega^* \end{bmatrix} + \begin{bmatrix} v_{ref}\cos(e_\theta) \\ v_{ref}\sin(e_\theta) \\ \dot{z}_{ref} \\ \omega_{ref} \end{bmatrix} \tag{18}$$

$\Delta \mathbf{U}^* = \begin{bmatrix} \Delta u_l^* & \Delta u_l^* & \Delta u_l^* & \Delta \omega^* \end{bmatrix}^T$ are the optimal compensation velocities for correction of the trajectory tracking error. Figure 3 shows the proposal of the control loop to track the trajectory of a UAV with Optima Gain.

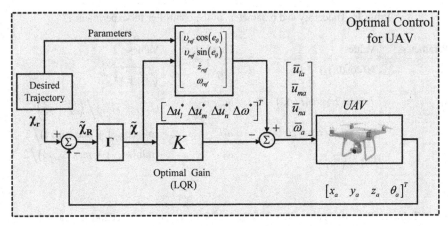

Fig. 3. Trajectory tracking control scheme with optimal gain.

4 Experimental Results

In order to validate the controller, two experimental tests are proposed with Phantom 3 Pro (see Fig. 4). The controller is processed in Matlab Software and the data is sent and received through the communication architecture developed in [10]. The proposed experiments are:

Fig. 4. Phantom 3 pro for experimental tests.

Experiment 1: Consists of following a defined trajectory and analyzing the behavior of the controller with different values of gain **Q** and **R** of the cost function proposed. The desired trajectory and parameters for the controller are defined in Table 1 for the experiment. The gains for experiment 1 are defined in Table 2. The proposed gains prioritize the error in x, y; The error in z corrects the desired position slowly.

Table 1. Trajectory and parameters of the controller for experiment 1.

Parameters	Values	Parameters	Values
x_r	$10\cos(0.1t) - 10$	υ_{ref}	$\sqrt{\dot{x}_r^2 + \dot{y}_r^2}$
y_r	$7\sin(0.2t)\cos(t/15) - 2$	ω_{ref}	$(\dot{x}_r\ddot{y}_r - \ddot{x}_r\dot{y}_r)\big/(\dot{x}_r^2 + \dot{y}_r^2)$
z_r	$0.5\sin(2t/5) + 20$	u_{lr}	$(\min(\upsilon_{ref}) + \max(\upsilon_{ref}))/2$
θ_r	$\tan^{-1}\left(\frac{\dot{y}_r}{\dot{x}_r}\right)$	ω_r	$(\min(\omega_{ref}) + \max(\omega_{ref}))/2$

Table 2. Proposed gain values for J and value of the performance index obtained

Gain	Value	Gain	Value	J
Q_1	$diag\begin{bmatrix} 10\ 5\ 5\ 0.1 \end{bmatrix}$	R_1	$diag\begin{bmatrix} 20\ 20\ 50\ 100 \end{bmatrix}$	1.1356×10^4
Q_2	$diag\begin{bmatrix} 0.1\ 0.1\ 10\ 0.1 \end{bmatrix}$	R_2	$diag\begin{bmatrix} 10\ 50\ 1000\ 100 \end{bmatrix}$	2.9186×10^5
Q_3	$diag\begin{bmatrix} 0.01\ 0.01\ 0.01\ 0.5 \end{bmatrix}$	R_3	$diag\begin{bmatrix} 5\ 5\ 5\ 5 \end{bmatrix}$	1.4341×10^3

Figure 5 indicates the behavior of the UAV from the starting point and along the trajectory executed, it is clearly seen how the evolution of the trajectory of the aerial robot resembles the desired trajectory $[x_r\ y_r\ z_r\ \theta_r]$, i.e., it meets the trajectory tracking objective according to the gains proposed in each test. Figure 6 shows the trajectory

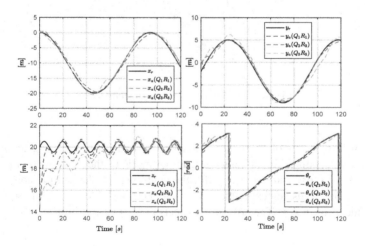

Fig. 5. Evolution of the trajectory described by the UAV.

tracking control errors, it is appreciated that the errors tend to be zero when the robot is on the proposed trajectory.

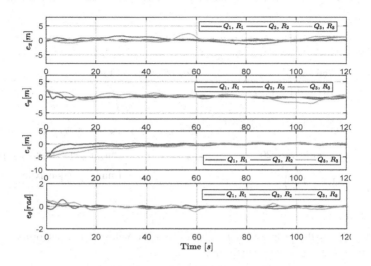

Fig. 6. Evolution of control errors.

Figure 7 shows the control actions injected into the aerial robot during the experimental test, you can see how the actions act according to the gains proposed in the performance index J.

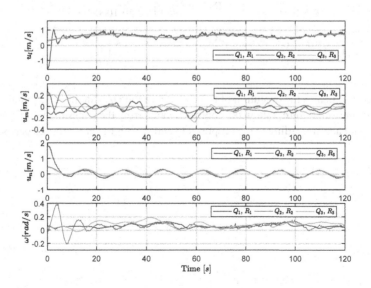

Fig. 7. Optimal control actions applied to the UAV.

Experiment 2: It consists in investigating what are the effects of the different gains $\mathbf{Q}$ and $\mathbf{R}$ that are placed in the functional J, $i.e.$, choose the appropriate values that best act for the proposed task. Table 3 presents the parameters proposed for experiment.

Table 3. Trajectory and parameters of the controller for experiment 2.

Parameters	Values	Parameters	Values
x_r	$10\sin(0.1t) + 10$	v_{ref}	$\sqrt{\dot{x}_r^2 + \dot{y}_r^2}$
y_r	$5\sin(0.1t)\cos(0.1t) - 6$	ω_{ref}	$(\dot{x}_r\ddot{y}_r - \ddot{x}_r\dot{y}_r)\big/\left(\dot{x}_r^2 + \dot{y}_r^2\right)$
z_r	$2\sin(0.1t) + 20$	u_{lr}	$\left(\min(v_{ref}) + \max(v_{ref})\right)/2$
θ_r	$\tan^{-1}\left(\frac{\dot{y}_r}{\dot{x}_r}\right)$	ω_r	$\left(\min(\omega_{ref}) + \max(\omega_{ref})\right)/2$

Table 4 shows the proposed gains values and the value of the performance index resulting from the experiment performed.

Table 4. Gains and value of the performance index obtained from experiment 2.

Gain	Value	Gain	Value	J
Q_1	$diag\begin{bmatrix} 10 & 5 & 5 & 0.1 \end{bmatrix}$	R_1	$diag\begin{bmatrix} 20 & 20 & 50 & 100 \end{bmatrix}$	4.6253×10^4
Q_2	$diag\begin{bmatrix} 0.1 & 0.1 & 10 & 0.1 \end{bmatrix}$	R_2	$diag\begin{bmatrix} 10 & 50 & 1000 & 100 \end{bmatrix}$	4.3345×10^4
Q_3	$diag\begin{bmatrix} 0.01 & 0.01 & 0.01 & 0.5 \end{bmatrix}$	R_3	$diag\begin{bmatrix} 5 & 5 & 5 & 5 \end{bmatrix}$	2.6360×10^4

Figure 8 indicates the evolution of the position of the UAV in the x, y, z plane during the task execution. Figure 9 show the control errors during the experiment, it is observed how the robot acts according to the proposed $\mathbf{Q}$ and $\mathbf{R}$ weight values for the performance index J.

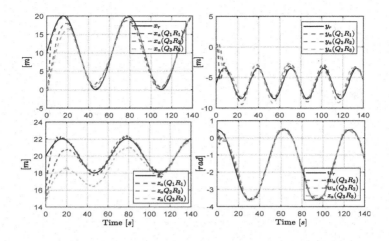

Fig. 8. Evolution of the trajectory described by the UAV.

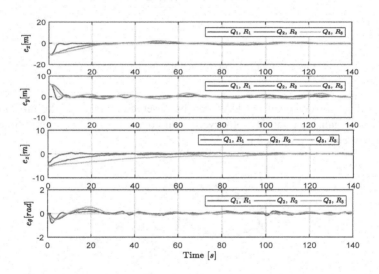

Fig. 9. Evolution of control errors e_z and e_θ.

Finally, Fig. 10 presents the control actions generated by the controller so that the UAV (Phantom 3 Pro) follows the desired trajectory.

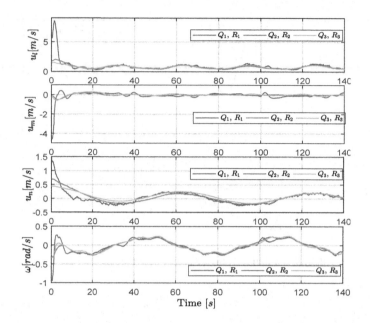

Fig. 10. Optimal control actions applied to the UAV.

The strobe movement of the UAV can be observed in Fig. 11, the movement is executed with gains Q_3 and R_3, observe how it corrects the error in an optimal way getting to execute the proposed task.

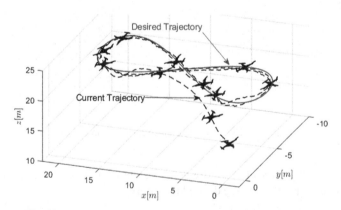

Fig. 11. Movement executed by the UAV during experiment 2.

5 Conclusions

The controller proposed in this work allows to track trajectories in an optimal way, that is, the corrections of the position errors at each instant of time act according to the weight values proposed in the performance index or functional cost J that are minimize through the LQR. With the experiments executed, it is verified that the proposed scheme satisfactorily meets the requirements of trajectory tracking, the result as optimal values are with Q_3 and R_3 in the two experiments these gains execute the optimal task a considerable value of the index performance, take into account that the error correction is done slowly but without applying too much effort in the UAV, that is, the control velocities are not high for correct the error.

Acknowledgments. The authors would like to thanks to the Corporación Ecuatoriana para el Desarrollo de la Investigación y Academia –CEDIA for the financing given to research, development, and innovation, through the CEPRA projects, especially the project CEPRA-XIII-2019-08; Sistema Colaborativo de Robots Aéreos para Manipular Cargas con Óptimo Consumo de Recursos; also to Universidad de las Fuerzas Armadas ESPE, Grupo de Investigación ARSI, and finally to Instituto de Automática de la Universidad Nacional de San Juan for the support and the theoric knowledge provided for the execution of this work.

References

1. Obradovic, R., Vasiljevic, I., Kovacevic, D., Marinkovic, Z., Farkas, R.: Drone aided inspection during bridge construction. In: 2019 Zooming Innovation in Consumer Technologies Conference (ZINC) (2019). https://doi.org/10.1109/zinc.2019.8769345
2. Castillo-Zamora, J.J., Escareno, J., Boussaada, I., Labbani, O., Camarillo, K.: Modeling and control of an aerial multi-cargo system: robust acquiring and transport operations. In: 2019 18th European Control Conference (ECC). Naples, Italy, pp. 1708–1713 (2019)
3. Acosta, A.G., et al.: E-tourism: governmental planning and management mechanism. In: De Paolis, L.T., Bourdot, P. (eds.) AVR 2018. LNCS, vol. 10850, pp. 162–170. Springer, Cham (2018). https://doi.org/10.1007/978-3-319-95270-3_11
4. Elfes, A., S., et al.: A semi-autonomous robotic airship for environmental monitoring missions. In: Proceedings 1998 IEEE International Conference on Robotics and Automation. vol.4, Leuven, Belgium, pp. 3449–3455 (1998). (Cat. No.98CH36146)
5. Andaluz, Víctor H., et al.: Autonomous monitoring of air quality through an unmanned aerial vehicle. In: Wotawa, F., Friedrich, G., Pill, I., Koitz-Hristov, R., Ali, M. (eds.) IEA/AIE 2019. LNCS (LNAI), vol. 11606, pp. 146–157. Springer, Cham (2019). https://doi.org/10.1007/978-3-030-22999-3_14
6. Hang, C., et al.: Three-dimensional fuzzy control of mini quadrotor UAV trajectory tracking under impact of wind disturbance. In: 2016 International Conference on Advanced Mechatronic Systems (ICAMechS) (2016)
7. Jurado, F., Lopez, S.: Continuous-time decentralized neural control of a quadrotor UAV. In: Artificial Neural Networks for Engineering Applications, pp. 39-53 (2019). https://doi.org/10.1016/b978-0-12-818247-5.00013-7
8. Pantoja, A., Yela, J.: Nonlinear control of a quadrotor for attitude stabilization. In: 2017 IEEE 3rd Colombian Conference on Automatic Control (CCAC) (2017). https://doi.org/10.1109/ccac.2017.8276467

9. Andaluz, V.H., Carvajal, C.P., Pérez, J.A., Proaño, L.E.: Kinematic nonlinear control of aerial mobile manipulators. In: Huang, Y., Wu, H., Liu, H., Yin, Z. (eds.) ICIRA 2017. LNCS (LNAI), vol. 10464, pp. 740–749. Springer, Cham (2017). https://doi.org/10.1007/978-3-319-65298-6_66

10. Andaluz, V., Gallardo, C., Chicaiza, F., Carvajal, C., et al.: Robot nonlinear control for unmanned aerial vehicles' multitasking. Assembly Autom. **38**(5), 645–660 (2018). https://doi.org/10.1108/AA-02-2018-036

11. Xiong, J., Zhang, G.: Global fast dynamic terminal sliding mode control for a quadrotor UAV. ISA Trans. **66**, 233–240 (2017). https://doi.org/10.1016/j.isatra.2016.09.019

12. Mathisen, S., Gryte, K., Johansen, T., Fossen, T.: Non-linear model predictive control for longitudinal and lateral guidance of a small fixed-wing UAV in precision deep stall landing. AIAA Infotech @ Aerospace (2016). https://doi.org/10.2514/6.2016-0512

13. Santos, M.C.P., Rosales, C.D., Sarapura, J.A., et al.: An adaptive dynamic controller for quadrotor to perform trajectory tracking tasks. J. Intell. Robot. Syst. **93**(1), 5–16 (2018). https://doi.org/10.1007/s10846-018-0799-3

14. Kumar, R., Dechering, M., Pai, A., et al.: Differential flatness based hybrid PID/LQR flight controller for complex trajectory tracking in quadcopter UAVs. In: 2017 IEEE National Aerospace and Electronics Conference (NAECON) (2017). https://doi.org/10.1109/naecon.2017.8268755

15. Rosales, C., Gandolfo, D., Scaglia, G., Jordan, M., Carelli, R.: Trajectory tracking of a mini four-rotor helicopter in dynamic environments - a linear algebra approach. Robotica **33**(8), 1628–1652 (2015)

16. Chen, Y., Luo, G., Mei, Y., et al.: UAV path planning using artificial potential field method updated by optimal control theory. Int. J. Syst. Sci. **47**(6), 1407–1420 (2014). https://doi.org/10.1080/00207721.2014.929191

17. Peng, H., Li, F., Liu, J., Ju, Z.: A symplectic instantaneous optimal control for robot trajectory tracking with differential-algebraic equation models. IEEE Trans. Ind. Electron. **67**(5), 3819–3829 (2019). https://doi.org/10.1109/TIE.2019.2916390

18. Cardenas Alzate, P., Velez, G., Mesa, F.: Optimum control using finite time quadratic linear regulator. Contemp. Eng. Sci. **11**, 4709–4716 (2018). https://doi.org/10.12988/ces.2018.89516

19. De Luca, A., Oriolo, G., Vendittelli, M.: Control of wheeled mobile robots: Experimental Overview. 181–226. Springer, Berlin, Heidelberg (2001). https://doi.org/10.1007/3-540-45000-9_8

Non-linear Control of Aerial Manipulator Robots Based on Numerical Methods

David F. Grijalva[✉], Jaime A. Alegría[✉], Víctor H. Andaluz[✉], and Cesar Naranjo[✉]

Universidad de las Fuerzas Armadas ESPE, Sangolquí, Ecuador
{dfgrijalva,jaalegria,vhandaluz1,canaranjo}@espe.edu.ec

Abstract. This work proposes a kinematic modelling and a non-linear kinematic controller for an autonomous aerial mobile manipulator robot that generates velocity commands for trajectory tracking problem. The kinematic modelling is considered using a hexarotor system and robotic arm. The stability and robustness of the entire control system are tested by this method. Finally, the experiment results are presented and discussed, and validate the proposed controller.

Keywords: Aerial manipulator · AMR · Control no lineal

1 Introduction

Robotics has greatly evolved and is now present in several areas of the industrial field, as well as the service robotics, where a wide study and research field exists due to its several applications, such as: robotic service assistant in nursing [1]; service robotics with social conscience for guiding and helping passengers in airports [2]; robotics for home assistance [3]; service robot used for preventing collisions [4]. Service robots may present unexpected behaviors that represent economic and safety risks, especially for the human staff around them [5], because of the wide operating field of service robots, some structures have been developed so that they can work in land, water and aerial environments, therefore, they can use wheels, legs, and propellers, as its application requires. For service robotics, one of the main workplaces are locations where there are only flat surfaces for movement, so in order to cover these locations, unmanned aerial vehicles are used (UAV) [6].

Unmanned aerial vehicles, also known as drones, are flying objects that are not manned by a pilot [7]. Aerial vehicles have been under research lately [8] generating several applications, like search and rescue operations, surveillance, handling or grabbing tasks, in which it is needed to include a robotic arm that can limit the vehicle from executing more complex and precise tasks.

The combination of mobile aerial systems with robotic arms is known as aerial mobile manipulators [9], which are a type of unmanned aerial vehicles with the ability to physically interact within an ideally unlimited workspace [6]. The most often used platforms for aerial mobile manipulators are helicopter-type or multirotor with their different varieties: quadrotor, hexarotor or octorotor, combined with robotic arms with

© Springer Nature Switzerland AG 2020
H. Fujita et al. (Eds.): IEA/AIE 2020, LNAI 12144, pp. 97–107, 2020.
https://doi.org/10.1007/978-3-030-55789-8_9

multiple freedom degrees [14]. The main applications of aerial mobile manipulators are: i) *Military field:* transportation of equipment tasks, search under rubble and rescue tasks [11]; ii) *Commercial field:* performing merchandise transportation tasks [10]; iii) Industrial field: welding, handling high-rise objects, equipment and light machinery moving, which require high precision for performing the task [9, 10, 12].

The aerial manipulators have recently covered a vast area of research, which has focused its attention on multiple research groups and international companies interested in such technology by boosting robotic systems with manipulation abilities as its main goal [10]; and especially focusing on studying: i) *Construction:* As the complexity of the task or application increases, the advances in mechanical design of the aerial manipulator must innovate and focus on designing channeled fans, boosting tiltable frames mechanisms, and studying the concept of the architecture of a tiltable rotor, which has been a very thoroughly studied topic to increase the flying time of such vehicles [15]; ii) *Energy consumption:* One of the most relevant parameters to be considered in these vehicles is the light weight and the inertia that the robotic arm must present due to the severe limitations of useful load and the convenience to extremely reduce the influence of the arm movement over the UAV's stability, apart from optimizing the tilting angles of the propeller depending on the application [14]; iii) *Modelling:* There are two criteria for modelling an aerial manipulator, which are a) Kinematic analysis, and b) Dynamic analysis. The kinematic analysis determines the movement and restrictions of the mobile manipulator. On the other hand, the dynamic analysis focuses on the study of several pairs and forces intervening in movement (inertia, centrifuge, coriolis, gravity, etc.), focused in using methods and proceedings based in the Newtonian and Lagrangian mechanics. This analysis is crucial for designing and assessing the mechanic structure of the aerial manipulator, as well as sizing the actuators and other parts [13, 15]; iv) *Control:* There are several types of controllers, but they are used depending on the aerial manipulator considerations. Such is the case of the included control that consider the UAV and the robotic arm as a whole system and in these controllers focused in one control-point only, (tip of the robotic arm) are used. There are also controllers focused on not included controllers which are characterized by considering 2 different systems, one for the UAV and one for the robotic arm, and must be applied to the work position, and later the robotic arm controller is used to perform the task [15]. However, it is still necessary to address the technological challenges before the use of this kind of technology can be considered reliable. Among these challenges, the ability to handle impacts during a task of a highly dynamic physical interaction is still an unexplored research topic [9].

This paper presents a non-linear control strategy for resolving the trajectory tracking problem of an Aerial manipulator robot that will be defined with the acronym ARM. Which is constituted by an hexacopter mounting a robotic arm of 3 degrees of freedom mounted on back of base. For the design of the controller, the kinematic model of the AMR is used which has as input the velocity and orientation, this controller is designed based on nine velocities commands of the AMR, six corresponding to the aerial platform: forward, lateral, up/downward and orientation, the last three are those who command the manipulator robot. It is also pointed out that the workspace has a single reference that is located in the operative end of the AMR R(x y z). The stability of the controller is analyzed by the Lyapunov's method and to validate the proposed control algorithm, experimental processes are presented and discussed in this paper.

The document is organized as described below. Section 1 describes the characteristics and applications of the aerial manipulators. Section 2 includes the movement characteristics of the hexarotor. Section 3 describes its kinematic model. Section 4 details the control to be implemented in the aerial manipulator, stability and robustness analysis. Section 5 includes the results obtained from the experimental tests. And, Sect. 6 presents the conclusions.

2 Motion Characteristics

An UAV is an unmanned aerial vehicle, which when combined with a robotic arm turns into an aerial manipulator, Fig. 1 shows the combination of a rotational-frames vehicle with an anthropomorphic-type robotic arm.

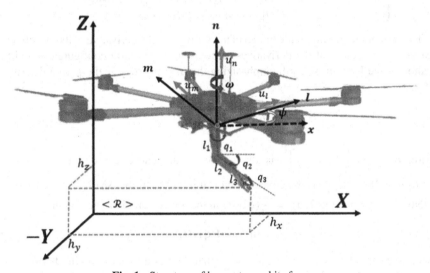

Fig. 1. Structure of hexarotor and its frames

Since this work is formed by a 3-DOF robotic arm on an hexarotor UAV, as shown in Fig. 1, in which the two principal forces for moving it are gravity and the rotors thrust, its movements are controlled by several effects, whether mechanical or aerodynamic. The main effects on the hexarotor are listed in Table 1.

3 AMR Model

The aerial manipulator robot configuration is defined by a vector $\mathbf{q} = \begin{bmatrix} \mathbf{q_h}^T & \mathbf{q_a}^T \end{bmatrix}^T$ where $\mathbf{q_h} = \begin{bmatrix} x_u & y_u & z_u & \psi \end{bmatrix}^T$ that represents the specific coordinates of the UAV, and $\mathbf{q_a} = \begin{bmatrix} q_1 & q_2 & q_3 \end{bmatrix}^T$ the specific coordinates of the robotic arm.

Table 1. The main effects on the acting hexarotor

Effects	Fountainhead
Aerodynamics effects	Rotating propellers
Inertiel counter torque	Velocity change of propellers
Effect of gravity	Position of the center of mass
Gyroscopic effects	Change in the direction of the drone
Friction effect	All drone movements

Where, the next system of equations represents the direct kinematic model of the ARM:

$$\begin{cases} h_x = x_u + l_2 \cos(q_1 + \psi) \cos(q_2) + l_3 \cos(q_1 + \psi) \cos(q_2 + q_3) \\ h_y = y_u + l_2 \sin(q_1 + \psi) \cos(q_2) + l_3 \sin(q_1 + \psi) \cos(q_2 + q_3) \\ \qquad h_z = z_u + l_1 + l_2 \sin(q_2) + l_3 \sin(q_2 + q_3) \end{cases} \quad (1)$$

The instantaneous kinematic model of an AMR gives the derivative of its end-effector location as a function of the derivatives of both the robotic arm configuration and the location of the UAV, it is worth emphasizing that it is practically the partial derivative of Eq. (1).

$$\dot{\mathbf{h}}(t) = \frac{df}{d\mathbf{q}}(\mathbf{q})\mathbf{v}(t)$$

where, $\dot{\mathbf{h}}(t) = \begin{bmatrix} \dot{h}_x & \dot{h}_y & \dot{h}_z \end{bmatrix}^T$ is the vector of the end-effector velocity, $\mathbf{v}(t) = \begin{bmatrix} u_l & u_m & u_n & \omega & \dot{q}_1 & \dot{q}_2 & \dot{q}_3 \end{bmatrix}^T$ is the control vector of mobility of the AMR.

Now, after replacing $\mathbf{J}(\mathbf{q}) = \frac{df}{d\mathbf{q}}(\mathbf{q})\mathbf{v}(t)$ in the above equation, we obtain

$$\dot{\mathbf{h}}(t) = \mathbf{J}(\mathbf{q})\mathbf{v}(t) \quad (2)$$

where, $\mathbf{J}(\mathbf{q}) \in \mathbb{R}^{3 \times 7}$ is the Jacobian matrix that defines a linear mapping between the vector of the AMR velocities $\mathbf{v}(t) \in \mathbb{R}^7$ and the vector of the end-effector velocity $\dot{\mathbf{h}}(t) \in \mathbb{R}^3$.

4   Control Algorithm: Numerical Methods

Through Euler's approximation to the kinematic model for AMR trajectory tracking, the following discrete kinematic model is obtained.

$$\mathbf{h}(k + 1) = \mathbf{h}(k) + T_0 \mathbf{J}(\mathbf{q}(k))\mathbf{v}_{ref}(k) \quad (3)$$

where, values of $\mathbf{h}$ at the discrete time $t = kT_0$ will be denoted as $\mathbf{h}(k)$, T_0 is the sample time, and $k \in \{1, 2, 3, 4, 5 \ldots\}$. Next by the Markov property and to adjusting the performance of the proposed control law, the states vector $\mathbf{h}(k + 1)$ is replaced by,

$$\mathbf{h}(k + 1) = \mathbf{h}_d(k + 1) - \mathbf{W}(\mathbf{h}_d(k) - \mathbf{h}(k)) \quad (4)$$

where, $\mathbf{W}$ is weight matrix of control errors defined by $\mathbf{h_d}(k) - \mathbf{h}(k)$.

For the design of the control law is used the Euler's approximation of the kinematic model of the ARM (2) and in turn the property of Markov (3), hence

$$\mathbf{h}_d(k+1) - \mathbf{W}(\mathbf{h}_d(k) - \mathbf{h}(k)) = \mathbf{h}(k) + T_0\mathbf{J}(\mathbf{q}(k))\mathbf{v}_{ref}(k)$$

$$\mathbf{J}(\mathbf{q}(k))\mathbf{v}_{ref}(k) = \frac{1}{T_0}(\mathbf{h}_d(k+1) - \mathbf{W}(\mathbf{h}_d(k) - \mathbf{h}(k))) \tag{5}$$

Equation (4) can be represented by, $\mathbf{Jv} = \mathbf{b}$, through the properties of Linear Algebra the following control law is proposed for trajectory tracking

$$\mathbf{v}_{ref} = \mathbf{J}^{\#}\mathbf{b}$$

where $\mathbf{J}^{\#}(\mathbf{q}(k)) = \mathbf{J}^T(\mathbf{q}(k))\big(\mathbf{J}(\mathbf{q}(k))\mathbf{J}^T(\mathbf{q}(k))\big)^{-1}$ represents the pseudoinverse matrix of $\mathbf{J}(\delta(k), \mathbf{q}(k))$, hence, the proposed control law is:

$$\mathbf{v}_{ref}(k) = \frac{1}{T_0}\mathbf{J}^{\#}(\mathbf{q}(k))(\mathbf{h}_d(k+1) - \mathbf{W}(\mathbf{h}_d(k) - \mathbf{h}(k)) - \mathbf{h}(k)) \tag{6}$$

where $\mathbf{v}_{ref}(k) = \begin{bmatrix} u_{l_{ref}}(k) & u_{m_{ref}}(k) & u_{n_{ref}}(k) & \omega_{r_{ref}}(k) & \dot{q}_{1_{ref}}(k) & \dot{q}_{2_{ref}}(k) & \dot{q}_{3_{ref}}(k) \end{bmatrix}^T$ is the maneuverability vector of AMR.

4.1 Stability Análisis

In order to evaluate the behavior of the AMR control errors, the stability analysis is performed, for which it is considered perfect velocity tracking, i.e. $\mathbf{v}_{ref}(k) \equiv \mathbf{v}(k)$. The behavior of control errors can be obtained by relating the dictated model of the AMR (3) and the proposed control law (6)

$$\frac{1}{T_0}(\mathbf{h}(k+1) - \mathbf{h}(k)) = \frac{1}{T_0}\mathbf{JJ}^{\#}(\mathbf{h}_d(k+1) - \mathbf{W}(\mathbf{h_d}(k) - \mathbf{h}(k)) - \mathbf{h}(k)).$$

Where $\mathbf{I} = \mathbf{JJ}^{\#}$ simplifying the terms, the closed-loop equation is

$$\mathbf{h}_d(k+1) - \mathbf{h}(k+1) = \mathbf{W}(\mathbf{h_d}(k) - \mathbf{h}(k)) \tag{7}$$

where the control error is defined by $\tilde{\mathbf{h}}(k) = \mathbf{h}_d(k) - \mathbf{h}(k)$ and $\tilde{\mathbf{h}}(k+1) = \mathbf{h}_d(k+1) - \mathbf{h}(k+1)$, therefore (7) can be rewritten as

$$\tilde{\mathbf{h}}(k+1) = \mathbf{W}\tilde{\mathbf{h}}(k). \tag{8}$$

In order to evaluate the evolution of the control error, the i-th control error is considered, $\tilde{h}_i(k+1)$ and $\tilde{h}_i(k)$ the weight matrix is defined as $\mathbf{W} = diag(w_{11}, w_{22}, w_{33})$

$$\tilde{h}_i(k+1) = w_{ii}\tilde{h}_i(k) \tag{9}$$

Table 2 represents the evolution of the i-th control error for different instants of time

If $k \to \infty$ then $\tilde{h}(\infty) = w_{ii}^{\infty}\tilde{h}(1)$, therefore so that the $\tilde{h}(\infty) \to 0$ values of the diagonal weight matrix must be between $0 < diag(w_{11}, w_{22}, w_{33}) < 1$. As described, it can be concluded that control errors have asymptotic stability, that is to say $\tilde{h}(k) \to 0$, when $k \to \infty$.

Table 2. Evolution of *it-th* control error.

k	$\tilde{h}_i(k+1)$	$w_{ii}\tilde{h}_i(k)$
1	$\tilde{h}_i(2)$	$w_{ii}\tilde{h}_i(1)$
2	$\tilde{h}_i(3)$	$w_{ii}\tilde{h}_i(2) = w_{ii}^2\tilde{h}_i(1)$
3	$\tilde{h}_i(4)$	$w_{ii}^3\tilde{h}_i(1)$
$\vdots$	$\vdots$	$\vdots$
n	$\tilde{h}_i(n+1)$	$w_{ii}^n\tilde{h}(1)$

4.2 Robustness Analysis

In order to the robustness analysis it is considered:

$$\mathbf{v}(k) = \mathbf{v}_{ref}(k) - \tilde{\mathbf{v}}(k)$$

The behavior of control errors can be obtained by relating the dictated model of the AMR (3), the proposed control law (6)

$$\frac{1}{T_0}(\mathbf{h}(k+1) - \mathbf{h}(k)) = \frac{1}{T_0}\mathbf{JJ}^{\#}\left(\mathbf{h}_d(k+1) - \mathbf{h}(k) - \mathbf{W}\left(\tilde{\mathbf{h}}(k)\right) - \mathbf{J}\tilde{\mathbf{v}}(k)\right)$$

where $\mathbf{I} = \mathbf{JJ}^{\#}$ simplifying the terms, the equation is

$$\mathbf{J}\tilde{\mathbf{v}}(k) = \mathbf{h}_d(k+1) - \mathbf{h}(k+1) - \mathbf{W}\left(\tilde{\mathbf{h}}(k)\right)$$

$$\tilde{\mathbf{h}}(k+1) = \mathbf{W}\left(\tilde{\mathbf{h}}(k)\right) + \mathbf{J}\tilde{\mathbf{v}}(k)$$

$$\tilde{\mathbf{h}}(n+1) = \mathbf{W}^n\left(\tilde{\mathbf{h}}(n)\right) + \mathbf{J}\tilde{\mathbf{v}}(n) \tag{10}$$

if $0 < \mathbf{W} < 1$ and $n \rightarrow \infty$

$$\therefore \left\|\tilde{\mathbf{h}}(n+1)\right\| < \|\mathbf{J}\tilde{\mathbf{v}}(n)\|.$$

5 Experimental Results

The experimental tests of the proposed control algorithm are performed on the Aerial Manipulator Robot consisting of a 3DOF robotic arm, a UAV (Matrice 600 pro) and the PC where the control actions are sent. Figure 2 shows the ARM used for the experimental tests.

A desired trajectory is established to verify the performance of the proposed control law. For the compliance of the trajectory the AMR PC sends the control actions that allow the AMR mobility and the tracking of the trajectory. It is necessary to know the initial conditions for the execution of the control law. The initial conditions are defined

Fig. 2. Aerial manipulator robot

by: $x_{UAV}(1)$, $y_{UAV}(1)$, $z_{UAV}(1)$, $q_1(1)$, $q_2(1)$, $q_3(1)$. The conditions can be defined from the controller or read the actual positions of the robotic arm and drone. Table 3 below shows the initial conditions and the desired values of the trajectory.

Figure 3, shows the desired and realized trajectory of the final effectory. It can be seen that the proposed controller has a good performance. The data of the trajectory made are real obtaining experimental tests.

The control errors are close to zero as shown in Fig. 4. This allows to say that the controller has a good performance, due to the fact that in the experimental tests in spite of the existence of perturbations the AMR fulfills the established trajectory.

The maneuverability commands shown in Fig. 5 are those that allow the movement of the UAV defined by u_l movement forward, u_m lateral move, u_n up and down motion and ψ orientation of UAV.

Table 3. Initial conditions and target trajectory values

Variables	Values	Variables	Values
$x_{UAV}(1)$	0 [m]	$q_3(1)$	0.4 [rad]
$y_{UAV}(1)$	0 [m]	$h_{xd}(t)$	$5\cos(0.05t) + 5$[m]
$z_{UAV}(1)$	0 [m]	$h_{yd}(t)$	$5\sin(0.05t) + 5$ [m]
$\psi_{UAV}(1)$	0 [m]	$h_{zd}(t)$	$\sin(0.3t) + 16$ [m]
$q_1(1)$	0.2 [rad]	$\psi_{zd}(t)$	$\tan^{-1}\left(\dot{h}_{yd} / \dot{h}_{xd}\right)$ [rad]
$q_2(1)$	0.8 [rad]		

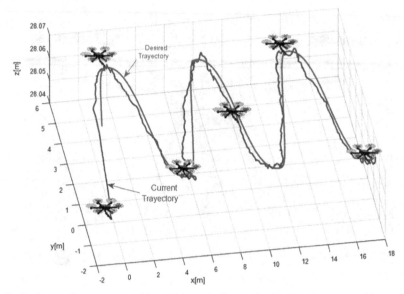

Fig. 3. Stroboscopic movement of the aerial manipulator robot in the trajectory tracking problem.

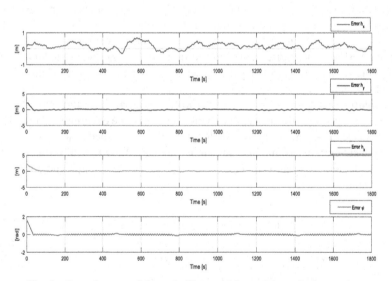

Fig. 4. Control errors of the end-effector of the aerial manipulator robot.

The robotic arm maneuverability commands are defined by q_1, q_2 y q_3 which will allow the end effector to comply with the established trajectory (Fig. 6).

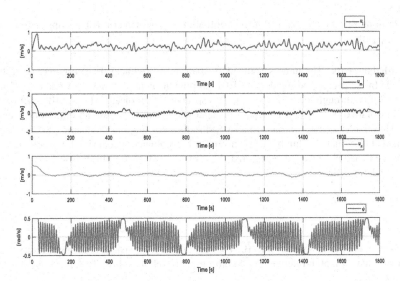

Fig. 5. Commands of maneuverability of aerial manipulator robot

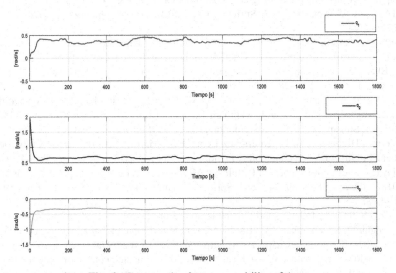

Fig. 6. Commands of maneuverability of Arm.

6 Conclusions

In this article, a non-linear controller based on numerical methods for trajectory tracking was proposed, through the kinematic analysis that allowed the implementation of an precise model of an aerial manipulator robot, the same one that was used for experimental tests. The easy implementation of the controller and its simplicity make it much more understandable and easy to use, in addition to presenting excellent stability and robustness, in analytical tests and experimental tests. The results of the experimental tests have demonstrated the capacity of the controller to perform control actions globally and asymptotically.

Acknowledgements. The authors would like to thank the Corporación Ecuatoria-na para el Desarrollo de la Investigación and Academia CEDIA for the financing given to research, development, and innovation, through the CEPRA projects, especially the project CEPRA-XIII-2019-08; Sistema colaborativo de robots aéreos para manejar cargas con un consumo óptimo de recursos; also to Universidad de las Fuerzas Armadas ESPE, Escuela Superior Politécnica de Chimborazo, Universidad Nacional de Chimborazo, Universidad tecnológica Indoamérica, Universidad internacional de Ecuador, Universidad central de Venezuela, and Grupo de Investigación ARSI, for the support to develop this work.

References

1. Baumgarten, S., Jacobs, T., Graf, B.: The robotic service assistant-relieving the nursing staff of workload. In: 50th International Symposium on Robotics, ISR 2018, pp. 1–4. VDE, June 2018
2. Triebel, R., et al.: SPENCER: a socially aware service robot for passenger guidance and help in busy airports. In: Wettergreen, D.S., Barfoot, T.D. (eds.) Field and Service Robotics. STAR, vol. 113, pp. 607–622. Springer, Cham (2016). https://doi.org/10.1007/978-3-319-27702-8_40
3. Cesta, A., Cortellessa, G., Orlandini, A., Sorrentino, A., Umbrico, A.: A semantic representation of sensor data to promote proactivity in home assistive robotics. In: Arai, K., Kapoor, S., Bhatia, R. (eds.) IntelliSys 2018. AISC, vol. 868, pp. 750–769. Springer, Cham (2019). https://doi.org/10.1007/978-3-030-01054-6_53
4. Mendes, M., Coimbra, A.P., Crisóstomo, M.M., Cruz, M.: Vision-based collision avoidance for service robot. In: Ao, S.-I., Gelman, L., Kim, H.K. (eds.) WCE 2017, pp. 233–248. Springer, Singapore (2019). https://doi.org/10.1007/978-981-13-0746-1_18
5. Guerrero-Higueras, Á.M., Rodríguez-Lera, F.J., Martín-Rico, F., Balsa-Comerón, J., Matellán-Olivera, V.: Accountability in mobile service robots. In: Fuentetaja Pizán, R., García Olaya, Á., Sesmero Lorente, M.P., Iglesias Martínez, J.A., Ledezma Espino, A. (eds.) WAF 2018. AISC, vol. 855, pp. 242–254. Springer, Cham (2019). https://doi.org/10.1007/978-3-319-99885-5_17
6. Ortiz, J.S., et al.: Modeling and kinematic nonlinear control of aerial mobile manipulators. In: Zeghloul, S., Romdhane, L., Laribi, M.A. (eds.) Computational Kinematics. MMS, vol. 50, pp. 87–95. Springer, Cham (2018). https://doi.org/10.1007/978-3-319-60867-9_11
7. Varela-Aldás, J., Andaluz, V.H., Chicaiza, F.A.: Modelling and control of a mobile manipulator for trajectory tracking. In: 2018 International Conference on Information Systems and Computer Science. INCISCOS, pp. 69–74. IEEE, November 2018
8. Škorput, P., Mandžuka, S., Gregurić, M., Vrančić, M.T.: Applying unmanned aerial vehicles (UAV) in traffic investigation process. In: Karabegović, I. (ed.) NT 2019. LNNS, vol. 76, pp. 401–405. Springer, Cham (2020). https://doi.org/10.1007/978-3-030-18072-0_46
9. Bartelds, T., Capra, A., Hamaza, S., Stramigioli, S., Fumagalli, M.: Compliant aerial manipulators: toward a new generation of aerial robotic workers. IEEE Robot. Autom. Lett. $\mathbf{1}$(1), 477–483 (2016)
10. Andaluz, V.H., Carvajal, C.P., Pérez, J.A., Proaño, L.E.: Kinematic nonlinear control of aerial mobile manipulators. In: Huang, Y., Wu, H., Liu, H., Yin, Z. (eds.) ICIRA 2017. LNCS (LNAI), vol. 10464, pp. 740–749. Springer, Cham (2017). https://doi.org/10.1007/978-3-319-65298-6_66

11. Suárez, A., Sanchez-Cuevas, P., Fernandez, M., Perez, M., Heredia, G., Ollero, A.: Lightweight and compliant long reach aerial manipulator for inspection operations. In: 2018 IEEE/RSJ International Conference on Intelligent Robots and Systems (IROS), pp. 6746–6752. IEEE, October 2018

12. Tognon, M., Franchi, A.: Dynamics, control, and estimation for aerial robots tethered by cables or bars. IEEE Trans. Robot. **33**(4), 834–845 (2017)

13. Molina, M.F., Ortiz, J.S.: Coordinated and cooperative control of heterogeneous mobile manipulators. In: Ge, S.S., et al. (eds.) ICSR 2018. LNCS (LNAI), vol. 11357, pp. 483–492. Springer, Cham (2018). https://doi.org/10.1007/978-3-030-05204-1_47

14. Andaluz, V.H., et al.: Nonlinear controller of quadcopters for agricultural monitoring. In: Bebis, G., Boyle, R., Parvin, B., Koracin, D., et al. (eds.) ISVC 2015. LNCS, vol. 9474, pp. 476–487. Springer, Cham (2015). https://doi.org/10.1007/978-3-319-27857-5_43

15. Rajappa, S., Ryll, M., Bülthoff, H.H., Franchi, A.: Modeling, control and design optimization for a fully-actuated hexarotor aerial vehicle with tilted propellers. In: 2015 IEEE International Conference on Robotics and Automation (ICRA), pp. 4006–4013. IEEE, May 2015

Non-linear 3D Visual Control for an Unmanned Aerial Vehicle

Daniel D. Guevara$^{(\boxtimes)}$ and Víctor H. Andaluz

Universidad de las Fuerzas Armadas ESPE, Sangolquí, Ecuador
{ddguevara1,vhandaluz1}@espe.edu.ec

Abstract. This document presents the development of a kinematic control with visual feedback based on images, to solve the objective tracking problem in the semi-structured 3D workspace in UAV. The design of the entire controller is based on a total Jacobian that contains the geometric, image, and object Jacobian. These Jacobians allow the calculation of the control errors that feedback the system through the characteristics of the image, allowing the stability and robustness of the system to be determined. The system implementation was developed by calibrating the UAV's vision sensor against a global reference system, using the Python programming language, which is the means of communication between the PC and the UAV, in the same way with MATLAB mathematical software.

Keywords: UAV visual control · Stability analysis and robustness

1 Introduction

Unmanned Aerial Vehicles (UAVs) play important roles in both, the consumer and commercial markets. Compared to manned aircrafts, UAVs are generally much more cost effective, smaller, able to approach dangerous areas and there-fore widely applied in search and rescue, emergency, delivery, infrastructure inspection and others [1–3]. A basic requirement for a UAV is autonomous and robust navigation and positioning, which will be carried out using vision-based control techniques. Computer vision is gaining importance in the field of mobile robots, currently used in the feedback control cycle, as a cheap, passive and in-formation-rich tool. The vision sensor, usually combined with an inertial measurement unit (IMU) provides solid information about an object, allowing autonomous positioning and navigation [4, 5].

When designing a Computer Vision system, a number of parameters are always taken into account that will be decisive for the appearance of the objects in the image to be the best for the subsequent analysis algorithms. However, for those Computer Vision algorithms that need to extract three-dimensional information from an image or image sequence or to establish the correspondence between two or more cameras, the calibration of the intrinsic and extrinsic parameters of the vision system is a fundamental step [6]. In this study, a camera with a large field of view is adopted to simplify the development and validation of the servo-visual control approach [7]. Currently there is a lot of work

© Springer Nature Switzerland AG 2020
H. Fujita et al. (Eds.): IEA/AIE 2020, LNAI 12144, pp. 108–115, 2020.
https://doi.org/10.1007/978-3-030-55789-8_10

related to visual servo control [8] based on path-following images [6] for manipulators [1], mobile robots and drones [9–11].

In the present research work it is proposed to implement a control with visual feedback in closed loop for the recognition and monitoring of 3D patterns by means of an unmanned aerial vehicle (UAV), based on the characteristics of movement of the aerial robot and the projection of the characteristic points of the image of the pattern to be followed. In addition, the stability and robustness of the proposed control scheme is mathematically analyzed in order to ensure that control errors tend to zero. Finally, the performance of the proposed controller in partially structured work environments was evaluated experimentally. It is important to emphasize that this research focuses on several important points such as pattern detection [6], mathematical modelling and control scheme presented in the Sects. 2 and 3. Inside the controller. It is worth mentioning that the models found will be used to determine the Jacobian matrix of the system, that is, the relationship between the Jacobian Image (vision sensor) and the Jacobian Geometry (UAV) presented in Sect. 2.

2 System Structure

The servo visual handheld camera control technique increases effective resolution and prevents obstruction of the target by the application in mobile robots such as UAVs. The system is composed of a visual sensor integrated into the UAV, so the geometric modelling of the robot is proposed as an integrated structure. The kinematic model of a UAV results in the location of the point of interest as a function of the location of the UAV. The kinematic model of a UAV gives the derivative of its location of the point of interest according to the location of the aerial mobile platform, $\dot{\mathbf{h}} = \partial f / \partial \mathbf{\theta}(\theta)\mathbf{v}$, where $\dot{\mathbf{h}} = \begin{bmatrix} \dot{hx} & \dot{hy} & \dot{hz} & \dot{\theta} \end{bmatrix}^T$ is the velocity vector at the point of interest, is the mobility control vector of the mobile air handler, in this case the dimension depends on the manoeuvring velocities of the UAV.

It is indispensable to enunciate the kinematic model of the UAV from the linear and angular velocity, in this way its movement can be initiated, in such a way the following kinematic model is obtained [1].

$$\dot{\mathbf{h}}(t) = \mathbf{J_g}(\theta)\mathbf{v}(t) \tag{1}$$

where, $\dot{\mathbf{h}} \in \mathbb{R}^n$ with $n = 4$ represents the vector of the velocities of the axis of the system $\sum_o$ and the angular velocity around the Z axis; the Jacobian matrix that defines a linear mapping between the vector of velocities $\mathbf{v}(t)$ of the UAV; and the maneuverability control of the UAV is defined as $\mathbf{v} \in \mathbb{R}^m$ with $m = 4$ represents the velocity vector of the UAV $\mathbf{v} = \begin{bmatrix} v_l & v_m & v_n & \omega \end{bmatrix}$, from the reference system $\sum_c$.

2.1 Mathematical Model of the Visual Sensor

The camera model used in this document is the perspective projection model or pinhole model. This implies a simplified model for an ideal viewing camera without distortion and without optical noise. The projection perspective with the Pinhole camera model is shown in Markus Richter [1], where f_c is the focal length, $^{w}\mathbf{p}_i \in \mathfrak{R}^3$ y $^{c}\mathbf{p}_i = \begin{bmatrix} ^{c}x_i & ^{c}y_i & ^{c}z_i \end{bmatrix}^T \in \mathfrak{R}^3$ are the 3D position vectors of the *ith* characteristic point of the target object in relation to $\sum_o$ y $\sum_c$, respectively, as shown in Fig. 1.

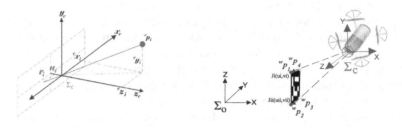

(a) Projection perspective of the UAV with the pattern to follow.   (b) Frame of reference for a UAV with visual feedback through selected points.

Fig. 1. Projection perspective of the characteristic image points.

3 Controller Design and Stability Analysis

A. Design Controller

This Section discusses the design of a visual controller based on image feature errors (image-based control) to enable a UAV with a built-in camera to perform a task of tracking moving objects in the 3D workspace, while image feature errors are defined as $\xi(t)$ converge asymptotically to zero. Therefore, the control objective can be defined as $\lim_{t\to\infty} \tilde{\xi}(t) = 0$. Depending on the frame of reference $\sum_o$ represents the framework of the world and $\sum_c$ is the frame of the camera, as shown in Fig. 2 using a transformation of coordinates, plus the relationship between them is expressed this way, $^{c}\mathbf{p}_i = {}^{c}\mathbf{R}_w \left(^{w}\mathbf{p}_i - {}^{c}\mathbf{p}_{Corg} \right)$. The perspective projection of the *ith* characteristic point in the plane of the image gives us the coordinate of the plane of the image $\xi_i = [u_i \quad v_i]^T \in \mathfrak{R}^2$ as

$$\xi_i \left(^{c}x_i, {}^{c}y_i, {}^{c}z_i \right) = -\frac{f_c}{^{c}z_i} \begin{bmatrix} ^{c}x_i \\ ^{c}y_i \end{bmatrix} \tag{2}$$

solving (2) and $^{c}\mathbf{p}_i$, it can be expressed as follows $\dot{\xi}_i$ in terms of UAV velocity as

$$\dot{\xi}_i = \mathbf{J}_{\mathbf{I}_i} \left(\xi_i, {}^{c}z_i \right) \begin{bmatrix} ^{c}\mathbf{R}_w & 0 \\ 0 & ^{c}\mathbf{R}_w \end{bmatrix} \mathbf{J}_{\mathbf{g}}(\theta)\mathbf{v} - \mathbf{J}_{\mathbf{o}_i} \left(\theta, {}^{c}\mathbf{p}_i \right) {}^{w}\dot{\mathbf{p}}_i \tag{3}$$

where $\mathbf{J}_{\mathbf{g}}(\theta)$ is the *Jacobian geometric* UAV defined in Eq. (1), $\mathbf{J}_i \left(\xi_i, {}^{c}z_i \right)$ is the *jacobian of image* defined by $\mathbf{J}_{\mathbf{I}_i}(\xi_i, {}^{c}z_i)$. In addition, $\mathbf{J}_{\mathbf{o}_i} \left(\theta, {}^{c}\mathbf{p}_i \right) {}^{w}\dot{\mathbf{p}}_i$ represents the movement of the *ith* characteristic point in the plane of the image, where $^{w}\dot{\mathbf{p}}_i$ is the velocity of the *ith* point in relation to $\sum_o$ and $\mathbf{J}_{\mathbf{o}_i}(\theta, {}^{c}\mathbf{p}_i)$ [1].

In applications where the object is in a 3D workspace, two or more image characteristics are needed for the visual servo control to be implemented [8, 12, 13]. To extend this model to r image points it is necessary to stack the vectors of the coordinate of the image plane, that is, $\xi = \begin{bmatrix} \xi_1^T & \xi_2^T & \xi_3^T & \cdots & \xi_r^T \end{bmatrix}^T \in \Re^{2r}$ and $^c\mathbf{p} = \begin{bmatrix} ^c\mathbf{p}_1 & ^c\mathbf{p}_2 & \cdots & ^c\mathbf{p}_r \end{bmatrix}^T \in \Re^{3r}$. It is assumed that multiple point entities are provided in a known object. From Eq. (3), for multiple point entities one can written

$$\dot{\xi} = \mathbf{J}(\theta, \xi, {}^c z)\mathbf{v} - \mathbf{J}_o(\theta, {}^c\mathbf{p})^w\dot{\mathbf{p}} \tag{4}$$

where

$$\mathbf{J}(\theta, \xi, {}^c z) = \mathbf{J_I}(\xi, {}^c z)\begin{bmatrix} {}^c\mathbf{R}_w & 0 \\ 0 & {}^c\mathbf{R}_w \end{bmatrix}\mathbf{J_g}(\theta); \qquad \mathbf{J_I}(\xi, {}^c z) = \begin{bmatrix} \mathbf{J}_1\left([u_1 \quad v_1]^T, {}^c z_1\right) \\ \vdots \\ \mathbf{J}_r\left([u_r \quad v_r]^T, {}^c z_r\right) \end{bmatrix};$$

$$\mathbf{J}_o(\theta, {}^c\mathbf{p}) = \begin{bmatrix} \dfrac{f_c}{{}^c z_1}\begin{bmatrix} 1 & 0 & -\frac{{}^c x_1}{{}^c z_1} \\ 0 & 1 & -\frac{{}^c y_1}{{}^c z_1} \end{bmatrix} & \cdots & \dfrac{f_c}{{}^c z_m}\begin{bmatrix} 1 & 0 & -\frac{{}^c x_r}{{}^c z_r} \\ 0 & 1 & -\frac{{}^c y_r}{{}^c z_r} \end{bmatrix} \end{bmatrix}^T {}^c\mathbf{R}_w$$

For the sake of simplicity, the following notation will now be used $\mathbf{J} = \mathbf{J}(\theta, \xi, {}^c z)$ and $\mathbf{J}_o = \mathbf{J}_o(\theta, {}^c\mathbf{p})$. This section, an image is captured at the desired reference position and the corresponding extracted characteristics represent the desired characteristics vector ξ_d. The control problem is to design a controller that calculates the velocities applied $\mathbf{v}_{ref}$ to move the UAV in such a way that the characteristics of the actual image reach the desired ones. The design of the kinematic controller is based on the kinematic model of the UAV and the projection model of the camera. Equation (4), $\mathbf{v}$ can be expressed in terms of $\dot{\xi}$ and $^w\dot{\mathbf{p}}$ using the pseudo-inverse matrix $\mathbf{J}$, i.e., $\mathbf{v} = \mathbf{J}^\#\left(\dot{\xi} + \mathbf{J}_o^w\dot{\mathbf{p}}\right)$ where, $\mathbf{J}^\# = \mathbf{W}^{-1}\mathbf{J}^T\left(\mathbf{J}\mathbf{W}^{-1}\mathbf{J}^T\right)^{-1}$, being $\mathbf{W}$ a positive defined matrix that weighs on the control actions of the system, $\mathbf{v} = \mathbf{W}^{-1}\mathbf{J}^T\left(\mathbf{J}\mathbf{W}^{-1}\mathbf{J}^T\right)^{-1}\left(\dot{\xi} + \mathbf{J}_o{}^w\dot{\mathbf{p}}\right)$. The controller is based on a simple solution, where you will reach your navigation target with as few movements as possible. The following control law is proposed for the visual control of the UAV,

$$\mathbf{v}_{ref} = \mathbf{J}^\#\left(\mathbf{J}_o{}^w\dot{\mathbf{p}} + \mathbf{L_K}tanh\left(\mathbf{L_K}^{-1}\mathbf{K}\tilde{\xi}\right)\right) \tag{5}$$

where, $\mathbf{J}_o{}^w\dot{\mathbf{p}}$ represents the velocity of the object to follow in the plane of the image, $\tilde{\xi}$ is the vector of control errors defined as $\tilde{\xi} = \xi_d - \xi$, $\mathbf{K} \in \Re^{2r}$, $\mathbf{L_K} \in \Re^{2r}$.

The mission of the UAV is based on the values of the image characteristics corresponding to the relative positions of the robot and the object in the image plane. In this approach, an image is captured at the desired reference position and the corresponding extracted characteristics represent the desired characteristic vector ξ_d. The problem with visual control is to design a controller that calculates the applied $\mathbf{v}_{ref}$ velocities to move the UAV in such a way that the real image characteristics reach the desired ones. To solve the problem of visual control a complete control scheme of the system is proposed as shown in Fig. 2. The error in the image is defined as $\tilde{\xi} = \xi_d - \xi$, it can be calculated at any time during the measurement and used to operate the mobile manipulator in a

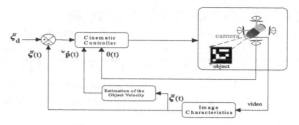

Fig. 2. Scheme of control of the complete system.

direction that reduces the error. Therefore, the purpose of the control is to ensure that $\lim_{t\to\infty} \tilde{\xi}(t) = 0 \in \Re^{2r}$.

How, the behavior of the control error $\tilde{\xi}$ is analyzed assuming a perfect follow up of the velocity $\mathbf{v} = \mathbf{v}_{ref}$, $\mathbf{v}_{ref} = [\begin{array}{cccc} v_l & v_m & v_n & \omega \end{array}]^T$. It must be taken into account that the vector of the desired image characteristic is constant, therefore, it can be concluded that $\dot{\tilde{\xi}} = -\dot{\xi}$. Now, substituting Eq. (5) in Eq. (4) gives the following closed-loop equation is obtained, $\dot{\tilde{\xi}} + \mathbf{L_K} \tanh\left(\mathbf{L_K^{-1} K}\tilde{\xi}\right) = 0$. For the stability analysis the following Lyapunov candidate function is considered $V\left(\tilde{\xi}\right) = \frac{1}{2}\tilde{\xi}^T\tilde{\xi}$. Its time derivative on the trajectories of the system is, $\dot{V}\left(\tilde{\xi}\right) = -\tilde{\xi}^T \mathbf{L_K} \tanh\left(\mathbf{L_K^{-1} K}\tilde{\xi}\right) < 0$ which implies that $\tilde{\xi}(t) \to \mathbf{0}$ asymptotically.

4   Robustness Analysis

In some studies of visual servo controller based on image do not choose to perform an analysis of stability and robustness [14, 15], in this document does not exclude all this analysis because it is a fundamental part of the controller to determine that it is valid, consistent and especially applicable experimentally in any mobile device, as in our case in a UAV [16].

The proposed controller presented above considers that the velocity of the object to follow $^w\dot{\mathbf{p}}$ is exactly known. However, this is not always possible in a real context. In practice, this velocity will be estimated using visual detection of the position of the object, $e.g.$ by means of a filter $\alpha - \beta$ [17]. This motivates to study the behavior of the image characteristic error $\tilde{\xi}$ considering the estimated velocity errors of the object to follow and also considering the assumption of a perfect velocity tracking. It defines the velocity estimation errors of the object in the image plane as $\varepsilon = \mathbf{Jo}\left(^w\hat{\dot{\mathbf{p}}} - {}^w\dot{\mathbf{p}}\right)$ where $^w\hat{\dot{\mathbf{p}}}$ y $\hat{\dot{\mathbf{p}}}$ are the actual and estimated velocities of the object, respectively. Hence $\dot{\tilde{\xi}} + \mathbf{L_K} \tanh\left(\mathbf{L_K^{-1} K}\tilde{\xi}\right) = 0$ can now written as

$$\dot{\tilde{\xi}} + \mathbf{L_K} \tanh\left(\mathbf{L_K^{-1} K}\tilde{\xi}\right) = \mathbf{J}\tilde{\mathbf{v}} + \varepsilon \tag{6}$$

Lyapunov candidate function $V\left(\tilde{\xi}\right) = \frac{1}{2}\tilde{\xi}^T\tilde{\xi}$ is considered again, which time derivative along the trajectories of the system Eq. (6) is $\dot{V}\left(\tilde{\xi}\right) = \tilde{\xi}^T(\mathbf{J}\tilde{\mathbf{v}} + \varepsilon) -$

$\mathbf{L_K} \tanh\left(\mathbf{L_K^{-1} K \tilde{\xi}}\right)$. A sufficient condition for $\dot{V}\left(\tilde{\xi}\right)$ to be negative definite is

$$\left|\tilde{\xi}^\mathbf{T} \mathbf{L_K} \tanh\left(\mathbf{L_K^{-1} K \tilde{\xi}}\right)\right| > \left|\tilde{\xi}^\mathbf{T}(\mathbf{J\tilde{v}} + \varepsilon)\right| \tag{7}$$

For large values of $\tilde{\xi}$, it can be considered that $\mathbf{L_K} \tanh\left(\mathbf{L_K^{-1} K \tilde{\xi}}\right) \approx \mathbf{L_K}$. Then, Eq. (7) can be expressed as $\|\mathbf{L_K}\| > \|\mathbf{J\tilde{v}} + \varepsilon\|$ thus, making the errors $\tilde{\xi}$ decrease. Now, for small values of $\tilde{\xi}$, $\mathbf{L_K} \tanh\left(\mathbf{L_K^{-1} K \tilde{\xi}}\right) \approx \mathbf{K \tilde{\xi}}$, thus Eq. (7) can be written as, $\left\|\tilde{\xi}\right\| > \|\mathbf{J\tilde{v}} + \varepsilon\| / \lambda_{\min}(\mathbf{K})$ thus, implying that the error $\tilde{\xi}$ is bounded by, $\left\|\tilde{\xi}\right\| \leq \|\mathbf{J\tilde{v}} + \varepsilon\| / \lambda_{\min}(\mathbf{K})$. Hence, it is concluded that the image feature error is ultimately bounded by the bound $\|\mathbf{J\tilde{v}} + \varepsilon\| / \lambda_{\min}(\mathbf{K})$ on a norm of the control error and the estimated velocity error the object to be followed.

5   Experimental Results

Both tests were carried out to evaluate the performance of the proposed controller, the first in real time using Python software as a means of communication between the PC with the Tello UAV of the DJI brand and the second test by simulation in the MATLAB mathematical software.

Experiment 1: The experimental tests are in a semi-structured environment or with little influence from wind, where the linear velocities have a very low margin of error, in the same way the angular velocity is stable most of the time. These control actions are observed in Fig. 3 and Fig. 4 shows that the control errors $\tilde{\xi}(t)$ are ultimately bounded with final values close to zero, i.e., achieving final feature errors max $\left(\left|\tilde{\xi}(t)\right|\right) < 40$ pixels with a sampling of 250 ms.

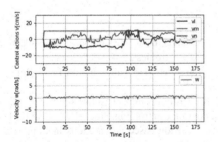
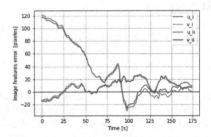

Fig. 3. Actions to control the linear velocities received by the UAV.

Fig. 4. Time evolution of control error $\tilde{\xi}(t)$.

Experiment 2: This experiment is a simulation result performed to evaluate the performance of the system when the target makes a movement in 3D space. In Fig. 5 it is observed that the control actions tend to zero in a very short period of time, as in Fig. 6 the control errors are minimal with a sampling of 100 ms, demonstrating the theory of control. Fig. 7 shows the strobe movement of the UAV with respect to the characteristic image points in the workspace (X, Y and Z).

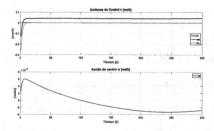
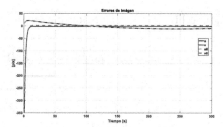

Fig. 5. Velocities commands to the UAV.   **Fig. 6.** Time evolution of control error $\tilde{\xi}(t)$.

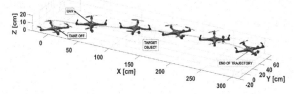

Fig. 7. Strobe movement of the UAV. UAV position and target position are displayed at the same instant. Five different moments of time are represented.

Acknowledgment. The authors would like to thank the Coorporación Ecuatoriana para el Desarrollo de la Investigación y Academia CEDIA for their contribution in innovation, through the CEPRA projects, especially the project CEPRA-XIII-2019-08; Sistema Colaborativo de Robots Aéreos para Manipular Cargas con Óptimo Consumo de Recursos; also the Universidad Nacional de Chimborazo, Universidad de las Fuerzas Armadas ESPE and the Research Group ARSI, for the support for the development of this work.

References

1. Ricardo, C., Salinas, L., Toibero, J.M., Roberti, F., Andaluz, V.: Visual control with adaptive dynamical compensation for 3D target tracking. Mechatronics **22**(4), 491 (2011)
2. Siegwart, R., Nourbakhsh, I.R., Scaramuzza, D.: Introduction to Autonomous Mobile Robots. MIT Press, Cambridge (2004)
3. Zhou, W., Li, B., Sun, J., Wen, C.Y., Chen, C.K.: Position control of a tail-sitter UAV using successive linearization based model predictive control. Control Eng. Pract. **91**, 104125 (2019)
4. Saripalli, S., Montgomery, J.F., Sukhatme, G.S.: Vision-based autonomous landing of an unmanned aerial vehicle. Cat. No. 02CH37292, Washington, DC, USA, 11–15 May 2002
5. Salazar, S., Romero, H., Lozano, R., Castillo, P.: Modeling and real-time stabilization of an aircraft having eight rotors. In: Valavanis, K.P., Oh, P., Piegl, L.A. (eds.) Unmanned Aircraft Systems. Springer, Dordrecht (2008). https://doi.org/10.1007/978-1-4020-9137-7_24
6. de la Escalera, A., Armingol, J.M., Pech, J.L., Gómez, J.J.: Detección Automática de un Patrón para la Calibración de Cámaras. Revista Iberoamericana de Automática e Informática Industrial RIAI **7**(4), 83–94 (2010)
7. Dong, G., Zhu, Z.H.: Kinematics-based incremental visual servo for robotic capture of non-cooperative target. Robot. Auton. Syst. **112**, 221–228 (2019)

8. Andaluz, V., Roberti, F., Carelli, R.: Robust control with redundancy resolution and dynamic compensation for mobile manipulators. In: IEEE-ICIT International Conference on Induatrial Technology, pp. 1449–1454 (2010)
9. Hamel, T., Mahony, R.: Image based visual servo control for a class of aerial robotic systems. Automatica **43**(11), 1975–1983 (2007)
10. Mehta, S.S., Ton, C., Rysz, M., Kan, Z., Doucette, E.A., Curtis, J.W.: New approach to visual servo control using terminal. J. Franklin Inst. **356**, 5001–5026 (2019)
11. Corke, P.I.: Visual control of robot manipulators – a review. In: World Scientific Series in Robotics and Intelligent Systems Visual Servoing, pp. 1–31 (1993)
12. Chaumette, F., Rives, P., Espiau, B.: Classification and realization of the different vision-based tasks. In: Visual Servoing: Real-Time Control of Robot Manipulators Based on Visual Sensory Feedback, pp. 199–228 (1993)
13. Hashimoto, K., Aoki, A., Noritsugu, T.: Visual servoing with redundant features. J. Robot. Soc. Jpn. **16**(3), 384–390 (1998)
14. Ceren, Z., Altuğ, E.: Image based and hybrid visual servo control of an unmanned aerial vehicle. J. Intell. Robot. Syst. **65**, 325–344 (2012). https://doi.org/10.1007/s10846-011-958 2-4T
15. Yüksel, T.: An intelligent visual servo control. Trans. Inst. Meas. Control **41**, 3–13 (2019)
16. Sarapura, J.A., Roberti, F., Toibero, J.M., Sebastián, J.M., Carelli, R.: Visual servo controllers for an UAV tracking vegetal paths. In: Sergiyenko, O., Flores-Fuentes, W., Mercorelli, P. (eds.) Machine Vision and Navigation, pp. 597–625. Springer, Cham (2020). https://doi.org/10.1007/978-3-030-22587-2_18
17. Kalata, P.R.: The tracking index: a generalized parameter for $\alpha-\beta$ and $\alpha-\beta-\gamma$ target trackers. IEEE Trans. Aerosp. Electron. Syst. **AES-20**(2), 174–182 (1984)

Construction and Control Aerial Manipulator Robot

Steeven J. Loor$^{(\boxtimes)}$, Alan R. Bejarano$^{(\boxtimes)}$, Franklin M. Silva$^{(\boxtimes)}$,
and Víctor H. Andaluz$^{(\boxtimes)}$

Universidad de Las Fuerzas Armadas ESPE, Sangolquí, Ecuador
`{sjloor,arbejarano2,fmsilva,vhandaluz1}@espe.edu.ec`

Abstract. This article presents the construction of an aerial manipulator robot composed of one or two robotic arms on an unmanned aerial vehicle, in order to execute control tasks in an autonomous or tele-operated manner. This aerial manipulator robot can work with one or two arms depending on the application requirements. The arms have been designed to serve several purposes: object manipulation and protect the actuating servos against direct impacts and overloads. Finally, a trajectory tracking algorithm is implemented and the simulation results are presented and discussed, which validate the controller and the proposed modeling.

Keywords: UAV · Robotic arms · Aerial manipulator robot

1 Introduction

During the last decades, research in robotics has been oriented to find solutions to the technical needs of applied robotics. Today, the creation of new needs and markets outside the traditional manufacturing robotics market (*e.g.,* cleaning, mine clearance, construction, shipbuilding, agriculture) and the world in which we live is demanding field and service robots to serve the new market and human social needs [1]. The International Federation of Robotics defines a service robot as a semi- or fully autonomous operated robot that performs useful services for humans and is not used for manufacturing [2]. They are those robotic systems that help people in their daily lives at work, at home, in leisure, and as part of the assistance to the disabled and the elderly. In other words, any robot used in the medical, health, military, domestic and educational industries is considered a service robot [3]. The UAVs have been widely used in military applications, Applications in agriculture and industrial environments are currently being exploited [5, 6, 7, 8]. There are several control strategies for UAVs some of which are energy optimization [9, 10] and fuzzy control.

The UAV industry is growing very rapidly due to the use of UAVs in commercial areas. However, unmanned aerial vehicles often do not meet the needs of complicated missions such as object handling, as they are only able to navigate in spaces that are difficult to access. For this reason, it is proposed to incorporate one or two robotic arms on an aerial platform (UAV). The combination of aerial mobile platforms with

© Springer Nature Switzerland AG 2020
H. Fujita et al. (Eds.): IEA/AIE 2020, LNAI 12144, pp. 116–123, 2020.
https://doi.org/10.1007/978-3-030-55789-8_11

robotic arms is known as aerial mobile manipulators [10]. Airborne manipulators have recently encompassed an extensive area of research, which has focused the attention of multiple research groups and international companies interested in such technology [11]. The objective of these projects is the development of robotic systems with handling capabilities. The most commonly used platforms for air manipulators are the helicopter or multirotor type, in its different variants: quadrotor, hexarotor or octrotor, combined with robotic arms with multiple degrees of freedom [12]. Given the versatility of this type of robotic system, different tasks can be performed, such as: welding [19]; transport of long objects [13]; handling of items placed at a height that is difficult to access [13]; construction [14]; transport of goods [15]; power-line applications [16]; hazardous tasks involving human beings, among others [19, 20, 21].

This application seeks to break the paradigm with respect to the manipulation of objects with the design and construction of robotic arms with 3° of freedom each one since at the moment there are manipulators with very simple arms with a single degree of freedom, in addition an algorithm of control in closed loop will be implemented to follow a task of control in trajectory and thus to be fortifying and improving the process of autonomous tasks in the different scopes leisure, work, military, among others for example cleaning, demining, construction. Construction and control are the main objectives of this work.

2   Aerial Manipulator Robot Construction

The airborne manipulator robot is a system conceived and designed to maintain its stability at different flight heights and under different weather conditions. Therefore, in order to comply with all the functional and safety requirements demanded for this type of equipment, a preliminary design of the drone and the robotic arm Fig. 1 was started and these were modified and optimized in geometry and dimensions during the design process, according to the maximum stresses generated in each component under critical operating conditions, until an adequate safety factor was reached and the definitive design of the air handling robot was obtained.

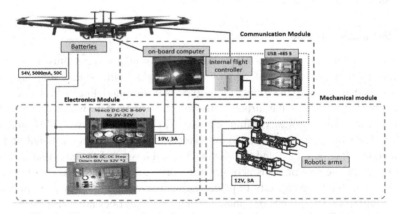

Fig. 1. Electronic, mechanical and communication connection diagram.

2.1 Electronics Module

For the implementation of the air handler, two voltage regulators are included that are supplied by the UAV batteries; one of 54 V 7A to 18 V 3A for the computer, and another of 54 V 6A to 12 V 6A. The addition of these electronic elements ensures the proper operation of both the computer and the robotic arm and prevents overvoltage and overcurrent in them, for which experimental tests are conducted on power consumption and battery life.

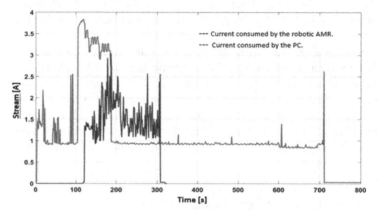

Fig. 2. Power tests of both regulators.

Additionally, an emergency circuit used to suspend erroneous arm actions is included, allowing power to be removed when it could become a hazard to the propellers of the Matrice 600pro, and achieving a safe scheduled landing, this in order to preserve the components and the UAV from an unexpected collision [17].

Finally, several power tests have been carried out to validate the correct operation of the implemented equipment. Figure 2 shows scenarios where both the computer and the robot arm have been subjected to maximum current tests *i.e.* the computer using all processor cores and the arm with a load of 2 kgs.

2.2 Mechanical Module

The main structure and the arms of the airborne manipulator robot must be analyzed to verify that they are capable of supporting the weight of the robotic arms and other components that make up the robot: a prototype is made (modeling) with its geometry and weight that simulates the airborne manipulator robot. Afterwards, the structural analysis is carried out by means of finite elements (FEM), with the use of CAD software. Finally, all the necessary mechanical components that make up the air handling robot are integrated. It should be noted that the mechanical design of the base that will serve as a support for the robotic arms is of great importance, and will be assembled on the UAV. This will have the capacity to place one or two robotic arms or just one so that the plate will have 3 couplings distributed evenly so that it does not affect the center of gravity of the UAV.

2.3 Bilateral Communication Module

The air handling system has a hardware remote control, the pilot has the option to intervene or not in basic maneuvers such as take-off and landing. A higher priority was given to pilot commands over autonomous control; in case of unforeseen situations, the pilot can avoid autonomous control and immediately operate the hardware controller.

This approach was implemented in a three-layer architecture in Robotic Operating System (ROS), this simplified its development and look to the future will make possible two things: the creation of a swarm of UAVs and their migration on board the UAV. Increasing the independence of the UAV from the ground control station (GCS).

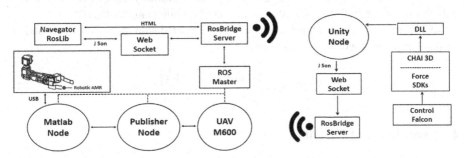

Fig. 3. Communication structure

The structure of this communication system in Fig. 3 uses the Onboard SDK, which has ROS compatibility and this is defined as a node, inside the UAV Matlab computer it runs as a node and with the help of a proprietary node of the Matlab libraries it will send the calculated data and then print the velocities. In the same way through a USB-485 serial communication the arm or arms are controlled depending on the type of configuration being used.

Besides, a ROS Bridge Server is initialized with the purpose that from the Master computer through a FALCOM Force SDK and with the CHAI 3D libraries, it can be read from the libraries to a Unity Node through shared memories (DLL), with this data in the Unity Node it will be written through the Web Socket that will be anchored to the ROS Bridge Server, and it will be possible to send the data from the FALCOM to Matlab that is running in the UAV computer.

3   Control Algorithm

The following is a closed-loop autonomous control scheme that validates the construction and the communication channel, which defines a desired task that through an addition node enters the non-linear controller, which sends the control actions to the AMR that is internally distributed in kinematic and dynamic;

PID feedback is performed in each of the segments to pass through the controller and enter a direct kinematics node which closes the loop and compensates errors by entering directly to the non-linear controller [18].

To carry out the experimental tests, it is necessary to implement a control law based on the kinematics of the aerial manipulator root described in the previous section, two types of control can be implemented by trajectory and position, in which the trajectory is defined as:

$$\mathbf{v}_{ref}(t) = \mathbf{J}^{\#}\left(\dot{\mathbf{h}}_{\mathbf{d}} + \mathbf{K_1}\tanh\left(\mathbf{K_2}\,\tilde{\mathbf{h}}\right)\right) \tag{1}$$

where $\dot{h}_d$ is the reference velocity input of the aerial mobile manipulator for the controller; $\mathbf{J}^{\#}$ is the matrix of pseudoinverse kinematics for the aerial mobile manipulator; while that $\mathbf{K_1} > 0$ and $\mathbf{K_2} > 0$ area gain matrix of the controller that weigh the control error respect to the inertial frame $< R >$; and the **tanh**(.) represents the function saturation of maniobrability velocities in the aerial mobile manipulator [18].

In addition, a position control can only be implemented with: $\dot{\mathbf{h}}_{\mathbf{d}} = 0$ so the equation for position control would look like this:

$$\mathbf{v}_{ref}(t) = \mathbf{J}^{\#}\left(\mathbf{K_1}\tanh\left(\mathbf{K_2}\,\tilde{\mathbf{h}}\right)\right) \tag{2}$$

4   Experimental Results

This section presents the built AMR, which is composed of a Hexarrotor Matrice-600 and the 3 DOF robot arm, in addition the communication described the results and the proposed controller. In Fig. 4 it is possible to observe the AMR manipulating an object, in this case a pin pon ball, the manipulation was made in experimental tests in laboratory and also in field tests, with this manipulation and flight tests the process of construction and implementation of the arm in the aerial platform is validated. Matlab node that is running on the UAV computer, a sampling period of 100 [ms] will be used).

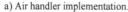
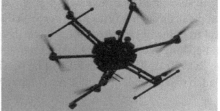

a) Air handler implementation.   b) Airborne robot with base deployment
                                           for greater arm work area

Fig. 4. AMR in operation and stand by.

Figure 5a, shows the desired trajectory and the current trajectory of the end-effector. It can be seen that the proposed controller presents a good performance. Figure 5c, shows the evolution of the tracking errors, which remain close to zero, while Fig. 5b and Fig. 5d show the control actions.

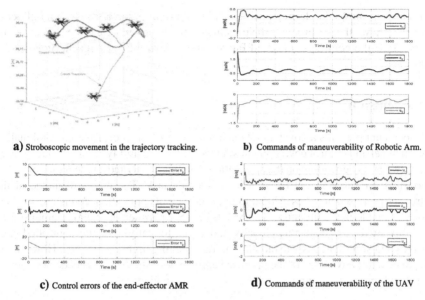

a) Stroboscopic movement in the trajectory tracking.

b) Commands of maneuverability of Robotic Arm.

c) Control errors of the end-effector AMR

d) Commands of maneuverability of the UAV

Fig. 5. Graphs obtained from the implemented system.

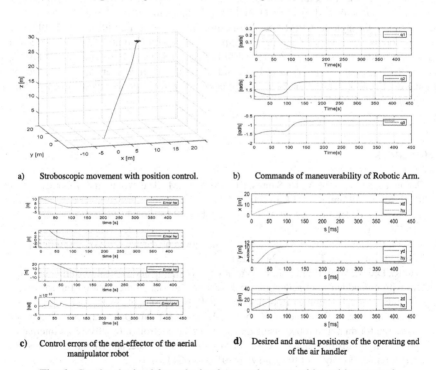

a)   Stroboscopic movement with position control.

b)   Commands of maneuverability of Robotic Arm.

c)   Control errors of the end-effector of the aerial manipulator robot

d)   Desired and actual positions of the operating end of the air handler

Fig. 6. Graphs obtained from the implemented system with position control.

Figure 6a, shows the desired position and the current trajectory of the end-effector. Figure 6c, shows the evolution of the tracking errors, which remain close to zero, while Fig. 6b show the control actions and Fig. 6d shows how the UAV reaches the desired positions.

5 Conclusion

This document presents a novel aerial manipulator robot that consists of a combination of mechanical and control development strategies, with the ability to have modular arms (one or two arms depending on the application that requires it) and a kinematic controller responsible for carrying out the trajectory tracking task. The main advantage of the control law proposed here lies in its simplicity and ease of application, compared to others already available in the literature. The results of the experimentation have demonstrated the ability of the controller to globally and asymptotically zero the controlled state variables and simultaneously avoid any saturation in the flight commands. However, in order to take full advantage of the proposed approach, further work is needed on motion control and trajectory planning of the position of the end effector.

Acknowledgements. The authors would like to thank the Corporación Ecuatoriana para el Desarrollo de la Investigación and Academia CEDIA for the financing given to research, development, and innovation, through the CEPRA projects, especially the project CEPRA-XIII-2019-08; Sistema colaborativo de robots Aéreos para Manipular Cargas con Optimo Consumo de Recursos; also to Universidad de las Fuerzas Armadas ESPE, Escuela Superior Politécnica de Chimborazo, Universidad Nacional de Chimborazo, Universidad tecnológica Indoamérica, Universidad internacional del Ecuador, Universidad central de Venezuela, and Grupo de Investigación ARSI, for the support to develop this work.

References

1. Garcia, E., Jimenez, M.A., De Santos, P.G., Armada, M.: The evolution of robotics research. IEEE Robot. Autom. Mag. **14**, 1 (2007)
2. International Federation of Robotics. Service robots: Provisional definition of service robots. http://www.ifr.org/service-robots/
3. Moradi, H., et al.: Service robotics (the rise and bloom of service robots) [tc spotlight]. IEEE Robot. Autom. Mag. **20**(3), 22–24 (2013)
4. Lee, K., Lee, J., Woo, B., Lee, J., Lee, Y., Ra, S.: Modeling and control of a articulated robot arm with embedded joint actuators. In: 2018 International Conference on Information and Communication Technology Robotics (ICT-ROBOT), Busan, pp. 1–4, (2018)
5. Thai, H.N., Phan, A.T., Nguyen, C.K., Ngo, Q.U., Dinh, P.T., Vo, Q.T.: Trajectory tracking control design for dual-arm robots using dynamic surface controller. In: 2019 First International Symposium on Instrumentation, Control, Artificial Intelligence, and Robotics (ICA-SYMP) (2019). https://doi.org/10.1109/ica-symp.2019.8646243
6. Ruan, L., et al.: Energy-efficient multi-UAV coverage deployment in UAV networks: a game-theoretic framework. China Commun. **15**(10), 194–209 (2018)
7. Pajares, G.: Overview and current status of remote sensing applications based on unmanned aerial vehicles (UAVs). Photogramm. Eng. Remote Sens. **81**(4), 281–330 (2015)

8. Ramon Soria, P., Arrue, B., Ollero, A.: Detection, location and grasping objects using a stereo sensor on UAV in outdoor environments. Sensors **17**(1), 103 (2017)

9. Suarez, A., Heredia, G., Ollero, A.: Lightweight compliant arm with compliant finger for aerial manipulation and grasping. In: 2016 IEEE/RSJ International Conference on Intelligent Robots and Systems (IROS) (2016)

10. Bartelds, T., Capra, A., Hamaza, S., Stramigioli, S., Fumagalli, M.: Compliant aerial manipulators: Toward a new generation of aerial robotic workers. IEEE Robot. Autom. Lett. **1**(1), 477–483 (2016)

11. Tognon, M., Franchi, A.: Dynamics, control, and estimation for aerial robots tethered by cables or bars. IEEE Trans. Rob. **33**(4), 834–845 (2017)

12. Suarez, A., Heredia, G., Ollero, A.: Lightweight compliant arm with compliant finger for aerial manipulation and grasping. In 2016 IEEE/RSJ International Conference on Intelligent Robots and Systems (IROS)

13. Suárez, A., Sanchez-Cuevas, P., Fernandez, M., Perez, M., Heredia, G., Ollero, A.: Lightweight and compliant long reach aerial manipulator for inspection operations. In: 2018 IEEE/RSJ International Conference on Intelligent Robots and Systems (IROS), pp. 6746–6752. IEEE, October 2018

14. Marquez, F., Maza, I., Ollero, A.: Comparacion de planificadores de caminos basados en muestro para un robot aereo equipado con brazo manipulador. Comité Español de Automática de la CEA-IFAC (2015)

15. Cano, R., Pérez, C., Pruaño, F., Ollero, A., Heredia, G.: Diseño Mecánico de un Manipulador Aéreo Ligero de 6 GDL para la Construcción de Estructuras de Barras. ARCAS (ICT-2011– 287617) del séptimo Programa Marco de la Comisión Europea y el proyecto CLEAR (DPI2011-28937-C02-01) (2013)

16. Kim, S., Choi, S., Kim, H.J.: Aerial manipulation using a quadrotor with a two DOF robotic arm. In: IEEE/RSJ International Conference on Intelligent Robots and Systems, Tokyo, pp. 4990–4995 (2013)

17. https://www.dji.com/matrice600-pro

18. Ortiz, J., Erazo, A., Carvajal, C., Pérez, J., Proaño, L., Silva M,F., Andaluz, V.: Modeling and kinematic nonlinear control of aerial mobile manipulators. In: Computational Kinematics, pp. 87–95 (2017)

19. Leica, P., Balseca, J., Cabascango, D., Chávez, D., Andaluz, G., Andaluz, V.H.: Controller based on null space and sliding mode (NSB-SMC) for bidirectional teleoperation of mobile robots formation in an environment with obstacles. In: 2019 IEEE Fourth Ecuador Technical Chapters Meeting (ETCM), Guayaquil, pp. 16 (2019)

20. Varela-Aldás, J., Andaluz, V.H., Chicaiza, F.A.: Modelling and control of a mobile manipulator for trajectory tracking. In: 2018 International Conference on Information Systems and Computer Science (INCISCOS), Quito (2018)

21. Andaluz, V., Rampinelli, V.T.L., Roberti, F., Carelli, R.: Coordinated cooperative control of mobile manipulators. In: 2011 IEEE International Conference on Industrial Technology, Auburn, AL, pp. 300–305 (2011)

Knowledge Based Systems

ConMerge – Arbitration of Constraint-Based Knowledge Bases

Mathias Uta$^{1(\boxtimes)}$ ⓘ and Alexander Felfernig$^{2(\boxtimes)}$ ⓘ

^{1} Siemens AG, Freyeslebenstr. 1, 91058 Erlangen, Germany
mathias.uta@siemens.com
^{2} Applied Software Engineering, IST, Graz University of Technology,
Inffeldgasse 16b/II, 8010 Graz, Austria
alexander.felfernig@tugraz.at

Abstract. Due to the increasing need to individualize mass products, product configurators are becoming more and more a manifest in the environment of *business to customer* retailers. Furthermore, technology-driven companies try to formalize expert knowledge to maintain their most valuable asset – their technological know-how. Consequently, insulated and diversified knowledge bases are created leading to complex challenges whenever knowledge needs to be consolidated. In this paper, we present the *ConMerge-Algorithm* which can integrate two constraint-based knowledge bases by applying redundancy detection and conflict detection. Based on detected conflicts, our algorithm applies resolution strategies and assures consistency of the resulting knowledge bases. Furthermore, the user can choose the operation mode of the algorithm: keeping all configuration solutions of each individual input knowledge base or only solutions which are valid in both original knowledge bases. With this method of knowledge base arbitration, the ability to consolidating distributed product configuration knowledge bases is provided.

Keywords: Constraint-based configuration · Arbitration · Merging knowledge bases

1 Introduction

Product configuration technologies have been used since the late 1970s [1]. While first mainly *business to business* applications were introduced [2], the new technology found its way as well into *business to custome*r approaches following the new ideas of mass customization formulated in the early 1990s [3]. Nowadays, product configuration is closely connected with the creation and customization of products and can be found on innumerable websites of online vendors as well as in company-internal applications to improve the engineering process or to support the sales department. Consequently, many different approaches of product knowledge formulization have been developed, leading to insulated systems. Issues occur whenever two or more knowledge bases must be integrated with each other. The necessity for this action can be triggered by various events. One scenario is the merging of two online retailers with two different product

H. Fujita et al. (Eds.): IEA/AIE 2020, LNAI 12144, pp. 127–139, 2020.
https://doi.org/10.1007/978-3-030-55789-8_12

configurators who want to maintain their entire portfolio on a newly created website. Another scenario is a consequence of digitalization efforts in companies by creating data and knowledge integrated systems as described in [4]. The maintenance of redundant data and knowledge is an unnecessary but increasingly observable challenge in companies. Attempts to consolidate existing software solutions are in most cases contempt to fail since no programmatic solution for knowledge base consolidation (also known as arbitration) is available. As a result, completely new solutions must be developed, accompanied by high costs and long implementation periods. In this paper, we provide a solution (the *ConMerge-Algorithm*) to merge constraint-based product configuration knowledge bases, a first step in the automatization of knowledge base arbitration.

The remainder of this paper is structured as follows. In Sect. 2, we introduce related work, followed by the presentation of a working example in Sect. 3. Based on this example, an understanding of the chosen constraint solver is given in Sect. 4. The *ConMerge-Algorithm* itself is introduced in Sect. 5. Finally, a performance analysis is presented in Sect. 6, continued by the résumé and an outlook on future work in Sect. 7.

2 Related Work

The main issues former publications of knowledge base arbitration dealt with is the versatility of knowledge bases – examples are [5, 6, 7]. As a result, inconsistencies occur after the merging step which must be resolved to assure a conflict-free knowledge base as required by belief revision or more general the minimal change principle defined in [8, 9]. Liberatore identified *three* possible mistakes of arbitrated knowledge bases [7]. They can be summarized as (1) "mistakes due to a wrong interpretation of variables (homonymies, synonymies, and subject misunderstanding); (2) mistakes due to a wrong interpretation of context (generalization, particularization, and extension); (3) mistakes due to a wrong interpretation of the logic (ambiguity and exclusion)".

In the domain of product configurators, in contrast to the general approach of knowledge base arbitration, the advantage of a much more restricted knowledge representation exists. Knowledge is represented as constraints following a defined syntax. The wrong interpretation of logic is consequently impossible. Nevertheless, some effort is needed to consolidate knowledge bases of the same product domain since various possibilities to describe the same product are used. A matching of variables is inevitable to reach knowledge base alignment and has already been researched in the domain of schema matching [10] as well as in numerous further publications which are summarized in [11]. Furthermore, a possibility to integrate two constraint-based knowledge bases by introducing a contextualization argument to each constraint has been presented in [12]. While this approach provides effective resolution strategies for identical constraints in both knowledge bases, a full integration of both knowledge bases is not accomplished and an unnecessary big amount of constraints is persisted.

The purpose of this paper is to provide a complete solution for conflicts created by the merging of knowledge bases in the sense of issue (2) defined by Liberatores [7]. In this context, we introduce the following assumptions:

1. The two knowledge bases A and B are individually consistent.
2. Variable names and domain definitions are the same in both knowledge bases.
3. Constraints in knowledge base A and B are based on the same knowledge representation with equivalent semantics, variables and domain definitions.

3 Working Example

To simplify and illustrate the discussion in the remainder of this paper, we introduce the following two knowledge bases as a working example.

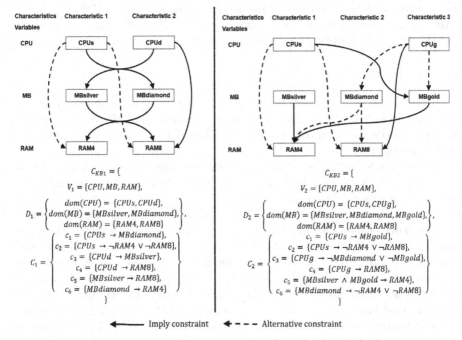

Fig. 1. Two consistent independent PC configuration knowledge bases

Both knowledge bases deal with the configuration of a fragment of a personal computer (PC), symbolizing the product configurators of two online resellers who perform a company merge. The PC consists of the component types *central processing unit* (CPU), *motherboard* (MB), and *random-access memory* (RAM). Each of these components is specified by two or more subtypes. The formal description and a graphical representation of the two knowledge bases is depicted in Fig. 1. Both knowledge bases consist, according to assumption 2, of the same variables but differ in their characteristics and their restrictions given by the constraints.

In accordance with [13], we describe the knowledge bases as a *constraint satisfaction problem* (CSP) defined by a triple (V, D, C) where V is a set of finite domain variables, D represents variable domains, and C represents a set of constraints defining restrictions on the possible variable value combinations. *"imply"*-constraints are used to describe a direct dependency of two variable domains and entail that no valid solution of the configuration can exist without the existence of the assigned variable domain. The second constraint type provides the possibility of further user choices if a variable domain is chosen. The choice of the trigger variable domain on the left side of the constraint explicitly allows *"alternative"* valid solutions containing the variable domains on the right side of the constraint. These two constraint types have been chosen under consideration of standard feature model representations [14]. They describe all positive restrictions of a CSP, making further constraint types in this constellation unnecessary in order to find a valid solution. To simplify the algorithm input for the user, a more abstract descriptions of the constraints can be provided to the algorithm instead of using propositional logic representations as shown in Fig. 1. These are for instance {"CPU", "CPUs", "imply", "MB", "MBdiamond"} to represent implications or {"CPU", "CPUs", "alternative", "RAM", "RAM4", "RAM8"} to achieve alternatives.

4 Constraint-Solver

Choco Solver [15] as a Java implementation and a wide-spread constraint programming system has been chosen for the *ConMerge* implementation. In the following, we provide an overview of the specialties of this constraint solver and the resulting design decisions for our merging algorithm.

To solve a CSP with *Choco Solver*, a model including the triple (V, D, C) must be created. Variables and their domains are declared and are represented in our implementation using the following syntax: {"intVar", "CPU", "CPUs", "CPUd"}. First, the *Choco Solver* internal variable type is declared, followed by the variable and all variable domains. Based on this semantic description, the CSP model is enriched with all relevant variables and variable domains. Additionally, a possibility to semantically describe the constraints to restrict the solution space of the model is provided by the *ConMerge*-Algorithm. As already indicated in the previous section, *"imply"*- and *"alternative"*-constraints are used to narrow the solution space of the CSP. The algorithm generically creates the model internal constraints necessary to solve the CSP. *ConMerge* converts the semantic descriptions into *If-Then*-relations which are translated by *Choco Solver* into several constraints following a syntax interpretable by the constraint solving routine. If the constraint $c_3 = \{CPUd \rightarrow MBsilver\}$ of C_{KB1} is taken as an example, the *Choco Solver* internal representation is: REIF_1 = [0,1] => ARITHM ([CPU = 1]), !REIF_1 = [0,1] => ARITHM ([CPU =/= 1]), REIF_2 = [0,1] => ARITHM ([MB = 3]), !REIF_2 = [0,1] => ARITHM ([MB =/= 3]), ARITHM ([prop(REIF_2.GEQ.REIF_1)]). This behavior leads to some restrictions for the utilized conflict solving method as explained in Sect. 5.1.

Unfortunately, the description of only positive constraints as input for *Choco Solver* will not lead to a restricted solution space since the solver treats unconnected variable domains as positive connections. Therefore, the *ConMerge-Algorithm* programmatical creates *"exclude"*-constraints for those variable domains, which are not explicitly declared as compatible by positive constraints (closed world assumption). On basis of C_{KB1}, the algorithm will create the following *"exclude"* constraints, leading to a knowledge base representation shown with Fig. 2.

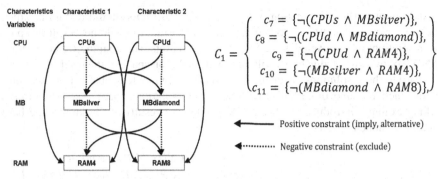

$$C_1 = \left\{ \begin{array}{c} c_7 = \{\neg(CPUs \land MBsilver)\}, \\ c_8 = \{\neg(CPUd \land MBdiamond)\}, \\ c_9 = \{\neg(CPUd \land RAM4)\}, \\ c_{10} = \{\neg(MBsilver \land RAM4)\}, \\ c_{11} = \{\neg(MBdiamond \land RAM8)\}, \end{array} \right\}$$

Fig. 2. PC configuration model – knowledge base CKB1 complemented with "exclude"-constraints

Complemented with these "exclude"-constraints, the constraint solver can determine the complete set of possible solutions for C_{KB1}:

$$S_1 = \left\{ \begin{array}{c} s_1 = \{CPUs, MBdiamond, RAM\,4\}, \\ s_2 = \{CPUd, MBsilver, RAM\,8\} \end{array} \right\}$$

The same process can be applied for C_{KB2} leading to the following set of solutions:

$$S_2 = \left\{ \begin{array}{c} s_1 = \{CPUs, MBgold, RAM\,4\}, \\ s_2 = \{CPUg, MBdiamond, RAM\,8\} \end{array} \right\}$$

Obviously, the explicit declaring of "exclude"-constraints, neglecting positive constraints, would also lead to these solutions, but would imply restrictions during the merging process which will be explained in the following section.

Finally, a further usability feature is provided to simplify the knowledge base descriptions. The user has the option to declare variables independent from each other, making it unnecessary to create "alternative"-constraints between variables which do not influence each other. For instance, the choice of a keyboard is independent from the choice of the central processing unit which can simply be described as {"Keyboard", "independent", "CPU"}.

5 The ConMerge-Algorithm

To perform the arbitration of the above introduced knowledge bases, we introduce the ConMerge-Algorithm. The algorithm solves conflicts occurring by merging two independently consistent knowledge bases and provides two options for integrating these two knowledge bases.

Basically, the arbitration of knowledge bases can, as well as every connection of two logical sentences, be described by logical operators. Therefore, the merging can be performed by connecting knowledge bases by "OR", "AND", "XOR", "NOR", etc. operations. Since in the configuration domain the solution space of the original knowledge bases is desired to be maintained, only the "OR" and "AND" connections remain as valid options. The combination of two positive constraints should remain positive, which excludes the "NOR" combination. Also, two negations could not result in a positive resolution, which excludes the "XOR" combination. The conjunctive connection ("AND") of two knowledge bases will narrow the solution space of the resulting knowledge base to these solutions occurring in both original knowledge bases (intersection). The disjunctive connection ("OR") will union both solution spaces. According to the working example introduced in Sect. 3, the following results of the knowledge base arbitration can be expected:

$$C_{KBMerge} = C_{KB1} \lor C_{KB2} \rightarrow S_{Merge} = \left\{ \begin{array}{l} s_1 = \{CPUs, MBdiamond, RAM\,4\}, \\ s_2 = \{CPUd, MBsilver, RAM\,8\}, \\ s_3 = \{CPUs, MBgold, RAM\,4\}, \\ s_4 = \{CPUg, MBdiamond, RAM\,8\} \end{array} \right\}$$

$$C_{KBMerge} = C_{KB1} \land C_{KB2} \rightarrow S_{Merge} = \{\}$$

5.1 The Algorithm

Based on the description method introduced in [16, 17], we now introduce the *ConMerge-Algorithm*. The input parameters for the algorithm are two knowledge bases C_{KB1} and C_{KB2} and the logical connection ("AND" or "OR") of these knowledge bases. The result is a consistent knowledge base C_{KB} merging the solution space of the original two knowledge bases.

```
Algorithm 1 - ConMerge(CKB1, CKB2, Mode)
{Input: CKB1/CKB2 - two consistent knowledge bases, Mode - desired
connection of knowledge bases (AND /OR)},
{CKB - merged knowledge base}, {ci - a single constraint}
CKB = CKB1 ∪ CKB2;
createExcludeConstraints(CKB);
For all ci of CKB do
  CKB ← solveConflict(Ci, CKB, Mode);
EndFor;
Return CKB;
```

In the first step of the algorithm, the union of the original knowledge bases is performed. Each part of the triple (V, D, C) describing the CSP of the first knowledge base is united with the corresponding part of the second knowledge base. As already indicated in Sect. 4, missing connections between variable domains must be considered incompatible. Therefore, the createExcludeConstraint() method is called in the next step of the algorithm to instantiate the missing "exclude"-constraints. This results in a knowledge base as depicted in Fig. 3 if the working example is executed.

The union of both knowledge base constraints obviously leads to conflicts and redundancies in the resulting knowledge base. For instance, $c_{10} = \{\neg(MBsilver \land RAM\,4)\}$ of C_{KB1} conflicts with the implication formulated in $c_5 = \{MBsilver \rightarrow RAM\,4\}$ of C_{KB2}.

To handle such situations, several publications in the configuration domain have been made in the past, including algorithms such as *Sequential* [18] and *CoreDiag* [19] for redundancy detection, *FastDiag* [20] for diagnosis tasks and *QuickXPlain* [21] for conflict detection. The most efficient algorithms of those named above rely mainly on the divide and conquer principle, meaning the analyzed constraint set is divided in half each time the algorithm detects that the first half is already inconsistent. Since this methodology promised the best option to solve the conflicts occurring in the arbitrated knowledge base, they have been extensively tested during the development of *ConMerge*. Unfortunately, *Choco Solver* showed some restrictions in combination with complex constraints and the divide and conquer principle. As already indicated in Sect. 4, complex constraints are *Choco Solver* internally instantiated as several single constraints, leading to a wrong behavior if the divide and conquer principle is applied. Constraints which must be handled as one set of constraints, are separated from each other with the consequence of a wrong diagnosis.

Analog to the conflict resolution algorithms in the configuration domain, ontological knowledge representation must handle similar problems [22, 23]. Changes in ontologies are performed by applying change kit algorithms to resolve occurring conflicts. Therefore, affected axioms are replaced and/or deleted to persist the consistency of the new ontology version. Based on this ontology approach we have decided to implement a comparable conflict solving algorithm for configuration knowledge bases. The algorithm does not operate on the level of instantiated constraints but on the semantical description level of the constraints (comparable to axioms in ontologies) we have introduced. In this *solveConflict*() algorithm two-dimensional string-arrays, where each string array represents one complex constraint, are analyzed with respect to redundancies and conflicts.

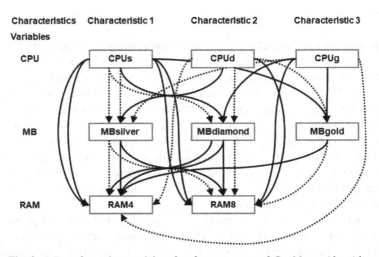

Fig. 3. PC configuration model – after first two steps of *ConMerge-Algorithm*

```
Algorithm 2 - solveConflict(Ci, CKB, Mode)
{Input: ci - input constraint, CKB - merged knowledge base, Mode -
desired connection of knowledge bases (AND /OR)}, {CREL - conflict-
ing constraints}, {c - a single constraint}
For all (c <> ci) of CKB do
//#1 determine all relevant constraints
  If (c.LeftSideOfConstraint == ci.LeftSideOfConstraint
  &&c.RightSideOfConstraint == ci.RightSideOfConstraint)
  Then
    CREL ← c;
  EnfIf;
EndFor;
For all c of CREL do
  If (c.Type == ci.Type) Then
  //#2 check redundancy
    CKB = CKB - c;
    Continue;
  EndIf;
  //#3 solve conflicts
  If (((c.Type == 'alternative' || c.Type == 'imply') &&
  Mode == 'OR') ||
  (c.Type == 'exclude' && Mode == 'AND')) Then
    CKB = CKB - ci;
    Break;
  Else
    CKB = CKB - c;
  EnfIf;
EndFor;
Return CKB;
```

Each constraint of the merged knowledge base is analyzed by applying the *solve-Conflict*()-method. In the first step, all relevant constraints are determined, namely those constraints which handle the same connection of two variable domains as the current analyzed (e.g. *MBsilver* and *RAM4* as mentioned above). In the second step, all relevant constraints are checked whether they are redundant to the current referenced constraint and deleted if this is the case. Finally, conflicts are solved under consideration of the connection mode the user has chosen for the knowledge base arbitration. In case of a conjunction, the *"imply"*-constraint in the example above needs to be deleted while in case of a disjunction the *"exclude"*-constraint must be deleted. By applying this procedure to all constraints, all conflicts and redundancies of the merged knowledge base are resolved, and a consistent condition is reached.

This approach explains the necessity for the explicit declaration of positive constraints indicated in Sect. 4. The *conflictsolve*()-algorithm would not determine a conflict if only the *"exclude"*-constraint would exist while the user expects the allowance of the combination "MBsilver" and "RAM4" according to C_{KB2}. Therefore, the user must describe the positive constraints as input for the algorithm while the negative constraints are generated programmatically.

5.2 Arbitration Result

In the following, the results based on our working example are presented and analyzed. For both arbitration modes (conjunction and disjunction), the merging of the variables and their domains lead to the same result since all options delivered by the configurators are maintained.

$$C_{KBMerge} = \{$$
$$V = \{CPU, MB, RAM\},$$
$$D = \left\{ \begin{array}{l} dom(CPU) = \{CPUs, CPUd, CPUg\}, \\ dom(MB) = \{MBsilver, MBdiamond, MBgold\}, \\ dom(RAM) = \{RAM\,4, RAM\,8\} \end{array} \right\}$$
$$\}$$

The resulting constraints differ based on the explanations given in the previous section. The remaining conjunction constraints are depicted in Fig. 4 and leading to the following solution space:

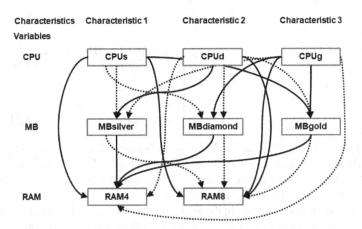

Fig. 4. PC configuration model – after conjunctive arbitration

$$C_{KBMerge} = C_{KB1} \wedge C_{KB2} \rightarrow S_{Merge} = \{s_1 = \{CPUs, MBgold, RAM\,4\}\}$$

Whereas an empty solution space was expected, since no solution existed which occurred in both original knowledge bases, *ConMerge* determined one solution. This can be explained by the new combinations resulting from the merge variable domain space. The determined solution was allowed in C_{KB2} while the combination of these domain variables was not explicitly restricted in C_{KB1} (the variable domain "MBgold" did not exist and the combination "CPUs" with "RAM4" was allowed). Consequently, the solution is not forbidden and therefore valid.

A similar phenomenon can be observed in the disjunctive combination of both knowledge bases shown in Fig. 5. As expected, all solutions originally existing in both separated knowledge bases were maintained after the merging. But additionally, again resulting from the new combinatoric given by the domain variable merge, a further solution is available.

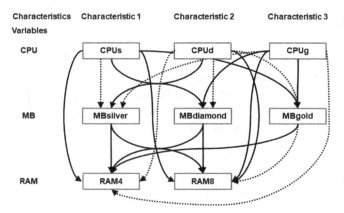

Fig. 5. PC configuration model – after disjunctive arbitration

$$C_{KBMerge} = C_{KB1} \vee C_{KB2} \rightarrow S_{Merge} = \left\{ \begin{array}{l} s_1 = \{CPUs, MBdiamond, RAM\,4\}, \\ s_2 = \{CPUd, MBsilver, RAM\,8\}, \\ s_3 = \{CPUs, MBgold, RAM\,4\}, \\ s_4 = \{CPUg, MBdiamond, RAM\,8\}, \\ s_5 = \{CPUs, MBdiamond, RAM\,8\} \end{array} \right\}$$

The combination "CPUs" and "MBdiamond" was explicitly allowed in C_{KB1} while it was forbidden in C_{KB2}. Contrary a solution including "MBdiamond" and "RAM8" was forbidden in C_{KB1} and allowed in C_{KB2}. Both conflicts were resolved in favor of the positive constraint resulting in the new solution.

Of course, both new solutions in the disjunction as well as in the conjunction need to be analyzed with respect to their real-world compatibility before releasing the merged knowledge base to the user. But nevertheless, *ConMerge* integrated two originally separated knowledge bases and even determined new options which might stayed hidden to the knowledge base engineers in a manual approach.

6 Performance Analysis

We now show some runtime evaluations on *ConMerge* demonstrating its applicability. Table 1 reflects the results of our analysis conducted with knowledge bases including an increasing number of variables and variable domains. The tested knowledge base pairs were created by a knowledge base randomizer. On the basis of the given number of variables, the randomizer generated between two and five variable domains per variable. All variable domains were connected by constraints. The type of constraint depended on the randomizer input parameters which have been chosen as 50% independent-, 25%

imply- and 25% alternative-constraints (chosen based on the expertise of knowledge base engineers). The tests have been executed on a standard desktop computer (Intel R CoreTM i5-6300 CPU Q9400 CPU with 2.4 GHz and 8 GB RAM) using a Java development environment.

Table 1. Performance overview

Average runtime in ms	Number of variables	Number of average original variable domains	Number of average original constraints
516	3	17	30
556	4	23	48
712	5	31	97
2347	10	62	489
23981	20	124	2210
799079	40	233	7636

Starting with an average runtime of round about 500 ms for very small knowledge bases, even knowledge bases with over 7500 constraints are merged in around 13 min. Nevertheless, a runtime improvement is desirable to assure applicability to even larger knowledge bases.

To measure the effectiveness of the merging procedure we introduce the merging rate as a key performance indicator in the following equation.

$$Merging\ rate = \frac{number\ of\ constraints_{KBmerge}}{number\ of\ constraints_{KB1} + number\ of\ constraints_{KB2}} \quad (1)$$

The measured knowledge base merging rate in our test scenario was 0.69. Hence, the number of constraints to be maintained by the knowledge engineers was significantly reduced and compactness of the configuration knowledge is conveyed. Furthermore, a significant increase of possible solutions can be observed in the merged knowledge base, if the conjunctive connection between both original knowledge bases is selected. Configurations previously forbidden and possibly hidden to the knowledge engineers are valid solutions after the arbitration operation, revealing new possibilities for users.

7   Conclusion and Future Work

In this paper, we have introduced an approach for an automated integration of configuration knowledge bases. The proposed solution helps to maintain the solution space of the two initially individual knowledge bases and contributes, in consequence, to the centralization of knowledge. Besides merging knowledge bases, *ConMerge* assures the compactness of the resulting knowledge base by minimalizing the amount of constraints.

With the application of the proposed algorithm, the arbitration of existing product configurators of the same domain gets possible. The potential to decrease implementation efforts if the knowledge of two configurators should be consolidated is provided. Nevertheless, knowledge engineers will still have effort in assuring the matching of the variables before the provided algorithm can be applied.

Future work will include research to implement an algorithm which goes beyond the introduced equal rating principle of knowledge bases (constraints of knowledge bases are valued as equally important). In real world applications, it is imaginable that one of the initial knowledge bases prevails the other one. Followed by this suggestion an adaption of *ConMerge* would be necessary. Finally, a more general solution to merge existing knowledge bases, which is not restricted to constraint-based knowledge, will be subject to further research and is inspired by the work of Liberatore in [6, 7] as well as Revesz in [5].

References

1. McDermott, J.: R1: a rule-based configurator of computer systems. Artif. Intell. **19**(1), 39–88 (1982)
2. Marcus, S., Stout, J., McDermott, J.: VT: an expert elevator designer that uses knowledge-based backtracking. AI Mag. **8**(4), 41–58 (1987)
3. Pine II, J.: Mass Customization: The New Frontier in Business Competition. Harvard Business School, Massachusetts, Boston (1992)
4. Uta, M., Felfernig, A.: Towards knowledge infrastructure for highly variant voltage transmission systems. In: Proceedings of the 20th Configuration Workshop, Graz, Austria, pp. 109–118 (2018)
5. Revesz, P. Z.: On the semantics of theory change: arbitration between old and new information. In: Proceedings of the Twelfth ACM SIGACT SIGMOD SIGART Symposium on Principles of Database Systems (PODS-93), pp. 71–82 (1993)
6. Liberatore, P., Schaerf, F.: Arbitration (or how to merge knowledgebases). IEEE Trans. Knowl. Data Eng. **10**(1), 76–90 (1998)
7. Libaratore, P.: Merging locally correct knowledgebases: a preliminary report. eprint (2002). arXiv:cs/0212053
8. Alchourrón, C.E., Gärdenfors, P., Makinson, D.: On the logic of theory change: partial meet contraction and revision functions. J. Symbolic Logic **50**, 510–530 (1985)
9. Gärdenfors, P.: Knowledge in Flux: Modeling the Dynamics of Epistemic States. Bradford Books MIT Press, Cambridge, MA (1988)
10. Hong-Hai, D.: Schema matching and mapping-based data integration, Leipzig (2017)
11. Christensen, P.: Data Matching. Springer, Heidelberg, Germany (2012)
12. Felfernig, A., Uta, M., Schenner, G., Spöcklberger, J.: Consistency-based merging of variability models. In: Proceedings of the 21st Configuration Workshop, Hamburg, (2019)
13. Hotz, L., Felfernig, A., Stumptner, M., Ryaboken, A., Bagley, C., Wolter, K.: Configuration knowledge representation and reasoning. In: Felfernig, A., Hotz, L., Bagley, C., Tiihonen, J. (eds.) Knowledge-Based configuration – From Research to BusinessCases. Morgan Kaufmann Publishers, Waltham, MA, pp. 41–72 (chapter 6) (2014)
14. Benavides, D., Ruiz-Cortes, A., Trinidad, P.: Automated reasoning on feature models. Comput. Sci. **3520**, 491–503 (2005)
15. Prud'homme, C., Fages, J.-G., Lorca, X.: Choco Documentation. TASC – LS2N CNRS UMR 6241, COSLING S.A.S (2017). http://www.choco-solver.org

16. Schubert, M., Felfernig, A., Mandl, M.: FastXplain: conflict detection for constraint-based recommendation problems. In: García-Pedrajas, N., Herrera, F., Fyfe, C., Benítez, J.M., Ali, M. (eds.) IEA/AIE 2010. LNCS (LNAI), vol. 6096, pp. 621–630. Springer, Heidelberg (2010). https://doi.org/10.1007/978-3-642-13022-9_62
17. Reiter, R.: A theory of diagnosis from first principles. Artif. Intell. **32**(1), 57–95 (1987)
18. Felfernig, A., Reiterer, S., Ninaus, G., Balzek, P.: Redundancy detection in configuration knowledge. In: Felfernig, A., Hotz, L., Bagley, C., Tiihonen, J. (eds.) Knowledge-Based configuration – From Research to BusinessCases. Morgan Kaufmann Publishers, Waltham, MA, pp. 157–166 (chapter 12) (2014)
19. Felfernig, A. Zehentner, C., Blazek, P.: CoreDiag: eliminating redundancy in constraint sets. In: 22nd International Workshop on Principles of Diagnosis (DX'2011), Murnau, Germany, pp. 219–224 (2011)
20. Felfernig, A., Reinfrank, F., Ninaus, G., Jeran, M.: Conflict detection and diagnosis in configuration. In: Felfernig, A., Hotz, L., Bagley, C., Tiihonen, J. (eds.) Knowledge-Based configuration – From Research to BusinessCases. Morgan Kaufmann Publishers, Waltham, MA, pp. 157–166 (chapter 7) (2014)
21. Junker, U.: QUICKXPLAIN: preferred explanations and relaxations for over-constraint problems. In: McGuinness, D.L., Ferguson, G. (eds.) 19th International Conferene on Artificial Intelligence (AAAI'04), pp. 167–172. AAAI Press, San Jose (2004)
22. Bayoudhi, L., Sassi, N., Jaziri, W.: Efficient management and storage of a multiversion OWL 2 DL domain ontology. Expert Syst. **36**(2), e12355 (2019)
23. Jaziri, W., Bayoudhi, L., Sassi, N.: A preventive approach for consistent OWL 2 DL ontology versions. Int. J. Semantic Web Inf. Syst. **15**(1), 76–101 (2019)

A Systematic Model to Model Transformation for Knowledge-Based Planning Generation Problems

Liwen Zhang[1,2]([⊠]), Franck Fontanili[2], Elyes Lamine[2], Christophe Bortolaso[1], Mustapha Derras[1], and Hervé Pingaud[3]

[1] Berger-Levrault, Jean Rostand 64, 31670 Labège, France
{liwen.zhang,christophe.bortolaso,
mustapha.derras}@berger-levrault.com, liwen.zhang@mines-albi.fr
[2] CGI, University of Toulouse-IMT Mines Albi, Campus Jarlard, 81000 Albi, France
{franck.fontanili,elyes.lamine}@mines-albi.fr
[3] CNRS LGC, University of Toulouse-INU Champollion, Place de Verdun, 81012 Albi, France
herve.pingaud@univ-jfc.fr

Abstract. The generation of an optimum planning problem and search for its solution are very complex regarding different business contexts. Generally, the problem is addressed by an optimization formulation and an optimizer is used to find a solution. This is a classical approach for the Operations Research (OR) community. However, business experts often need to express specific requirements and planning goals mathematically on a case by case basis. They also need to compute a planning result within various business constraints. In this paper, we try to support these experts during this preliminary problem design phase using a model driven engineering framework. An OR model could be generated from the knowledge included in a business conceptual model. A model to model transformation is described to support this goal. The Traveling Salesman Problem is used as a simple case study that allows explanation of our model transformation rules. We implemented our approach in the ADOxx meta-modelling Platform.

Keywords: Planning problem · Conceptual modeling · Operations Research · Model driven engineering · Model to model transformation · ADOxx · CPLEX

1 Introduction

Currently, in our connected world, there is a very wide demand from many organizations to regularly route and schedule people and goods. However, depending on the management culture of a given organization, planning will not be approached with the same spirit. A decision maker will certainly appreciate using a customized Human Machine Interface (HMI) in an efficient and comprehensive manner to obtain correct business-oriented knowledge to implement its own target planning needs.

Knowing this context, under the assumption of varying demands and specificities of service deliveries of each organization, it is necessary to embrace a broad set of

© Springer Nature Switzerland AG 2020
H. Fujita et al. (Eds.): IEA/AIE 2020, LNAI 12144, pp. 140–152, 2020.
https://doi.org/10.1007/978-3-030-55789-8_13

planning problems including a variety of mathematical formulations. An adapted mathematical formulation is a preliminary condition to call an optimizer with an embedded mathematical modelling language (i.e. CPLEX, Gurobi, LocalSolver), that will compute a solution with respect to an objective function and a set of constraints. For sake of efficiency as well as agility, a decision maker will appreciate the use of a strong decision-making support system in the complex combination of business constraints with which he must operate. Furthermore, the intrinsic nature of the demands for planning within a special application area (i.e. healthcare management, supply chain management, cooperation management), will tend to evolve considering the better control of system states obtained from the intensive use of information and communication facilities. This will lead to new pressure on decision makers as they face more complex situations and starve to more frequently formulate effective responses. It will pave the way to achieve business objectives while respecting the constraints inherent to the reality of an operational field that is ever changing. However, we think there is a lack of a strong decision support system which better captures knowledge using models about the system under study, whatever the specificity of the planning problem.

Let us illustrate this, making reference to the traditional planning optimization process that is often addressed as a mathematical problem in optimizing the planning of staff. The subject has been widely studied by the Operations Research (OR) community. It is often titled in the literature under the heading VRP (Vehicle Routing Problem) [1]. The mathematical formulation is two part, either mono-criteria or multi-criteria. In addition, the planning model can also be derived from a large number of variants depending on the different business constraints to be satisfied (i.e. means of transportation, geographical features, user preferences). As an illustration, a rise of activity on the operational field will inevitably reveal new characteristics to be taken into account and enforce a revision of the problem to be addressed. Facing this diversity in terms of operational planning model, the decision maker in the organization is not always able to react and change the model expeditiously. Decision makers do not necessarily have the expertise to consider the best or impact of the revised planning model. They are, however, business experts with sufficient knowledge to master business-oriented processes and content in order to apply them in the required business application. So, we must master a transition from their knowledge and business expertise to an ability to perform planning generations. We translate this need into a rationale of interactions between a real business-oriented model on the one hand, and the OR decision-making support model on the other. To do so, we use the best available knowledge management methods as part of a model driven engineering (MDE) approach. Our research aims to build a decision support system platform for helping these businessmen make ad-hoc planning decisions individually.

In this paper, based on a very simplified VRP problem, the Travelling Salesman Problem (TSP) [2], we show how the whole model transformations are put into practice following a three step generation process. i) Two models are considered at the two parts of the transformation. A given business-oriented conceptual model is a starting point, the source of transformation into a graph-based OR model that is a target. The knowledge transformation has been performed within the ADOxx meta-modeling platform developed by the OMILAB community [3]. ii) The graph-based OR model will provide the basis to parameterize a mathematical model predefined in the optimizer CPLEX. iii) CPLEX will compute the TSP result.

The following sections are divided into 6 parts. The overview of our approach is illustrated in Sect. 2. In the Sect. 3, we review the related work on MDE as well as MTL (Model Transformation Language) which will support an automatic model to model transformation. The Sect. 4 details the two meta-models of step i) of our generation process, as well as the transformation rules. Next, we explain the other remaining steps using the simple TSP case study. We make the assumption here that these final steps are not at the center of our concerns and we have effectively simplified our approach in consequence. The last section gives some conclusions and draws perspectives for future progress.

2   Approach Based on Model Driven Engineering (MDE)

Our MDE decision support platform is based on a series of three modules forming a consistent architecture. Together, they will deliver a generative solution for the TSP. Figure 1 shows an overview of the linear process that goes through the three modules. We firstly focus on the interaction between the business-oriented conceptual model (BCM) and the graph-based operation research model (GORM). Then, the GORM built from the BCM allow us to parameterize a predefined TSP solution-oriented mathematical model written in the OPL language and by doing so, prepare it to be submitted to the IBM CPLEX optimizer to generate a result.

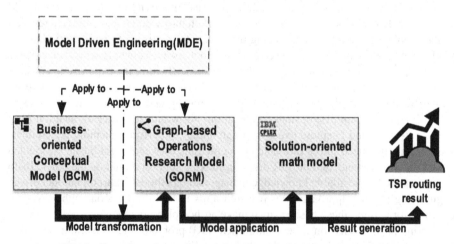

Fig. 1. Overview of our approach based on Model Driven Engineering

The first module (BCM) consists of embedding a domain specific modeling language (DSML) for defining the key concepts in the TSP, based on a business-oriented vocabulary. This conceptual modeling [4] uses a syntax that is meaningful for the business expert and shall provide him the means to easily specify the requirements of her/his TSP. It allows us to give a clear and specific definition of:

- Human support in the analysis of organizational and technical relationships.
- Creation of models from graphs or diagrams.
- Facultative definition of formal semantics.
- Improvement of human understanding and model mechanical processing.

BCM allows the user to characterize the knowledge required in the TSP formulation phase, and to identify the concepts associated with it such as the objective function. It could be the minimization of total salesman travel distances.

In the second module (GORM), as mentioned previously, OR knowledge is made available in order to translate business requirements into a mathematical formulation [5]. For the TSP case study, it will mainly consist in setting up a graph based representation for which the nodes and the edges will be symbols associated with variables and associations between them [6], matching ultimately the TSP's mathematical formulation that IBM CPLEX OPL model is waiting for.

The set of 2 first modules structure the model-driven engineering approach [7]. For MDE, models are considered as primary artifacts flowing throughout an engineering life cycle, during which it ensures model conformity to languages as well as ensuring model to model transformation (M to M). In order to do so, the keys required are a source meta-model and a target meta-model, which specify the concepts, and therefore the primitives of the languages related to their modelling. Figure 2 shows the relationship between the real world, the model and the meta-model. The M to M transformation is realized by respecting the mapping rules defined in the two meta-models. In our case, BCM is as the source model and conforms to a DSML meta-model. The GORM is the target model and conforms to the OR graph language meta-model. In the following, we will call them respectively BCMM and GORMM.

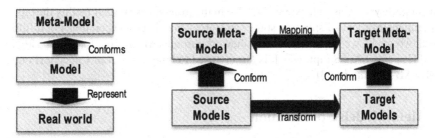

Fig. 2. Relationships between real world, model, meta-model and meta-models based M to M transformation Adapted from Wang et al. [8]

When the GORM is derived correctly from the source BCM, all the necessary parameters have been mandatorily translated into mathematical elements in the graph model.

3 Related Work

In recent years, model transformation methods have been extensively studied. Czarnecki, K., Helsen, S [9] gave us a clear view about the classification of general existing approaches in model transformation. For a simple use case based on UML/MOF class diagram [10] in conjunction with OCL constraint [11], the model transformation is addressed by an ordinary and basic model, from the entity-relationship model to the relational model. Moreover, as model validation is an essential phase after model transformation, a specific constraint and query language (i.e. OCL) is needed to control the distinct properties during the transformation [12]. In terms of application in the industry, interoperability problems linked to collaboration network configurations have been largely addressed. For example, we cite the mediation information system engineering (MISE) approach based on MDE to design a mediation information system (MIS), with the aim of improving interoperability in emerging collaborative situations [13]. Before achieving this goal, the design of the collaboration network could also be based on MDE using added knowledge given by use of ontologies [14]. Facing the improvement of the cooperation capabilities, model-to-model mapping is applied to detect the meanings and relations between different core business enterprise models automatically [8]. For another applicative domain such as healthcare system, a prototype based on MDE is introduced for realizing the mobile device interface code generation from the visual care plan modeling [15]. In regard to our work, some approaches are presented to perform the graph transformation though a simple visual model [16]. However, we have not found related research referring to the application of MDE for model transformation between a business context model and graph-OR model.

As mentioned above, the main ingredient for supporting model transformation is Model Transformation Language (MTL). ATL is largely used in the related research works [17]. Furthermore, the recent research of Burgueño, L et al. shows a survey about the MTL in practice [18], as well as the current and future state of MTL against the evolution of artificial intelligence. One objective of our work is to describe the transformation process from BCM to GORM. We will do it not only in the backend of our support system, as a black box, but also in the frontend as an HMI facility during our experiment in order to display our source and target model during operation. For fulfilling these two goals, we implement our framework through ADOxx platform embedded in an MTL named AdoScript, which is largely used to build a Proof of Concept (PoC) in OMILAB community [3].

4 Meta-Models Specification

4.1 Meta-Model: BCMM

The knowledge (business objects and their relationships) conveyed by TSP must be based on a meta-model whose content will set requirements for structuring data. The main concepts of a TSP consist of:

- **Salesman:** the person who will execute the city tour, characterized by her/his *name*.
- **City:** the town entity visited by the salesman, characterized by an *identifier* (*id*).
- **Travelling:** the routing itinerary between each pair of cities, characterized by distance length in kilometers. The distance of a two-way journey for a pair of cities might be different.

In terms of business constraints (**BuConst**), the constraint concept must guarantee that the salesman visits each city only once and finishes the tour in the starting city (the journey is a cycle). An enumeration list characterizes this concept within the *"type"* attribute, which includes the three types of constraints assimilated to options:

- Each city reached from exactly one other city by the salesman.
- From each city, there is only a departure to another city.
- There is only one tour covering all the cities to be visited.

Finally, the **Goal** concept is introduced to specify our objective function, the minimization of the salesman's total routing distance during the chosen visits. It is also explained by a *"type"* attribute. Figure 3 illustrates the BCMM, the goal is to provide the user with a platform for collecting the necessary data in a user-friendly way, though conceptual knowledge modeling covering all dimensions of the TSP.

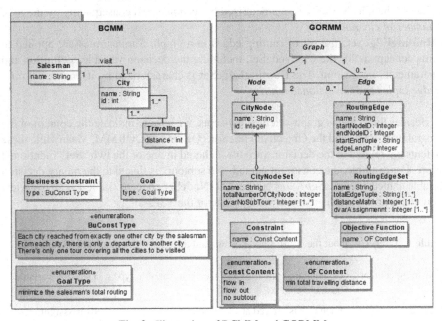

Fig. 3. Illustration of BCMM and GORMM

4.2 Meta-Model: GORMM

In OR, TSP is regarded as a typical combinatorial optimization problem in NP-optimization. This problem has always been represented using a directed-graph, with the form $G = (V, E)$ where V is the set of nodes and E is the set of edges [6]. Moreover, as the meta-model for the graph structure has yet to be addressed by Enríquez et al. [19], we extend this proposal to fit our problem. The explicit concepts in our GORMM are the followings:

- **CityNode:** the city node inherited the meta-class **Node** that makes up a graph. A graph can contain zero or more city nodes, and a city node is part of a graph. **CityNode** characterized by *name* and *identifier (id)*.
- **RoutingEdge:** the routing edge inherited the meta-class **Edge** of a graph. A graph can contain zero or more routing edges and a routing edge is part of a graph. Two types of edge exist: income or outcome edge in respect of each city node, due to the intrinsic directed-graph structure. A **RoutingEdge** is characterized by its *name, start node id* (i.e. 3*), end node id* (i.e. 1), *start-end tuple* (i.e. $< 3,1 >$) and *edge length* associated to a distance in a given unit (i.e. 8). The *start-end tuple* attribute must fit with the required input in the CPLEX environment.
- **CityNodeSet:** the set of city nodes in a graph. This is the key concept in the combinatorial optimization, because the decision variable will be formally designated in the field of concept attributes, which will be defined in accordance with the constraint or objective functions. The set of nodes is a class whose attributes will become parameters (as opposed to variables) of the mathematical model afterwards (i.e. number of cities). The **CityNodeSet** is characterized by its *name*, and consequently by the *total number of city node*.
- **RoutingEdgeSet:** the set of routing edges in a graph. Similar rules are applied to this concept as for the CityNodeSet, including the decision variables definition and parameters integration. The **RoutingEdgeSet** is characterized by its *name*, the *total edge tuples* and the *distance matrix*.

Besides these mains graph-oriented concepts, we have to indicate the content of the constraints (**Const**) and the Objective Function (**OF**) in our GORMM. Meanwhile these contents are related to the decision variable gathered in one of the two "**Set**" mentioned before. For example, the constraint named "No subtour" is linked to the binary decision variable named "*dvarNoSubTour*" built in **CityNodeSet** concept. Table 1 is an overview about the relevant mechanisms to set the stage of the TSP.

Table 1. Overview about the constraints' content and the related decision variable under TSP

Constraint content	Constraint explication in business context	Decision variable type	Decision variable nomination
Flow in	Each city is arrived at from exactly one other city	Boolean	dvarAssignmennt
Flow out	From each city there is a departure to exactly one other city		
No subtour	Only one tour covering all the cities to be visited covering all cities	Positive integer	dvarNoSubTour

Figure 3 illustrates the GORMM. The goal of this meta-model is to represent the graph-oriented knowledge in 3 aspects:

- Visualization of the structuring graph-oriented knowledge.
- Generation of the input data for our mathematical IBM CPLEX optimizer.
- Definition of the decision variable based on the constraint content.

5   Implementation

In this section, we introduce the methodology for processing the Model to Model transformation from BCM (source model) to GORM (target model).

The translation from one model to the other is divided into two main parts: class (concept) mapping and attribute mapping, which are performed simultaneously. The transformation process operates progressively and can be described in 4 steps. Figure 4 shows what happens at the model level on both sides subject to inference of the class mapping rules at the meta level.

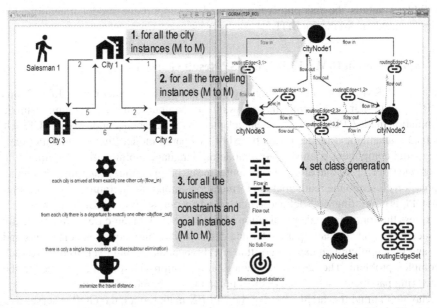

Fig. 4. M to M transformation steps in model level by one TSP Use Case: 3 cities to be visited

1. At first, we execute the mapping from **City** to **CityNode**, the corresponding attribute value such as identifier in city class will be assigned directly in the identification of **CityNode**.
2. The second step pertains to city interrelation building. The **Travelling** class (an arrow represents the relation in BCM) specifies the distance between each pair of

cities. In the GORM, the relation class will be endowed a new pattern [**flow in** (relation) – **RoutingEdge** (class) – **flow out** (relation)]. This is for explaining the constraint named "flow in" and "flow out" regarding each **CityNode** in the result generation phase. As for the attributes, we focus not only on distance, but also on the identification of the incoming and outgoing **CityNode,** referring to each routing edge.

3. Next, the business constraints and goal are translated. Through this operation, on the mathematical side, we provide an indicator to indicate the relative equations in mathematic model.

4. Finally, the **Set** class generation is done. It is not a translation, but it depends on the result of steps 1 and 2. The purpose is to determine the defined set parameter value, which will be the input data of our IBM CPLEX model. To progress, the operation must consist of managing data in a new structure gathering together the attribute value in homogeneous classes, such as the whole set of distances defined in a routing edge matrix. This is realized by "browsing" functions which collect a singular value (i.e. a distance) as input and store it in a collection in output (i.e. distance matrix).

Table 2 details the mapping rules applied to support the model transformation process, based on AdoScript MTL.

5.1 Application in IBM CPLEX for TSP Result Generation

Once the GORM is generated by applying the explicit rules enumerated in Table 2, this GORM provides an architecture that allows us to manipulate a predefined mathematic formulation in the CPLEX environment, adapted to the TSP being addressed. A CPLEX oriented mathematical model is composed of two parts: model (.mod) and data (.dat). The model component is embedded a modeling language Optimization Programming Language (OPL) [20] that allows to compute linear (or quadratic) programs, through an expression closed to the mathematical formulation among the literature of OR community. OPL is promoting well-defined syntaxes and operators within CPLEX environment, to express the combinatorial optimization problem (i.e. TSP in our case) mathematically by specifying relevant data (parameter and decision variable), equation constraints and OF, and then applying an exact algorithm (i.e. branch and bound [21]) to resolve such a complex problem. The data component is aimed to provide the initial values to the structuring parameters which are defined in the model component.

Having in mind this architecture of the CPLEX math model, the GORM serves as a source for a new translation going to the TSP expression into a mathematical optimization model. Figure 5 shows the overview of the roadmap concerning the application of GORM in CPLEX model. With respect to this figure, the GORM is processing the following knowledge:

1. Indication of decision variable. The attribute with prefix "dvar" of "set" object in GORM indicates the needed decision variables in respect of TSP.

Table 2. Mapping rules between BCMM and GORMM

Source		Mapping rules base on AdoScript	Target	
Concept	Attribute		Attribute	Concept

City	idOfCity	For srObj **in** srObjs {
		   CC "**Core**" getClass (City)
		   CC "**Core**" getAttribut (srObj)of(City)
		   SET aVal = "idOfCity" (srObj)of(City)
		   CC "**Core**" getClass (CityNode)
		   CC "**Core**" createObj (tarObj=srObj)of(CityNode)
		   CC "**Core**" set(aVal)in"id" (tarObj)of(CityNode) }
			id	**CityNode**

City	idOfCity	For srObj **in** srObjs {
		   SET conns = getConnecctors(srObj)of(City)
		   SET aValStart = "idOfCity" (srObj)of(City)
		For conn **in** conns {
		   CC "**Core**" getAttribut (conn)of(Travelling)
		   SET aValDist= "distance" (conn)of(Travelling)
		   SET srEndP = getEndPoint(conn)
		   CC "**Core**" getAttribut (srEndP)of(City)
		   SET aValEnd = "idOfCity" (srEndP)of(City)
		   CC "**Core**" getClass (RoutingEdge)
		   CC "**Core**" createObj (rTarObj)of(RoutingEdge)
		   CC "**Core**" createConnector
		     From (tarObj= srObj) To (rTarObj)
		   CC "**Core**" createConnector
		     From (rTarObj) To (tarObj =srEndP)
		   CC "**Core**" set(aValStart)in"startNodeId"
		(rTarObj)of(RoutingEdge)
			startNodeId
			
			endNodeId
			
			startEndTuple
				RoutingEdge

Travelling	distance	CC "**Core**" set(aValEnd)in"endNodeId"
		(rTarObj)of(RoutingEdge)
		   CC "**Core**" set(<aValStart ,aValEnd>)in"startEndTuple"
		(rTarObj)of(RoutingEdge)
		   CC "**Core**" set(aValDist)in"edgeLength"
		(rTarObj)of(RoutingEdge) }}
			edgeLength	

BuConst	type	For srObj **in** srObjs {
		   CC "**Core**" getClass (BuConst)
		   CC "**Core**" getAttribut (srObj)of(BuConst)
		   SET aValConst = "type" (srObj)of(BuConst)
		   CC "**Core**" getClass (Goal)
		   CC "**Core**" getAttribut (srObj)of(Goal)
		   SET aValGoal = "type" (srObj)of(Goal)
			name	**Const**

Goal	type	CC "**Core**" getClass (Const)
		   CC "**Core**" createObj (tarObj=srObj)of(Const)
		   CC "**Core**" set(aValConst)in"name" (tarObj)of(Const)
		   CC "**Core**" getClass (OF)
		   CC "**Core**" createObj (tarObj=srObj)of(OF)
		   CC "**Core**" set(aValGoal)in"name" (tarObj)of(OF) }
			name	**OF**

		CC "**Core**" getClass (CityNode)
		   CC "**Core**" getAllObjs(tarObjs)of(CityNode)
		   SET aNum = countTheNumber.(tarObjs)of(CityNode)
		   CC "**Core**" createObj(tarObj)of(CityNodeSet)
		   CC "**Core**" set(aNum)in"numOfCityNode"
		(tarObj)of(CityNodeSet)
			numOfCityNode	**CityNodeSet**

		CC "**Core**" getClass (RoutingEdge)
		   CC "**Core**" getAllObjs(tarObjs)of(RoutingEdge)
		   SET aEdgeTuple = append."startEndTuple"
		(tarObjs)of(RoutingEdge)
		   SET aDistMat = append."edgeLength" (tarObjs)of(RoutingEdge)
		   CC "**Core**" createObj(tarObj)of(RoutingEdgeSet)
		   CC "**Core**" set(aEdgeTuple)in"edgeTuple"
		(tarObj)of(RoutingEdgeSet)
		   CC "**Core**" set(aDistMat)in"distanceMatrix"
		(tarObj)of(RoutingEdgeSet)
			edgeTuple
			
			distanceMatrix
				RoutingEdge Set

2. Indication of data declaration. Some objects' attributes in GORM can be considered as a pointer, in order to identify the type's declaration of relative data in CPLEX math model.

3. Input data generation. After declaring data types, this stage allows us to assign the value to the related parameter with a specific format: parameter name – value in data part (.dat) of math model in CPLEX.

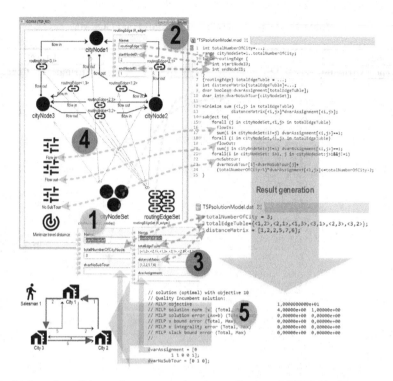

Fig. 5. Overview of the roadmap concerning the application of GORM in CPLEX model

4. Constraint and objective indication. The value of "*name*" attribute in constraint and OF object in GORM, draws a link to the respective equation in the CPLEX model. This mapping enables the computation module to process the algorithm and find the optimum value of the indicated decision variables.
5. The last part is a configuration for providing the decision-making support to the user, by properly returning the results from the optimizer, ideally in the BCM. These values announce the best routing result for TSP regarding the conceptual modelling. We emphasize the flexibility that comes with the modification of the TSP's dimension, such as the increase of the number of cities to be visited on the modelling component of BCM.

6   Conclusions and Perspectives

This research work addresses a framework to solve the classic combinatorial optimization problem of TSP based on MDE. We emphasized the M to M transformation between BCM and GORM by adopting the MTL AdoScript as a component of the ADOxx platform for supporting this transformation, as well as the visualization of the two models. Furthermore, we specified the knowledge to be handled during the main steps for processing the wanted transformation: 2 meta-models and the explicit mapping rules between meta-models at an upper level. During the framework design and experimentation, we decided the final target model to be an IBM CPLEX solution-oriented one, which facilitates the generation of a TSP solution for a given set of data.

We are not sure at this time we have reached the best transformation model level, so we cannot affirm that we have captured all the existing knowledge about how the experts resolve the planning problem. But we think that our MDE framework could be a basis for the continuous improvement in the future on that point, which leads to a better optimized model to model transformation.

In the future, at first, it will be interesting to make progress at the end of the MDE chain. We intend to develop a meta-model adapted to OPL in CPLEX, for supporting the M to M transformation from GORM to this math model in CPLEX. Then we wish to extend this approach not only for TSP, but also to other Vehicle Routing Problems (VRP), and test its added value for a decision maker facing more and more complex OR problems. Starting from this prototype, the following question is to demonstrate how we could handle a broad variety of planning problems with less effort. Considering the TSP experimentation, for example, we work on the assessment of our tool's effectiveness when many problem configurations are considered. It means that the OF or some constraints have to be systematically modified through model transformation, subject to a change in specifications made at the business level. This is the price to be paid for quality proof of our innovation.

References

1. Toth, P., Vigo, D.: The vehicle routing problem. SIAM (2002)
2. Dantzig, G.B., Ramser, J.H.: The truck dispatching problem. Manag. Sci. **6**, 80–91 (1959)
3. Fill, H.-G., Karagiannis, D.: On the conceptualisation of modelling methods using the ADOxx Meta modelling platform. Enterp. Model. Inf. Syst. Architect. (EMISAJ) **8**, 4–25 (2013). https://doi.org/10.18417/emisa.8.1.1
4. Fill, H.-G.: Semantic evaluation of business processes using SeMFIS. Domain-Specific Conceptual Modeling, pp. 149–170. Springer, Cham (2016). https://doi.org/10.1007/978-3-319-39417-6_7
5. Hastings, K.J.: Introduction to the Mathematics of Operations Research with Mathematica®. CRC Press, CRC Press (2018)
6. Bondy, J.A., Murty, U.S.R.: Graph theory with applications (1976)
7. MOF, O.: OMG Meta Object Facility (MOF) Specification v1. 4. OMG Document formal/02-04-03. http://www.omg.org/cgibin/apps/doc (2002)
8. Wang, T., Truptil, S., Benaben, F.: An automatic model-to-model mapping and transformation methodology to serve model-based systems engineering. Inf. Syst. E-Bus. Manag. **15**(2), 323–376 (2016). https://doi.org/10.1007/s10257-016-0321-z
9. Czarnecki, K., Helsen, S.: Classification of model transformation approaches. In: Proceedings of the 2nd OOPSLA Workshop on Generative Techniques in the Context of the Model Driven Architecture, pp. 1–17, USA (2003)
10. Loecher, S., Ocke, S.: A metamodel-based OCL-compiler for UML and MOF. Electron. Notes Theor. Comput. Sci. **102**, 43–61 (2004). https://doi.org/10.1016/j.entcs.2003.09.003

11. Bézivin, J., Büttner, F., Gogolla, M., Jouault, F., Kurtev, I., Lindow, A.: Model transformations? transformation models! In: Nierstrasz, O., Whittle, J., Harel, D., Reggio, G. (eds.) MODELS 2006. LNCS, vol. 4199, pp. 440–453. Springer, Heidelberg (2006). https://doi.org/10.1007/11880240_31

12. Lengyel, L., Levendovszky, T., Mezei, G., Forstner, B., Charaf, H.: Metamodel-based model transformation with aspect-oriented constraints. Electron. Notes Theor. Comput. Sci. **152**, 111–123 (2006). https://doi.org/10.1016/j.entcs.2005.10.020

13. Mu, W., Benaben, F., Boissel-Dallier, N., Pingaud, H.: collaborative knowledge framework for mediation information system engineering. Sci. Program. **2017**, 1–18 (2017). https://doi.org/10.1155/2017/9026387

14. Wang, T., Montarnal, A., Truptil, S., Benaben, F., Lauras, M., Lamothe, J.: A semantic-checking based model-driven approach to serve multi-organization collaboration. Procedia Comput. Sci. **126**, 136–145 (2018). https://doi.org/10.1016/j.procs.2018.07.217

15. Khambati, A., Grundy, J., Warren, J., Hosking, J.: Model-driven development of mobile personal health care applications. In: Proceedings of the 2008 23rd IEEE/ACM International Conference on Automated Software Engineering. pp. 467–470. IEEE Computer Society, Washington, DC, USA (2008). https://doi.org/10.1109/ASE.2008.75

16. Taentzer, G., et al.: Model transformation by graph transformation: a comparative study (2005)

17. Jouault, F., Allilaire, F., Bézivin, J., Kurtev, I.: ATL: a model transformation tool. Sci. Comput. Program. **72**, 31–39 (2008). https://doi.org/10.1016/j.scico.2007.08.002

18. Burgueño, L., Cabot, J., Gérard, S.: The future of model transformation languages: an open community discussion. JOT **18**(7), 1 (2019). https://doi.org/10.5381/jot.2019.18.3.a7

19. Enríquez, J., Domínguez-Mayo, F., Escalona, M., García, J.G., Lee, V., Goto, M.: Entity identity reconciliation based big data federation-a MDE approach. In: International Conference on Information Systems Development (ISD) (2015)

20. Kordafahar, N.M., Rafeh, R.: On the optimization of cplex models. Int. Res. J. Appl. Basic Sci. **4**, 2810–2816 (2013)

21. Lawler, E.L., Wood, D.E.: Branch-and-bound methods: a survey. Oper. Res. **14**, 699–719 (1966)

Innovative Applications of Intelligent Systems

Mathematical Expression Retrieval in PDFs from the Web Using Mathematical Term Queries

Kuniko Yamada$^{(\boxtimes)}$ and Harumi Murakami

Osaka City University, 3-3-138, Sugimoto, Sumiyoshi, Osaka 558-8585, Japan
d18ud512@eb.osaka-cu.ac.jp, harumi@osaka-cu.ac.jp

Abstract. Since mathematical expressions on the web are not annotated with natural language, searching for expressions by conventional search engines is difficult. Our method performs web searches using a mathematical term as a query and extracts expressions related to it from the obtained PDF files. We convert the PDF to TeX, create images from the mathematical descriptions in TeX and obtain image feature quantities. The expressions are discriminated by a support vector machine (SVM) using the feature quantities. Our experimental results show that eliminating slide-derived PDF files effectively improves F-measure and the mean reciprocal rank (MRR) is best when using both PDFs and HTML.

Keywords: Mathematical expression retrieval · PDF file · Web search

1 Introduction

There is much useful mathematical information on the web, especially in PDF files that contain reliable information. However, it is difficult to search for mathematical expressions in PDF files efficiently. Our method performs an ordinary text search using a mathematical term as a query and presents mathematical expressions related to an input query from the PDF files. After converting the PDF to TeX, we create images from the mathematical descriptions in TeX and obtain the image feature quantities. Our method measures the relevance between a query and a mathematical expression from the following viewpoints: the expression is in a separate line; the query is in the neighborhood and it has appropriate image feature quantities when converting it to an image; and it appears in the first part of the PDF file.

2 Approach

2.1 Mathematical Expression

Mathematical documents have features which differ from ordinary documents. The writing style for mathematical expressions in a document is: (a) variables

© Springer Nature Switzerland AG 2020
H. Fujita et al. (Eds.): IEA/AIE 2020, LNAI 12144, pp. 155–161, 2020.
https://doi.org/10.1007/978-3-030-55789-8_14

and signs, or an additional expression are on a line and are expressed with math notations instead of characters because the mathematical font is special; (b) an important theorem or formula is on a separate line, even though it is part of a sentence; and (c) when a theorem is derived or a calculation example is shown, its expression is very long. In such a case, the expression is usually not important.

2.2   Method

Our method is illustrated in Fig. 1. This approach does not depend on the language; however, we conducted our original study in Japanese, so we translated the data into English for ease of explanation. In our previous study [1], we proposed a method which retrieves mathematical expression images on the web using mathematical terms as queries. Here, we propose a method which retrieves mathematical expressions in PDF files. In Fig. 1, highlighted parts ((2) and (3)) are the new proposal and the others are adjusted to suit the TeX documents.

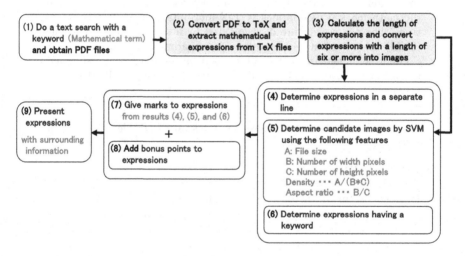

Fig. 1. Overview

(1) We perform a web search using mathematical terms as a query and capture the top 100. From the search result page returned by the search engine, we obtain PDF files and the boldface parts of the snippets.

(2) Preliminary experiment results showed that PDF files derived from slides had poor accuracy. In order to eliminate these, we obtain the aspect ratio of the PDF files from "MediaBox" and "Rotate" in the PDF page object, and remove landscape-oriented files beforehand. Then, we covert the PDF files to TeX files using InftyReader (http://www.inftyproject.org/en/index.html).

(3) In order to eliminate variables and fragments of expressions, we define and calculate the length of a mathematical expression according to the following procedure. (a) TeX commands are extracted from dataset D_1 (described later) and the command is given a 1 if it is a constituent element of an expression; otherwise, it is given a 0. For example, commands with 1 are "\frac", "\partial", and "\alpha", and commands with 0 are "\mathrm" and "\overline". (b) These TeX commands are checked to determine whether they match part of the expression. The number of commands with 1 is length1 and all matched commands are deleted. The number of the characters in the rest of the expression is length2. The length of the expression is then the sum of length1 and length2. Since the length of $y = f(x)$ or $E = mc^2$ (E = mc^2 in TeX notation) is 6, we define 6 or more as a mathematical expression.

(4) When a mathematical expression part in TeX is in the display math mode, the expression is given a 1. When an expression part is in the inline mode, the expression is given a 0.

(5) Mathematical expressions converted to images are classified by SVM. We convert an expression to a PNG format image and use the SVM classifier developed by LIBSVM [2] from dataset D_2 (described later). We previously developed this classifier in [1]. The feature quantities are shown in Fig. 1(5). If an output of the SVM is positive, the expression is given a 1 and, if not, a 0.

(6) We search for the keyword (query) as a window size that is set from -200 to $+200$ characters surrounding a mathematical expression in TeX. If it exists, the expression is given a 1; if not, a 0. However, when there is no expression with a keyword, we search again for an alternative keyword which is the boldface part of the snippet.

(7) Each expression is given a score using (4) to (6). At this point, three points is a full score.

(8) The first order of expressions is returned by the search engine and the appearance order in the PDF file. When expressions are ranked in descending order of their score, incorrect expressions are still highly ranked because their order is affected by the ranking of the original web pages. As important information often appears in the first part of a document, we propose a bonus point system to solve this problem. This system gives an additional point to an expression when it first obtains the full score in the appearance order in the PDF. With this system, the possibility increases that the correct expression in each PDF rises to a higher order. From the above, $score(i_k)$ is given to each expression i_k by Eq. (1) in [1].

$$score(i_k) = x_{line} + x_{key} + x_{svm} + x_{bo}, \qquad (1)$$

when $x_{line} = 1$, an expression is in a separate line; when $x_{key} = 1$, an expression has a keyword; when $x_{svm} = 1$, an expression is discriminated to be positive by the SVM; and when $x_{bo} = 1$, an expression has a bonus point.

(9) Our system ranks expressions according to $score(i_k)$, and obtains the top ten. Finally, the top ten expressions with their surrounding information are displayed. In order to show collateral conditions and commentary on the expressions, 200 characters before and after the expression are also displayed. Figure 2 shows a screen display example. Here, the top one with "kernel trick" as the keyword was retrieved correctly. This expression is extracted from the end of the second page of the original five-page PDF document. This PDF ranks 12th in the original ranking returned by the search engine.

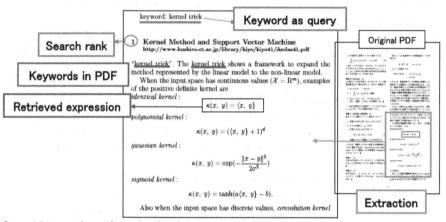

Some right parentheses changed to brackets as a result of conversion mistakes.

Fig. 2. Display example of keyword: kernel trick

3 Experiments

3.1 Dataset

In [1], we randomly selected keywords from the index of "Pattern Recognition and Machine Learning [3] (Japanese version)" by Bishop and conducted a web search using keywords as queries. We obtained the top 100 results for each keyword. We created dataset D_1 using keywords 1 to 30, D_2 using keywords 31 to 60, and D_3 using keywords 61 to 70 for evaluation. Table 1 shows the contents of D_3. The expressions in D_3 were judged manually.

Table 1. Contents of dataset D_3

PDF			HTML			Other
Expression		File	Image		File	
Correct	Total		Correct	Total		
Total 398	29,708	297 (included 60 slide-derived PDFs)	181	6,030	554	149

Keywords: positive definite matrix, multi modality, equality constraint, Newton-Raphson method, root-mean-square error, Lagrange multiplier, sum rule of probability, kernel trick, uniform sampling, kinetic energy

Other: includes broken links, other file formats, PDF that cannot be converted to TeX, etc.

3.2 Experiment1

The aim of this experiment was to examine the effect of eliminating slide-derived PDF files. Using F-measure, MRR (Eq. (2)), and mean average precision (MAP), we evaluated the output results of the top ten. MAP was obtained by calculating the macro mean of the Average Precision (AP) (Eq. (2)).

$$\text{F-measure} = \frac{2 \cdot \text{Precision} \cdot \text{Recall}}{\text{Precision} + \text{Recall}}, \text{MRR} = \frac{1}{n}\sum_{k=1}^{n}\frac{1}{r'_k}, \text{AP} = \frac{1}{\min(n,c)}\sum_{s}I(s)\text{Prec}(s), \quad (2)$$

where $\text{Precision} = \frac{r}{n}, \text{Recall} = \frac{r}{c}$, r is the number of correct expressions of top n, c is the total number of the correct expressions, r'_k is the rank of the correct expression at the top of the k-th keyword, $I(s)$ is the flag indicating whether the expression at s-th is correct or not, and $\text{Prec}(s)$ is precision at s-th.

In Condition A, our method is applied to all PDF files in D_3 and, in Condition B, it is applied to the PDF files which were not derived from slides in D_3. The results show that the F-measure and MRR improved by eliminating slide-derived PDF files, as illustrated in Fig. 3 and Table 2.

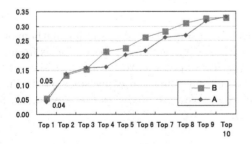

Table 2. Results of Experiment1 (MRR, MAP)

	A	B
MRR	0.50	**0.56**
MAP	**0.17**	**0.17**

Fig. 3. Results of Experiment1 (F-measure)

3.3 Experiment2

From the results of Experiment 1, we used the PDF files with the slide-derived PDFs eliminated. The aim of this experiment was to compare three conditions: Condition B uses PDF (same as in Experiment 1); C uses HTML (same as in [1]); and D uses both PDF and HTML, and these are ranked again after restoring to the original ranking by the search engine. As shown in Fig. 4 and Table 3, almost all values demonstrate that C is better than B and D; however D is the best in MRR.

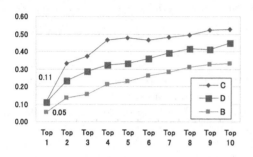

Table 3. Results of Experiment2 (MRR, MAP)

	B	C	D
MRR	0.56	0.75	**0.77**
MAP	0.17	**0.34**	0.29

Fig. 4. Results of Experiment2 (F-measure)

3.4 Discussion

PDF files have many correct expressions (see Table 1) and clear sources. 99.4% of D_3 was obtained from organizations such as universities, academic societies, research institutes, and enterprises. To use these PDF files more effectively, our future tasks are to: (1) increase evaluation data to verify experimental results in more detail; (2) develop a new classifier to improve accuracy; and (3) perform preprocessing, other than the length, to improve accuracy.

4 Related Work

Mathematical expression retrieval usually means retrieval of similar mathematical expressions and a search scope within a particular database. For example, Zanibbi et al. conduct similar mathematical expression retrieval which searches in the dataset developed from arXiv using mathematical expressions written in LaTeX and those converted into images [4]. We were unable to find any research using a dataset from PDFs on the web.

5 Conclusions

We proposed a method to perform web searches using a mathematical term as a query and extracted mathematical expressions related to it from the obtained PDF. Our experiments demonstrate the usefulness of using PDFs other than slide-derived PDFs as well as using both PDF and HTML in MRR.

References

1. Yamada, K., Ueda, H., Murakami, H., Oka, I.: Mathematical expression image retrieval on web using mathematical terms as queries. Trans. Jpn. Soc. Artif. Intell. **33**(4), A-H91_1-13 (2018)
2. Chang, C.-C., Lin, C.-J.: LIBSVM: a library for support vector machines. ACM Trans. Intell. Syst. Technol. **2**, 27:1–27:27 (2011)
3. Bishop, C.M.: Pattern Recognition and Machine Learning. Springer, New York (2006)
4. Zanibbi, R., Yuan, B.: Keyword and image-based retrieval for mathematical expressions. In: Proceedings of Document Recognition and Retrieval XVIII, pp. 011–019 (2011)

Automatic Identification of Account Sharing for Video Streaming Services

Wei Zhang$^{(\boxtimes)}$ and Chris Challis

Adobe Inc., McLean, USA
wzhang@adobe.com

Abstract. According to multiple studies, account sharing is common among subscribers of video streaming services. This leads to huge revenue loss for service providers. Although they have strong financial interests to address the problem, service providers face multiple challenges when trying to identify shared accounts. On one hand, the huge volume of unstructured and noisy data makes it hard to manually process data. On the other hand, it is legitimate for family members to share an account, from anywhere and use many devices as they want. Only these accounts which are shared outside of the household are against policies. In this paper, we propose an efficient solution to address the account sharing problem. Based on a massive volume of session data, our solution builds user profile through accumulating and representing the geolocation and device usage information. Then we estimate the account sharing risk by analyzing the usage pattern of each account. The proposed solution can identify a significant number of shared accounts and help service providers to recoup a huge amount of lost revenue.

1 Introduction

Video streaming is a massive industry and keeps growing. Account sharing is a major problem faced by streaming service providers. According to a poll by Consumer Reports in 2015 [2], 46% of respondents who use a streaming service share their account with someone outside their households. An earlier poll by Thomson Reuters in 2014 found that 15% to 20% millennials shared their accounts [1]. More recent, a study by CNBC in 2018 found that an estimated 35% of millennials share passwords for streaming services [7]. Consequently, the streaming industry loses huge potential revenue due to account sharing. The loss could be hundreds of millions of dollars annually for Netflix alone [7].

Although streaming service providers have strong financial interests in addressing the problem, they face multiple challenges when trying to identify shared accounts. 1. Huge volume of data: leading providers have millions of users and billions of sessions each month. 2. Unstructured and noisy data: session logs are plain text with lots of noise: missing information, numerical error, etc. 3. Perhaps the trickiest problem is that it is perfectly legitimate to share within a household. Family members can share one account from anywhere (e.g. home, office, school or on travel) and use any device. Only accounts which are shared across household are against policies and should be pursued.

H. Fujita et al. (Eds.): IEA/AIE 2020, LNAI 12144, pp. 162–173, 2020.
https://doi.org/10.1007/978-3-030-55789-8_15

Currently, service providers choose not to manually identify/label account sharing because it is too costly and prone to error. Note *account sharing* in this paper refers to against-policy sharing unless we explicitly note otherwise.

1.1   Problem Definition

In this paper, we focus on the video streaming service in the TV Everywhere ecosystem, although our solution can be easily adapted to other services. TV Everywhere is also known as authenticated streaming service, where subscribers are authenticated and authorized to stream video from Multichannel Video Programming Distributors (MVPDs). Major MVPDs in USA all have millions of subscribers. For example, both *AT&T* and Comcast have 20+ million subscribers [5]. The TV Everywhere ecosystem has access to session logs of all subscribers. Each session log contains a slew of information such as account ID, location and some content information, as shown in Fig. 1.

```
2018-04-02 01:00:55.911 [cex-38] INFO  com.adobe.tve.metrics.MetricsLogger.onlyMetrics  - [METRICS]
event=authzg&mvpd=Charter_Direct&prog=ESPN&uid=130d92fdc8e2eb8e906dfafe93b0785f&plainUid=AQICi2beoJu07syR9xJRvCcxTBq
PPZMKuQymHesNIM0bEi7BoGI1uTLy5KWPE1Vr8Xxg3qEWiEwdjXM%3D&ip=174.110.163.2&deviceId=%3CsimpleTokenFingerprint+xmlns%3D
%22http%3A%2F%2Ftve.adobe.com%2Fsdk%2Ftokens%2Fsimple%22%3Eba0458ac546100d7259410da848aa4321ac62875%3C%2FsimpleToken
Fingerprint%3E&arch=3.0&latency=325&res=%3Crss+version%3D%222.0%22+xmlns%3Amedia%3D%22http%3A%2F%2Fsearch.yahoo.com%
2Fmrss%2F%22%3E%3Cchannel%3E%3Ctitle%3E%3C%21%5BCDATA%5Bespn1%5D%5D%3E%3C%2Ftitle%3E%3Citem%3E%3Ctitle%3E%3C%21%5BCD
ATA%5Bst.+Louis+Cardinals+vs.+New+York+Mets%3Ere-
air%29%5D%5D%3E%3C%2Ftitle%3E%3Cguid%3E%3C%21%5BCDATA%5Bespn1%2FSt.+Louis+Cardinals+vs.+New+York+Mets%3Ere-
air%29%2F3292023%5D%5D%3E%3C%2Fguid%3E%3Cmedia%3Arating+scheme%3D%22urn%3Av-
chip%22%3E%3C%21%5BCDATA%5BG%5D%5D%3E%3C%2Fmedia%3Arating%3E%3C%2Fitem%3E%3C%2Fchannel%3E%3C%2Frss%3E&ttl=86400&devi
ce=dmr&clientType=Clientless&clientVersion=v1&os=RokuOS&pht=SetTopBox&dmd=Digital+Video+player&dhwmd=Digital+Video+p
layer&dhwvn=Roku&dhwmf=Roku&dosnm=Roku+OS&dosfm=Roku+OS&dosvn=Roku&ddsw=0&ddsh=0&ddsp=0&userAgent=Roku%2FDVP-
8.0+%28288.00E04128A%29&cdt=roku&country=usa&region=sc&city=westcolumbia&lat=33.989&long=-81.1001&postalCode=29169&c
onnType=cable
```

Fig. 1. An example session log. Each session log contains information such as the subscriber's ID (obfuscated), the device used for connection, the GPS location, etc.

We propose to utilize the session information for creating a service which automatically identifies shared accounts. MVPDs can benefit from the service in two ways: (1) the opportunity of growing revenue remarkably: conversion of shared accounts to regular paid accounts, even just a small fraction of them, will bring in millions of dollars because of their large customer bases; (2) limiting the number of shared accounts also translates to server/network load reduction, which leads to significant cost cutting.

One big challenge for this service is that there is no ground truth label, since the labeling cost is prohibitive as we discussed in Sect. 1. This certainly constrains our choice of algorithms. More importantly, a big question arises: without any ground truth to compare against, how can we justify the results of the service? This question is *crucial* for demonstrating the value of the service due to the consequences of regulating account sharing: treating normal accounts as shared accounts (false alarm) will annoy their subscribers and lead to potential loss of business. Classifying shared accounts as regular accounts, on the other hand, means the solution brings no value to them. We believe that only an

explainable and **presentable** solution can address this challenge. Whether an account is identified as shared or not, it is paramount that MVPDs can easily understand the reason so to trust the results.

2   Existing Work

The streaming industry has tried to restrain account sharing by adding constraints to user accounts: (1) limiting the number of concurrent streaming; (2) asking users to register a limited number of devices to their accounts. However, the first approach can adversely affect concurrent streaming by family members, who are entitled to do so. In addition, limiting concurrent streaming can be circumvented by sharing accounts at different time periods. Having to register a limited number of devices will hurt customer experience and is not desirable either: first, it is a hassle to do the registration; second, customers are having more and more devices and they can easily hit the limit. In addition, an account owner can sell/give the "registered" device to other people, so they can use the device for streaming and easily defeat the policy.

In the academic community, there has been considerable research [4, 8–11] on modeling user behavior from session logs, mainly for improving recommendations. They mostly focus on identifying multiple users by the content that they watched. The techniques that were used are: collaborative filtering [8], subspace clustering [10], graph partitions [9] and topic modeling [11]. [6, 10] are the very few paper which attempt to determine whether an account is shared by multiple users. However, multiple users sharing one account are not necessarily against-policy. In fact, more often than not, they are shared within a household. Therefore, they cannot solve the account sharing problem.

3   Our Solution

The account sharing detection service must accommodate all variations that a normal account could have, so regular users are not impacted. After all, good user experience is the key for MVPDs to maintaining and growing their customer bases. This means that it needs to handle the following scenarios and label them as normal accounts: 1. a big family with a large number of concurrent sessions, since everyone likes the freedom of choosing his/her own content; 2. ever-growing number of devices in a household as new devices are being added all the time; 3. family members commuting to places such as school/office/mall, or traveling to other states and stream video anywhere they want.

Although the problem is complicated, the following assumptions usually hold. (1). Even though they share the account, users outside of the household (against-policy sharing)) are unlikely to share devices with account holders, since they live in different places. They might use devices which used to be owned by the account holder, through sale/gift, but the devices are transferred and not shared, i.e., the account owner is unlikely to use it again. (2). Non-family users are likely to stream videos from separate locations, not the home of account holders.

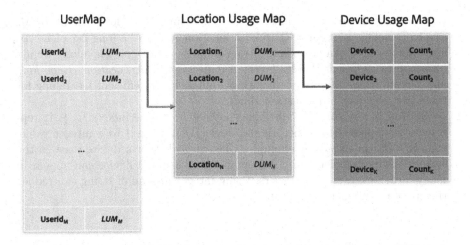

Fig. 2. The data structures used in the algorithms.

Otherwise, they are more like a part of the household and virtually impossible to be identified. Based on these analysis, we propose an novel approach to estimate a sharing score of each account. It utilizes both geolocation and device information to address the problem. Algorithm 2 describes how the sharing score is estimated. It depends on Algorithm 1 to process data and construct efficient retrievable user profiles. Note the GPS coordinates in the session logs are usually noisy; we need to mitigate the problem when processing raw log files as described in Algorithm 1. Without this process, multiple geolocations might be associated with an account even if the owner only streams in her/his home, because the GPS coordinates of difference sessions can be different due to noise.

We first introduce the notation and data structures shown in Fig. 2.

1. *Device Usage Map (DUM)* represents a distribution of device usages. It is a hashmap $\mathcal{M} = \{(\mathcal{D}^0 : \mathcal{C}^0), (\mathcal{D}^1 : \mathcal{C}^1), \ldots, (\mathcal{D}^K : \mathcal{C}^K)\}$, K is the number of devices used in the location. $\mathcal{D}^k$ is the kth device in the list, $\mathcal{C}^k$ is the count (histogram) of usages for the kth device.
2. *Location Usage Map (LUM)* is a hashmap in the form $\{(\mathcal{L}_0 : \mathcal{M}_0), (\mathcal{L}_1 : \mathcal{M}_1), \ldots, (\mathcal{L}_N : \mathcal{M}_N)\}$, where N is the number of locations associated with the account. $\mathcal{L}_i$ is the ith GPS coordinates, $\mathcal{M}_i$ is the *Device Usage Map* associated with $\mathcal{L}_i$. For instance, a *LUM* with only one element: $\{(40.2814, -111.698) : \{(4277780 : 16),\quad (11090085 : 2)\}\}$. It means that 2 devices have been used at location $(40.2814, -111.698)$: device #4277780 was used 16 times (appears in 16 sessions), device #11090085 was used twice.
3. *userMap* stores the profile of all users; each entry contains an userID and a list of *Location Usage Maps* associated with the account. It stores all locations and devices that are associated with the account, as well as the relationship between locations and devices (which devices are used in which locations). The relationship can be represented as a 2D (location-device) matrix, but

the matrix will be very sparse: an account can have a long list of *LUMs* due to traveling or account sharing; a location can have a long list of *DUMs* since a household can have an arbitrary number of devices. Therefore, we use hashmap to represent these 3 levels of maps: *userMap*, *Location Usage Map* and *Device Usage Map*, so to make our approach very efficient (searching for their keys happens in a constant time).

4. *deviceMap* is a hashmap: { (original deviceID : device index), ... }. It represents a mapping from the original deviceID (a string) to a integer value, e.g., 4277780 in the above example. Checking whether a device exist in the system is super efficient using hashmap. In addition, a device can appear in many locations, even across users. Using integer instead of string can reduce the space requirement significantly.

Algorithm 1. Build user profiles (location and device graph).

---

1. Initialization: deviceMap $\leftarrow$ {}, userMap $\leftarrow$ {}, $nDevices \leftarrow 0$
2. Go through all session logs, one session at a time, to build userMap:
   - (a) Extract usage information, e.g., deviceID and GPS coordinates $\mathcal{X}$.
   - (b) **if** deviceID $\in$ deviceMap **then**
             Get its device index k.
          else
             Add { deviceID : nDevices } to deviceMap
             $k \leftarrow nDevices$, $nDevices \leftarrow nDevices + 1$
   - (c) **if** userID $\in$ userMap **then**
             Merge to the existing entry in userMap:
             if $\mathcal{X} \in$ the associated *LUM*, say $\mathcal{X} = \mathcal{L}_j$ **then**
                if $k \in \mathcal{M}_j$ **then**
                   Increase usage count in the corresponding entry in $\mathcal{M}_j$
                else
                   Add a new entry $\{k : 1\}$ to $\mathcal{M}_j$
             else
                Add { $\mathcal{X} : \{k : 1\}$} to the *LUM*
          else
             Create an entry in userMap for this new account
3. Go through each user in the userMap, combine neighboring *Location Usage Maps*:
   - (a) Sort *LUMs* by their number of device usages, in descending order.
   - (b) $i \leftarrow 0$, $P \leftarrow$ number of *LUMs*
          while $i \neq P$ **do**
             for all $(\mathcal{L}_j : \mathcal{M}_j), j > i$ **do**
                if $dist(\mathcal{L}_i : \mathcal{L}_j) < \sigma$ **then**
                   Merge $\mathcal{M}_j$ into $\mathcal{M}_i$ (combine their device usage)
                   Delete $(\mathcal{L}_j : \mathcal{M}_j)$
                   P $\leftarrow P - 1$
             $i \leftarrow i + 1$

Assuming the GPS noise follows a zero mean distribution, the observed coordinates will be centered around the true coordinates. The step 3 in Algorithm 1 is essentially doing non-maximum suppression, it combines neighboring Location Usage Maps into one single usage map in the center location[1]. σ is set to be 50 m in our experiment. The average GPS accuracy is about 7.8 m [3], well within the σ range. So we can almost guarantee to handle GPS noise. This σ setting is also fine enough to identify account sharing across street. If the GPS noise do not follow a zero-mean distribution, the coordinates will be shifted by the non-zero mean. However, it will not affect the scoring algorithm since Algorithm 2 is only based on the usage pattern, not the exact location.

Algorithm 2. Sharing score estimation for each user in userMap.

1. Given $\left[(\mathcal{L}_0 : \mathcal{M}_0), (\mathcal{L}_1 : \mathcal{M}_1), \ldots, (\mathcal{L}_N : \mathcal{M}_N)\right]$, Set the *base location* $\mathcal{L} \leftarrow \mathcal{L}_0$.
2. Initiate a "registered" device map $\mathcal{M} \leftarrow \mathcal{M}_0$. Initiate risk score $R \leftarrow 1$.
3. For $i = 1, 2, \ldots, N$:
   (a) Calculate distance weight: $W_d \leftarrow log_\alpha(max(dist, \alpha))$, where $dist \leftarrow |\mathcal{L}_i - \mathcal{L}|$.
   (b) Let $\mathcal{M}_i$ be $\{(\mathcal{D}_i^0 : \mathcal{C}_i^0), (\mathcal{D}_i^1 : \mathcal{C}_i^1), \ldots, (\mathcal{D}_i^J : \mathcal{C}_i^J)\}$, where J is the number of devices associated with the ith location. $\mathcal{D}_i^j$ is the jth device ID, $\mathcal{C}_i^j$ is the count of usages for the jth device.

   for $j = 0, 1, \ldots, J$ **do**
       $R_j \leftarrow 0$
       if $\mathcal{D}_i^j \in \mathcal{M}$, say $\mathcal{D}_i^j = \mathcal{D}^k$ **then**
           $r \leftarrow \frac{\mathcal{C}_i^j}{\mathcal{C}^k}$
           $R_j \leftarrow R_j + \frac{1}{e^{-(r-\beta)/3} + 1}$
           $\mathcal{C}^k \leftarrow \mathcal{C}^k + \mathcal{C}_i^j$
       else
           $R_j \leftarrow R_j + 1$
           Add $(\mathcal{D}_i^j, \mathcal{C}_i^j)$ to $\mathcal{M}$

   (c) $R \leftarrow R + R_j * W_d$
4. $W_t \leftarrow log_\gamma(max(t, \gamma))$, t is the number of devices used by the account.
5. $R \leftarrow R * W_t$
6. Sharing score $S \leftarrow 2 * \frac{1}{e^{-(R-1)} + 1} - 1$

The userMap generated from Algorithm 1 captures the usage pattern of all accounts: one entry for an account. Each entry contains a list of Location Usage Maps, in descending order of their device usages. We use Google Map API for visualizing the usage pattern, so people can easily see why an account is labeled as normal or abusive sharing. Some screen copies of the interactive map are shown in the result section, such as Table 1, Table 4 and Table 5. Each location associated with the account is tagged with a red balloon. A red circle is centered at the root of each balloon, representing the usage in that location. The bigger

---

[1] In our implementation, we use kernel density estimation to in stead of the simple histogram count. So it will be more robust to GPS drift.

the circle, the larger the number of usages (sum of all device usages in that locations). Note the circle size is not linearly proportional to the number of usages. Because the range of usage numbers is very wide, from 1 to multiple thousands. Consequently, a large circle will make all other circles too small to see. Instead, the size is based on the natural logarithm of the numerical value, so that we can see the difference in usages across different locations. The location with the most usages is called the *base location* of the account, presumably it is the owners home place. The base location is important for the scoring Algorithm 2. A regular account is more likely to have a dominant base location, because that is the place where most household members enjoy the streaming service. For a heavily shared account, the usage pattern is more distributed.

Algorithm 2 estimates the score (risk) of an account being shared, by checking the device usage of all locations other than the base. If a device is not in the list of "registered" devices, i.e., it is a new device never used in the base location, it is more likely to be used by outsiders (users not belonging to the household). If it appears in the list, but used much more often in other locations than the base, it is also possibly an outsider's device (e.g. a friend who visits occasionally), although the probability is much lower than an "unregistered" device. This is captured in (5) in Algorithm 2. We believe that the risk of sharing is fairly low when the non-base usage is not significantly higher than the usages at the base location. β is set to be 20 in our experiments. That is, when the usage in other locations is 20 times as high, we increase R_j by 0.5. Higher R_j leads to higher R, which ultimately leads to a higher sharing score. $r - \beta$ is divided by 3 so the logistic curve does not saturate too quickly.

We also take the distribution of locations into account when estimating sharing scores. The idea is that uses far apart are more likely to be due to account sharing. Household members may go to office or school and stream video every day, but are less likely to go to the other side of the country. The distance weight W_d is introduced for this purpose. The minimum distance α is set to be 50 so the distance weight will have no effect in Algorithm 2(3c) for usages within 50 miles, while usage in locations which are hundreds of miles away will be penalized and lead to higher scores.

Even if all streaming sessions appear to happen in the base location, it is still possible that the account is shared since people can fake their location by geo-spoofing. For example, geo-spoofing has been used by *Pokemon Go* players to "go" to places without physically being there. It is not a widespread practice yet, but we should be ready to tackle it. The device weight W_t in Algorithm 2 is designed for this purpose. The higher the number of devices, the higher the weight. So even if all steaming sessions share the same location, the score will still be higher if there is an extremely large number of devices. This is our first attempt to tackle the geo-spoofing problem, so we are relatively generous on the parameter setting, with $\gamma = 20$. For example, based on the current settings, if 400 devices are used in an account, the weight will equal to 2. If the account's *Location Usage Maps* has only one location (all sessions happen in one location), the final score will be 0.46. If the number of devices is less than γ, the weight is always 1. For accounts with just one location and less than 20 devices, their R value in Algorithm 2(5) are always 1. Consequently, they get score 0, which signifies an unquestionably safe account. Even with this generous setting, we identified some potentially shared accounts where all sessions appeared to be in one location; an example is shown in Table 2. The worst case has 6,829 devices under one account, clearly an shared account using geo-spoofing. It gets a score of 0.75, not extreme but high enough to be identified.

4   Experiments Result

We use a three-month session log of TV everywhere system for testing the proposed solution. The number of users in this data is 30,620,878. The total number of session records in the data is 1,032,254,858 and the size of this data is 1.01 terabyte. Using 0.5 as the threshold for sharing score, we identified about 6.45% accounts as shared, with very high confidence. Those accounts are usually quite obvious to be shared as what we show in Table 4 and Table 5. Using a threshold of 0.05, we identified about 15.66% accounts as shared. Some manual verification might be need for the these results. Nevertheless, based on random check, we can see the identified accounts are indeed likely to have been shared. About 70% of users stream videos from just one location using less than 10 devices. They are all labeled as regular/non-sharing accounts as their sharing scores are 0.

4.1   Multiple Devices in One Location

Two cases are shown in Table 1 and 2. Although both have only one location for all sessions, the case in Table 2 is more likely to be a shared account due to an extraordinary number of devices being used. The accounts which have zero sharing scores all have the same visualization (one location with few devices).

4.2   Multiple Devices Multiple Locations

The case in Table 3 represents an interesting pattern: many locations and few devices. We call it a traveling account. The account has two devices (#747409 and #868586) associated with it. The base location is $(40.7046, -73.9216)$ where device #747409 was used 15 times and device #868586 was used 6 times. Both devices were used in other locations, suggesting that they were taken to travel around. The account sharing score is relatively low for this case (only 0.01) and it is labeled as a safe account. The score is not 0 though, because it is possible that a friend visited the account holder's base location (probably home) multiple times with device #868586, thus he/she got the device "registered" to the account and lowered the account sharing score. Nevertheless, the probability is very low in comparison with other shared accounts. Note the sharing score is updated monthly with incoming data, so next month device #868586 will not be "registered" with the account if the friend no longer brings it to the account holder's base location. As a result, the sharing score would be much

Table 1. This account has been used by 27 devices in total. However, they are all used in the base location and the number of devices is not super high, so it is still likely a legitimated account with score 0.05.

Usage visualization	Location Usage Map
	{(39.9194, -75.4205): {11687811: 2, 2217860: 7, 12263382: 5, 1016583: 57, 77: 10, 578254: 3, 17937: 20, 10821077: 4, 3304278: 3, 11051844: 2, 12391413: 4, 10117597: 2, 8262432: 3, 2583585: 22, 6699239: 1, 6699240: 5, 1317930: 25, 2012653: 35, 3827630: 1, 11688048: 4, 3908979: 1, 12391412: 1, 1015669: 11, 5412342: 12, 2622200: 16, 12721209: 2, 10457717: 2}}

Table 2. All sessions happens in just one location for this account. However, the account has been used by many more devices, 72 in total. The number of devices suggests that geo-spoofing might be used here. The score is 0.21.

Usage visualization	Location Usage Map
	{(39.0329, -77.4866): {6435225: 4, 7796609: 2, 13210498: 2, 5182979: 4, 8522116: 2, 8911041: 3, 11916204: 2, 12181898: 1, 13113230: 2, 12885648: 2, 6745537: 2, 9590035: 1, 9984404: 2, 6644503: 1, 7834904: 2, 9573145: 2, 8482497: 2, 2710447: 2, 9993242: 2, 7778113: 2, 7055009: 3, 13225123: 2, 8222628: 1, 6520613: 2, 7564072: 3, 3754666: 2, 7653420: 2, 11134208: 2, 8911023: 2, 2710448: 2, 12746418: 2, 12634526: 2, 7379895: 4, ... } }

Table 3. This account has been used in many locations. The location with most usage is labeled bold. However, the account is identified as a regular account because few devices are used. It is most likely family members traveling around.

Usage visualization	Location Usage Map
	{(40.6797, -73.9503): {747409: 2}, (40.859, -73.8908): {747409: 1}, (40.8371, -73.8807): {868586: 1}, (40.679, -73.9618): {747409: 1}, (40.809, -73.9168): {868586: 1}, (40.8276, -73.896): {747409: 1}, **(40.7046, -73.9216): {747409: 15, 868586: 6}**, (40.8187, -73.8572): {747409: 3}, (40.728, -73.9493): {747409: 1}, (40.6936, -73.9265): {747409: 1}, (40.6471, -73.9549): {747409: 1}, (40.7903, -73.9468): {747409: 7}, (40.8202, -73.9202): {747409: 3}, (40.8464, -73.9027): {747409: 4}, (40.7758, -73.8749): {747409: 5}, (40.7608, -73.9457): {747409: 1}, ... }

higher according to Algorithm 2, as the device is not used in the base location. This would cause the account to be labeled as shared, which is the desired result. Therefore, even if users know about how we identify shared accounts, it is not easy for them to game the system: they have to pay regular visits to account holders' base location, in order to "register" their devices to the account. Otherwise, they will be identified.

4.3   Identified Shared Accounts

See Table 4 and Table 5 for some typical cases.

Table 4. A typical shared account: used in many locations, without a clear base location (many locations have similar numbers of usages); 39 devices have been used for streaming with this account.

Usage visualization	Location Usage Map
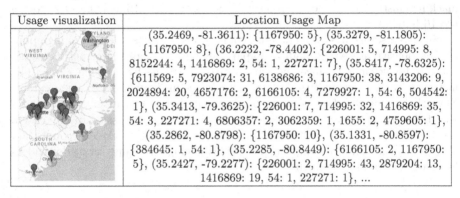	(35.2469, -81.3611): {1167950: 5}, (35.3279, -81.1805): {1167950: 8}, (36.2232, -78.4402): {226001: 5, 714995: 8, 8152244: 4, 1416869: 2, 54: 1, 227271: 7}, (35.8417, -78.6325): {611569: 5, 7923074: 31, 6138686: 3, 1167950: 38, 3143206: 9, 2024894: 20, 4657176: 2, 6166105: 4, 7279927: 1, 54: 6, 504542: 1}, (35.3413, -79.3625): {226001: 7, 714995: 32, 1416869: 35, 54: 3, 227271: 4, 6806357: 2, 3062359: 1, 1655: 2, 4759605: 1}, (35.2862, -80.8798): {1167950: 10}, (35.1331, -80.8597): {384645: 1, 54: 1}, (35.2285, -80.8449): {6166105: 2, 1167950: 5}, (35.2427, -79.2277): {226001: 2, 714995: 43, 2879204: 13, 1416869: 19, 54: 1, 227271: 1}, ...

Table 5. A wildly shared account: used in numerous locations; 33,909 devices have been used for streaming with this account in the 3-month period.

Usage visualization	Location Usage Map
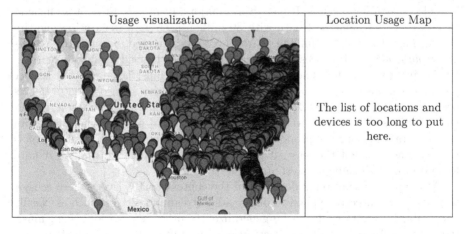	The list of locations and devices is too long to put here.

4.4 Discussion

As we have argued, the solution has to be **explainable** and **presentable** so people can understand and trust it. This has been a design principle for our solution. As we have demonstrated, our identification results can be illustrated intuitively and digested easily. This surely helps our solution and results to be more trustworthy.

Both Algorithm 1 and Algorithm 2 are naturally parallelizable: we can easily split the computation by grouping account userIDs. In our implementation, we divide the work by the first character of userIds, i.e., [0–9, a–f], so the work was split into 16 batches. Thus we don't need to have a huge hashmap of all users, instead we work with 1/16 of them at each batch. This greatly reduced the memory requirement for our implementation. We use only one workstation for this experiment. It can finish all jobs in a week, which is enough for (the currently designed) monthly sharing score update. In the future, we can use a machine cluster to scale for more users if necessary, e.g., 16 machines to process the 16 batches.

Since our solution is based on GPS coordinates, it is possible that in high population density areas, e.g. high rise apartment, people can share their accounts without being caught, since they are indistinguishable by position alone. The fact that we also consider the number of devices mitigates this to some extent. Nevertheless, more information such as the number of concurrent sessions and user behavior analysis will be needed to better address the issue. In any cased, the fact that we can identify over 6% accounts as reliably shared accounts can already have a significant impact, potentially saving streaming service providers hundreds of millions of dollars.

5 Conclusion

In this paper, we have proposed a novel solution for identifying shared accounts for video streaming services. It has several major advantages. First, it is efficient; we can process 3 months of data with 30 million users in a week using one single machine, with over 2 million shared accounts detected. Second, the results are *explainable*. Each processed subscriber, whether labeled as shared or regular, is giving an intuitive and interactive web-based illustration, so that service providers can understand and trust the results. Note that our solution preserves privacy: we obfuscate deviceID using an integer when showing the result, so service providers do not need to worry about violation of privacy when using our solution. In addition, the proposed solution handles noise in geolocation information. Last but not the least, it guards against geo-spoofing, making it hard for subscribers to game the system.

Although our solution is designed in the context of TV Everywhere ecosystem, it can be directly applied to other video streaming services such as Netflix and Hulu. In addition, it can be generalized to other applications, e.g., music streaming and more broadly, subscription-based software, e.g., Photoshop.

References

1. Thomson Reuters poll (2014). https://fingfx.thomsonreuters.com/gfx/rngs/USA-TELEVISION-PASSWORDS-POLL/010041YS48H/index.html
2. Consumer Reports poll (2015). https://www.consumerreports.org/cro/magazine/2015/01/share-logins-streaming-services/index.htm
3. GPS accuracy (2019). https://www.gps.gov/systems/gps/performance/accuracy/
4. Bajaj, P., Shekhar, S.: Experience individualization on online TV platforms through persona-based account decomposition. In: 24th ACM International Conference on Multimedia, pp. 252–256. ACM (2016)
5. Farrell, M.: Top 25 MVPDs (2018). https://www.multichannel.com/news/top-25-mvpds-411157
6. Jiang, J.Y., Li, C.T., Chen, Y., Wang, W.: Identifying users behind shared accounts in online streaming services. In: The 41st International ACM SIGIR Conference on Research & Development in Information Retrieval, pp. 65–74. ACM (2018)
7. Salinas, S.: Millennials are going to extreme lengths to share streaming passwords (2018). https://www.cnbc.com
8. Verstrepen, K., Goethals, B.: Top-n recommendation for shared accounts. In: 9th ACM Conference on Recommender Systems, pp. 59–66. ACM (2015)
9. Wang, Z., Yang, Y., He, L., Gu, J.: User identification within a shared account: improving IP-TV recommender performance. In: Manolopoulos, Y., Trajcevski, G., Kon-Popovska, M. (eds.) ADBIS 2014. LNCS, vol. 8716, pp. 219–233. Springer, Cham (2014). https://doi.org/10.1007/978-3-319-10933-6_17
10. Zhang, A., Fawaz, N., Ioannidis, S., Montanari, A.: Guess who rated this movie: identifying users through subspace clustering. In: Proceedings of the Twenty-Eighth Conference on Uncertainty in Artificial Intelligence (2012)
11. Zhao, Y., Cao, J., Tan, Y.: Passenger prediction in shared accounts for flight service recommendation. In: Wang, G., Han, Y., Martínez Pérez, G. (eds.) APSCC 2016. LNCS, vol. 10065, pp. 159–172. Springer, Cham (2016). https://doi.org/10.1007/978-3-319-49178-3_12

A Model for Predicting Terrorist Network Lethality and Cohesiveness

Botambu Collins, Dinh Tuyen Hoang, and Dosam Hwang[(⊠)]

Department of Computer Engineering, Yeungnam University,
Gyeongsan, South Korea
botambucollins@gmail.com, hoangdinhtuyen@gmail.com, dosamhwang@gmail.com

Abstract. The recurrent nature of terrorist attacks in recent times has ushered in a new wave of study in the field of terrorism known as social network analysis (SNA). Terrorist groups operate as a social network in a stealthy manner to enhance their survival thereby making sure their activities are unperturbed. Graphical modeling of the terrorist network in order to uncloak their target is an NP-complete problem. In this study, we assume that the terrorist group usually confine themselves in a neighborhood that is conducive or regarded as a safe haven. The conducive nature of the environment ramified the terrorist action process from areas of low action (soft target) to areas of high action (maximally lethal target). It is the desired of every terrorist to be maximally lethal, therefore curbing terrorist activities becomes a crucial issue. This study explores the 9/11 event and examines the M-19 network that carried out the attack. We use a combination of methods to rank vertices in the network to show their role. Using the 4-centrality measures, we rank the vertices to get the most connected nodes in the network. The betweenness centrality and traversing graph shortest path are calculated as well. We opine that in curbing terrorist network activities, the removal of a vertex with high centrality measures is a condition sine qua non. We further examine the lethality of the network as well as network bonding to get the degree of cohesiveness which is a crucial determinant for network survival. The experimental results show that M-01 is the most important vertex in the network and that timing is a salient feature in unveiling network lethality. We posit that more study is required to include the global network that was involved in the 9/11 event.

Keywords: Modeling Terrorist Network · Network bonding · Modeling lethality · Terrorist network cohesiveness · 4-Centrality measures

1 Introduction

The recurrent nature of terrorist attacks has now emerged as a core concern across the globe [5]. The world is now faced with a new form of terror induced by non-state actors in the furtherance of their ideologies [11]. Terrorism has not only plagued our society into chaos, but it has also meted untold suffering

© Springer Nature Switzerland AG 2020
H. Fujita et al. (Eds.): IEA/AIE 2020, LNAI 12144, pp. 174–185, 2020.
https://doi.org/10.1007/978-3-030-55789-8_16

to mankind as well [5,8,11]. The traditional method of countering terrorism has yielded little fruit due to the perennial nature of terrorist attacks in recent times. Consequently, terrorism has registered an exponential spike in global attacks spanning from 1970 to 2016 with a whopping figure over 170,000 cases as indicated by Global Terrorism Database (GTD). As a result, many studies have been carried in order to find ways of curbing terrorist activities. Terrorism is an evolutionary processes [10], and the actions of terrorists are predicated on the supposition that when a group successfully carries out an attack, other group copy and employ the same tactics. Therefore, in order to perturb terrorist activities, we must apply models that are dynamic and evolutional as well. The use of social network analysis (SNA) emerges as a novel approach to give an insight into how terrorist organization operates [3,6,8]. Understanding the nature in which a terrorist group operates is a fundamental step in countering terrorism [2]. Scholars involved in this approach has often alluded that the modeling of the terrorist network is an NP-complete problem [11]. Terrorist network (TN) is different from a social network (SN) [2,3,7,9], unlike SN, TN operate a subnetwork outside of its main network also referred to as cell [2], TN operates in a hierarchical structure [3,7] with a well-defined leadership style with each vertex assigned to a well-defined task [7]. TN operates with a finite set of goals and objectives. The first author to mapped out the 9/11 network was Valdis Krebs in [6], his work became the baseline for the study of the 9/11 terror network [3,9]. Other authors engaged in graph modeling of terrorist networks citing 9/11 as a case study will consider the baseline provided by Krebs in [6].

The terrorist network is a covert social network [2] with a hierarchical structure that mimics an organizational network in its operation [5,7]. Although usually classified under social network analysis, the terrorist network operates in a stealthy style that can be modeled in a graph [6]. As mentioned above the graphical modeling of a terrorist network is considered the problem of NP-Complete. This is because there is no clear approach to perturbing terrorist activities [6,8], and gaining access to terrorist data is inherently a challenging and complex issue [11]. Intuitively, gathering intelligence to curb the action of a given terrorist network is known only after the action had occurred such as the 9/11 hijackers [6]. It is, therefore, necessary to employ a brute force and a heuristic approach in uncloaking terrorist networks.

Terrorists usually operate in a network of secrecy [6] to ensure their activities are unperturbed thereby gaining the necessary skills required to be maximally lethal. We assume that the terrorist action takes the process of ramification with a cascading effect (see network lethality section). The prevalence of maximal lethal attacks is as a result of a conducive environment [11]. Terrorist only operate in a neighborhood that it sees as a safe haven with conducive environmental features capable of concealing their operations as well as enhances its efficacy. We consider terrorist network (TN) to be a covert network graph $TN = G$ with a set of vertices V (terrorist) with edges E linking each other such that $TN = G(VE)$ which indicates the interaction within the network such that VE shows there exists a relationship. Our study is motivated base on the premise that terrorism is a dynamic and evolutional process such that, when a group successfully carries out a lethal operation, another group will copy the same tactic. We assume that since the M-19 attack was a success, other terrorist organizations will use the same pattern of operation and hence, a vivid study of the M-19 network and operational model will give an insight into preventing future attacks. The rest of this paper is organized as follows: Sect. 2 is the related work, while Sect. 3 methodology and Sect. 4 is our case study and methodology where we succinctly give an account of the M-19 network that was responsible for the 9/11 event, we model the lethality of the M-19 network and show the network bonding also referred to as network cohesiveness. We discuss our results in Sect. 4 while the conclusion and future directions in Sect. 5.

2   Related Work

Studies on terrorism have experienced a research boom in recent years especially with respect to gaining insight into a terrorist network. In [10], the author explains the causes of the demise of a terrorist network and posited that the rationale behind the demised of certain terror networks is splintering. Some terrorist splint from the main group to join another group after analyzing the cost and benefits. Also, terrorist seeks to foster an ideology and usually terrorist group with similar or the same ideological pattern merge together to enhance their survival rate. The discrepancy between individual thought may facilitate a vertex to splint from a network. Large groups with a higher span will attract small groups thereby leading to their demise [10]. The application of the probabilistic models to curb terrorist threats using the combination of the Hidden Markov Model HMM and Bayesian Networks BN was examined in [11], the authors explain the chain of events that takes place before an attack such as planning, target identification, recruitment and gathering financial resources. Citing examples of the pattern of the event surrounding an attack, the author alluded that "a terrorist withdraws money from a bank, buys chemical which can be used to make a bomb, purchase a flight ticket into the USA, and carry out an attack in the USA such as the 9/11" it took nearly four years just to plan the 9/11 and about one year to plan the Bali and Madrid bombing [11]. Therefore uncloaking these events will yield fruit in preventing such an attack from occurring in the future. The author in [3] sees the modeling of a terror network to be a complex and challenging task that requires a heuristic approach. The author purposes the application of network analysis, agent-based simulation, and NK-Boolean fitness landscapes to measure the complex and dynamic nature of a covert terror network. He contends that a terror network is a hierarchical structure with disinformation being an important tactic employ by the terrorist. Examining the behavioral pattern to understand the correlated nature of agents in a network was coined by [1], the author examines the opaque nature of terrorist networks by vividly studying the interactive relationship that exists within the terrorist in the network and contends that the finance manager plays a pivotal role in the network. The terror network mimics real organization structure with the main leader giving out command and distributing tasks, while the financial operator is in charge of paying out salaries [1]. Concealed links in the network could be unveiled through past actions and interpersonal relationship analyses which is useful in constructing and developing a terrorist network. Although the authors employ centrality measures in their study, they failed to calculate these measures and only give a narrative critique. Furthermore, contrary to our work, no real case study was used in this work and instead they use only 3-centrality measures whereas we used 4-centrality measures which include eigenvector centrality with M-19 network being our case study. Analogous to this is the work of Bo Li and colleagues in [7], the authors championed the agent-based terrorist network modeling and predict the role that each agent plays in the network for it to remain active. The authors maintain that the role of the agents is not mutually exclusive with the outcome of the network performance. The terrorist network is different from a social network with its structure mimicking the form of a real organization with a well-defined leadership structure and chain of command [7]. In [12], the author examine the terrorist network to solve the person's successor problem. The person successor problem here refers to who will replace a given vertex in case such a vertex is disconnected from the network. The study focuses on gaining insight into how a terrorist network will reshape itself in the advent of the removal of a vertex. When a key vertex or member of a network is removed, it exerts existential pressures that cause network malfunction. Predicting which node to remove from a network that will perturb network activity is essential in limiting the lethality of the terrorist network [12]. Our work differs in that, we focus on the M-19 network responsible for the tragic events of September 11th, 2001. Instead of predicting what will happen if a vertex is removed from the network, rather, we focused on uncloaking the entire network which is crucial in counterterrorism.

3   Methodology

The study of the terrorist network is an amorphous one in which graph networks are modeled base on the information which is uncovered usually after the attack has occurred. In the case of the M-19, the 19 vertices (men) responsible for the attack were only uncovered after the attack and further investigation traced a linked network across other parts of the globe. We use the dataset from [4] and a modified version of the M-19 graph modeled by [6]. We then affix a code (M-01 to M-19) to the 19 individuals that were responsible for the September 11th, 2001 planes hijacking in the USA as seen in Table 1. We use BFS to traverse the graph to uncover the geodesics which is the first step in calculating betweenness centrality. We then model our link prediction using the 4-centrality measures where we rank the various vertices to know which vertex is most important based on the graph. We represented the various nodes on a matrix and iterate them up to k times where $K = 200$. We then applied the Girvan–Newman algorithm to get the betweenness centrality. It should be noted that each of the centrality measures was modeled using different algorithms. Our algorithms revealed that

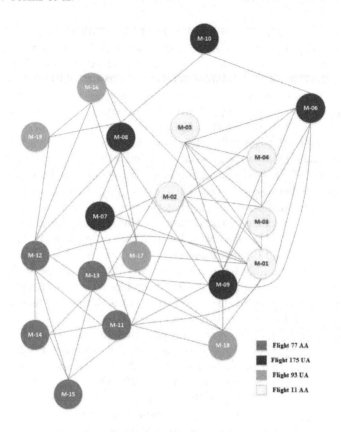

Fig. 1. M-19 network

M-01 that is Mohammed Atta is the most linked node in the network such that his removal will cause network malfunction as seen in Fig. 1. We calculated the lethality of the network as well as network bonding.

3.1   The M-19 Network

The M-19 network composed of the nineteen individuals responsible for the tragic events of September 11th, 2001. Under the auspices of al Qaeda leader Osama Bin Laden where four US commercial airlines were hijacked.

The terrorist network action takes 5 processes; *identification, recruitment, training, indoctrination, and deployment to a target mission.* Once indoctrination of a terrorist is complete, a target is identified and then design in the form of a mission where a set of vertices are sent to construct a subgroup or cell. This cell has a direct link to the main network. The M-19 contains a network of 19 vertices (terrorists) who form a cell of 5nodes each based on a target mission as seen in Fig. 1. The different 4 subcells can be seen in Fig. 2.

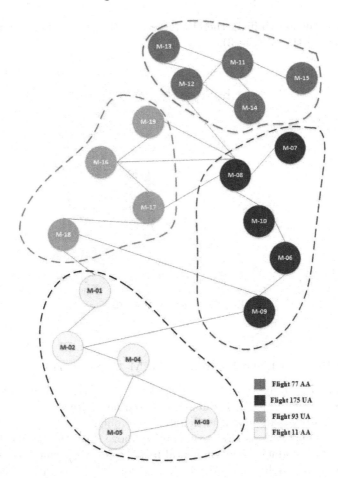

Fig. 2. Showing 4 cells of the M-19 network

Out of the 4 hijacked airplanes, only one cell has 4 vertices with other 3 having equal vertices of 5 each. Figure 2 shows the different 4 subnetwork or cells. We split the M-19 network into cells of 4 based on the 4 hijacked planes, 3 cells made their target except for the 4th cell with only 4 vertices which ended up crashing in a field in Shanksville, Pennsylvania. This shows that the number of vertices in a cell is very important. Because the 4th cell was short of one vertex, the passengers were able to fight back prompting the pilot to crash in a different destination other than the original target. Another salient feature uncover in this study is timing, timing is a crucial factor for the lethality of the network, the flight were just a few minutes away from each other, this was a well-calculated game in the sense that, should a plane crash and one hour later another one follow, there will be high possibility that all flights will be grounded, hence, before flights were grounded, all the target flights were already up in the sky and three already hit the target, the 4th one did not hit the target, there was a delay in the fourth flight thereby increasing the interval between the three cells and the fourth cell (Fig. 3).

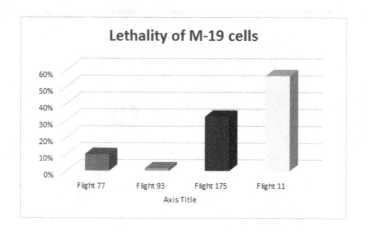

Fig. 3. Lethality of M-19

3.2   Lethality and Cohesiveness of the M-19 Network

Terrorist Network Lethality (TNL) is defined as the maximum damage a given TN can cause on its target. It is calculated based on the severity of the network [9,12]; prior history is imminent to denote its formula.

The terrorist network as already explained above takes the ramification process trigger by a conducive environmental factor ($ConE$) or enabling event [11] thereby, ramifying itself from areas of low action LA (soft Target denoted as S_iT_i) to the areas of high action Ha_i (Maximally Lethal Target denoted as ($Max_{L_i \in T_i}$). The ramification process is denoted as;

$$Ram = ConE(LA..S_iT_i => Max_{L_i \in T_i}Ta(HA)) \qquad (1)$$

Where Ram is the ramification process from areas of low action $ConE$ is the safe haven or conducive environment which enhances the process. Where S_iT_i is the soft Target when the environment is unconducive for the terrorist to be lethal. $Max_{L_i \in T_i}$ is the maximally Lethal Target which is trigger by conducive environmental factors. Ha_i is the area of high action.

In this light, we allude that the M-19 enjoys a conducive environment in the United States (USA) which maximizes their lethality. Equation (1) above explains the process for the network to be lethal. For the Lethality itself, the formula is as follows:

$$LT(Severity)PT_i = \frac{1}{n} \sum_{CaS+n \in T_{1i},...,T_{ni}} \% \frac{TotalAction_i(A_i) + Target(T_i)}{n(numberofplannedTarget(PT_i))}$$

$$(2)$$

where LT be lethality which represents the severity and all the planned target of the network. PT_i is the planned target or desired target, we hypothesized that terrorists set out a finite target to be met. Therefore, the probability for all desired PT_i to be met is $\frac{1}{PT_i}$ or $\frac{1}{n}$.

$CaS + n \in T_{1i}, ..., T_{ni}$ denote the casualty representing the summation of different targets such as $T_{1i}, T_{2i}, ..., T_{4i}$, which depicts the 4 hijacked planes by the M-19 network.

Finally, the lethality will be derived simply by the summation of all the casualties reached from all given targets, plus the percentage of all actions carried out plus the target met. This is further divided by the intended target or planned target. We give the notion that a terrorist may inflict casualties but not necessarily reaching its desired target. To get the lethality, we must know the level of network cohesiveness (NC) or group bonding. We hypothesize the efficacy of the network is guaranteed by the level of the network cohesiveness. For a network to be maximally lethal, there must be high group bonding. Given a Terrorist Network $TN = G(V_i \in NC)$

For $V_i \in NC$: Let R_i be a set of requirements for a vertex to join the network.

Let T_iT_i be the training time requirements for a vertex to be sent on a mission.

Let $fb(g, b)$ be a set of factor binding V_i with G.

Let $Lp(g, b)$ be the level of punishment that V_i will receive for disconnecting from G.

Let dp be the probability for the vertex to disconnect from the network.

Let w_i be the probability that V_i will disconnect at will $w_i = 1$.

Let S_i be the probability that V_i will disconnect but not at will (possible arrest by police) $S_i = 0$.

The degree of network cohesiveness is derived as follows:

$$NC(g, b) = d\frac{T_iT_i}{R_i} \sum_{dp(w_iS_i)} fb(sn) \tag{3}$$

In calculating the NC, we made 4 assumptions which guide our formulation as seen below;

Assumption 1

Given a Terrorist Network $TN = G$ with a network cohesiveness (NC) such that the nc prevents a certain vertex V_i from disconnecting from the network such that the damages to be incurred by the vertex as a result of disconnecting from the network outweighs the benefits. If $NC > g = 1$ otherwise if $NC < g = 0$. In other words, in order to guarantee group loyalty herein refers to as network cohesiveness, a given vertex is held of something more precious than his life such as immediate family members in order to guarantee that the vertex does not disconnect from the network (betray the network) The formula for this is outlined as:

$$NC(betrayal) = \frac{T_iT_i}{R_i} \sum_{ai1,ai2,...,ai5(loyalty)} fb(n(LP)) \tag{4}$$

where $ai1, ai2, ..., ai5$ be a set of binding factors that guarantees loyalty.

Assumption 2

Suppose a Vertex V_i is a suicide bomber and is tasked to carry out a Target T_i. Once the task is carried out the Vertex is automatically deleted from the network. Hence, compensation to family members or nodes connected to the vertex is done in order to enhance network bonding and network cohesiveness.

Let mb_w be a set of family relations binding V_i to G.

Let V_i d be the disconnected vertex.

Let $V_i r$ be a replacement vertex such that $V_i r$ must match the features of $V_i d$.

$$mb_w \begin{cases} if\ V_i d \notin G, \\ input(V_i r), if\ V_i d \in V_i r, \\ 1, if\ w \in N, \\ else : trainnewnode \end{cases} \tag{5}$$

We hypothesized that compensation is done before a vertex is deployed to a mission. Once the above conditions are made, a replacing vertex is prepared for future targets.

Assumption 3

Suppose a Vertex V_i is a suicide bomber and is tasked to carry out a Target T_i and the Vertex disconnect from the network without completing the task, the vertex will be deleted from the network including nodes connected to him which guaranteed network bonding and cohesion.

$$Ti(0, 1) = dp(w_i S_i) => V_{iR}n(disconnect) \tag{6}$$

In Eq. (6) we believed that once a vertex disconnect from G whether at will w_i or not S_i, such a vertex is deleted from the network and every E relating to the node is blocked. We give $Ti = 0$ if he is not and $Ti = 1$ if he is disconnected. Therefore a replacing vertex is prepared for future targets.

Assumption 4

Suppose a vertex (V) is a suicide bomber and is task to carry out a Target Ti and the vertex is disconnected from the network without his desired (such as the case of arrest or killed by FBI), the nodes connected to the vertex is not removed from the network rather, the nodes are maintained and compensation as well as more roles given to them. Hence, a replacement vertex is being prepared. The outcome of this assumption is the same as Eq. (6) above.

3.3   4-Centrality Measures

Measures of centrality are important elements in terrorist network modeling especially when gaining an insight into the role of vertices in the network. In [2], the author contends that in order to curb a terrorist network, the elimination of vertices with high centrality measures is a condition sine qua non.

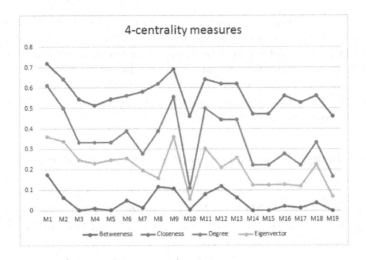

Fig. 4. Showing 4-centrality measures

Eigenvector Centrality. This indicates the extent to which a vertex is connected to influence other vertices in the network. We used the PageRank algorithm which is an ideal algorithm to calculate the eigenvector centrality measures.

Degree of Centrality. This is the number of edges connected to a vertex such it indicates how useful such a vertex is, with the eigenvector centrality, the number edges connected to a vertex does not imply such a vertex is useful in the network rather the number of important edges connected to a given vertex gives it meaningful position in the network.

Closeness Centrality. The mean of the shortest path to all other vertices in the network is known as closeness centrality. Closeness centrality depicts how a vertex can quickly access information in the network.

Betweenness Centrality. The Betweenness centrality here is the extent to which a vertex lies on the geodesics path between other vertices. The betweenness of edges is calculated based on the number of the shortest path. Here we used the Girvan-Newman Algorithm, which visits every vertex once and computes the number of shortest paths from M-01 to M-19 which go through each of the edges within the M-19 network.

4 Discussion of Results

In Fig. 1 we unveil the entire M-19 network and predict the link importance using the 4-centrality measures. The four centrality measure as seen in Fig. 4 indicates that M-01 (Mohammed Atta) is the central figure in the 9/11 case. He got the highest betweenness centrality, closeness, eigenvector as well as the degree of centrality which indicates that his removal could have prone the network to failure. In terms of network lethality, the M-19 so far has the highest lethality rate ever recorded in the history of terrorism. The entire network recorded 2996 fatalities which is highest so far with a single strike from any terrorist network. Flight 11 had the highest fatality rate with 56% follow by flight 175 with 32%. Flight 77 had a 10% fatality rate and with flight 93 recording just 1%. This is because flight 93 did not meet up its target and the 1% came from the 44 casualties on board the plane (Table 1).

Table 1. Results of 4-centrality measures of M-19 network

Vertices	Code	Betweenness	Closeness	Degree	Eigenvector
Mohamed Atta	**M1**	**0.175**	**0.720**	**0.611**	**0.362**
Abdul Aziz al Omari	M2	0.063	0.643	0.500	0.339
Wail al Shehri	M3	0.002	0.545	0.333	0.247
Waleed al Shehri	M4	0.009	0.514	0.333	0.230
Satam al Suqami	M5	0.002	0.545	0.333	0.247
Fayez Banihammad	M6	0.048	0.563	0.389	0.254
Ahmed al Ghamdi	M7	0.014	0.581	0.278	0.195
Hamza al Ghamdi	M8	0.116	0.621	0.389	0.157
Marwan al Shehhi	M9	0.109	0.692	0.556	0.361
Mohand al Shehri	M10	0.005	0.462	0.111	0.057
Hani Hanjour	M11	0.080	0.643	0.500	0.303
Nawaf al Hazmi	M12	0.119	0.621	0.444	0.209
Salem al Hazmi	M13	0.064	0.621	0.444	0.259
Khalid al Mihdhar	M14	0.000	0.474	0.222	0.123
Majed Moqed	M15	0.000	0.474	0.222	0.123
Saeed al Ghamdi	M16	0.021	0.563	0.278	0.126
Ahmad al Haznawi	M17	0.012	0.529	0.222	0.120
Ziad Jarrah	M18	0.037	0.563	0.333	0.225
Ahmed al Nami	M19	0.000	0.462	0.167	0.068

5   Conclusion and Future Work

A terrorist network is organized in a complex and stealthy manner such that tracking and detecting their action is very difficult. For this to happen, there must be a high degree of network cohesiveness (NC) and for such a group to command respect and gain media attention, it must maintain maximal lethality. We examine the Al-Qaeda link network which carried out the 9/11 attack in the USA under the auspices of Osama Bin Laden. A terrorist group understands that they are being hunted and hence would confine itself in a network of secrecy. A neighborhood where people around will accept and ordained their action. That is to say a conducive environment or safe haven. Although terrorists are sets of individuals, they don't operate in isolation rather their modus operandi takes the form of an orderly hierarchical organizational network having a chain of command. Our findings revealed that Mohammed Atta plays the center role in the network in which his removal could have cause network dysfunction which is in line with Carley et al. in [2] who contends that an important aspect in perturbing a terror network activities is to remove a vertex with the highest degree of centrality, betweenness centrality which will intend cause network malfunction. Mohammed Atta enjoys all these features which make our results more accurate.

References

1. Berzinji, A., Kaati, L., Rezine, A.: Detecting key players in terrorist networks. In: 2012 European Intelligence and Security Informatics Conference, pp. 297–302. IEEE (2012)
2. Carley, K.M., Reminga, J., Kamneva, N.: Destabilizing terrorist networks. In: NAACSOS Conference Proceedings, Pittsburgh, PA (2003)
3. Fellman, P.V.: The complexity of terrorist networks. In: 2008 12th International Conference Information Visualisation, pp. 338–340. IEEE (2008)
4. Kean, T.: The 9/11 commission report: final report of the national commission on terrorist attacks upon the United States. Government Printing Office (2011)
5. Keller, J.P., Desouza, K.C., Lin, Y.: Dismantling terrorist networks: evaluating strategic options using agent-based modeling. Technol. Forecast. Soc. Chang. **77**(7), 1014–1036 (2010)
6. Krebs, V.: Uncloaking terrorist networks. First Monday **7**(4), 1–7 (2002)
7. Li, B., Sun, D., Zhu, R., Li, Z.: Agent based modeling on organizational dynamics of terrorist network. Discrete Dyn. Nat. Soc. **2015**, 1–17 (2015)
8. Li, G., Hu, J., Song, Y., Yang, Y., Li, H.J.: Analysis of the terrorist organization alliance network based on complex network theory. IEEE Access **7**, 103854–103862 (2019)
9. Malang, K., Wang, S., Phaphuangwittayakul, A., Lv, Y., Yuan, H., Zhang, X.: Identifying influential nodes of global terrorism network: a comparison for skeleton network extraction. Phys. A **545**, 123769 (2019)
10. Mehta, D.M.: The transformation and demise of terrorist organizations: causes and categories. Honors Undergraduate Theses (2016)
11. Singh, S., Tu, H., Allanach, J., Areta, J., Willett, P., Pattipati, K.: Modeling threats. IEEE Potentials **23**(3), 18–21 (2004)
12. Spezzano, F., Subrahmanian, V., Mannes, A.: Stone: shaping terrorist organizational network efficiency. In: Proceedings of the 2013 IEEE/ACM International Conference on Advances in Social Networks Analysis and Mining, pp. 348–355. ACM (2013)

S2RSCS: An Efficient Scientific Submission Recommendation System for Computer Science

Son T. Huynh[1,2,3], Phong T. Huynh[1,2,3], Dac H. Nguyen[3,4],
Dinh V. Cuong[1,2,3], and Binh T. Nguyen[1,2,3(✉)]

[1] University of Science, Ho Chi Minh City, Vietnam
ngtbinh@hcmus.edu.vn
[2] Vietnam National University, Ho Chi Minh City, Vietnam
[3] AISIA Research Lab, Ho Chi Minh City, Vietnam
[4] John Von Neumann Institute, Ho Chi Minh City, Vietnam

Abstract. With the increasing number of scientific publications as well as conferences and journals, it is often hard for researchers (especially newcomers) to find a suitable venue to present their studies. A submission recommendation system would be hugely helpful to assist authors in deciding where they can submit their work. In this paper, we propose a novel approach for a Scientific Submission Recommendation System for Computer Science (S2RSCS) by using the necessary information from the title, the abstract, and the list of keywords in given paper submission. By using tf-idf, the chi-square statistics, and the one-hot encoding technique, we consider different schemes for feature selection, which can be extracted from the title, the abstract, and keywords, to generate various groups of features. We investigate two machine-learning models, including Logistic Linear Regression (LLR) and Multi-layer Perceptrons (MLP), for constructing an appropriate recommendation engine. The experimental results show that using keywords can help to increase the performance of the recommendation model significantly. Prominently, the proposed methods outperform the previous work [1] for different groups of features in terms of top-3 accuracy. These results can give a promising contribution to the current research of the paper recommendation topic.

Keywords: Paper recommendation · tf-idf · Logistic regression · Multi-layer perceptrons

1 Introduction

Recommendation systems have been extensively studied during the last decades. There have been a massive number of applications for recommendation systems in various areas, and they have been playing a vital role in this internet era [2,3]. Many big-name companies (Google, Facebook, Amazon, eBay, Spotify, Netflix)

© Springer Nature Switzerland AG 2020
H. Fujita et al. (Eds.): IEA/AIE 2020, LNAI 12144, pp. 186–198, 2020.
https://doi.org/10.1007/978-3-030-55789-8_17

have been employing recommendation systems (items, books, movies, songs, or products) to boost their businesses as efficiently as possible [4]. In academics, recommendation systems have been investigated to accelerate research processes, for instance, suggesting collaborations [5,6], papers [7], citations [8,9], or publication venues [1,10–13].

One of the most fundamental jobs of a researcher is to present their fascinating findings to the scientific community. But as science grows, the number of publication venues (conferences, journals, workshops, etc.) is increasing dramatically over each year. Up to now, the total number of computer science conferences and journals has been above 13,000 [1]. Each venue even has its topics and the corresponding quality ranking. Understandably, a researcher may feel overwhelmed when trying to look for the most suitable site to publish a given work. Primarily, it can also occur for researchers who want to explore new domains, work on interdisciplinary projects, or not have much experience. As a consequence, it is useful to construct such a paper recommendation system for researchers. Recently, there are various studies related to this topic [1,10–13]. Some of them are available for public usage [1,10].

To best of our knowledge, there are still few works in building a paper recommendation system for the field of computer science. The current state-of-the-art public platforms have not got much high performance and even had challenges for improvement. Wang and co-workers [1] use the abstract from each paper submission for training a recommendation model to suggest the top relevant journals or conferences. Typically, they apply the Chi-square, the term frequency-inverse document frequency (tf-idf), and the linear logistic regression for extracting potential features of the problem and learning a suitable model. The experimental results show that the recommendation system can achieve an accuracy of 61.37%. Similarly, Feng and colleagues [10] use pre-trained word2vec embeddings and convolutional neural networks to transform each abstract into a more meaningful vector space. By collecting about 880,165 papers from 1130 different journals in PubMed Central, they extract all abstracts from these papers and create an experimental dataset for proposing a publication recommendation system, namely Pubmender. After that, a fully connected softmax model can be used for constructing the corresponding recommendation engine. Pradhan and Pal [11] present a more comprehensive scholarly venue recommendation system that is brilliantly exploiting the content of papers and social network analysis. Related to the social network analysis, they consider the following features: centrality measure calculation, citation and co-citation analysis, topic modeling based contextual similarity, and key-route identification based primary path analysis of a bibliographic citation network. With these fruitful factors, the experimental results show that their DISCOVER system can outperform state-of-the-art recommendation techniques in terms of various metrics, including precision@k, nDCG@k, accuracy, MRR, diversity, stability, average venue quality, and F-scores.

In this paper, we investigate the paper recommendation problem based on the useful information from a paper submission in computer science, including the title, the abstract, and the list of keywords. The main goal of the problem is to recommend the list of top N relevant journals or conferences for the paper submission so that researchers can quickly get more insights and make a better decision for submitting their work with the highest chance. For building the paper recommendation engine, we use the same dataset, which was introduced by Wang et al. [1]. This dataset has 9347 documents in the training dataset and 4665 documents in the testing dataset, which are classified into 65 different categories (journals or conferences). It is worth noting that in that work, they only use the abstract for learning essential features of the problem and the softmax regression for training an appropriate recommendation model. In this work, we aim at using all existent information of one paper submission (including the title, the abstract, and the list of keywords) for improving the performance of the paper recommendation model with this dataset. As no keyword is available, we decide to collect all existing keywords from the corresponding link of each publication (among all documents of the original dataset). Afterward, for essential computing features of the recommendation engine, we use tf-idf, Chi-square statistics, and one-hot encoding. We apply the logistic regression and multi-layer perceptrons (MLP) for learning an appropriate model and compare the performance of proposed models with different combinations of features (derived from the title, the abstract, and the list of keywords). The experimental results show that by using all features and the multi-layer perceptrons, we can achieve the highest accuracy (top 3) is about 89.07%, which significantly outperforms the most recent work in this dataset.

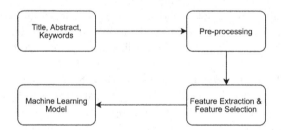

Fig. 1. The workflow of the paper recommendation problem.

2   Our Proposed Methods

In this paper, we aim to investigate an efficient paper recommendation system for researchers. This system can help scientists to gain more useful information by recommending the top N journals or conferences in Computer Science with given a title, an abstract, and a list of relevant keywords. It is worth noting that such a system is crucial and becomes an indispensable tool in the scientific community. In what follows, we present our proposed approach for the paper recommendation. Figure 1 depicts the workflow of our problem, including three primary steps: data processing, feature extraction/selection, and machine-learning modeling.

2.1 Data Processing

Data processing is a critical step in natural language processing (NLP). It helps to transform human language contents into a form that machine learning models can easily understand and perform well. In this problem, for each document (title or abstract), we intend to summarize which words have a significant impact on the prediction. Then, we focus on detecting top words that can be regarded as the main characteristics of the document. Before training or testing an appropriate model, we propose the following pre-processing procedure. First, we remove all numbers, English stopwords, and punctuation marks by using the Python library, NLTK[1]. Also, in each title or abstract, words are usually written in different forms (e.g., noun, verb, adjective, adverb, etc.) As we aim at measuring the frequency of the occurrence of words correctly, we only use the root of each word without lowercase, uppercase, or any other forms of that word. Typically, we apply these following steps to pre-process title and abstract before using them as inputs: (1) lowercase all characters; (2) remove punctuation marks; (3) remove numbers and stop words; (4) stem the whole texts by using the function Porter Stemmer (**NLTK package**[2]). As mentioned previously, after this processing step, we use Chi-square statistics and the tf-idf technique to extract influenced words as potential features (Fig. 2).

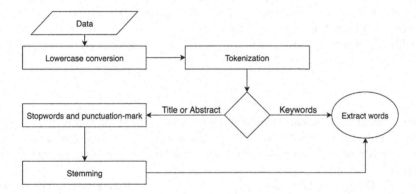

Fig. 2. A proposed data processing step for the paper recommendation problem.

2.2 Chi-Square Statistics

The chi-square test is ubiquitously used in statistics to test the independence of two events. Given a data of two variables, one can get an observed count vector o, the expected count vector e, and d.f is degrees of freedom. Typically,

---

[1] https://gist.github.com/sebleier/554280.

[2] https://www.nltk.org/.

chi-squared χ can measure the deviation between two events by the following formula:

$$\chi_v^2 = \sum \frac{(o_i - e_i)^2}{e_i}$$

Similar to [1], we want to estimate the dependence of words compared to categories. Assume that X is the number of documents including a word w which belongs to a category c, Y is the number of documents including the word w which does not belong to the category c, Z is the number of documents in the category c which does not include the word w, and T is the number of documents in other categories and without the word w. The $\chi(w, c)^2$ can be determined as

$$\chi(w, c)^2 = \frac{(XT - YZ)^2(X + Y + Z + T)}{(X + Y)(Z + T)(X + Z)(Y + Z)}$$

2.3 Term Frequency - Inverse Document Frequency

Term frequency - Inverse document frequency (tf-idf) is a numerical statistic that reflects how important one term (which is a word in this problem) is to a document among all documents (categories) in one collection. It is a common technique used in text mining. This weight is also utilized to evaluate the importance of a word in a text. Importantly, a high value can indicate a high level of importance, and it depends on the number of times words appearing in the text but offset by the frequency of the word in the data set. Term frequency (denoted as TF) is the number of times words appear in the text. As texts can have different lengths, some words may appear more times in a longer document than a shorter one.

$$tf(w, d) = \frac{f(w, d)}{\max(\{f(w, d) : w \in d\})},$$

where $tf(w, d)$ is the term frequency of the word w in the document d, $f(w, d)$ is the number of the word w existing in the document d, $\max(f(w, d) : w \in d)$ is the number of occurrences of the word w with the most occurrences in the document d.

Inverse Document Frequency (denoted as IDF) helps to evaluate the importance of a word. When calculating TF, all words are considered to have equal importance. However, there exist several words (such as "is," "of," and "that") often appearing so many times, but the corresponding importance is quite low. For this reason, one prefers using IDF values which can be computed as:

$$idf(w, D) = \log\left(\frac{|D|}{|\{d \in D : w \in d)\}|}\right),$$

where $idf(w, D)$ is the value IDF of the word w in all documents, D is the set of documents, and $|d \in D : w \in d)|$ is the number of documents including the word w in D. Finally, the tf-idf value of a word in a text can be determined as the product of the corresponding TF and IDF values of that word:

$$\text{tf-idf}(w, d, D) = tf(w, d) \times idf(w, D).$$

2.4   Feature Extraction

In this section, we present the feature extraction step for the paper recommendation problem. As the input data include the paper title, the abstract, and the list of keywords for each paper submission, we aim at using the Chi-square statistics, tf-idf, and the one-hot encoding method to derive potential features.

It is crucial to note that the paper title and the list of keywords provided by authors usually have specific information, which may have a high impact on the prediction results. For this reason, in this paper, we also consider different scenarios by combining different types of features extracted from the paper title, the abstract, and the list of keywords. In experiments, we study the following seven kinds of features: title (T), abstract (A), keywords (K), title + abstract (TA), title + keywords (TK), abstract + keywords (AK), and title + abstract + keywords (TAK), and compare the performance of each proposed model.

Now, we describe how we can extract useful factors from the title, the abstract, and the list of keywords. As the number of words in the title is limited, it is usually quite short. Meanwhile, the number of words for an abstract is often less than 250, and the number of keywords given by authors is from three to five. Consequently, one needs to have different ways to derive useful features from three of them. In this work, we apply the tf-idf formula to find the top valued words from the title and utilize both tf-idf and chi-square statistics to obtain features from the abstract.

Feature Selection from the Abstract.   Typically, each abstract may have many unnecessary words for the recommendation problem. It causes the number of unique words that may be huge, which leads to high computing costs as well as time complexity. Before applying tf-idf, we use the Chi-squared statistics to select top m words for each category in C categories by calculating the value $\chi^2(W^i, c_i)$, where

$$W^i = \{w_1^i, w_2^i, ...w_t^i\},$$

is list of all the words in category c_i, and

$$\chi^2(W^i, c_i) = [\chi^2(w_1^i, c_i), \chi^2(w_2^i, c_i), \ldots, \chi^2(w_t^i, c_i)].$$

After that, we sort this vector in the descending order and choose the top M words having the largest chi-square values in the category c_i. By this way, one can obtain the following list of vectors $\{FD^1, FD^2, ..., FD^{M_c}\}$, where FD^i is the vector contains top words that depend on category c_i, and M_c is the number of categories. Next, we combine all vectors FD_i and remove duplicates to gain the following set of vectors $W = \{w_1, w_2, ..., w_{N_{F_D}}\}$, where N_{F_D} is the number of elements in the final vector-feature space.

Finally, we use tf-idf to digitize the list of feature vectors W as follows. For each document i, we define the corresponding feature vector f_i with tf-idf(w_j, d_i) is tf-idf value of the feature word w_j in the document i. Then, we obtain $F = [f_1, f_2, ..., f_N]$, where N is the number of documents in all data:

$$f_i = [\text{tf-idf}(w_1, d_i), \text{tf-idf}(W_2, d_i), \ldots, \text{tf-idf}(W_{N_{F_D}}, d_i)].$$

Feature Selection from Keywords. As authors provide keywords, when computing more features from the list of keywords, one only needs to lowercase them. We denote $Td^i(K^i, c_i)$ is the frequency that each keyword in K^i appears in the category i, where

$$K^i = \{k_1^i, k_2^i, ...k_t^i\}$$

is the list of all keywords in the category c_i, and t is the number of keywords in the category c_i. By this way, one can obtain the following list of vectors $[Td(k_1^i, c_i), Td(k_2^i, c_i), ...Td(k_t^i, c_i)]$. Next, we sort all values in this vector in the descending order and choose the top H high-valued keywords. Similar to what have done for the feature selection with the abstract, one can obtain $\{FK^1, FK^2, ..., FK^{M_c}\}$, where FK^i is the vector containing the top keywords that depend on the category c_i, and vector $K = \{k_1, k_2, ..., k_{N_{T_D}}\}$, where N_{T_D} is the number of elements in the final vector feature space. Instead of using tf-idf, we use one-hot-encoding for feature extraction. It means one can get

$$V = [v_1, v_2, ..., v_N],$$

where N is the number of documents in all data and v_i is a vector which can be determined by: $v_i[j] = 1$ if the keyword k_j is in the document i; otherwise, $v_i[j] = 0$.

2.5   Predictive Models

After extracting the final feature vector from the input data of one paper submission, this vector can be fed into a machine-learning model to make predictions. In this study, we investigate two different models, consisting of the logistic regression and the multi-layer perceptrons (MLP).

Mathematically a logistic model can estimate the conditional probability $P(c|p)$ of a given paper submission p belonging to a venue c. It can statistically score the matching relationship between p and c. Hence, these such scores can be used to extract the top N relevant venues for the paper p. Assume that $v = (v_1, ..., v_k)$ is the feature vector computed from p. Then, the linear logistic regression model can calculate the corresponding score as follows:

$$\text{logits} = vW + b,$$
$$\text{scores} = \sigma(\text{logits}),$$

where W is a matrix of size $k \times C$, C is the number of categories (venues), b is a vector $1 \times C$, and σ is the softmax function determined by

$$\sigma(z)_i = \frac{e^{z_i}}{\sum_{j=0}^{C} e^{z_j}}, i \in \overline{1, C}.$$

Similarly, MLP has one fully connected hidden layer, and it can be formulated as

$$\text{logits} = \max(vW_1 + b_1, 0)W_2 + b_2,$$
$$\text{scores} = \sigma(\text{logits}),$$

where W_1, W_2 are matrices of size $k \times h_d$, $h_d \times C$, b_1 and b_2 are vectors of size $1 \times h_d$, $1 \times C$ respectively. Note that max is an element-wise function and we use $h_d = 100$ in our experiments.

We train our model by minimizing the cross-entropy loss with the tolerance $tol = 1e - 4$. The parameters W, b, W_1, W_2, b_2, and b_2 can be learned during the training process. Also, we apply the L2 weight decay for a better generalization. All hyper-parameters, such as the weight decay rate, the learning rate, the number of iterations, etc. are moderately tuned to achieve the best performance. Indeed, we also consider other traditional machine learning methods in experiments, i.e., random forests, support vector machines, XGBoost, and so on, but the corresponding performances are less or equal. We measure the performance of different models by using the top 3 accuracy, which is comparable to the experimental results shown in [1].

3 Experiments

In this paper, we run all experiments on a computer with Intel(R) Core(TM) i7 2 CPUs running at 2.4 GHz with 8 GB of RAM and an Nvidia GeForce RTX 2080Ti GPU. During the implementation, we use **pandas** and **numpy** packages to read and process data faster. In the data processing step, we utilize the following packages: **textblob, nltk,** and **re**. For the modeling procedure, we apply the **scikit-learn**[3] package for all calculations and training models like the Logistic Linear Regression and MLP models. Also, **scikit-learn** provides a powerful tool that helps us in tuning the hyper-parameters of each model. Finally, we use **pickle** to store models' weights for predicting purpose in the future without training.

3.1 Datasets

For comparing the performance of our approaches with different groups of features and machine-learning models, we use the same dataset as investigated by Wang et al. [1]. This dataset has 9347 scientific papers (academic journal articles or proceedings papers) for the training dataset and 4665 documents for the testing dataset. However, there is a lack of keywords for each document in this dataset, as authors only focus on investigating the paper recommendation problem by using the abstract. For this reason, we need to crawl the corresponding list of keywords of each paper from the corresponding publication website.

After this step, we can create a new dataset for training a suitable model for the problem. Importantly, among 9347 documents in the training dataset, there are only 8789 documents containing keywords. Meanwhile, in the testing dataset, there exist only 4453 documents having keywords among 4665 original ones. Table 1 shows an example of one title, one abstract, and the list of keywords in one paper submission.

---

[3] https://scikit-learn.org/stable/.

Table 1. An example for a title, an abstract, and a list of keywords.

	Examples
Title	Fighting malicious code: an eternal struggle
Abstract	Despite many years of research and significant commercial investment, the malware problem is far from being solved (or even reasonably well contained)...
Keywords	Security and privacy, Intrusion/anomaly detection and malware mitigation, Systems security, Operating systems security

Table 2. An example for the data processing procedure.

	Original	After pre-processed
Title	"Fighting malicious code: an eternal struggle"	"fight malici code etern struggl"
	"Letting the puss in boots sweat: detecting fake access points using dependency of clock skews on temperature"	"let puss boot sweat detect fake access point use depend clock skew temperatur"
Abstract	"Despite many years of research and significant commercial investment, the malware problem is far from being solved (or even reasonably well contained)..."	"despit year research signific commerci invest malwar problem far solv reason contain week mainstream press publish articl incid million credit card leak larg ..."
Keywords	"PAC-Bayes bound, support vector machine, generalization capability prediction, classification"	"pac-bayes bound", "support vector machine", "generalization capability prediction", "classification"
	"Security and privacy, Intrusion/ anomaly detection and malware mitigation, Systems security, Operating systems security"	"security and privacy", "intrusion/ anomaly detection and malware mitigation", "systems security", "operating systems security"

In summary, we aim at doing experiments with seven different groups of features (title (T), abstract (A), keywords (K), title + abstract (TA), title + keywords (TK), abstract + keywords (AK), and title + abstract + keywords (TAK)), and compare two machine learning models (logistic regression and multi-layer perceptrons).

3.2 Experimental Results

For a given paper submission, the input data include a title, an abstract, and a list of keywords. By applying all necessary processing steps, the input data can be transformed as depicted at Table 2. To ensure we get the best performance for each set of features on each model, we tune hyper-parameters and choose different optimizers, which are "sag", "saga", "lbfgs", and "newton-cg" for Logistic Linear Regression model. We find that $C = 1.0$ and solver $=$ "newton-cg" as the best parameters for this model. Other details can be found in Sect. 2.5.

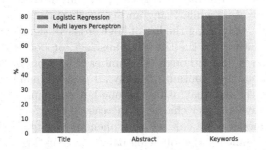

Fig. 3. The experimental results between LLR and MLP models for three groups of features: T, A, and K.

Moreover, during the training process, we apply different feature selection schemes for each group of features (among our seven proposed groups) to maximize the performance of each learning model. Table 3 presents the best feature selection scheme for those. Subsequently, by using those tuned hyper-parameters and thresholds, we reach to the final experimental results, as shown in Figs. 3 and 4. More detailed results can be found at Table 4.

Table 3. The best feature selection scheme tuned for different groups of features during the training process with two learning models in experiments.

	Feature Selection Scheme(Title, Abstract, Keywords)
T	We use all words in the title
A	We choose top 50 words for extracting feature vectors
K	We only choose keywords appearing more than five times in each category
TA	We only choose the top 50 words for both title and abstract
TK	We use all words in the title and only choose keywords appearing more than five times in each category
AK	We only choose keywords appearing more than five times in each category and the top 50 words for abstracts
TAK	We only choose keywords appearing more than five times in each category and the top 50 words for abstracts

From Table 4, one can see that the MLP model can achieve better results than the Logistic Linear Regression (LLR) model for all seven different groups of features in terms of the top-3 accuracy. For both two models, using the combination of all features extracted from the title, the abstract, and the list of keywords can obtain the highest performance in comparisons with smaller groups of features. Especially, comparing with the highest performance (63.1% in the top-3 accuracy) in the previous work [1], our LLR model can get a better result, 67.01%. It is worth noting that in this case, we only use the top 50 words for extracting feature vectors while Wang and co-workers decide using the top 200 words. Interestingly, only using the list of keywords can significantly outperform three scenarios when we use the title, the abstract, or both title and abstract. One possible reason is that although the number of keywords given in one paper submission is limited, they contain precisely all possibly main topics of the scientific works. It turns out that using the one-hot encoding technique to mark all keywords can have a better chance to increase the performance of the paper recommendation model. One can see this phenomenon quite clearly when we compare the performance of the following pairs: T versus TK, A versus AK, and TA versus TAK, for both LLR and MLP models in Table 4.

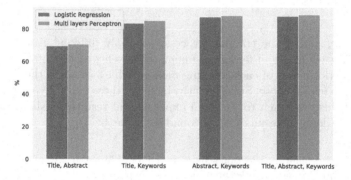

Fig. 4. The experimental results between LLR and MLP models for four groups of features: TA, TK, AK, and TAK.

Table 4. The top 3 accuracy of two models among seven different features.

Features	T	A	K	TA	TK	AK	TAK
LLR	51%	67.01%	80.32%	69.71%	84.07%	87.87%	88.60%
MLP	**55.78%**	**69.88%**	**80.86%**	**71.12%**	**85.32%**	**88.47%**	**89.07%**

Finally, the highest performance in our proposed methods is **89.07%** for the top-3 accuracy when using the group of features **TAK** and the MLP model. It is a promising results when we gain **25.97%** higher in comparison with the state-of-the-art methods in the same dataset.

4 Conclusion and Further Works

We have proposed a novel approach for the paper recommendation problem by using the necessary information from the title, the abstract, and the list of keywords in one paper submission. By using tf-idf, the Chi-square statistics, and the one-hot encoding technique, we have presented different schemes for feature selection from the title, the abstract, and keywords. By comparing seven groups of features as well as using two well-known machine learning models (LLR and MLP), the experimental results have shown that using keywords can help to improve the performance of the recommendation model significantly. Notably, the proposed methods outperform the previous work [1] for six different groups of features (A, K, TK, AK, TA, TAK) even when using LLR or MLP model. These results can give a promising contribution to the current research of the paper recommendation topic. In the future, we aim to apply new trending methods for pre-processing texts (BERT, ALBERT, or ELMO) and use other deep learning techniques to improve the performance of our proposed methods.

Acknowledgement. We would like to thank The National Foundation for Science and Technology Development (NAFOSTED), University of Science, Inspectorio Research Lab, and AISIA Research Lab for supporting us throughout this paper.

References

1. Wang, D., Liang, Y., Xu, D., Feng, X., Guan, R.: A content-based recommender system for computer science publications. Knowl.-Based Syst. **157**, 1–9 (2018)
2. Khusro, S., Ali, Z., Ullah, I.: Recommender systems: issues, challenges, and research opportunities. Information Science and Applications (ICISA) 2016. LNEE, vol. 376, pp. 1179–1189. Springer, Singapore (2016). https://doi.org/10.1007/978-981-10-0557-2_112
3. Mohamed, M., Khafagy, M., Ibrahim, M.: Recommender systems challenges and solutions survey (2019)
4. Amatriain, X., Basilico, J.: Recommender systems in industry: a netflix case study. In: Ricci, F., Rokach, L., Shapira, B. (eds.) Recommender Systems Handbook, pp. 385–419. Springer, Boston, MA (2015). https://doi.org/10.1007/978-1-4899-7637-6_11
5. Chaiwanarom, P., Lursinsap, C.: Collaborator recommendation in interdisciplinary computer science using degrees of collaborative forces, temporal evolution of research interest, and comparative seniority status. Knowl.-Based Syst. **75**, 161–172 (2015)
6. Liu, Z., Xie, X., Chen, L.: Context-aware academic collaborator recommendation. In: Proceedings of the 24th ACM SIGKDD International Conference on Knowledge Discovery & Data Mining, pp. 1870–1879 (2018)
7. Bai, X., Wang, M., Lee, I., Yang, Z., Kong, X., Xia, F.: Scientific paper recommendation: a survey. IEEE Access **7**, 9324–9339 (2019)
8. Bhagavatula, C., Feldman, S., Power, R., Ammar, W.: Content-Based Citation Recommendation (2018). arXiv e-prints, arXiv:1802.08301

9. Jeong, C., Jang, S., Shin, H., Park, E., Choi, S.: A context-aware citation recommendation model with BERT and graph convolutional networks (2019). arXiv e-prints, arXiv:1903.06464
10. Feng, X., et al.: The deep learning-based recommender system "pubmender" for choosing a biomedical publication venue: Development and validation study. J. Med. Internet Res. **21**, e12957 (2019)
11. Pradhan, T., Pal, S.: A hybrid personalized scholarly venue recommender system integrating social network analysis and contextual similarity. Fut. Gener. Comput. Syst. **110**, 1139–1166 (2019)
12. Medvet, E., Bartoli, A., Piccinin, G.: Publication venue recommendation based on paper abstract. In: 2014 IEEE 26th International Conference on Tools with Artificial Intelligence, pp. 1004–1010 (2014)
13. Safa, R., Mirroshandel, S., Javadi, S., Azizi, M.: Venue recommendation based on paper's title and co-authors network **6** (2018)

Improved Grey Model by Dragonfly Algorithm for Chinese Tourism Demand Forecasting

Jinran Wu[1] and Zhe Ding[2]([⊠])

[1] School of Mathematical Sciences, Queensland University of Technology,
Brisbane, QLD 4001, Australia
jinran.wu@hdr.qut.edu.au
[2] School of Computer Science, Queensland University of Technology,
Brisbane, QLD 4001, Australia
zhe.ding@hdr.qut.edu.au

Abstract. For Chinese tourism demand forecasting, we present a novel hybrid framework, a rolling grey model optimized by dragonfly algorithm (RGM-DA). In our framework, a rolling grey model is deployed to forecast the following demand, while the weight parameter in grey model is optimized by the dragonfly algorithm. Using the Experimental data from National Bureau of Statistics of China during 1994–2015, it shows our proposed framework is superior to all considered benchmark models with higher accuracy. Moreover, our proposed framework is a promising tool for short time series modelling.

Keywords: Tourism forecasting · Grey model · Dragonfly Algorithm · Short time series

1 Introduction

In recent decades, tourism has become a populated area in China, especially powered by the souring disposable personal income of Chinese residents [1]. Such huge Chinese tourism industry is usually divided into three major segments: international inbound tourism, domestic tourism, and outbound tourism. Among these three major segments, domestic tourism plays the major role. However, there is a huge gap between the souring demand from Chinese residents and existing tourism infrastructures, such as transportation, accommodation, catering, entertainment, and retailing sectors. Therefore, an accurate tourism (population) demand forecasting system is vital to improve the quality of tourism production and service in China.

Time-series forecasting is often implemented in tourism demand. Relevant time series forecasting models can be classified into two categories: linear or nonlinear methods. Linear methods (e.g. traditional approaches for tourism demand

This work was supported by the ARC Centre of Excellence for Mathematical and Statistical Frontiers.

forecasting) have satisfactory popularity in tourism management [2,3]. However, tourism-demand time series is an open, non-linear, dynamic, and complex system. Thus, traditional linear methods are hard to establish great models, which guarantee great ability of achieving the relationship between input and output variables [4]. Therefore, artificial intelligence techniques are developed in the non-linear dynamic system modelling [5,6], requiring greater amount of data for the model training. Nevertheless, Chinese domestic tourism demand is absence of sufficient data over a short time, thus the mentioned methods cannot achieve a high accuracy for this forecasting. Fortunately, Deng proposes Grey Theory to model the uncertain systems with limited data [7], like the Chinese domestic tourism forecasting system mentioned above. Now, GM(1, 1) has been the most popular Grey forecasting model because of its simplicity [8,9].

However, traditional grey models do not remove previous samples, so the accuracy of prediction may decrease because of useless previous samples. To overcome this problem, the rolling mechanism is introduced, updating new sample and removing previous sample [10]. Meanwhile, the weight parameter in grey model determines the prediction performance. As a result, meta-heuristic optimizations are developed to consolidate the weight parameter in grey model. The hybrid framework, combining grey mode, rolling mechanism and optimization algorithm, has been developed in the uncertain system modelling [11,12].

Considering the literature above, we propose a new RGM-DA framework, combining grey forecasting model, rolling mechanism, and dragonfly algorithm, for Chinese domestic tourism demand forecasting. The simulation is performed based on the data set from National Bureau of Statistics of China during 1994–2015, shown in Fig. 1.

This paper consists of four parts: i). the GM(1, 1), Rolling GM(1, 1) and dragonfly algorithm (DA) are introduced for 1-step ahead Chinese domestic tourism demand forecasting; ii). the data of this research and the process of hybrid model RGM-DA is illustrated in details; iii). the third part talks about the evaluation criteria and results of the experiment; and iv). the paper is concluded.

2   Methodology

2.1   GM(1,1)

GM(1, 1) is a forecasting technique, dealing with uncertain and imperfect information [7]. The basic procedure of building GM(1, 1) is summarized as,

Step 1: assume the the original non-negative data sequence is given as,

$$\{x^{(0)}(1), x^{(0)}(2), \ldots x^{(0)}(n)\}, \tag{1}$$

Step 2: construct the one order accumulated generating operation sequence,

$$\{x^{(1)}(1), x^{(1)}(2), \ldots x^{(1)}(n)\}, \tag{2}$$

with $x^{(1)}(l) = \sum_{i=1}^{l} x^{(0)}(l)$, $l = 1, 2, \ldots, n$.

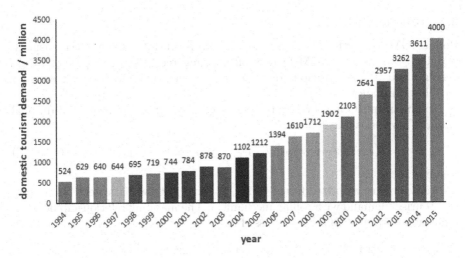

Fig. 1. Domestic tourism demand in China.

Step 3: the basic form of GM(1, 1) is shown as,

$$x^{(0)}(l) + a \cdot z^{(1)}(l) = b, \tag{3}$$

where $z^{(1)}(l)$ is calculates by,

$$z^{(1)}(l) = \lambda \cdot x^{(1)}(l) + (1 - \lambda) \cdot x^{(1)}(l - 1), \tag{4}$$

with $\lambda = 0.5$.

Step 4: calculate the values a and b by least square method as,

$$[a, b]^T = (B^T B)^{-1} B^T Y, \tag{5}$$

with $Y = \begin{bmatrix} x^{(0)}(2) \\ x^{(0)}(3) \\ \vdots \\ x^{(0)}(n) \end{bmatrix}$, and $B = \begin{bmatrix} -z^{(1)}(2) & 1 \\ -z^{(1)}(3) & 1 \\ \vdots & \vdots \\ -z^{(1)}(n) & 1 \end{bmatrix}$.

Step 5: the solution of Eq. (3) ix expressed as,

$$\hat{x}^{(1)}(l+1) = [x^{(0)}(l) - \frac{b}{a}] \cdot \exp^{-a \cdot l} + \frac{b}{a}, \tag{6}$$

where $l = 0, 1, \ldots, n$.

Step 6: based on the inverse accumulated generating operator, the prediction can be given,

$$\hat{x}^{(0)}(l) = \begin{cases} \hat{x}^{(1)}(1), & l = 1, \\ \hat{x}^{(1)}(l) - \hat{x}^{(1)}(l - 1), & l = 2, 3, \ldots, n. \end{cases} \tag{7}$$

2.2   Rolling GM(1,1)

Rolling GM(1, 1) (RGM) is developed, incorporating rolling mechanism in GM(1,1). Specifically, RGM focuses on utilizing recent data to predict future demand and remove previous data. Thus, the new modification can enhance model performance.

To illustrate RGM clearly, in this study case, input and output of traditional GM(1, 1) are shown as follows,

$$(x_{1994}, x_{1995}, x_{1996}, x_{1997}, \ldots, x_{1997+p}) \rightarrow (x_{1998+p}), \tag{8}$$

where $p = 1, 2, \ldots, 17$.

And compare with traditional GM(1, 1), we select five data elements to forecast tourism demand with rolling mechanism. The mapping relationship between input and output of RGM is expressed as,

$$(x_{1993+j}, x_{1994+j}, x_{1995+j}, x_{1996+j}, x_{1997+j}) \rightarrow (x_{1998+j}), \tag{9}$$

where $j = 1, 2, \ldots, 17$.

2.3   Dragonfly Algorithm

To optimize weight parameter λ of RGM, dragonfly algorithm is introduced to improve tourism demand forecasting [13]. In static swarm, dragonflies make small groups and fly in a small area to hunt other flying preys. Meanwhile, in dynamic swarms, a massive number of dragonflies make the swarm for migrating in one direction over long distances [14]. Based on the behaviours of dragonfly, three principles of DA are proposed as [13]:

a. separation, which refers to the static collision avoidance of the individuals from other individuals in the neighbourhood;
b. alignment, which indicates velocity matching of individuals to that of other individuals in neighbourhood;
c. cohesion, which refers to the tendency of individuals towards the center of the mass of the neighbourhood;
d. all of the individuals should be attracted towards food sources and distracted outward enemies.

Thus, the behaviours of dragonfly are mathematically modelled as, Firstly, the separation is calculated as,

$$S_i = -\sum_{j=1}^{N} X - X_j, \tag{10}$$

where X is the position of the current dragonfly, and X_j stands for the position j−th neighbouring dragonfly. (N is the number of neighbouring dragonflies.) Secondly, the alignment is given as,

$$A_i = \frac{1}{N} \sum_{j=1}^{N} V_j, \tag{11}$$

where V_j represents the velocity of j–th neighbouring dragonfly. Thirdly, the cohesion is shown as,

$$C_i = \frac{1}{N} \sum_{j=1}^{N} X_j - X, \tag{12}$$

Then, attraction towards a food source is expressed as,

$$F_i = X^+ - X, \tag{13}$$

where X^+ stands for the position of the food source. Finally, distraction outwards an enemy is calculated as,

$$E_i = X^- + X, \tag{14}$$

where X^- is the position of the enemy.

Combining these five corrective patterns above, the step vector of the dragonfly is defined as,

$$\Delta X_{t+1} = (s \cdot S_i + a \cdot A_i + c \cdot C_i + f \cdot F_i + e \cdot E_i) + \omega \cdot \Delta X_t, \tag{15}$$

where s, a, c, f, e and ω stand for separation weights, alignment weight, cohesion weight, food factor, enemy factor and interior weight, respectively. S_i, A_i, C_i, F_i and E_i are the separation, alignment, cohesion, food source and position of enemy of the i–th dragonfly, respectively. (t is the iteration counter.)

Therefore, position vectors are calculated as,

$$X_{t+1} = X_t + \Delta X_{t+1}, \tag{16}$$

where X_t stands for the position at t, and X_{t+1} is the position of dragonfly updated at $t + 1$.

Furthermore, for improving the randomness, stochastic behaviour and exploration of dragonflies, the Levy flight is employed to deal with problem of no neighbouring solutions. In the condition, the position is updated using,

$$X_{t+1} = X_t + Levy(d) \otimes X_t, \tag{17}$$

where Levy flight provides a random walk, and the random step length is draw from a Levy distribution $Levy(d) : u = t^{-d}$ ($d = 1.5$).

3   Process of RGM-DA

This section mainly introduces the data for Chinese domestic tourism demand forecasting and the process of implementing RGM-DA framework.

First, Chinese domestic tourism demand data set is shown in Fig. 1.

Next, we note the λ-GM(1, 1) as $f(\lambda)$, where λ is optimized by DA instead of the empirical setting.

Then, the framework of RGM-DA is designated in Algorithm 1.

Algorithm 1. Framework of RGM-DA

Input: Recent five tourism demand
Output: Tourism demand prediction
Initialization: Randomize parameter λ
Counter $n = 0$
while $n \leq Max\_Iteration$ **do**
  Calculate GM(1,1)
  Calculate ARE of fitted values
  Run DA and update λ
  $n = n + 1$
end while
Achieve the best λ
Return the following demand

At last, for our demand forecasting, in RGM-DA training, for example, suppose $p = 1$ in the formula (7) with DA, the fitness function F of RGM-DA can be shown as,

$$F = \frac{1}{5} \sum_{i=1994}^{1998} |\frac{x_i - \hat{x}_i}{x_i}|, \tag{18}$$

where x_i is predicted using $f(\lambda)$. When F is converged, the optimal weight parameter λ^* can be achieved. From $f(\lambda^*)$, the prediction in 1999 can be gotten. Furthermore, the process of RGM-DA for Chinese domestic tourism demand forecasting from 1999 to 2005 is shown in Fig. 2.

4   Forecasting Performance Evaluation

4.1   Evaluation Criteria

To measure the prediction accuracy of different models, three indexes, mean relative error (MRE), mean absolute error (MAE), and root mean square error (RMSE), are applied.

These evaluation criteria are defined as,

$$\text{MRE} = \frac{1}{n} \sum_{t=1}^{n} |\frac{\hat{y}_t - y_t}{y_t}|, \tag{19}$$

$$\text{MAE} = \sum_{t=1}^{n} \frac{|\hat{y}_t - y_t|}{n}, \tag{20}$$

$$\text{RMSE} = \sqrt{\frac{\sum_{t=1}^{n}(\hat{y}_t - y_t)^2}{n}}, \tag{21}$$

Meanwhile, the absolute relative error, denoted as ARE, is applied to evaluate models in our experiments, which can reflect the error obtained by different models in each year.

$$\text{ARE} = |\frac{\hat{y}_t - y_t}{y_t}|, \tag{22}$$

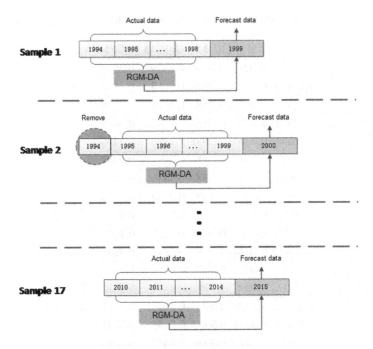

Fig. 2. Process of RGM-DA for demand forecasting.

with the observation y_t, and the prediction $\hat{y}_t$ at time t.

4.2 Results of the Experiment

The detailed procedures of forecasting with rolling mechanism is displayed in
Fig. 2. After 17 times rolling, the weight parameter λ^* can be achieved, listed in
Table 1. Then, the corresponding weight is updated to obtain the prediction by
different models, recorded in Table 2.

Firstly, for GM, it has a fixed λ and retains the previous data. In this model,
the domestic tourism demand sequence form X_{1994} to X_{1997+p} is selected as the
input variables, and then the prediction value X_{1998+p} can be obtained where
$p = 1, 2, \ldots, 17$.

Secondly, for RGM, the forecasting principle is shown in Fig. 2. This model
considers the impact of recent data but has a fixed λ through the whole prediction
period. It selects the last five domestic tourism demand data
$(X_{1993+j}, X_{1994+j}, X_{1995+j}, X_{1996+j}, X_{1997+j})$ as input variables, and the output
variable is X_{1998+j}, where $j = 1, 2, \ldots, 17$.

Thirdly, for GM-DA, the weight parameter λ of GM are optimally determined
by DA. This model adjusts λ at each rolling stage, and it does not remove the
previous data. The optimal weight λ^* are listed in Table 1, and the forecasting
results are shown in Table 2.

Table 1. Optimal parameter value λ^* of different models

Year	GM	RGM	GM-DA	RGM-DA
1999	0.5000	0.5000	0.4339	0.4340
2000	0.5000	0.5000	0.6819	0.5167
2001	0.5000	0.5000	0.5621	0.3658
2002	0.5000	0.5000	0.5134	0.4618
2003	0.5000	0.5000	0.2573	0.3400
2004	0.5000	0.5000	0.3891	0.2907
2005	0.5000	0.5000	0.3394	0.6884
2006	0.5000	0.5000	0.2730	0.5150
2007	0.5000	0.5000	0.6099	0.4741
2008	0.5000	0.5000	0.4555	0.4709
2009	0.5000	0.5000	0.6264	0.3869
2010	0.5000	0.5000	0.6071	0.4964
2011	0.5000	0.5000	0.6308	0.4812
2012	0.5000	0.5000	0.5461	0.4814
2013	0.5000	0.5000	0.5468	0.4735
2014	0.5000	0.5000	0.6343	0.4163
2015	0.5000	0.5000	0.6528	0.4992

Finally, for RGM-DA, it introduces DA on the basis of RGM. Compared with others three models above, this model not only considers the recent data but also adjusts λ in each rolling step.

Furthermore, the values of ARE with four forecasting models are listed in Table 2. RGM-DA obtains the smallest gap between the actual value and prediction value in most years. Besides that, with the data point increases, the gap between the actual value and prediction value by these model which without rolling mechanism becomes lager and larger. From the last three lines of Table 2, RGM-DA has the best forecasting performance due to its smallest Minimum ARE, Average ARE and the second smallest Maximum ARE.

The predictions from 1999 to 2015, utilizing the different models above, are illustrated in Fig. 3. Among all benchmark models, the fitted curve of the proposed RGM-DA framework approximates to the real curve best, indicating that RGM-DA has the highest accuracy for domestic tourism demand forecasting.

Table 3 shows the results of evaluation criterion among GM, RGM, GM-DA and RGM-DA. It is demonstrated that RGM-DA with the MRE value 0.0368, compared with that of GM, RGM and GM-DA is 0.0847, 0.0416 and 0.0373, respectively, achieves much better performance than the other three models. Similarly, MAE and RMSE of RGM-DA are the smallest among these four different forecasting models. In a word, all the indicators indicate that the proposed model RGM-DA has the highest accuracy. Furthermore, comparing GM

Table 2. Comparison between observations and predictions

Year	Observe	GM		RGM		GM-DA		RGM-DA	
		Predict.	ARE	Predict.	ARE	Predict.	ARE	Predict.	ARE
1999	719	705	0.0199	705	0.0199	703	0.0222	703	0.0222
2000	744	740	0.0060	750	0.0080	745	0.0017	751	0.0088
2001	784	770	0.0183	784	0.0006	772	0.0155	779	0.0067
2002	878	808	0.0802	812	0.0756	808	0.0796	810	0.0772
2003	870	881	0.0129	921	0.0590	867	0.0031	909	0.0444
2004	1102	918	0.1673	942	0.1451	911	0.1733	928	0.1575
2005	1212	1056	0.1289	1177	0.0286	1040	0.1422	1212	0.0002
2006	1394	1199	0.1397	1368	0.0185	1166	0.1634	1372	0.0157
2007	1610	1379	0.1434	1621	0.0067	1403	0.1284	1611	0.0006
2008	1712	1597	0.0670	1817	0.0612	1584	0.0750	1806	0.0549
2009	1902	1789	0.0595	1954	0.0272	1839	0.0332	1916	0.0076
2010	2103	1998	0.0499	2098	0.0022	2050	0.0254	2097	0.0027
2011	2641	2224	0.1579	2291	0.1326	2299	0.1293	2285	0.1347
2012	2957	2574	0.1294	2969	0.0042	2610	0.1173	2956	0.0003
2013	3262	2933	0.1009	3474	0.0650	2979	0.0867	3450	0.0577
2014	3611	3304	0.0851	3798	0.0518	3469	0.0393	3730	0.0330
2015	4000	3706	0.0736	4004	0.0009	3929	0.0179	4003	0.0008
ARE	Minimum		0.0060		0.0060		0.0017		0.0002
	Maximum		0.1673		0.1451		0.1733		0.1575
	Mean		0.0847		0.0416		0.0737		0.0368

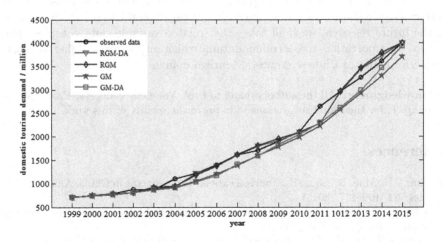

Fig. 3. Predictions for 1-step ahead tourism demand.

and GM-DA, the GM-DA shows better forecasting performance than GM, which indicates the newly proposed algorithm DA is a effective optimization tool for domestic tourism demand forecasting. Additionally, comparing GM and RGM, we can find the grey model with rolling mechanism shows better forecasting performance than that without rolling mechanism. This is because the rolling mechanism utilizes the most recent data as input variables, which can be considered as the latest characteristics and development trend of the forecasting object.

Table 3. Indicators of four benchmark models

Indicators	GM	RGM	GM-DA	RGM-DA
MRE	0.0847	0.0416	0.0737	**0.0368**
MAE	13.1609	8.7328	11.7058	**8.0887**
RMSE	217.2131	121.5844	176.8440	**114.3492**

5 Conclusion

In this paper, the proposed RGM-DA framework, where GM's weight parameter is optimized by DA with rolling mechanism for Chinese domestic tourism demand forecasting, is effective and accurate in the following two aspects. On one hand, deploying DA to optimize the parameter λ can enhance the forecasting accuracy. On the other hand, the introduction of rolling mechanism can also make the forecasting value much closer to the actual value. The primary reason is that the forecasting model with rolling mechanism can use more up-to-date data, and remove the previous data.

From the study above, the proposed RGM-DA framework can achieve great performance in short time series forecasting. It can extend into other study areas. In the future research, we shall take other related variables into consideration, such as transportation and accommodation, which may cause a higher forecasting performance of Chinese domestic tourism demand.

Acknowledgment. All the authors thank to Prof. You-Gan Wang and Prof. Yu-Chu Tian, QUT, for their kind suggestions to improve the quality of this work.

References

1. Lin, V.S., Mao, R., Song, H.: Tourism expenditure patterns in China. Ann. Tourism Res. **54**, 100–117 (2015)
2. Chu, F.-L.: Forecasting tourism demand with ARMA-based methods. Tourism Manag. **30**(5), 740–751 (2009)
3. Witt, S.F., Martin, C.A.: Forecasting future trends in European tourist demand. Tourist Rev. **40**(4), 12–20 (1985)

4. Wu, J., Cui, Z., Chen, Y., Kong, D., Wang, Y.-G.: A new hybrid model to predict the electrical load in five states of Australia. Energy **166**, 598–609 (2019)

5. Hong, W.-C., et al.: SVR with hybrid chaotic genetic algorithms for tourism demand forecasting. Appl. Soft Comput. **11**(2), 1881–1890 (2011)

6. Chen, K.-Y.: Combining linear and nonlinear model in forecasting tourism demand. Exp. Syst. Appl. **38**(8), 10368–10376 (2011)

7. Julong, D.: Introduction to grey system theory. J. Grey Syst. **1**(1), 1–24 (1989)

8. Lin, C.-S., Liou, F.-M., Huang, C.-P.: Grey forecasting model for CO2 emissions: a Taiwan study. Appl. Energy **88**(11), 3816–3820 (2011)

9. Dhouib, A., Trabelsi, A., Kolski, C., Neji, M.: A multi-criteria decision support framework for interactive adaptive systems evaluation. In: Benferhat, S., Tabia, K., Ali, M. (eds.) IEA/AIE 2017. LNCS (LNAI), vol. 10350, pp. 371–382. Springer, Cham (2017). https://doi.org/10.1007/978-3-319-60042-0_42

10. Kusakci, A.O., Ayvaz, B.: Electrical energy consumption forecasting for Turkey using grey forecasting technics with rolling mechanism. In: 2015 2nd International Conference on Knowledge-Based Engineering and Innovation (KBEI). IEEE (2015)

11. Liu, L., et al.: A rolling grey model optimized by particle swarm optimization in economic prediction. Comput. Intell. **32**(3), 391–419 (2016)

12. Zhao, Z., et al.: Using a grey model optimized by differential evolution algorithm to forecast the per capita annual net income of rural households in China. Omega **40**(5), 525–532 (2012)

13. Mirjalili, S.: Dragonfly algorithm: a new meta-heuristic optimization technique for solving single-objective, discrete, and multi-objective problems. Neural Comput. Appl. **27**(4), 1053–1073 (2015). https://doi.org/10.1007/s00521-015-1920-1

14. Russell, R.W., et al.: Massive swarm migrations of dragonflies (Odonata) in eastern North America. Am. Midland Nat. **140**(2), 325–342 (1998)

Variable Transformation to a 2 × 2 Domain Space for Edge Matching Puzzles

Thomas Aspinall⬤, Adrian Gepp⁽✉⁾⬤, Geoff Harris⬤,
and Bruce James Vanstone⬤

Bond Business School, Bond University, Gold Coast, Australia
{taspinal,adgepp,gharris,bvanston}@bond.edu.au

Abstract. The Eternity II (E2) challenge is a well-known instance of
the set of Edge Matching Puzzles (EMP), which are examples of combinatorial problem spaces of the worst-case complexity. Transformation
of the domain space to consider pieces at the 2 × 2 level increases the
total number of elements but is shown to result in orders of magnitude
smaller search spaces. While the original domain space has uniform cardinality, the transformed space exhibits statistically exploitable features.
Two heuristics are proposed and compared to both the original search
space and the raw transformed search space. The efficacy of the two
heuristics is empirically demonstrated. An explanation of how the mapping results in an overall decrease in the number of nodes in the solution
search space of the transformed problem is outlined.

Keywords: Worst-case complexity · Transformed search space · Edge
Matching Puzzle · EMP · Eternity II (E2) challenge

1 Introduction

A new methodology for pre-processing EMPs, in which the variables and domain
elements are transformed into a different EMP problem, is presented. The use of
heuristics to exploit the structure of the transformed domains are investigated.

EMPs belong to the NP-complete (NP-C) problem set. Depending on the
objective function chosen, solution optimization is an NP-hard problem [1]. This
work focuses upon two-set Framed Generic Edge Matching Puzzles (GEMP-F)
[2] that adhere to the structural characteristics of the Eternity II problem (E2-
style). The E2-Style GEMP-F has an $n \times n$ board with n^2 slots and four-edged
pieces and possess a distinguishing border pattern and two sets of patterns for
edge and inner matches [2]. Each tile must be placed such that pairwise adjacent tiles have matching patterns. E2-style GEMP-F are designed to eliminate
statistical information inherent within the characteristics of both the pieces and
distribution of edge patterns. This feature is designed to minimize the ability
of Solution Algorithms to reorder domain variables via some form of metric
function in order to improve search space traversal efficiency.

H. Fujita et al. (Eds.): IEA/AIE 2020, LNAI 12144, pp. 210–221, 2020.
https://doi.org/10.1007/978-3-030-55789-8_19

This paper presents a transformation of the variables such that the resulting domains do exhibit statistical weaknesses not apparent in the original E2-style GEMP-F problem sets. This involves considering all valid 2 × 2 tiling combinations that can be generated from four single tiles within a problem instance. The transformed EMP problems have several properties that make each instance easier to solve. First, transforming these problem instances results in a collapse of the depth of the search space tree by a factor of four. Second, unlike the original variables, the domains of the transformed variables have statistically exploitable features allowing heuristic search tree trimming thus enabling directed searching towards solutions. The results of this study demonstrate that performing the domain transformation and reordering tiles according to the presented heuristics does reduce the median number of nodes traversed to approach a global solution. This approach effectively solves the smallest sized puzzles, with the impact of the transformation becoming less pronounced for larger puzzle sizes as the total number of variables within the domain space increases.

After examining the E2-style GEMP-F literature (Sect. 2), the impacts of variable transformation on the EMP problem space and its generation of statistically exploitable features are presented (Sect. 3). Two efficient search space traversal heuristics are outlined (Sect. 4). The results of applying these heuristics to problem instances are presented and discussed (Sect. 5). Future research avenues are outlined in Sect. 6.

2 Literature Review

E2-style problems were developed following the claimant of the £1 million cash prize offered for the solution of the first Eternity puzzle [3]. The winners of the original Eternity prize were contracted to collaboratively develop the Eternity II puzzle and were likely given the task of eliminating the combinatorial flaws used to solve the original puzzle. The result was an GEMP-F, of size $n = 16$, possessing structural characteristics that cannot be as readily used to reduce the problem search space. Harris et al. [4] describes a process for generating E2-style GEMP-F instances. The 3-partition problem has been proven to be NP-Hard [5], with E2-style puzzles the worst-case instances of NP-hard problems [2].

Strategies for solving E2-style GEMP-F have used algorithms belonging to methodologies such as Evolutionary Algorithms [6], Boolean satisfiability [2,7, 8] and constraint satisfaction [9,10]. These algorithms employed heuristics and meta-heuristics, e.g. tabu search, look-ahead, back jumping and arc consistency, to reduce total search space size and optimize the scores.

Extending the Eternity II challenge, the 3rd International Conference on Metaheuristics and Nature Inspired Computing (META'10, 2010), provided four online downloadable E2-style GEMP-F instances of size $n = \{10, 12, 14, 16\}$ [11]. The winning approach of the META'10 challenge, which coincidentally did not fully solve any of the problem instances, was the application of a two-phase guide-and-observe hyper-heuristic solution algorithm with a score of 461/480 on the size $n = 16$ problem instance [12]. The heuristics utilized involved swap and rotate moves at both the single, double and triple tile level. The work of Wauters et al. [13] considered the number of inner matches for all 4×4 regions in the puzzle, an extension of the 3×3 objective that provided promising results. These experiments showed that by first optimizing an objective besides the raw score of the puzzle and switching to this objective, a higher average score is reached [13]. The work of Salassa et al. [14] applied a hybrid local search approach through both mixed-integer linear programming and Max-Clique formulation, building complete solutions by constructing optimal sub-regions across the problem space. The application of both greedy and backtracking constructive heuristics resulted in maximum scores of 459/480 for the commercial E2 instance and the META'10 size $n = 16$ problem instance [14].

Recently it was shown that the computational overheads involved in heuristics that reduce search space size are greater than those for brute force implementations imposed upon a static variable instantiation scheme [4]. This is due to E2-style EMP's design having no statistically exploitable weaknesses. The Zero-Look Ahead (ZLA) algorithm was shown to outperform all other Solution Algorithms (in 40 of the 48 benchmarks) and was up to 3 orders of magnitude faster than previous published solvers [4].

Many variable transformation techniques have been used while the area of NP-Complete problems has been studied, e.g. variable transformation for Latin Hyper-cubes [15] and in published attempts at solving the real-world Car Sequencing Challenge, with instances provided by Renault [16]. Given the widespread opinion that $P \neq NP$ (e.g. [17]) each transformation is a problem specific creation. To date, there have been no published variable domain transformations for the E2-style EMP. This paper develops such a transformation specifically for the E2 problem and which can be applied to all other EMPs.

3   Variable Transformation

The domain elements of the variables for E2-style GEMP-F problems are single tiles, each being square, non-symmetrically patterned and unique. Each tile rotation (C4 geometric symmetry) represents a valid placement into the variables of the problem space. The elements of an EMP are each piece in each possible rotation, providing the total variable space. Satisfaction of the constraints of each placement determines if the final configuration (all variables instantiated as shown in Fig. 1) represents a valid solution or not to the problem. Consider all 2×2 tiling combinations to be generated from 4 single tiles (Fig. 1). A 2×2 piece is the valid aggregation of four individual pieces all with matching inner edges. The resultant 2×2 combination is an individual piece (domain element) of a higher order.

The characteristics of 2×2 pieces used when further solving EMP are its outer edge patterns and the four individual pieces making up the transformed variable (Fig. 1). If all possible 2×2 combinations are generated then the original single tile $n \times n$ problem can be transformed into an $\frac{n}{2} \times \frac{n}{2}$ problem (assuming n is even – a requirement for an EMP problem to belong to the E2 class).

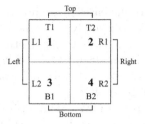

Fig. 1. 2 × 2 tiling generated from four single tiles.

The number of variables has been reduced from n^2 to $\frac{n^2}{4}$. As the combinatorial problem space grows with n, this reduction in the number of levels in the search space tree is associated with a combinatorial decrease in the number of nodes in the transformed search space.

Note that transforming the variables of an EMP to the 2 × 2 domain space results in more variables available for assignment than there are positions on the board. The transformed problem instance thus involves the assignment of a subset of these variables to the domains of a 2 × 2 piece board of size $= \frac{n}{2}$. A constraint upon the variables of an EMP is that each single piece is used once only. As 2 × 2 pieces comprise four individual pieces, when a 2 × 2 piece is placed onto the board, all other 2 × 2 pieces that feature any identical individual pieces are removed from the variable domain space. This 2 × 2 domain transformation produces an important effect upon piece types, i.e. corner, edge and inner pieces. Piece types at the 2 × 2 level feature a combination of one, two or all three piece-types at the individual level (see Fig. 2).

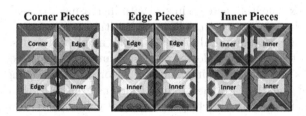

Fig. 2. The three 2 × 2 piece types.

The distinction between solving for the frame and inner sections of the board is blended at the 2 × 2 level as assigning frame pieces onto the board reduces available inner pieces and vice versa. This characteristic of 2 × 2 pieces results in a collapse of the branching factor of the expanded search space as 2 × 2 pieces are assigned. This leads to faster backtracking of Solution Algorithms, increasing the speed of search space traversal.

Although the domain transformation results in a collapse of the number of levels of the search space, the number of branches from each node in the transformed 2×2 tiling space increases in conjunction with the increased number of elements. Sadly, no known analytic method can determine, a priori, if the trade-off between: (a) the decrease in tree depth and (b) the increase in the average branching factor at each node in the transformed tree, results in a significant reduction in search space size. Instead empirical examination of randomly generated E2-style GEMP-F is required.

Pre-processing pieces by performing a domain transformation is classified as pre-processing as the propagation of variables within CSP Solution Algorithms need be applied only once per problem instance. Processing time for the pre-processing of pieces before they are utilized in a Solution Algorithm is, therefore, not considered in evaluating the efficiency of Solution Algorithms.

Table 1. GEMP-F 2×2 frequency of Piece types and pattern IDs.

Puzzle size	Corner pieces	Edge pieces	Inner pieces	Total patterns
4	180	0	0	32
6	727	5,138	6,866	73
8	2,138	26,570	77,108	113
META'10-10	2,280	58,388	342,046	180
META'10-12	3,536	136,832	1,033,686	240
META'10-14	3,512	165,592	1,888,540	376
META'10-16	5,180	290,864	4,079,776	459
Eternity II-16	5,824	292,012	4,059,952	459

The number of elements of 2×2-EMP elements demonstrates neither a combinatorial nor exponential increase in size with increasing n (Table 1), leading to the possibility in net increase in Solution Algorithm performance as a result of the corresponding combinatorial decrease in the number of nodes in the search space. Efficient traversal and indexing of these elements will be necessary to effectively utilize 2×2 elements within a Solution Algorithm. For solving 2×2-EMP's to be more efficient than for the original (untransformed) EMP, the benefit of an increase in efficiency of search space traversal will have to outweigh the cost of the increase in total variable domain sizes.

3.1   Effect upon Edge Pattern Distribution

One of the properties of E2-style GEMP-F is a uniform distribution of edge patterns. That is, there is no statistical information about the distribution that could be used to determine the order in which to instantiate variables (assuming a SAT/CSP methodology). The pattern distribution of the original Eternity II puzzle is shown in Fig. 3.

The consequence of this uniformity of pattern distribution is to maximise entropy thereby making variable selection heuristics and guided search techniques ineffective at both reducing and identifying solution rich regions of the

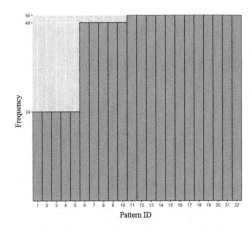

Fig. 3. The Eternity II puzzle pattern ID distribution.

search space. In terms of statistical frequency, there is no reason to target pieces of one pattern over another. If the pattern distribution was not uniform, domain minimizing heuristics can be used to select pieces that are the most constrained (have the least number of patterns of the puzzle). These are the pieces most likely to collapse the search space and instigate backtracking. This is very similar in methodology to the first Eternity puzzle solver [3]. Nonetheless, the pattern distribution for the transformed variable space exhibits statistical features that can be exploited. Figure 4 shows the distribution of the Eternity II puzzle in its 2 × 2 space.

The most important result of the transformation to the 2 × 2 phase space is the effect this transformation has on the pattern frequency distribution of the 2 × 2 elements. The difficulty of the commercial E2 instance was optimized through careful assignment of both the number of patterns and the distribution they possess. The patterns were not however optimized in any way at the 2 × 2 level as is dramatically shown in Fig. 4.

The pre-processing of E2-Style GEMP-F by generating 2 × 2 elements introduces statistical information within the pattern distributions and variable domains. Thus, by definition, pre-processing elements has transformed the problem instance into one that is no longer an E2-style problem. The non-uniform distribution of the edge patterns in the transformed puzzle enables the use of heuristics and CSP techniques, such as arc consistency and dynamic ordering of variable instantiation, to reduce the search space compared to a pure brute force approach. This systemic process of traversing the search space, as opposed to randomly instantiating elements and traversing through root nodes, results in an increase in Solution Algorithm performance.

At the 2 × 2 level the presence of more elements than variables on the board might allow statistical information to be used to distinguish between solution and non-solution pieces. If this is possible, then there may be ways to reorder

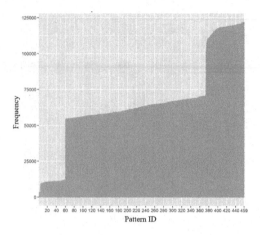

Fig. 4. The Eternity II puzzle 2×2 pattern ID distribution.

the variables of the elements to search elements that feature in a solution rich search space. Furthermore, as 2×2 pieces (elements) are placed onto the board, the domains and subsequent pattern distribution of remaining domain elements change, since piece placement in the 2×2 problem results in immediate domain size reduction across all related non-instantiated domains. This suggests two domain variable ordering heuristics, which are now proposed and discussed.

4   Domain Variable Ordering Heuristics

Two variable selection heuristics are proposed to filter the problem space to search in regions that contain a statistically higher likelihood of containing a global solution. These heuristics do so by increasing the rate at which solution 2×2 pieces that are a subset of a global solution are analysed by a 2×2 Solution Algorithm. A domain characteristic of 2×2-EMP is that it possesses more 2×2 piece elements than there are positions upon the board for variable assignment. Thus a subset of these 2×2 elements belong in a global solution, and, therefore, belong to a solution rich region of the search space. If the benefit of a more efficient search space traversal outweighs the computational cost of possessing all 2×2 pieces within the domain space, then the net Solution Algorithm performance will increase. This is as a result of pre-processing elements to the 2×2 space and running variable selection heuristics upon them.

The two variable selection heuristics are inspired by Bayesian statistical theory and attempt to determine the probabilistic likelihood that a 2×2 element matches a given position. The goal of reordering the variable elements by these metrics is to rank them by this probabilistic likelihood, to be used in the filtering of domains. Call the two heuristics H1 and H2, with H2 being an iterative improvement upon H1. The heuristics construct metrics equal to the product of

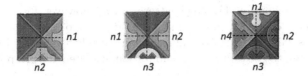

Fig. 5. Domain ordering metrics per 2 × 2 piece type.

the domain sizes of each border edge, i.e. the number of 2 × 2 pieces that can form a valid placement. Figure 5 displays the edge matches per 2 × 2 piece type.

Heuristic 1 (H1): the product of the domain sizes of each border edge.

$$\prod n_i \qquad (1)$$

Then index the 2 × 2 pieces according to their metric to enable the piece with the highest metric value to be chosen first during the traversal of the search tree. Eliminate any 2 × 2 piece for which the metric is 0. The higher the value of H1, the more likely a 2 × 2 element can be placed into a partial solution, as the metric has calculated more adjacent matches to this element than others. Figure 6 provides an example of the application of H1 with valid matches.

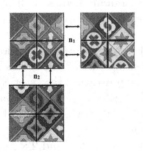

Fig. 6. Heuristic 1 applied to a 2 × 2 corner piece.

Heuristic 2 (H2): the product of the domain sizes of each border edge such that pieces with common single tiles are not counted twice.

$$\prod (n_i - x_i) \qquad (2)$$

Index the 2 × 2 pieces by metric to enable the piece with the highest metric value to be chosen first during the traversal of the search tree. Eliminate any 2 × 2 piece for which the metric is 0. The additional condition of Heuristic 2 is that the elements that make up matching 2 × 2 pieces must be unique. Figure 7 provides an example of valid and invalid pairings of 2 × 2 pieces. Although the edge patterns for n_1 form a valid match of these 2 × 2 pieces, the individual elements that make up these 2 × 2 pieces are not unique, resulting in an invalid match of these 2×2 pieces. The edge patterns for n_2, however, form a valid match with no duplicate elements, thus forming a valid match of these 2×2 pieces. The additional constraint provided by H2 results in a product function that is always lower than H1 at the cost of more computationally intensive pre-processing.

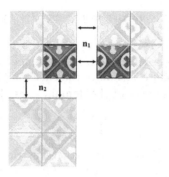

Fig. 7. Example of valid (n_2) and invalid (n_1) pairing.

5   Results and Discussion

The solution generation algorithm applied at both the single and 2×2 tile level was a brute force approach featuring no forward checking, back jumping or k-arc consistency, as first proposed in the work of Harris et al. [4]. This algorithm has been empirically determined to outperform all previous results that utilized the aforementioned domain trimming tactics [4]; it was argued that this is a consequence of E2-style GEMP-F possessing zero statistically exploitable features. For this initial work on the transformation of the variable space, it was determined to utilize this particular Solution Algorithm for both the untransformed and transformed problem spaces.

Solution Algorithms for the single tile and 2×2 tile approaches were implemented in Pascal on a current generation Intel$^{TM}Core^{TM}$i9-8950HK CPU @ 2.90 GHz. The heuristics were applied to border 2×2 pieces within this work and were encoded separately to the Solution Algorithm, computed as the 2×2 input pieces were generated from the corresponding single piece element file. Consequently, no runtime search time was required for the various heuristics; it was entirely absorbed in the input file generation phase. These implementations of the Solution Algorithm output the number of nodes traversed before a global solution was generated for each problem instance presented to it. This enabled a direct comparison between the median number of nodes (as suggested by Ansótegui [7]) traversed to find a solution as a function of E2 puzzle size n.

Sets of 20 randomly generated single tile E2-style GEMP-F instances were created of puzzle sizes $n = \{4, 6, 8\}$ with differing border and inner colour combinations (see Table 2). Table 2 presents the results of running the Solution Algorithm at both the single and 2×2 tile level, with two variable domain heuristics applied at the 2×2 level.

Table 2. Comparison of the effect of nodes traversed by puzzle size, domain transformation and domain ordering metrics.

Puzzle size	4	4	6	6	6	8	8
Border:Inner	3:3	3:4	3:5	3:6	3:7	3:6	3:7
Single tiles	88	63	305,428	179,379	25,421	14,012,064	542,359,516
2 × 2 tiles	14	6	103,648	94,499	8,800	11,151,356	399,094,862
Heuristic 1	10	5	91,730	101,713	4,789	4,967,412	433,661,359
Heuristic 2	7	4	83,689	96,052	5,052	6,309,757	393,869,489

The results above clearly demonstrate the efficacy of the heuristics, particularly when applied to lower puzzle sizes. Indeed, we can see that the size $n = 4$ puzzles are for all intents and purposes a solved problem: the reordering of the tiles according to either heuristic when approached as a search space spanning problem solves the puzzle. In all of the other puzzles we can see that the inclusion of additional constraints to the heuristics leads to a reduction in the total number of nodes traversed per solution.

The general trend of the above results indicates that the impact of the transformation was less dramatic for larger puzzle sizes. In this work, the heuristics were only applied to the border pieces and although the border pieces increase linearly with puzzle size n, the number of inner pieces increases at least quadratically as a function of n. It is thus not too surprising to note that for this initial study, applying the heuristics to solely the border pieces has its greatest impact when there are no inner pieces (size $n = 4$) and the least impact for the largest puzzle instances (size $n = 8$).

6   Conclusion and Future Work

This paper has presented a transformation of the variable and domain space for E2-style GEMP-F which is shown to exhibit statistical structure which should be able to be used to reduce the size of the search space. Two heuristics are proposed to exploit this observed structure, albeit of just the border pieces, and an empirical study demonstrates their effectiveness. It is noted that the impact of the heuristics is smaller for larger n, a consequence of the border pieces increasing linearly in n whilst the inner pieces grow quadratically. Given the results above, and taking into account the factorial increase in nodes per solution as a function of n, it is not surprising that in the literature to date there has been no report of solving a single $n = 10$ instance of the E2-style GEMP-F. Given this is the first attempt to transform the E2 problem into a problem with statistically exploitable features, these initial results appear very promising. It is fair to assume that similar heuristics, developed specifically for the inner pieces, would reduce the nodes traversed in especially the $n = 8$ puzzles, bringing the overall effectiveness more into line with the observed results of the $n = 4$ and $n = 6$ puzzles.

The potential research tree is vast, especially as pre-processing elements to a 2×2 phase space on Eternity II has not been explored previously. This includes extensions of the present empirical analysis; development of a 2×2 Solution Algorithm; development of further 2×2 variable selection metrics, especially for inner pieces, and the construction of analytical proofs of the empirically observed characteristics of EMP in this research. Calculating multiple global solutions to an EMP instance may provide associated insights into why specific 2×2 elements are filtered to the top of the reordered variables. If it were observed that the metric filtered solution pieces across multiple solutions that the EMP possesses, it would be further evidence of changes to solution density of the search space at the upper percentiles of the reordered variables.

The value of pre-processing 2×2 elements of an EMP as well as the scaling ability of the Solution Algorithm could be empirically verified through the development of a 2×2 Solution Algorithm. This Solution Algorithm could be used to analyse net change in search space efficiency and solving time as a result of transforming the domains of EMP's and the reordering according to suggested variable selection heuristics. This would require an efficient indexation of the large numbers of 2×2 variables as well as update the variable space to eliminate invalid elements as pieces featuring identical individual pieces were placed into a solution. This would also provide evidence as to whether or not the transformed problem space features puzzle hardness equal to an E2-style GEMP-F.

The creation of new domain reordering heuristics is suggested to further determine the likelihood that an element belongs to a certain position on the board. Performing Monte-Carlo simulations as the 2×2 Solution Algorithm traverses the search space of an EMP could be used in an attempt to extract statistical information inherent within the domain variables of elements. This would determine the probabilistic likelihood that an element is a subset of a global solution in each position on the board of the EMP. The implementation of a 2×2 Solution Algorithm, using Monte-Carlo simulations to generate this new variable selection metric, is the next natural step recommended to continue this research.

Finally, it is proposed that the use of a transformation to 3×3 pieces be investigated and applied to $n = 6$ puzzles. This may very well, in conjunction with the heuristics proposed above, reduce the $n = 6$ puzzle into a solved problem, much as has been demonstrated here for the $n = 4$ puzzle sizes.

References

1. Demaine, E.D., Demaine, M.L.: Jigsaw puzzles, edge matching, and polyomino packing: connections and complexity. Graphs Combinator. **23**, 195–208 (2007). https://doi.org/10.1007/s00373-007-0713-4
2. Ansótegui, C., Bonet, M.L., Levy, J., Manya, F.: Measuring the hardness of sat instances. In: Twenty-Third AAAI Conference on Artificial Intelligence, pp. 222–228 (2008)
3. Selby, A., Riordan, O.: Eternity - description of method (2000). http://www.archduke.org/eternity/method/desc.html

4. Harris, G., Vanstone, B.J., Gepp, A.: Automatically generating and solving Eternity II style puzzles. In: Mouhoub, M., Sadaoui, S., Ait Mohamed, O., Ali, M. (eds.) IEA/AIE 2018. LNCS (LNAI), vol. 10868, pp. 626–632. Springer, Cham (2018). https://doi.org/10.1007/978-3-319-92058-0_60

5. Garey, M.R., Johnson, D.S.: Computers and Intractability: A Guide to the Theory of NP-Completeness. Series of Books in the Mathematical Sciences. W. H. Freeman, New York (1979)

6. Muñoz, J., Gutierrez, G., Sanchis, A.: Evolutionary genetic algorithms in a constraint satisfaction problem: puzzle Eternity II. In: Cabestany, J., Sandoval, F., Prieto, A., Corchado, J.M. (eds.) IWANN 2009. LNCS, vol. 5517, pp. 720–727. Springer, Heidelberg (2009). https://doi.org/10.1007/978-3-642-02478-8_90

7. Ansótegui, C., Béjar, R., Fernández, C., Mateu, C.: On the hardness of solving edge matching puzzles as SAT or CSP problems. Constraints 18, 7–37 (2013). https://doi.org/10.1007/s10601-012-9128-9

8. Heule, M.J.H.: Solving edge-matching problems with satisfiability solvers. In: Kullmann, O. (ed.) International Conference on Theory and Applications of Satisfiability Testing, pp. 69–82 (2009)

9. Schaus, P., Deville, Y.: Hybridization of CP and VLNS for eternity II. In: Journées Francophones de Programmation par Contraintes (JFPC 2008) (2008)

10. Bourreau, E., Benoist, T.: Fast global filtering for Eternity II. Constraint Program. Lett. (CPL) 3, 036–049 (2008)

11. Meta 2010 problem instances (2010). http://web.ntnu.edu.tw/~tcchiang/publications/META2010-Instances.rar

12. Vancroonenburg, W., Wauters, T., Vanden Berghe, G.: A two phase hyper-heuristic approach for solving the Eternity-II puzzle. In: Proceedings of the 3rd International Conference on Metaheuristics and Nature Inspired Computing (META), vol. 10 (2010)

13. Wauters, T., Vancroonenburg, W., Berghe, G.V.: A guide-and-observe hyper-heuristic approach to the Eternity II puzzle. J. Math. Model. Algor. 11(3), 217–233 (2012). https://doi.org/10.1007/s10852-012-9178-4

14. Salassa, F., Vancroonenburg, W., Wauters, T., Croce, F.D., Berghe, G.V.: MILP and max-clique based heuristics for the eternity II puzzle. CoRR abs/1709.00252 (2017)

15. Liefvendahl, M., Stocki, R.: A study on algorithms for optimization of Latin hypercubes. J. Stat. Plan. Infer. 136(9), 3231–3247 (2006)

16. Solnon, C., Cung, V.D., Nguyen, A., Artigues, C.: The car sequencing problem: overview of state-of-the-art methods and industrial case-study of the ROADEF'2005 challenge problem. Eur. J. Oper. Res. 191(3), 912–927 (2008)

17. Aaronson, S.: Guest column: NP-complete problems and physical reality. ACM SIGACT News 36(1), 30–52 (2005)

Industrial Applications

Using Deep Learning Techniques to Detect Rice Diseases from Images of Rice Fields

Kantip Kiratiratanapruk[1]([✉]) [iD], Pitchayagan Temniranrat[1] [iD],
Apichon Kitvimonrat[1] [iD], Wasin Sinthupinyo[1] [iD], and Sujin Patarapuwadol[2] [iD]

[1] National Electronics and Computer Technology Center (NECTEC),
Pathum Thani, Thailand
kantip.kir@nectec.or.th

[2] Department of Plant Pathology, Faculty of Agriculture at Kamphaeng Saen,
Kasetsart University, Nakhon Pathom, Thailand
agrsujp@ku.ac.th

Abstract. Rice is a staple food feeding more than half of the world's population. Rice disease is one of the major problems affecting rice production. Machine Vision Technology has been used to help develop agricultural production, both in terms of quality and quantity. In this study, convolutional neural network (CNN) was applied to detect and identify diseases in images. We studied 6 varieties of major rice diseases, including blast, bacterial leaf blight, brown spot, narrow brown spot, bacterial leaf streak and rice ragged stunt virus disease. Our studied used well known pre-trained models namely Faster R-CNN, RetinaNet, YOLOv3 and Mask RCNN, and compared their detection performance. The database of rice diseases used in our study contained photographs of rice leaves taken from fields of planting areas. The images were taken under natural uncontrolled environment. We conducted experiments to train and test each model using a total of 6,330 images. The experimental results showed that YOLOv3 provided the best performance in term of mean average precision (mAP) at 79.19% in the detection and classification of rice leaf diseases. The precision obtained from Mask R-CNN, Faster R-CNN, and RetinaNet was at 75.92%, 70.96%, and 36.11%, respectively.

Keywords: Image processing · Rice disease · Deep learning · Object detection and classification

1 Introduction

Rice is an important economic crop in Thailand and is the main food source for over half of the world population. Recent problem of weather anomalies is one of the factors causing plant disease epidemics, which affect rice production. Therefore, rice disease diagnosis technology is one of the important tools that can

© Springer Nature Switzerland AG 2020
H. Fujita et al. (Eds.): IEA/AIE 2020, LNAI 12144, pp. 225–237, 2020.
https://doi.org/10.1007/978-3-030-55789-8_20

help in controlling the epidemic areas and preventing diseases from spreading and causing severe damage. Plant disease diagnosis usually requires a specialist of plant diseases who can analyze the signs and symptoms, physical character- istics of the diseases or requires chemical tests in laboratories. The method is accurate, however, there are many limitations including time-consumption, cost, complexity and inconvenience. In the past decade, computers vision, artificial intelligence, and machine learning techniques are widely used in various fields with good solutions. Recently, deep learning techniques have become a technol- ogy that is more attractive than tradition machine learning techniques because they have several advantages of being able to learn to extract features directly from the input image data. The traditional machine learning techniques required basic knowledge from experts, pre-processing steps, feature extraction and algo- rithms implementation, in order to separate certain features such as color and shape. Conversely, deep learning techniques can enable computers to learn the appropriate characteristics from direct image input. Deep learning has network structure connection in many layers in which each layer is capable of extract- ing features from low level to high level. Among various network architectures used in deep learning, convolutional neural networks (CNN) are used extensively in the field of image recognition. CNN has succeeded in detecting any objects that the class has learned beforehand. Some CNN network architectures have a potential to detect and classify objects on various backgrounds without any necessity to rely on segmenting the object area from the background.

There are about 20 diseases in rice [1,2], which are caused by infection from fungus, virus or bacteria. The symptoms of the disease can show up in various parts of the plant, such as leaves, stalks, stems and roots. In this study, we focused on rice diseases occurred in leaves because it was where the diseases usually appeared. There are many obvious wounds that make it easier to collect than the other locations. In this research, we presented the diagnosis of rice leaf diseases using deep learning techniques. There are various works related to rice plant disease diagnosis as the following. Sukhvir Kaur et al. [3] reviewed the liter- ature on agriculture application, especially with different plants. They described advantages and disadvantages that obtained from each of those studied papers, including the direction of research that should be developed in the future. Vimal K. Shrivastava et al. [4] used convolutional neural network (CNN) based transfer learning technique for rice plant disease classification. Their experiment evalu- ated a total of 619 rice plants from four classes, including three for diseases in leaves and one for healthy leaves. The study showed results by dividing between training and testing sets at various sizes to assess the best performance. V. Vanitha [5] identified four types of rice diseases the same as the study by Vimal K. Shrivastava et al. [4] above. Their experiment tested three training meth- ods and compared their performances from three CNN architectures including VGG16, ResNet50, and InceptionV3. Muhammad Hammad et al. [6] reviewed CNN techniques along with the data visualization techniques. They summarized CNN architectures in the most frequently used methods for detecting and clas- sifying plant disease symptoms. Srdjan Sladojevic et al. [7] presented soybean diseases identification method using CNN technique based on a transfer learning approach for three disease types. Their experiment showed the performance rate of the model when modifying various hyper parameters in pretrained AlexNet and GoogleNet. Prakruti V. Bhatt et al. [8] used YOLO based detection methods with different feature extraction architectures to identify and localize the disease or pest in tea plantations in uncontrolled conditions. Their experiment showed that detection using YOLOv3 can achieves mean average precision (mAP) of about 86% evaluated on a total of 250 images for each pest with different resolutions, quality and brightness. In almost literature, the analyzed images have a simple background or the analysis only focus on the limit area of interest. The work on practical scenarios are still relatively few. The CNN techniques applied to solve these problems only focus on object classification. In this paper, we focus rice plant disease in the practical environment. The difficulty of the rice plant is the clump consisting of many leaves that are long and narrow shape. Therefore, it is not easy to capture a single rice leaf or to avoid the background. We have previously tested these images with the CNN techniques based object classification method and found that the results were not satisfactory. In this paper, we applied the CNN techniques based object detection method that can support both object localization and classification.

In the last few years, it has been shown that CNN technique is one of the best detection and classification techniques. Therefore, in this study, we used four popular and effective models of CNN techniques including Faster R-CNN [10], RetinaNet [13], YOLOv3 [11], Mask R-CNN [12] to identify locations of the disease in rice leaves and to classify the disease type. There are six most common disease types of rice in Thailand including blast, bacterial leaf blight (blight), brown spot, narrow brown spot, bacterial leaf streak and rice ragged stunt virus (RRSV). Over 6,000 images collected from various plant sources were used. The images of rice diseases were taken from various uncontrolled environments in the actual rice fields in Thailand, which had a variety of weather conditions of tropical climate. In addition, our study was not limited to any particular stages, symptoms, or severity of the diseases. The objective of this work was to evaluate the feasibility and performances of the four CNN models on the efficiency of diagnosis to identify diseases on rice leaves. In the future, we plan to develop a rice disease diagnostic system and create a mobile application as an alternative tool to the traditional method of using plant disease specialists.

2   Material Collection and Preparation

Over 6,300 images in this study were collected from photographs of rice leaves in the actual rice fields under variety of weather conditions and background compositions. However, from the photography technical point of view, it was a challenge to keep the whole image of narrow and long shape objects, such as rice leaves, in focus. Some wounds on some leaves might not appear in the pictures because they blended into the background environments. In some cases, it was necessary to hold the leaves in a position and angle to show the wounds clearly. Various stages of infection in each disease could also change the physical characteristics of the wound. This variation caused derivation in the data characteristics. A smartphone camera was used as a photographic tool to obtain the database images because the device was widely and conveniently used, and in order to control consistency of image quality to the actual conditions of common usage.

The physical appearances of the 6 disease types are shown in Fig. 1. It can be seen that each type of wound is different in color, size, shape and texture. In data preparation, we put the label on each diseased leaf area as a polygon shape area using LabelMe Image Annotation Tool [9]. The purpose of using the polygon label was to identify the leaf region without background mixing. After we had identified disease areas, labeled, and classified which disease was the infected area, the image and labeling information were re-examined by the plant disease specialists to ensure that those images were identified correctly according to the type of disease. This part was quite tedious and required many trained people and plant experts to avoid the risk of making a wrong judgment by one person. The obtained regions could easily be generated by masks in training Mask RCNN model. It also was time-saving because it was easy to convert polygon coordinates to bounding boxes, which was an input for the three models: Faster R-CNN, RetinaNet, YOLOv3.

(a) blast   (b) blight   (c) brown spot

(d) narrow brown spot   (e) bacterial leaf streak   (f) RRSV

Fig. 1. Photographs of the six rice diseases: (a) blast; (b) blight; (c) brown spot; (d) narrow brown spot; (e) bacterial leaf streak; and (f) rice ragged stunt virus (RRSV) disease

3   Methodology

The object detection process in images was to locate an object of interest, which could be one or more objects in one image, and to classify its type. We used four common object detection models in our experiment: Faster R-CNN, RetinaNet, YOLOv3, and Mask R-CNN. It was well known that training deep learning network from scratch required a big data set and was a time consuming process. In this work, we used transfer learning technique with initialized weights from a pre-trained model, which had already been trained on the famous data set, such as ImageNet [14]. The pre-trained weight was retrained with our target data set in order to update the weight according to the type to be classified while the learning parameters were adjusted to fit our purposes and decreased training time.

Additionally, over-fitting was another problem needed to be considered during the CNN training. It caused the model well learned on the training data set without generalizing new data that was not in the training set. The augmentation technique was a solution to reduce over-fitting problem. In this work, an augmentation technique was applied to increase the amount of training data by using several transformations such as random, flipping, scaling, rotation, translation, brightness, contrast, and hue. It increased the model learning of sample data patterns during the training process.

3.1   Faster R-CNN and Mask R-CNN

There were three versions of the R-CNN family. In the early version, the model used selective search to extract region proposals and fed to CNN computation. With the structure designed at that time, CNN had to calculate the feature map repeatedly in the processing, which increased the training time. There had been a continuous improvement to become the Faster R-CNN that could process in a shorter time than before. Mask R-CNN was the object detection in R-CNN family, which was capable of marking specific pixels in the image. It was an extension of the Faster R-CNN supported both bounding boxes and pixel segmentation. This extension added branches for the prediction of object masks parallel with the existing branches that only had box boundary recognition.

3.2   RetinaNet

RetinaNet model was evolved to deal with the issue of imbalances class between foreground and background. The model focused on the overwhelming sample of the background instead of the foreground. RetinaNet model was designed to have a structure consisting of a feature Pyramid Network (FPN) and Focal loss. FPN produces features maps of different scales on multiple levels in the network to enable multi-scale object detection, while focal loss optimizes the weight between easy and hard training samples.

3.3 YOLOv3

YOLO (You Only Look Once) family is a great innovation. Unlike other models, it is capable of performing the detection by predicting the bounding boxes and the class probabilities for each object with only a single pass to the network. This ability makes YOLO extremely outstanding in real-time processing applications. YOLO model relied on dividing an image into a grid of boxes. YOLOv2 proposed anchor boxes for better bounding boxes prediction and implemented to generalize the image size. YOLOv3 improved bounding boxes to predict results at different scales and increased the depth of convolutional layers in the Darknet architecture network.

4   Experimental Results

Our work was implemented by Keras Python library with Tensorflow backend for deep learning techniques. The training process of the four detection models was the task, which was computationally intensive and time consuming. Three computers were used to support the operation, each of which had different specifications as follows: 1. Intel Core i7 3.2 GHz CPUs, 16 GB Ram with NVIDIA GeForce GTX 1080; 2. Google Colaboratory with Tesla K80 GPU acceleration; 3. NVIDIA DGX-1 with Dual 20-Core Intel Xeon E5-2698 v4 2.2 GHz CPU, 512 GB 2,133 MHz, and 8X NVIDIA® Tesla® V100 GPU. Our dataset used to evaluate the performance of the models consisted of 6 diseases, blast, bacterial leaf blight (Blight), brown spot (BSP), narrow brown spot (NBS), bacterial leaf streak (Streak) and rice ragged stunt virus disease (RRSV). To reduce impact from unbalance class problem, number of bounding boxes of wounded leaf samples from the entire annotated data was random at equally proportion in each class as presented in the left hand column of Table 1, while number of images belonging to those bounding boxes were presented in the right hand columns. This Table showed a proportion of the data consisting of 3 sets: training, validation, and testing set. The number of bounding box used for training and validation were divided into 80:20% approximately, while testing images was about 120 images for each class.

In the Table 2, model configuration setting in training of each model is summarized in the left three columns. For the two right hand columns, numbers of epoch and its corresponding training time obtained when the model training gave a validation loss with constant decreasing. The efficiency of the model to localize object and classify their type was evaluated in terms of average precision (AP) and precision-recall graph [15]. These terms were derived from Intersection over Union (IoU), which was an overlapping ratio between our predicted boundary results and the ground truth. IoU threshold was commonly defined as 0.5. If the IoU of a predicted result was higher than the threshold, the result was considered to be a correct detection, or else it was considered to be a false detection. These parameters were used to calculate the precision and recall curve. In this experiment, we computed the AP for each class and averaged them (mAP).

Table 1. Dataset of six rice diseases in our experiment.

Class	No. of label box			No. of image data			
Name	Train	Validate	Total	Train	Validate	Test	Total
Blast	873	217	1090	805	200	120	1125
Blight	881	214	1095	866	205	120	1191
BSP	873	216	1089	513	131	120	764
NBS	874	214	1088	822	183	120	1125
Streak	874	215	1089	829	204	120	1153
RRSV	873	214	1087	682	170	120	972
Total	5248	1290	6538	4517	1093	720	6330
Total (%)	80.27	19.73	100	71.36	17.27	11.37	100

Table 2. Model configuration settings.

Computer	Model	Backbone	Augmentation	Epoch	Training time (d:hr:min)
Intel Core i7 GeForce GTX1080	Faster R-CNN	InceptionV2	Fl, Sc, Hu, Co, Br	50	0:21:39
	Faster R-CNN	ResNet101	Fl, Sc, Hu, Co, Br	50	0:20:49
	Mask R-CNN	ResNet101	$Fl, Sc, Sh, Tr,$ Ro, Co, Bl	160	5:08:24
	RetinaNet	ResNet50	Fl, Sc, Sh, Tr, Ro	50	3:18:41
DGX-1	RetinaNet	ResNet101	Fl, Sc, Sh, Tr, Ro	50	2:18:30
Colab Tesla K80	YOLOv3	DarkNet53	Ro, Co, Br, Hu	23,000	2:09:30

Here we defined augmentation parameters Fl = Flip, Sc = Scaling, Hu = Hue, Co = Contrast, Br = Brightness, Sh = Shear, Tr = Translation, Ro = Rotation, and Bl = Blur.

Figure 2 and Table 3 showed AP and mAP performances on six disease types obtained from the four models. We found that YOLOv3 model had the highest precision of five diseases except of the RRSV. Mask R-CNN and Faster R-CNN models had slightly lower precisions than YOLOv3 and RetinaNet model had the lowest performance.

Considering precision-recall graphs shown in Fig. 3, the average precision of YOLOv3 model was high for both precision and recall rates in four disease types including streak, blight, blast and NBS. In the same group of diseases, Mask R-CNN and Faster R-CNN models had recall rates about the same as the YOLOv3 model. However, their average precisions were lower than the YOLO. Both models could be a larger part of all positive results but some of its predicted labels were incorrect when compared to the ground truth labels. For RetinaNet model, the recall was very low when compared with the three other models. The model had only a few prediction results. From the comparison of six types of

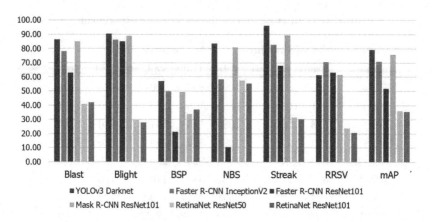

Fig. 2. Bar plots showing average precision (AP) of the four models on detecting six rice diseases.

Table 3. Average precision (AP) of the four models on detecting six rice diseases.

	Average Precision (AP)					
Model	Faster R-CNN		RetinaNet		YOLOv3	Mask R-CNN
Backbone	Inception V2	ResNet 101	ResNet 50	ResNet 101	Darknet	ResNet101
Blast	77.98	63.05	40.89	41.87	86.46	84.95
Blight	86.22	84.94	29.75	27.84	90.33	88.83
BSP	49.67	21.36	33.90	36.97	56.88	49.37
NBS	58.41	10.48	57.30	55.46	83.72	81.07
Streak	82.70	68.00	31.65	30.21	96.33	89.55
RRSV	70.75	63.31	24.00	20.51	61.44	61.76
mAP	70.96	51.86	36.11	35.48	79.19	75.92

diseases, the performance of precision-recall graphs of BSP and RRSV diseases (see in Fig. 3(c) and (f)) was slightly lower than the other four diseases. The reason might be because RRSV had more physical shape varieties than other diseases. In BSP disease, we found that there was missed detection result due to usual occurrence at the early stage of wound.

The sample images of six types of diseases in our dataset and the detection and classification results of the four different models were presented in Fig. 4. This Figure showed the comparison results from each model. All models provided the coordinates of bounding box corresponding to the locations and its sizes where leaf diseases appeared. The color of bounding box results representing each type of disease was defined as the following, blast (red), blight (green), brown spot (pink), narrow brown spot (vivid green), streak (yellow), and RRSV (blue). It was noticed that images from YOLOv3 were found to be mostly accurate while other models contained some errors. In the Mask R-CNN mode with using pixel

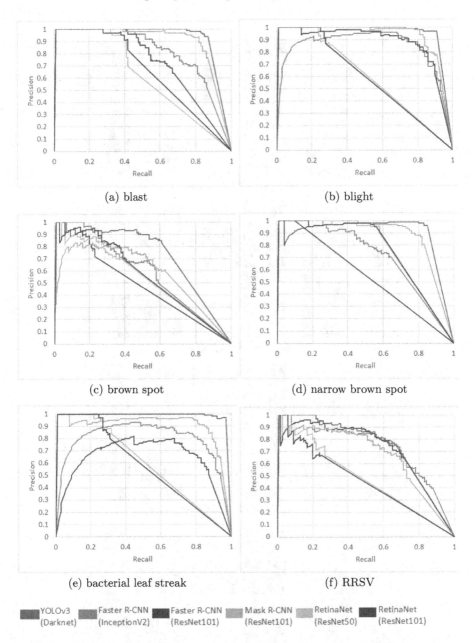

Fig. 3. Illustration of precision-recall curves for six diseases, (a) blast; (b) blight; (c) brown spot; (d) narrow brown spot; (e) bacterial leaf streak; and (f) RRSV

masking labeling data, the results showed reasonable performance to identify the region, which was consistent with the edge of the leaf disease.

(a) Faster R-CNN model

(b) RetinaNet model

(c) YOLOv3 model

(d) Mask R-CNN model

Fig. 4. Comparison of prediction results from the four models, (a) Faster R-CNN (b) RetinaNet (c) YOLOv3 and; (d) Mask R-CNN model (Color figure online)

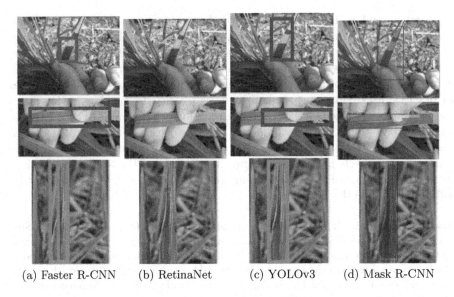

(a) Faster R-CNN   (b) RetinaNet   (c) YOLOv3   (d) Mask R-CNN

Fig. 5. Comparison of prediction results from the four models

Figure 5 showed false predictions. We found that it occurred from the complexity of the data, caused by two main factors. 1. images taken too far distance resulting in unclear wound positions; 2. stage of the severity of the disease that made the wound physical appear differently and more varieties. For example, in the middle row of Fig. 5, the camera was not close up and the wound was in the initial stages showing the image of a small wound compared to the image size. This might cause the bounding box results to miss classification and was not consistent with the type of disease appeared in the image. In addition, images of some disease stages in our dataset may have relatively few data for the effective model training. It resulted in the model not performing well in those cases.

5   Conclusion

This paper showed the results of the study of using CNN to diagnose 6 rice leaf diseases, consisting of blast, blight, brown spot, narrow brown spot, streak, and RRSV. The disease images in our dataset were taken under the challenges of various weather conditions based on actual usage, without limiting it to be only a close-up view of the wounds and the severity of the diseases. In our experiment, the four popular CNN models, Faster R-CNN, RetinaNet, YOLOv3, and Mask R-CNN, were studied for performance comparison. The experiment was evaluated in term of mAP, precision and recall, on over 6,300 images. The best model was YOLOv3 having high precision of almost 80%. It indicated that using the CNN model has a great potential to be used for rice disease identification. The results of this study will be used to improve the technique in future efforts.

For further studies, in order to improve the performance, we plan to extend the investigation to cover many other parameters available in CNN training. The size and quality of the training dataset is an important factor to build the reliability of the model performance. We also plan to gather additional rice disease types for diagnosing rice diseases in order to cover enough disease types for practical uses. Moreover, we plan to develop a rice disease diagnostic system and develop a user friendly mobile application for rice diagnosis. The users only need a mobile phone to take a photo of the rice disease sample and send it to the diagnostic system. The system also reconfirms the detected results with the plant disease experts in order to ensure the accuracy of the result. In addition, all the information will be shared among related persons and the reliable advice will be inferred to the farmers.

Acknowledgments. This study was supported by grants from Innovation for Sustainable Agriculture (ISA), National Science and Technology Development Agency, Thailand. We would like to thanks team from Kamphaeng Saen, Kasetsart University, who is encourage support data and provides useful knowledge in plant disease diagnosis. We would also like to thank Dr. Supapan Seraphin and Dr. Suwich Kunaruttanapruk who helped improving the quality of this paper.

References

1. Thailand Rice Department Homepage. http://www.ricethailand.go.th/rkb3/Disease.htm. Accessed 2016
2. IRRI Rice Knowledge Bank Homepage. http://www.knowledgebank.irri.org/step-by-step-production/growth/pests-and-diseases/diseases
3. Kaur, S., Pandey, S., Goel, S.: Plants disease identification and classification through leaf images: a survey. Arch. Computat. Methods Eng. **26**(2), 507–530 (2018). https://doi.org/10.1007/s11831-018-9255-6
4. Shrivastava, V.K., Pradhan, M.K., Minz, S., Thakur, M.P.: Rice plant disease classification using transfer learning of deep convolution neural network, pp. 631–635 (2019)
5. Vanitha, V.: Rice disease detection using deep learning. Int. J. Recent Technol. Eng. (IJRTE) **7**, 534–542 (2019)
6. Saleem, M.H., Potgieter, J., Arif, K.M.: Plant disease detection and classification by deep learning. Plants (Basel) **8**(11) (2019). https://doi.org/10.3390/plants8110468
7. Sladojevic, S., Arsenovic, M., Anderla, A., Culibrk, D., Stefanovic, D.: Deep neural networks based recognition of plant diseases by leaf image classification. Comput. Intell. Neurosci. (2016). https://doi.org/10.1155/2016/3289801
8. Bhatt, P.V., Sarangi, S., Pappula, S.: Detection of diseases and pests on images captured in uncontrolled conditions from tea plantations. In: SPIE Proceedings, Autonomous Air and Ground Sensing Systems for Agricultural Optimization and Phenotyping IV, vol. 11008 (2019). https://doi.org/10.1117/12.2518868
9. LabelMe: Image Polygonal Annotation with Python. https://github.com/wkentaro/labelme. Accessed Nov 2019
10. Ren, S., He, K., Girshick, R.B., Sun, J.: Faster R-CNN: towards real-time object detection with region proposal networks. IEEE Trans. Pattern Anal. Mach. Intell. **39**, 1137–1149 (2015)

11. Redmon, J., Farhadi, A.: YOLOv3: an incremental improvement. arXiv (2018)
12. He, K., Gkioxari, G., Dollár, P., Girshick, R.B.: Mask R-CNN. In: IEEE International Conference on Computer Vision (ICCV), pp. 2980–2988 (2017)
13. Lin, T.-Y., Goyal, P., Girshick, R.B., He, K., Dollár, P.: Focal loss for dense object detection. arXiv (2017)
14. Deng, J., Dong, W., Socher, R., Li, L.-J., Li, K., Fei-Fei, L.: ImageNet: a large-scale hierarchical image database. In: IEEE Conference on Computer Vision and Pattern Recognition (CVPR), pp. 248–255 (2009). https://doi.org/10.1109/CVPR.2009.5206848
15. Everingham, M., Van Gool, L., Williams, C.K.I., Winn, J., Zisserman, A.: The PASCAL visual object classes (VOC) challenge. Int. J. Comput. Vis. **88**, 303–338 (2010). https://doi.org/10.1007/s11263-009-0275-4

Machine Learning for Water Supply Supervision

Thomas Schranz, Gerald Schweiger, Siegfried Pabst, and Franz Wotawa[✉]

Graz University of Technology, Graz, Austria
{thomas.schranz,gerald.schweiger,siegfired.pabst}@tugraz.at,
wotawa@ist.tugraz.at

Abstract. In an industrial setting water supply systems can be complex. Constructing physical models for fault diagnosis or prediction requires extensive knowledge about the system's components and characteristics. Through advances in embedded computing, consumption meter data is often readily available. This data can be used to construct black box models that describe system behavior and highlight irregularities such as leakages. In this paper we discuss the application of artificial intelligence to the task of identifying irregular consumption patterns. We describe and evaluate data models based on neural networks and decision trees that were used for consumption prediction in buildings at the Graz University of Technology.

Keywords: Machine learning · Fault diagnosis · Data science

1 Introduction

The World Economic Forum lists water crises among the top global risks in terms of potential impact on society [18]. Mekonnen et al. [13] conducted a study that considers the large inter-annual variations in water consumption and availability. This study finds that more than four billion people experience severe water shortages for at least one month per year. A shortage of unpolluted, drinkable water threatens biodiversity and the livelihood of entire communities. Water consumption habits play a key role in managing water scarcity. A study by Vörösmarty et al. [17] claims that rising water demands even exceed the impact of greenhouse warming on the health of water systems. It stands to reason that changing consumption habits can help mitigate water crises. One step can be an efficient means to detect and mitigate leakages.

Building complexes are streaked with water pipes and sewers. Opposed to residential buildings, industrial structures typically have little to no occupancy. Thus, leakages from broken pipes may go undetected. This entails two major problems: i.) water is wasted and ii.) the spillage can damage equipment. The equipment can be expensive and the damage considerable. Investing in automated tools to monitor the water supply system can provide positive financial returns.

© Springer Nature Switzerland AG 2020
H. Fujita et al. (Eds.): IEA/AIE 2020, LNAI 12144, pp. 238–249, 2020.
https://doi.org/10.1007/978-3-030-55789-8_21

Recent advances in the field of embedded computing and cloud services allow us to collect minute data on the states of supply systems. Martin Rätz et al. [14] have highlighted the capabilities of data science methods such as machine learning algorithms in building energy optimization and control. It stands to reason that these black box models have the potential to address the issue of water leakage detection in complex supply systems. A first step towards this direction is to develop models that predict regular water consumption.

In this paper we discuss the application of artificial intelligence (AI) to the task of detecting irregular water consumption patterns. We describe and evaluate models based on neural networks and decision trees that were used to predict consumption in buildings at the Graz University of Technology.

1.1 Machine Learning in Prediction and Fault Detection

Su et al. [15] recently performed a comprehensive analysis of neural networks in short-term prediction of photo voltaic (PV) power generation. Edwards et al. [6] expanded on studies that applied machine learning to forecast electrical consumption in commercial buildings, such as the one by Karatasou et al. [12]. Idowu et al. [9] evaluated data-driven methods used to analyze and predict heat load in district heating systems.

For complex systems such as PV, there exist accurate and reliable fault diagnosis models based on machine learning techniques such as kernel extreme learning [3], neural networks and fuzzy logic systems [4]. In water leakage detection both convolutional neural networks and support vector machines were successfully used by Kang et al. [11] and Javadiha et al. [10]. Their main takeaway is that black box approaches are well suited for prediction. Especially in complex systems with non-linear output characteristics, constructing physical models is infeasible. Data models provide a viable alternative.

2   Method

In the following, we outline the requirements on the AI and explain why we chose a black box model over a physical model. We describe the dataset and why it was necessary to build a regression model in order to ensure unbiased classification into regular and faulty behavior. Furthermore, we address the issues of: i) preprocessing, ii) training data selection, iii) feature construction and selection and iv) encoding. The flow graph in Fig. 1 visualizes the feature engineering and encoding process used to obtain the input feature vector. The subsequent sections contain a description of the models we built and the metrics we used for evaluation.

2.1 Use Case Description

The Buildings and Technical Support department (BATS) at TU Graz monitors and maintains infrastructure across three campuses and cumulative floor space of

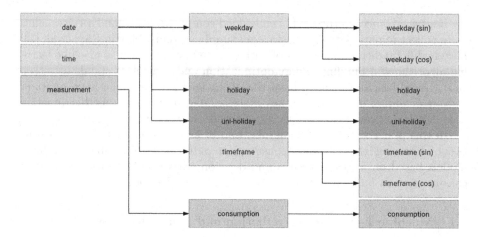

Fig. 1. Feature construction and selection. The flow graph shows the transformation of a data entry into a feature vector.

240,000 m^2 [1]. Among other tasks, the department manages ventilation and air conditioning (HVAC), water supply and drainage. During its operation BATS has accumulated large amounts of data. One group of datasets contains the measurements from water consumption meters on all three campuses.

Because of the water supply system's complexity it is impossible to monitor all consumers manually, let alone ensure 24/7 coverage. Especially during holidays, leakages are likely to go unnoticed for days. It stood to reason that the monitoring process should be automated. However, as pointed out by Rätz et al. [14] physically modelling existing systems requires detailed information on the system's components and its interconnection, which is not always available. In addition, it is difficult to properly describe the habits of occupants with forward modelling [7] because of the inherent complexity of human behavior. Instead, developing a data model appeared to be a more reasonable approach. Water leakages cause irregularities in the consumption patterns, hence it should be possible to train a machine learning model that can be used to detect them.

2.2 Data

The consumption meters record measurements marked with a date and a timestamp. The data entries do not provide any information on whether the measurement was recorded during a leakage or during normal system behavior. Tagging the entries would have required the input of an human expert. However, talking to operators from the BATS department revealed that the number of leakages was, at least in the scope of machine learning, comparatively low. For the consumption meter that provided the test data there was only one instance of a known leakage in three years. Thus, there was simply too little data representing leakages to build an unbiased classification model. Instead, we decided to train a regression model that forecasts the expected consumption value. We would then detect leakages through the difference between projection and actual measurement. This, however, raises the issue of comparing trajectory similarities. Toothey and Duckham [16] reviewed common measures. Simple approaches such as comparing ordered value pairs struggle with outliers. Sophisticated methods such as Dynamic time warping (DTW) depend on choosing an appropriate distance function. Combining approaches into an adequate method is an ongoing development.

2.3 Preprocessing

The dataset provides consumption meter readings taken every 15 min. These readings were used to calculate the consumption in liters per time interval. A resolution of 15 min captures short term influences, such as peaks caused by the toilet flushing system. Yet, for the use case described here it was clearly desirable to model the overall consumption pattern and ignore peaks. Downsampling is a means to increase prediction accuracy in such a situation [14]. Experiments without resampling (i.e. on 15 min intervals) and with aggregates over one, two, four, eight and twelve hours were performed. It was an objective to provide fast real-time feedback on the state of the water supply system while maintaining sufficient prediction accuracy so we finally settled on one hour intervals.

2.4 Period Selection

To correctly train a regression model on the intended behavior it is necessary to remove all data considered irregular. To do so all entries recorded during known leakages were dropped. Other instances of irregular behavior, for example the consumption patterns during conferences or public events were removed as well. Proper sample selection significantly improves the data model, but applying advance techniques, such as numeric wrapper methods or Voronoi algorithms were beyond the scope of this paper. Instead, the selection process was guided by the experts from the BATS department. For evaluation purposes irregular data was reused to verify that any model trained on the regular patterns did not incidentally fit the irregular patterns as well.

2.5 Feature Construction and Selection

Reducing the input along the feature axis has benefits such as lower computational complexity. Especially in regression, however, the inherent weighing process of a neural network may outperform manual feature selection [14], which is not to say that it does not have potential. In a setting such as the one described here, where only two features are available, selection seems to be crucial. Using the time only, for example provided insufficient information to the model, resulting in low accuracy on the training set. Using date and time, on the other hand, caused overfitting on the date. It was necessary to construct additional features using knowledge about the water supply system and the consumption trajectories.

Plotting the hourly consumption in liters showed recurrent patterns that differ in amplitude depending on the weekday. The consumption trajectory for a week in October 2016 can be seen in Fig. 2. Because of its apparent influence on the consumption's magnitude it seemed reasonable to use *weekday* as a feature in the data model.

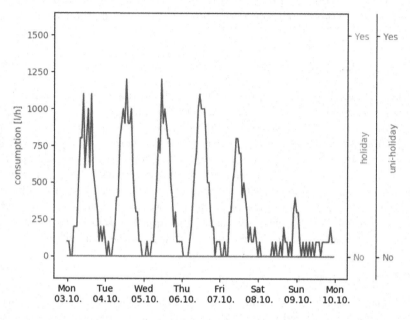

Fig. 2. Consumption trajectory in liters per hour over a week in October 2016. October is part of the winter semester (i.e. courses were held). No holidays during this week.

Other than that, the plots showed significant drops in amplitude on holidays whenever they occur between Monday and Friday. Figure 3 shows the influence of the Austrian National Day, a public holiday, on the consumption value. Consequently, we used *holiday* as a binary input component in the feature vector.

While the weekly consumption patterns are generally similar, the amplitudes vary depending on whether classes are held or not. In Austria the academic year is divided into semesters, a winter semester (from October to February) and a summer semester (from March until the end of June). The data shows that consumption increases by approximately 200 to 400 liters per hour around peak times during semesters. Similar to the *holiday* feature, *uni-holiday* was used as a binary component that captures if a measurement was taken during or outside the semester.

2.6   Encoding

The features *time frame* and *weekday* assume recurrent values; meaning that they run from zero to a maximum value periodically. While it is obvious to

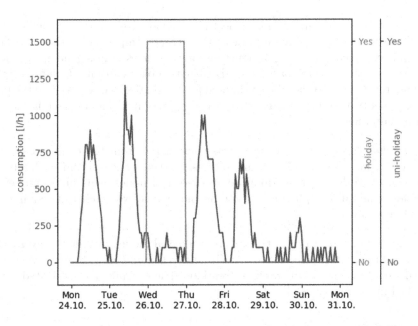

Fig. 3. Consumption trajectory in liters per hour over a week in October 2016. There is a noticeable drop in consumption on the Austrian National Day (October 26th).

a human that the difference between weekday six (Sunday) and weekday zero (Monday) is the same as the difference between weekday two (Wednesday) and three (Thursday), this fact is not properly represented numerically if the *weekday* values run from zero to six. In order to capture the cyclic nature of the features *time frame* and *weekday*, they were encoded with variable pairs, following the approach used by Drezga and Rahman [5]:

$$c_{\sin}(v) = \sin\left(\frac{2\pi v}{T}\right) \tag{1}$$

$$c_{\cos}(v) = \cos\left(\frac{2\pi v}{T}\right), \tag{2}$$

where v is the variable value and T is the period. The *time frame* value is calculated from the timestamp's hour, multiplied by 100 plus the minute normalized to 100 time units. Hence T is set to 2,400 for the *time frame* and 7 for the *weekday*.

2.7   Algorithms

Neural networks are commonly used for prediction. However, performance is subject to hyperparameter settings and choosing the optimal ones can be non-trivial. Training and validation is time-consuming. Decision tree induction is commonly used in data mining to obtain a predictive model from observations.

Decision tree algorithms require selecting comparatively few parameters. On the downside, trees cannot be grown to arbitrary complexity so they can lack generalization accuracy [8]. Constructing multiple trees in independent, random subspaces of the feature space expands the ensemble's capacity. Combining the outputs results in what is called a random forest. We decided to compare the performance of two models: an ensemble-learning-version of decision tree induction and a neural network.

The random forest model we used was constructed from ten estimators, with the mean squared error (MSE) measure as splitting criterion and no limits on maximum tree depth or maximum number of leaf nodes. Minimum number of samples required for splitting an internal node was two, each leaf required at least one sample.

The neural network was constructed as sequential model with dense layers. We compared the performance of models with different numbers of layers, layer sizes, batch sizes and number of training iterations. The neurons were activated using rectified linear functions (ReLU).

2.8   Performance Metrics

The specifics of the classification, i.e. the threshold on the difference between projections and actual measurements have to be developed in cooperation with the BATS operators. This process is subject to further research. Besides, because we lacked sufficient data representing leakages, it was difficult to validate the models on classification metrics. Consequently, we decided to evaluate the models based on the error between their predictions on unseen input data and the actual measurements corresponding to these inputs. Higher accuracy on regular data and lower accuracy on irregular data was desirable. In [14] the coefficient of determination (R^2), the root-mean-square error (rMSE) and mean absolute error (MAE) are listed as typical error metrics for regression problems. Ahmad et al. [2] used the rMSE and R^2 as the primary metrics to compare random forests and neural networks for building energy consumption prediction. Because their setup was similar to ours we chose to evaluate the models on the same metrics.

3   Results

The data tuning framework and the models were implemented in Python 3. We used TensorFlow's[1] implementation of the Keras[2] API for the neural network and sklearn's[3] implementation of the random forest classifier. For data handling and transformations on the input data we used pandas[4] and numpy[5]. Preprocessing,

---

[1] https://www.tensorflow.org/.
[2] https://keras.io/.
[3] https://scikit-learn.org/stable/.
[4] https://pandas.pydata.org/.
[5] https://numpy.org/.

training, hyperparameter tuning and prediction took place on a machine running Arch Linux, kernel version 4.19, with 32 GB of RAM and with 8 CPU cores.

The models were trained on measurements recorded between August 1, 2016 and December 31, 2017. We used a randomly resampled split of 80% training data (9,697 samples) and 20% test data (2,242 samples). To validate generalization performance we compared predictions and readings for inputs from between January 1, 2018 and June 30, 2019. For the neural network we performed a hyperparameter search with two-fold cross validation on the training data. As scoring method we used the R^2 value. Table 1 shows the candidate values used in the hyperparameter search. We used grid search to test all possible value combinations. Table 2 shows the three setups that performed best. One of the setups showed up twice in the top-three results. We decided to benchmark it against the random forest model using the subsamples (test), the regular data (validation) and the irregular data (leak) collected between January 10, 2018 and March 7, 2018 (1,354 samples).

Table 1. Values used in hyperparameter search.

Hyperparameter	Values
Hidden layers	1, 2, 3
Neurons	24, 48, 96, 128, 192, 384
Epochs	24, 48, 96, 192, 384, 512
Batch size	32, 64, 128, 512

Table 2. Top-three results from hyperparameter search. R^2 was calculated on a independent validation set, generated from the training data.

Layers	Neurons	Epochs	Batch size	R^2 (split)
2	128	240	384	0.85003
2	384	80	192	0.84971
2	128	480	384	0.84965

We decided to compare the metrics between days with regular and irregular behavior. The correlation (R^2) between prediction and actual measurement was above 0.7 and below 0 for regular and irregular measurements, respectively, on all test days. Finding a threshold for the classification is ongoing, however, a negative R^2 value seems to be a reasonable first indicator for a leak. Developing a method that can detect divergences quickly, i.e. at least within a few hours, is a primary objective.

Table 3. Root mean squared error (rMSE) and the R^2 performance metrics, calculated on training data, randomly resampled test data, separate validation data and data recorded during a leakage. Last row shows average training time in milliseconds.

	Random forest	Neural network
R^2 (training)	0.8590	0.8556
R^2 (test)	0.8441	0.8500
R^2 (validation)	0.7733	0.7763
R^2 (leak)	−4.8426	−4.8404
rMSE (training)	117.52	118.92
rMSE (test)	127.31	124.87
rMSE (validation)	148.22	147.22
rMSE (leak)	777.16	777.02
Average training time	500 ms	15000 ms

Results as seen in Table 3 show that the neural network slightly outperforms the random forest in terms of accuracy on the test and validation set. However, the margin is extremely narrow. A more significant difference, can be seen in computational complexity. While training the random forest model took 500 ms on average, training the neural network was significantly slower with an average runtime of 15 s. Figure 4 shows the values predicted by the neural network projected against actual consumption during regular operation. Figure 5 shows how significantly the prediction diverges from actual consumption when a leakage occurs.

While promising the results show potential for improvement, especially when looking at how well the models generalize. Errors are significantly lower on entries that were randomly selected from the same time frame as the training data (2016 to 2017) opposed to examples from a later time (2018 or 2019). The two most obvious approaches to improving accuracy concern feature engineering and model selection. For example, the model does not take into account variance in occupancy during a semester or that some holidays, such as Christmas, are traditionally less busy than others. This provides the opportunity to improve feature construction and selection. It would, for instance, be possible to develop a feature that represents the time of year in order to capture long-term consumption patterns. This feature, however, would have to be constructed carefully so that the model does not overfit on patterns in the training data. While general purpose approaches such as decision trees and neural networks seem to perform well, there exist methods more specifically tailored to temporal behaviors, such as recurrent neural networks or Markov chains. In combination with features capturing long-term consumption tendencies, these methods may perform even better than the ones described here.

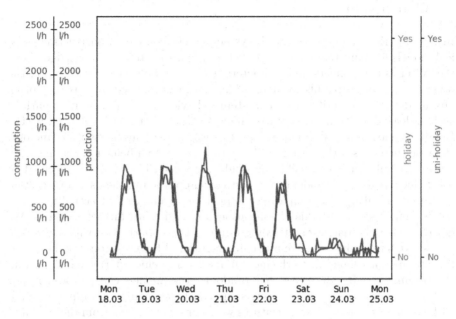

Fig. 4. Prediction and actual measurements during regular system operation, the prediction and actual values match closely.

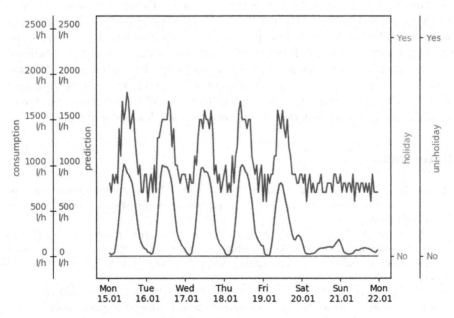

Fig. 5. Prediction and actual measurements during a leakage. Values constantly differ by approximately 600 l/h, which corresponds to the amount of water spilled.

4   Conclusion

In an industrial setting water supply systems can be complex. Constructing physical models for fault diagnosis or prediction requires extensive knowledge about
the system's components and characteristics. Consumption meter data, on the
other hand is often readily available. This data can be used to construct black
box models that describe regular system behavior and highlight irregularities
such as leakages. We developed two models based on the data collected by the
Buildings and Technical Support department at Graz University of Technology.
Because there was little data describing irregular system behavior, classification
between regular/irregular behavior would have been biased. So we decided to
construct a regression model to forecast consumption and then set up a system
that detects leakages by comparing the projected value against actual measurements. The specifics of this classification are subject to further research. We
downsampled the data from 15-minute-intervals to hourly intervals to average
over short-term influences. We manually engineered features from the dates and
timestamps corresponding to the measurements and removed irregular data from
the training set. Results show that random forest models can perform on par
with neural networks while requiring significantly less computation power. Even
with a simple setup, where the features were constructed only from the time and
date of a measurement performance was adequate. However, there is potential for
improvement, especially in terms of generalization. Errors are consistently lower
on entries that were randomly selected from the same time frame as the training
data opposed to samples from other time frames. To do so it would be necessary to develop features that capture long-term patterns throughout the year
without overfitting. While general purpose approaches such as decision trees and
neural networks perform well, selecting other methods more specifically tailored
to temporal behaviors, such as recurrent neural networks or Markov chains, may
improve the monitoring system's performance. However, we are confident that
in the setting described here, any type of data model is more than capable of
handling the task and that there is no need for complex physical models.

References

1. Buildings and Technical Support - TU Graz. https://www.tugraz.at/en/tu-
   graz/organisational-structure/service-departments-and-staff-units/buildings-and-
   technical-support/
2. Ahmad, M.W., Mourshed, M., Rezgui, Y.: Trees vs neurons: comparison between
   random forest and ANN for high-resolution prediction of building energy consumption. Energy Build. **147**, 77–89 (2017). https://doi.org/10.1016/j.enbuild.2017.04.
   038
3. Chen, Z., Wu, L., Cheng, S., Lin, P., Wu, Y., Lin, W.: Intelligent fault diagnosis
   of photovoltaic arrays based on optimized kernel extreme learning machine and
   I-V characteristics. Appl. Energy **204**, 912–931 (2017). https://doi.org/10.1016/j.
   apenergy.2017.05.034

4. Dhimish, M., Holmes, V., Mehrdadi, B., Dales, M.: Comparing Mamdani Sugeno fuzzy logic and RBF ANN network for PV fault detection. Renewable Energy **117**, 257–274 (2018). https://doi.org/10.1016/j.renene.2017.10.066

5. Drezga, I., Rahman, S.: Input variable selection for ANN-based short-term load forecasting. IEEE Trans. Power Syst. **13**(4), 1238–1244 (1998). https://doi.org/10.1109/59.736244

6. Edwards, R.E., New, J., Parker, L.E.: Predicting future hourly residential electrical consumption: a machine learning case study. Energy Build. **49**, 591–603 (2012). https://doi.org/10.1016/j.enbuild.2012.03.010

7. Fumo, N.: A review on the basics of building energy estimation (2014). https://doi.org/10.1016/j.rser.2013.11.040

8. Ho, T.K.: Random decision forests. In: Proceedings of the International Conference on Document Analysis and Recognition, ICDAR, vol. 1, pp. 278–282. IEEE Computer Society (1995). https://doi.org/10.1109/ICDAR.1995.598994

9. Idowu, S., Saguna, S., Åhlund, C., Schelén, O.: Applied machine learning: forecasting heat load in district heating system. Energy Build. **133**, 478–488 (2016). https://doi.org/10.1016/j.enbuild.2016.09.068

10. Javadiha, M., Blesa, J., Soldevila, A., Puig, V.: Leak localization in water distribution networks using deep learning, pp. 1426–1431. Institute of Electrical and Electronics Engineers (IEEE), September 2019. https://doi.org/10.1109/codit.2019.8820627

11. Kang, J., Park, Y.J., Lee, J., Wang, S.H., Eom, D.S.: Novel leakage detection by ensemble CNN-SVM and graph-based localization in water distribution systems. IEEE Trans. Industr. Electron. **65**(5), 4279–4289 (2018). https://doi.org/10.1109/TIE.2017.2764861

12. Karatasou, S., Santamouris, M., Geros, V.: Modeling and predicting building's energy use with artificial neural networks: methods and results. Energy Build. **38**(8), 949–958 (2006). https://doi.org/10.1016/j.enbuild.2005.11.005

13. Mekonnen, M.M., Hoekstra, A.Y.: Sustainability: four billion people facing severe water scarcity. Sci. Adv. **2**(2) (2016). https://doi.org/10.1126/sciadv.1500323

14. Rätz, M., Javadi, A.P., Baranski, M., Finkbeiner, K., Müller, D.: Automated data-driven modeling of building energy systems via machine learning algorithms. Energy Build. **202**, 109384 (2019). https://doi.org/10.1016/j.enbuild.2019.109384

15. Su, D., Batzelis, E., Pal, B.: Machine learning algorithms in forecasting of photovoltaic power generation, pp. 1–6. Institute of Electrical and Electronics Engineers (IEEE), September 2019. https://doi.org/10.1109/sest.2019.8849106

16. Toohey, K., Duckham, M.: Trajectory similarity measures. SIGSPATIAL Spec. **7**(1), 43–50 (2015). https://doi.org/10.1145/2782759.2782767

17. Vörösmarty, C.J., Green, P., Salisbury, J., Lammers, R.B.: Global water resources: vulnerability from climate change and population growth. Science **289**(5477), 284–288 (2000). https://doi.org/10.1126/science.289.5477.284

18. World Economic Forum: The Global Risks Report 2019 14th Edition Insight Report (2019). http://wef.ch/risks2019

An Enhanced Whale Optimization Algorithm for the Two-Dimensional Irregular Strip Packing Problem

Qiang Liu, Zehui Huang, Hao Zhang, and Lijun Wei$^{(\boxtimes)}$

Guangdong Provincial Key Laboratory of Computer Integrated Manufacturing System, State Key Laboratory of Precision Electronic Manufacturing Technology and Equipment, Guangdong University of Technology, Guangzhou 510006, China
liuqiang@gdut.edu.cn, hzh3386@163.com, zhanghao0815@126.com, villagerwei@gmail.com

Abstract. In this paper, we propose an enhanced whale optimization algorithm for the two-dimensional strip packing problem, which requires cutting a given set of polygons from a sheet with fixed-width such that the used length of the sheet is minimized. Based on the original whale swarm algorithm, this algorithm introduces strategies such as adaptive weighting factors, local perturbation, and global beat variation, which can better balance the global optimization and local optimization search capabilities of the whale algorithm. The proposed algorithm was tested using the standard test case and compared with other algorithms in the literature. The results show that the proposed algorithm can improve the best-known solution for some instances.

Keywords: Irregular packing · Whale optimization algorithm · Local search

1 Introduction

The two-dimensional irregular strip packing problem (2DIRSP), also known as the nesting problem, is an NP-hard combinatorial optimization problem, which has been widely used in steel material cutting, textile, paper, and other materials industries. It requires that a given set of two-dimensional polygons be placed in a rectangular container. The polygons must not overlap each other, and cannot exceed the boundaries of the container, where each polygon is not necessarily convex. The container has a fixed width, and its length can be changed. Our goal is to find a set of valid polygon packing so that the length of the container is the smallest. This problem has several variations depending on the rotation of the polygon, and we deal with the case that allows rotation at a limited number of angles.

Researchers have tried many solutions to the 2DIRSP. Meta-heuristics and hybrid algorithms are often used to solve this problem. Hopper and Turton [8]

© Springer Nature Switzerland AG 2020
H. Fujita et al. (Eds.): IEA/AIE 2020, LNAI 12144, pp. 250–261, 2020.
https://doi.org/10.1007/978-3-030-55789-8_22

reviewed these methods, and a summary of geometric methods was published by Bennell and Oliveira [2]. In recent years, with the continuous deepening of research, many breakthroughs have been made in this issue. Among them, The typical methods include the 2DNest algorithm of Egeblad et al. [6], the ILSQN algorithm of Imamichi et al. [9], the ELS algorithm of Stephen C.H. Leung et al. [10], and the GCS algorithm of Ahmed Elkeran [7].

2   Related Description of the 2DIRSP

2.1   Mathematical Model of 2DIRSP

Given a rectangular container $C(W, L)$ of fixed width W and unlimited length L, a polygon set $P = \{P_1, P_2...P_n\}$ of size n, and a set O of allowed rotation angles. The goal of the two-dimensional irregular strip packing problem is to seek an optimal packing to maximize the utilization rate of raw materials. To formally describe the 2DIRSP, we adopt the relevant definitions proposed by Stephen C.H. Leung et al. [10].

Definition 1. point. *A point p on the plane is defined as a binary group: $p = (p_x, p_y) \in R^2$.*

Definition 2. line segment. *Line segment s is a set of points defined by its two endpoints, p_o and p_t: $s = \{p \in R^2 | p = p_o + t \times (p_t - p_o), t \in [0, 1], t \in R\}$.*

Definition 3. polygon. *A polygon can be defined as a finite set S consisting of line segments, where the line segments are the edges of the polygon, the endpoints of the line segments are the vertices of the polygon, and the number of vertices is equal to the number of edges.*

Definition 4. $\oplus$. *Given a polygon P and a translation vector $v = (v_x, v_y)$ represented by a binary group, the $\oplus$ operation can be defined as: $p \oplus v = \{(p'_x + v_x, p'_y + v_y) | p' \in P\}$, where $\oplus$ represents the translation of the polygon along the vector.*

Definition 5. rotation operation. *Given the polygon P and rotation angle r, the rotation operation is defined as: $p(r) = \{(p'_x \times \cos(r) + p'_y \times \sin(r), -p'_x \times \sin(r) + p'_y \times \cos(r)) | p' \in P\}$.*

The 2DIRSP can be formally defined as:

$$\text{Minimize } L \tag{1}$$

$$\text{s.t. } (P_i(r_i) \oplus v_i) \cap (P_j(r_j) \oplus v_j) = \varnothing, 1 \leq i, j \leq n, \tag{2}$$

$$(P_i(r_i) \oplus v_i) \subseteq C(W, L), 1 \leq i \leq n \tag{3}$$

$$r_i \in O, 1 \leq i \leq n \tag{4}$$

$$v_i \in R^2, 1 \leq i \leq n \tag{5}$$

where vector $V = \{v_1, v_2, ..., v_n\}$ is the position of the reference point of each piece and $R = \{r_1, r_2, ..., r_n\}$ is the rotation angle. A packing scheme can be represented by (V, R). After the lower left corner of the rectangular strip is fixed on the origin of coordinates (as shown in Fig. 1), the length L of the rectangular strip can be given by the following formula: $L = L(V, R) = \max\{p_x | (p_x, p_y) \in P_i(r_i) \oplus v_i, P_i \in P, 1 \le i \le n\}$.

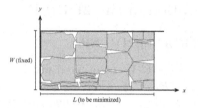

Fig. 1. A legal packing scheme

2.2   Dealing with Geometry

A basic geometric operation in 2DIRSP is to determine whether two polygons overlap. In this paper, we use the no-fit polygon technology to solve the intersection test and overlap calculation problem, as shown in Fig. 2(a). The no-fit polygon has been used by [6,9,10] and [7] for solving 2DIRSP. For any two polygons (including polygons and rectangular containers), we need to build their no-fit polygons. For the all allowable rotation angles, all the no-fit polygons can be calculated and saved in the pre-processing phase. The specific algorithm to build the no-fit polygon can be found from [1,4], and [3].

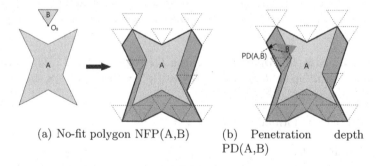

(a) No-fit polygon NFP(A,B)   (b)   Penetration   depth
                                           PD(A,B)

Fig. 2. The no-fit polygon for overlap calculation [7]

2.3   Overlap Minimization Problem

The overlap minimization problem is a subproblem of nesting problem, which must be solved to get a valid packing. Define function $f_{ij}(V, R)$ to represent the amount of overlap between two polygons $P_i(r_i) \oplus v_i$ and $P_j(r_j) \oplus v_j$, and define function $g_i(V, R)$ to represent the amount of overlap between polygons $P_i(r_i) \oplus v_i$ and rectangular strip $C(W, L)$. The overlap minimization problem can be formally described as: Minimize $F(V, R) = \sum_{1 \leq i \leq j \leq n} w_{ij} \times f_{ij}(V, R) + \sum_{1 \leq i \leq n} g_i(V, R)$, where w_{ij} is the penalty weights. The goal of the overlap minimization problem is to find a solution (V, R) to minimize the total amount of overlap. When the value $F(V, R)$ is less than ε which is close to 0, we think we have obtained a legitimate packing scheme (V, R).

Penetration depth (also known as cross depth) is an important concept used for robotics and computer vision [5]. The penetration depth of two overlapping polygons $PD(A, B)$ can be defined as the minimum translation distance to separate them. Its formal definition is as follows: $PD(A, B) = \min\{\|v\| \| (A \cap (B \oplus v) = \varnothing), v \in R^2\}$, where the symbol $\|.\|$ represents the Euclidean normal form.

The value of $PD(A, B)$ can be calculated by the no-fit polygon generated by A and B. If the reference point O_B is inside, $PD(A, B)$ is equal to the minimum distance from O_B to the boundary of $NFP(A, B)$ (as shown in Fig. 2(b)). If O_B is outside the no-fit polygon, $PD(A, B)$ is equal to 0. Therefore, the value of penetration depth can be used to evaluate the overlap of two polygons.

For polygons A and B, this paper takes the square of the penetration depth as the overlap value between them, and uses a function to represent the overlap value, as follows: $Overlap(A, B) = \|PD(A, B)\|^2$.

3   Enhanced Whale Optimization Algorithm

Inspired by the observation of the foraging behavior of humpback whales, Mirjalili and Lewis proposed a new swarm intelligence optimization algorithm – whale optimization algorithm (WOA) [11]. The algorithm mainly simulates the hunting behavior of whales to solve the optimization problem, which has the advantages of simple principles, unique search mechanism, and easy programming. In terms of function optimization, it has been proved that it is better than PSO and gravity search algorithm in solving accuracy and convergence speed performance [11]. However, as a new member of the swarm optimization algorithm, the whale algorithm also has some common problems of the swarm optimization algorithm, such as unbalanced global and local optimization ability, easy to fall into local optimal solution, slow convergence speed, and low convergence accuracy. To overcome these issues in the whale optimization algorithm, an enhanced whale optimization algorithm (PEWOA) is proposed. In Sect. 4, the algorithm is applied to solve the 2DIRSP.

WOA algorithm is mainly divided into three stages: search foraging, contraction surrounding, and spiral updating position. The switching of these three stages is mainly determined by the values of probability factor p and coefficient $|A|$. To overcome the shortcomings of the basic WOA algorithm, this section proposes an enhanced whale optimization algorithm (PEWOA), which makes targeted improvements in the three stages of the basic WOA algorithm.

3.1   Nonlinear Time-Varying Adaptive Weights

In the basic whale optimization algorithm, when $|A| \geq 1$, the whale group enters the search and foraging stage; when $|A| < 1$, the whale group enters the contraction surrounding stage. The size of A is determined by the control parameter a. In the basic WOA algorithm, the value of a decreases linearly from 2 to 0 as the number of iterations increases, which makes the global and local search of the algorithm insufficient. Therefore, a weight factor w_1, which varies with the number of iterations, is introduced to control the size change of a. The weight factor w_1 is defined as follows: $w_1 = 0.5 \times (1 + \cos(\frac{\pi t}{MaxGen}))$, where $MaxGen$ is maximal number of iterations.

The weight factor w_1 proposed in this section decreases slowly at the beginning of iteration, and the algorithm can maintain better global exploration ability. After a certain number of iterations, the weight decreases rapidly, so that the algorithm can search the optimal solution more precisely in the local development stage. Therefore, after introducing the new nonlinear time-varying adaptive weight factors w_1, a and the $X(t + 1)$ of the contraction surrounding stage are updated as follows:

$$a = 2 \times w_1, X(t + 1) = X_{best}(t) - A \times |C \times X_{best}(t) - X(t)|,$$

where the values of A and C are the same as those in WOA, $A = 2 \times a \times r_1 - a$, $C = 2 \times r_2$, r_1 and r_2 are both random numbers between $[0, 1]$.

3.2   Differential Variation Perturbation Factor

During the contraction surrounding phase of the whale group, whale individuals are mainly based on the position information of the current best whale individuals and approach them to update their positions. In the whole solution process of the basic whale algorithm, new feasible solutions are constantly generated around the optimal solution, which will lead to continuous loss of populational diversity and precocious convergence. This paper refers to the mutation operator idea of the differential evolutionary algorithm, and introduces a differential mutation perturbation factor d in the contraction surrounding stage, which is defined as follows: $d = a \times (X_{best}(t) - X(t))$. The a in the equation is the same as that mentioned above, and the $X(t+1)$ introducing the differential variation perturbation factor is updated as: $X(t+1) = X_{best}(t) - A \times |C \times X_{best}(t) - X(t)| + d$. During the period of whale's contraction and surrounding, the introduction of differential variation perturbation factor can make it easier for individual whales to jump out of the local optimal, increase the diversity of population to a certain extent, and improve the solution accuracy of local optimization.

3.3   Poor Individual Variation

For each whale in the population, we selected some individuals with poor fitness for random variation. The coordinates of the selected whale individuals were randomly assigned again, and the individuals that received the mutation entered the next generation of the search regardless of their fitness. The purpose of introducing a mutation strategy is to make some individuals with worse fitness in the population more likely to be eliminated and, at the same time, retain some good individuals to enter the next generation of search. After the mutation of these poor individuals, the diversity of the population can be increased, which will enable the improved whale algorithm to conduct global search more fully.

3.4   Improved Spiral Update Mode

In basic whale optimization algorithm of spiral update location phase, each individual whales to current best whales, at this time to take the position of the update methods is the logarithmic spiral updates, literature [13] points out that the logarithmic spiral search way has some limitations when stepping spiral spacing than the search scope, will make the algorithm cannot traversal search the whole solution space, thus lost some potentially feasible solution. The Archimedes spiral search mode is used here instead of the log-spiral update mode of the original algorithm, so it is further improved as follows: $X(t+1) = |X_{best}(t) - X(t)| \times e^{bl} \times \cos(2\pi l) + X_{best}(t)$.

4   PEWOA Algorithm for 2DIRSP

In this section, an enhanced whale optimization algorithm (PEWOA) is proposed for the two-dimensional irregular strip packing problem. We first give the main framework of the EWS algorithm (Sect. 4.1) and then goes on to introduce the specific algorithms of the main modules.

4.1   Framework of the EWS Algorithm

EWS adopts three stages to solve the irregular strip packing problem.

First stage: a fast bottom-left heuristic algorithm (BL-Heuristic) is applied to obtain an initial legal packing scheme.

Second stage: in the legal packing, reduce or increase the length of the rectangular strip, and use layout adjustment(Layout-Adjustment) to search for a better packing plan.

Third stage: when the stop condition is reached, a global compression algorithm (Global-Adjustment) is used to improve further the solution obtained in stage 2. After the completion of stage 3, the whole solution method is completed.

The second stage is the core of the whole method, and PEWOA is called in the Layout-Adjustment algorithm, which was used to minimize the overlap of the rectangular strip of fixed length. The whole algorithm flow is as follows:

Algorithm 1. The EWS algorithm for 2DIRSP

1: Using the BL-Heuristic algorithm to obtain an initial legal packing scheme (V, R)
2: $l_{best} = L(V, R)$
3: $l_{best} = (1 - r_{dec}) \times l_{best}$
4: **if** $polygon.number < 30$ **then**
5:    $p_s = 1.0$
6: **else**
7:    $p_s = 0.3$
8: **end if**
9: **while** time limit not exceeded **do**
10:    (V, R)=Layout-Adjustment$(P, (V, R), C(W, L), R)$
11:    **if** $F(V, R) \leq \varepsilon$ **then**
12:       $(V_{best}, R_{best}) = (V, R)$
13:       $l_{best} = L(V_{best}, R_{best})$
14:       $l = (1 - r_{dec}) \times l_{best}$
15:    **else**
16:       $l = (1 + r_{inc}) \times l_{best}$
17:       **if** $l > l_{best}$ **then**
18:          $(V, R) = (V_{best}, R_{best})$
19:          $l = (1 - r_{dec}) \times l_{best}$
20:       **end if**
21:    **end if**
22: **end while**
23: (V_{best}, R_{best})=Global-Adjustment$(P, (V_{best}, R_{best}), C(W, L), R)$
24: **return** (V_{best}, R_{best})

4.2   Bottom-Left Heuristic

To generate an initial legal layout, this paper uses the bottom-left heuristic method(BL-Heuristic) proposed by Gomes and Oliveira (2000) [12]. The polygon parts were sorted according to the area, and the polygons with the large area were first placed on the rectangular strip. According to the bottom-left heuristic strategy, all waiting parts were placed one by one.

4.3   Layout-Adjustment Algorithm

The Layout-Adjustment algorithm adopts an iterative approach. In each step of the iteration, there are two strategies, one is to select a polygon randomly on the rectangle and move it, the other is to select two polygons randomly and exchange them, and the exchange strategy consists of two moving steps. When placing a moving polygon, we tried to move polygons to a position with minimal overlap, using the enhanced whale optimization algorithm (PEWOA) to find this placement. The process of the Layout-Adjustment algorithm is as Algorithm 2. Adaptive penalty weights w_{ij} are used to avoid the Layout-Adjustment algorithm falling into local optimality, and the rules proposed by Umetani et al. [14] are used for updating. After each move and placement of a polygon, if it still overlaps with other polygons on the rectangle strip, then

Algorithm 2. The Layout-Adjustment Algorithm

1: $step = 1$
2: Initializes penalty weights $w_{ij} = 1$
3: **while** $step < n_{max}$ **do**
4:    **if** $randNum(0,1) > p_s$ **then**
5:       Select a polygon $P_i \in P$ randomly and move it
6:       (V', R')=PEWOA$(P_i, P_i, (V, R), C(W, L), R)$
7:    **else**
8:       Select two polygons $P_i \in P$ and $P_j \in P$ randomly and swap them
9:       (V', R')=PEWOA$(P_i, P_j, (V, R), C(W, L), R)$
10:      (V', R')=PEWOA$(P_j, P_j, (V, R), C(W, L), R)$
11:    **end if**
12:    (V'', R'')=Separate$((V', R'), C(W, L))$
13:    **if** $F(V'', R'') \leq \varepsilon$ **then**
14:       **return** (V'', R'')
15:    **else if** $F(V'', R'') < F(V, R)$ **then**
16:       $(V, R) = (V'', R'')$
17:      $step = 0$
18:    **end if**
19:    Increase the value of w_{ij} according to rule below
20:    $step = step + 1$
21: **end while**
22: **return** (V, R)

---

the w_{ij} between these polygons will be added according to the following rules: $w_{ij} = w_{ij} + \frac{f_{ij}(V_i, V_j, R_i, R_j)}{max(f_{kl}(V_k, V_l, R_k, R_l), 1 \leq k < l \leq n)}$.

Select a polygon P_i from set P randomly and move it to a new position, where P_i will be placed with less total overlap. During this process, PEWOA is used to find this position. The specific algorithm flow of PEWOA is shown as Algorithm 3, where $C_i^t.fitness$ represents the fitness of the i-th individual of population in the t generation. By initializing a whale population n_c, each whale individual represents the coordinates of a certain position. According to the fitness evaluation, the better the fitness is, the smaller the overlap value will be when the whale individual is placed on the position. After the evolutionary iteration of the $MaxGen$ generation, the coordinates of the best-fit whale individuals are the placement of polygons.

4.4 Global-Adjustment Algorithm

At the end of the second phase, EWS obtains a current optimal solution. Here, we use a global compression algorithm proposed by Stephen C.H. Leung et al. [10] to improve the packing further and try to obtain a higher utilization. The idea of this global compression algorithm is to move polygonal parts as far to the right as possible, making the required container length shorter.

Algorithm 3. PEWOA

1: **for** each $r \in R$ **do**
2:     initializing the whale populations $C_i(i = 1, 2, ...n_c)$
3:     **for** $i = 1$ to n_c **do**
4:         $C_i = randNum(container.minX, container.maxX),$
                $randNum(container.minY, container.maxY)$
5:     calculate fitness of C_i and record the best fitness
6:     **end for**
7:     **while** $t < MaxGen$ **do**
8:         Calculate the weight factor w_1 and convergence factor a, and update the values of A and C
9:         **if** $t \ mod \ 3 == 0$ **then**
10:             **if** $|A| > 1$ **then**
11:                 **for** $i = 1$ to n_c **do**
12:                     Update the current location of individual whale and its fitness according to the search and foraging method
13:                 **end for**
14:             **else**
15:                 **for** $i = 1$ to n_c **do**
16:                     Update the value of the perturbation factor d
17:                     Update the current location of individual whale and its fitness according to the contraction surrounding method
18:                 **end for**
19:             **end if**
20:         **else**
21:             **for** $i = 1$ to n_c **do**
22:                 Update the current location of individual whale and its fitness according to the improved spiral update method
23:             **end for**
24:         **end if**
25:         Calculate the total fitness $newSum$ in this generation
26:         Find the individuals with the best fitness of this generation, update and save their location and fitness
27:         **for** $i = 1$ to n_c **do**
28:             Take the random number between $(0.2w_1, w_1)$ and assign the run-out factor p_2
29:             **if** $C_i^t.fitness \times p_2 > newSum/n_c$ **then**
30:                 $C_i = randNum(container.minX, container.maxX),$
                        $randNum(container.minY, container.maxY)$
31:             **end if**
32:         **end for**
33:         Update t to $t + 1$
34:     **end while**
35:     Save the best position and fitness of individual whales at the current rotation angle r
36: **end for**
37: Compare the best whale individuals corresponding to each r and choose the one with the best fitness as the location of P_i, and place P_i
38: **return** (V, R)

---

5   Computational Results

This section reports the results on the computational experiments of the algorithm EWS and previously published algorithms using the same framework. Nesting problem instances available in the literature were used to evaluate the performance of the EWS algorithm. They are available online at EURO Special Interest Group on Cutting and Packing (ESICUP) website (http://www.apdio.pt/esicup). The proposed algorithm has been implemented in C++ and compiled by Visual Studio 2010. The working environment is a PC with a Core i7 3.10 GigaHertz processor and 32 Gigabytes memory utilizing only one core to run the program. For each instance, our approach is ran ten times by setting the random seed from 1 to 10.

The values of parameters r_{dec}, r_{inc} and n_{max} are set to 0.04, 0.01 and 200 as recommended by Imamichi et al. [9]. For other main EWS parameters, they are set as follows $Pa = 0.5$, $n_c = 20$ and $MaxGen = 20$. The quality of the solution is judged by the fill ratio $\sum_{i=1}^{n} Area(P_i)/Area(Container)$.

Table 1. Comparison of four algorithms' best results and computational times.

Instance	EWS i7 3.1 GHz		GCS i7 2.2 GHz		ELS Pentium4 2.4 GHz		ILSQN Xeon 2.8 GHz	
	Best %	Time(s)	Best %	Time(s)	Best %	Time(s)	Best %	Time(s)
Albano	88.90	1202	**89.58**	1200	88.48	1203	88.16	1200
Dagli	89.06	1202	**89.51**	1200	88.11	1205	87.40	1200
Dighe1	**100.00**	600	**100.00**	600	**100.00**	601	99.89	600
Dighe2	**100.00**	600	**100.00**	600	**100.00**	600	99.99	600
Fu	92.12	600	**92.41**	600	91.94	600	90.67	600
Jakobs1	**89.10**	601	**89.10**	600	**89.10**	603	86.89	600
Jakobs2	86.19	601	**87.73**	600	83.92	602	82.51	600
Mao	84.97	1200	**85.44**	1200	84.33	1204	83.44	1200
Marques	89.94	1201	**90.59**	1200	89.73	1204	89.03	1200
Shapes0	68.00	1206	**68.79**	1200	67.63	1207	68.44	1200
Shapes1	76.00	1206	**76.73**	1200	75.29	1212	73.84	1200
Shapes2	83.89	1202	**84.84**	1200	84.23	1205	84.25	1200
Shirts	88.15	1217	**88.96**	1200	88.40	1293	88.78	1200
Swim	**76.13**	1217	75.94	1200	75.43	1246	75.29	1200
Trousers	89.85	1212	**91.00**	1200	89.63	1237	89.79	1200

Tables 1 and 2 show the comparison of EWS algorithm's best and average solutions with other existing algorithms using the same framework. To the best of our knowledge, the competitive published results are found in Ahmed Elkeran (2013) (GCS) [7], Leung et al. (2012) (ELS) [10], Imamichi et al. (2009) (ILSQN) [9]. All the algorithms are all conducted ten runs for each instance, and the computational times are listed in seconds. For all the 15 instances, EWS got four best results, with the Swim instance updating the best know results. It

Table 2. Comparison of four Algorithms' average results and computational times.

| Instance | EWS i7 3.1 GHz | | GCS i7 2.2 GHz | | ELS Pentium4 2.4 GHz | | ILSQN Xeon 2.8 GHz | |
	Avg %	Time(s)	Avg %	Time(s)	Avg %	Time(s)	Avg %	Time(s)
Albano	**87.54**	1202	87.47	1200	87.38	1203	87.14	1200
Dagli	**87.18**	1202	87.06	1200	86.27	1205	85.80	1200
Dighe1	93.43	600	**100.00**	600	91.61	601	90.49	600
Dighe2	**100.00**	600	**100.00**	600	**100.00**	600	84.21	600
Fu	**90.96**	600	90.68	600	90.00	600	87.57	600
Jakobs1	**89.00**	601	88.90	600	88.35	603	84.78	600
Jakobs2	**82.10**	601	81.14	600	80.97	602	80.50	600
Mao	**83.14**	1200	82.93	1200	82.57	1204	81.31	1200
Marques	89.07	1201	**89.40**	1200	88.32	1204	86.81	1200
Shapes0	67.01	1206	**67.26**	1200	66.85	1207	66.49	1200
Shapes1	74.12	1206	73.79	1200	**74.24**	1212	72.83	1200
Shapes2	**82.94**	1202	82.40	1200	82.55	1205	81.72	1200
Shirts	87.16	1217	87.59	1200	87.20	1293	**88.12**	1200
Swim	**74.69**	1217	74.49	1200	74.10	1246	74.62	1200
Trousers	88.66	1212	**89.02**	1200	88.29	1237	88.69	1200

achieved the same results as GCS and ELS for Jakobs1 instance; however, the obtained layouts are different. On the other hand, EWS updates average results for nine instances.

6   Conclusion

In this paper, a new algorithm (EWS) is presented for solving the 2DIRSP. The use of the enhanced whale optimization algorithm (PEWOA) has proved to be a competitive alternative to other published nesting algorithms. Computational experiments show that the implementation is robust and also reasonably fast. Since the whale optimization algorithm was proposed in 2016, this paper firstly applied the optimization algorithm to solve the 2DIRSP problem, and obtained good optimization results after some improvements and enhancement strategies. It shows that the whale optimization algorithm also has excellent research potential and application prospect in dealing with nonlinear and multi-peak optimization problems. Compared with other optimization algorithms that have been widely used to solve 2DIRSP problems, such as simulated annealing algorithm, genetic algorithm, particle swarm optimization algorithm, etc., PEWOA has a significant effect on the improvement of layout utilization and has strong competitiveness. If the EWS algorithm containing PEWOA is applied to the nesting and cutting of cloth, leather, plate, and other manufacturing industries on a large scale, the improved utilization rate of raw materials will save significant material costs for enterprises and bring considerable economic benefits. For future work, it is proposed to further study the efficiency of PEWOA in large-scale layout cases.

References

1. Bennell, J.A., Dowsland, K.A., Dowsland, W.B.: The irregular cutting-stock problem - a new procedure for deriving the no-fit polygon. Comput. Oper. Res. **28**(3), 271–287 (2001)
2. Bennell, J.A., Oliveira, J.F.: The geometry of nesting problems: a tutorial. Eur. J. Oper. Res. **184**(2), 397–415 (2008)
3. Bennell, J.A., Song, X.: A comprehensive and robust procedure for obtaining the nofit polygon using Minkowski sums. Comput. Oper. Res. **35**(1), 267–281 (2008)
4. Burke, E., Hellier, R., Kendall, G., Whitwell, G.: Complete and robust no-fit polygon generation for the irregular stock cutting problem. Eur. J. Oper. Res. **179**(1), 27–49 (2007)
5. Dobkin, D., Hershberger, J., Kirkpatrick, D., Suri, S.: Computing the intersection-depth of polyhedra. Algorithmica **9**(6), 518–533 (1993). https://doi.org/10.1007/BF01190153
6. Egeblad, J., Nielsen, B.K., Odgaard, A.: Fast neighborhood search for two- and three-dimensional nesting problems. Eur. J. Oper. Res. **183**(3), 1249–1266 (2007)
7. Elkeran, A.: A new approach for sheet nesting problem using guided cuckoo search and pairwise clustering. Eur. J. Oper. Res. **231**(3), 757–769 (2013)
8. Hopper, E., Turton, B.C.H.: A review of the application of meta-heuristic algorithms to 2D strip packing problems. Artif. Intell. Rev. **16**(4), 257–300 (2001). https://doi.org/10.1023/A:1012590107280
9. Imamichi, T., Yagiura, M., Nagamochi, H.: An iterated local search algorithm based on nonlinear programming for the irregular strip packing problem. Discrete Optim. **6**(4), 345–361 (2009)
10. Leung, S.C., Lin, Y., Zhang, D.: Extended local search algorithm based on nonlinear programming for two-dimensional irregular strip packing problem. Comput. Oper. Res. **39**(3), 678–686 (2012)
11. Mirjalili, S., Lewis, A.: The whale optimization algorithm. Adv. Eng. Softw. **95**, 51–67 (2016)
12. Oliveira, J.F., Gomes, A.M., Ferreira, J.S.: TOPOS – a new constructive algorithm for nesting problems. OR Spektrum **22**(2), 263–284 (2000). https://doi.org/10.1007/s002910050105
13. Sun, W., Wang, J., Wei, X.: An improved whale optimization algorithm based on different searching paths and perceptual disturbance. Symmetry **10**(6), 210 (2018)
14. Umetani, S., Yagiura, M., Imahori, S., Imamichi, T., Nonobe, K., Ibaraki, T.: Solving the irregular strip packing problem via guided local search for overlap minimization. Int. Trans. Oper. Res. **16**(6), 661–683 (2009)

A Heuristic Approach to the Three Dimensional Strip Packing Problem Considering Practical Constraints

Qiang Liu, Dehao Lin, Hao Zhang, and Lijun Wei[(✉)]

Guangdong Provincial Key Laboratory of Computer Integrated Manufacturing System, State Key Laboratory of Precision Electronic Manufacturing Technology and Equipment, Guangdong University of Technology, Guangzhou 510006, China
liuqiang@gdut.edu.cn, benben0605@gmail.com, villagerwei@gmail.com

Abstract. This paper studies the three dimensional strip packing problem (3DSPP) with practical constraints. The objective is to pack a set of three-dimensional items into a three-dimensional container with one open dimension such that the used length of the open dimension is minimized. Several practical constraints, including the *orientation, support, multiple-drop, moving* and *maximum length overlap*, are considered. An open space heuristic is proposed to decode a given sequence of items into a feasible solution, and a random local search is used to improve the solution. Computational experiments on 30 practical instances are used to test the performance of our approach.

Keywords: Greedy heuristic · 3D packing · Practical 3D packing

1 Introduction

Sinotrans logistics LTD., one of the biggest international logistics companies in China, receives orders from different customers and helps the customers to transport the cargos from China to different places all over the world. The cargos from different customers will be first stored in a different warehouse before loading into the container sent to the destination by the trip. To save costs, the company will load the cargos having the same destination into the same container. One of the tasks the company encountered is to determine whether a given set of cargos can be loaded into a given container. If the answer is yes, they want the occupied space of these cargos is minimized such that the left space can be used by further coming cargos. We model this problem as the three dimensional strip packing problem (3DSPP) with practical constraints.

The three-dimensional strip packing problem(3DSP), which is an NP-Hard problem, is relatively less well-studied. 3DSP occurs in many fields of application, such as in transportation and logistics. Recent literature on the problem examines metaheuristic approaches in the majority. Two metaheuristics for 3DSP were introduced in the early work by Bortfeldt and Gehring [2], namely a tabu

© Springer Nature Switzerland AG 2020
H. Fujita et al. (Eds.): IEA/AIE 2020, LNAI 12144, pp. 262–267, 2020.
https://doi.org/10.1007/978-3-030-55789-8_23

search algorithm and genetic algorithm. Bortfeldt and Mack [4], extended the branch-and-bound approach for the single container loading problem proposed by Pisinger [7] to the 3DSP. Inspired by them, our approach follows the wall-building strategy. Zhang et al. [10] use open space to examine the feasibility of the solutions of vehicle routing problems under three-dimensional loading constraints. Our approach uses open space to meet Moving constraint(see below).

In our problem, we are given a set $O = \{O_1, \cdots, O_K\}$ of orders. Each order $O_i(1 \leq i \leq K)$ contains a set $O_i = \{O_{i1}, \cdots, O_{in_i}\}$ of three-dimensional items (represent the cargo) and there is a warehouse s_i representing the warehouse the items in this order located at. The dimension of the item O_{ij} is $l_{ij} \times w_{ij} \times h_{ij}$, and the quantity of item O_{ij} is m_{ij}. The objective is to orthogonally load all the items without overlap into a three-dimensional container of width W and height H such that the used length L of the loading is minimized. All the cargos in the same order are stored in the same warehouse and must be loaded into the container in successive order. Besides, a feasible loading plan must satisfy the following constraints. **Loading constraint:** the items should be loaded into the container with their edges parallel to the sides of the container, and no overlap exists between any pair of items. **Orientation constraint:** for each dimension $k(k = 1, 2, 3)$ of each item O_{ij}, there is a flag f_{ijk} indicating whether this dimension can be the one parallel to the height of the container. **Support constraint:** the base of any item must be fully supported by the bottom of the container or other items. **Multiple-drop constraint:** the warehouse is visited once in increasing order, and all the cargos located in this warehouse must be loaded. The items in the same order must be loaded in successive order. **Moving constraint:** when loading an item Oij, we assume that the item can only be moved directly from the rear to its position, which requires that no already packed items are placed between Oij and the rear of the vehicle. **Maximum length overlap constraint:** assumed that the used length of already packed items is L_{i-1} before loading items in order O_i, it is required that the x position of all the items in O_i must not be smaller than $L_{i-1} - dep$, where dep is the maximal allowed overlap. The 3DSPP has been studied by [1,3] and [8]. However, to the best of our knowledge, only the orientation constraint is considered in previous literature.

2 Open Space Heuristic for the 3DSPP with Practical Constraints

In this section, we propose a new greedy heuristic OPENSPACEHEURISTIC for the 3DSPP. This heuristic takes as input a sequence of boxes $B = (b_1, b_2, \ldots, b_k)$, container of three dimensions $L \times W \times H$ as the input, the used length of the final loading plan is returned final. The free space is represented by the open space. In any step of the construction process of our solution, the state of our partial packing can be described by: **R:** a list of open spaces representing the free space in the container. **B:** a list of candidate boxes. **PB:** a list of placed boxes and their locations. The process of our heuristic to construct a loading plan can be described as follows: (1) start from the list of free space consists of the whole container and the list of boxes. (2) select the first box b in list B, and check the free space in the list according to front-bottom-left strategy until this placement satisfies all the constraints. (3) place b at the selected free space, update the state including R, B, and PB. (4) repeat step (2)–(3) until all the boxes are loaded. (5) return the used length of the loading plan.

We use the *open space* to represent the free space in the container. The open space is introduced by [10] for solving the sub loading problem of vehicle routing problem with loading constraints, later adapted by [9] to solve the 2D strip packing problem with unloading constraints. It is similar to the *maximal space* concept to represent the free space in the container [5,6]. Concerning *maximal space*, when a box is placed at a corner of the container, the remaining free space can be represented by three over-lapping cuboids, and each cuboid is the largest rectangular box that is interior-disjoint with the placed box. Such a cuboid is called a residual space. Due to the overlapping nature of the maximal space representation, when a box is placed at some corner of residual space, it may overlap with other residual spaces. After a box is placed, these overlapped residual spaces must be updated. To ensure the *moving* constraint of the solution, we also use *open space* concept to represent the free space in the container. Workers may not be able to reach the space at specific loading steps if we only use *maximal space* strategy. As shown in Fig. 1, workers can not reach r' which will be selected if use *maximal space*. The *open space* is similar to the *maximal space*. The *maximal space* is the largest rectangular cuboid that is interior-disjoint with the placed items. The main difference is that one of the planes of open space must be placed in the container's rear door, but no such constraint is required for maximal space [10].

2.1   Corner of Selected Free Space

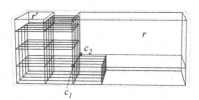

Fig. 1. Candidate corners of selected free space r

It is well known that place boxes in the corners of the container tends to result in better utilization of space. In the practical loading process, the space close to the front of container will be preferred to be loaded, like wall-building loading [3,7]. Hence, we use front-bottom-left strategy to select free space to placed selected box until satisfies support constraint. A free space r has eight corner, two corners of them close to front-bottom c_1, c_2 as candidates for box to place. Our heuristic checks in order whether c_1, c_2 meet the support constraint through calculating support rate. r will be selected if one of candidate corners meet the constraint.

3   Random Local Search

Algorithm 1. Random local search for 3DSPP

1: sort the input box B according to our sorting rule mentioned above consists of three steps
2: $minLength \leftarrow$ OpenSpaceHeuristic(B, W, H)
3: **while** time limit is not exceeded **do**
4:     $r \leftarrow$ randomly choose between 0 and 1
5:     **if** r == 0 **then**
6:         choose on warehouse s randomly
7:         randomly shuffle the sequence of the order(with items in this order) belonging to s and let the new resulting boxes sequence be B'
8:     **end if**
9:     **if** r == 1 **then**
10:        randomly choose one order o
11:        randomly shuffle the sequence of the items in o and let the new resulting boxes sequence be B'
12:     **end if**
13:     $length \leftarrow$ OpenSpaceHeuristic(B', W, H)
14:     **if** $length <= minLength$ **then**
15:        $minLength \leftarrow length$
16:        $B \leftarrow B'$
17:     **end if**
18: **end while**
19: **return** $minLength$

The open space heuristic constructs a solution for a given list of boxes B. To improve the solution further, a random local search approach is introduced to find more solutions. The overall approach is given as Algorithm 1. We first sort the input box according to the following rules. Firstly, the order is sorted by increasing of the warehouse id. Secondly, the order belongs to the same warehouse is sorted by the decreasing total items volume. Thirdly, the items in the same order are sorted by decreasing volume. Two operators are used to construct a new boxes list. In the first operator, one of the warehouses is selected, and the loading sequence of orders belong to this warehouse is shuffled. In the second operator, one of the order is randomly selected, and the items in this order are randomly shuffled.

Table 1. Results on the practical instances

Inst.	K	N	W	H	L	T	fr(%)	Inst.	K	N	W	H	L	T	fr(%)
1	1	145	234	238	348	500	75.72	16	11	190	234	238	654	500	79.51
2	6	217	235	268	861	500	86.04	17	8	385	235	268	1265	501	81.12
3	5	105	234	238	445	500	78.21	18	10	487	227	254	731	502	83.12
4	7	725	233	238	1094	502	93.82	19	8	757	235	268	1264	500	85.36
5	5	2388	233	238	948	510	93.97	20	9	889	227	254	1217	519	85.78
6	7	170	234	238	426	500	79.09	21	10	335	227	254	1164	501	81.23
7	9	810	227	254	1223	501	88.8	22	10	653	235	268	974	509	84.14
8	7	334	234	238	404	501	83.97	23	10	924	235	268	1159	522	83.62
9	6	352	234	238	452	500	85.99	24	10	380	235	268	1104	501	85.76
10	9	314	234	238	602	500	86.12	25	11	655	227	254	1216	507	81.58
11	6	255	227	254	1177	500	83.5	26	8	206	234	238	537	500	84.36
12	8	515	235	268	1071	503	86.88	27	9	429	235	268	1289	500	79.05
13	10	357	227	254	1141	501	86.29	28	7	823	235	268	1223	501	85.74
14	10	712	235	268	1382	504	76.46	29	13	994	235	268	1256	503	84.77
15	11	1216	235	268	649	531	78.41	30	12	521	235	268	1108	500	82.8

4 Computational Experiments

The algorithms proposed were implemented in Java using Sun Oracle JDK 1.7.0. All experiments were tested on a computer with an Inter (R) Xeon(R) CPU E5-2620 v4 @2.10 Ghz, with 16 GB RAM. Test instances are derived from 30 practical instances provided by Sinotrans Logistics LTD. Each instance includes: (1)the width W and height H of the container; (2) the items to be loaded, including order and warehouse to which each item belongs. The time limit is set to 500 s, which is not only the time that customers can accept but also the time to find better results. The maximum length overlap is set to 100 cm. The results are shown as Table 1, where K is the number of orders, N is the total number of boxes, and $fr(\%)$ is the volume utilization of the container. K and N are used to reflect the heterogeneity of the instances. $fr(\%)$ is defined as: $\frac{\sum_{i=1}^{K} \sum_{j=1}^{n_i} m_{ij} \cdot l_{ij} \cdot w_{ij} \cdot h_{ij}}{minLength \cdot W \cdot H} \times 100$. The column L equal to $minLength$ return by Algorithm 1 and column T is the time used for each instance. We can see from this table that the utilization of most of the instances is above 80%, and two of the instances reach 90%. Figure 2 is plotted automatically according to the output result of our algorithm, shows the loading plans for two different instances.

5 Conclusions

In this paper, we propose a random local search approach to the 3DSPP with practical constraints. It makes use of an open space based heuristic to construct a solution for a given sequence. Two operators are introduced to generate a

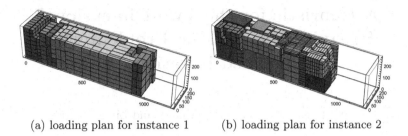

(a) loading plan for instance 1   (b) loading plan for instance 2

Fig. 2. Loading plans for two instances

neighbor sequence. A reality set of instances are used to test our approach. For further work, we will explore the effect of different constraint deeply. In our algorithm, the method of packing plan evaluation may lead to a longer time. Our approach will attempt other evaluations, for instance, evaluate the used length after placing a certain number of orders.

References

1. Allen, S.D., Burke, E.K.E., Kendall, G.: A hybrid placement strategy for the three-dimensional strip packing problem. Eur. J. Oper. Res. **209**(3), 219–227 (2011)
2. Bortfeldt, A., Gehring, H.: Two metaheuristics for strip packing problems. In: Proceedings band der 5th International Conference of the Decision Sciences Institute, 141(2), pp. 1153–1156 (1999)
3. Bortfeldt, A., Gehring, H.: A hybrid genetic algorithm for the container loading problem. Eur. J. Oper. Res. **131**(1), 143–161 (2001)
4. Bortfeldt, A., Mack, D.: A heuristic for the three-dimensional strip packing problem. Eur. J. Oper. Res. **183**(3), 1267–1279 (2007)
5. Parreño, F., Alvarez-Valdes, R., Oliveira, J.F., Tamarit, J.M.: Neighborhood structures for the container loading problem: a VNS implementation. J. Heuristics **16**(1), 1–22 (2010)
6. Parreño, F., Alvarez-Valdes, R., Tamarit, J.M., Oliveira, J.F.: A maximal-space algorithm for the container loading problem. INFORMS J. Comput. **20**(3), 412–422 (2008)
7. Pisinger, D.: Heuristics for the container loading problem. Eur. J. Oper. Res. **141**(2), 382–392 (2002)
8. Wei, L., Oon, W.-C., Zhu, W., Lim, A.: A reference length approach for the 3D strip packing problem. Eur. J. Oper. Res. **220**(1), 37–47 (2012)
9. Wei, L., Wang, Y., Cheng, H., Huang, J.: An open space based heuristic for the 2D strip packing problem with unloading constraints. Appl. Math. Model. **70**, 67–81 (2019)
10. Zhang, Z., Wei, L., Lim, A.: An evolutionary local search for the capacitated vehicle routing problem minimizing fuel consumption under three-dimensional loading constraints. Transp. Res. Part B Methodological **82**, 20–35 (2015)

A Heuristic for the Two-Dimensional Irregular Bin Packing Problem with Limited Rotations

Qiang Liu, Jiawei Zeng, Hao Zhang, and Lijun Wei[✉]

Guangdong Provincial Key Laboratory of Computer Integrated Manufacturing System, State Key Laboratory of Precision Electronic Manufacturing Technology and Equipment, Guangdong University of Technology, Guangzhou 510006, China
liuqiang@gdut.edu.cn, jwzeng1994@gmail.com, zhanghao0815@126.com,
villagerwei@gmail.com

Abstract. In this paper, we propose a heuristic for the two-dimensional irregular bin packing problems (2DIRBPP) with limited rotations. To solve the 2DIRBPP, we use the First Fit Decreasing (FFD) strategy to assign the pieces to the bins, then use the Bottom-Left algorithm and the pieces exchange method to place pieces into each bin, and finally perform a local search to improve the solution. The results show that our approach is competitive with all existing approaches on several sets of standard benchmark.

Keywords: Cutting and packing · Irregular bin packing · Heuristic

1 Introduction

The two-dimensional irregular bin packing problem (2DIRBPP) is fundamental in cutting and packing literature, which is widely used in industrial production. In the furniture plate processing, shipbuilding industry, automobile industry, sheet metal processing industry, clothing production, leather processing, etc., it is necessary to deal with cutting and packing problems. Although the packing problem in different industrial fields has various restrictions and constraints, the common basic problem is to find an effective method for placing the required part pieces into the raw material bins, which makes the area utilization of the raw material bins high to save as much material as possible.

The most studied irregular packing problem in literature is the two-dimensional strip packing problem, also known as nesting problem, and many algorithms have achieved excellent results. Imamichi et al. [5] proposed an algorithm to separate overlapping polygons based on nonlinear programming, and an algorithm to exchange two polygons in one bin in order to find their new positions in the bin with minimal overlap, and then formed an iterative local search algorithm to solve the problem of irregular strip packaging. Leung et al. [8] extended the iterative local search algorithm of Imamichi et al. [5], added

© Springer Nature Switzerland AG 2020
H. Fujita et al. (Eds.): IEA/AIE 2020, LNAI 12144, pp. 268–279, 2020.
https://doi.org/10.1007/978-3-030-55789-8_24

a tabu search algorithm to avoid local minima, and proposed a compression algorithm to improve the results. Based on Imamichi et al. [5] and Leung et al. [8], Elkeran [4] further added a cuckoo search to improve the search scope and efficiency, and introduced pair clustering to combine paired pieces together, reducing the complexity of nesting problems.

Due to the complexity of geometry calculation and the limitation of computer functions, there are few literatures on 2DIRBPP compared with 2D strip packing problem. Terashima-marin et al. [13] introduced a hyper-heuristic method based on a genetic algorithm, which can solve 2d regular and irregular bin packing problems. The hyper-heuristic algorithm combines two methods: the Bottom-Left algorithm of Jakobs [6] and the improved bottom left algorithm of Liu et al. [9]. Lopez-camacho et al. [10] used Djang and Finch heuristics (DJD) to solve two-dimensional irregular packing problems and achieved excellent results on instance types of convex polygons. On this basis, lopez-camacho et al. [11] solved non-convex polygon instances and reported the percentage of problems solved using more, the same or fewer bins.

In other literatures, 2DIRBPP is considered for piece rotation. Pieces rotation is generally divided into no rotation allowed, 4 rotations ($90°, 180°, 270°, 360°$), and free rotations. Martinez-sykora et al. [12] proposed multiple variants of a construction algorithm, which considered free rotation of pieces and could solve 2DIRBPP. Several integer programming models are used to determine the assignment of pieces to bins, and then a mixed-integer programming model is used to place pieces into each bin. Abeysooriya et al. [1] proposed a heuristic method based on the Jostle algorithm. Different from common algorithms, it solved both the allocation and placement of pieces together. At the same time, two strategies for dealing with pieces rotation are proposed, including four rotations and free rotations.

This paper describes a constructive algorithm that considers three aspects of the packing problem: the assignment of pieces to bins, the placement of assigned pieces on each bin, and the local search between two bins. At present, there are two commonly used placement strategy of pieces in the bin: the feasible solution based placement and overlapping removal based placement. The feasible solution based placement is usually used in the irregular strip packing problem, which does not allow overlapping and takes a short time. The overlapping removal based placement is usually used in irregular bin packing problem, which allows overlapping and achieves excellent results. We want to use the overlapping removal based placement in irregular bin packing problem. So the placement strategy of pieces in the bin is adapted from the placement method in the strip packing problem of Imamichi et al. [5] and Leung et al. [8], so that two pieces could be exchanged in one bin to find their new positions with minimal overlap. The local search referred to the method in martinez-sykora et al. [12], and carried out one-to-many exchange of pieces between two bins to improve the utilization rate.

The paper is organized as follows. We briefly introduce the two-dimensional irregular bin packing problem and give some mathematical definitions in Sect. 2. In Sect. 3, we describe our constructive approach to generate the initial solution. The local search algorithm to further improve the solution is introduced in Sect. 4. In Sect. 5, we conduct experimental tests and give the result. Finally, we summarize our conclusions in Sect. 6.

2   Problem Description

Let $P = \{p_1, ..., p_n\}$ be the set of n pieces, represented as simple polygons, to be placed into identical rectangular bins of dimensions $W \times L$, with no limit on the number of bins. For each piece p_j, the corresponding area is a_j, and the rotation angle is O $(90°, 180°, 270°, 360°)$. The task is to pack all the pieces into the *bins* without overlap such that the used number of *bins* is minimized.

A solution to the described problem is given by a set of N bins $B = \{b_1, ..., b_N\}$. Each bin $b_i(P_i, R_i, V_i)$ represents the packing result of one bin, $i = 1, ..., N$. We denote $P_i = \{p_1, ..., p_{n_i}\}$ as the set of packed pieces in bin b_i and n_i is the number of placed pieces in bin b_i, $P_i \subset P$, $P = \cup_{i=1}^{N} P_i$. A vector $R_i = \{r_1, ..., r_{n_i}\}$ is the set of rotation angles of packed pieces in bin b_i, $r_j \in O, j = 1, ..., n_i$. $V_i = \{v_1, ..., v_{n_i}\}$ is the set of positions of packed pieces in bin b_i and for a position $v_j = (x_j, y_j), j = 1, ..., n_i$, x_j and y_j are the x-coordinate and y-coordinate of the reference point of each piece p_j. We define the reference point of the piece p_j as the bottom-left corner of the enclosing rectangle of the piece when in rotation angle r_j.

The common evaluation measure is the number of bins used, N. But it cannot distinguish the solutions with the same number of bins. In order to compare more accurately with other algorithms, the evaluation measure F is used: $F = \frac{\sum_{i=1}^{N} U_i^2}{N}$. And U_i is utilization ratio of each bin b_i, defined as: $U_i = \frac{\sum_{j=1}^{n_i} a_j}{LW}$. This measure rewards solutions that highly utilize bins while trying to empty one weakest bin, which supports the search in reducing the number of bins.

3   Initial Solution Generation

In this section, we present a new constructive heuristic for 2DIRBPP to generate the initial solution. It consists of two main parts: the assignment of pieces to a bin (Sect. 3.1) and how to place the assigned pieces into a single bin (Sect. 3.2).

3.1   Assignment Strategy

Our assignment strategy adopts the fast and well known First Fit Decreasing Algorithm (FFD) (see Johnson et al. [7]). FFD directly assigns pieces into a bin in a given sorted order following a First Fit Decreasing strategy. The pieces are placed into the bins using the procedure described in Sect. 3.2. The structure is as follows:

(1) Sort all the pieces in order of decreasing area, set $i = 1, j = 1$, open bin b_i.
(2) For each b_k in open bins, try to pack piece p_j into the open bin b_k using the placement method described in Sect. 3.2.
(3) If a feasible solution is found, set $j = j + 1$, if $j \leq n$, go to (2), else return; otherwise, $i = i + 1$, open new bin b_i and place p_j at b_i, $j = j + 1$, go to (2).

3.2   Placement of a Single Bin

We use the overlap minimization algorithm to place the assigned pieces into a single bin. The reference algorithm is shown in the Fig. 1 (see Imamichi et al. [5]). We simplify the operation of reducing and increasing length in the algorithm. The main process is shown in Fig. 2 and Algorithm 1. The classical **Bottom-Left(BL)** strategy (Sect. 3.2.2) is used to obtain an **initial position** for each piece, where the candidate positions of the pieces is calculated through **No-fit-polygon(NFP)** (Sect. 3.2.1). Note that the initial position may not be infeasible as the used length may be larger than that of the bin. To get a feasible solution, we iteratively use the **swap two pieces** (Sect. 3.2.3) operators and the **separation algorithm** (Sect. 3.2.4) to minimize the overlap between bin and pieces, where the number of iterations is controlled by the parameter N_{mo}. A feasible solution to place all the pieces is found if the final overlap is zero.

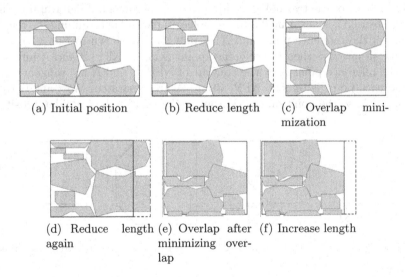

(a) Initial position   (b) Reduce length   (c) Overlap minimization

(d) Reduce length again   (e) Overlap after minimizing overlap   (f) Increase length

Fig. 1. Referenced placement policy

3.2.1   No-fit-polygon
No-fit-polygon (NFP) is shown in Fig. 3. Given two polygons A and B, fix polygon A and slide polygon B around the outer contour of polygon A. During the sliding process, B keeps contact with A but does not overlap. The trace of the

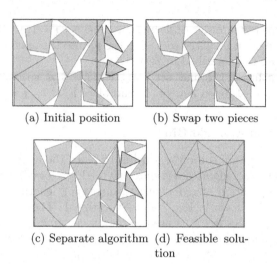

(a) Initial position   (b) Swap two pieces

(c) Separate algorithm   (d) Feasible solution

Fig. 2. Placement policy

reference point on B is the contour of the NFP. The algorithm to build NFP can be seen in Zhang et al. [14], Bennell et al. [2] and Burke et al. [3].

When NFP of the two polygons are given, the problem of determining their relative position is transformed into a simpler problem of determining the relative position of the reference point and NFP polygon. As shown in Fig. 3(d), there are three situations: when the reference point is inside NFP, B and A intersect; when the reference point is at the contour of NFP, B and A are tangent; when the reference point is outside NFP, B and A are separated. We want B to be tangent to A, so the point on the contour of the NFP is the candidate position.

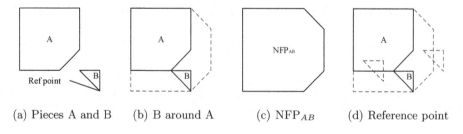

(a) Pieces A and B   (b) B around A   (c) NFP$_{AB}$ (d) Reference point

Fig. 3. NFP

3.2.2 Bottom-Left Strategy to Generate the Initial Position

We use the Bottom-Left strategy to select the initial position. The method is shown as Algorithm 2. The pieces in P_i are placed one by one according to the given order. For each piece, the vertices and the intersections of the NFP is considered as the candidate positions and the bottom-left point without overlap

---

Algorithm 1. Place assigned P_i into b_i and get final position V_i and R_i

---

1: $(V_i, R_i) = \text{InitialPosition}(P_i, b_i)$
2: **if** $\text{Overlap}(V_i, R_i) == 0$ **then**
3:     **return** True
4: **end if**
5: **for** $n = 1$ to N_{mo} **do**
6:     Randomly choose p_j and p_k from P_i
7:     $(V_i^1, R_i^1) = \text{Swap}(p_j, p_k, V_i, R_i, b_i)$
8:     $(V_i^2, R_i^2) = \text{Separate}(V_i^1, R_i^1)$
9:     **if** $\text{Overlap}(V_i^2, R_i^2) < \text{Overlap}(V_i, R_i)$ **then**
10:       $(V_i, R_i) = (V_i^2, R_i^2)$
11:     **end if**
12:     **if** $\text{Overlap}(V_i, R_i) == 0$ **then**
13:       **return** TRUE
14:     **end if**
15: **end for**
16: **return** False

Algorithm 2. Initial Position of P_i in b_i

1: **for** $j = 1$ to n_j **do**
2:     **for** each $r_j \in O$ **do**
3:       **for** $k = 1$ to $j - 1$ **do**
4:         Gets the vertices and intersections(denote S) of $\text{NFP}(p_j, p_k)$
5:       **end for**
6:     **end for**
7:     $(V_i, R_i)_j = (v_j, r_j) = \text{BottomLeft}(S, P_i, b_i)$
8: **end for**
9: **return** (V_i, R_i)

(excluding the overlap between the piece and the right side of the bin) is selected finally.

3.2.3 Swap Two Polygons Operator

When the overlap of the initial position obtained by the Bottom-Left strategy is not zero, the swap algorithm and separation algorithm is called to minimize the overlap. The algorithm for swapping two polygons p_j and p_k is constructed with moving polygon. It is shown as in Algorithms 3 and 4 (see Leung et al. [8]). Here the overlap is calculated by the function:

$$H(V_i, R_i) = \text{Overlap } (V_i, R_i) + \text{Overlap } (p_j, p_j^1)$$

where function Overlap (V_i, R_i) is as described in Sect. 3.2.4, and p_j^1 is denoted as the polygon after p_j moves. The function Overlap (p_j, p_j^1) is the overlap between p_j and p_j^1.

Algorithm 3. Swap two polygons p_j and p_k in b_i

1: $(V_i, R_i)_j = (v_j, r_j) = (+\infty, +\infty)$
2: $(V_i, R_i) = \text{Move}(p_k, V_i, R_i, b_i)$
3: $(V_i, R_i) = \text{Move}(p_j, V_i, R_i, b_i)$
4: **return** (V_i, R_i)

Algorithm 4. Move polygon p_j in b_i

1: $\text{minOverlap} = +\infty$
2: **for** each $r_j \in O$ **do**
3:    **for** $k = 1$ **to** n_i **do**
4:       Gets the vertices and midpoint(denote S) of $\text{NFP}(p_j, p_k)$
5:    **end for**
6:    Gets the vertices and midpoint(add to S) of $\text{NFP}(p_j, b_i)$
7:    **for** each point $s \in S$ **do**
8:       $(V_i^1, R_i^1) = (V_i, R_i)$
9:       $(V_i, R_i)_j = (v_j, r_j) = (s, r_j)$
10:      **if** $\text{H}(V_i, R_i) < \text{minOverlap}$ **then**
11:         $\text{minOverlap} = \text{H}(V_i, R_i)$
12:      **else**
13:         $(V_i, R_i) = (V_i^1, R_i^1)$
14:      **end if**
15:    **end for**
16: **end for**
17: **return** (V_i, R_i)

3.2.4 Separation Algorithm

There will be overlap after swapping pieces, and the separation algorithm can reduce the overlap. The separation algorithm is to reposition the pieces to get the new position with the minimum overlap. Given a placement (V_i, R_i) in b_i, we denote $f_{jk}(V_i, R_i)$ as the measure of the overlap between polygons p_j and p_k, and $g_j(V_i, R_i, b_i)$ as the measure of the overlap between polygon p_j and b_i. The overlap minimization problem in b_i is formulated by the following equation:

Minimize Overlap $(V_i, R_i) = \sum_{1 \leq j < k \leq ni} f_{jk}(V_i, R_i) + \sum_{1 \leq j \leq n} g_j(V_i, R_i, b_i)$

By fixing the rotation R_i, it can be transformed as follow:

Minimize Overlap $(V_i) = \sum_{1 \leq j < k \leq ni} f_{jk}(V_i) + \sum_{1 \leq j \leq n} g_j(V_i, b_i)$

The unconstrained nonlinear programming model is used to solve the problem. There exist many different separation algorithms. In this paper, we adopt the separation algorithm proposed by Imamichi et al. [5].

4 Local Search for 2DIRBPP

The utilization rate of the initial solution maybe not high, local search is introduced to improve the utilization rate further. Given a solution containing a set of bins $B = \{b_1, ..., b_N\}$ sorted by decreasing utilization, i.e, $U_1 \geq U_2 \geq \cdots \geq U_N$. The proposed local search procedure considers all pairs of bins, b_i and b_j, with

$i < j$, and attempts to move pieces one by one from b_j into b_i. If all pieces from two bins fit into only one bin, the new solution is accepted. Moreover, the moving is also accepted if U_i is increased, and consequently, U_j is decreased. The new solution will have the same number of bins but a higher F. It is shown as in Algorithm 5 (see martinez-sykora et al. [12]).

Algorithm 5. Local search between b_i and b_j

1: $b_i^1 = b_i, b_j^1 = b_j$
2: Sort pieces in b_i^1 by increasing area
3: **for** each piece p_j in b_i^1 **do**
4:     $area_{out} = area_{p_j}, area_{in} = 0, unplaced = \varnothing$
5:     Remove p_j from b_i^1
6:     placeList: Sort pieces in b_j^1 by non-increasing area
7:     Add p_j to the end of placeList
8:     **for** each piece p_k in b_j^1 **do**
9:         Solve placement policy to know whether p_k can be placed into b_i^1
10:        **if** p_k can be placed **then**
11:            add p_k to b_i^1
12:            $area_{in} += area_{p_k}$
13:        **else**
14:            add p_k to $unplaced$
15:        **end if**
16:    **end for**
17:    **if** $area_{in} > area_{out}$ **then**
18:        **if** $unplaced = \varnothing$ **then**
19:            $b_j^1 = \varnothing$
20:            **return** b_i^1 and b_j^1
21:        **else**
22:            $b_j^1 = \varnothing$
23:            Solve placement policy to know whether all $unplaced$ can be placed into b_j^1
24:            **if** all $unplaced$ can be placed **then**
25:                add all $unplaced$ to b_j^1
26:                **return** b_i^1 and b_j^1
27:            **else**
28:                $b_i^1 = b_i, b_j^1 = b_j$
29:            **end if**
30:        **end if**
31:    **else**
32:        $b_i^1 = b_i, b_j^1 = b_j$
33:    **end if**
34: **end for**

---

5   Computational Experiments

Our algorithms were implemented in C++ and compiled by Visual Studio 2012, and no multi-threading was explicitly utilized. The computational environment is PC, Intel i7 3.60 GHz CPU, 8 G RAM.

5.1   Data Instances

We use two sets of benchmarks instances to test the performance of our approach: nesting instances commonly used in strip packing problems and instances of jigsaw puzzles. Both instances can be obtained from the ESICUP web site https://www.euro-online.org/websites/esicup/.

Nesting instances are derived from 23 strip packing data instances, including convex and non-convex pieces. The goal of the strip packing is to minimize the length with the fixed width, so we need to define the length and width of the bin to make it as an instance of the bin packing problem. Martinez-sykora et al. [12] defined three bin sizes for each instance. Let d_{max} be the maximum length or width across all the pieces in their initial orientation for a given instance, the three bin sizes are set as follows. The length and width of the small bin in the first type (SB) are set to $1.1d_{max}$, and they are set to $1.5d_{max}$ in the second type (MB). While in the last type (LB), the length and width of the bin are set to $2d_{max}$ (LB).

The jigsaw puzzle instances were proposed by lopez-camacho et al. [10] and consist of JP1 and JP2 instance sets. They are cut down from many pieces of the whole bin, and the utilization rate can reach 100%. JP1 instance set has 18 classes with 30 instances in each class, resulting in a total of 540 instances that the pieces are all convex polygons. JP1 instance set has 16 classes with 30 instances in each class, resulting in a total of 480 instances that the pieces have convex polygons or non-convex polygons.

5.2   Experimental results

There are two main experimental parameters: times of swapping two polygons N_{mo} and running time t. When pieces numbers n is less than 45, N_{mo} is 100; When n is greater than or equal to 45, N_{mo} is 20. The running time t is 10 minutes. For each instance, our approach is ran 10 times by setting the random seed from 1 to 10. Tables 1 and 2 are the average results of the two data sets after 10 runs, where IJS1 and IJS3 are the approach by Abeysooriya et al. [1] and LS is our approach. Better results are highlighted as bold. The suffix FR represents the free rotation of the pieces, and the $4R$ represents only 4 rotation angles $(90°, 180°, 270°, 360°)$ is allowed. The column F (see Sect. 2) and T is the averaged objective value and the time used for each instance.

We can see from Table 1 that our approach improves the solution for many of the instances for the nesting instances, and the averaged F value is better than that of Abeysooriya et al. [1] in all the subset. For the artificial jigsaw puzzle instances, there is still some gap between our result and the existing solution. This is that because the pieces of jigsaw puzzle instances are cut from the whole bin, the best utilization rate is 100%. Therefore, the utilization of the final solution significantly depends on the assignment strategy. Since we only use the First Fit Decreasing strategy to assign the piece to the bin, it is hard to reach the assignment with 100% utilization.

Table 1. Results for nesting instances

Instance	LB				MB				SB			
	IJS1-FR		LS-4R		IJS1-FR		LS-4R		IJS1-FR		LS-4R	
	F	T	F	T	F	T	F	T	F	T	F	T
albano	0.388	163	**0.413**	720	**0.532**	178	0.52	612	**0.605**	185	0.604	618
shapes2	0.451	110	**0.473**	618	0.339	124	**0.351**	277	0.3	134	**0.301**	44
trousers	0.462	754	0.44	1017	**0.574**	815	0.567	625	**0.684**	839	0.678	649
shapes0	0.294	549	**0.462**	1062	0.268	593	**0.367**	621	0.244	617	**0.33**	601
shapes1	0.318	748	**0.462**	1057	**0.369**	823	0.367	616	0.32	922	**0.33**	616
shirts	**0.652**	1118	0.642	717	**0.664**	1230	0.554	616	0.611	1279	**0.666**	602
dighe2	**0.26**	35	**0.26**	2	**0.322**	42	0.321	120	0.394	46	**0.395**	105
dighe1	**0.329**	35	**0.329**	4	**0.365**	39	0.352	633	0.427	42	**0.449**	178
fu	0.505	8	**0.514**	130	0.431	9	**0.478**	21	**0.448**	9	0.366	8
han	**0.412**	144	0.32	739	**0.402**	162	0.4	619	**0.446**	168	0.432	614
jakobs1	**0.616**	92	0.602	678	**0.579**	109	0.565	628	**0.433**	119	0.372	276
jakobs2	0.448	97	**0.463**	667	0.438	97	**0.446**	619	**0.418**	102	0.411	572
mao	**0.61**	177	**0.61**	130	**0.501**	193	0.489	632	**0.472**	191	0.451	600
poly1a	**0.368**	42	**0.368**	5	**0.349**	46	0.343	612	**0.454**	48	0.45	609
poly2a	0.386	129	**0.394**	728	0.366	151	**0.517**	681	**0.452**	151	0.45	602
poly3a	**0.416**	247	0.399	613	**0.442**	274	0.441	601	**0.451**	304	0.448	603
poly4a	**0.418**	421	**0.418**	629	0.437	442	**0.517**	602	**0.409**	437	0.408	603
poly5a	0.408	590	**0.437**	603	0.465	613	**0.473**	624	0.413	693	**0.452**	602
poly2b	**0.459**	126	0.454	618	0.402	154	**0.414**	651	**0.419**	154	0.417	606
poly3b	**0.418**	240	0.393	619	**0.462**	292	0.438	612	**0.456**	304	0.446	600
poly4b	0.406	466	**0.408**	849	**0.483**	568	0.467	602	0.428	563	**0.429**	602
poly5b	0.473	702	**0.475**	1374	0.479	835	**0.48**	898	0.435	935	**0.479**	602
swim	**0.408**	3447	0.347	1112	0.368	3689	**0.379**	1074	0.337	3689	**0.395**	621
Avg	0.431	454	**0.438**	639	0.436	499	**0.445**	591	0.437	519	**0.442**	501

Table 2. Results for jigsaw puzzle instances

(a) Results for JP1

	IJS3-4R(F value)	LS-4R(F value)
TA	**0.632**	0.616
TB	**0.88**	0.879
TC	**0.797**	0.751
TD	**0.61**	0.578
TE	**0.54**	0.5
TF	**0.548**	0.512
TG	**0.819**	0.809
TH	**0.883**	**0.883**
TI	**0.697**	0.658
TJ	**0.698**	0.669
TK	**0.768**	0.754
TL	**0.63**	0.591
TM	**0.723**	0.675
TN	**0.549**	0.5
TO	**0.937**	0.876
TP	**0.717**	0.68
TQ	0.956	**0.997**
TR	**0.786**	0.739
Avg.	**0.732**	0.704
Avg.time	75	465

(b) Results for JP2

	IJS3-4R(F value)	LS-4R(F value)
TA	**0.604**	0.593
TB	**0.804**	0.762
TC	**0.692**	0.666
TF	**0.525**	0.496
TH	**0.809**	0.783
TL	**0.588**	0.573
TM	0.627	**0.634**
TO	**0.826**	0.813
TS	**0.642**	0.57
TT	**0.959**	0.876
TU	**0.823**	0.797
TV	**0.993**	0.879
TW	**0.843**	0.675
TX	**0.649**	0.595
TY	**0.735**	0.689
TZ	**0.839**	0.811
Avg.	**0.747**	0.701
Avg.time	58	504

6   Conclusions

In this paper, we present an approach for solving the 2DIRBPP. The assignment method, swap strategy, and separation algorithm are used to obtain the initial solution. Local searches between the bins are used to improve utilization and get better solutions. Compared with the methods in existing literature, the results are competitive.

References

1. Abeysooriya, R.P., Bennell, J.A., Martinez-Sykora, A.: Jostle heuristics for the 2d-irregular shapes bin packing problems with free rotation. Int. J. Prod. Econ. **195**, 12–26 (2018)
2. Bennell, J.A., Dowsland, K.A., Dowsland, W.B.: The irregular cutting-stock problem–a new procedure for deriving the no-fit polygon. Comput. Oper. Res. **28**(3), 271–287 (2001)
3. Burke, E.K., Hellier, R.S., Kendall, G., Whitwell, G.: Complete and robust no-fit polygon generation for the irregular stock cutting problem. Eur. J. Oper. Res. **179**(1), 27–49 (2007)
4. Elkeran, A.: A new approach for sheet nesting problem using guided cuckoo search and pairwise clustering. Eur. J. Oper. Res. **231**(3), 757–769 (2013)
5. Imamichi, T., Yagiura, M., Nagamochi, H.: An iterated local search algorithm based on nonlinear programming for the irregular strip packing problem. Discrete Optim. **6**(4), 345–361 (2009)

6. Jakobs, S.: On genetic algorithms for the packing of polygons. Eur. J. Oper. Res. **88**(1), 165–181 (1996)
7. Johnson, D.S., Demers, A., Ullman, J.D., Garey, M.R., Graham, R.L.: Worst-case performance bounds for simple one-dimensional packing algorithms. SIAM J. Comput. **3**(4), 299–325 (1974)
8. Leung, S.C., Lin, Y., Zhang, D.: Extended local search algorithm based on non-linear programming for two-dimensional irregular strip packing problem. Comput. Oper. Res. **39**(3), 678–686 (2012)
9. Liu, D., Teng, H.: An improved BL-algorithm for genetic algorithm of the orthogonal packing of rectangles. Eur. J. Oper. Res. **112**(2), 413–420 (1999)
10. López-Camacho, E., Ochoa, G., Terashima-Marín, H., Burke, E.K.: An effective heuristic for the two-dimensional irregular bin packing problem. Ann. Oper. Res. **206**(1), 241–264 (2013)
11. López-Camacho, E., Terashima-Marin, H., Ross, P., Ochoa, G.: A unified hyper-heuristic framework for solving bin packing problems. Expert Syst. Appl. **41**(15), 6876–6889 (2014)
12. Martinez-Sykora, A., Alvarez-Valdés, R., Bennell, J.A., Ruiz, R., Tamarit, J.M.: Matheuristics for the irregular bin packing problem with free rotations. Eur. J. Oper. Res. **258**(2), 440–455 (2017)
13. Terashima-Marín, H., Ross, P., Farías-Zárate, C., López-Camacho, E., Valenzuela-Rendón, M.: Generalized hyper-heuristics for solving 2D regular and irregular packing problems. Ann. Oper. Res. **179**(1), 369–392 (2010)
14. Zhang, D., Chen, J., Liu, Y., Chen, H.: Discrete no-fit polygon, a simple structure for the 2-D irregular packing problem. J. Softw. **20**(6), 1511–1520 (2009)

Faster R-CNN Based Fault Detection in Industrial Images

Faisal Saeed[1], Anand Paul[1], and Seungmin Rho[2(✉)]

[1] Department of Computer Science and Engineering, Kyungpook National University,
Daegu, South Korea
bscsfaisal821@gmail.com, paul.editor@gmail.com
[2] Department of Software, Sejong University, Seoul, South Korea
smrho@sejong.edu

Abstract. Industry 4.0 requires smart environment to find defects or faults in their products. A defective product in the market can impact negatively on the overall image of the industry. Thus, there is continuous struggle for industrial environment to reduce impulsive downtime, concert deprivation and safety risks. Defect detection in industrial products using the images is very hot topic in era of current research. Machine learning provides various solution but most of the time such solutions are not suitable for environment where product is on conveyor belt and traveling from one point to another. To detect fault using industrial images, we proposed a method which is based on Faster R-CNN which is suitable for smart environment as it can the product efficiently. We simulated our environment using python language and proposed model has almost 99% accuracy. To make our proposed scheme adaptable for the industry 4.0, we also developed an android application which make it easy to interact with the model and industry can train this model according to their needs. Android application is able to take pictures of defective product and feed it to model which improve accuracy and eventually reduces time identify defective product.

Keywords: Industrial images · Defect detection · Fault identification · Convolution neural networks · Fast R-CNN · RPN

1 Introduction

In present era, the industrial sector is growing in an increasing pace. Therefore, it is very much essential for every industry to manufacture and produce the quality products without any faults such as missing of hardware parts i.e. screws, in the manufactured products i.e. computer hardware. However, the problem of product rejection rates are still existing due to faults or missing parts in the manufactured products. There are many reasons for product rejection during quality checking process in the industry. Some of the reasons for rejection are there might be system misconfiguration, human error [1] etc. Therefore, the need for an accurate fault detection system or missing parts detection system is inevitable for the industry to manufacture and deliver the

© Springer Nature Switzerland AG 2020
H. Fujita et al. (Eds.): IEA/AIE 2020, LNAI 12144, pp. 280–287, 2020.
https://doi.org/10.1007/978-3-030-55789-8_25

quality products to the end user or customers. Faster R-CNN model is used in this paper. CNN models have advantages in learning the appropriate features from the captured images by themselves but CNN read the features from whole image which increase the computational time. To avoid this issue, we have used Faster R-CNN in us proposed framework. The proposed Faster R-CNN model in this paper can learn appropriate features from the images captured using a smartphone and sent for testing using our proposed model to extract the features. The complete analysis report is delivered to the GUI interface of the mobile application to verify the faults during quality checking process so that product rejection rates can be reduced effectively which pave a way for delivering high-quality products to the customers. Thus, the productivity in the industry improves as the quality of process is improved.

2 Related Work

Object detection is becoming famous in recent years and evolve with time. Many techniques were introduced to detect object with higher efficiency. Region proposal networks (RPN) open ways for researchers to explore more and make object detection better than ever. RPN used along with Multi feature concatenation can reduce computational time and study of [2] shows that authors were able to reduce pattern computational time up to 66% by reducing parameters of zf-based network in real time. Definition of object can vary from domain to domain e.g. text can be represented as object as [3] take text as object to use RPN to detect text after which localization or classification can be applied and they were able to reduce false rejection by 15%. Such techniques are very good where written text is hard to read for computer or text from the old age scripts where humans cannot find the writing patterns, such models proven to be effective for identifying text. signal to noise ratio in x-ray images are very low and guide wire have some similarity so, [4] used this information along with RPN on 22 different images and their model average precision is 89.2% with 40fps speed. Such study shows that RPN not only increase the accuracy but also decrease the computational time which is very crucial in industry 4.0. Industry cannot slow down their production rate and also cannot send defective product in the market thus RPN is suitable for such environments. Robot-assisted surgery (RAS) is also gaining popularity and require real time feedback from real-time images. Study by [5] uses deep neural networks for region proposal and detection for RAS and achieve average precision about 91% with mean computational time 0.1 s. Deep convolution networks are also complex in term of learning process which make them hard to retune again and again. Detector proposed by [6] streamline learning process of CNN and when combined with SPP, it accelerates detection by integrating several learning step into single algorithm. This result in high accuracy also, with low computational time that is what Industry 4.0 needs in this fast pace of production. A variant of R-CNN is proposed by [7] where they generate mask with high quality segmentation and by training their model with 5fps they outperformed other R-CNN models. To detect sealed and unsealed cracks with complex road background, [8] used Fast-R-CNN (a variant of R-CNN) which detects the cracks with higher efficiency even when the background is complex to analyse. An. Faster R-CNN combined with techniques like calibration of parameters, hard negative mining, feature concatenation [9] shows good benchmark as compared to previously existing machine learning models. Refinement according to models and techniques are studied by [11] where they find that models can be modified to adopt according to the research problem and can make it faster in the object detection results.

3   Proposed Framework

Our proposed model is composed of Faster R-CNN. The basic Idea of using Faster R-CNN for fault detection in the image like the missing screw, misplace or missing labels is to test only the region of interests where these types of object exists. Faster R-CNN, is based of two modules, the first module is called RPN a full deep convolutional neural network to produced RoIs, and the second module that faster R-CNN is using is the Fast R-CNN detector which uses RoIs as an input to classify. The detail of Faster R-CNN is in the coming section. Figure 2 is the description of our proposed model. We used three industrial cameras to capture the images. As shown in the figure the products are on conveyor belt coming in front of cameras where in first turn three pictures captured and then after rotating the object remaining two sides are captured. Then these images send to local server where our faster R-CNN model works to detect fault in the captured images. In addition, we also made GUI mobile application shown in Fig. 1. In this GUI based android app, we can capture images using mobile camera and after capturing the image the can be sent to the server where we already have working Faster R-CNN model to detect the fault in the image.

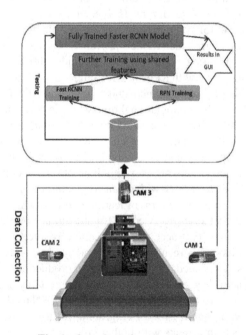

Fig. 1. Overview of proposed model

3.1 RPN

To make the segmentations of the image RPN uses already trained CNN model [10] than other methods which are category-independent e.g. edge boxes [3] and selective search [4]. The input to RPN is an image of arbitrary size it gives number of region of proposals in the rectangular form as an output. As convolutional neural networks are very flexible, because to this reason the training of RPN can be done using specific class as in our case classes will be screw, labels, missing screws and labels.

3.2 Fast R-CNN

Fast R-CNN is the fast region-based CNN model commonly used for the object detection. The Fast R-CNN takes an image and its corresponding region proproposals as an input. After that it uses to classifiers to produce box bonding and class score layers. The input image is first going to VGG-16 network where it processes and generates the feature maps.

3.3 Faster R-CNN

Faster R-CNN uses RPN for region proposal and as a detector network, it uses Fast R-CNN. Refer to this proposed idea, during the training phase both networks Fast-R-CNN and RPN act as one combine network. During the training of Fast R-CNN detector the region proposals are pass forwarded and treated as pre-computed and fixed in every SGD repetition. For both R-CNN and RPN the losses are cumulated during the backward signal propagation in shared layers, while in normal the backward signals are propagated in usual manner. The given solution is highly recommended for simple implementation and easy to use. The main flaws in this theme is network coordinate boxes' responses in the proposal which limits the derivative too. As for as our given solution we solve this problem and empirically made the results close to ideal product. Up to 25–50% of training time saving are reduced as compare to conventional training. The said solution is comprised in our implementation.

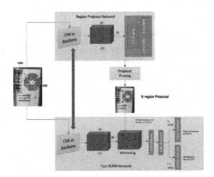

Fig. 2. Faster RCNN architecture

4 Proposed Framework

We implemented our Faster R-CNN model in python and the training and testing process of these models is made on a CPU i.e. Intel R CoreTM i5-3570 CPU @ 3.40 GHz 3.80 GHz having 64-bit windows operating system, 64 GB RAM and GTX 970 graphics card. To train and test our Faster R-CNN model we used our own generated dataset. Total 40520 images are used in which 19251 images are used for training 1543 images for validation and remaining are used for testing. Initially the learning rate was 0.01 but we used step decay learning process so, every after 1000 iterations it decreases by a factor 0.5.

4.1 Results

To train and test our proposed model we used the dataset that we generate. Our dataset includes screw and label images as well as missing screw and missing, misplaced or wrong placed images. During the training and testing of our model, we calculate the training and testing accuracies. Figure 3 is showing the combined training and testing accuracy of our models. We compute it against two different data sets. The results show that our model have good accuracy against these datasets. Our model accuracy is almost 99%. We also computed training and testing loss of our model. Figure 4 is the illustration of training and testing loss of model. We calculated results for precision p, recall r, accuracy a, are shown in Table 1. We also compared our model ROIs formation with other models. Figure 5 is showing the comparison of our model. These results show that our trained model has near 99% _re detection accuracy. We tested our complete proposed model on one of industrial product step by step. First, we capture the original image with mobile camera in mobile GUI. We send this image to the server where our Faster-R-CNN model is working. Where this picture is tested and sent back the mobile GUI with tested results. Figure 6 shows mobile GUI and its tested results.

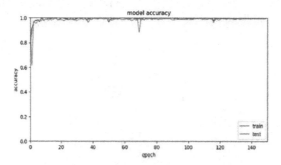

Fig. 3. Model accuracy

Table 1. Table of precision, recall and accuracy

	Precision	Recall	Accuracy
Screw	88.9	94.35	98.92
Label	92.3	95.35	99.40

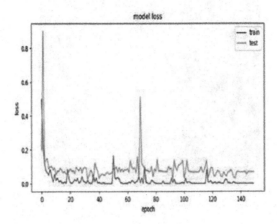

Fig. 4. Model accuracy

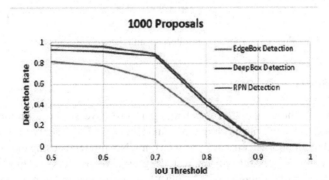

Fig. 5. Comparison analysis of our model

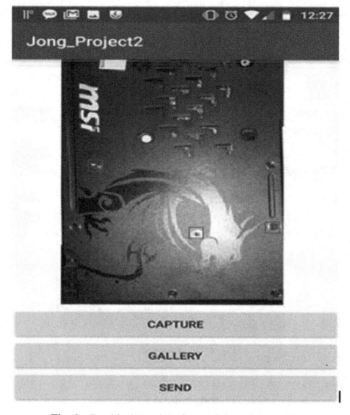

Fig. 6. Graphical user interface and detection results.

5   Conclusion

Fault detection in the images is very challenging task. We solved a problem that is, industrial products some time missing screw and labels etc. For that, we proposed Faster R-CNN based model for detection. We trained our CNN model by using our own dataset that we also created. We test the only interested regions. We saw our CNN model has good accuracy against testing dataset. We test our proposed model in the real environment. The accuracy of our model almost 99%. In future we are going to extend our idea.

Acknowledgement. This work was supported by the National Research Foundation of Korea (NRF) grant funded by the Korea government (MSIT) (NRF-2019R1F1A1060668).

References

1. Cognex. https://www.cognex.com/en-cz/applications/customer-stories/automotive/vision-system-pays-for-itself-in-one-week-by-preventing-two
2. Shih, K.-H., Chiu, C.-T., Lin, J.-A., Bu, Y.-Y.: Real-time object detection with reduced region proposal network via multi-feature concatenation. IEEE Trans. Neural Networks Learn. Syst. **31**, 2164–2173 (2019)
3. Hou, J., Shi, Y., Ostendorf, M., Hwang, M.-Y., Xie, L.: Real-time object detection with reduced region proposal network via multi-feature concatenation region proposal network based small-footprint keyword spotting. IEEE Signal Process. Lett. **31**, 2164–2173 (2019)
4. Wang, L., Xie, X.L., Bian, G.B., Hou, Z.G., Cheng, X.R., Prasong, P.: Guide-wire detection using region proposal network for X-ray image-guided navigation. In: Proceedings of the International Joint Conference on Neural Networks (2017)
5. Sarikaya, D., Corso, J.J., Guru, K.A.: Detection and localization of robotic tools in robot-assisted surgery videos using deep neural networks for region proposal and detection. IEEE Trans. Med. Imaging **PP**(99), 1(2017)
6. Lenc, K., Vedaldi, A.: R-CNN minus R. In: Proceedings of the British Machine Vision Conference 2015, British Machine Vision Association (2015)
7. He, K., Gkioxari, G., Dollar, P., Girshick, R.: Mask R-CNN. In: Proceedings of the IEEE International Conference on Computer Vision (2017)
8. Huyan, J., Li, W., Tighe, S., Zhai, J., Xu, Z., Chen, Y.: Detection of sealed and unsealed cracks with complex backgrounds using deep convolutional neural network. Autom. Constr. (2019)
9. Sun, X., Wu, P., Hoi, S.C.H.: Convolutional neural net- works (CNN), Face detection, Faster RCNN, Feature concatenation, Hard negative mining, Multi-scale training. Neurocomputing (2018)
10. Long, J., Shelhamer, E., Darrell T.: Fully convolutional networks for semantic segmentation. In: Proceedings of the IEEE Conference on Computer Vision and Pattern Recognition, Boston (2015)
11. Roh, M.C., Lee, J.Y.: Re_ning faster-RCNN for accurate object detection. In: Proceedings of the 15th IAPR International Conference on Machine Vision Applications, MVA 2017 (2017)

Estimation of Cable Lines Insulating Materials Resource Using Multistage Neural Network Forecasting Method

Nikolay K. Poluyanovich[✉], Mikhail Yu Medvedev, Marina N. Dubyago, Nikolay V. Azarov, and Alexander V. Ogrenichev

Southern Federal University, Taganrog 347900, Russian Federation
nik1-58@mail.ru

Abstract. The use of artificial neural networks (ANN) for predicting the temperature regime of a power cable line (PCL) core is considered. The relevance of the task of creating neural networks for assessing the throughput, calculating and forecasting the PCL core temperature in real time based on the data from a temperature monitoring system, taking into account changes in the current load of the line and external conditions of heat removal, is substantiated. When analyzing the data, it was determined that the maximum deviation of the neural network (NN) data from the data of the training sample was less than 3% and 5.6% for experimental data, which is an acceptable result. It was established that ANN can be used to make a forecast of the cable core temperature regime with a given accuracy of the core temperature. The comparison of the predicted values with the actual ones allows us to talk about the adequacy of the selected network model and its applicability in practice for reliable operation of the cable power supply system of consumers. The model allows evaluating the insulation current state and predicting the PCL residual life. The results analysis showed that the more aged PCL insulation material, the greater the temperature difference between the original and aged sample. This is due to the loss of electrical insulation properties of the material due to the accumulation of the destroyed structure fragments, containing, in an increasing amount, inclusions of pure carbon and other conductive inclusions.

Keywords: Artificial intelligence · Neural networks · Forecasting · Insulating materials · Thermofluctuation processes · Reliability of energy supply systems

1 Introduction and Statement of the Problem

During operation, PCL insulation is exposed to thermal, electrical, chemical, mechanical and other types of influences, resulting in a change in its electrical properties and characteristics of insulating materials. In the insulation defect places (gas and solid inclusions) the partial discharges (PD) are formed, the development of which leads to an insulation breakdown.

© Springer Nature Switzerland AG 2020
H. Fujita et al. (Eds.): IEA/AIE 2020, LNAI 12144, pp. 288–296, 2020.
https://doi.org/10.1007/978-3-030-55789-8_26

The main factor that has an adverse effect on the electrical insulation state during operation of the power cable under load is heating.

If the core temperature approaches the maximum permissible, then the process of intense insulation thermal deterioration begins, its thermal aging, which leads to thermal breakdown of the insulation. As the heating temperature of the core increases, the specific thermal resistance of the insulating material decreases.

With an increase in the insulation temperature, chemical reactions in the insulation materials accelerated, they enhanced by internal inhomogeneities, environment exposure, absorbed moisture. This reduces the insulation electrical properties and leads to the insulation gap breakdown or mechanical destruction of the insulation.

One of the problem solutions is to monitor the cable temperature during operation [1]. A promising area is the creation of the CL fault diagnosis infrastructure based on the use of artificial neural networks. A method for recognizing partial discharge models based on an extended neural network for high-voltage power devices was proposed in [2]. Signal analysis and the formation of partial discharge identification signs in high-voltage equipment were proposed in [3, 4]. The application of predictive modular neural networks for short-term forecasting was considered at [5].

The intellectualization of the monitoring system for cable systems is based on modern technology for distributed temperature measurement of optical fiber (DistributedFiber-OpticSensing). This technology allows temperature measurements along the entire length of the high-voltage cable [6] in real time [7]. The objectives of the study are:
- development of an intelligent PCL core temperature forecasting system for planning power grid operating modes in order to increase the reliability and energy efficiency of their interaction with the integrated power system.

1.1 Development of a PCL Thermal Substitution Scheme

The resource of insulating materials depends on such parameters as temperature, current, humidity [8], therefore, studies of the temperature field in the PCL section [9] are necessary, as the main factor, Fig. 1. The low thermal conductivity of the cable insulation, Fig. 1, leads to a high temperature gradient [10], therefore, the temperature of the most heated section in the cable section (near the core) differs significantly from the measured temperature, and in transient conditions this difference can increase several times. Therefore, the urgent task of the creating neural networks for assessing the throughput, calculating and forecasting the PCL core temperature in real time based on data from a temperature monitoring system, taking into account changes in the current load of the line and external conditions of heat removal.

Based on the PCL core temperature prediction, the initial and optimal modes of electric power systems are calculated, their reliability, efficiency, quality of the supplied electricity, etc. are evaluated.

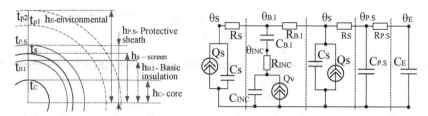

Fig. 1. Cable cross section and its equivalent thermal circuit

2 ANN Analysis, Type Selection and Development

To determine the neural network input variables when solving problems of forecasting the PCL core temperature, a model was used that describes the time changes of the actual temperature values, which in general is represented by a nonlinear function:

$$Y_t = f(X_{t-n}, T_{t-n}, N_t) + \varepsilon_t \tag{1}$$

where X_t - the actual core temperature value at time t; t - the current time; W_{t-n} - previous observations of core temperature, T_{t-n} - previous observations of ambient temperature; n - the data retrospective index; I_t - the value of the effective current ($I_{min} < I < I_{max}$); ε_t - a random component representing unobservable factors affecting core temperatures.

In the framework of the work, four types of neural networks were analyzed (Table 1). To determine the effectiveness of the studied neural networks, the standard error was used, averaged over the number of output variables of the neural network and calculated based on the predicted and real values of the test sample by the formula:

$$E = \frac{1}{N * K} \sum_{i=1}^{K} \sum_{j=1}^{N} \left(y_{ij}^{real} - y_{ij}^{predict} \right)^2 \tag{2}$$

where i - the value of the i-th output variable of the neural network for the j-th training or test example; - the predicted value of the i-th output variable of the neural network for the j-th training or test example; N - the number of examples in a training or test sample; K - the number of output variables of the neural network. The solution of the problem, the creation of ANN model, is determined by the choice of its optimal configuration, taking into account the results of the calculation of the mean square error. The Levenberg-Markard training algorithm and cascade network with direct signal propagation and back propagation of error with 10 neurons were selected. The results of comparing the types of neural network are presented in Table 1.

Table 1. Comparison of neural networks types.

№ п.п	Neural networks type	Average forecast error	
		ε, °C	ε, %
1	Multilayered perceptron (MLP)	0.9	2.3%
2	Layer reccurent	1.2	2,5%
3	Cascade forward backpropagation	1.63	4.2%
4	Elman backpropagation	2.14	5.3%
5	Backpropagation with delay	2.75	7.4%

Thus, of the considered neural networks, the multilayered perceptron has the highest accuracy (Table 1). To solve the problem of PCL resource forecasting, a network with direct data distribution and feedback error Feed-forwardbackprop was chosen, because Networks of this type, together with an activation function in the form of a hyperbolic tangent $(f(x) = (ex - e - x)/(ex + e - x))$, are to some extent a universal structure for approximation and prediction problems. The network was trained every time from scratch, the weights obtained during the initial training of the network are not used for retraining.

3 Implementation of Neural Network Models for PCL Core Temperature Prediction

Thermofluctuation processes in the power cable insulating material (without damage) were investigated for APvPu g-1x240/25-10 with a real diagram of the current value of the cable core current and the developed forecast model using artificial neural networks (ANN) on deep retrospective temperature data θp.c, θm.i (Fig. 1) of the studied cable for 2015–2019. The value of the transmitted current by PCL was varied discretely, and for each value of the core current, 400 points of the casing temperature θp.c, and the main insulation θm.i were taken. The article presents prediction cable core temperature results (θc.c.), using ANN, for one studied sample of the cable. The average error in prediction temperature of the cable core θc.c. for various values of the core current ($440 \leq$ Ic.c. ≤ 660, A) does not exceed 5 °C, which indicates the possibility of using the artificial neural network method for forecasting the temperature of the cable core by temperature on the surface θp.c. Graphs of the actual temperature of the studied samples and graphs based on the data of the training sample of the neural network were constructed (Fig. 2).

Thus, a neural network was developed to determine the temperature regime of a cable core of a power cable. When analyzing the data, it was determined that the maximum deviation of the data received by the neural network from the data of the training sample was less than 3% and for experimental data −5.6%, which is an acceptable result.

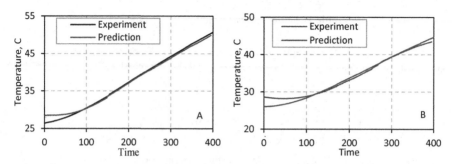

Fig. 2. Graph of the actual temperature, samples No. 1–2, and forecast, built on the basis of the data of the training sample and the neural network.

Further, the studied cable samples were artificially aged in a heating cabinet at a constant temperature, but with different exposure time intervals. Sample No. 1 was in the heating cabinet for 8 h, sample No. 2 was aged for 16 h, and sample No. 3 was 24 h. Table 2 shows the data on both unaged cable samples (training set) and artificially aged (control set).

Table 2. Results of an experimental study of a control sampling of aged samples.

Initial parameters			Temperature calculation, t °C			Predicted error	
№ sample		Time $t_{CT,}$, hour	Experiment$T_{av,}$ °C	Forecast$T_{av,}$ °C	$\Delta t_{exp}/\Delta t_{for}t$ °C	t °C	%
Training	1	–	33,97	34,185	–	−0,354	1,021
	2	–	40,34	40,348	–	−0,613	1,526
	3	–	38,28	39,602	–	−1,322	3,339
Control	1	8	35,65	38,623	1,68/4,36	−2,967	3,36
	2	16	42,59	42,375	2,25/3,42	1,542	4,517
	3	24	41,43	41,413	3,15/3,83	1,952	4,852

An analysis of the results showed that the longer the aging time, the greater the temperature difference between the original and aged samples. We believe that this is due to the loss of material electrical insulating properties due to the accumulation of destroyed structure fragments containing, in an increasing amount, pure carbon inclusions and other conductive inclusions. The polyethylene incomplete oxidation to produce carbon:

$$[-CH2 - CH2]\,n + n\,O2 \rightarrow 2n\,C + 2nH2O$$

4   Multistage Forecasting Method of Thermofluctuation Processes

It is proposed to solve the problem of forecasting thermofluctuation processes using non-traditional methods, namely, using models based on expert systems and artificial neural networks [11–16]. They do not require building a model of the object, their performance is not lost with incomplete input information, they are resistant to interference, and have a high speed.

Short-term forecasting of a parameter is performed by the step size of the sliding window in the training sample [17]. If a long-term forecast is necessary, a sequential iterative forecast is applied for the step value that is a multiple of the step of the training sample. Moreover, the quality of a long-term prediction usually worsens with an increase in the number of iterations of the forecast (for example, with an increase in the time period of the forecast) [18–20, 21, 22].

To solve the problem of worsening the long-term prediction, it was proposed to use, as an assessment of the quality of training, not the deviation in the magnitude of the forecast for one step of the sliding window on the training sample, but the total deviation of the forecast for all values of the training sample, and with obtaining new values based on the forecast obtained on the previous one step.

The proposed neural network algorithm for predicting the characteristics of electrical insulation was tested on a control sample of experimental data on which training of an artificial neural network was not carried out. The forecast results (Table 3) showed the effectiveness of the selected model.

To build a multi-stage forecast of a multi-dimensional series of input:

$$X_n = \theta_{p.c.}(n), I_c, N_k \tag{3}$$

and output:

$$Y_n = \theta_{p.c.}(n+1), \theta_{p.c}(n+2) \tag{4}$$

variables, where θp.c. is the temperature of the protective casing, I_c is the operating current of the core, N_k is the number of the sample, θp.c.(n + 1) is the predicted temperature of the casing in the next step, are formed so that the width of the sliding window was equal to one step, and the forecast was two-stage. Having formed the data in this way, we obtain a transformed multidimensional time series. The algorithm iteratively makes a prediction based on the data of the training sample X(n) and is adjusted by updating the weight coefficients. Training stops after an acceptable level of efficiency is achieved (minimum forecast error ε). The error values for a multi-stage temperature forecast of the conductors of various cable samples are shown in Table 3, Fig. 3.

Table 3. Results of multi-stage cable temperature forecasting.

Cable sample number	The average forecast error	
	ε, °C,	ε, %
sample № 5	1.6	2.3%
sample № 6	1.8	2.5%

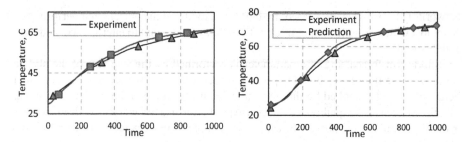

Fig. 3. Graph of predicted and experimental temperatures of samples No. 5–6.

4.1 Experimental Investigations of the Developed NN

The calculation of the forecast and the assessment of the long-term permissible values of the core cable heating in the dynamically changing load mode, was carried out in a wide range of cable core currents (400, A $\leq$ Ic $\leq$ 650, A). In Fig. 4, the synthesized two-network NN architecture with a sequentially distributed structure is presented.

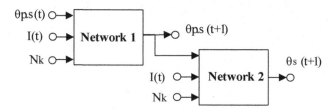

Fig. 4. Neural network architecture for predicting thermofluctuation processes.

Figure 5 shows graphical dependences of the cable core temperature on different values of the core current for 10 kV cables with insulation from SPE.

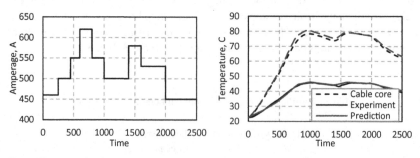

Fig. 5. a) diagram of the cable test current; b) a graph of the predicted and experimental cable temperatures.

5 Conclusion

Error estimation of the forecasted and experimental temperatures of cable No. 5 showed that the average forecast error ε for the training sample was 2.6 °C and for experimental data 5.6°. The introduction of systems using temperature monitoring will make it possible to practically monitor online the status of each cable, including its actual throughput, and this means that you can pass significantly more current without exceeding the permissible cable temperature. A comprehensive solution for temperature monitoring will help quickly monitor the current state of cable lines and optimize using of their real bandwidth. The main application area of the developed neural network for determining the temperature condition of a cable core is to assess the current state of insulation and predict its residual life.

This work was supported by a grant: Development of theoretical foundations and methods for constructing intelligent multi-connected systems for controlling the processes of energy production, transportation, distribution and consumption, No. VnGr-07/2017-15.

References

1. Boltze, M., Markalous, S.M., Bolliger, A.: On-line partial discharge monitoring and diagnosis at power cables. In: Doble Engineering Company – 76th Annual International Doble Client Conference, pp. 1–21 (2009)
2. Chen, H.-C., Gu, F.-C., Wang, M.-H.: A novel extension neural network based partial discharge pattern recognition method for high-voltage power apparatus. Expert Syst. Appl. **39**(3), 3423–3431 (2012)
3. Perpiñán, O.: Signal analysis and feature generation for pattern identification of partial discharges in high-voltage equipment. O. Perpiñán и др. Electr. Power Syst. Res. **95**, 56–65 (2013)
4. Dubyago, M., Poluyanovich, N.: Partial discharge signal selection method for interference diagnostics of insulating materials. In: Conference Proceedings - 2019 Radiation and Scattering of Electromagnetic Waves, RSEMW 2019 8792693, pp. 124–127. IEEE Xplore Digital Library (2019) https://ieeexplore.ieee.org/xpl/conhome/8784099/proceeding
5. Gross, G., Galiana, F.D.: Short-term load forecasting. Proc. IEEE **75**(12), 1558–1571 (1987)
6. Ukil, A., Braendle, H., Krippner, P.: Distributed temperature sensing: review of technology and applications. Sens. J. **12**(5), 885–892 (2012)
7. Deng, J., Xiao, H., Huo, W., et al.: Optical fiber sensor-based detection of partial discharges in power transformers. Optics Laser Technol. **33**(5), 305–311 (2001)
8. Dubyago, M.N., Poluyanovich, N.K.: Thermal processes of the insulating materials in problems of nondestructive diagnostics of the main and power supply systems. EAI Endorsed Trans. Energy Web **5**(16), e3 (2018)
9. Terracciano, M., Purushothaman, S.: Thermal analysis of cables in unfilled troughs: investigation of the IEC standard and a methodical approach for cable rating. IEEE Trans. Power Deliv. **27**(3), 1423–1431 (2012)
10. Dubyago, M.N., Poluyanovich, N.K, Burkov, D.V., Tibeyko, I.A.: Dependence of dielectric loss tangent on the parameters of multilayer insulation materials cable, pp. 1003–1007. Taylor & Francis Group, London (2015)
11. Gross, G., Galiana, F.D.: Short term load forecasting. Proc. IEEE **75**(12), 1558–1573 (1987)

12. Chen, S.T., David, C.Y., Moghaddamjo, A.R.: Weather sensitive short-term load forecasting using non fully connected artificial neural network. IEEE Trans. Power Syst. **7**(3), 1098–1105 (1992)
13. Dash, P.K., Ramakrishna, G., Liew, A.C., Rahman, S.: Fuzzy neural networks for time-series forecasting of electric load. IEE Proc. Gener. Transm. Distrib. **142**(5), 535–544 (1995)
14. Hsy, Y., Ho, K.: Fuzzy expert systems: an application to short term load forecasting. IEE Proc. C. **139**(6), 471–477 (1992)
15. Lee, K.Y., Park, J.H.: Short-term load forecasting using an artificial neural network. IEEE Trans. Power Syst. **7**(1), 124–130 (1992)
16. Meldorf, M., Kilter, J., Pajo, R.: Comprehensive modelling of load. In: CIGRE Regional Meeting, 18–20 June 2007, Tallinn, Estonia, pp. 145–150 (2007)
17. Gross, G., Galiana, F.D.: Short-term load forecasting. Proc. IEEE **75**(12), 1558–1571 (1987)
18. Chen, S.T., David, C.Y., Moghaddamjo, A.R.: Weather sensitive short-term load forecasting using non fully connected artificial neural network. IEEE Trans. Power Syst. **7**(3), 1098–1105 (1992)
19. Dash, P.K., Ramakrishna, G., Liew, A.C., Rahman, S.: Fuzzy neural networks for time-series forecasting of electric load. IEE Proc. Gener. Transm. Distrib. **142**(5), 535–544 (1995)
20. Meldorf, M., Kilter, J., Pajo, R.: Comprehensive modelling of load. In: CIGRE Regional Meeting, 18–20 June 2007, Tallinn, Estonia, pp. 145–150 (2007)

Networking Applications

User Grouping and Power Allocation in NOMA Systems: A Reinforcement Learning-Based Solution

Rebekka Olsson Omslandseter[1(✉)], Lei Jiao[1(✉)], Yuanwei Liu[2(✉)], and B. John Oommen[1,3(✉)]

[1] University of Agder, Grimstad, Norway
{rebekka.o.omslandseter,lei.jiao}@uia.no
[2] Queen Mary University of London, London, UK
yuanwei.liu@qmul.ac.uk
[3] Carleton University, Ottawa, Canada
oommen@scs.carleton.ca

Abstract. In this paper, we present a pioneering solution to the problem of user grouping and power allocation in Non-Orthogonal Multiple Access (NOMA) systems. There are two fundamentally salient and difficult issues associated with NOMA systems. The first involves the task of grouping users together into the pre-specified time slots. The subsequent second phase augments this with the solution of determining how much power should be allocated to the respective users. We resolve this with the first reported Reinforcement Learning (RL)-based solution, which attempts to solve the partitioning phase of this issue. In particular, we invoke the Object Migration Automata (OMA) and one of its variants to resolve the user grouping problem for NOMA systems in stochastic environments. Thereafter, we use the consequent groupings to infer the power allocation based on a greedy heuristic. Our simulation results confirm that our solution is able to resolve the issue accurately, and in a very time-efficient manner.

Keywords: Learning Automata · Non-Orthogonal Multiple Access · Object Migration Automata · Object Partitioning

1 Introduction

The Non-Orthogonal Multiple Access (NOMA) paradigm has been proposed and promoted as a promising technique to meet the future requirements of wireless capacity [6]. With NOMA, the diversity of the users' channels and power is exploited through Successive Interference Cancellation (SIC) techniques in receivers [1]. This technology introduces questions concerning users who are ideally supposed to be grouped together, so as to obtain the maximum capacity gain. Additionally, the power level of the signal intended for each user is a crucial component for the successful SIC in NOMA operations. Consequently, it is an accepted fact that the performance of NOMA is highly dependent on both the grouping of the users and the power allocation.

© Springer Nature Switzerland AG 2020
H. Fujita et al. (Eds.): IEA/AIE 2020, LNAI 12144, pp. 299–311, 2020.
https://doi.org/10.1007/978-3-030-55789-8_27

The user grouping and power allocation problems in NOMA systems are, in general, intricate. First of all, the user grouping problem, in and of itself, introduces a combinatorially difficult task, and is infeasible as the number of users increases. This is further complicated by the channel conditions, and the random nature of the users' behaviors in communication scenarios. For this reason, the foundation for grouping, and consequently, for power allocation, can change rapidly. It is, therefore, necessary for a modern communication system to accommodate and adapt to such changes.

The user grouping in NOMA systems, is akin to a classic problem, i.e., to the Object Partitioning Problem (OPP). The OPP concerns grouping "objects" into sub-collections, with the goal of optimizing a related objective function, so as to obtain an optimal grouping [3]. Our goal is utilize Machine Learning (ML) techniques to solve this, and in particular, the ever-increasing domain of Reinforcement Learning (RL), and its subdomain, of Learning Automata (LA). When it concerns RL-based solutions for the OPP, the literature reports many recent studies to solve Equi-Partitioning Problems (EPPs). EPPs are a sub-class of the OPP, where all the groups are constrained to be of equal size. Among these ML solutions, the Enhanced Object Migration Automata (EOMA) performs well for solving different variants of EPPs [2]. They can effectively handle the stochastic behavior of the users, and are thus powerful in highly dynamic environments, similar to those encountered in the user grouping phase in NOMA systems.

Moving now to the second phase, the task of allocating power to the different users of a group in NOMA systems, further complicates the NOMA operation. However, a crucial observation is that the problem resembles a similar well-known problem in combinatorial optimization, i.e., the *Knapsack Problem* (KP). KPs, and their variants, have been studied for decades [4], and numerous solutions to such problems have been proposed. Among the numerous solutions, it is well known that many fundamental issues can be resolved by invoking a greedy solution to the KP. This is because a greedy solution can be exquisite to a highly complex problem, and can quickly utilize a relation among the items, to yield a near-optimal allocation of the resources based on this relation. The power allocation problem in NOMA systems can be modeled as a variation of a KP, and this can yield a near-optimal solution based on such a greedy heuristic.

In this paper, we concentrate on the problem's stochastic nature and propose an adaptive RL-based solution. More specifically, by invoking a technique within the OMA paradigm, we see that partitioning problems can be solved even in highly stochastic environments. They thus constitute valuable methods for handling the behavior of components in a NOMA system. In particular, we shall show that such methods are compelling in resolving the task of grouping the users. Indeed, even though the number of possible groupings can be exponentially large, the OMA-based scheme yields a remarkably accurate result within *a few hundred iterations*. This constitutes the first phase of our solution. It is pertinent to mention that the solution is unique, and *that we are not aware of any analogous RL-based solution for this phase of NOMA systems.*

The second phase groups users with different channel behaviors, and allocates power to the respective users. Here, we observe that the power allocation problem can be mapped onto a variation of a KP. Although the types of reported KPs are numerous, our specific problem is more analogous to a *linear* KP. By observing this, we are able to resolve the power allocation by solving a linear (in the number of users) number of algebraic equations, all of which are also *algebraically linear*. This two-step solution constitutes a straightforward, but comprehensive strategy. Neither of them, individually or together, has been considered in the prior literature.

The paper is organized as follows. In Sect. 2, we depict the configuration of the adopted system. Then, in Sect. 3, we formulate and analyze the optimization problem. Section 4 details the proposed solution for the optimization problem. We briefly present numerical results in Sect. 5, and conclude the paper in Sect. 6.

2 System Description

Consider a simplified single-carrier down-link cellular system that consists of one base station (BS) and K users that are to be divided into N groups for NOMA operation. NOMA is applied to each group, but different groups are assigned to orthogonal resources. For example, one BS assigns a single frequency band to the K users. The users are to be grouped in N groups, each of which occupies a time slot. User k is denoted by U_k where $k \in \mathcal{K} = \{1, 2, \ldots, K\}$. Similarly, the set of groups are denoted by $\mathcal{G} = \{g_n\}$, $n \in \mathcal{N} = \{1, 2, \ldots, N\}$, where g_n is the set of users inside the n-th group. The groups are mutually exclusive and collectively exhaustive, and thus, $g_n \cap g_o = \emptyset$ with $n \neq o$. When a User U_k belongs to Group n, we use the notation $U_{n,k}$ to refer to this user and its group. We adopt the simplified notation U_k to refer to a user when the user's group is trivial or undetermined. Thus, if we have 4 users in the system, User 1 and User 3 could belong to Group 1, and User 2 and User 4 belong to group 2. In this case, when we want to refer to User 1 without its group, we use U_1. Likewise, when we want to mention User 4 belonging to Group 2, we apply $U_{2,4}$. For mobility, the users are expected to move within a defined area. The user behavior in a university or an office building are examples of where the user behavior coincides with our mobility model.

2.1 Channel Model

The channel model coefficient for U_k is denoted by $h_k(t)$ and refers to the channel fading between the BS and U_k along time. The channel coefficient is generated based on the well-recognized mobile channel model, which statistically follows a Rayleigh distribution [8]. The parameters of the channel configuration will be detailed in the section describing the numerical results. Note that the LA solution to be proposed can handle a non-stationary stochastic process, and the solution proposed in this work is distribution-independent. Therefore, the current Rayleigh distribution can be replaced by any other channel model, based on the application scenario and environment.

2.2 Signal Model

Based on the NOMA concept, the BS sends different messages to the users of a group in a single time slot via the same frequency band. Consequently, the received signal y_k at time t for $U_{n,k}$ can be expressed as

$$y_k(t) = \sqrt{p_{n,k}} h_k(t) s_k + \sum_{e=1}^{|g_n|-1} \sqrt{p_{n,e}} h_k(t) s_e + n_k, \tag{1}$$

where e is the index of the users in the set $g_n \setminus U_{n,k}$, which is the complementary set of $U_{n,k}$ in g_n. $|g_n|$ returns the number of users in g_n. The received signal $y_k(t)$ has three parts, including the signal intended for $U_{n,k}$, the signal from all users other than $U_{n,k}$ in the same group, and the additive white Gaussian noise (AWGN) $n_k \sim C\mathcal{N}(0, \sigma_k^2)$ [10]. The transmitted signal intended for $U_{n,k}$ and $U_{n,e}$ is given by s_k and $s_e \sim C\mathcal{N}(0, 1)$ respectively. $p_{n,k}$ is the allocated power for $U_{n,k}$, and the total power budget for group g_n is given by P_n.

The BS' signals are decoded at the users through SIC by means of channel coefficients in an ascending order [9]. As a result, through SIC, a user with a good channel quality can remove the interference from the users of poor channel quality, while users of poor channel quality decode their signals without applying SIC. Hence, for the User $U_{n,k}$, successful SIC is applied when $|h_{n,w}(t)|^2 \leq |h_{n,k}(t)|^2$ fulfills, where w is the index of the users that have lower channel coefficients than User k in the user Group g_n.

3   Problem Formulation

In this section, we formulate the problem to be solved. The problem is divided into two sub-problems. Specifically, in the first problem, we cluster the users into categories based on the time average of the channel coefficients. In the second step, we group the users based on the learned categories and solve the resultant power allocation problem.

3.1   Problem Formulation for the Clustering Phase

To initiate discussions, we emphasize that the channel coefficients of the users in a group need to be as different as possible so as to achieve successful NOMA operation. To group the users with different coefficients, we thus first cluster the users with *similar* coefficients, and then select one user from each cluster to formulate the groups. The first problem, the clustering problem, is formulated in this subsection. The problem for user grouping, together with power allocation, is formulated later.

The criterion that we have used for clustering the users is the *time average* of the channel coefficients, $\overline{h_k(t)}$. The reason motivating this is because the user grouping is computationally relatively costly, and the fact that the environment may change rapidly, i.e., $h(t)$ might change after channel sounding. If we cluster the users according to the time average, we can reduce the computational cost, and at the same time, capture the advantages of employing NOMA statistically.

We consider clustering users to clusters of the same size, where the number of the clusters is $L_c = K/N$, and where L_c and N are integers[1]. Let q_c be the set of users in Cluster c, where $c \in [1, 2, \ldots, L_c]$ is the index of the cluster. Clearly, for the clustering problem, the difference of coefficients in each cluster needs to be minimized, and the problem can be formulated as

---

[1] In reality, if L_c is not an integer, we can add dummy users to the system so as to satisfy this constraint. *Dummy users* are virtual users that are not part of the real network scenario, but are needed for constituting equal-sized partitions in the clustering phase.

$$\min_{\{\varphi_{c,k}\}} \sum_{c=1}^{L_c} \sum_{k=1}^{K} \varphi_{c,k} |\overline{h_k(t)} - E_c|, \tag{2a}$$

$$\text{s.t.} \sum_{c=1}^{L_c} \sum_{k=1}^{K} \varphi_{c,k} = K, \ c \in C, k \in K, \tag{2b}$$

$$\sum_{k=1}^{K} \varphi_{c,k} = N, \ \forall c, \tag{2c}$$

where $\varphi_{c,k}$ is an indicator function showing the relationship of users and clusters, as $\varphi_{c,k} = 1$ when U_k belongs to Cluster c, and 0 otherwise. Additionally, the mean value of the channel fading in each cluster is denoted by the parameters E_c and δ, which are given by $E_c = \frac{1}{\delta} \sum_{k=1}^{K} \overline{h_k(t)} \varphi_{c,k}$, and $\delta = \sum_{k=1}^{K} \varphi_{c,k}$ respectively. To explain the above equation, we mention that Eq. (2a) states the objective function. Specifically, for all the clusters, we want to minimize the difference of channel coefficients between the users within each cluster. Equations (2b) and (2c), state a description of the variable $\varphi_{c,k}$ for User k in c. Hence, the sum of the variables $\varphi_{c,k}$ needs to be equal to the number of users, meaning that all the users need to be a part of a single cluster, and in each cluster, there needs to be an equal number of users.

The result of the clustering problem, i.e., the $\{\varphi_{c,k}\}$ that minimizes the objective function, formulates L_c sets of users, each of which has exactly N users.

3.2 Problem Formulation for the Power Allocation

From the output of the clustering, we know which users that are similar. Thus, when we take one user from each cluster and construct N groups, the size of each group is L_c. Without loss of generality, we can assume that the average channel coefficients are sorted in ascending order, i.e., $\overline{h_1(t)} \leq \overline{h_2(t)} \leq \ldots \leq \overline{h_K(t)}$. If we consider user grouping and power allocation based on average channel coefficients, the reduces to:

$$\max_{\{g_n\}, \{p_{n,k}\}} R = \sum_{k=1}^{K} b \log_2 \left(1 + \frac{p_{n,k} |\overline{h_{n,k}}|^2}{I_{n,k} + \sigma^2} \right) \tag{3a}$$

$$\text{s.t.} \ g_n \cap g_o = \emptyset, \ n \neq o, \ n, o \in \mathcal{N}, \tag{3b}$$

$$\sum_{j, \forall U_j \in g_n} p_{n,j} \leq P_n, \ n \in \mathcal{N}, \tag{3c}$$

$$R_{n,j}(t) \geq R_{QoS}, \ j \in \mathcal{K}, n \in \mathcal{N}, \tag{3d}$$

$$\overline{h_i(t)} > \overline{h_j(t)}, \ \forall i > j, \ i, j \in \mathcal{K}, \tag{3e}$$

$$|g_n \cap q_c| = 1, \ \forall c, \forall n, \tag{3f}$$

$$\sum_{j, \forall U_j \in g_n} \tau_{n,j} = L_c, \ \forall n, \tag{3g}$$

$$\sum_{n, \forall n \in \mathcal{N}} \sum_{j, \forall U_j \in g_n} \tau_{n,j} = NL_c. \tag{3h}$$

In Eq. (3a), $I_{n,k} = \sum_{j, \forall j > k, \{U_j, U_k\} \in g_n} |\overline{h_k}|^2 p_{n,j}$ is the interference to User k in Group n. In Eq. (3b), we state that the groups need to be disjoint. Hence, one user can only be in one group. In (3c), we address the constraint for the power budget. The QoS constraint is given in Eq.(3d), where $R_{n,j}(t)$ is the achievable data rate for User j in Group n, and Eq. (3e) gives the SIC constraint. The constraint in Eq. (3f) specifies that only a single user is selected to formulate a group from each cluster. In Eq. (3g), we introduce an indicator $\tau_{n,k}$, stating whether U_k is in Group n, as $\tau_{n,k} = 1$ when U_k belongs to Group n, and 0 otherwise. Furthermore, all users should belong to a certain group, which is given in Eq. (3h). Table 1 summarizes the notation.

Table 1. Summary of notations

Notation	Description	Notation	Description
$h, h(t)$	Channel coefficient, and h for t	$p_{n,k}$	Allocated power for $U_{n,k}$
$h_k, h_{n,k}$	h for U_k and $U_{n,k}$	n_k, σ^2	AWGN at U_k and Gaussian noise
$\overline{h_k(t)}, \overline{h_{n,k}(t)}$	The mean of h for U_k	$h_k(t), h_{n,k}(t)$	h for U_k and $U_{n,k}$ at t
K	Total number of users	$I_{n,k}$	Interference from other users to $U_{n,k}$
N	Total number of groups	E_c	Mean of channel fading in q_c
$\mathcal{K}$	Set of user indexes	R	Total data rate (capacity)
$\mathcal{N}$	Set of group indexes	S	Number of states per action
$\mathcal{G}$	Set of groups	R_{QoS}	Minimum required data rate for a user
g_n	Set of users inside the n-th group	v_U, v_L	Mobility factor and speed of light
$|g_n|$	Number of users inside g_n	f_c, f_d	Carrier and Doppler frequency
$U_k, U_{n,k}$	User k and user k in group n	L_c	Number of clusters
$g_n \backslash U_{n,k}$	The complementary set of users in set g_n	q_c	Set of users in cluster c
$\emptyset$	Empty set	C	Set of clusters
$y_k, y_k(t)$	Signal from BS at U_k and U_k at time t	$\varphi_{c,k}$	Indicator of whether U_k is in cluster c
s_k	Transmitted signal intended for U_k	δ	Number of $\varphi_{c,k} = 1$ for a cluster
P_n	Power budget for g_n	b	Channel bandwidth
$\tau_{n,k}$	Indicator of whether U_k is in group n	r_k	Rank of U_k
Δ_t	Time period for average of h	Υ_k	Ranking category of user k
ε_k	Index of the current state of user k	Θ_k	Cluster of U_k
$Q = (U_a, U_b)$	Input *query* of users to the EOMA	$W, Mbps$	Watt and Megabits per second
r	Rank	c	Index of the set of clusters

4 Solution to User Grouping and Power Allocation

The problem of grouping and power allocation in NOMA systems is two-pronged. Therefore, in Sect. 4.1, we only consider the first issue of the two, namely the grouping of users. We will show that our solution can handle the stochastic nature of the channel coefficients of the users, while also being able to follow changes in their channel behaviors over time. This will ensure that the system will be able to follow the nature of the channels in a manner that is similar to what we will expect in a real system. Thereafter, in Sect. 4.2, we will present our solution to the power allocation problem. Once the groups have been established in Sect. 4.1, we can utilize these groups to allocate power among them either instantaneously, or over a time interval using a greedy solution to the problem.

4.1 Clustering Through EOMA

The family of OMA algorithms are based on *tabula rasa* Reinforcement Learning. Without any prior knowledge of the system parameters, the channels, or the clusters, (as in our case), the OMA self-learns by observing, over time, the Environment that it interacts with. For our problem, the communication system constitutes the Environment, which can be observed by the OMA through, e.g., channel sounding. By gaining knowledge from the system behavior and incrementally improving through each interaction with the Environment, the OMA algorithms are compelling mechanisms for solving complex and stochastic problems. In the OMA, the users of our system need to be represented as abstract objects. Therefore, as far as the OMA is concerned, the users are called "objects". The OMA algorithms require a number of states per action, indicated by S. For the LA, an action is a solution that the algorithm can converge to. In our system, the actions are the different clusters that the objects may belong to. Hence, based on the current state of an object, we know that object's action, which is equal to its current cluster in our system. Therefore, each object, or user in our case, has a given state indicated by $\varepsilon_k = \{1, 2, ..., SL_c\}$, where ε_k denotes the current state of U_k, S is the number of states per action, and L_c is the number of clusters. Clearly, because we have L_c clusters, the total number of possible states is SL_c. To indicate the set of users inside

---

Algorithm 1. Clustering of Users

Require: $h_k(t)$ for all users K
  while not converged **do** // Converged if all users are in the two innermost states of any action
    for all K **do**
      Rank the users from 1 to K // 1 is given to the user with lowest h (K to the highest)
    end for
    for $\frac{K}{N}$ pairs (U_a, U_b) of K **do**   // The pairs are chosen uniformly from all possible pairs
      if $\Upsilon_a = \Upsilon_b$ **then**   // If U_a and U_b have the same ranking category
        if $\Theta_a = \Theta_b$ **then**   // If U_a and U_b are clustered together in the EOMA
          Process Reward
        else   // If U_a and U_b are not clustered together in the EOMA
          Process Penalty
        end if
      end if
    end for
  end while   // Convergence has been reached

Cluster c, where $c \in [1, 2, \ldots, L_c]$, we have q_c. The cluster for a given User, k, is represented by Θ_k, where the set of clusters is denoted by C and $\Theta_k \in C = \{q_1, q_2, \ldots, q_{L_c}\}$.

The states are central to the OMA algorithms, and the objects are moved in and out of states as they are penalized or rewarded in the Reinforcement Learning process. When all objects have reached the two innermost states of an action, we say that the algorithm has converged. When convergence is attained, we consider the solution that the EOMA algorithm has found to be sufficiently accurate. In the EOMA, the numbering of the states follows a certain pattern. By way of example, consider a case of three possible clusters: the first cluster of the EOMA has the states numbered from 1 to S, where the innermost state is 1, the second innermost state is 2, and the boundary state is S. The second cluster has the innermost state $S + 1$ and the second innermost state $S + 2$, while the boundary state is $2S$. Likewise, for the third cluster, the numbering will be $2S + 1$ for the innermost and $2S + 2$ for the second innermost state, while $3S$ is the boundary state.

Algorithm 1 presents the overall operation for the clustering of the users. The functionality for reward and penalty, as the EOMA interacts with the NOMA system, are given in Algorithms 2 and 3 respectively. In the algorithms, we consider the operation in relation to a pair of users U_a and U_b, and so $Q = \{(U_a, U_b)\}$. The EOMA considers users in pairs (called *queries*, denoted by Q). Through the information contained in their pairwise ranking, we obtain a clustering of the users into the different channel categories. For each time instant, Δ_t, the BS obtains values of $h_k(t)$ through channel sounding, and we use the average of Δ_t samples as the input to the EOMA $(\overline{h_k(t)})$. The BS then ranks the users, indicated by $r_k = \{1, 2, \ldots, K\}$, where each U_k is given a single value of r_k for each Δ_t. For the ranks, $r_k = 1$ is given to the user that has the lowest channel coefficient compared to the total number of users, and $r_k = K$ is given to the user with the highest channel coefficient of the users. The others are filled in between them with ranks from worst to best. Furthermore, the values of these ranks corresponds to ranking categories, denoted by Υ_k for U_k, where $\Upsilon_k = \{r \in [1, N] = 1, r \in [1 + N, 2N] = 2, r \in [1 + 2N, 3N] = 3, \ldots, r \in [K - N + 1, K] = L_c\}$. In this way, even if the users have similar channel conditions, they will be compared, and the solution can work on finding the current best categorization of the K users for the given communication scenario. As depicted in Algorithm 1, we check the users' ranking categories in a pairwise manner. If the users in a pair (query) are in the same ranking category, they will be sent as a query to the EOMA algorithm. The EOMA algorithm will then work on putting the users that are queried together in the same cluster, which, in the end, will yield clusters of users with similar channel coefficients. More specifically, if two users have the same ranking category, they are sent as a query to the EOMA and the LA is rewarded if these two users are clustered together (penalized if they are not together).

When the algorithm converges to obtain the groups that are needed for the power allocation, we rank the users within each cluster based on $\overline{h_k(t)}$ that was obtained in the clustering process, and then formulate the groups that consist of one user from each cluster with the same rank.

Algorithm 2. Process Reward

Require: $Q = (U_a, U_b)$ // A query (Q), consisting of U_a and U_b
Require: The state of U_a (ε_a) and U_b (ε_b)
   if $\varepsilon_a \bmod S \neq 1$ **then**   // U_a not in innermost state
     $\varepsilon_a = \varepsilon_a - 1$ // Move U_a towards innermost state
   end if
   if $\varepsilon_b \bmod S \neq 1$ **then**   // U_b not in innermost state
     $\varepsilon_b = \varepsilon_b - 1$ // Move U_b towards innermost state
   end if
   return The next states of U_a and U_b

4.2 Power Allocation Through a Greedy Solution

Once the grouping of the users has been established, we can allocate power to different users in such a way that the joint data rate (R) is maximized. There are numerous ways of power allocation in various communication scenarios [5, 10]. The power allocation can be replaced by any other algorithm and will not change the nature of the Reinforcement Learning procedure. However, in this paper, we will consider the problem of power allocation as a variation of the KP, and solve it through a greedy solution.

Algorithm 3. Process Penalty

Require: $Q = (U_a, U_b)$ // A query (Q), consisting of U_a and U_b
Require: The state of U_a (ε_a) and U_b (ε_b)
   if $\varepsilon_a \bmod S \neq 0$ and $\varepsilon_b \bmod S \neq 0$ **then**   // Neither of the users are in boundary states
     $\varepsilon_a = \varepsilon_a + 1, \varepsilon_b = \varepsilon_b + 1$ // Move U_a and U_b towards boundary state
   else if $\varepsilon_a \bmod S \neq 0$ and $\varepsilon_b \bmod S = 0$ **then**   // U_b in boundary state but not U_a
     $\varepsilon_a = \varepsilon_a + 1, temp = \varepsilon_b$
     $x =$ unaccessed user in cluster of U_a which is closest to boundary state
     $\varepsilon_x = temp, \varepsilon_b = \varepsilon_a$
   else if $\varepsilon_b \bmod S \neq 0$ and $\varepsilon_a \bmod S = 0$ **then**   // U_a in boundary state but not U_b
     $\varepsilon_b = \varepsilon_b + 1, temp = \varepsilon_a$
     $x =$ unaccessed user in cluster of U_b closest to boundary state
     $\varepsilon_x = temp, \varepsilon_a = \varepsilon_b$
   else   // Both users are in boundary states
     $\varepsilon_y = \varepsilon_{\{a \text{ or } b\}}$ // y equals a or b with equal probability, and y is the staying user
     $\varepsilon_z = \varepsilon_{\{a \text{ or } b\}}$ // z is the moving user, and is a if b was chosen as y (b if a chosen)
     $temp = \varepsilon_z$
     $x =$ unaccessed user in cluster of U_y closest to boundary state
     $\varepsilon_x = temp$
     $\varepsilon_z = \varepsilon_y$ // Move U_z to cluster of U_y
   end if
   return The next states of U_a and U_b

Our aim for the greedy solution is that of maximizing the total data rate of the system. Thus, the weakest user will always be limited to the minimum required data rate. The heuristic involves allocating the majority of the power to the users with higher values of h, and this will result in a higher sum rate for the system. Consequently, the stronger users are benefited more from the greedy solution than those with weaker channel coefficients. However, the weak users' required data rate is ensured and can be adjusted to the given scenario. The formal algorithms are not explicitly given here in the interest of brevity, and due to space limitations. They are included in [7].

5   Numerical Results

The techniques explained above have been extensively tested for numerous numbers of users, power settings etc., and we give here the results of the experiments. In the interest of brevity, and due to space limitations, the results presented are a very brief summary of the results that we have obtained. More detailed results are included in [7] and in the Doctoral Thesis of the First Author.

We employed Matlab for simulating the values of the channel coefficient, h. Additionally, we invoked a Python script for simulating the LA solution to the user grouping and the greedy solution to power allocation. The numerical results for the power allocation solution are based on the results obtained from the EOMA clustering and grouping. For the simulations, we used a carrier frequency of 5.7 GHz and an underlying Rayleigh distribution for the corresponding values of $h(t)$. For the mobility in our model, we utilized a moving pace corresponding to the movement inside an office building, i.e., $v_U = 2$ km/h. We sampled the values of h according to $\frac{1}{2f_d}$, where f_d is the Doppler frequency and f_c is the carrier frequency. The Doppler frequency can be expressed as $f_d = f_c(\frac{v_U}{v_L})$ and v_L is the speed of light. Therefore, in the following figures, we use "Sample Number" as the notation on the X-axis. Figure 1 illustrates the snap-shot of h values for four users and the principle for the simulation when the number of users increased.

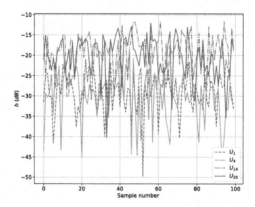

Fig. 1. Example of the simulated $h(t)$ for four different users. In the interest of clarity, and to avoid confusion, we did not plot all the 20 users.

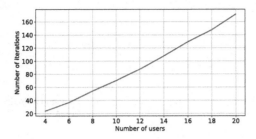

Fig. 2. A plot of the average number of iterations needed before convergence, as a function of number of users, where there were two users in each group. This was obtained by executing 100 independent experiments.

For evaluating the simulation for the clustering phase, we recorded whether or not the LA were able to determine the clusters that corresponded to the minimized difference between the users in a cluster, based on the users' mean values of h in the simulations. Remarkably, in the simulation, the EOMA yielded a 100 % accuracy in which the learned clustering was identical to the unknown underlying clustering in every single run for the example provided with -10 dB difference between values of h within the different clusters. This occurred for groups of sizes $4, 6, 8, 10, 12, 14, 16, 18$ and 20, where the number of users in a group was equal to two. The difference between the users can be replaced by any "equivalent metric", and it should be mentioned that these values were only generated for testing the solution, since in a real scenario, the "True partitioning" is always unknown. The number of iterations that it took the EOMA to achieve 100% accuracy for the different number of users is depicted in Fig. 2. Notably, the EOMA retains its extremely high accuracy as the number of users increased, and yielded 100% accuracy both for 4 users as well as 20 users.

The simulation for the greedy solution to the power allocation phase, was carried out based on the groups established in the LA solution. Again, in the interest of brevity, we only report the results for the cases with 20 users in total and 2 users in each group, and more extensive results are included in [7] and in the Doctoral Thesis of the First Author. The optimal power allocation for 2 users in a group was obtained when we gave the minimum required power to the user with smaller h value and then allocated the rest to the user with lager h value. The optimality of the greedy algorithm for the 2 user-group case was verified by an alternate independent exhaustive search. For illustrating the advantages of our NOMA greedy solution when NOMA was employed, we compared it with the data rate that would be achieved with orthogonal multiple access[2]. In Fig. 3, we depict the results obtained for the greedy NOMA solution together with the orthogonal multiple access, for an average over $\Delta t = 5$ samples of h. Further, with regard to the parameters used, the data rate for the simulations depicted in Fig. 3 was based on the following configuration: The minimum required data rate was configured

---

[2] The achieved data rate for User k in Group n in orthogonal multiple access is given by $R_{n,k} = \frac{1}{2} \log_2 \left(1 + \frac{P_n |h_{n,k}|^2}{\sigma^2}\right)$. The factor $\frac{1}{2}$ is due to the multiplexing loss when 2 users share the orthogonal resource.

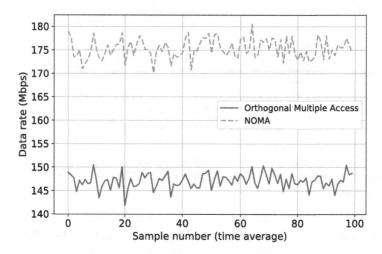

Fig. 3. Data rate for orthogonal multiple access compared to NOMA for time averages of h. Based on averages over 500 samples of h.

to 2.0 Mbps, the noise to 10^{-8} W, the bandwidth to 1 MHz, and the power level for all groups to 0.125 W. As illustrated, the simulation results obtained show that the greedy solution to the power allocation is higher than the data rate achieved with orthogonal multiple access. From the graph in Fig. 3 we see that the average difference between the orthogonal multiple access and NOMA was approximately 28.17 Mbps.

6   Conclusions

In this paper, we have proposed a novel solution to the user grouping and power allocation problems in NOMA systems, where we have considered the stochastic nature of the users' channel coefficients. The grouping has been achieved by using the *tabula rasa* Reinforcement Learning technique of the EOMA, and the simulation results presented demonstrate that a 100% accuracy for finding clusters of similar $h(t)$ over time can be obtained within a limited number of iterations. With respect to power allocation, we proposed a greedy solution, and again the simulation results confirm the advantages of the NOMA solution. Our solutions offer flexibility, as both the grouping and the power allocation phases, can be used as stand-alone components of a NOMA system.

References

1. Cui, J., Ding, Z., Fan, P., Al-Dhahir, N.: Unsupervised machine learning-based user clustering in millimeter-wave-NOMA systems. IEEE Trans. Wirel. Commun. **17**(11), 7425–7440 (2018)
2. Gale, W., Das, S., Yu, C.T.: Improvements to an algorithm for equipartitioning. IEEE Trans. Comput. **39**(5), 706–710 (1990)

3. Glimsdal, S., Granmo, O.: A novel bayesian network based scheme for finding the optimal solution to stochastic online equi-partitioning problems. In: 2014 13th International Conference on Machine Learning and Applications, pp. 594–599, December 2014
4. Kellerer, H., Pferschy, U., Pisinger, D.: Knapsack Problems. Springer, Berlin (2004)
5. Liu, Y., Elkashlan, M., Ding, Z., Karagiannidis, G.K.: Fairness of user clustering in MIMO non-orthogonal multiple access systems. IEEE Commun. Lett. **20**(7), 1465–1468 (2016)
6. Liu, Y., Qin, Z., Elkashlan, M., Ding, Z., Nallanathan, A., Hanzo, L.: Nonorthogonal multiple access for 5G and beyond. Proc. IEEE **105**(12), 2347–2381 (2017)
7. Omslandseter, R.O., Jiao, L., Liu, Y., Oommen, B.J.: An Efficient and Fast Reinforcement Learning-Based Solution to the Problem of User Grouping and Power Allocation in NOMA Systems. Unabridged version of this paper. To be submitted for publication
8. Pätzold, M.: Mobile Radio Channels, 2nd edn. Wiley, Chichester (2012)
9. Pischella, M., Le Ruyet, D.: Noma-relevant clustering and resource allocation for proportional fair uplink communications. IEEE Wirel. Commun. Lett. **8**(3), 873–876 (2019)
10. Xing, H., Liu, Y., Nallanathan, A., Ding, Z., Poor, H.V.: Optimal throughput fairness tradeoffs for downlink non-orthogonal multiple access over fading channels. IEEE Trans. Wirel. Commun. **17**(6), 3556–3571 (2018)

Deep Learning for QoS-Aware Resource Allocation in Cognitive Radio Networks

Jerzy Martyna[✉]

Institute of Computer Science, Faculty of Mathematics and Computer Science, Jagiellonian University, ul. Prof. S. Lojasiewicza 6, 30-348 Cracow, Poland
jerzy.martyna@uj.edu.pl

Abstract. This paper focuses on the application of deep learning (DL) to obtain solutions for radio resource allocation problems in cognitive radio networks (CRNs). In the proposed approach, a deep neural network (DNN) as a DL model is proposed which can decide the transmit power without any help from other nodes. The resource allocation policies have been shown in the context of effective capacity theory. The numerical results demonstrate that the proposed model outperforms the scheme in terms of radio resource utilization efficiency. Simulation results also support the effectiveness on the delay guarantee performance.

Keywords: Deep learning · Convolutional Neural Network · Support vector machines · Cognitive radio network · Statistical QoS guarantee · Effective capacity

1 Introduction

Machine learning (ML) was introduced in the late 1950s as a technique useful in AI [1]. Over time, ML has allowed the development of dozens of algorithms and techniques that have become not only popular in many fields of science and technology, but also extremely stable and effective in solving problems. It is recalled that these tasks included classification, regression and density estimation, optimisation in a variety of application areas such as spam detection, automatic speech recognition, computer vision, face detection, hacker attack detection, network management, etc. This also applies to wireless computer networks, including cognitive radio networks (CRNs) [2] and future next generation networks [3].

Recently, deep learning (DL) or deep neural network with multi layers has become very popular in the literature due to the huge number of problems solved in a wide range of areas with performance improvement, pattern recognition [4], natural language processing [5], computer vision [6], speech recognition [7], and so on. One of the most popular deep neural networks is the Convolutional Neural Network (CNN) [8]. It takes its name from mathematical linear operation between matrixes called convolution. CNNs have multiple layers; including a convolutional layer, a non-linearity layer, a pooling layer and a fully-connected

© Springer Nature Switzerland AG 2020
H. Fujita et al. (Eds.): IEA/AIE 2020, LNAI 12144, pp. 312–323, 2020.
https://doi.org/10.1007/978-3-030-55789-8_28

layer. The convolutional and fully-connected layers have parameters, whereas pooling and non-linearity layers do not. The CNN has an excellent performance in machine learning problems.

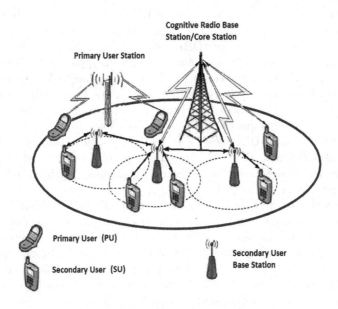

Fig. 1. Basic architecture of downlink/uplink cognitive radio network.

The CRN mentioned above is based on the concept of software radio (SR) introduced by J. A. Mitola [9]. The basic architecture of this system is shown in Fig. 1. This concept requires cognitive radio (CR) devices, and allows the use of an underutilised spectrum. The CR nodes also allow the spectrum to be shared with licensed primary user (PU) nodes and their transmission characteristics adapted according to the instantaneous behaviour of licensed users. CR networks based on CR devices can improve the overall spectral utilisation. However, the maximisation of the throughput of the network leads to the undesirable interference power of the secondary users (SUs) into PUs [10]. From the PU's perspective, the transmission of the SU is allowed as long as the interference caused by the SU does not affect its quality of service (QoS). On the other hand, the SU should transmit until it reaches a higher transmission rate without causing too much interference to the PU. Therefore, the key is to define the channel capacity of the SU link with the influence of the interference present.

The QoS in CR networks has been the subject of a number of papers. The cross-layer methods in CRN proposed to allocate the resource with heterogeneous QoS requirements have been proposed in [11]. To satisfy the delay requirement in the cognitive radio system, the authors of the paper [12] proposed a novel component carrier configuration and switching scheme for real-time traffic. A comprehensive analytical framework based on queueing theory is proposed to analyse

QoS of SUs in [13]. A two-level quality protocol for provisioning media access control for CRN has been proposed by V. Mishra et al. [14]. Nevertheless, this work solves the problem of resource allocation with QoS requirements in CRN networks using advanced AI techniques.

Recently, genetic algorithm was used by L. Zhu et al. [15] for optimising spectrum utilisation while providing a fairness guarantee between users in CRNs. The deep reinforcement learning (DRL) method to solving the resource allocation problem in cloud radio access networks has been applied by Z. Xu et al. [16]. The same approach for power control and spectrum sharing has been proposed by X. Li [17]. In turn, deep learning methods for the radio resource allocation problems in multi-cell networks have been presented by K.I. Ahmed et al. [18].

This paper first explores the concept of a deep neural network (DNN) with SVM architecture as a DL model. For achieving optimal subband and power allocation for CRN with the goal of maximizing the total network throughput, a supervised DL model has been developed. A learning algorithm has been prepared specifically for this model.

Major contribution of the paper can be summarized as follows: (1) Propose the QoS-aware resource allocation scheme in cognitive radio networks (CRNs) based on Convolutional Neural Network (CNN) and Support Vector Machine (SVM) as a deep learning (DL) model. Previous systems cannot explore such QoS-aware resource allocation in CRNs using the DL model. (2) Explore an adaptive structure of CNN for the proposed radio resource allocation scheme. (3) Develop the Stochastic Gradient Scheme for training together both CNN and SVM. (4) The proposed system performs well any priori information.

The rest of this paper is organised as follows. The effective capacity for QoS guarantees of CRN is described in Sect. 2. Section 3 provides the details of the convolutional neural network. Section 4 is devoted to the SVM classifier used here as a decision element. In Sect. 5, the training algorithm is presented. In Sect. 6, the simulation results are presented and the conclusions are drawn in Sect. 7.

2   Fulfilling the QoS Requirements

All real-time multimedia applications used in wireless communication require a guaranteed bandwidth. This is because if a real-time packet violates its delay, it will be discarded. The solution to this problem has been made possible by the concept of effective capacity developed by C.-S. Chang [19] to provide the statistical QoS guarantee in general real-time wireless communication. As has been shown in [19] for a queueing system with a stationary ergodic arrival and service process, the queue length process $Q(t)$ converges to a random variable $Q(\infty)$ so that

$$- \lim_{x \to \infty} \frac{\log(Pr\{Q(\infty) > x\})}{x} = \theta \tag{1}$$

where θ and $\theta > 0$ characterises the degree of statistical QoS guarantees that can be provided by the system. Analytically, the effective capacity can be formally defined as [20]:

$$E_C(\theta) \overset{\triangle}{=} = -\frac{\Lambda_c(-\theta)}{\theta} = -\lim_{t \to \infty} \frac{1}{\theta t} \log\left(E[e^{-\theta S[t]}]\right) \tag{2}$$

where $S[t] \overset{\triangle}{=} \sum_{i=1}^{t} R[t]$ is the partial sum of the discrete-time stationary and ergodic service process $\{R[i], i = 1, 2, \ldots\}$ and

$$\Lambda_c(\theta) = \lim_{t \to \infty} \frac{1}{t} \log\left(E[e^{\theta S[t]}]\right) \tag{3}$$

is a convex function differentiable for all real θ.

Thus, the probability that the packet delay violates the delay requirement is given by

$$\Pr\{Delay > d_{max}\} \approx e^{-\theta \delta d_{max}} \tag{4}$$

where d_{max} is the delay requirement, δ is a constant jointly determined by the arrival process and their service process.

3   Convolutional Neural Network

A basic DL model is a multi-layered feed-forward neural network. This neural network enables the implementation of a deep neural network (DNN), which is composed of a convolutional neural network (CNN) providing two basic operations: convolution and sub-sampling (e.g. pooling operations). The CNN network has the following layers:

1) *Convolutional layer*, which performs a 2D input images x and a bank of filters w and produces another set of images h. The input-output relationships are represented by a connection table (CT). The mapping, called *scaled hyperbolic tangent*, $h_j = \sum_{i,k \in CT_{i,k}} (x_i * w_k)$ is similar to the case of MLP network.
2) *Sub-sampling layer*, which implements the so called "max-pooling operation". This operation leads to faster convergence to and improved generalisation.
3) *Fully-connected layer*. This layer combines the outputs of the previous layer into a 1-dimensional feature vector, which can be treated as a *scaled hyperbolic tangent activation function*.
4) *Output layer*, which acts as a linear classifier operating on the 1-dimensional feature vectors computed from the fully-connected layer.

In a typical application of CNN, gradient descent approaches are used for minimising the training error. This is because all of the neurons of the CNN layer are differentiable. Nevertheless, this approach cannot be applied here, because in this case the CNN network cooperates with the SVM classifier, which plays the role of the decision-making element. Therefore, a specially developed training algorithm has been used here that allows you to teach CNN and SVM in a combined way.

4   Support Vector Machine Classifier

Support vector machines (SVMs) demonstrated excellent performance in hyperspectral data classification in terms of accuracy [21]. Therefore, SVM was chosen here as the primary classifier to data classes in the feature space.

Given a training set of data set $\{(x_1, y_1), (x_2, y_2), \ldots, (x_n, y_n)\}$, where $x_i \in \mathbb{R}^n$ is the i-th input pattern and $y_i \in \mathbb{R}$ is the i-th output pattern, and a non-linear mapping $\phi(.)$, the SVM method solves

$$\min_{w, \xi, b} \left\{ \frac{1}{2} \parallel w \parallel^2 + C \sum_i \xi_i \right\} \tag{5}$$

constrained to

$$y_i \Big(\langle \phi(x_i), w \rangle + b \Big) \geq 1 - \xi_i, \quad \forall i = 1, \ldots, n \tag{6}$$

$$\xi_i \geq 0, \quad \forall i = 1, \ldots, n \tag{7}$$

where w and b define a linear classifier in the feature space. The regularisation parameter C controls the generalisation capabilities of the classifier and it must be selected by the user, ξ_i are positive slack variables enabling permitted errors to be dealt with.

Primal function given by Eq. (5) is usually solved by Langrangian sual problem. It requires a solution of

$$\max_{\alpha_i} \left\{ \sum_i \alpha_i - \frac{1}{2} \sum_{i,j} \alpha_i \alpha_j y_i y_j \langle \phi(x_i), \phi(x_j) \rangle \right\} \tag{8}$$

constrained to $0 \leq \alpha_i \leq C$ and $\sum_i \alpha_i y_i = 0$ are Langrange multipliers corresponding to constraints in Eq. (6). It allows to define a kernel function K, namely

$$K(x_i, x_j) = \langle \phi(x_i), \phi(x_j) \rangle \tag{9}$$

Then, a non-linear SVM classifier can be constructed using only the kernel function, without having to consider the mapping function ϕ. Thus, the decision function implemented by the SVM classifier for any test vector $\mathbf{x}$ is given by

$$f(\mathbf{x}) = sgn \Big(\sum_{i=1}^n y_i \alpha_i K(x_i, \mathbf{x}) + b \Big) \tag{10}$$

where b can be computed from α_i that are neither 0 nor C.

5   Training Algorithm of CNN and SVM

To achieve the desired operation of a hybrid system built from CNN and SVM for resource allocation with QoS guarantees, the following cost function should be minimised

$$\min_w \left\{ \frac{\lambda}{2} \parallel \mathbf{w} \parallel^2 + \frac{1}{m} \sum_{i=1}^m l(x_i, y_i; \mathbf{w}) \right\} \tag{11}$$

Algorithm 1 Hybrid System Training

1: **procedure** SVRG TRAINING OF CNN AND SVM
2:     **Input parameters:** m, η, λ, T;
3:     **Initialise w** such that $\| \mathbf{w} \| \leq \frac{1}{\sqrt{\lambda}}$
4:     **for** $t \leftarrow 1, T$ **do**
5:         Acquire m samples: $\{A_t(s) \mid s = 1, \ldots, m\}$
6:         $A_t^+ = \{(x, y) \in A_t : 1 - y(w_t \cdot x) < 1\}$
7:         $\eta_t = \frac{1}{\lambda t}$
8:         $w_t = (1 - \eta_t \lambda) w_t + \frac{\eta_t}{m} \sum_{(x,y) \in A_t^+} yx$
9:         **for** $s \leftarrow 1, m$ **do**
10:             Randomly pick $s_t \in \{1, \ldots, m\}$
11:             $w_s = w_{s-1} - \eta_t (\nabla \psi_{s_t}(w_{s-1}) - \nabla \psi_{s_t}(w_s) + \widetilde{\mu})^T$
12:         **end for**
13:         $w_t = \min \left\{ 1, \frac{1/\sqrt{\lambda}}{\|w_m\|} \right\}$
14:     **end for**
15: **end procedure**

Fig. 2. Pseudo-code of SVRG training algorithm.

where $l(x_i, y_i; \mathbf{w}) = \max\{0, 1 - y_i f(x_i)\}$. λ is a regularisation parameter that is used to scaling of w.

The operation of the training algorithm developed here is based on the Stochastic Gradient Descent (SGD) method using predictive variance reduction [22]. This method was deliberately chosen because the SGD method leads to slower convergence.

According to the stochastic variance reduction gradient (SVRG) [22], a new auxiliary function is defined

$$\psi_i(\widetilde{w}) = \psi_i(w) - (\nabla \psi_i(\widetilde{w}) - \widetilde{\mu})^T w \tag{12}$$

where $\widetilde{\mu}$ is the average gradient, namely

$$\widetilde{\mu} = \nabla P(\widetilde{w}) = \frac{1}{m} \sum_{i=1}^{m} \nabla \psi_i(\widetilde{w}) \tag{13}$$

Details of the CNN and SVM training algorithm are shown in Fig. 2. The operation of this algorithm is as follows. The *SVRG Training of CNN and SVM* procedure starts its operation from reading such parameters as: m - the number of training samples to use in each so-called mini-batch of training data, T - the number of iterations to perform, η - the learning rate, λ - the regularisation parameter. The weight vector w is initially set to any vector whose norm satisfies the condition $\| \mathbf{w} \| \leq \frac{1}{\sqrt{\lambda}}$. In addition, at each iteration t of the algorithm, the required set of sample $A_t(s) \mid s = 1, 2, \ldots, m$ is required. The following action is performed during each iteration:

$$w_{t+1} = (1 - \eta_t \lambda) w_t + \frac{\eta_t}{m} \sum_{(x,y) \in A_t^+} yx \tag{14}$$

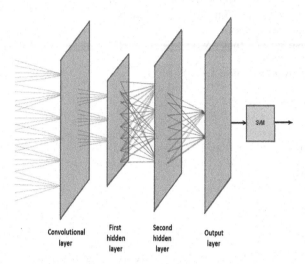

Convolutional Neural Network

Fig. 3. The whole process of DL system consisting of CNN and SVM.

Then, in the internal loop of the SVRG training algorithm, actions are performed that cause a sharp decrease in the value of the learning rate parameter. The sub-gradient projection of vector $\mathbf{w_t}$ is calculated each time. The optimal solution is obtained by minimising the updated vector $\mathbf{w}_t$.

6   Simulation Results

In this section, simulation results are given to evaluate the performance of the DL system used to resource allocation with QoS guarantees in CRN networks.

For the needs of the simulation, a moderate scale CRN network consisting of 50 SUs nodes uniformly distributed in a 80 m × 80 m square area and 2 PUs deployed in the network was established. Each PU can operate on one channel such that SUs can only access unused channels for data communication. It means it is free or occupied at any time depending on PU activity. The maximum false alarm constraint is set to $\alpha = 0.1$, as recommended by the IEEE 802.22 standard [23].

The DL system has been implemented in the Matlab language. The architecture of the DL system consists of the four layers of the CNN system described above and a decision system based on the SVM classifier (see Fig. 3). The inputs of the CNN are the instantaneous transmission rates, current SINR level, and the value of the PU signal. The activation function for the hidden layers and the output layer is ReLU, namely $y = max$, where x and y denote the input and output of the neural unit, respectively. In the SVM, the RBF kernel with regularisation parameter C equal to 100 for all experiments has been applied.

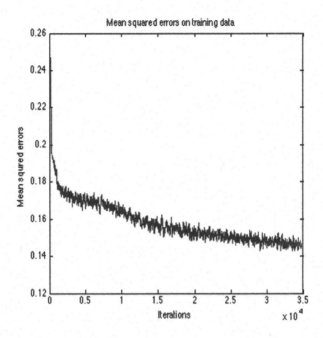

Fig. 4. The mean square errors of training process of DL system.

The σ_{min} has been set as the average minimal distance between any two training samples. The output of the DL system provides the resource allocation strategy, i.e. gives the effective capacity depending on the QoS exponent θ and the probability that the delay exceeds the delay requirement.

The training process for the proposed DL system was obtained using the given training procedure and the conventional resource allocation in CRN presented in [24]. The training data are divided into 100 mini-batches and each mini-batch consists of 1000 samples. The input vectors of the training process concerned the effective capacity under varying interference power. The output data was optimal transmit power. In the training process, the mean squared error minimisation criterion was applied.

Figure 4 shows the mean square error between the optimal transmit power obtained using the method [24] and the resource allocation achieved thanks to the method of DL system. It can be seen that the mean square error decreases and approaches 0 with the increase in training steps in the training process.

Figure 5 shows the effective capacity of the SUs in the CRN obtained using the DL system versus the value of the QoS exponent. It is visible from the graph in the figure that the effective capacity decreases with an increased system load, given by ρ.

Figure 6 shows the simulation results achieved by use the DL system of the delay bound violation probability of voice stream. The graph in the figure

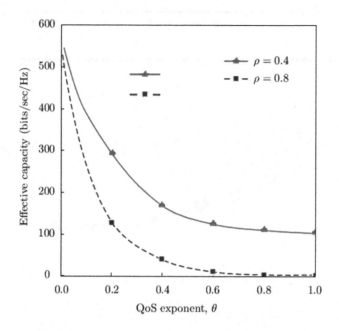

Fig. 5. The effective capacity of the SUs versus QoS exponent θ.

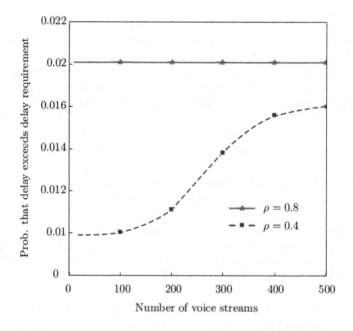

Fig. 6. The probability of the delay bound violation of voice streams.

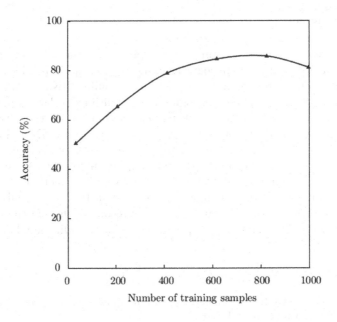

Fig. 7. Test accuracy versus number of training samples.

Table 1. Comparative results of radio resource allocation using DL methods.

Publication	Model	Objective	Accuracy
Xu *et al.* [16]	Q-learning	Minimise the total power consumption with QoS constraints	–
Ahmed *et al.* [18]	Deep neural network	Radio resource allocation	80%
Li *et al.* [17]	Deep neural network	Power control and spectrum sharing	–
Proposed method	Convolutional neural network and SVM	QoS-aware resource allocation	85%

demonstrates that the DL system restricts bound violation probability below the required value.

Figure 7 shows the test accuracy versus to the number of training samples. In the figure we observe that the test accuracy increases slowly with the training data. However, the increasing number of training samples leads to overfit and, as a result, reduces test accuracy.

A summary of approaches to the problem of resource allocation solved using DL methods is provided in Table 1 for comparison the results obtained in terms of the resources management objectives applied, as well as the methods used and their accuracy.

7 Conclusion

The paper presents a deep learning (DL) approach for the sub-band allocation problem with QoS guarantees in cognitive radio networks (CRNs). The use of this approach gives an advantage in helping design and build CRN networks that will meet the ever-growing demand for high-quality multimedia data transmissions. CRN networks must employ an advanced QoS mechanism for their users.

The optimum configuration of the DL system based on CNN and SVM was demonstrated, e.g. the number of hidden layers and size input samples are a basic challenge. For this purpose, a training algorithm for the DL system was developed. It is also important to develop effective offline methods to generate optimal/almost optimal data samples for DL system training. Finally, the use of DL system in simulation scenarios and their results confirmed the usefulness of the introduced DL system for its practical applications in the CRN network.

References

1. Ayodele, T.O.: Introduction to machine learning. In: Zhang, Y. (ed.) New Advances in Machine Learning. InTech (2010)
2. Karunaratne, S., Gacanin, H.: An Overview of Machine Learning Approaches in Wireless Mesh Network (2018). https://arxiv.org/ftp/arxiv/papers/1806/1806.10523.pdf
3. Jiang, C., Zhang, H., Ren, Y., Han, Z., Chen, K.-C., Hanzo, L.: Machine learning paradigms for next-generation wireless networks. IEEE Wirel. Commun. 24(1), 98–105 (2016)
4. Zilong, H., Tang, J., Wang, Z., Zhang, K., Sun, Q.: Deep learning for image-based cancer detection and diagnosis - a survey. Pattern Recogn. 83(11), 123–149 (2018)
5. Cho, K., Van Merriënboer, B., Gulcehre, C.: Learning phrase representations using RNN encoder-decoder for statistical machine translation (2014). http://arxiv.org/abs/1406.1078
6. Krizhevsky, A., Sutskever, I., Hinton, G.E.: ImageNet classification with deep convolutional neural networks. In: Proceedings of the Advances in Neural Information Processing Systems, pp. 1097–1105 (2012)
7. Weng, C., Yu, D., Watanabe, S., Juang, B.-H.F.: Recurrent deep neural networks for robust speech recognition. In: Proceedings of the ICASSP, pp. 5532–5536, Florence, Italy, May 2014
8. Schmidhuber, J.: Deep learning in neural networks: an overview. Neural Networks 61, 85–117 (2015). https://arxiv.org/abs/1404.7828
9. Mitola, J., Maguire, G.Q.: Cognitive radio: making software radios more personal. IEEE Pers. Commun. 6(4), 13–18 (1999)
10. Haykin, S., Radio, C.: Brain-empowered wireless communications. IEEE J. Sel. Areas Commun. 23(2), 201–220 (2005)
11. Chen, Y., Feng, Z., Chen, X.: Cross-layer resource allocation with heterogeneous QoS requirements in cognitive radio networks. In: IEEE Wireless Communications and Networking Conference, pp. 96–101 (2011)
12. Fu, Y., Ma, L., Xu, Y.: A novel component carrier configuration and switching scheme for real-time traffic in a cognitive-radio-based spectrum aggregation system. Sensors 9, 23706–23726 (2015)

13. Doost, R., Naderi, M.Y., Chowdhury, K.R.: Spectrum allocation and QoS provisioning framework for cognitive radio with heterogeneous service classes. IEEE Trans. Wirel. Commun. **13**(7), 3938–3950 (2014)
14. Mishra, V., Tong, L., Chan, S., Mathew, J.: TQCR-media access control: two-level quality of service provisioning media access control protocol for cognitive radio network. IET Networks **2**, 74–81 (2014)
15. Zhu, L., Xu, Y., Chen, J., Li, Z.: The design of scheduling algorithm for cognitive radio networks based on genetic algorithm. In: IEEE International Conference on Computational Intelligence Communication Technology (CICT), pp. 459–464 (2015)
16. Xu, Z., Wang, Y., Wang, J., Cursoy, M.C.: A deep reinforcement learning based framework for power efficient resource allocation in cloud RANs. In: IEEE International Conference on Communication (ICC), pp. 1–6 (2017)
17. Li, X., Fang, J., Cheng, W., Duan, H., Chen, Z., Li, H.: Intelligent power control for spectrum sharing in cognitive radios: a deep reinforcement learning approach. IEEE Access **6**, 25463–25473 (2018)
18. Ahmed, K.I., Tabassum, H., Hossain, E.: Deep Learning for Radio Response Allocation in Multi-Cell Networks (2018). https://arxiv.org/pdf/1808.00667.pdf
19. Chang, C.-S.: Stability, queue length, and delay of deterministic and stochastic queueing networks. IEEE Trans. Autom. Control **39**(5), 913–931 (1994)
20. Wu, D., Negi, R.: Effective capacity: a wireless link model for support of quality of service. IEEE Trans. Wirel. Commun. **12**(4), 630–643 (2003)
21. Schölkopf, B., Smola, A.: Learning with Kernels - Support Vector Machines, Regularization, Optimization and Beyond. Cambridge, Massachussets. London, England: The MIT Press Series (2001)
22. Johnson, R., Zhang, T.: Accelerating stochastic gradient descent using predictive variance reduction. Adv. Neural Inf. Process. **26** 315–323 (2013). http://papers.nips.cc/paper/4937-accelerating-stochastic-gradient-descent-using-predictive-variance-reduction.pdf
23. IEEE 802.22, Cognitive Wireless Regional Area Network (2011). https://standards.ieee.org/about/get/802/802.22.html
24. Kang, X., Liang, Y.C., Nallanathan, A., Garg, H.K., Zhang, R.: Optimal power allocation for fading channels in cognitive radio networks: ergodic capacity and outage capacity. IEEE Trans. Wirel. Commun. **8**(2), 940–950 (2009)

Social Network Analysis

Self-understanding Support Tool Using Twitter Sentiment Analysis

Harumi Murakami[(⊠)], Naoya Ejima, and Naoto Kumagai

Osaka City University, Sugimoto, Sumiyoshi, Osaka 558-8585, Japan
harumi@osaka-cu.ac.jp
http://murakami.media.osaka-cu.ac.jp/

Abstract. Self-understanding, which is important for such daily aspects as decision making, requires grasping one's strengths and weaknesses, likes and dislikes, joys and disappointments. However, gathering such useful data for self-understanding is difficult. The aim of this research is to develop a system to support self-understanding using sentiment analysis of Twitter posts. We developed a tool that displays bar graphs, tweet log lists, pie charts, and word lists that indicate the positive/negative emotions of a user's tweets. Preliminary experiments revealed that our approach and tool are promising for self-understanding, understanding emotions, and recalling memories.

Keywords: Sentiment analysis · Twitter · Graph · Self-understanding

1 Introduction

Self-understanding is important for various daily elements such as decision making. It requires grasping one's strengths and weaknesses, likes and dislikes, joys and disappointments. However, it is difficult to gather such useful data for self-understanding. On the other hand, the spread of smartphones, microblogging, and social networking services simplifies the storage of personal information. Among such services, Twitter posts (called tweets) often include a user's personal experiences and emotions to them. We believe that extracting and visualizing the emotions included in tweets will increase self-awareness. The aim of this research is to develop a system that supports self-understanding using the sentiment analysis of a user's tweets.

We conducted our research using Japanese tweets and translated the examples in this paper from Japanese into English for publication.

2 Approach

We visualize the emotions included in user tweets for self-understanding. The Google Natural Language API sentiment analysis (Google SA) is used to get a score and a magnitude for each tweet. We display the following: several types of

© Springer Nature Switzerland AG 2020
H. Fujita et al. (Eds.): IEA/AIE 2020, LNAI 12144, pp. 327–332, 2020.
https://doi.org/10.1007/978-3-030-55789-8_29

bar graphs that show the transition of emotions, a list of tweets with emotional scores, a pie chart showing the proportion of emotions, and word lists included in positive and negative tweets.

Google SA calculates a score and a magnitude for a given text. The score of the sentiment ranges between –1.0 (negative) and 1.0 (positive) and corresponds to the overall emotional leaning of the text. The magnitude indicates the overall strength of the emotion within the given text that ranges between 0.0 and infinity.

2.1 Bar Graphs

We developed two kinds of bar graphs with different units: tweets and periods. The bar graph by tweet displays the scores of tweets. We implemented two periods: day and month and developed three bar graphs with different values: (a) maximal values of the scores (both positive and negative); (b) an average value of the scores; and (c) a modified average value using the scores and magnitudes. In addition, there are two modes for bar graphs, log lists, and pie charts: (1) positive and negative and (2) positive, negative, and neutral. In mode (1), a score less than 0 is negative, 0 is neutral, and greater than 0 is positive. In mode (2), a score less than –0.25 is negative, –0.25 to 0.25 is neutral, greater than 0.25 is positive, identical as Google SA. We combined three types (a)–(c) and two modes (1)–(2) in the bar graphs. For example, bar graph (a)–2 displays the maximal positive, negative, positive-side-neutral, and negative-side-neutral scores using three colors.

Graphs (a) and (b) have individual advantages and disadvantages. Graph (a) emphasizes the highest valued tweets but hide other tweets in the designated period. Graph (b) summarizes the values in one value, but offsets the plusses and minuses and the values tend to be small, complicating the detection of tweets with high emotion content. To overcome the disadvantage of graph (b), we designed graph (c), which emphasizes stronger emotions using a modified average value (an *E-score*):

$$E\text{-}score = \begin{cases} \frac{\sum_{i=1}^{n} |Score_i \times Magnitude_i|}{\sum_{i=1}^{n} Magnitude_i} & \text{if } \sum_{i=1}^{n} Score_i \geq 0, \\ -\frac{\sum_{i=1}^{n} |Score_i \times Magnitude_i|}{\sum_{i=1}^{n} Magnitude_i} & \text{otherwise,} \end{cases} \tag{1}$$

where $Score_i$ is a Google's score for tweet t_i, $Magnitude_i$ is the magnitude for tweet t_i, and n is the number of tweets except for those whose scores are 0.

2.2 Tweet Log Lists, Pie Charts, and Word Lists

Tweets can be sorted by date or by score. A pie chart shows the ratio of the positive/negative/neutral scores of the tweets.

The tool displays the top five most frequent word lists extracted from the positive and negative tweets. First, we performed morphological analysis (MeCab was used for Japanese) and assigned the score of one tweet to the words contained in it. For each positive and negative category, the summation of the word's score is divided by its frequency.

2.3   Example

Figure 1 shows an example of a user's monthly usage. After he chose the period (1/Jun/2018–30/Jun/2018) and selected "Per day: A" with "displaying neutral," a bar graph (a)–2, a tweet log list, a pie chart, and positive/negative word lists are displayed. In this period, he enjoyed gaming and watching soccer and baseball. The pie chart shows that 42.9% (9/21) of the tweets were positive, 19% were negative (4/21), and 38.1% (8/21) were neutral. On June 18, 2018 (JST), he posted a very positive tweet about Mexico's FIFA World Cup victory and a very negative tweet about a big earthquake (2018 Osaka earthquake). The positive word list includes *Japan, Mexico, Fujinami* (baseball player), and *this season*, all of which are related to soccer and baseball. The negative word list includes *aftershocks*, related to the earthquake, and *Columbia*, related to soccer.

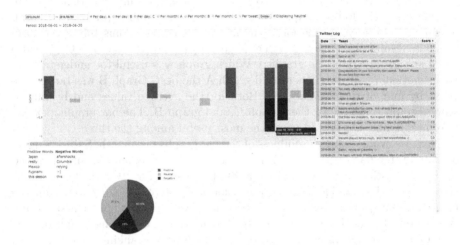

Note: The colors (red: positive, blue: negative, and green: neutral) are based on pre-liminary investigation for Japanese people. Although we chose these colors based on Japanese preferences, they can be tailored to satisfy specific cultural requirements.

Fig. 1. Example of screen with bar graph (a)–2 per day.

3   Preliminary Experiments

We conducted preliminary experiments to evaluate our approach and the tool and recruited subjects who satisfied the following conditions: (a) have used Twitter more than one year, and (b) post more than 15 tweets per month. The subjects were four Japanese males (average age, 22.0 years) in Experiment 1 and five Japanese males (average age, 22.8 years) in Experiments 2 and 3.

3.1 Experiment 1: Google's Scores

We verified the validity of Google's scores for the tweets. The subjects evaluated their emotions of 50 recent tweets by five values (1: very negative; 5: very positive). Two types of correlation analysis (Kendall and Spearman) were performed between Google's scores and manual scores. The average values were over 0.2 in both types of analysis (0.242 in Kendall and 0.313 in Spearman). We believe Google's scores for tweets are promising for our approach.

3.2 Experiment 2: Bar Graphs

We evaluated the usefulness of daily bar graphs. In Experiment 2, not the whole tool but only bar graphs only were displayed. After the subjects described their positive and negative experiences for a month, they operated three bar graphs, (a)–2, (b)–2, and (c)–2, to describe those monthly experiences again and answered the following questions: "Was the bar graph A (or B or C) useful for understanding your emotions?" by five values (1: very unuseful; 5: very useful). The results of the average values were 4.2 (A), 3.6 (B), and 3.8 (C). We also asked subjects to rank the graphs: "Which graph was useful for understanding your emotions?" All subjects ranked graph A the best. Graph B was ranked the worst by four of the five subjects. Both for the average values and by ranked scores, graph A was best, then graph C, and graph B. The bar graphs per day were useful, and bar graph A was the most useful among the three.

Next we evaluated the usefulness of the bar graphs per month. Subjects again operated three bar graphs, (a)–2, (b)–2, and (c)–2, for six months and answered the same questions as for the daily graphs. The results for "Was the bar graph A (B or C) useful for understanding your emotions?" were 3.4 (A), 2.0 (B), and 3.0 (C). Three subjects ranked graph C first and two ranked graph A first. Graph B was ranked worst by all the subjects. Among the three graphs, graph (a) was best for the average values and graph (c) was best for ranking. Bar graph (b) per month is not very useful.

From the results of Experiment 2, we identified the usefulness of bar graph A per day and per month. Since graph C is superior to graph B, it compensates for the latter's disadvantage.

3.3 Experiment 3: Tool

Finally, we evaluated the utility of our entire tool. After using it, our subjects described the most positive and negative events for a month and answered the questions shown in Table 1.

The average values exceeded 4.0 except for Q3 and Q4. The bar graphs and the log lists were useful, but the words were less useful. The following are some comments from the subjects: "I had fun during the experiment," "It was fun because I recalled some good memories," "I recollected past memories." "It might be useful for looking back at the end of the year." We believe that our approach and tool are promising for self-awareness and emotions and remembering.

Table 1. Results in experiment 3: tool

Questions	Mean	SD
Q1 Was the bar graph useful for understanding your emotions?	4.0	0.00
Q2 Was the tweet log useful for understanding your emotions?	4.4	0.55
Q3 Were the words useful for understanding your emotions?	2.8	0.84
Q4 Was the pie chart useful for understanding your emotions?	3.0	0.70
Q5 Was the system useful for understanding your emotions?	4.4	0.55
Q6 Was the system useful for recalling your memories?	4.4	0.55
Q7 Was the system useful for understanding yourself?	4.0	0.71
Q8 Was using the system fun?	4.8	0.45
Q9 Do you want to use the system in the future?	4.6	0.55

4   Related Work

This research is a part of our work on storing personal information. Using Twitter data, Murakami et al. [1] developed a knowledge-space browser that displays a network and another work [2,3] developed a tag browser that displays a tag cloud to support human memory recollection.

Much research assigns sentiment values to texts, some of which are publicly available. Google SA is the most popular such service. Our preliminary investigation tried some services and research and decided to use it for getting initial scores. We presented three types of scoring for the bar graphs: maximal, average, and modified average.

Other research also did sentiment analysis and visualization from tweets. Kumamoto et al. [4] proposed a web application system for visualizing Twitter users based on temporal changes in the impressions from tweets. Mitsui et al. [5] developed an emotion lifelog system called Emote that classifies the emotions in Twitter and Facebook texts into six categories and visualizes them on a calendar. Wang et al. [6] presented a system for the real-time analysis of public sentiment toward presidential candidates in the 2012 U.S. election from Twitter using graph representations. Applications are available for sentiment analysis and visualization from tweets, such as sentiment viz [7]. Our research's purpose, method of calculation, and interface are different.

The following are the three main contributions of this paper: (a) presented an approach for using sentiment analysis from tweets for self-understanding; (b) presented a prototype tool; and (c) described how our approach and tool are promising for understanding self and emotions and recalling memories.

Giachanou and Crestani [8] surveyed sentiment analysis on Twitter and discussed fields related to sentiment analysis on it. They didn't address our task: self-understanding.

5 Summary

We developed a tool that displays bar graphs, tweet log lists, pie charts, and word lists using Twitter sentiment analysis for self-understanding support. Preliminary experiments revealed that our approach and tool are promising for self-understanding, understanding emotions, and recalling memories. We need in the future to improve the algorithms that calculate the modified average values and extract positive/negative words and the tool, especially such interfaces as selecting bar graphs. We also need to evaluate the tool with more subjects and both genders.

References

1. Murakami, H., Mitsuhashi, K., Senba, K.: Creating user's knowledge space from various information usages to support human recollection. In: Jiang, H., Ding, W., Ali, M., Wu, X. (eds.) IEA/AIE 2012. LNCS (LNAI), vol. 7345, pp. 596–605. Springer, Heidelberg (2012). https://doi.org/10.1007/978-3-642-31087-4_61
2. Matsumoto, M., Matsuura, S., Mitsuhashi, K., Murakami H.: Supporting human recollection of the impressive events using the number of photos. In: Proceedings of the 6th International Conference on Agents and Artificial Intelligence - Volume 1: ICAART, pp. 538–543 (2014)
3. Murakami, H., Murakami, R.: A system using tag cloud for recalling personal memories. In: Wotawa, F., Friedrich, G., Pill, I., Koitz-Hristov, R., Ali, M. (eds.) IEA/AIE 2019. LNCS (LNAI), vol. 11606, pp. 398–405. Springer, Cham (2019). https://doi.org/10.1007/978-3-030-22999-3_35
4. Kumamoto, T., Wada, H., Suzuki, T.: Visualizing temporal changes in impressions from tweets. In: Proceedings of the iiWAS 2014, pp. 116–125. ACM Press, New York (2014). https://doi.org/10.1145/2684200.2684279
5. Mitsui, T., Ito, T., Nakanishi, H., Hamakawa, R.: Emote: emotion lifelog system using SNS. In: IPSJ SIG Technical Report, 2014-EC-32, vol. 1, pp. 1–6 (2014). (in Japanese)
6. Wang, H., Can, D., Kazemzadeh, A., Bar, F., Narayanan, S.: A system for real-time Twitter sentiment analysis of 2012 U.S. presidential election cycle. In: Proceedings of the ACL 2012, pp. 115–120 (2012)
7. Healey, C.: Tweet Sentiment Visualization App. https://www.csc2.ncsu.edu/faculty/healey/tweet_viz/tweet_app/
8. Giachanou, A., Crestani, F.: Like it or not: a survey of twitter sentiment analysis methods. ACM Comput.Surv. 49(2), 1–41 (2016). https://doi.org/10.1145/2938640. ACM Press, New York

Integrating Crowdsourcing and Active Learning for Classification of Work-Life Events from Tweets

Yunpeng Zhao[1], Mattia Prosperi[1], Tianchen Lyu[1], Yi Guo[1], Le Zhou[2], and Jiang Bian[1(✉)]

[1] University of Florida, Gainesville, FL 32611, USA
{yup111,m.prosperi,paint.94,yiguo,bianjiang}@ufl.edu
[2] University of Minnesota, Twin Cities, MN 55455, USA
zhoul@umn.edu

Abstract. Social media, especially Twitter, is being increasingly used for research with predictive analytics. In social media studies, natural language processing (NLP) techniques are used in conjunction with expert-based, manual and qualitative analyses. However, social media data are unstructured and must undergo complex manipulation for research use. The manual annotation is the most resource and time-consuming process that multiple expert raters have to reach consensus on every item, but is essential to create gold-standard datasets for training NLP-based machine learning classifiers. To reduce the burden of the manual annotation, yet maintaining its reliability, we devised a crowdsourcing pipeline combined with active learning strategies. We demonstrated its effectiveness through a case study that identifies job loss events from individual tweets. We used Amazon Mechanical Turk platform to recruit annotators from the Internet and designed a number of quality control measures to assure annotation accuracy. We evaluated 4 different active learning strategies (i.e., least confident, entropy, vote entropy, and Kullback-Leibler divergence). The active learning strategies aim at reducing the number of tweets needed to reach a desired performance of automated classification. Results show that crowdsourcing is useful to create high-quality annotations and active learning helps in reducing the number of required tweets, although there was no substantial difference among the strategies tested.

Keywords: Social media · Crowdsourcing · Active learning

1 Introduction

Micro-blogging social media platforms have become very popular in recent years. One of the most popular platforms is Twitter, which allows users to broadcast short texts (i.e., 140 characters initially, and 280 characters in a recent platform update) in real time with almost no restrictions on content. Twitter is a source of people's attitudes, opinions, and thoughts toward the things that happen in

© Springer Nature Switzerland AG 2020
H. Fujita et al. (Eds.): IEA/AIE 2020, LNAI 12144, pp. 333–344, 2020.
https://doi.org/10.1007/978-3-030-55789-8_30

their daily life. Twitter data are publicly accessible through Twitter application programming interface (API); and there are several tools to download and process these data. Twitter is being increasingly used as a valuable instrument for surveillance research and predictive analytics in many fields including epidemiology, psychology, and social sciences. For example, Bian et al. explored the relation between promotional information and laypeople's discussion on Twitter by using topic modeling and sentiment analysis [1]. Zhao et al. assessed the mental health signals among sexual and gender minorities using Twitter data [2]. Twitter data can be used to study and predict population-level targets, such as disease incidence [3], political trends [4], earthquake detection [5], and crime perdition [6], and individual-level outcomes or life events, such as job loss [7], depression [8], and adverse events [9]. Since tweets are unstructured textual data, natural language processing (NLP) and machine learning, especially deep learning nowadays, are often used for preprocessing and analytics. However, for many studies [10–12], especially those that analyze individual-level targets, manual annotations of several thousands of tweets, often by experts, is needed to create gold-standard training datasets, to be fed to the NLP and machine learning tools for subsequent, reliable automated processing of millions of tweets. Manual annotation is obviously labor intense and time consuming.

Crowdsourcing can scale up manual labor by distributing tasks to a large set of workers working in parallel instead of a single people working serially [13]. Commercial platforms such as Amazon's Mechanical Turk (MTurk, https:// www.mturk.com/), make it easy to recruit a large crowd of people working remotely to perform time consuming manual tasks such as entity resolution [14,15], image or sentiment annotation [16,17]. The annotation tasks published on MTurk can be done on a piecework basis and, given the very large pool of workers usually available (even by selecting a subset of those who have, say, a college degree), the tasks can be done almost immediately. However, any crowdsourcing service that solely relies on human workers will eventually be expensive when large datasets are needed, that is often the case when creating training datasets for NLP and deep learning tasks. Therefore, reducing the training dataset size (without losing performance and quality) would also improve efficiency while contain costs.

Query optimization techniques (e.g., active learning) can reduce the number of tweets that need to be labeled, while yielding comparable performance for the downstream machine learning tasks [18–20]. Active learning algorithms have been widely applied in various areas including NLP [21] and image processing [22]. In a pool-based active learning scenario, data samples for training a machine learning algorithm (e.g., a classifier for identifying job loss events) are drawn from a pool of unlabeled data according to some forms of informativeness measure (a.k.a. active learning strategies [23]), and then the most informative instances are selected to be annotated. For a classification task, in essence, an active learning strategy should be able to pick the "best" samples to be labelled that will improve the classification performance the most.

In this study, we integrated active learning into a crowdsourcing pipeline for the classification of life events based on individual tweets. We analyzed the quality of crowdsourcing annotations and then experimented with different machine/deep learning classifiers combined with different active learning strategies to answer the following two research questions (RQs):

- **RQ1.** How does (1) the amount of time that a human worker spends on and (2) the number of workers assigned to each annotation task impact the quality of an-notation results?
- **RQ2.** Which active learning strategy is most efficient and cost-effective to build event classification models using Twitter data?

2   Methods

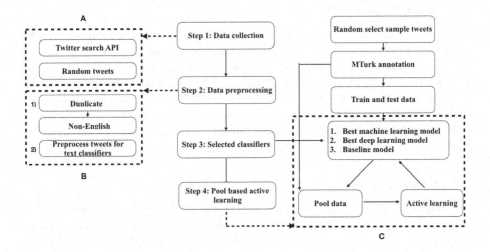

Fig. 1. The workflow of our Twitter analysis pipeline.

We first collected tweets based on a list of job loss-related keywords. We then randomly selected a set of sample tweets and had these tweets annotated (i.e., whether the tweet is a job loss event) using the Amazon MTurk platform. With these annotated tweets, we then evaluated 4 different active learning strategies (i.e., least confident, entropy, vote entropy, and Kullback-Leibler (KL) divergence) through simulations.

2.1   Data Collection

Our data were collected from two data sources based on a list of job loss-related keywords. The keywords were developed using a snowball sampling process, where we started with an initial list of 8 keywords that indicates a job-loss event (e.g., "got fired" and "lost my job"). Using these keywords, we then queried (1) Twitter's own search engine (i.e., https://twitter.com/search-home?lang=en), and (2) a database of public random tweets that we have collected using the Twitter steaming application programming interface (API) from January 1, 2013 to December 30, 2017, to identify job loss-related tweets. We then manually reviewed a sample of randomly selected tweets to discover new job loss-related keywords. We repeated the search then review process iteratively until no new keywords were found. Through this process, we found 33 keywords from the historical random tweet database and 57 keywords through Twitter web search. We then (1) not only collected tweets based on the over-all of 68 unique keywords from the historical random tweet database, but also (2) crawled new Twitter data using Twitter search API from December 10, 2018 to December 26, 2018 (17 days).

2.2 Data Preprocessing

We preprocessed the collected data to eliminate tweets that were (1) duplicated or (2) not written in English. For building classifiers, we preprocessed the tweets following the preprocessing steps used by GloVe [24] with minor modifications as follows: (1) all hashtags (e.g., "#gotfired") were replaced with "<hashtag> PHRASE" (e.g.,, "<hashtag> gotfired"); (2) user mentions (e.g., "@Rob_Bradley") were replaced with "<user>"; (3) web links (eg, "https:// t.co/fMmFWAHEuM") were replaced with "<url>"; and (4) all emojis were replaced with "<emoji>."

2.3 Classifier Selection

Machine learning and deep learning have been wildly used in classification of tweets tasks. We evaluated 8 different classifiers: 4 traditional machine learning models (i.e., logistic regress [LR], Naïve Bayes [NB], random forest [RF], and support vector machine [SVM]) and 4 deep learning models (i.e., convolutional neural network [CNN], recurrent neural network [RNN], long short-term memory [LSTM] RNN, and gated recurrent unit [GRU] RNN). 3,000 tweets out of 7,220 Amazon MTurk annotated dataset was used for classifier training (n = 2,000) and testing (n = 1,000). The rest of MTurk annotated dataset were used for the subsequent active learning experiments. Each classifier was trained 10 times and 95 confidence intervals (CI) for mean value were reported. We explored two language models as the features for the classifiers (i.e., n-gram and word-embedding). All the machine learning classifiers were developed with n-gram features; while we used both n-gram and word-embedding features on the CNN classifier to test which feature set is more suitable for deep learning classifiers. CNN classifier with word embedding features had a better performance which is consistent with other studies [25,26] We then selected one machine learning and one deep learning classifiers based on the prediction performance (i.e., F-score). Logistic regression was used as the baseline classifier.

2.4   Pool-Based Active Learning

In pool-based sampling for active learning, instances are drawn from a pool of samples according to some sort of informativeness measure, and then the most informative instances are selected to be annotated. This is the most common scenario in active learning studies [27]. The informativeness measures of the pool instances are called active learning strategies (or query strategies). We evaluated 4 active learning strategies (i.e., least confident, entropy, vote entropy and KL divergence). Figure 1C shows the workflow of our pool-based active learning experiments: for a given active learning strategy and classifiers trained with an initial set of training data (1) the classifiers make predictions of the remaining to-be-labelled dataset; (2) a set of samples is selected using the specific active learning strategy and annotated by human reviewers; (3) the classifiers are retrained with the newly annotated set of tweets. We repeated this process iteratively until the pool of data exhausts. For the least confident and entropy active learning strategies, we used the best performed machine learning classifier and the best performed deep learning classifier plus the baseline classifier (LR). Note that vote entropy and KL divergence are query-by-committee strategies, which were tested upon three deep learning classifiers (i.e., CNN, RNN and LSTM) and three machine learning classifiers (i.e., LR, RF, and SVM) as two separate committees, respectively.

3   Results

3.1   Data Collection

Our data came from two different sources as shown in Table 1. First, we collected 2,803,164 tweets using the Twitter search API [28] from December 10, 2018 to December 26, 2018 base on a list of job loss-related keywords ($n = 68$). After filtering out duplicates and non-English tweets, 1,952,079 tweets were left. Second, we used the same list of keywords to identify relevant tweets from a database of historical random public tweets we collected from January 1, 2013 to December 30, 2017. We found 1,733,905 relevant tweets from this database. Due to the different mechanisms behind the two Twitter APIs (i.e., streaming API vs. search API), the volumes of the tweets from the two data sources were significantly different. For the Twitter search API, users can retrieve most of the public tweets related to the provided keywords within 10 to 14 days before the time of data collection; while the Twitter streaming API returns a random sample (i.e., roughly 1% to 20% varying across the years) of all public tweets at the time and covers a wide range of topics. After integrating the tweets from the two data sources, there were 3,685,984 unique tweets.

3.2   RQ1. How Does (1) the Amount of Time that a Human Worker Spends on and (2) the Number of Workers Assigned to Each Annotation Task Impact the Quality of Annotation Results?

We randomly selected 7,220 tweets from our Twitter data based on keyword distributions and had those tweets annotated using workers recruited through

Table 1. Descriptive statistics of job loss-related tweets from our two data sources.

Data source	Year	# of tweets	# of English tweets
Historical random public tweet database	2013	434, 624	293, 664
	2014	468, 432	279, 876
	2015	401, 861	228, 301
	2016	591, 948	322, 459
	2017	1, 299, 006	609, 605
Collected through Twitter search API	2018	2, 803, 164	1, 952, 079
	Total	5, 999, 035	3, 685, 984

Amazon MTurk. Each tweet was also annotated by an expert annotator (i.e., one of the authors). We treated the consensus answer of the crowdsourcing workers (i.e., at least 5 annotators for each tweet assignment) and the expert annotator as the gold-standard. Using control tweets is a common strategy to identify workers who cheat (e.g., randomly select an answer without reading the instructions and/or tweets) on annotation tasks. We introduced two control tweets in each annotation assignment, where each annotation assignment contains a total of 12 tweets (including the 2 control tweets). Only responses with the two control tweets answered corrected were considered valid responses and the worker would receive the 10 cents incentive.

The amount of time that a worker spends on a task is another factor associated with annotation quality. We measured the time that one spent on clicking through the annotation task without thinking about the content and repeated the experiment five times. The mean amount time spent on the task is 57.01 (95% CI [47.19, 66.43]) seconds. Thus, responses with less than 47 s were considered invalid regardless how the control tweets were answered.

We then did two experiments to explore the relation between the amount of time that workers spend on annotation tasks and annotation quality. Figure 2A shows annotation quality by selecting different amounts of lower cut-off time (i.e., only considering assignments where workers spent more time than the cut-off time as valid responses), which tests whether the annotation is of low quality when workers spent more time on the task. The performance of the crowdsourcing workers was measured by the agreement (i.e., Cohan's kappa) between labels from each crowdsourcing worker and the gold-standard labels. Figure 2B shows annotation quality by selecting different upper cut-off time (i.e., keep assignments whose time consumption were less than the cut-off time), which tests whether the annotation is of low quality when workers spent less time on the task. As shown in Fig. 2A and B, it does not affect the annotation quality when a worker spent more time on the task; while, the annotaion quality is significantly lower if the worker spent less than 90 s on the task.

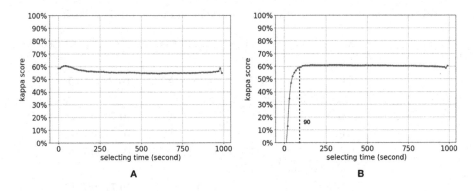

Fig. 2. Annotation quality by selecting different cut-off time.

We also tested the annotation reliability (i.e., Fleiss' Kappa score) between using 3 workers vs. using 5 workers. The Fleiss' kappa score of 3 workers is 0.53 (95% CI [0.46, 0.61]. The Fleiss' kappa score of 5 workers is 0.56 (95% CI [0.51, 0.61]. Thus, using 3 workers vs. 5 workers does not make any difference on the annotation reliability, while it is obviously cheaper to use only 3 workers.

3.3 RQ2. Which Active Learning Strategy is Most Efficient and Cost-Effective to Build Event Classification Models Using Twitter Data?

We randomly selected 3,000 tweets from the 7,220 MTurk annotated dataset to build the initial classifiers. Two thousands out of 3,000 tweets were used to train the classifiers and the rest 1,000 tweets were used as independent test dataset to benchmark their performance. We explored 4 machine learning classifiers (i.e., Logistic Regression [LR], Naïve Bayes [NB], Random Forest [RF], and Support Vector Machine [SVM]) and 4 deep learning classifiers (i.e., Convolutional Neural Network [CNN], Recurrent Neural Network [RNN], Long Short-Term Memory [LSTM], and Gated Recurrent Unit [GRU]). Each classifier was trained 10 times. The performance was measured in terms of precision, recall, and F-score. 95% confidence intervals (CIs) of the mean F-score across the ten runs were also reported. Table 2 shows the performance of classifiers. We chose logistic regression as the baseline model. RF and CNN were chosen for subsequent active learning experiments, since they outperformed other machine learning and deep learning classifiers.

We implemented a pool-based active learning pipeline to test which classifier and active learning strategy is most efficient to build up an event classification classifier of Twitter data. We queried the top 300 most "informative" tweets from the rest of the pool (i.e., excluding the tweets used for training the classifiers) at each iteration. Table 3 shows the active learning and classifier combinations that we evaluated. The performance of the classifiers was measured by F-score. Figure 3 shows the results of the different active learning strategies

Table 2. The performance of machine learning and deep learning classifiers.

Feature	Model name	Precision	Recall	F-score	95% CIs of F-score
N-gram	Baseline				
	LR	**0.75**	**0.75**	**0.74**	**(0.74, 0.75)**
	Machine learning				
	NB	0.73	0.73	0.72	(0.72, 0.73)
	RF	0.74	0.74	0.74	(0.74, 0.75)
	SVM	0.73	0.73	0.72	(0.71, 0.73)
	Deep learning				
	CNN	0.71	0.71	0.72	(0.71, 0.73)
Word-embedding	**CNN**	**0.79**	**0.79**	**0.79**	**(0.77, 0.80)**
	RNN	0.75	0.75	0.74	(0.74, 0.75)
	LSTM	0.72	0.72	0.71	(0.69, 0.72)
	LSTM (GRU)	0.75	0.75	0.74	(0.72, 0.76)

combined with LR (i.e., the baseline), RF (i.e., the best performed machine learning model), and CNN (i.e., the best performed deep learning model). For both machine learning models (i.e., LR and RF), using the entropy strategy can reach the optimal performance the quickest (i.e., the least amount of tweets). While, the least confident algorithm does not have any clear advantages compared with random selection. For deep learning model (i.e., CNN), none of the active learning strategies tested are useful to improve the CNN classifier's performance. Figure 4 shows the results of query-by-committee algorithms (i.e., vote entropy and KL divergence) combined with machine learning and deep learning ensemble classifiers. Query-by-committee algorithms are slightly better than random selection when it applied to machine learning ensemble classifier. However, query-by-committee algorithms are not useful for the deep learning ensemble classifier.

Table 3. The active learning strategy and classifier combinations tested.

Model	Query strategies
LR	Random query, least confident, entropy
RF	Random query, least confident, entropy
CNN	Random query, least confident, entropy
Ensemble[1]	Vote entropy, KL divergence

[1] machine learning ensemble classifier: LR, RF, and SVM., Deep learning ensemble classifier: CNN, RNN, and LSTM

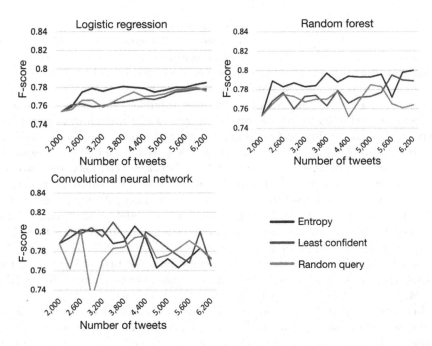

Fig. 3. The performance of active learning strategies combined with linear regression, random forest, and convolutional neural network.

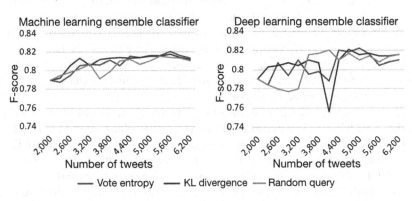

Fig. 4. The performance of active learning strategies combined with machine learning and deep learning ensemble classifiers.

4 Discussion

The goal of our study was to test the feasibility of building classifiers by using crowdsourcing and active learning strategies. We had 7,220 sample job loss-related tweets annotated using Amazon MTurk, tested 8 classification models, and evaluated 4 active learning strategies to answer our two RQs.

The key benefit of crowdsourcing is to have a large number of workers available to carry out tasks on a piecework basis. This means that it is likely to get the crowd to start work on tasks almost immediately and be able to have a large number of tasks completed quickly. However, even welltrained workers are only human and can make mistakes. Our first RQ was to find an optimal and economical way to get reliable annotations from crowdsourcing. Beyond using control tweets, we tested different cut-off time to assess how the amount of time workers spent on the task would affect annotation quality. We found that the annotation quality is low if the tasks were finished within 90 s. We also found that the annotation quality is not affected by the number of workers (i.e., between 3 worker group vs 5 worker group), which was also demonstrated by Mozafari et al. [29].

In second RQ, we aimed to find which active learning strategy is most efficient and cost-effective to build event classification models using Twitter data. We started with selecting representative machine learning and deep learning classifiers. Among the 4 machine learning classifiers (i.e., LR, NB, RF, and SVM), LR and RF classifiers have the best performance on the task of identifying job loss events from tweets. Among the 4 deep learning methods (i.e., CNN, RNN, LSTM, LSTM with GRU), CNN has the best performance.

In active learning, the learning algorithm is set to proactively select a subset of available examples to be manually labeled next from a pool of yet unlabeled instances. The fundamental idea behind the concept is that a machine learning algorithm could potentially achieve a better accuracy quicker and using fewer training data if it were allowed to choose the most informative data it wants to learn from. In our experiment, we found that the entropy algorithm is the best way to build machine learning models fast and efficiently. Vote entropy and KL divergence, the query-by-committee active learning methods are helpful for the training of machine learning ensemble classifiers. However, all the active learning strategies we tested do not work well with deep learning model (i.e., CNN) or deep learning-based ensemble classifier.

We also recognize the limitations of our study. First, we only tested 5 classifiers (i.e., LR, RF, CNN, a machine learning ensemble classifier, and a deep learning classifier) and 4 active learning strategies (i.e., least confident, entropy, vote entropy, KL divergence). Other state-of-art methods for building tweet classifiers (e.g., BERT [30]) and other active learning strategies (e.g., variance reduction [31]) are worth exploring. Second, other crowdsourcing quality control methods such as using prequalification questions to identify high-quality workers also warrant further investigations. Third, the crowdsourcing and active learning pipeline can potentially be applied to other data and tasks. However, more experiments are needed to test the feasibility. Fourth, the current study only focused on which active learning strategy is most efficient and cost-effective to build event classification models using crowdsourcing labels. Other research questions such as how the correctness of the crowdsourced labels would impact classifier performance warrant future investigations.

In sum, our study demonstrated that crowdsourcing with active learning is a possible way to build up machine learning classifiers efficiently. However, active learning strategies do not benefit deep learning classifiers in our study.

Acknowledgement. This study was supported by NSF Award #1734134.

References

1. Bian, J., et al.: Using social media data to understand the impact of promotional information on laypeople's discussions: a case study of lynch syndrome. J. Med. Internet Res. **19**(12), e414 (2017)
2. Zhao, Y., et al.: Assessing mental health signals among sexual and gender minorities using twitter data. In: 2018 IEEE International Conference on Healthcare Informatics Workshop (ICHI-W), pp. 51–52. IEEE, New York (2018)
3. Eichstaedt, J.C., et al.: Psychological language on twitter predicts county-level heart disease mortality. Psychol. Sci. **26**(2), 159–169 (2015)
4. Gayo-Avello, D.: A meta-analysis of state-of-the-art electoral prediction from twitter data. Soc. Sci. Comput. Rev. **31**(6), 649–679 (2013)
5. Sakaki, T., Okazaki, M., Matsuo, Y.: Earthquake shakes Twitter users: real-time event detection by social sensors. In: Proceedings of the 19th international conference on World wide web - WWW 2010, p. 851. ACM Press, Raleigh (2010)
6. Wang, X., Gerber, M.S., Brown, D.E.: Automatic crime prediction using events extracted from twitter posts. In: Yang, S.J., Greenberg, A.M., Endsley, M. (eds.) SBP 2012. LNCS, vol. 7227, pp. 231–238. Springer, Heidelberg (2012). https://doi.org/10.1007/978-3-642-29047-3_28
7. Du, X., Bian, J., Prosperi, M.: An operational deep learning pipeline for classifying life events from individual tweets. In: Lossio-Ventura, J.A., Muñante, D., Alatrista-Salas, H. (eds.) SIMBig 2018. CCIS, vol. 898, pp. 54–66. Springer, Cham (2019). https://doi.org/10.1007/978-3-030-11680-4_7
8. Leis, A., Ronzano, F., Mayer, M.A., Furlong, L.I., Sanz, F.: Detecting signs of depression in tweets in Spanish: behavioral and linguistic analysis. J. Med. Internet Res. **21**(6), e14199 (2019)
9. Wang, J., Zhao, L., Ye, Y., Zhang, Y.: Adverse event detection by integrating twitter data and VAERS. J. Biomed. Seman. **9**(1), 1–10 (2018)
10. Finin, T., Murnane, W., Karandikar, A., Keller, N., Martineau, J., Dredze, M.: Annotating named entities in twitter data with crowdsourcing. In: Proceedings of the NAACL HLT 2010 Workshop on Creating Speech and Language Data with Amazon's Mechanical Turk, pp. 80–88. Association for Computational Linguistics, Los Angeles (2010)
11. Mozetič, I., Grčar, M., Smailović, J.: Multilingual twitter sentiment classification: the role of human annotators. PLOS ONE **11**(5), e0155036 (2016)
12. Stowe, K., et al.: Developing and evaluating annotation procedures for twitter data during hazard events. In: Proceedings of the Joint Workshop on Linguistic Annotation, Multiword Expressions and Constructions (LAW-MWE-CxG-2018), pp. 133–143. Association for Computational Linguistics, Santa Fe (2018)
13. Carenini, G., Cheung, J.C.K.: Extractive vs. NLG-based abstractive summarization of evaluative text: the effect of corpus controversiality. In: Proceedings of the Fifth International Natural Language Generation Conference, pp. 33–41. Association for Computational Linguistics, Salt Fork (2008)
14. Arasu, A., Götz, M., Kaushik, R.: On active learning of record matching packages. In: Proceedings of the 2010 International Conference on Management of Data - SIGMOD 2010, p. 783. ACM Press, Indianapolis (2010)

15. Bellare, K., Iyengar, S., Parameswaran, A.G., Rastogi, V.: Active sampling for entity matching. In: Proceedings of the 18th ACM SIGKDD International Conference on Knowledge Discovery and Data Mining - KDD 2012, p. 1131. ACM Press, Beijing (2012)

16. Vijayanarasimhan, S., Grauman, K.: Cost-sensitive active visual category learning. Int. J. Comput. Vis. **91**(1), 24–44 (2011)

17. Pak, A., Paroubek, P.: Twitter as a corpus for sentiment analysis and opinion mining. In: LREC (2010)

18. Marcus, A., Karger, D., Madden, S., Miller, R., Oh, S.: Counting with the crowd. Proc. VLDB Endowment **6**(2), 109–120 (2012)

19. Franklin, M.J., Kossmann, D., Kraska, T., Ramesh, S., Xin, R.: CrowdDB: answering queries with crowdsourcing. In: Proceedings of the 2011 International Conference on Management of Data - SIGMOD 2011, p. 61. ACM Press, Athens (2011)

20. Parameswaran, A.G., Garcia-Molina, H., Park, H., Polyzotis, N., Ramesh, A., Widom, J.: CrowdScreen: algorithms for filtering data with humans. In: Proceedings of the 2012 International Conference on Management of Data - SIGMOD 2012, p. 361. ACM Press, Scottsdale (2012)

21. Tang, M., Luo, X., Roukos, S.: Active learning for statistical natural language parsing. In: Proceedings of the 40th Annual Meeting of the Association for Computational Linguistics, pp. 120–127. Association for Computational Linguistics, Philadelphia (2002)

22. Wang, K., Zhang, D., Li, Y., Zhang, R., Lin, L.: Cost-effective active learning for deep image classification. IEEE Trans. Circ. Syst. Video Technol. **27**(12), 2591–2600 (2017). arXiv: 1701.03551

23. Settles, B.: Active Learning Literature Survey. Computer Sciences Technical Report 1648, University of Wisconsin-Madison (2009)

24. Pennington, J., Socher, R., Manning, C.D.: GloVe: global vectors for word representation. In: Empirical Methods in Natural Language Processing (EMNLP), pp. 1532–1543 (2014)

25. Le, H.P., Le, A.C.: A comparative study of neural network models for sentence classification. In: 2018 5th NAFOSTED Conference on Information and Computer Science (NICS), pp. 360–365 (2018)

26. Badjatiya, P., Gupta, S., Gupta, M., Varma, V.: Deep learning for hate speech detection in tweets. In: Proceedings of the 26th International Conference on World Wide Web Companion - WWW 2017 Companion, pp. 759–760. ACM Press, Perth (2017)

27. Min, X., Shi, Y., Cui, L., Yu, H., Miao, Y.: Efficient crowd-powered active learning for reliable review evaluation. In: Proceedings of the 2nd International Conference on Crowd Science and Engineering - ICCSE 2017, pp. 136–143. ACM Press, Beijing (2017)

28. Twitter, I.: Twitter developer API reference index (2020)

29. Mozafari, B., Sarkar, P., Franklin, M., Jordan, M., Madden, S.: Scaling up crowdsourcing to very large datasets: a case for active learning. Proc. VLDB Endowment **8**(2), 125–136 (2014)

30. Devlin, J., Chang, M.W., Lee, K., Toutanova, K.: BERT: pre-training of deep bidirectional transformers for language understanding (2018). arXiv preprint arXiv:1810.04805

31. Yang, Y., Loog, M.: A variance maximization criterion for active learning. Pattern Recogn. **78**, 358–370 (2018)

Many-to-One Stable Matching
for Prediction in Social Networks

Ke Dong[1,2], Zengchang Qin[1,3(✉)], and Tao Wan[4]

[1] Intelligent Computing and Machine Learning Lab, School of ASEE,
Beihang University, Beijing, China
zcqin@buaa.edu.cn
[2] Dynamics System Lab, University of Toronto, Toronto, Canada
[3] AI Research, Codemao, Shenzhen, China
[4] School of Biological Science and Medical Engineering,
Beihang University, Beijing, China
taowan@buaa.edu.cn

Abstract. Stable matching investigates how to pair elements of two
disjoint sets with the purpose to achieve a matching that satisfies all
participants based on their preference lists. In this paper, we consider the
case of matching with incomplete information in a social network where
agents are not fully connected. A new many-to-one matching algorithm is
proposed based on the classical Gale-Shapley algorithm with constraints
of given network topology. In simulated experiments, we find that the
matching outcomes in scale-free networks yield the best average utility
with least connective costs compared to other structured networks in one-
to-one problems. But in many-to-one matching cases, network structure
has no significant influence on matching utilities. We also apply the new
matching model to a real-world social network matching problem and
we find a significant increase of accuracy in matching pair prediction
comparing to classical methods.

Keywords: Stable matching · One-to-one matching · Many-to-one
matching · Gale-Shapley algorithm

1 Introduction

The presence of game theory in computer science has become impossible to ignore
in recent years. It is an interesting historic fact that both modern computer
science and game theory originated in Princeton under the influence of John von
Neumann. The emerging field of *Algorithmic Game Theory* [20] has attracted
much attention as an interaction between computer science, artificial intelligence
and game theory. *Stable matching* is an interesting topic in game theory, it
mainly concerns about how to pair members of two disjoint sets to satisfy all
participants based on their preferences. This problem is first put forward in [6]
as the *stable marriage* problem, where a set of men and a set of women need to
be matched. Each person ranks people in the opposite sex in accordance to his

© Springer Nature Switzerland AG 2020
H. Fujita et al. (Eds.): IEA/AIE 2020, LNAI 12144, pp. 345–356, 2020.
https://doi.org/10.1007/978-3-030-55789-8_31

or her preferences for a marriage partner. After receiving everyone's *preference list*, a centralized matching system gives matching results where there is no man and woman that both prefer each other than their current partners, and this is called stable matching. Gale and Shapley showed that every instance of such stable marriage problem allows at least one stable matching with an $O(n^2)$ algorithm named after them [6].

Besides the marriage problem, the stable matching in two-sided markets has many other real-world matching scenarios like hospital-resident (H-R) [18], firm-worker and university-student [10]. These problems can be roughly divided into two categories, namely *one-to-one* matching and *many-to-one* matching, according to whether multiple agents can be matched to a same agent in the other side. Roth[1] pointed out that one-to-one problems are not equivalent to many-to-one problems from the point of dominant strategies and side-optimal matching mechanisms [17]. Recently, the Knaster-Tarski fixed point theory is used to explain the stable marriage problem [4]. The stable marriage problem with unequal number of men and women is investigated by [14]. Roth [15] also introduced the incomplete preference list, where women can regard some men as unacceptable and prefer to remain single than married to them. In the research of [7] and [8], indifferent preference is allowed, and there would still be a stable matching result for every marriage problem. Generally, these literature mainly focus on weakening the assumption of complete and strict preference list given the classical matching problems, yet throw little light on the assumption of complete accessibility information caused by social networks. Accessibility between different agents could be influenced by the actions of his or her neighbors. This is well known as *six degree of separation* and it has been applied in the studies of network structure or dynamics [22].

Social networks are often modeled as graphs, where nodes (vertices) represent individuals and links (edges) represent connections between the individuals. Jackson *et al.* [9] developed a general framework for games including the marriage matching problems played in networks. But they mainly focus on how the matching results can influence the topology of networks, where networks are generated dynamically in their framework. Different from their works, in this research, we hope to understand how the topology of social networks can influence the choices that agents make in matching and if such matching pairs can be predicted in a social network.

2   Stable Matching

In setting of one-to-one matching like the marriage problem, there are finite sets $M = \{m_1, m_2, ..., m_{N_m}\}$ of men, and $W = \{w_1, w_2, ..., w_{N_w}\}$ of women. The preference of each man m_i over set W is represented by a vector $P^M(m_i) = [p^m_{i,1}, p^m_{i,2}, ..., p^m_{i,N_w}]$, where $p^m_{i,j} \in [1, N_w]$ means woman w_j is the $p^m_{i,j}$-th choice of man m_i. Similarly the women w_j's preference vector is

---

[1] A. Roth and L. Shapley shared Nobel Prize in Economics in 2012 for the theory of stable allocations and the practice of market design.

$P^W(w_j) = [p^w_{1,j}, p^w_{2,j}, ..., p^w_{N_m,j}]$. $w >_m w'$ denotes that a man m prefers w to w'. The preference matrix for man and woman can be represented by

$$\begin{aligned}
\mathbf{P}^M &= [P^M(m_1); P^M(m_2); \ldots; P^M(m_{N_w})] \\
\mathbf{P}^W &= [P^W(w_1); P^W(w_2); \ldots; P^W(w_{N_m})]
\end{aligned} \tag{1}$$

For simplicity, we assume $N_m = N_w$ and only strict preference (i.e. no ties) is allowed in this research. Then we introduce definitions of matching functions μ and matching utility u.

Definition 1. *[19] A matching μ is a function from the set of $M \cup W$ onto itself of order of two ($\mu^2(x) = x$) such that if $\mu(m) \neq m$, then $\mu(m)$ is in W and if $\mu(w) \neq w$, then $\mu(w)$ is in M. $\mu(x)$ is referred as the partner of x.*

The matching function μ is bidirectional and symmetrical, and $\mu(x) = x$ means element x remains unmatched. (w, m) is a *blocking pair* [6] if $w >_m \mu(m)$ and $m >_w \mu(w)$, namely both w and m prefer each other than their current partner. A *stable* matching is the matching result without blocking pairs. To better evaluate the matching results, the utility of agent is defined as follows.

Definition 2. *[13] The utility of an agent measures how well his/her preference list is meet in the final matching result.*

$$\begin{cases}
u(m) = 10\frac{N_w + 1 - P^m_{m,\mu(m)}}{N_w}, & u(m) \in [0, 10] \\
u(w) = 10\frac{N_m + 1 - P^w_{\mu(w),w}}{N_m}, & u(w) \in [0, 10]
\end{cases} \tag{2}$$

Zero utility is assigned to unmatched elements, and the average utility of a matching across all agents ($N = N_m + N_w$) is defined by: $U = \frac{1}{N}\sum_{i=1}^{N} u_i$.

Now, we consider the many-to-one matching problem in the scenario of hospital-resident problem [18]. Say there are finite sets $H = \{h_1, h_2, .., h_{N_h}\}$ of hospitals, and $R = \{r_1, r_2, ..., r_{N_r}\}$ of residents. Each resident can be matched to at most one hospital, but one hospital, say h_i, has a quota q_i indicating the maximum number of positions it may fill. Similar to the one-to-one problem, each hospital has preference over individual residents and each resident has preference over individual hospitals. The preference matrix for hospitals and residents can be denoted as

$$\begin{aligned}
\mathbf{P}^H &= [P^H(h_1); P^H(h_2); \ldots; P^H(h_{N_h})] \\
\mathbf{P}^R &= [P^R(r_1); P^R(r_2); \ldots; P^R(r_{N_r})],
\end{aligned} \tag{3}$$

where $P^H(h_i)$ and $P^R(r_i)$ are the preference vector of hospital h_i and r_i. The matching function γ is defined by the following definition.

Definition 3. *[19] A matching γ is a function from the set of $H \cup R$ onto the set of subsets of $H \cup R$, namely $Pwr(H \cup R)$, such that:*

$$\begin{aligned}
\gamma(h_i) &= \begin{cases} \{h_i\}, \text{No residents are admitted to } h_i \\ \{r_{i1}, \ldots, r_{is}\}, \text{residents admitted to } h_i \end{cases} \\
\gamma(r_j) &= \begin{cases} \{r_j\}, \text{No hospital has admitted } r_j \\ \{h_i\}, \text{Hospital } h_i \text{ has admitted } r_j \end{cases}
\end{aligned} \tag{4}$$

The definition of utility is slightly modified since hospitals can admit more than one residents.

Definition 4. *The utility of an agent in the hospital-resident problem:*

$$
\begin{cases}
u(h) = \frac{10}{|\gamma(h)|} \sum_{r \in \gamma(h)} \frac{N_r + 1 - P^h_{h,\gamma(r)}}{N_r}, & u(h) \in [0, 10] \\
u(r) = 10 \times \frac{N_h + 1 - P^r_{\gamma(r),r}}{N_h}, & u(r) \in [0, 10]
\end{cases}
\tag{5}
$$

Similarly, the average utility of the whole matching problem is $U = \frac{1}{N_h + N_r} \sum_i u_i$.

3 Stable Matching in Networks

3.1 Structured Networks

Most social, biological, and technological networks display substantial non-trivial topological features, with patterns of connection between their elements that are neither purely regular nor purely random. Complex networks have always been an important research field, there are various kinds of network structures. In this paper, we only considers the following four types of well-studied networks: Nearest-neighbor coupled network (NCN model) [1], Random network (Erods-Renyi model) [3], Small World network (Watts-Strogatz model) [23] and Scale-free network (Barabasi-Albert model) [2] (Fig. 1).

Fig. 1. Four types of networks with $N = 20$ nodes and $\overline{k} = 4$. From left to right: (a) Nearest-neighbor coupled (NCN) model; (b) Erods-Renyi (ER) model; (c) Watts-Strogatz (WS) model; (d) Barabasi-Albert (BA) model.

These structures are representative and often used together in other studies including [11] and [12]. The topology of the network decides the dynamics of the network, which could be characterized by two parameters: *degree distribution* $p(k)$ and *average path length* (APL) [21]. The degree distribution $p(k)$ is the probability distribution of node degrees over the whole network. And the average degree is $\overline{k} = \left(\sum_{i=1}^{N} k_i \right) / N$. The APL is defined as the average value of Dijkstra distance between any two nodes, namely $\overline{d} = \frac{\sum_{i>j} d(i,j)}{N(N-1)/2}$, where the Dijkstra distance refers to the shortest path connecting two nodes using the Dijkstra algorithm [13].

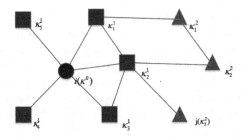

Fig. 2. An example of shortest path from node i to j in a given network. The nodes are colored based on the distance to the starting node i (κ^0). (Color figure online)

In real social networks, regional limitation and attenuation of information flow force people to develop neighborhoods to obtain indirect information. It also divides agents into groups implicitly and it is always costly to interact with agents far away. Thus it is necessary to characterize this phenomenon for the matching problem.

Definition 5 (D-neighborhood). *D-neighborhood are the nodes within the maximum permissible contact range D. For a given depth D and agent i and j, the function $\Delta(i, j|D)$ can determine whether j is in i's D-neighborhood:*

$$\Delta(i, j|D) = \begin{cases} 1 & d(i, j) \leq D \\ 0 & d(i, j) > D \end{cases} \tag{6}$$

$\Delta(i, j|D)$ is symmetric since $d(i, j) = d(j, i)$. $\sum_j \Delta(i, j|D)$ calculates the number of D-neighbors of the node i in a given network. Given $d(i, j) = l$ ($0 \leq l \leq D$), the shortest path from the starting node i (denoted by κ^0) to the ending node j (denoted by κ^l) can be denoted by: $\langle i(\kappa^0), \kappa^1, \kappa^2, \ldots, \kappa^{l-1}, j(\kappa^l) \rangle$

Figure 2 shows an example with the starting node i (green circle). The nodes within one unit distance are in blue square and the nodes within two units distance are in red triangle. The shortest path between node i and j is $\langle i(\kappa^0), \kappa_2^1, j(\kappa_3^2) \rangle$ with $d(i, j) = 2$, where κ_s^d represents the s-th node (clockwise) in the set of nodes with d units distance to the starting node. In order to find such a shortest path in the whole network, firstly κ_2^1 has to be selected from five blue nodes: $P(\kappa_2^1) = 1/5$. Then the next node has to be chosen from three red nodes with two units distance to the starting node i (κ^0), but only two of them are connected directly to κ_2^1. So the probability to choose κ_3^2 is $P(\kappa_3^2) = 1/2$. Thus, the probability of a node appearing in the shortest path can be calculated by:

$$P(\kappa^s) = \frac{1}{\sum_t \Delta(\kappa^{s-1}, \kappa_t^s|1)} \qquad s.t. \quad : P(\kappa^0) = 1 \tag{7}$$

Definition 6 (connective cost). *Connective cost $c_{i,j}$ measures the cost for agent i to know j through an intermediate path in a given network.*

$$c_{i,j} = \prod_{d=1}^{l} \log\left(\frac{1}{P(\kappa^d)}\right) * \exp(d) \tag{8}$$

Two factors are considered in the connective cost: (1) $c_{i,j} \propto \frac{1}{P(\kappa^d)}$, the lower probability a node has, the larger cost for it to get connected. (2) $c_{i,j} \propto \exp(d)$ implies that, the increase of cost grows exponentially with respect to the increase of depth. Logarithm is used here to re-scale the exponential cost when the network gets really large. The average connective cost between all matching pairs in a network with N nodes is: $C = \sum_{ij} \frac{c_{i,j}}{N/2}$, where C is the average connective cost of a network, and $N/2$ is the maximum number of potential matched pairs.

Definition 7 (network connectivity). *Network connectivity ϕ of a network refers to the proportion of the number of paths whose lengths are less than the maximum depth (D) to the number of all possible paths in the network.*

$$\phi = \frac{count(d \leq D)}{N(N-1)/2} \tag{9}$$

The parameter ϕ reveals the extent to which agents are connected with other nodes in the network. Intuitively, ϕ may be negatively correlated with the APL of networks, since with the increase of APL, it would be harder for two nodes to connect each other. For example, in the ER model, the distribution of the shortest path length follows the Poisson distribution [5]: Then, the connectivity of ER can be represented by:

$$\phi = \int_0^D \frac{\lambda^l e^{-\lambda}}{l!} dl \tag{10}$$

where λ is APL and D is the maximum depth. Thus, in the ER model, ϕ is negatively correlated with APL.

3.2 Many-to-One Matching Algorithm

Based on the *deferred acceptance mechanism* [6,16] in classical stable matching, we can propose the matching algorithm in networks, by using the hospital-resident problem as an example: in round one, each resident proposes to its most preferred accessible hospital, then each hospital rejects all but the most preferred residents to fulfill its quota. In each subsequent round, unmatched residents continue to propose to its most preferred accessible hospital which the resident has never applied before, and then each hospital compares the new applicants with admitted residents to select its most preferred residents. Such procedure ends when there is no free residents or free residents who have applied to all the accessible hospitals. The pseudo-code is shown in Algorithm 1. Note that *accessible* implies the existence of incomplete preference list. Agents would only give preferences to nodes within its D-neighborhood. In the implementation, we would assign the lowest preference to the nodes not in the D-neighborhood to ensure that no agent will be matched with them.

Algorithm 1. Many-to-One Stable Matching Algorithm in Networks.

Inputs: Network G, Maximum depth D, Preference matrix $\mathbf{P^H}, \mathbf{P^R}$ and
        Quota vector $Q = \{q_1, q_2, ... q_{N_h}\}$

Outputs: Matching outcome $\gamma : H \cup R \to Pwr\{H \cup R\}$

while for $r_j \in R$, if r_j is free and has never applied to at least one hospitals before **Do**
    $h_i \leftarrow r_j$'s top hospital in its preference list that it has never applied before
    if h_i still has at least one empty position
    then (h_i, r_j) become a match
    else $r_{j'} \leftarrow$ The resident with the lowest preference score in $\gamma(h_i)$
      if h_i prefers $r_{j'}$ to r_j
      then r_j stays free and apply to the next ranking hospital $h_{i'}$
        if $h_{i'}$ is beyond the D-neighbourhood of r_j
        then r_j stays free in this round
      else (h_i, r_j) become a match
        $r_{j'}$ becomes free

4 Experimental Studies

4.1 Simulated Results

In our experiments, only small networks with $N \leq 100$ are considered because of the time complexity. In implementation of marriage problem, nodes labeled as either man or woman, they can make acquaintance in a D-neighborhood [13]. However, this is not the case for many-to-one matching scenarios like the hospital-resident problem, because hospitals would not make friends with residents, i.e. the connections within hospitals have different meanings from those within residents. The former can be recognized as hospital cooperation and supervision, while the latter represents the friendship or acquaintance relationship between residents. Therefore, in this paper, we generate hospital networks and resident networks independently, and then connect hospital nodes and resident nodes randomly. Therefore, there are three different kinds of social connections in the hospital-resident problem: 1) resident-resident: this represents co-operations between residents, mainly sharing job information about hospitals. 2) resident-hospital: this means that the resident submits an application to the hospital. 3) hospital-hospital: this represents co-operations between hospitals, i.e. connected hospitals would share applicant information with each other.

In our simulated experiments, considering the fact that the WS model is an intermediate form of the NCN model and the ER model, and the BA model is totally different with the other three networks for its preferential attachment mechanism, we only use the BA model and the WS model to build the hospital network. The hospital network has $N_h = 30$ nodes with the average degree $\overline{k_h} = 2$, and the resident network has $N_r = 70$ nodes. The sum of job quotas $Q = \sum_i^{N_h} q_i$ equals to the number of resident nodes, namely $Q = N_r$, and these quotas are distributed to hospitals in proportion to their degrees. When the hospital network is modeled as BA network, its relationship of utility U against resident node degree $\overline{k_r}$ and a fixed maximum depth $D = 3$ is shown in Fig. 3:

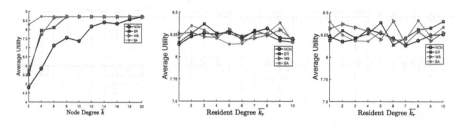

Fig. 3. Average utility of four networks with increasing node degree $\overline{k_r}$. Left-hand: one-to-one matching; middle: many-to-one matching with $D = 3$ in the BA model; right-hand: many-to-one matching with $D = 3$ in the WS model.

we can see that in many-to-one matching, there is no significant difference of utility by using BA (middle) and WS (right) to model resident networks, but in the one-to-one problem [13], the BA model (left) has the highest utilities among all models.

Intuitively, in the hospital-resident problem or other many-to-one matching scenarios, most of the applicants tend to submit massive applications to their target hospitals. These are the random connection between hospital nodes and resident nodes in our simulated experiments, which adds additional social connections to the original structured networks. Besides, hospitals are connected to many resident nodes, making them act like the central nodes in the BA model. Recall that the existence of central nodes would make nodes have accesses to more nodes, which is proven by the BA model in one-to-one matching problem. These two factors make many-to-one matching in social network more like a classical stable matching problem with complete accessibility information and weaken the influence of different network topology.

4.2   Experiments on Real-World Data

In this section, we apply the newly proposed matching algorithm on the real-world social network in order to predict the matching results. All code and datasets are available for free[2].

Data Resource. ResearchGate[3] is a social network for academic professors, researchers, students and engineers. In ResearchGate, researchers can use the platform to upload their own publications, ask and answer research-related questions, and follow other researchers to receive their publication updates. Research-Gate also collects the co-author relationships between different researchers based on the publications. This kind of typical social links of a researcher is partially shown in Fig. 4. A researcher may follow others and she/he also owns followers. We are interested in predicting the co-author relationship between students

---

[2] https://github.com/KevinDong0810/Stable-Matching-In-Social-Network.

[3] https://www.researchgate.net/about.

Fig. 4. An example of basic relations of ResearchGate. Given a particular researcher (e.g. Prof Kewu Peng), his top co-authors, followers and followings are listed. An co-author could be either his followings or his followers.

and professors based on their following-follower relationship in ResearchGate through the new proposed matching algorithm, and results are validated by true co-author relationships collected from ResearchGate.net.

We connect two nodes in the network when the corresponding people follow each other and they form a social network G. We then use the RG score, an metric for measuring researcher's scientific reputation by ResearchGate, to form the preference matrix $\mathbf{P}$. That is to say, people with higher RG scores would always have higher priority in other people's preference list. The quota of each professor is given according to the number of his/her student co-authors in ResearchGate. The problem is stated as follows: There is a finite set $T = \{t_1, t_2, ..., t_{N_t}\}$ of professors, and a finite set $S = \{s_1, s_2, ..., s_{N_s}\}$ of students. Each student can publish a paper with at most one professor, but one professor, say t_i, has a quota q_i of student co-authors, that makes a typical many-to-one stable matching problem. Use the preference matrix $\mathbf{P}^T, \mathbf{P}^S$ and the network $G = (V, E)$ to predict the matching results (co-authorship) between elements of T and S.

Data Processing. Social network prediction has always been a hot topic, especially in WWW and e-commerce. However, most research focus on predicting missing links in homogeneous networks. What we are interested is *matching* in social networks, we can hardly find any free data for academic research as far as we know, that is the reason we build our own data set based on Research-Gate to share with the community. We start from twelve professors who have close collaboration with each other in the department of electric and electrical engineering of Tsinghua University. All of the followings and followers of these twelve professors are collected. After deleting incomplete profiles, there are 559 people left. To distinguish with the matching members shown later, we refer these 559 people as *community members*. Then all of the followings, followers

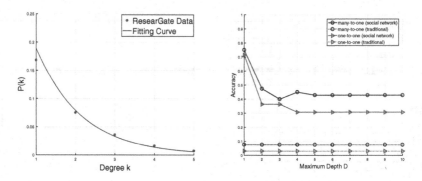

Fig. 5. Left-hand: probability distribution for k in ResearchGate network ($k \leq 6$). Right-hand: experimental results for one/many-to-one algorithms.

and co-authors of these 559 members are collected from the website, but only those social links between these 559 community members are kept and formed connections in the network G correspondingly. G has 363 social connections or edges[4], and the average degree $\overline{k} = 0.649$, which shows that G is a sparse graph. The average clustering coefficient is 0.049 [1]. The probability distribution of k is shown in Fig. 5. The maximum k is 15, but when $k \geq 6$, $P(k) \approx 0$. $P(k)$ decreases exponentially with k. The fitting formula is $P(k) = 0.4527e^{-0.873k}$, with $R^2 = 0.9866$. According to the research of [1] and [23], this network resembles the characteristics of the WS model.

Recall that we are interested in the matching results between professors and students. But there are people who do not fall into this category among the 559 community members. Thus, we first pick out professors and students according to their titles posted online, and then pick out those professors and students who have co-authored with people in the other side manually. Finally we get 99 people in total, 33 professors and 66 students. We refer them as *matching members*. According to the assumption that one student can only co-author with at most one professor, only the professor with the most co-authored papers is remained if the student has more than one professor cu-authors. For example, there are two pairs $(t_1, s), (t_2, s)$, but s has published more papers with t_1 than papers with t_2, then (t_2, s) is deleted. By doing so, we have 66 co-author pairs, which can be used as the ground-truth for obtaining the accuracy rate.

Experiment Results. The network matching algorithm (Alg.1) is applied to predict the possible matching results for 99 matching members using the inputs introduced above. For comparison, we also conduct the traditional Gale-Shapley algorithm [6]. We give the same preference list $\mathbf{P}^T, \mathbf{P}^S$ and quota vector $\mathbf{Q}$, collected from ResearchGate.net, to these members, but connect every two people

---

[4] Although there are 5115 social links between community members, but only when two members follow each other would a connection be formed in G. Thus there are only 363 edges in G.

in the matching pool according to the assumption of full accessibility. The predicted co-author pairs of the two algorithms are verified with the true co-author pairs. And the accuracy rate ρ is gained through

$$\rho = \frac{Correct\ Co\text{-}author\ Pair}{True\ Co\text{-}author\ Pair}$$

The accuracy rate ρ against different maximum depth D (larger D implies much larger connected neighborhood) are shown in the right-hand figure of Fig. 5.

It can be seen that the accuracy rate of our algorithm decreases with the increase of D. Intuitively, with the increase of D, the true student co-authors have to compete with more candidates and some of them may fail in the matching competition. This results in the decrease of accuracy rate. The accuracy rate reaches to its peak 0.75 when $D = 1$ and stabilizes at 0.43 when $D \geq 5$. In contrast, the accuracy rate of classical algorithm equals to 0.08 (D dose not have a influence on the fully-connected graph). For the one-to-one algorithm, we take the same procedures, except that we have to force every professor to admit at most one student. As we can see from Fig. 5, many-to-one matching in social networks has the best results. Such significant increase of accuracy rate by using our proposed algorithm convincingly demonstrates that the information of social networks should be taken into consideration when applying stable matching algorithms into real life.

5 Conclusions

In this paper, we proposed many-to-one stable matching algorithms by considering incomplete information in social networks, where agents in both sides are not fully connected to each other. Four types of well-used network structures are tested, *D-neighbourhood* and *connective cost* are defined to measure matching utilities. We found that the BA model has the most desired *average utility* with less connective costs in one-to-one matching. But in the many-to-one problem, the superiority of BA model disappears and four types of network structures' performance are nearly the same. This is because connections largely increase the connectivity and weaken the influence of network topologies. We also tested the new model on a real-world network matching problem. Experimental results show that the new matching model can achieve the best accuracy of 0.75 while traditional model has only 0.03. In our future work, we will investigate stable matching in large-scale social networks.

References

1. Albert, R., Barabási, A.L.: Statistical mechanics of complex networks. Rev. Mod. Phys. **74**, 47–97 (2002)
2. Barabási, A.L., Albert, R.: Emergence of scaling in random networks. Science **286**(5439), 509–512 (1999)

3. Erdős, P., Rényi, A.: On the evolution of random graphs. In: Publication of the Mathematical Institute of the Hungarian Academy of Sciences, pp. 17–61 (1960)

4. Fleiner, T.: A fixed-point approach to stable matchings and some applications. Math. Oper. Res. **28**(1), 103–126 (2003)

5. Fronczak, A., Fronczak, P., Hołyst, J.A.: Average path length in random networks. Phys. Rev. E **70**(5), 056110 (2004)

6. Gale, D., Shapley, L.S.: College admissions and the stability of marriage. Am. Math. Monthly **69**(1), 9–15 (1962)

7. Gusfield, D., Irving, R.W.: The Stable Marriage Problem: Structure and Algorithms. MIT Press, Cambridge (1989)

8. Irving, R.W.: Stable marriage and indifference. Discrete Appl. Math. **48**(3), 261–272 (1994)

9. Jackson, M.O.: Allocation rules for network games. Games Econ. Behav. **51**(1), 128–154 (2005)

10. Kelso Jr, A.S., Crawford, V.P.: Job matching, coalition formation, and gross substitutes. Econometrica: J. Econometric Soc. **50**(6), 1483–1504 (1982)

11. Li, Z., Chang, Y.-H., Maheswaran, R.: Graph formation effects on social welfare and inequality in a networked resource game. In: Greenberg, A.M., Kennedy, W.G., Bos, N.D. (eds.) SBP 2013. LNCS, vol. 7812, pp. 221–230. Springer, Heidelberg (2013). https://doi.org/10.1007/978-3-642-37210-0_24

12. Li, Z., Qin, Z.: Impact of social network structure on social welfare and inequality. In: Pedrycz, W., Chen, S.-M. (eds.) Social Networks: A Framework of Computational Intelligence. SCI, vol. 526, pp. 123–144. Springer, Cham (2014). https://doi.org/10.1007/978-3-319-02993-1_7

13. Ling, Y., Wan, T., Qin, Z.: Stable matching in structured networks. In: Ohwada, H., Yoshida, K. (eds.) PKAW 2016. LNCS (LNAI), vol. 9806, pp. 271–280. Springer, Cham (2016). https://doi.org/10.1007/978-3-319-42706-5_21

14. McVitie, D.G., Wilson, L.B.: Stable marriage assignment for unequal sets. BIT Numer. Math. **10**(3), 295–309 (1970)

15. Roth, A.E.: The economics of matching: stability and incentives. Math. Oper. Res. **7**(4), 617–628 (1982)

16. Roth, A.E.: The evolution of the labor market for medical interns and residents: a case study in game theory. J. Polit. Econ. **92**(6), 991–1016 (1984)

17. Roth, A.E.: The college admissions problem is not equivalent to the marriage problem. J. Econ. Theor. **36**(2), 277–288 (1985)

18. Roth, A.E.: On the allocation of residents to rural hospitals: a general property of two-sided matching markets. Econometrica **54**(2), 425–427 (1986)

19. Roth, A.E., Sotomayor, M.: Two-sided matching. In: Handbook of Game Theory with Economic Applications, vol. 1, chap. 16, pp. 485–541. Elsevier (1992)

20. Shoham, Y.: Computer science and game theory. Commun. ACM **51**(8), 75–79 (2008)

21. Strogatz, S.H.: Exploring complex networks. Nature **410**(6825), 268–76 (2001)

22. Travers, J., Milgram, S.: The small world problem. Phychology Today, pp. 61–67 (1967)

23. Watts, D.J., Strogatz, S.H.: Collective dynamics of 'small-world' networks. Nature **393**(6684), 440–2 (1998)

A Framework for Detecting User's Psychological Tendencies on Twitter Based on Tweets Sentiment Analysis

Huyen Trang Phan[1], Van Cuong Tran[2], Ngoc Thanh Nguyen[3], and Dosam Hwang[1(✉)]

[1] Department of Computer Engineering, Yeungnam University, Gyeongsan, Republic of Korea
huyentrangtin@gmail.com, dosamhwang@gmail.com
[2] Faculty of Engineering and Information Technology, Quang Binh University, Dong Hoi, Vietnam
vancuongqbuni@gmail.com
[3] Faculty of Computer Science and Management, Wroclaw University of Science and Technology, Wroclaw, Poland
Ngoc-Thanh.Nguyen@pwr.edu.pl

Abstract. The more societies develop, people have less time for interacting face-to-face with each other. Therefore, more and more users express their opinions on many topics on Twitter. The sentiments contained in these opinions are becoming a valuable source of data for politicians, researchers, producers, and celebrities. Many studies have used this data source to solve a variety of practical problems. However, most of the previous studies only focused on using the sentiment in tweets to address the issues regarding commercial without considering the negative aspects related to user psychology, such as psychological disorders, cyberbullying, antisocial behaviors, depression, and negative thoughts. These problems have a significant effect on users and societies. This paper proposes a method to detect the psychological tendency that hides insides one person and to give the causations that lead to this psychological tendency based on analyzing sentiment of tweets by combining the feature ensemble model and the convolutional neural network model. The results prove the efficacy of the proposed approach in terms of the F_1 score and received information.

Keywords: Sentiment analysis · User psychology · User detection

1 Introduction

Nowadays, along with the development of technologies, people want to have an intermediary tool to help them can exchange, share information, or even talk to each other without any concern for space distance as well as time. The social networks can help them do these without taking much time. Therefore, the number of users on social networks is increasing daily. Over time, social networks

© Springer Nature Switzerland AG 2020
H. Fujita et al. (Eds.): IEA/AIE 2020, LNAI 12144, pp. 357–372, 2020.
https://doi.org/10.1007/978-3-030-55789-8_32

have gradually replaced traditional communication with online communication between individuals, communities, and societies globally. Social networks become close friends with many people, where they can confide their thoughts anytime and anywhere. Besides, some people have a habit of posting everything from happiness to sadness, from joy to unhappiness, from love to hate. They post without regard for how others will react to the contents which they have posted. Currently, one of the most popular social networks is Twitter. The number of Twitter's users has increased quickly, with approximately 500 million tweets published daily in 2014, the last time stats were released[1]. According to statistics[2], the number of active Twitter users per month is 326 million worldwide for Q3, 2018. This is a source of available information that could provide many benefits.

Many researchers tried to solve many practical problems through tweet sentiment analysis (TSA) [13]. In which, most of the studies have focused on using TSA to determine critic opinions about a given product [17], to track the shifting attitudes of the general public toward a political candidate [15]. Other researchers have used the sentiment in tweets for the market analysis [5]. Besides, there are a few studies that applied TSA to build recommendation systems [11], decision support systems [20], prediction systems. From the above analysis, we can see that most of the previous research focused only on using tweets sentiment analysis to solve commercial problems. Very few studies have concerned about using TSA to detect the psychological tendencies of users on Twitter that aim early to identify users who have signs of depression, excessive stress, or pessimism about life. Early intervention of such problems can prevent negative effects on users and society. Additionally, the causations that lead to the psychological tendencies of users have not considered. These drawbacks motivated us to propose a method to detect psychological tendencies that hide inside users and causations of these psychological tendencies based on TSA by the combination of a feature ensemble model and convolutional neural network (CNN) model. In the proposed method, first, a set of features related to syntactic, lexical, semantic, position and sentiment score of the words is extracted. Second, the tweet embeddings are created by using the feature ensemble model. Third, the sentiments of tweets are classified as positive, neutral, and negative by using the CNN model. Finally, the psychological tendencies of users are detected as negative or positive based on the TSA results, and the causations of these psychological tendencies are analyzed.

The remainder of the paper is organized as follows. In Sect. 2, we summarize existing work related to approaches for sentiment analysis. The research problem is described in Sect. 3 and the proposed method is presented in Sect. 4. The experimental results and evaluations are shown in Sect. 5. Finally, the conclusion and a discussion of future work are presented in Sect. 6.

---

[1] http://www.internetlivestats.com/twitter-statistics/.

[2] https://zephoria.com/twitter-statistics-top-ten/.

2   Related Works

Many researchers have used TSA to solve practical problems. In which, most studies have applied the sentiment in tweets to analyze the commercial market or determine the user's opinions and attitudes about products, brands, or celebrities. Jain *et al.* [5] predicted the prices of Bitcoin and Litecoin with a multilinear regression model of tweets tagged with a cryptocurrency name (Bitcoin, Litecoin) and concurrent price data (from Coindesk[3]). Significant features were explored and mapped with the prices to build prediction curves. These prediction curves can predict cryptocurrency prices in the near future. Yussupova *et al.* [20] built a novel domain-independent decision support model to determine customer satisfaction that was based on consumer reviews posted on the Internet in natural language. The efficacy of the approach was evaluated quantitatively and qualitatively on two datasets related to hotels and banks. Phan *et al.* [14] proposed to support decision-making by combining sentiment analysis and data mining. The results proved the performance of this approach for decision-making based on customer satisfaction on Twitter. Only a few of the following researchers have sought to uncover issues related to the user's behavior by using the contents on Twitter. Yang *et al.* [19] studied the effect of climate and seasonality on the prevalence of depression in Twitter users using text mining and geospatial analysis. McManus *et al.* [10] detected Twitter users with schizophrenia by leveraging a large corpus of tweets and machine learning. Features selected from the tweets, such as emoticons, time of tweets, and dictionary terms were used to train and validate several machine learning models. This enabled outreach to undiagnosed individuals, improved physician diagnoses, and addressed the stigmatization of schizophrenia. O'Dea *et al.* [12] created direct or indirect textual or audio-visual references related to suicidality based on detecting the level of concern for an individual's tweets. The feasibility of this automated prediction was examined using recall and precision metrics. Generally, many social problems can be detected early if the psychological tendencies of users are known. However, there are few studies using sentiment analysis of tweets to detect psychological tendencies of users, as well as having not considered the causations that lead to these psychological tendencies.

3   Research Problem

3.1   Sentiment Score of a Word in Lexicons

Let $\mathcal{F}$ be a set of fundamental sentiment words that include positive and negative words. These words are used to express emotional states of users, such as "angry," "sad," and "happy." $\mathcal{F}$ was collected from SentiWordNet (SWN) proposed by Baccianella *et al.* [2]. Let $\mathcal{F}s$ be a set of fuzzy semantic words that consist of intensifier and diminisher words. Intensifier words are positioned before the fundamental sentiment words and can increase the polarity of these

---

[3] https://www.coindesk.com/price/bitcoin.

words, e.g., "too", "so", "overly". Diminisher words are positioned before the fundamental sentiment words and can decrease the polarity of these words, e.g., "quite", "fairly", "slightly". $\mathcal{F}s$ was created by using the English modifier words proposed by Benzinger[4]. Let $\mathcal{N}$ be a set of negation words that are positioned before the fundamental sentiment words, and change the polarity of these words, e.g., "not", "n't". $\mathcal{N}$ selected from [8].

Given a set of tweets $\mathcal{T}$.

For $t \in \mathcal{T}$: let $\mathcal{W}$ be a set of words existing in t.

For $w \in \mathcal{W}$: let S be a set of synsets of word w.

For $sn \in S$: let $\mathcal{P}$ be a set of Part-Of-Speech (POS) tags of words in $\mathcal{W}$, let $\mathcal{P}s$ be the positivity score assigned by SWN to synset sn, let $\mathcal{N}s$ be the negativity score assigned by SWN to synset sn. In which, $\mathcal{P}s, \mathcal{N}s \in [0.0, 1.0]$ and $\mathcal{N}s + \mathcal{P}s \leq 1.0$.

For $p \in \mathcal{P}$: let $Sp(w, p)$ be a positive score of word w corresponding to synsets, let $Sn(w, p)$ be a negative score of word w corresponding to synsets.

$$Sp(w, p) = \frac{1}{m} \sum_{sn \in Sp} \mathcal{P}s(sn) \tag{1}$$

$$Sn(w, p) = \frac{1}{m} \sum_{sn \in Sp} \mathcal{N}s(sn) \tag{2}$$

where m represents the number of synsets of word w.

For $(f \in \mathcal{F})$ and $(p \in \mathcal{P} :)$ let $Sc(f, p)$ be the sentiment score of word f and $Sc(f, p)$ is computed based on $Sp(f, p)$, $Sn(f, p)$ as follows:

$$Sc(f, p) = Sp(f, p) - Sn(f, p) \tag{3}$$

Next, for $n \in \mathcal{N}$: let $Sc(n)$ be the score of negation word n and $Sc(n)$ is determined based on [8]. The score of negation words is shown in Table 1:

Table 1. The score for negation words

On positive words							On negative words	
Word	Score	Word	Score	Word	Score	Word	Score
Will not be	−1.066	Was no	−0.939	Would not	−0.869	Will not	0.878
Cannot	−1.030	Will not	−0.935	Had no	−0.862	Does not	0.823
Did not	−0.978	Was not	−0.928	Would not be	−0.848	Was not	0.786
Not very	−0.961	Have no	−0.917	May not	−0.758	no	0.768
Not	−0.959	Does not	−0.907	Nothing	−0.755	Not	0.735
No	−0.948	Could not	−0.893	Never	−0.650		

---

[4] (https://archive.org/stream/intensifiersincu00benz/intensifiersincu00benz_djvu.txt).

Then, the sentiment score of the fuzzy semantic words are determined as follows: For $fs \in \mathcal{Fs}$: let $Sc(fs)$ be a sentiment score of fs. The sentiment score of fundamental words was in the range [–0.75,0.75]; therefore the value of fuzzy semantic words was in the range [–0.25,0.25] [18]. In this study, we used the numeric values offered by [1,6], and [16] to assign the score for intensifier and diminisher word lists. Numeric scores to each fuzzy semantic word were normalized to fit with our proposal by mapping from range [–100%,+100%] to range [–0.25,0.25]. The score of the fuzzy semantic words is shown in Table 2.

Table 2. The score for some fuzzy semantic words

Intensifier words				Diminisher words				
Word	List 1	List 2	Normalized	Score	Word	List 1	List 2	Normalized	Score
Really	+15%	+15%	+15%	0.04	Slightly	–50%	–40%	–45%	–0.11
Very	+25%	+50%	+38%	0.09	Somewhat	–30%		–30%	–0.08
Most		+90%	+95%	0.24	Quite		–20%	–20%	–0.05
Totally		+70%	+70%	0.18	A little		–40%	–40%	–0.1
Too		+45%	+45%	0.11	A bit		-35%	-35%	–0.09

List 1 is the score extracted from paper [6,16],
List 2 is the score extracted from paper [1].

3.2   Definitions Related to Sentiment of Tweets

Given a set of users $\mathcal{U}$: For $u \in \mathcal{U}$: let $\mathcal{T}_u$ be a set of tweets which user u posted on Twitter, $\mathcal{L} = \{$"*Positive*", "*Neutral*", "*Negative*"$\}$ be a set of sentiment labels, $\mathcal{Lb} : \mathcal{T}_u \to \mathcal{L}$ be a mapping function from a tweet t to a label $\mathcal{Lb}(t)$ determining the sentiment of tweet t.

Definition 1. *A negative tweet is a tweet that assigned the sentiment label as "Negative." It means that in this tweet, a user expressed a not good emotion or opinion towards a specific subject. The negative tweet is denoted by* $t_n = \{t | t \in \mathcal{T}_u \wedge \mathcal{Lb}(t) =$ *"Negative"*$\}$.

Definition 2. *A neutral tweet is a tweet that assigned the sentiment label as "Neutral." It means that in this tweet, a user expressed a hesitation of emotion or opinion (not good as well as not bad) towards a specific subject. The neutral tweet is denoted by* $t_{ne} = \{t | t \in \mathcal{T}_u \wedge \mathcal{Lb}(t) =$ *"Neutral"*$\}$.

Definition 3. *A positive tweet is a tweet that assigned the sentiment label as "Positive." It means that in this tweet, a user expressed a great emotion or opinion towards a specific subject. The positive tweet is denoted by* $t_p = \{t | t \in \mathcal{T}_u \wedge \mathcal{Lb}(t) =$ *"Positive"*$\}$.

3.3   Definitions Related to User Psychological Tendencies

For $t \in T_u$: let T_{t_p} be a set of positive tweets posted by user u, and T_{t_n} be a set of negative tweets posted by user u.

Definition 4. *The negative ratio, denoted by R_u, is the ratio of negative tweets in T_u. The negative ratio is defined as follows:*

$$R_u = \frac{Card\left(T_{t_n}\right)}{Card\left(T_u\right)} \tag{4}$$

Definition 5. *The positive ratio, denoted by R_p, is the ratio of positive tweets in T_u. The positive ratio is defined as follows:*

$$R_p = \frac{Card\left(T_{t_p}\right)}{Card\left(T_u\right)} \tag{5}$$

Definition 6. *A user with a negative psychological tendency, referred to as a negative user, denoted by u_n, is a user who tends to post negative tweets. The negative user is determined as follows:*

$$u_n = \{u | u \in U \wedge R_u > \alpha \wedge R_p < \gamma\} \tag{6}$$

Definition 7. *A user with positive psychological tendency, referred to as a positive user, denoted by u_p, is a user who tends to post positive tweets. The positive user is defined as follows:*

$$u_p = \{u | u \in U \wedge R_p > \gamma \wedge R_u < \alpha\} \tag{7}$$

Definition 8. *Let $V_n = \bigcup_{w \in t, t \in T_{t_n}} (w)$ be a dictionary of set T_{t_n}. A cause that leads to a user's negative psychological tendency, denoted by c_n, is a noun in V_n with a high frequency of occurrence. The cause c_n is defined as follows:*

$$c_n = \{w | w \in V_n \wedge tag(w) = \text{'}noun\text{'} \wedge Ra(w) \leq \beta\} \tag{8}$$

where $Ra(w)$ is a mapping function from a word to a numeric indicating a frequency of this word.

Definition 9. *Let $V_p = \bigcup_{w \in t, t \in T_{t_p}} (w)$ be a dictionary of set T_{t_p}. A cause that leads to a user's positive psychological tendency, denoted by c_p, is a noun in V_p with a high frequency of occurrence. The cause c_p is defined as follows:*

$$c_p = \{w | w \in V_p \wedge tag(w) = \text{'}noun\text{'} \wedge Ra(w) \leq \beta\} \tag{9}$$

In which, α, γ, β are the thresholds for selecting the negative ratio, positive ratio, and number of causes of a user's psychological tendencies, respectively.

3.4 Research Question

This study addresses the research question: *How can we build a framework for detecting a user's psychological tendencies on Twitter by using tweets sentiment analysis?*. This question is partitioned into the two following sub-questions:

The first question: *How can we detect users with negative and positive psychological tendencies from their sentiment?*

The second question: *What are the possible causes of a user's negative or positive psychological tendencies?*

4 Proposed Framework

In this section, we present a methodology to implement our proposal. The workflow of the method is shown in Fig. 1.

Given a set of users $\mathcal{U}$: For $u \in \mathcal{U}$: let $\mathcal{T}_u$ be a set of tweets which user u posted on Twitter. The proposed method has five steps: First, vectors are created for lexicon, word-type, sentiment score, semantics, and position of each word in a tweet. Second, all extracted vectors are combined into a word embedding vector. Third, the CNN model is used to classify the sentiment of tweets in $\mathcal{T}_u$ as either negative, neutral, or positive. Fourth, user u is classified as having either positive or negative psychological tendencies. Finally, frequent nouns found in tweets are sorted to predict the causations of positive or negative psychological tendencies. These steps are explained in more detail in the following sub-sections.

4.1 Creating Tweet Embeddings

A tweet embedding is created by concatenating feature vectors of each word, i.e., lexical (l2v), word-type (sy2v), sentiment score (pl2v), position (ps2v), and semantic (se2v) [4] .

The lexical vector of a word is created based on n-grams of this word. The n-grams of a word include 1-gram, 2-grams, and 3-grams started by this word. The lexical vector is illustrated in Table 3.

Table 3. Creating lexical vector of a word

Words	1-gram	2-grams				3-grams			
w_i	w_i	$w_i w_1$	$w_i w_2$	...	$w_i w_n$	$w_i w_1 w_2$	$w_i w_1 w_3$	...	$w_i w_{n-1} w_n$
tf-idf(w_i)	t_i	t_{i1}	t_{i2}	...	t_{in}	t_{i12}	t_{i13}	...	t_{in-1n}
l2v(w_i)	$[t_i,$	$t_{i1},$	$t_{i1},$	...,	$t_{in},$	$t_{i12},$	$t_{i13},$		$t_{in-1n}]$

The semantic vector of a word is generated by using the 300-dimensional pre-trained word embeddings from Glove[5]. If a word exists in the Glove dataset,

---

[5] http://nlp.stanford.edu/projects/glove/.

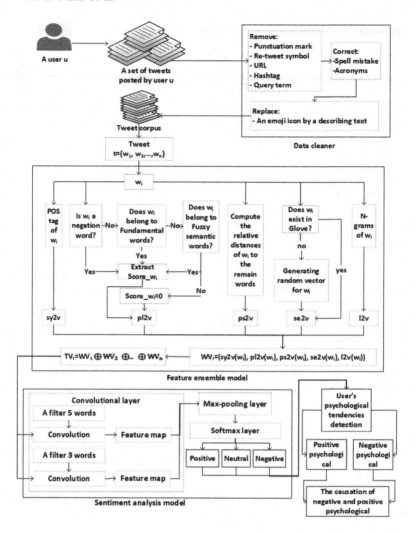

Fig. 1. The workflow of proposed method.

its vector will be extracted from Glove embeddings. Otherwise, if a word doesn't exist in the Glove dataset, its vector will be generated randomly.

The word-type vector of a word is created based on POS tag of this word in tweet t. The NLTK toolkit [9] is used to annotate the POS tags of words. The word-type vector is shown in Table 4.

The sentiment score vector of a word is built based on the score feature of kinds of words, such as the score of positive, negative, negation, intensifier, and diminisher words. This feature is extracted by using a window of 1 to 3 words before a sentiment word and search for these kinds of words. The appearance of these words in the tweet and their scores become features and feature values, respectively. The sentiment score vector is presented in Table 5.

Table 4. Creating word-type vector of a word

Words	CD	DT	IN	JJ	TO	MD	NN	PRP	SYM	TO	VB	WRB
w_i	0	0	0	0	0	0	0	1	0	0	0	0
sy2v(w_i)	$[0,0,0,0,0,0,0,0,0,0,1,...,0]$											

Table 5. Creating sentiment score vector of a word

Word	Negation word	Fundamental word	Fuzzy semantic word
w_i	score(w_i) $= Sc(n)$	score(w_i) $= Sc(f,p)$	score(w_i) $= Sc(fs)$
pl2v(w_i)	[score(w_i)]		

The position vector of a word is created by computing the relative distances of the current word to the remaining words in a tweet and shown in Table 6.

Table 6. Creating position vector of a word

Words	Distance to				ps2v(w_i)
	w_0	w_1	...	w_n	
w_i	d_{i0}	d_{i1}	...	d_{in}	$[d_{i0}, d_{i1}, ..., d_{in}]$

The relative distances of the current word to the remaining words in tweet t are encoded into vectors d_i by using a position embedding table (Initialized randomly).

4.2   Analyzing Sentiment of Tweets

The CNN model is used to analyze the sentiment of tweets. The CNN has become a significant Deep learning model using in the NLP field since the research of Collobert *et al.* [3] are published and then Kim [7] who applied the success of CNN in sentiment analysis. The sentiment analysis model is built as the following steps:

Tweet Embeddings Layer: Each tweet is represented by a vector $T2V$:

$$T2V_{1:n} \in \mathcal{R}^{d \times n} \wedge T2V_{1:n} = v_1 \oplus v_2 \oplus v_3 \oplus ... \oplus v_n \tag{10}$$

where $\oplus$ is the concatenation operator, d is the dimension of v_i, v_i is i-th word vector, $v_i = l2v(w_i) \oplus sy2v(w_i) \oplus se2v(w_i) \oplus pl2v(w_i) \oplus ps2v(w_i)$.

Convolutional Layer: This layer create a feature map (c) from the tweet embedding layer by filtering important features with a sliding window of length

q words from i to $i + q - 1$. Each sliding of the window creates a new feature vector as follows:

$$c_i = ReLU(\mathcal{M}.T2V_{i:i+q-1} + b) \tag{11}$$

where $ReLU$ is a rectified linear function. b is a bias term. $\mathcal{M} \in \mathcal{R}^{h \times qd}$ is a transition matrix created for each filter, h is the number of hidden units in the convolutional layer. Therefore, when a tweet is slided completely, the features map is generated as follows:

$$c = [c_1, c_2, .., c_{d-q+1}], c \in \mathcal{R}^{d-q+1} \tag{12}$$

Max-Pooling Layer: This layer is to reduce the dimensionality of the feature map by taking the maximum value $\hat{c} = max(c)$ as the feature corresponding to each filter. Assume that we use m filters, after this step the obtained new feature is $\hat{c} = [\hat{c}_1, \hat{c}_2, .., \hat{c}_m]$. Then this vector is fed into next layer.

Softmax Layer: This layer uses a fully connected layer to adjust the sentiment characteristic of the input layer, and predict tweet sentiment polarity by using Softmax function as follows:

$$y = softmax(\mathcal{M}\hat{c} + b) \tag{13}$$

where $\mathcal{M}$ is a transition matrix of Softmax layer.

In this study, the CNN has two convolutional layers, one max pooling layer, and one softmax output layer. The detail of the hyperparameters of the CNN model is presented in Table 7.

Table 7. Hyperparameters for CNN model

Hyperparameters	Values
# hidden layer	2
Comma-separated filter sizes	5, 3
# filters	128
l_2 regularization	0.0001
# epochs	100
Dropout keep probability	0.1

4.3   Detecting User's Psychological Tendencies

A user's psychological tendencies are determined by computing the degree of negative and positive sentiments of a user's tweets. The steps to identify a user's psychological tendencies and the causes are shown in Algorithm 1.

Algorithm 1. User's psychological tendencies detection

Input:

1: $\mathcal{T}_u = \{t_1, t_2, ..., t_n\}$, a set of tweets which posted bu user u

Output: u is negative user or u is positive user

2: **for** $t \in \mathcal{T}_u$ **do**

3:    **if** $\mathcal{L}b(t) = $ "*Positive*" **then**

4:       insert t into $\mathcal{T}_{t_p}$

5:    **else if** $\mathcal{L}b(t) = $ "*Negative*" **then**

6:       insert t into $\mathcal{T}_{t_n}$

7:    **end if**

8: **end for**

9: $\mathcal{R}_u = \dfrac{Card\left(\mathcal{T}_{t_n}\right)}{Card\left(\mathcal{T}_u\right)}; \quad \mathcal{R}_p = \dfrac{Card\left(\mathcal{T}_{t_p}\right)}{Card\left(\mathcal{T}_u\right)}$

10: **if** $\mathcal{R}_u > \alpha$ and $\mathcal{R}_p < \gamma$ **then**

11:    u is a negative user

12:    **for** $t \in \mathcal{T}_{t_n}$ **do**

13:       $\mathcal{V}_n = \{\bigcup(w)|w \in t\}$

14:    **end for**

15:    **for** $w \in \mathcal{V}_n$ **do**

16:       $c_n = \mathcal{R}a(w)$

17:       **if** $c_n > \beta$ and $tag(w) = $ "*noun*" **then**

18:          insert w into C_n

19:       **end if**

20:    **end for**

21: **else if** $\mathcal{R}_p > \alpha$ and $\wedge \mathcal{R}_u < \gamma$ **then**

22:    u is a positive user

23:    **for** $t \in \mathcal{T}_{t_p}$ **do**

24:       $\mathcal{V}_p = \{\bigcup(w)|w \in t\}$

25:    **end for**

26:    **for** $w \in \mathcal{V}_p$ **do**

27:       $c_p = \mathcal{R}a(w)$

28:       **if** $c_p > \beta$ and $tag(w) = $ "*noun*" **then**

29:          insert w into C_p

30:       **end if**

31:    **end for**

32: **end if**

33: **return** $(u_n$ and $C_n)$ or $(u_p$ and $C_p)$

5   Experiment

5.1   Data Acquisition

The model was trained with 7368 English tweets from [14]. The training data had 2880 positive tweets, 1883 neutral tweets, and 2605 negative tweets. The Python package Tweepy[6] was then used to collect tweets from nine users on Twitter. Only English tweets were collected. The tweets of nine users are used

---

[6] https://pypi.org/project/tweepy/.

as testing data. Tweets were cleaned by removing punctuation marks, re-tweet symbols, URLs, hashtags, and query term. Emoji icons were replaced with text that describes an emoji by using the Python emoji package[7]. Accuracy can be affected by acronyms and spelling mistakes are common because tweets are informal. The Python-based Aspell library[8] was used to correct spelling. Tweets in the training and testing data were labeled as either *Positive, Neutral*, and *Negative* by three annotators. We annotated the testing set as the gold standard to assess the performance. The detail of the testing set is presented in Table 8.

Table 8. Tweets statistics by annotators.

User	User-1	User-2	User-3	User-4	User-5	User-6	User-7	User-8	User-9
$\#t$	685	379	493	103	768	1098	1450	674	60
$\#t_p$	315	165	93	41	325	674	730	267	31
$\#t_{ne}$	295	141	88	17	237	245	429	95	17
$\#t_n$	75	73	312	45	206	179	291	312	12

5.2   Evaluation Results

Metrics used to assess the proposed method include *Precision, Recall*, and $\mathcal{F}_1$. The values of *Precision, Recall*, and $\mathcal{F}_1$ are computed as follows:

$$Precision = \frac{TP}{TP + FP}; \quad (14) \qquad Recall = \frac{TP}{TP + FN}; \quad (15) \qquad \mathcal{F}_1 = 2 \times \frac{Precision \times Recall}{Precision + Recall} \quad (16)$$

where, TP (True Positive) is the number of exactly classified items, FP (False Positive) is the number of misclassified items, TN (True Negative) is the number of exactly classified non-items, FN (False Negative) is the number of misclassified non-items.

5.3   Results and Discussion

The average performance of the sentiment analysis in the tweets is shown in Table 9.

Looking at Table 9, it can be seen that for tweets of nine users, the positive class has performed better than the negative and neutral classes. Specifically, there are eight cases the highest performance falls in the positive sentiment; meanwhile, there is only one case that falls in the negative sentiment. Intuitively, the performance was low because there were fewer tweets with negative

---

[7] https://pypi.org/project/emoji/.
[8] https://pypi.org/project/aspell-python-py2/.

Table 9. The average performance of the sentiment analysis step.

	P	Ne	N
Precision	0.85	0.71	0.69
Recall	0.78	0.78	0.83
F₁	0.81	0.73	0.74

P: Positive, N: Negative, Ne: Neutral

sentiment. We believe that more data and data that is balanced can significantly improve performance.

In addition, to prove the performance of the TSA step, we implemented three TSA experiments on the same dataset. The first time, the traditional lexicon-based approach (LBA) is performed. The second time, the scaling method (SM [21]) is employed, and the third time, our method is used. The results are shown in Table 10.

Table 10. Performance comparison in terms of sentiment analysis.

Method	*Precision*	*Recall*	*F₁*
LBA	0.70	0.72	0.71
SM	0.73	0.77	0.75
Our method	0.75	0.79	0.76

Table 10 shows our method had the better results compared with the other methods in terms of sentiment analysis. According to our assessment, this performance was achieved because the CNN model is a good algorithm in analyzing sentiment. Furthermore, the feature ensemble model also has a significant impact on improving the accuracy of the CNN model. Additionally, we get tweets statistics as Table 11.

According to Table 11, it can be observed that when $\alpha = 0.5$ and $\gamma = 0.25$, user-3 and user-6 will be negative and positive users, respectively. We will focus on analyzing the causes of the negative and positive psychological tendencies of two users.

For user-3, we can see that the number of negative tweets accounts for 53.35%, which is higher than their positive and neutral tweets. This suggests that user-3 often has negative thoughts or is pessimistic. Such users can spread their negativity to others. Therefore, if detected early, it can prevent their state from worsening, as the worsened state may even affect society. This can be indicative of stress, depression, or a suicidal tendency.

For user-6, the number of positive tweets accounts for 55.01%, which is higher than their negative and neutral tweets. This suggests that user-6 often has positive thoughts or is optimistic. Such users can spread their positivity to others. Therefore, we can use these people to propagate positive messages.

Table 11. Tweets statistics by proposed method (%).

User	$Card(\mathcal{T}_u)$	$Card(\mathcal{T}_{t_p})$	$\mathcal{R}_p$	$Card(\mathcal{T}_{t_n})$	$\mathcal{R}_u$
User-1	685	272	39.71	164	23.94
User-2	379	160	42.22	101	26.65
User-3	493	121	24.54	263	53.35
User-4	103	36	34.95	40	38.84
User-5	768	294	38.28	235	30.60
User-6	1098	604	55.01	248	22.59
User-7	1450	594	40.97	395	27.24
User-8	674	240	35.61	287	42.58
User-9	60	27	45.00	17	28.33

$\mathcal{R}_u$ and $\mathcal{R}_p$ are computed by using Eq. 4 and Eq. 5.

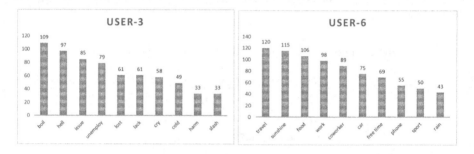

Fig. 2. Ten words with high-frequency of user-3 and user-6

In Fig. 2, for $\beta = 10$, there are ten words with POS tag as "noun" and their frequency will be listed. These words are considered as causes that lead to the psychological tendencies of the user. For example, ten causes lead to the psychological tendency of the user-3 that is *a boil, hell, issue, unemploy, lost, lack, cry, cold, harm, slash*. Meanwhile, ten causes lead to the psychological tendency of the user-6 that is *travel, sunshine, food, coworker, car, free-time, phone, sport, rain*.

6   Conclusion and Future Work

This study presents a framework for detecting user's psychological tendencies on Twitter based on tweet sentiment analysis. The experiments revealed that the proposed model classified Twitter users according to their psychological tendencies with significant performance. This model provides a set of words that describe the causes that lead to psychological tendencies. The main limitation of the proposed approach is identifying a suitable method for comparison in terms of detecting user's psychological tendencies on Twitter. So, the performance of the model has not been compared to other methods. Our future work will compare the model with other methods.

Acknowledgment. This research was supported by the National Research Foundation of Korea (NRF) grant funded by the BK21PLUS Program (22A20130012009).

References

1. Asghar, M.Z., Khan, A., Ahmad, S., Qasim, M., Khan, I.A.: Lexicon-enhanced sentiment analysis framework using rule-based classification scheme. PloS one **12**(2), e0171649 (2017)
2. Baccianella, S., Esuli, A., Sebastiani, F.: Sentiwordnet 3.0:an enhanced lexical resource for sentiment analysis and opinion mining. In: Lrec, vol. 10, pp. 2200–2204 (2010)
3. Collobert, R., Weston, J.: A unified architecture for natural language processing: deep neural networks with multitask learning. In: Proceedings of the 25th International Conference on Machine Learning, pp. 160–167. ACM (2008)
4. Phan, H.T., Tran, V.C., Nguyen, N.T., Hwang, D.: Improving the performance of sentiment analysis of tweets containing fuzzy sentiment using the feature ensemble model. IEEE Access **8**, 14630–14641 (2020). https://doi.org/10.1109/ACCESS.2019.2963702
5. Jain, A., Tripathi, S., DharDwivedi, H., Saxena, P.: Forecasting price of cryptocurrencies using tweets sentiment analysis. In: proceedings of the 2018 Eleventh International Conference on Contemporary Computing (IC3), pp. 1–7. IEEE (2018)
6. Kennedy, A., Inkpen, D.: Sentiment classification of movie reviews using contextual valence shifters. Comput. Intell. **22**(2), 110–125 (2006)
7. Kim, Y.: Convolutional neural networks for sentence classification (2014). arXiv preprint arXiv:1408.5882
8. Kiritchenko, S., Mohammad, S.M.: The effect of negators, modals, and degree adverbs on sentiment composition (2017). arXiv preprint arXiv:1712.01794
9. Loper, E., Bird, S.: NLTK: the natural language toolkit (2002). arXiv preprint cs/0205028
10. McManus, K., Mallory, E.K., Goldfeder, R.L., Haynes, W.A., Tatum, J.D.: Mining Twitter data to improve detection of schizophrenia. AMIA Summits Transl. Sci. Proc. **2015**, 122 (2015)
11. Nabil, S., Elbouhdidi, J., Yassin, M.: Recommendation system based on data analysis-application on tweets sentiment analysis. In: Proceedings of the 2018 IEEE 5th International Congress on Information Science and Technology (CiSt), pp. 155–160. IEEE (2018)
12. O'Dea, B., Wan, S., Batterham, P.J., Calear, A.L., Paris, C., Christensen, H.: Detecting suicidality on Twitter. Internet Interventions **2**(2), 183–188 (2015)
13. Phan, H.T., Nguyen, N.T., Hwang, D.: A tweet summarization method based on maximal association rules. In: Nguyen, N.T., Pimenidis, E., Khan, Z., Trawiński, B. (eds.) ICCCI 2018. LNCS (LNAI), vol. 11055, pp. 373–382. Springer, Cham (2018). https://doi.org/10.1007/978-3-319-98443-8_34
14. Phan, H.T., Tran, V.C., Nguyen, N.T., Hwang, D.: Decision-making support method based on sentiment analysis of objects and binary decision tree mining. In: Wotawa, F., Friedrich, G., Pill, I., Koitz-Hristov, R., Ali, M. (eds.) IEA/AIE 2019. LNCS (LNAI), vol. 11606, pp. 753–767. Springer, Cham (2019). https://doi.org/10.1007/978-3-030-22999-3_64

15. Subramaniyaswamy, V., Logesh, R., Abejith, M., Umasankar, S., Umamakeswari, A.: Sentiment analysis of tweets for estimating criticality and security of events. J. Organ. End User Comput. (JOEUC) **29**(4), 51–71 (2017)
16. Taboada, M., Brooke, J., Tofiloski, M., Voll, K., Stede, M.: Lexicon-based methods for sentiment analysis. Comput. Linguist. **37**(2), 267–307 (2011)
17. Tomihira, T., Otsuka, A., Yamashita, A., Satoh, T.: What does your tweet emotion mean?: neural emoji prediction for sentiment analysis. In: Proceedings of the 20th International Conference on Information Integration and Web-based Applications & Services, pp. 289–296. ACM (2018)
18. Trang Phan, H., Nguyen, N.T., Tran, V.C., Hwang, D.: A sentiment analysis method of objects by integrating sentiments from tweets. J. Intell. Fuzzy Syst. (Preprint) **37**(6), 1–13 (2019)
19. Yang, W., Lan, M., Shen, Y.: Effect of climate and seasonality on depressed mood among twitter users. Appl. Geogr. **63**, 184–191 (2015)
20. Yussupova, N., Boyko, M., Bogdanova, D., Hilbert, A.: A decision support approach based on sentiment analysis combined with data mining for customer satisfaction research. Int. J. Adv. Intell. Syst. **8**(1&2), 145–158 (2015). Published by IARIA
21. Dinakar, S., Andhale, P., Rege, M.: Sentiment analysis of social network content. In: Proceedings of the 2015 IEEE International Conference on Information Reuse and Integration, San Francisco, CA, pp. 189–192 (2015). https://doi.org/10.1109/IRI.2015.37

Automatic Fake News Detection by Exploiting User's Assessments on Social Networks: A Case Study of Twitter

Van Cuong Tran[1(✉)], Van Du Nguyen[2], and Ngoc Thanh Nguyen[3,4]

[1] Faculty of Engineering and Information Technology, Quang Binh University,
Dong Hoi, Vietnam
vancuongqbuni@gmail.com

[2] Faculty of Information Technology, Nong Lam University,
Ho Chi Minh City, Vietnam
nvdu@hcmuaf.edu.vn

[3] Department of Applied Informatics,
Wroclaw University of Science and Technology, Wrocław, Poland
Ngoc-Thanh.Nguyen@pwr.edu.pl

[4] Faculty of Information Technology, Nguyen Tat Thanh University,
Ho Chi Minh City, Vietnam

Abstract. Nowadays, social media has been becoming the main news source for millions of people all over the world. Users easily can create and share their information on social platforms. Information on social media can spread rapidly in the community. However, the spreading of misleading information is a critical issue. There are much intentionally written to mislead the readers, that are called fake news. The fake news represents the most forms of false or unverified information. The extensive spread of fake news has negative impacts on society. Detecting and blocking early fake news is very essential to avoid the negative effect on the community. In this paper, we exploit the news content, the wisdom of crowds in the social interaction and the user's credibility characteristics to automatically detect fake news on Twitter. First, the user profile is exploited to measure the credibility level. Second, the users' interactions for a post such as Comment, Favorite, Retweet are collected to determine the user's opinion and exhortation level. Finally, a Support Vector Machine (SVM) model with the Radial Basis Function (RBF) kernel is applied to determine the authenticity of the news. Experiments conducted on a Twitter dataset and demonstrated the effectiveness of the proposed method.

Keywords: Fake news detection · User's credibility · Social interaction · User's opinion

1 Introduction

Traditional information channels such as newspapers, television has become less prominent in recent years. Nowadays, social networks have become a popular

© Springer Nature Switzerland AG 2020
H. Fujita et al. (Eds.): IEA/AIE 2020, LNAI 12144, pp. 373–384, 2020.
https://doi.org/10.1007/978-3-030-55789-8_33

source for society and play a crucial role in spread the news in the community. In 2018, about 68% of American adults got news on social media, while it was only 62% in 2016[1]. Social network services such as Twitter, Facebook has rapidly developed and attracted millions of users who publish and share the most up-to-date information, emergent social events, and personal opinions everyday. Taking Twitter as an example, it has more than 313 million monthly active users and 500 million tweets are sent per day[2]. In Twitter, users can publish short messages called a tweet. Tweets can read by anyone and reply from other users that leads to real-time conversations and rapidly spread information. However, the quality of news on social networks is a critical issue for users. Because of the easy news publishing, there is a lot of fake news posting on social networks to mislead the readers. Fake news represents the most forms of false or unverified information. The extensive spread of fake news has negative impacts on society and political instability. Furthermore, misleading or wrong information has a higher potential to become viral and lead to negative discussions [2,18]. Detecting and blocking early fake news is very essential to avoid the negative effect on the community.

The fake news detection is a research topic that has attracted much more attention from researchers in recent years. The number of papers concerning fake news has been published from 2006 to 2018 and indexed in the Scopus database grown rapidly, especially in the last two years [2]. Many different aspects of fake news have been analyzed by researchers in the literature. The most popular approach uses content to classify news [1,7,9,12]. The language features are analyzed to determine fake or real news. Generally, this approach is appropriate for formal and long news from traditional news on websites in which the writing styles misunderstanding readers are quite clear. However, the fake news on social networks is quite different from formal news since the short texts, creative lexical variations. Therefore, the approach combining the content and context-based features has considered much for fake news detection on social networks [13,14]. Machine learning algorithms have applied extremely usefully to solve the fake news detection task [2]. SVM-based approaches have made more prominent use of content and context-based features. Features are combined to train the classifiers [3,8]. Random forest, logistic regression, conditional random field, and hidden Markov models field have been employed in a number of fake news detection studies and achieved high accuracy in experiments [4,6]. An approach exploiting the wisdom of crowds is also especially considered with the data from social networks. The users' interactions can be used to identify the authenticity of news [5,17]. The combination of features significantly improves the accuracy of fake news detection problem.

A tweet is too short in order to analyze content for detecting fake news. The user's understanding presenting via interactions is useful information that should be exploited in fake news problems. The wisdom of crowds, especially high credibility users, could improve the performance of fake news detection on social

---

[1] https://www.journalism.org/2018/09/10/news-use-across-social-media-platforms-2018/.

[2] https://about.twitter.com/company.

networks. In this study, we present a framework to detect fake news on social networks based on exploiting news content, user profile, and social interactions related to the news. To deal with the weakness of analyzing short text such as tweets, the context characteristics of the social network are considered to support the fake news detection. User assessments have an important role to determine the authenticity of the news. The user's understanding and opinion are expressed via the comments, favorites, and sharing the news. These data are exploited to determine the user's viewpoint of the news. Besides, the trust factor of the news sources is also considered based on the user profile. The features of the user profile are analyzed to determine the credibility level. Finally, a Support Vector Machine (SVM) model with the Radial Basis Function (RBF) kernel is applied to classify tweets into two classes, those are fake news and real news. The experiments were conducted on real data collected on Twitter to evaluate the performance of the proposed method. The experiments reveal that the method is effective and significantly improves the performance of fake news detection problem.

The organization of this paper is as follows. Section 2 briefly presents related works. The problem statement and the method are presented in Sect. 3. The experiments and results are presented in the subsequent section. The last is the conclusion.

2 Related Work

Fake news detection is a research topic that has been attracted much more attention from researchers in recent years. Bondielli and Marcelloni surveyed the different approaches to detect fake news proposed in the recent literature [2]. A diversity of techniques proposed to solve the fake news detection problem. The approach based on analyzing news content is first attention, especially for detecting fake news in traditional news. The feature such as vocabulary, writing style are considered to determine misleading information [1,7,9]. Pérez-Rosas et al. proposed a method to identify the fake content in online news [7]. They conducted experimental analyses on the collected data to identify linguistic properties that express in fake content. The model relying on the linguistic feature was built and they achieved the accuracy up to 78%. Afroz et al. argue that some linguistic features change when the author hides the writing style [1]. They proposed a method to detect stylistic deception in documents. The experiment results on a large feature set show that the F-measure was 96.6%.

The linguistic feature is appropriate to detect fake news in regular documents. However, the news on social networks is the usual difference from traditional news on websites. The writing style is usually not formal and users often write to mislead readers by mimicking true news. Existing methods focusing on analyzing news content is often not effective and reduce performance when applying to the news on social media. Therefore, exploring the auxiliary information to improve the detection performance is should be concerned. Existing research mainly considers two types of data sources: news content and social context. Social context during spreading news on social media forms the tri-relationship those are the relationship among publishers, news pieces, and users which has the potential to improve fake news detection [14]. Kai Shu et al. used a tri-relationship embedding framework to model publisher-news relations and user-news interactions simultaneously for fake news classification. Besides, Kai Shu et al. also proposed exploited three dimensions in news spread ecosystem on social media those are a content dimension, a social dimension, and temporal dimension [11]. The authors introduced popular network types for studying fake news and proposed how to use these networks to detect and mitigate fake news on social media. User social engagements are auxiliary information to improve fake news detection. User social engagements can be the correlation between user profiles on social media and fake news. Kai Shu and Huan Liu [13] constructed a real dataset measuring the trust level of the user on fake news. Two representative user groups are experienced users who are able to recognize fake news as false and naive users who are gullible to believe fake news. Existing methods only rely on one of three general characteristics of fake news, those are text content, user responses, and source users achieved limited results [10]. Natali Ruchansky et al. proposed a combination model of these characteristics for a more accurate and automated prediction. The experiment results on real datasets demonstrated the effectiveness when comparing the accuracy to existing models.

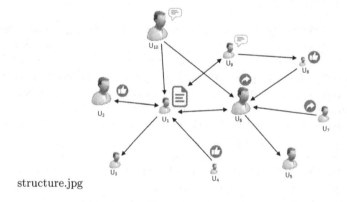

structure.jpg

Fig. 1. The illustration of users' relationships on Twitter

Jin et al. proposed exploiting the wisdom of crowds to improve news verification by mining conflicting viewpoints in microblogs [5]. The conflicting viewpoints in tweets are discovered by a topic model and then, a credibility propagation network of tweets linked with supporting or opposing relations is built to deduct the news' credibility. The experiments on real-world datasets showed that the proposed approach significantly outperformed the baselines. Like action from users on social networks could reflect their opinion toward the news. Tacchini proposed to use two classification techniques relying on logistic regression and a novel adaptation of boolean crowdsourcing algorithms based on users' like [15]. The method applied to Facebook posts and obtain a high accuracy classification. The crowd on social media can assess the veracity of news by posting their responses or reporting. Wei proposed to combine crowd judgment with machine intelligence to detect fake news [17]. The human and machine judgment were extracted from data sources and then an unsupervised Bayesian aggregation model is applied to these judgments to obtain the final prediction model. The method demonstrated the effectiveness of exploiting the complementary value of human intelligence contained in crowd judgment to help detect fake news.

In this paper, we propose a fake news detection method by using a set of features such as news content, social interactions, and user's credibility. In the next sections, we present the proposed method and experiment results.

3   Method

3.1   Problem Statement

When a user posts news on Twitter, that news spreads to a group of users who are his/her followers. Users express their viewpoints on the news by Favorite, Retweet, or Comment. For example, Fig. 1 illustrates a simple graph of relationships between users on Twitter. In this graph, user U_1 has five followers those are user $U_2, U_4, U_6, U_9, U_{10}$. Once user U_1 posts news, his/her followers will assess and express their viewpoints on that news. Intuitively, each user has a credibility level to the community/group (e.g., politicians, journalists, influencers are

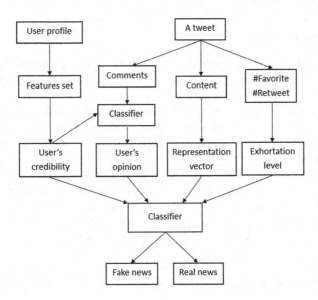

Fig. 2. A framework for detecting fake news based on content, social interactions, and user's credibility

high credibility, and new users having few friends are low credibility). In this example, the high credibility users are drawn to a bigger size and a less size in the opposite case.

Given a target tweet T posted by user U. That news is spread to the group of users that is U's follower. In general, when followers read the news either they can leave a comment representing their opinion or they can express their viewpoint by Favorite, Retweet or nothing. Assume that each user has a credibility level determined relying on his/her profile and we collected a set of social interactions related to that news. Let's determine the authenticity of the news if it is fake or real news by using the news content, social interactions, and user's credibility features.

In fact, a tweet is a short text and a lack of context to clearly understand its content. In some cases, the reader has to have wide knowledge to determine the authenticity of news. Therefore, detecting fake news only relying on content features usually obtains low accuracy, especially short text like tweets on Twitter. In this study, the fake news detection problem is solved by integrating features which are content, social interactions, and user's credibility. The framework of the proposed method to detect fake news is described in Fig. 2. It contains the following steps:

- First, users' data on Twitter are collected and exploited to measure the credibility level.
- Second, a target tweet is converted to a vector of real numbers.
- Third, the user's interactions for the tweet such as comments, favorites, retweets are collected to determine the user's opinions and exhortation level.
- Finally, a classification model is applied to these features to determine the authenticity of the news.

3.2   Feature Extraction

News Content Feature. Normally, a simple method representing a short text with lots of misspellings like tweets into a vector is based on the particular vectors of feature words of text. The feature words are extracted by removing stop words, a modal, symbols, a determiner, a list item marker, etc. First, word vectors of text using a Word2vec model are constructed and then added up the word vectors to compose a tweet vector [16]. In this study, we use the implementation of Word2vec from the gensim package[3] and only feature words of a tweet are converted into vectors of real numbers. A word vector is a k-dimensional vector denoted $W = (w_1, w_2, ..., w_k)$. We use a linear combination of word vectors for generating tweet vectors. A tweet vector (V) is formed as follows:

$$V = \sum_{i=1}^{n} W_i = (\sum_{i=1}^{n} w_1, \sum_{i=1}^{n} w_2, ..., \sum_{i=1}^{n} w_k),$$ (1)

where n is the number of feature words of the tweet.

---

[3] https://radimrehurek.com/gensim/models/word2vec.html.

User's Credibility Feature. Based on user profile and historical data, we can extract information to determine the user's credibility level. Some significant information should be considered such as joined date, the number of followings, the number of followers, a total of favorites, a total of tweets, a total of comments, a total of retweets, etc. A vector of a user's characteristics is a 5-dimensional vector represented as follows:

$$U = \{t, fo, fa, co, re\}, \tag{2}$$

where t is the number of months in that time the account is created, fo is the ratio of the number of followers to the number of followings, fa is the ratio of the number of favorites to the number of tweets, co is the ratio of the number of comments to the number of tweets, re is the ratio of the number of retweets to the number of tweets.

User's Opinion Feature. In social networks, users express viewpoints to the news by comments. They can leave comments for friends' news and content of comments expresses their understanding and opinion about news content. Tweets are short and lack context so the users generally comment relying on their knowledge. Analyzing the comments from users is useful to detect fake news. Users have a wide knowledge and can analyze related problems to understand the news content. They can recognize the authenticity of news by leaving comments expressing their viewpoints. Therefore, we can exploit the wisdom of crowds that is the user's comments to detect fake news.

First, followers' comments related to a target tweet are crawled. Then, data is cleaned to eliminate inessential elements. Vectors of the comments are constructed as the vectors of tweets. A vector of a user's opinion is formed by concatenating the vector of the user's comment and the vector of the user's credibility. The integration of the user's comment and their credibility can increase the efficiency of exploiting comments. To classify opinions, a multiSVM model[4] is used to classify the comments into three categories such as *positive, neutral*, and *negative*. In which *negative* presents a fake news assessment and *positive* presents a real news assessment. The comments do not express the authenticity of news classified into *neutral*. In fact that each tweet has more comments from different users. Therefore, the users' opinions for a tweet is determined based on a combination of opinions. A vector of the users' opinions is defined as follows:

$$O = \{pos, neu, neg\}, \tag{3}$$

where pos is the percentage of positive comments, neu is the percentage of neutral comments, neg is the percentage of negative comments.

---

[4] https://scikit-learn.org.

Exhortation Feature. In social networks, the users usually express their viewpoints toward news by leaving comments. Analyzing comments to know users' opinions is valuable for fake news detection problem. Besides, the user also encourages the news by favorite or retweet action. In general, once users interest the news they will click on Favorite to identify their particular interests. They can also re-post a tweet on their timeline, called Retweet. By this way, users can spread the news to wider groups that they are a member of those groups. In general, the favorite and retweet actions present the user's agreement to the news. Since the user's tweets will only present on the timeline of followers. Therefore, in this work, we consider the ratio of the number of favorites to the number of followers (fav) and the ratio of the number of retweets to the number of followers (ret). The exhortation of the news is described as:

$$E = \{fav, ret\}. \tag{4}$$

Algorithm 1. Constructing a feature vector

Input: A target tweet T
Output: A feature vector F

1: Preprocessing T;
2: Converting T to a vector, V;
3: Determining a set of users related to T (i.e., poster, followers who have interacted with T);
4: Crawling user profile and historical data of users;
5: Extracting features to determine the user's credibility, U;
6: Collecting social interaction data related to T (i.e., comments, #favorite, #retweet);
7: Determining the user's opinion, O;
8: Determining the exhortation level, E;
9: $F \leftarrow V \& U \& O \& E$;//Integrating features
10: **return** F.

3.3 Classification Performance

In this study, we use a set of features to classify news. The news content and data related to users who interact with the news and their assessments are collect to construct the feature set as described in Sect. 3.2. The news content, user's credibility, user's opinion, and exhortation feature are integrated to generate the feature vector. The steps of constructing the feature vector are described in Algorithm 1. A classification model is applied to the feature vector to classify the news. In this work, we use a supervised learning method to implement the classification task that is the SVM model with the RBF kernel from Scikit-learn (See footnote 4). The model's parameters are adjusted to achieve the best performance. The classifier's output is two classes (i.e., real news and fake news). In this study, in order to assess the effectiveness of proposed features, we conduct experiments the fake news detection based on particular features and the integrated features also. The experimental results are presented in the next section.

4   Experiments

4.1   Evaluation Metrics

To evaluate the performance of proposed method, we used the precision and recall measurement. Precision (P) and Recall (R) are calculated as follows:

$$P = \frac{|TruePositive|}{|TruePositive| + |FalsePositive|} \tag{5}$$

$$R = \frac{|TruePositive|}{|TruePositive| + |FalseNegative|} \tag{6}$$

where True Positive is a set of detected fake news which are actually fake news, False Positive is a set of detected fake news which are actually real news, False Negative is a set of detected real news which are actually fake news. The F_1 score that combines precision and recall is the harmonic mean of precision and recall, defined as follows:

$$F_1 = 2 \times \frac{P \times R}{P + R} \tag{7}$$

4.2   Data Collection

In order to construct experimental data, first, we collected a set of fake news and a set of real news from the website www.politifact.com. Then, using Twitter's search API to crawl tweets related to that news that is spreading on Twitter. In fact that there are many tweets mentioned the content of each news. Therefore, we just randomly selected up to 10 tweets for each news. Finally, 255 fake news and 260 real news were selected for experiments. For each tweet, user profile and historical data of related users who have published and interacted news are also collected to construct features as described in Sect. 3.2. To train classifiers for classifying opinions, we annotated 1486 comments. The number of negative, neutral, positive comments are 435, 521, 530, respectively. In this study, we used the available tool Twitter4J[5] (i.e., an unofficial Java library for the Twitter API) to crawl data. Because of the limitation of Twitter API for crawling data, we only constructed features related to the user profile and historical data collected during one recent year. Then, data are cleaned to eliminate elements that are not necessary. Elements such as hashtags, mentions, hyperlinks, symbols, and emoticons are removed from tweets to create the content of tweets.

---

[5] http://twitter4j.org.

4.3 Results and Discussions

With extracted features as mentioned in Sect. 3.2, we used the SVM model with the RBF kernel to train classifiers. In order to evaluating the effect of proposal, the different subsets of features were implemented with a 5-fold cross validation strategy. A summary of experimental results is shown in the Table 1.

Table 1. The experimental results for different subsets of features.

Features	P	R	F_1
Content	50.0	56.9	53.2
User's opinion	70.5	60.8	65.3
Content+User's credibility	52.7	56.9	54.7
Content+Exhortation	49.2	58.8	53.6
Content+User's opinion	72.3	66.7	69.4
User's opinion+User's credibility	71.4	68.6	70.0
User's opinion+Exhortation	66.7	62.7	66.6
Content+User's opinion+User's credibility	75.0	70.6	72.7
User's opinion+User's credibility+Exhortation	73.5	70.6	72.0
All features	**75.5**	**72.5**	**74.0**

First, we used two particular features that were content-based and user's opinion-based feature. The performance of the system was only based on the user's opinion better than content-based by 12.1%. This result shows the effect of the user's opinions discovery in fake news detection. Detecting fake news by mining the user's assessments is better than news content-based. We also used two separate features that were user's credibility and exhortation feature to train classifiers. However, the accuracy of the classifiers was very low. That demonstrates that using these features particularly is ineffective. Then, we combined these features with the news content and the system performance also improved not much. Intuitively, based on news content and exhortation level is hard to determine the authenticity of the news. The main reason is that user's information having a favorite or retweet action can not be collected on Twitter. The experiment on combining users' opinions and their credibility level achieved good results. It was better than the user's opinion only by 4.7%. Especially, the combination of the user's opinion and content feature achieved results better than content-based only by 16.2%.

We conducted the experiments with a combination of more than two features. The obtained classifiers produced better results. The F_1 score is 74.0% when combining all features. It was better than content-based, user's opinion feature by 20.8% and 8.7%, respectively. Obviously, compared to one or two features of experimental results, the precision, and recall of fake news detection improve significantly. In general, the precision and recall of combining all features exceeded those of other feature subsets. The user's credibility and exhortation features showed effectiveness when incorporated with many other features. The experiments with the combination of three features obtained results that were not different from each other much. The F_1 score is less than that of all features about 2.0%.

5   Conclusions

In this paper, we proposed using a set of features to detect fake news on Twitter. We exploited the news content, social interactions, and the user's credibility to automatically detect fake news. The experiments on the SVM model with the proposed feature set demonstrated the effectiveness of the method. The best performance was 74.0% in the F_1 score when combining all features. The user's opinion proved the effectiveness in fake news detection on social networks. Exploiting the user's assessment and their profile significantly improved the performance of the classifier. The proposed method efficiently mined the wisdom of crowds to discover the fake news.

References

1. Afroz, S., Brennan, M., Greenstadt, R.: Detecting hoaxes, frauds, and deception in writing style online. In: 2012 IEEE Symposium on Security and Privacy, pp. 461–475. IEEE (2012)
2. Bondielli, A., Marcelloni, F.: A survey on fake news and rumour detection techniques. Inf. Sci. **497**, 38–55 (2019)
3. Della Vedova, M.L., Tacchini, E., Moret, S., Ballarin, G., DiPierro, M., de Alfaro, L.: Automatic online fake news detection combining content and social signals. In: 2018 22nd Conference of Open Innovations Association (FRUCT), pp. 272–279. IEEE (2018)
4. Hardalov, M., Koychev, I., Nakov, P.: In search of credible news. In: Dichev, C., Agre, G. (eds.) AIMSA 2016. LNCS (LNAI), vol. 9883, pp. 172–180. Springer, Cham (2016). https://doi.org/10.1007/978-3-319-44748-3_17
5. Jin, Z., Cao, J., Zhang, Y., Luo, J.: News verification by exploiting conflicting social viewpoints in microblogs. In: Thirtieth AAAI Conference on Artificial Intelligence (2016)
6. Kwon, S., Cha, M., Jung, K., Chen, W., Wang, Y.: Prominent features of rumor propagation in online social media. In: 2013 IEEE 13th International Conference on Data Mining, pp. 1103–1108. IEEE (2013)
7. Pérez-Rosas, V., Kleinberg, B., Lefevre, A., Mihalcea, R.: Automatic detection of fake news. arXiv preprint arXiv:1708.07104 (2017)
8. Qin, Y., Wurzer, D., Lavrenko, V., Tang, C.: Spotting rumors via novelty detection. arXiv preprint arXiv:1611.06322 (2016)
9. Rashkin, H., Choi, E., Jang, J.Y., Volkova, S., Choi, Y.: Truth of varying shades: analyzing language in fake news and political fact-checking. In: Proceedings of the 2017 Conference on Empirical Methods in Natural Language Processing, pp. 2931–2937 (2017)

10. Ruchansky, N., Seo, S., Liu, Y.: CSI: a hybrid deep model for fake news detection. In: Proceedings of the 2017 ACM on Conference on Information and Knowledge Management, pp. 797–806. ACM (2017)

11. Shu, K., Bernard, H.R., Liu, H.: Studying fake news via network analysis: detection and mitigation. In: Agarwal, N., Dokoohaki, N., Tokdemir, S. (eds.) Emerging Research Challenges and Opportunities in Computational Social Network Analysis and Mining. LNSN, pp. 43–65. Springer, Cham (2019). https://doi.org/10.1007/978-3-319-94105-9_3

12. Shu, K., Sliva, A., Wang, S., Tang, J., Liu, H.: Fake news detection on social media: a data mining perspective. ACM SIGKDD Explor. Newslett. **19**(1), 22–36 (2017)

13. Shu, K., Wang, S., Liu, H.: Understanding user profiles on social media for fake news detection. In: 2018 IEEE Conference on Multimedia Information Processing and Retrieval (MIPR), pp. 430–435. IEEE (2018)

14. Shu, K., Wang, S., Liu, H.: Beyond news contents: the role of social context for fake news detection. In: Proceedings of the Twelfth ACM International Conference on Web Search and Data Mining, pp. 312–320. ACM (2019)

15. Tacchini, E., Ballarin, G., Della Vedova, M.L., Moret, S., de Alfaro, L.: Some like it hoax: automated fake news detection in social networks. arXiv preprint arXiv:1704.07506 (2017)

16. Tran, V.C., Hwang, D., Nguyen, N.T.: Hashtag recommendation approach based on content and user characteristics. Cybern. Syst. **49**(5–6), 368–383 (2018)

17. Wei, X., Zhang, Z., Zeng, D.D.: Combining crowd and machine intelligence to detect false news in social media. Available at SSRN 3355763 (2019)

18. Zollo, F., et al.: Emotional dynamics in the age of misinformation. PLoS ONE **10**(9), e0138740 (2015)

Financial Applications and Blockchain

Deep Reinforcement Learning for Foreign Exchange Trading

Yun-Cheng Tsai[1]([✉]), Chun-Chieh Wang[2], Fu-Min Szu[3], and Kuan-Jen Wang[4]

[1] School of Big Data Management, Soochow University, Taipei, Taiwan
pecutsai@gm.scu.edu.tw
[2] Department of Computer Science, National Chengchi University, Taipei, Taiwan
[3] Department of Electrical Engineering, National Taiwan University, Taipei, Taiwan
[4] Department of Mathematics, National Taiwan University, Taipei, Taiwan

Abstract. We optimized the Sure-Fire statistical arbitrage policy, set three different actions, encoded the continuous price over some time into a heat-map view of the Gramian Angular Field (GAF), and compared the Deep Q Learning (DQN) and Proximal Policy Optimization (PPO) algorithms. To test feasibility, we analyzed three currency pairs, namely EUR/USD, GBP/USD, and AUD/USD. We trained the data in units of four hours from 1 August 2018 to 30 November 2018 and tested model performance using data between 1 December 2018 and 31 December 2018. The test results of the various models indicated that favorable investment performance achieves as long as the model can handle complex and random processes, and the state can describe the environment, validating the feasibility of reinforcement learning in the development of trading strategies.

Keywords: Gramian Angular Field (GAF) · Deep Q Learning (DQN) · Proximal Policy Optimization (PPO) · Reinforcement learning · Foreign exchange trading

1 Introduction

We plan to use deep-enhanced learning to mimic how humans make decisions, using the state of the current environment to execute actions and obtain rewards from the environment. Moreover, people's actions impact the environment, causing the situation to enter a new state. To check the feasibility of this approach, we adopted the method of training four-hour units of EUR/USD, GBP/USD, and AUD/USD data between 1 August 2018 and 30 November 2018. We then applied the trained data to the period between 1 December 2018 and 31 December 2018 to validate the system performance.

2 Preliminary

In the paper, we adopted the Sure-Fire arbitrage strategy, which is a variant of the Martingale. It involves increasing bets after every loss so that the first

© Springer Nature Switzerland AG 2020
H. Fujita et al. (Eds.): IEA/AIE 2020, LNAI 12144, pp. 387–392, 2020.
https://doi.org/10.1007/978-3-030-55789-8_34

win recovers all previous losses plus a small profit. After entering the market and initiating trading, the investor uses the margin between the stop-loss and stop-gain prices as the raised margin. As long as the price fluctuates within the increased margin and touches on the risen price, the Sure-Fire Strategy urges investors to continue growing the stakes until they surpass the margin to profit. The most prominent shortcoming of Martingale is the lack of stakes or funds when the favorable odds are low. Therefore, we applied reinforcement learning to optimize the Sure-Fire Strategy. Data train to obtain the trading behavior with the minimum number of raises to achieve the maximum winning odds. A detailed transactions are illustrated from Figs. 1, 2 and Fig. 3. In Sure-Fire arbitrage strategy 1, we purchase one unit at any price and set a stop-gain price of $+k$ and a stop-loss price of $-2k$. At the same time, we select an amount with a difference of $-k$ to the buy price and $+k$ to the stop-loss price and set a backhand limit order for three units. Backhand refers to engaging in the opposite behavior. The backhand of buying is selling, and the backhand of selling is buying. A limit order refers to the automatic acquisition of corresponding units. In Sure-Fire arbitrage strategy 2, we place an additional backhand limit order, where the buy price is $+k$ to the selling price, and $-k$ to the stop-loss price when a limit order is triggered, and three units successfully sold backhand. We set the stop-gain point as the difference of $+k$ and the stop-loss point as the difference of $-2k$, after which an additional six units buy. In Sure-Fire arbitrage strategy 3, the limit order triggers in the third transaction. The final price exceeded the stop-gain amount of the first transaction, the stop-loss price of the second transaction, and the stop-gain price of the third transaction. In this instance, the purchase is complete. The calculation in the right block shows that the profit is $+1k$.

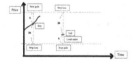

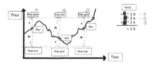

Fig. 1. Sure-Fire arbitrage strategy 1.

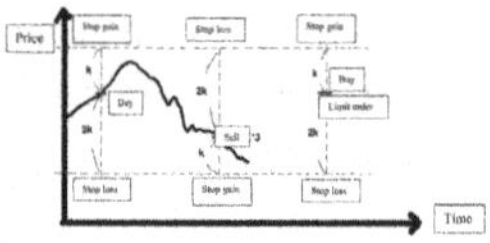

Fig. 2. Sure-Fire arbitrage strategy 2.

Fig. 3. Sure-Fire arbitrage strategy 3.

2.1   Deep Q Network (DQN)

The Deep Q Network (DQN) is a deep, reinforcement learning framework [1]. For the selection of actions, if the greedy method alone is applied to select the action with the highest expected return every round, then the chance of selecting other actions would be lost. Therefore, DQN adopts the ϵ-greedy exploration method. Specifically, each time an action is selected, the system provides a small probability (ϵ) for exploration, in which the best action abandoned and other new actions executed. The probability of exploration increases concurrently with the ϵ value, making the agent more likely to try a new state or action. This process improves the learning effects. The disadvantage is that the convergence time increases exponentially with the duration of exploration.

2.2 Proximal Policy Optimization (PPO)

Proximal policy optimization (PPO) is a modified version of the policy gradient method [2]. It used to address the refresh rate problem of policy functions. The algorithms used in the reinforcement learning method can be categorized into dynamic programming (DP), the Monte Carlo method and temporal-difference (TD) [3]. The policy gradient method uses the Monte Carlo method, wherein samples collected for learning. The advantage of this method is that it can apply to unknown environments. The policy gradient method abandons value functions. Instead, it directly uses rewards to update the policy function, which outputs the probability of taking specific action in an individual state. Thus, it can effectively facilitate decision making for non-discrete actions [4].

2.3 Gramian Angular Field (GAF)

Foreign exchange prices can be viewed as a time series. Therefore, it is necessary to consider prices over time rather than adopting a set of opening and closing prices as a single state. The paper uses the Gramian Angular Field method, which is a time series coding method consisting of many Gram matrices. A Gram matrix is a useful tool in linear algebra and geometry. It is often used to calculate the linear correlation of a set of vectors [5]. In Fig. 4, the left image is a heat-map view of the GAF, and the right image is the corresponding numerical line chart. The chart shows high values.

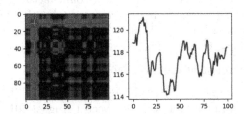

Fig. 4. GAF results: Heat-map and numerical line chart.

3   Methodology

Our reinforcement learning trading system designs as follows:

1. State Design: States are derived from an agent's observations of the environment. They are used to describe or represent environments. The agents then perform actions corresponding to the perceived state. Therefore, defining the state is key to learning performance. In the paper, we use the sliding window method on the "opening price, highest point, lowest point, and closing price" of the foreign exchange state in units of four hours to obtain sets of 12. After GAF encoding, the data are inputted as a state with dimension $12 \times 12 \times 4$.
2. Action Design: The purpose of the paper is to optimize the Sure-Fire policy. There are two optimization objectives. The first is to reduce overall buy-ins and the second is to set favorable stop-loss and stop-gain points. The former is to avoid the lack of funding while the latter is to prevent the fluctuation of prices and unnecessary placement of investments. After pragmatic consideration, three discrete actions are developed:
   - Upper limit to additional buy-ins after entering the market: $\{1, 2, 3\}$.
   - First buy or sell after entering the market: $\{BUY, SELL\}$.
   - Stop-gain (stop-loss is double stop-gain): $\{20, 25, 30\}$.
3. Reward Design: Reward is defined as follows:

$$\text{Reward} = \text{Profit} \times \text{Discount},$$

where profit is the net income of the current transaction and the discount is

$$1.0 - 0.1 \times \text{(number of additional buy-ins)}.$$

This calculation is a variant of the system quality number (SQN). SQNs are one of the indicators used to score trading strategies. The formula is:

(expected profit/standard deviation)×(square root of the number of transactions).

The discount factor is a variable that decreases concurrently with the increase in transactions. The reason for adding a discount factor to profit is to inform the agent that the reward decreased concurrently with the increase in the number of additional buy-ins. That is, risk increases concurrently with the number of additional buy-ins. Therefore, excessive buy-ins should be avoided.

Oue experiment environment designs as follows:

1. Experiment Data: we used the EUR/USD, GBP/USD, and AUD/USD exchange data in units of four hours between 1 August 2018 and 31 December 2018 as a reference for the environment design and performance calculation of the system. The period between 1 August 2018 and 30 November 2018 is the training period for the agent. In this period, the agent repeatedly applied the data to learn, eventually obtaining an optimal investment strategy. The period between 1 December 2018 and 31 December 2018 is the agent's performance evaluation period. The agent uses a trading strategy to form decisions. The system accumulates the rewards obtained during the evaluation period, which served as a reference for performance.

2. Trade Environment Settings: EUR/USD, GBP/USD, and AUD/USD are adopted as investment targets. The smallest unit is 1 pip (0.00001 of the price). The default earnings for an action is set to at least 20 pips, which would be greater than the slip price and transaction fee. Therefore, the slip price and transaction fees are not taken into account. Whenever a transaction is tested, the closing price is used as the data point rather than considering whether the transactions are triggered at a high point or low point in the data (that is, of the four values for each day-opening prices, high point, low point, and closing price-the final one is chosen). Time scales smaller than one day rarely contained stop-loss and stop-gain points.
3. Experiment Model Design: a constant agent is added to the experiment model, serving as a reference value for performance. The input of the three different algorithms is a GAF-encoded time series of dimension $12 \times 12 \times 4$. A CNN is used as the policy network or Q-network.

4   Experiment Results and Discussions

In Table 2, PEU and DEU profits are lower than those of CEU. The accumulated return trajectories for PEU and DEU are almost identical, with the sole differences being the slope. The lines for accumulated returns of PEU and DEU are smoother than CEU. The max drawdown is also much lower, which shows that net profits for PEU and DEU are lower only because no wrong decisions made in the more conservative markup models. The max drawdown of PEU is almost half that of CEU. A more economical risk-orientated strategy for stable returns very possibly discovered for PEU and DEU. The returns for the three are about the same, but the PGU profit factor is much higher than that of the other two, and the PGU max drawdown is the lowest of the three. It can see that PGU using the PPO trade algorithm is the most consistently profitable. Interestingly enough, the trading decisions made in DGU are almost all reached later than those in PGU. PGU more clearly captured profit maximization properties than DGU during the training period, resulting in a situation in which PGU did not experience the same delays as DGU. PAU made trading decisions that maximized profitability, with every evaluation being far better than that of CAU. Though the max drawdown of PAU is approximately 1.25 times that of DAU, returns are almost 1.5 times those of DAU, while the profit factor remained the highest of the three. Besides the unexpectedly high PGU earnings, there is not much out of the ordinary in the accumulated returns line graphs for DAU and

Table 1. Model code information

Model	Algorithm	Currencies	Model	Algorithm	Currencies	Model	Algorithm	Currencies
CEU	Constant	EUR/USD	DEU	DQN	EUR/USD	PEU	PPO	EUR/USD
CGU	Constant	GBP/USD	DGU	DQN	GBP/USD	PGU	PPO	GBP/USD
CAU	Constant	AUD/USD	DAU	DQN	EUR/USD	PAU	PPO	EUR/USD

CAU. DAU continually made sound trading decisions, and its max drawdown is the lowest of the three (Table 1).

Table 2. Comparing DQN model with PPO model trading performance. Using EUR/USD 4 h data from 2018/12/01 to 2018/12/31 at 1300 episode.

Model code	Net profit	Profit factor	Max draw-down
CEU	753	3.67	−5.18%
DEU	486	3.09	−3.07%
PEU	615	3.82	−2.92%
CGU	753	3.24	−5.58%
DGU	717	3.10	−4.63%
PGU	763	4.47	−4.33%
CAU	351	1.62	−14.53%
DAU	402	2.32	−5.97%
PAU	597	2.85	−7.62%

5   Conclusion

We used a GAF-encoded time series as the state, two different algorithms (DQN and PPO), and three different currency pairs (EUR/USD, GBP/USD, AUD/USD) to create six different trading models. From the above comparisons that using PPO's reinforcement learning algorithm to optimize forex trading strategies is quite feasible, with PPO performance being better than DQN. The results of the various models indicated that favorable investment performance achieved as long as the model can handle complex and random processes, and the state can describe the environment, validating the feasibility of reinforcement learning in the development of trading strategies. We found that the most challenging aspect is the definition of reward.

References

1. Hausknecht, M., Stone. P.: Deep recurrent q-learning for partially observable mdps. In: 2015 AAAI Fall Symposium Series (2015)
2. Schulman, J., Wolski, F., Dhariwal, P., Radford, A., Klimov, O.: Proximal policy optimization algorithms (2017). arXiv preprint arXiv:1707.06347
3. Sutton, R.S., Barto, A.G.: Reinforcement Learning: An Introduction. MIT Press, Cambridge (2018)
4. Sutton, R.S., McAllester, D.A., Singh, S.P., Mansour, Y.: Policy gradient methods for reinforcement learning with function approximation. In: Advances in Neural Information Processing Systems, pp. 1057–1063 (2000)
5. Tsai, Y.C., Chen, J.H., Wang, C.C.: Encoding candlesticks as images for patterns classification using convolutional neural networks (2019). arXiv preprint arXiv:1901.05237

Human-Centred Automated Reasoning for Regulatory Reporting via Knowledge-Driven Computing

Dilhan J. Thilakarathne[✉], Newres Al Haider, and Joost Bosman

ING Tech R&D, ING Bank, Haarlerbergweg 13-19, 1101 Amsterdam, CH, The Netherlands
Dilhan.Thilakarathne@ing.com

Abstract. The rise in both the complexity and volume of regulations in the regulatory landscape have contributed to an increase in the awareness of the level of automation necessary for becoming fully compliant. Nevertheless, the question of how exactly to become fully compliant by adhering to all necessary laws and regulations remains. This paper presents a human-centred, knowledge-driven approach to regulatory reporting. A given regulation is represented in a controlled natural language form including its metadata and bindings, with the assistance of subject matter experts. This representation of a semi-formal controlled natural language translates into a self-executable formal representation via a context-free grammar. A meta reasoner with the knowledge to execute the given self-executable formal representation while generating regulatory reports including explanations on derived results has been developed. Finally, the proposed approach has been implemented as a prototype and validated in the realm of financial regulation: Money Market Statistical Reporting Regulation.

Keywords: Regulations · Compliance · RegTech · Human-centred · Reasoning · Knowledge systems · Explainability · Rule-based systems · Finance · Banking

1 Introduction and Background

The Global Financial Crisis in 2008 changed the way in which financial markets, services and institutions manage financial risk. One necessity in financial risk management is understanding the importance of fully adhering to a process of compliance with its concealed complexity [3, 5]. Compliance adherence involves a significant allocation of both time and expertise; therefore, the associated cost has led to an increase in compliance budgets due to both a rise in the complexity and volume of regulations [5]. According to Duff & Phelps, financial institutes already spend a significant amount of money during the process of becoming compliant and expect a substantial increase in future compliance costs: 24% of participants expect to spend more than 5% of their annual revenue on compliance by 2023, while 10% expect to spend more than 10% by 2023 [10]. Given the increasing complexity and volume of regulations, it is clearly necessary to provide a cost-effective approach that is easy to explain [3, 5, 10, 14].

© Springer Nature Switzerland AG 2020
H. Fujita et al. (Eds.): IEA/AIE 2020, LNAI 12144, pp. 393–406, 2020.
https://doi.org/10.1007/978-3-030-55789-8_35

Regulatory technology (RegTech) is the centrifugal force both in re-conceptualizing the financial regulation process and achieving principal regulatory objectives [3]. According to Arner and colleagues, RegTech is '*a contraction of the terms regulatory and technology, and it comprises the use of technology, particularly information technology (IT), in the context of regulatory monitoring, reporting, and compliance*' [1, p. 373]. Given that many objectives are associated with RegTech-driven solutions in the financial domain, an appealing objective is to ensure the prudential safety and soundness of financial markets. To achieve this, the wisest approach is to enhance the transparency of the process with the support of an explainable causal reasoning methodology [8]. In this way, Subject Matter Experts (SMEs) can understand the causation behind derived conclusions (whether compliant or non-compliant) not only to gain a retrospective interpretation of precisely how compliant something is/was but also to unambiguously explain it to regulatory bodies. Furthermore, '*right to explanation*' has become a social convention (e.g., GDPR); therefore, explainability in the execution of financial tasks is necessary even from the customer/user point of view [6].

Another key goal in RegTech-driven regulatory compliance solutions is reducing outlay while managing compliance obligations via more efficient compliance system automation [3, 8, 10, 14, 27]. The efficacy of regulatory compliance is associated with three perspectives: corrective, detective and preventative steps [27]. The ideal RegTech approach has the power to reduce the cost associated with each of these perspectives. In this paper, the scope of the proposed approach is limited to before-the-fact detection (under detective perspective) with a focus mainly on regulatory compliance reporting.

This paper presents a human-centred approach for regulatory reporting through the use of knowledge-driven computing techniques, that aims to achieve the objectives of being both explainable and capable of automating arbitrary compliance regulations.

In this process the regulatory text is translated into a semi-formal representation, which is labelled Controlled Natural Language (CNL), that can be easily understood by SMEs. This CNL contains a hierarchical tree structure (root, intermediate and leaf) where leaf nodes only contain atomic facts and top-level nodes represent aggregated information. This CNL representation is further enriched by meta-data which is translated into a more formal rule set where individual rules are self-executable. A self-executable rule set is interpreted via a symbolic reasoner implemented on top of actual data that is required to validate its compliant status. The interpreter provides a final status (compliant or non-compliant) which always has a detailed explanation behind its derivation, along with extra information via metadata. The proposed approach is validated using a section of Money Market Statistical Reporting (MMSR) regulation.

The rest of the paper is structured as follows: Sect. 2 covers the related work and highlights key approaches in automating the regulatory compliance process; Sect. 3 describes the approach with relevant insight; Sect. 4 presents an example based on the MMSR document which systematically explains how the proposed approach can be used in a practical way; Sect. 5 concludes and draws possible links to future work.

2 Related Work

There have been significant contributions to regulatory compliance automation with many adopted approaches in several domains: construction [4, 9, 21], health [29], finance

[11, 12, 18, 26] and business processes (see [16]). Nevertheless, the research community is still far from fully overcoming the associated issues, though there are some key areas for further development [3, 5, 10, 16].

A known approach for automating regulatory compliance is to represent it as an (logical-) object model together with domain-specific parameters [4, 9, 17, 21]. This object model approach has been applied to many building regulations (see [17]); Malsane and colleagues have extended this generic approach to filter the rules before interpreting them [21]. In the knowledge formalisation, regulations are interpreted into a set of computer-implementable rules which are used to create an object model that can be logically validated. This representation is then used as the basis to create an Industry Foundation Class compliant data model. This model includes a hierarchical structure that helps to rationalise terminological ambiguities that exist in the semantic representation and uses the derived objects and associated relations as rules to validate compliance needs [21]. Furthermore, the data model can be converted to a machine-executable rule format: Semantic Web Rule Language (SWRL) that provides binary results [4].

SWRL together with ontologies is used for regulatory purposes in different domains as ontological informatics systems [11, 12, 29]. In the pharmaceutical domain, a key role is played by a domain-specific reasoner and a rule engine that manually parses a pharmaceutical regulation text into OWL axioms and SWRL rules [29]. The purpose of having a (OWL-) reasoner is to identify logical inconsistencies while a rule engine (e.g. Java Jess) interprets and enforces the (SWRL-) rules imposed by the SMEs. Finally these two results are combined to provide a structural incorporation to facilitate user interaction via Protégé. Instead of translating regulatory text into SWRL, different control languages such as SBVR have been used to address the same problem but are more focused on (financial-) business vocabularies as ontologies (e.g., FIRO, FIBO) together with formal reasoning and verification techniques.

Semantics of Business Vocabulary and Business Rules (SBVR) is one of the leading control languages in automating compliance applications [11, 18, 26]. Roychoudhury and colleagues presented an approach that translates legal text into SBVR via a control language called 'Structured English' (StEng) [26]. The purpose of introducing StEng in this approach is to partly eliminate the inherent complexity of understanding the legal text representation in SBVR by SMEs. StEng provides a format which is comparatively easier for SMEs to understand so they can easily intervene in this semi-automated task to validate and improve the transformation from legal text to an StEng [26]. Once there is StEng representation, it is translated into a regulatory model in SBVR and then further translated into formal logic that enables validation of a given data set via inferencing engines like DROOLs or DR-Prolog [25]. There is an extra step in this approach where a conceptual data model—DDL (i.e., data required for checking)—is generated from SBVR as a basis for mapping enterprise schema to the conceptual schema by a database expert, as it is required to populate the fact base in a DB in this approach.

Abdellatif and colleagues used SBVR StEng notation to translate legal text into SBVR with POS tagging and dependency parsing in NLP to automate this step [15]. Mercury, proposes a language to capture regulations for the purpose of compliance checking that consist of two main parts: Mercury StEng and Mercury Markup Language [7]. Mercury StEng is an extension of a subset of SBVR that is workable enough for

SMEs to directly represent and maintains the information available in regulatory text. This allows it to eliminate the complexity and verboseness of standard SBVR-driven logical formulation together with the introduction of Mercury Markup Language. This language contains a vocabulary and a rulebook which both help to map into FIRO ontology. Nevertheless, the main difficulty in translating legal text to SBVR is a complex semantic analysis of the English language, which is a challenging task for SMEs [7, 28].

3   Approach

A human-centred approach motivated by knowledge-driven computing techniques is used in this paper. Figure 1 presents the overall approach and highlights the end-to-end process including specific human interventions. The first step in knowledge acquisition takes place when a selected regulatory text is (manually-) translated into the proposed CNL representation. This representation is more human-oriented than machine-oriented [28] with the purpose of facilitating Subject Matter Experts (SMEs) to easily represent the given clauses.

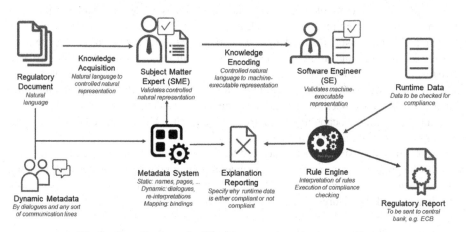

Fig. 1. Overall approach of human-centred regulatory compliance

A unique feature of this knowledge-acquisition phase is embedding metadata into the CNL. This includes not only the causal information of a clause in a regulatory text but it also includes the associated descriptive information including conversations about different possible interpretations; both of these result in more meaningful explanations in regulatory reporting. This CNL includes specific grammar and vocabulary; therefore, performing knowledge encoding (in this phase translating CNL into a Machine-Executable Representation) is achieved via a parser that translates CNLs context-free grammar (CFG) into a machine-executable representation form. This CNL contains a hierarchical tree structure (root, intermediate and leaf) where leaf nodes only contain atomic facts and inner nodes represent aggregated information.

This executable representation is validated by software engineers (SEs) with necessary interventions including the addition of correct functions from a pre-developed regulatory library for any missing or incorrect references provided for the calculation of the atomic facts represented in the leaf nodes. A specific declarative rule engine is implemented that interprets rules represented in machine-executable form together with given runtime data to generate a regulatory report that includes statuses and explanations. These explanations are enriched with metadata binding to facilitate clearer and unambiguous elaboration of the scenario. More detailed information about each part is explained in following subsections.

3.1 Process of Regulatory Text into CNL Representation

Even though laws and regulations are supposed to be unambiguous, ambiguities are prevalent in regulatory texts due to the inherent complexity of natural languages and their underlying processes [22]. As a result of this complexity, the key step in automating the regulatory compliance process is to extract these rules into a more formal structure, to make the rules more explicit and eliminate ambiguities.

As the first step of this process, SMEs need to translate a given regulatory text into a CNL, which is a semi-formal representation of a natural text (e.g., English text). Unsurprisingly, if the CNL can facilitate a closer expressivity in line with its selected natural text to non-technical specialists while enabling accurate and efficient processing to machines, then such a CNL is an ideal representation for this task [13]. Nevertheless, such an ideal representation is difficult to obtain practically; therefore, this approach uses a human-oriented CNL which is enriched with strong readability and comprehensibility features that are essential for SMEs when they need to manually transform regulatory text into CNL text [28].

For this transformation a specific CFG-based CNL is created as in Table 1. The root term of this grammar is the '*specification*' which holds a set of rules. Each rule contains a head (unique string), state, conditions (unique strings) and meta knowledge to elaborate a selected clause or phrase in the given regulatory text. Specifically, if the given conditions (in here conditions are conjunctions of individual conditions) are true then the head holds the value of the state which is a boolean value. Furthermore, conditions are constraints that should be satisfied to conclude the state of the head as in formal logic. Finally, meta knowledge consists of extra information to improve the meaning and understanding of the formation of the rule which provides clear insight into the derivation of content. As per the CFG both head and condition hold a statement, therefore we maintain a chain of rules within this specific hierarchy.

In the process of transforming regulatory text into its mapping form of CNL representation, a top-down approach is used in which the division of aggregated relationships in a compound clause contributes to dividing the full text into a set of rules. In other terms, structure of the 'specification' (specifically a single rule of the specification) is a tree and is composed of a root-node, intermediate-nodes and leaf-nodes. A root-node represents the aggregated goal of the regulatory document. There is commonly one single root for the document but it is possible to have many root nodes if the goals are mutually exclusive. The intermediate-nodes break the overall regulation into sub-clauses. Furthermore, each intermediate-node may further sub-divide into sub-intermediate-nodes depending on its level of aggregation and complexity. Finally, any element which does not require

Table 1. CFG for proposed CNL representation

specification	::= rule⁺
rule	::= *rule* head *is* state *if following conditions hold:* { conditions } *with meta knowledge* { know }
head	::= statement
state	::= *true* | *false*
conditions	::= condition⁺
condition	::= statement | (*not* statement) | (statement op condition⁺)
op	::= *and* | *or*
statement	::= ^[a-zA-Z]+([a-zA-Z0-9\s]+.*)
know	::= *static data* { pair⁺ } | *dynamic data* { pair* } | *mapping data* { mapping* }
pair	::= key : value ;
mapping	::= key : value | { variable [; variable ...] }
key	::= *string*
value	::= *string*
variable	::= *string*

any further sub-division, is labelled as a leaf-node, which carries the real atomic computation information to derive its ultimate runtime value. In order to evaluate whether a regulation holds or not, the computation at the leaf nodes can be executed with the appropriate data, and the result of which can be aggregated up through the links with the intermediate nodes to the root.

A unique feature of this knowledge-acquisition phase is embedding metadata into the CNL. In each leaf-node, SME has the option to include static, dynamic and mapping data (static and dynamic data can be used with all other nodes too). Static data includes direct reference information from the regulatory text such as the URL, title, version, page number, source text, etc. Dynamic data includes information that is used to finalise meaning and understanding of a selected clause/phrase with other colleagues, authorities or even legal experts (due to ambiguities and the level of clarity in the text). This data includes emails, written confirmations, approvals, references for voice data stored via phone calls, etc. Finally mapping data is used specifically as an aid for rule engine via machine-executable representation. As leaf nodes are atomic computation units, each node's data should translate into a computable function at the machine-executable level which requires considerable parameter-binding information. These mapping data include expected function names from a pre-developed regulatory library and relevant environment variables that should be used with a function to derive a value. The completion of mapping data is optional to SME unless they have a technical background; otherwise this is completed by SEs.

3.2 Process of CNL into Machine-Executable Representation

Having a CNL representation of a given regulatory text provides many benefits beyond just an unambiguous interpretation, though it cannot be directly executed on a rule engine. Therefore, the main second step in the regulatory compliance process is to translate CNL into a machine-readable form. This specific form should be able to directly execute on a rule engine. The rule engine, SWI Prolog-based meta interpreter [31], takes two inputs: machine-executable representation and runtime data. The structure of the representation is selected to directly encode CNL data to make this step fully-automated. Nevertheless, due to the possibility of missing information in leaf nodes—specifically mapping data— SEs should intervene in this step to validate the given library mappings and/or to inject correct mapping to derive expected results. Due to the availability of meta knowledge, this step is a comparably easy task for SEs and their scope is mainly limited to individual leaf nodes. The structure of machine-executable representation is presented in Table 2.

Table 2. Structure of machine-executable representation

```
rule(
    type(< root | intermediate | leaf >),
    rule_header(< HEADER NAME >),
    rule_conditions([
        ( < ?not > < CONDITION 1> ) operator ( < ?not > < CONDITION 2> ) operator
        ... ( < ?not > < CONDITION n> )]),
    meta_knowledge([
        static_data([
            url(< URL >), title(< TITLE >), ver(< VERSION >), page_num(< NUM >), ... ]),
        dynamic_data([...]),
        mapping_data([
            prolog_call(< FUN NAME, VAR1, VAR2 >),
            get_in_xml_path(< KEY1 >, < VAR1 >),
            get_in_xml_path(< KEY2 >, < VAR2 >)])])])).
```

Due to the fact that the CNL version is already validated by the SME, the logical distribution of information in it is already accurate; therefore, the most important micro-step here is to validate mapping for functions in the auto-generated machine-executable representation. As a result, this approach is more of a human-centred automated process in which experts encode knowledge but the system automates the execution. This ensures that the interpretation responsibility is not taken away from the subject matter experts and SEs in the pipeline.

3.3 Rule Engine and Explanations

In a regulatory compliance system, a rule engine (also called reasoner or inference engine) plays a key role by executing formal knowledge extracted from a given regulatory text with runtime data [4, 12, 25]. Nevertheless, the best rule engine for a given domain is subjective and in most situations this is determined by the features required and computational complexity of the problem. Declarative rule engines are well-known in knowledge-based systems, especially with explanations [23, 24, 30]. Therefore, a backward reasoning rule engine is developed with SWI Prolog to draw logical conclusions based on the given rules and runtime data.

This engine is a reference implementation inspired by a meta-interpreter explained in the Art of Prolog [31] to reason and derive new facts to validate the state of compliance. Furthermore, this uses backward reasoning (starts with the desired conclusion and performs backward to find supporting facts) as the reasoning strategy and provides proof-tracing as a reasoning feature for the purpose of explainability [23, 31]. Additionally, this rule engine consists of a pre-built regulatory function library to facilitate various functionalities required to execute the rules defined in CNL (leaf node rules). This library can be extended by SEs to include missing behaviours; it is also possible to combine existing functions to implement custom functions.

An important feature of a regulatory compliance system (especially in a regulatory reporting system) is the explanation it gives [23]. Explanation is an inherent feature of rule engine and proof-tracing, which explains the steps involved in logical reasoning and is used as the insight into this system. As the system uses a backward reasoning strategy (goal to facts) it has a systematic way of exploring the search space that provides insight into how the system reaches given conclusions [31]. A rule engine collects this information in tandem with the reasoning steps and returns meaningful systematic explanation data. This systematic data is encoded as a proof tree and included together with regulatory reports to label the given runtime data as compliant or not.

4   Use Case on Money Market Statistical Reporting

To demonstrate this approach, a portion of the Money Market Statistical Reporting (MMSR) regulations[1] (version 3.0) relating to the unsecured market segment (one from four main areas) has been implemented. For this purpose two reference documents are used: 'reporting instructions' (v3.0) and 'questions and answers' (v3.0). A prototype system has been developed with all major components of the presented approach in conjunction with the support of Subject Matter Experts (SMEs) and SEs responsible for implementing the MMSR regulations at ING Bank Netherlands. SMEs manually translate selected text into the CNL representation. Table 3 presents a sample of CNL text generated without including the metadata that was extracted from the MMSR 'Reporting Instructions' document version 3.0 in page numbers 32–33.

Encoding knowledge directly from a gigantic regulation (e.g., MMSR) into its CNL form may lead to many errors unless a significant patience and attention were given. To avoid such errors, a web based prototypical tool was developed for the SMEs to interact with the encoded CNL representation. The tool helps SMEs to easily interact with CNL data to validate the correctness, accuracy and consistency of (manually-) encoded rules. The given tool includes two options to navigate the given CNL: 1) via an accordion menu (left side list view) and 2) via a deep hierarchy tree. These views provide a greater complexity decomposition to SMEs to perform their tasks and this is found to be further empowered by the attached meta data information specifically on validating the correctness of rules. Figure 2 highlights some key features of the prototypical tool.

CNL representation includes all the meta knowledge except for mapping data which is handled by SEs. SEs use machine-executable representation which is derived from

---

[1] https://www.ecb.europa.eu/stats/financial_markets_and_interest_rates/money_market/html/index.en.html.

Table 3. A sample CNL data from machine-executable representation

Regulatory Text from MMSR 'reporting instructions' (v3.0) pp. 32-33: *"The MMSR message is sent by the reporting agents to the relevant NCB or to the ECB to report all unsecured transactions covering:* *• borrowing via unsecured deposits and call accounts, excluding current accounts, as defined in the table below, in euro with a maturity of up to and including one year (defined as transactions with a maturity date of not more than 397 days after the settlement date) by the reporting agent from financial corporations (except central banks where the transaction is not for investment purposes), general government or from non-financial corporations classified as 'wholesale' under the Basel III LCR framework ..."*

Mapping CNL:
rule is-an-unsecured-money-market-transaction-to-be-reported **is** true **if following conditions hold: {**
   is-the-transaction-either-borrowing-via-unsecured-deposit-and-call-account **and**
   (**not** transaction-is-a-current-account) **and**
   is-a-euro-transaction **and**
   has-the-transaction-a-maturity-of-no-more-than-one-year **and**
   is-the-transaction-with-a-valid-borrowing-entity-for-unsecured-market-segment **} with meta knowledge {** Source: "Money Market Statistical Reporting – Reporting Instructions for the Electronic Transactions", Author: "ECB", Version: "3.0", Section: "4.1", Page:"32", Text:"", ... **}**
rule is-the-transaction-either-borrowing-via-unsecured-deposit-or-call-account **is** true **if following conditions hold: {**
   is-the-transaction-a-borrowing-via-unsecured-deposit **or**
   is-the-transaction-a-borrowing-via-unsecured-call-account **} with meta knowledge { ... }**
rule is-the-transaction-a-borrowing-via-unsecured-deposit **is** true **if following conditions hold: {**
   is-the-deposit-redeemable-at-notice **or**
   has-the-deposit-a-maturity-of-no-more-than-one-year **} with meta knowledge { ... }**

CNL via a given CFG and they inject all mappings for leaf nodes to easily execute them on the rule engine. As explained in Sect. 3 a rule engine has been developed with SWI Prolog which accepts machine-executable representation as an input and internally represents it as a rule set to use in regulatory compliance validation. This rule set together with given runtime data are used as knowledge by the rule engine to systematically validate whether the given data set is compliant with the rule set. In addition to providing a compliant status, the rule engine provides an explanation to prove its derivation. This explanation is enriched with metadata so that it will develop the correct awareness of validation in problematic situations.

4.1 Validation of the Approach

The above mentioned prototypical system is validated by a team of three members who are responsible for implementing and managing the MMSR regulations at ING Bank Netherlands. They selected unsecured market segment of MMSR regulation for this experiment and three hours of workshop was conducted to train them about the features

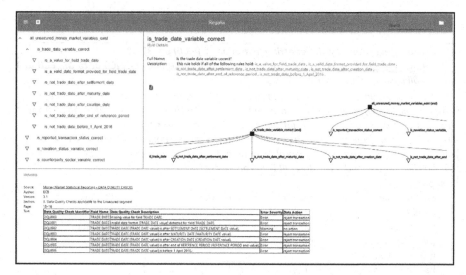

Fig. 2. User interface of developed prototype to navigate CNL data including the metadata

of the implemented prototypical system together with a description of the study's purpose and tasks it includes. In this workshop, they achieved all the core knowledge necessary to execute the experiment: translating the unsecured market segment of MMSR regulation into its CNL representation and encoding those CNL data into the proposed machine understandable representation with leaf node bindings through the provided regulatory library.

After the workshop, they work fully independently on the given task and 23 concepts of the unsecured market segment of MMSR regulation are used where each concept includes at least three leaf nodes (with many intermediate nodes) and a sufficiently complex rule set is used. Team processes the data available in the regulation as blocks and translated them into CNL manually in incremental steps where in each step they validate the correctness, accuracy and consistency of translated information. Having successfully translated the regulatory text into CNL, the next step that executed is encoding the CNL into machine understandable representation with binding leaf node functions from a regulatory library. Finally, twenty data sets (all data sets are synthetic data) are used for this experiment. Each set has labelled information and all sets are classified with an explanation that has also been identified as accurate. The quality of the generated explanations and outputs are validated by SME, which identifies whether outputs are accurate and includes interesting information as explanations. The compliance report generated over all 20 sample data sets which were accurate to the extent of 100%. Three sample explanation outputs are provided in an external appendix with the file name: 'SampleOutputs.pdf'[2]. Furthermore, a significant time is saved due to the introduction of CNL when encoding regulatory text into machine-executable representation.

This being an human centred approach, as the evaluation method psychometric evaluation of an after-scenario questionnaire is used as presented by James Lewis [20].

---

[2] https://github.com/DilhanAMS/conferences/tree/master/iea-aie2020.

Therefore, three main questions are considered: *1) Overall, we are satisfied with the ease of completing the tasks in this scenario, 2) Overall, we are satisfied with the amount of time it took to complete the tasks in this scenario and 3) Overall, we are satisfied with the support information when completing the tasks.* We collected the accumulative decision from the group rather than individually due to the fact that this experiment was motivated as a human centred collaborative task. We used the 7-point scale provided by James with the terms "Strongly agree" for 1 and "Strongly disagree" for 7. The team gave 2, 2, and 5 respectively for above mentioned 3 questions as a summary with some interesting feedback comments. With the above cumulative results (from 3 experts) and the feedback given, they have recognised a significant value in this approach specially in the CNL representation. They have found that CNL as an interesting representation with the feature of embedded meta data to represent rules in a less ambiguous form. Furthermore, they have stated that dynamic metadata part is very interesting in specific situations where they need to explain the reasoning behind their interpretations or decisions to an agent. Also, they have found that the time they took for this is relatively comparable with manual process, but in the feedback they have highlighted that with more experience on this approach they can reduce the required time at least by 20% (mainly due to the fact that regulatory library and limited errors in machine readable representation). The main remark for lower value on the question: support information when completing the tasks, is that the proposed prototypical tool still expect them to manually translate regulatory text into CNL and they strongly expect some degree of automation. They expect new features in the user interface such as highlight key phrases of the text together with auto suggested mapping CNL (irrespective of its accuracy).

5 Discussion and Conclusion

This paper presents a human-centred automated reasoning approach for regulatory reporting. The interpretation of legal concepts and the underlying system of rules are highly-debatable in legal philosophy [2]. This approach combines the legal expertise required to interpret the regulatory text and the reasoning techniques used in knowledge systems for inferencing. As a human-centred approach, the legal expertise required to interpret the regulatory text is achieved through the introduction of a CNL representation to Subject Matter Expert (SME). Because this CNL is more human-oriented, it can be easily followed by SME to encode legal text directly into this form. Via this approach, the system facilitates discussions of the different possible interpretations and allows them to use the final agreed-upon interpretation with supportive meta knowledge. Such interpretations can be used as part of explanations to auditors and/or authorities.

The designed CNL has a specific CFG and therefore it has encoded into a specific machine-executable representation that has been pre-identified. This executable form needs to be validated by SEs to have the correct bindings to functions in order to be executed. The results show full explanations relating to the CNL in order to be easily understood by the SMEs. Finally, the system provides the status of the given data set with explanations on why it is compliant or not.

The proposed approach has been validated through three experts responsible for MMSR regulation at ING. They have examined the approach with a sub-section selected from the MMSR regulation report and yielded interesting results which were confirmed by SME too. This includes 20 synthetic transaction data sets with pre-labelled compliant status and the results that were obtained are 100% accurate. The main limitation that encountered was the inability of auto generating at least partly correct CNL from regulatory text. This will be a key focus on future work.

Another key part of the future work of this research is to encode the full MMSR document and validate this approach with a considerably larger transaction data set. This work mainly applies to regulatory reporting, though the proposed approach is generic enough to apply to challenges in many parts of regulatory compliance.

This approach is an extension from previous work [1] on using a CNL based approach for implementing regulations using Software Contracts. The approach presented in this paper has a more structural solution to the issue of expressing the data flow in a more explainable manner, which was a limitation in the previous work.

Many different techniques have been used to translate regulatory text into SBVR, most of which use an intermediate representation language such as StEng, SBVR StEng and Mercury StEng as presented in related work. The main strength of the approach of this paper is that it allows to express arbitrary compliance rules while using a very straightforward CNL that is both easy to express for SMEs and easy to translate into an executable process. For example, the methodologies via StEng [15, 26] the translation of English legal text to SBVR has been noted to be a challenging task for SME [7, 28]. Both the SBVR StEng notation [25, 26] and the Mercury StEng [7] contain the features of SBVR and depend on a specific pre-developed ontology in the encoding process. Nevertheless, the use of ontologies to help the encoding process is a feature [15, 19, 28] that could be used to extend the current approach.

The object model approaches have a hierarchical structure that is similar to the one in the proposed work. However, the SWRL rules that are used in these approaches have a limited expressivity (see [11, 12, 24, 29]) in relation to a full logic-programming language implementation such as SWI Prolog [23, 31]. This limits them in scenarios where arbitrarily complex compliance rules have to be used. These approaches have a good integration with Semantic Web based technologies through their object model. In future work, the meta-knowledge in the approach can be extended to make use of Semantic Web technologies, which allows for an easy integration of domain specific ontologies like FIBO and FIRO [11, 12, 29].

Another way to extend the approach is to increase the expressivity of the intermediate nodes in the tree structure by adding modal logic operators. A key feature of this methodology is using bindings with a pre-developed regulatory library. Although initially the creation of such bindings is a manual process, learning techniques could be applied to automatically help to derive this information.

Acknowledgment. We would like to thank Robbert Haring, Vaibhav Saxena, Priyanka Kumar, John Marapengopie, Tim Slingsby for useful discussions and their expertise.

References

1. Al Haider, N., et al.: Solving financial regulatory compliance using software contracts. Presented at the 22nd International Conference on Applications of Declarative Programming and Knowledge Management (2019)
2. Araszkiewicz, M.: Analogy, similarity and factors. In: Proceedings of the 13th International Conference on Artificial Intelligence and Law - ICAIL 2011, Pittsburgh, Pennsylvania, pp. 101–105. ACM Press (2011)
3. Arner, D.W., et al.: FinTech, RegTech and the reconceptualization of financial regulation. Northwest. J. Int. Law Bus. **37**(3), 371–414 (2016)
4. Beach, T.H., et al.: A rule-based semantic approach for automated regulatory compliance in the construction sector. Expert Syst. Appl. **42**(12), 5219–5231 (2015)
5. Butler, T., O'Brien, L.: Understanding RegTech for digital regulatory compliance. In: Lynn, T., Mooney, J.G., Rosati, P., Cummins, M. (eds.) Disrupting Finance. PSDBET, pp. 85–102. Springer, Cham (2019). https://doi.org/10.1007/978-3-030-02330-0_6
6. Casey, B., et al.: Rethinking explainable machines: the GDPR's "right to explanation" debate and the rise of algorithmic audits in enterprise. Berkeley Technol. Law J. **34**, 145–189 (2019)
7. Ceci, M., et al.: Making sense of regulations with SBVR. In: RuleML (2016)
8. Chartis Research, IBM: AI in RegTech: A Quiet Upheaval (2018)
9. Dimyadi, J., Amor, R.: Regulatory knowledge representation for automated compliance audit of BIM-based models (2013)
10. Duff, Phelps: Global Regulatory Outlook 2018: Viewpoint 2018 Opinions on Global Financial Services Regulation and Industry Developments for the Year Ahead (2018)
11. Elgammal, A., Butler, T.: Towards a framework for semantically-enabled compliance management in financial services. In: Toumani, F., et al. (eds.) ICSOC 2014. LNCS, vol. 8954, pp. 171–184. Springer, Cham (2015). https://doi.org/10.1007/978-3-319-22885-3_15
12. Ford, R., et al.: Automating financial regulatory compliance using ontology+rules and sunflower. In: Proceedings of the 12th International Conference on Semantic Systems, pp. 113–120. ACM, New York (2016)
13. Fuchs, N.E., Schwitter, R.: Specifying logic programs in controlled natural language. In: Workshop on Computational Logic for Natural Language Processing, Edinburgh, p. 16 (1995)
14. Gurkan, C.: Banks' interactions with listed non-financial firms as a determinant of corporate governance in banking: an agency theory analysis. In: Díaz Díaz, B., Idowu, S.O., Molyneux, P. (eds.) Corporate Governance in Banking and Investor Protection. CSEG, pp. 21–35. Springer, Cham (2018). https://doi.org/10.1007/978-3-319-70007-6_2
15. Haj, A., Balouki, Y., Gadi, T.: Automatic extraction of SBVR based business vocabulary from natural language business rules. In: Khoukhi, F., Bahaj, M., Ezziyyani, M. (eds.) AIT2S 2018. LNNS, vol. 66, pp. 170–182. Springer, Cham (2019). https://doi.org/10.1007/978-3-030-11914-0_19
16. Hashmi, M., et al.: Are we done with business process compliance: state of the art and challenges ahead. Knowl. Inf. Syst. **57**(1), 79–133 (2018)
17. Ismail, A.S., et al.: A review on BIM-based automated code compliance checking system. In: 2017 International Conference on Research and Innovation in Information Systems (ICRIIS), Langkawi, Malaysia, pp. 1–6. IEEE (2017)
18. Kholkar, D., et al.: Towards automated generation of regulation rule bases using MDA. In: Proceedings of the 5th International Conference on Model-Driven Engineering and Software Development, Porto, Portugal, pp. 617–628. SCITEPRESS - Science and Technology Publications (2017)
19. Kuhn, T.: A survey and classification of controlled natural languages. Comput. Linguist. **40**(1), 121–170 (2014)

20. Lewis, J.R.: Psychometric evaluation of an after-scenario questionnaire for computer usability studies: the ASQ. SIGCHI Bull. **23**(1), 78–81 (1990)
21. Malsane, S., et al.: Development of an object model for automated compliance checking. Autom. Constr. **49**, 51–58 (2015)
22. Massey, A.K., et al.: Identifying and classifying ambiguity for regulatory requirements. Presented at the August 2014
23. Merritt, D.: Building Expert Systems in Prolog. Springer, New York (2012). https://doi.org/10.1007/978-1-4613-8911-8
24. Rattanasawad, T., et al.: A comparative study of rule-based inference engines for the semantic web. IEICE Trans. Inf. Syst. **E101.D**(1), 82–89 (2018)
25. Roychoudhury, S., Sunkle, S., Choudhary, N., Kholkar, D., Kulkarni, V.: A case study on modeling and validating financial regulations using (semi-) automated compliance framework. In: Buchmann, R.A., Karagiannis, D., Kirikova, M. (eds.) PoEM 2018. LNBIP, vol. 335, pp. 288–302. Springer, Cham (2018). https://doi.org/10.1007/978-3-030-02302-7_18
26. Roychoudhury, S., et al.: From natural language to SBVR model authoring using structured english for compliance checking. In: 2017 IEEE 21st International Enterprise Distributed Object Computing Conference (EDOC), pp. 73–78 (2017)
27. Sadiq, S., Governatori, G.: Managing regulatory compliance in business processes. In: vom Brocke, J., Rosemann, M. (eds.) Handbook on Business Process Management 2. IHIS, pp. 265–288. Springer, Heidelberg (2015). https://doi.org/10.1007/978-3-642-45103-4_11
28. Schwitter, R.: Controlled natural languages for knowledge representation. In: 23rd International Conference on Computational Linguistics, Beijing, China, pp. 1113–1121 (2010)
29. Sesen, M.B., et al.: An ontological framework for automated regulatory compliance in pharmaceutical manufacturing. Comput. Chem. Eng. **34**(7), 1155–1169 (2010)
30. Stefik, M.: Introduction to Knowledge Systems. Morgan Kaufmann Publishers, San Francisco (1995)
31. Sterling, L., Shapiro, E.Y.: The Art of Prolog: Advanced Programming Techniques. MIT Press, Cambridge (1994)

Security of Blockchain Distributed Ledger Consensus Mechanism in Context of the Sybil Attack

Michal Kedziora[✉], Patryk Kozlowski, and Piotr Jozwiak

Faculty of Computer Science and Management,
Wroclaw University of Science and Technology, Wroclaw, Poland
michal.kedziora@pwr.edu.pl

Abstract. In this paper, the susceptibility of blockchain network to the Sybil attack in which the creation of false identities is used was checked. We performed series of simulations using implemented environment, the cryptocurrency process was simulated using the proof of work consensus mechanism and limited network parameters, such as network computing power expressed in the number of computed hash results, difficulty in extracting the block and the number of users participating in the block in progress. The first of the examined situations was the one in which network nodes have an evenly distributed power allocation and the distribution of block output by individual units is checked. Another tested situation concerned moment of random allocation of computing power to each of the nodes in the network. Research was concluded with statistical analysis of simulation cases of the block mining process due to several parameters.

Keywords: Blockchain networks · Sybil attack · Consensus mechanisms

1 Introduction

Mechanisms underlying the blockchain, such as consensus algorithms are not 100% resistant to the malicious activity of cybercriminals, which is also acknowledged by researchers, indicating the possibility of various attacks: attack 51% [6], nothing at stake [8], long range attack, or deanonymisation of cryptocurrency network users on the basis of tracking their transactions [9–11]. In this work we focus on vectors of attacks on the mechanism of achieving a proof of work consensus, such as a Sybil attack. There is a problem of unambiguous user identification in distributed computer networks. In order to protect and mitigate the threat from malicious users, these networks, such as peer-to-peer blockchain networks, rely on the existence of many independent, often redundant units. Many systems duplicate the task performed between their nodes, and some of them fragment them to protect against data leaks. In each of these cases, the

© Springer Nature Switzerland AG 2020
H. Fujita et al. (Eds.): IEA/AIE 2020, LNAI 12144, pp. 407–418, 2020.
https://doi.org/10.1007/978-3-030-55789-8_36

use of redundancy requires making sure that two seemingly different units are indeed different [17,18]. When a local node has no real knowledge of a remote node, it can only be seen as a certain, abstract identifier, and the system must be sure that each of these identifiers corresponds to another real node. Otherwise, when a local node selects a subset from the pool of all nodes, it may turn out that several of the selected identifiers belong to the same node [1]. The situation in which the counterfeit and many identifiers are seized is a Sybil attack.

2   Sybil Attacks

In the Sybil attack one can distinguish three orthogonal dimensions [2]. First is Communication between nodes, second is Impersonation of identity and third one is Simultanism. Sybil attack due to impersonation of identity can take two forms: First one is when in some cases, the adversary can arbitrarily create new identities without incurring additional costs. For example, in a situation where each node is represented by a 32-bit identifier, the attacker can assign a random, 32-bit value to the node [19,20]. Second form is if there is a node identity verification mechanism, the adversary can not fake the identity. For example, the namespace is deliberately limited to prevent a Sybil attack. In this situation, the attacker must steal the identity of the authentic node, which may go unnoticed if the attacked node is temporarily shut down or destroyed [15,16].

Relying on redundancy of nodes in a distributed network, may prove to be fatal if the identifiers are not controlled in an explicit manner, through a central authorization or secret unit. Such systems are susceptible to a Sybil attack, where a small number of nodes can counterfeit multiple identities to compromise the network [1]. In the case of a network of cryptocurrencies, there could be situations in which the attacker tries to impersonate several users, thus trying to achieve benefits that are not achievable when honestly following the protocol of a given cryptocurrency.

The development of the Sybil attack in blockchain networks is an eclipse attack. It consists in deliberately cutting off the network node from the possibility of connecting with honest nodes [3]. The weakness of the implementation of the Bitcoin peer-to-peer network is used, which consists in the openness in the selection of nodes between which the connection is created. In the Bitcoin network, each node creates connections randomly to already existing nodes. This is based on information about IP addresses and no cryptographic mechanisms are used to protect this process. In an eclipse attack, the adversary takes over the monopoly on the outgoing and incoming connection, thus isolating it from the actual cryptocurrency network. In this way, the attacker can influence what information is sent to the victim-node, and thus force him to waste his computing power on pointless work for a block that is not currently being processed by the rest of the network or used to work on its own block, effectively taking over the victim's computing power [3].

In addition to the ability to influence the perception of the block chain by a isolated node, there are also a number of other potential applications, including it can be used as the basis for carrying out other attacks to which they belong:

- Create races between competing blocks. At the moment of the branching of the block, only one of the competing blocks will be finally incorporated into the public chain, and the remaining ones will be orphaned and no reward will be given for their extraction. An attacker whose aim is several miners can freely transfer any blocks to the affected nodes, which means they can waste their power.
- Turning off some of the power of the network. Isolation of part of the miners cuts them off from the rest of the functioning part, which makes it easier to carry out a 51% attack. In order to hide the natural fluctuations in the computing power of the network, cutting miners could be carried out gradually or only temporarily.
- Double spending without transaction confirmation. In the attack of double spending with the lack of confirmation, there is an exchange of goods before confirmation by inserting them in a blockchain. In order to carry out such an attack, the adversary sends one transaction T to the seller, who is cut off from the network by him and the other T' to the rest of the network. The seller gives good to the attacker based on the transaction T received, while the transaction actually confirmed by the network is T', as a result of which the attacker receives the good without spending anything.

The eclipse attack is most efficiently carried out on nodes with a public IP address. For Bitcoin, it consists of filling the table with its IP addresses. Then, in the new table, it saves "junk" IP addresses that do not actually belong to network nodes, because unlike addresses of the adversary, they are not a valuable resource.

Table 1. Values of Shapiro-Wilk and Kolmogorov-Smirnov tests with the Liliefors correction for a network with an even power distribution between nodes

The length of theblockchain	p-value	Shapiro-Wilk test	p-value	Kolmogorov - Smirnov test
100	0.85586	2.238e-05	0.1946	5.938e-05
500	0.96191	0.43	0.12447	0.06
1000	0.97705	0.46	0.08621	0.41
2000	0.96337	0.44	0.10847	0.15
3500	0.98556	0.71	0.091	0.64

The attack is carried out until the node is restarted (there is a 25% chance that such a node will restart in the next 10 h). After restarting, there is a high probability that all incoming and outgoing connections will be established to IP addresses controlled by the adversary [3].

3 Feasibility of Sybil Attacks in Blockchain

3.1 Description of the Research Environment

A simulation environment in C# 6.0 was prepared for the implementation of the research Microsoft .NET version 4.5 platform, based on the technical specification the implementation of the Bitcoin client [4]. Each of the network nodes performed actual work in the search for the value of the hash function corresponding to the corresponding one assumptions about the difficulty of finding a block, and its amount could be freely chosen because of the multi-threaded architecture in which each node worked regardless of others [13]. Connections between nodes in the graph were created in a way random, and each of the graphs formed from n to n, where n is the number of users of the network connection settings [12,14]. All statistical tests and charts are generated in the R-Studio version 1.0.136 using the language R. In the assumption of a proof of work algorithm, finding computational proof should be expensive, whereas its verification should be a non-calculated process. Based on these assumptions, the creator of Bitcoin decided to apply a one-way SHA256 hash function, for which even a small change in the input data will result in a significant change in the result, while knowing the input data, it is easy to verify the obtained result.

Table 2. Values of Shapiro-Wilk and Kolmogorov-Smirnov tests with the Liliefors amendment for a network with random distribution of each node

The length of the blockchain	p-value	Shapiro-Wilk test	p-value	Kolmogorov - Smirnov test
100	0.83754	2.238e-05	0.0841	3.672e-05
500	0.90764	0.002	0.13685	0.02
1000	0.91129	0.002	0.14085	0.01
2000	0.91253	0.002	0.15417	0.02
3500	0.92541	0.003	0.13877	0.03

1. Each node involved in the mining of the block collects a list of publicly announced transactions waiting for approval and, according to the rules defined by it, selects those that can be included in the newly created block, (e.g. assigning the transaction priority to the amount of the fee) for its processing
2. Before the selected transactions are attached to a block, it is necessary to verify them. It is checked whether the input data in the transaction was previously used to protect against double issue
3. Before creating the block, find the latest block of the longest existing chain, e.g. one that required the greatest amount of computing power. The result of the hash function of the block found is appended to the header of the newly created block

4. Finding the right nonce value such that after substituting for Equation 1, the found value will be less than the set T block
5. When the proof of work for the block is found by another node and broadcast in the network, each of the remaining nodes verifies the correctness of the transactions it contains and adds to its copy of the block chain

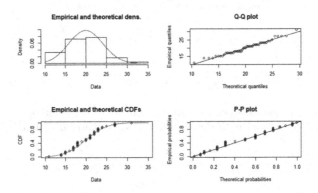

Fig. 1. Histogram, quantile-quantile graph, cumulative distribution and probability-probability plot for a 1000-length chain with an even power distribution.

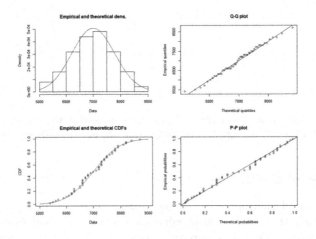

Fig. 2. Histogram, quantile-quantile chart, cumulative distribution and probability-probability plot for a 3500 length chain with an even power distribution

In order to find the right value, none of the miners searches the solution space consisting of 2256 numbers and checks the value of the hash function for each one, comparing its result with the block's difficulty value, and if it is smaller, the miner announces a new block. In the case when after searching the whole space, the appropriate value of the hash function was not found, the current block is abandoned and a new one is created with a changed transaction that awards a profit for finding a correct solution [7].

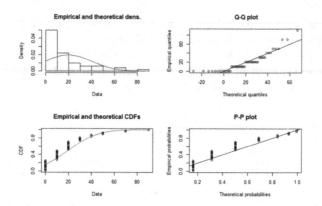

Fig. 3. Histogram, quantile-quantile graph, cumulative distribution and probability-probability plot for a 100-length chain with a random power distribution.

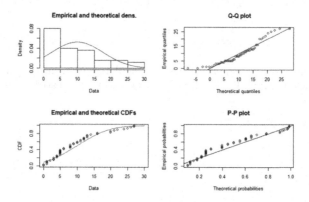

Fig. 4. Histogram, quantile-quantile graph, cumulative distribution and probability-probability plot for a 500-length chain with a random power distribution.

3.2 Course of Simulations

Using the created environment, the cryptocurrency extraction process can be simulated using the proof of work consensus mechanism and limited network parameters, such as network computing power expressed in the number of computed hash results, difficulty in extracting the block and the number of users participating in the block in progress. The first of the examined situations is the one in which network nodes have an evenly distributed power allocation and the distribution of block output by individual units is checked.

The following parameters were constant for all simulations: $n = 50$, $d = 3$, $P = 1500$, where: n - is number of users, d - is mining difficulty expressed in the number of initial zeros of the result of the hash function and P - is computational power of the network in the number of hashes/second, while the length of the chain was changing, after which the network stopped working.

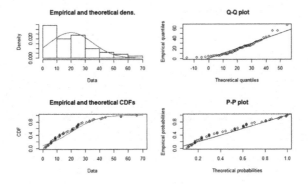

Fig. 5. Histogram, quantile-quantile graph, cumulative distribution and probability-probability plot for a 1000-length chain with a random power distribution.

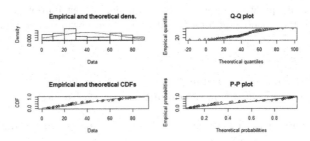

Fig. 6. Histogram, quantile-quantile graph, cumulative distribution and probability-probability plot for a 2000-length chain with a random power distribution.

It was decided to do this selection of parameters due to the fact that for given difficulty D and power P in the applied model, the situation of branching of the chain was relatively rare, which could influence the obtained results due to the block propagation mechanism and the solution of such branching. In the course of making the block mining extracts the number of n = 50 nodes was the smallest, for which in each case each node extracted a non-zero number of blocks, and in addition is the sample size for which the statistical test values can be found in the mathematical tables. The greater number of nodes also influenced the simulation performance due to the multithreaded architecture of the resulting model and the need to synchronize the data blocks between them. Another of the simulated situations that also occurs in real networks is when nodes have uneven power distribution, and also this situation was tested during research.

3.3 Analysis of Results

Statistical analysis was carried out for the process of extracting blocks using the proof of work algorithm with the assumption of two scenarios:

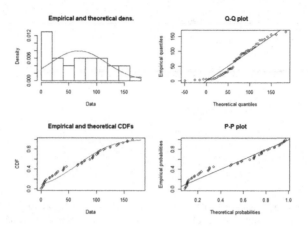

Fig. 7. Histogram, quantile-quantile graph, cumulative distribution and probability-probability plot for a 3500-length chain with a random power distribution.

- An even distribution of computing power between nodes.
- Random, unequal allocation of computing power of nodes.

For each of them, statistical tests have been carried out on testing hypotheses of statistical normality of the data distribution:

- Shapiro-Wilk Test.
- Kolmogorov-Smirnov Test.

They let us determine whether there are grounds for a given process to be called stochastic, with the Shapiro-Wilk test being more powerful. In order to rule out the occurrence of the first type of error, the significance level $\alpha = 0.05$ was adopted for all tests. In addition, it was decided to visualize the results obtained in order to compare them with theoretical runs for quantile-quantile, probability-probability, distribution, and histogram diagrams. They allow not only to assess whether the data belong to or not to the normal distribution, but also to notice how the trends change depending on the power of the network.

For the first of the scenarios studied, which is a theoretical situation, because in the history of the existence of cryptocurrencies no such was recorded. In the case of chain length 100 (Fig. 1), it is not possible to speak of the origin of random distribution data, as the results of statistical tests reject the null hypothesis. In addition, a large part of the nodes extracted a small number of blocks (from 0 to 3), which translates into a strong histogram concentration in a narrow range. With the increase in the number of blocks, an increase in the value of statistical tests, as well as the correlation of density diagrams, probability-probability and quantile-quantile to their theoretical trends is noticeable. For chains of 500, 1000, 2000 and 3500, it is noticeable to expand the range in which the major part of the blocks was extracted, which translates into an increase in the variance of the set. In addition, the clustering of points on the discriminate, probability-probability and quantile-quantile charts is reduced. The above tests do not allow to reject the hypothesis that the process of extracting blocks (digging) using the proof of work consensus mechanism and an even distribution of computing power between all participating nodes is a stochastic process with a normal distribution. The test results collected in Table 1 allow to trace how well individual simulations for different chain lengths were matched to the normal distribution estimated for parameters from the sample, such as variance, expected value and standard deviation. Another situation occurs at the moment of random allocation of computing power to each of the nodes in the network. For each of the tested chain lengths, one can notice a strong concentration of points on the histogram in the left part of the graph, provides it's a large number of nodes digs a small part of the blocks.

In addition, for a chain of 3500, a very even distribution of blocks is visible in the histogram. The results of statistical tests performed for chains with the following lengths: 100, 500, 1000, 2000 and 3500, (Figs. 1, 2, 3, 4, 5 and 6) order the rejection of the zero hypothesis concerning the normal distribution of the variable, which is the number of extracted blocks. In addition, quantile-quantile charts allow to assess to what extent the data from the sample deviate from the normal distribution. For a chain of 100 reasons for deviations in a much larger number of nodes extracting individual blocks compared to the estimated normal distribution for a given sample, which is visible in the histogram and in higher probabilities in the probability-probability diagram. For a chain length of 500 (Fig. 7), the reason for the deviation from the normal distribution is the re-existence of more nodes extracting a small number of blocks than for the estimate of the normal distribution of data from the sample, which can be seen both in the histogram and distribution - number of nodes extracting blocks from 5 to 15 is greater than theoretically predicted. For a length chain of 1000 (Fig. 8), the non-normal distribution is again caused by a larger number of nodes extracting blocks between 10 and 25 blocks and smaller extracting up to 10 blocks, and there are individual nodes that extract more blocks.

In the case of a length chain 2000 with a random power distribution (Fig. 8), the deviation from the normal distribution is caused by a smaller than predicted number of nodes selecting the most common values - between 30 and 60 blocks and more nodes extracting about 30 blocks. In the quantile-quantile chart, it is possible to observe in this area the shift of empirical values below the simple one defining theoretical quantiles. For a chain length of 3500, deviations from the normal distribution should be seen similarly as in previous situations, in a larger than the theoretical number of nodes extracting small amounts of blocks, here it is from 0 to 60 blocks, which is noticeable both in the quantile chart, probability - probability and cumulative distribution in the form of deviation from theoretical transitions. Additionally, the results of statistical tests collected in Table 2 allow to show how much individual simulations for different chain lengths deviated from the normal distribution estimated for the parameters from the sample.

The above situations can be explained by drawing a relatively large number of nodes with low computing power, so their chance of finding a block was

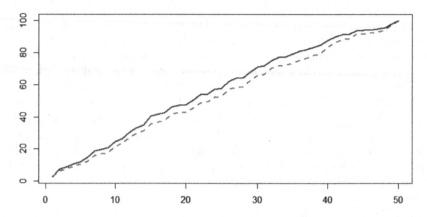

Fig. 8. Chart showing the dependence of node power in the network and mining of blocks, depending on the percentage of power in the network.

smaller. Despite the lack of data belonging to the random power distribution, the relationship between the relative share of power in the network and the relative number of extracted blocks was noticed. The r-Pearson coefficient for this data has the value r = 0.95, which indicates a very strong positive correlation between two random variables. The existence of the above correlation and the fact of a normal block extraction distribution in case of an even distribution of power between nodes, clearly shows that the execution of Sybil attack in the network of cryptocurrencies based on the mechanism of achieving a proof of work consensus is impossible to impersonate the network participants without an additional contribution of computing power, because regardless of the number of identity, the relative share of the attacker's computing power will remain unchanged. The susceptibility of cryptocurrencies based on the proof-of-concept mechanism to the Sybil attack was checked and confirmed. This was done by statistical analysis of the block mining process due to several parameters mentioned below. The architecture of cryptocurrency networks based on the proof of work mechanism should prove resistant to this type of attack.

4   Conclusion and Future Work

The purpose of the work was to analyze to what extent Sybil attack risks may affect the integrity of the cryptocurrency digging process in blockchain network. This goal was achieved by carrying out simulations using appropriate models, which were presented in the paper. Particularly noteworthy, however, are threats that are not widely considered, and were the subject of research, and to which attacks include: Sybil attack type. In the case of a Sybil attack, the mere creation of "virtual miners" who have no computing power has no impact on network security, so using such nodes to carry out an eclipse attack can give the adversary the intended effects in the form. In the case when the attacker would increase its share in the network's computing power, there is no question of carrying out a Sybil attack, as it is contrary to the idea of attack. In addition, the adversary would not act maliciously, and would become a full-fledged miner who could benefit from regular mining. While the use of "virtual miners" with zero computing power in order to increase the chance of obtaining a disproportionately higher number of blocks than it results from the computing power of the adversary, there is no chance of success, the Sybil attack using such virtual nodes it can contribute to increasing the chances of using selfish or stubborn mining strategies. Virtual nodes participate in the block propagation process and by introducing additional number of dishonest nodes into the network, cryptocurrencies to which network participants connect, with their appropriate modification in the scope of block propagation behavior, can bring disproportionately greater benefits to the attacker [5]. Future work will be focus on applying results of this research into selfish/stubborn mining attack techniques.

Acknowledgement. This work was partially supported by the European Union's Horizon 2020 research and innovation programme under the Marie Sklodowska-Curie grant agreement No. 691152 (RENOIR) and the Polish Ministry of Science and Higher Education fund for supporting internationally co-financed projects in 2016–2019 (agreement no. 3628/H2020/2016/2).

References

1. Douceur, J.R.: The Sybil attack. In: Druschel, P., Kaashoek, F., Rowstron, A. (eds.) IPTPS 2002. LNCS, vol. 2429, pp. 251–260. Springer, Heidelberg (2002). https://doi.org/10.1007/3-540-45748-8_24. ISBN 3-540-44179-4, http://dl.acm.org/citation.cfm?id=646334.687813

2. Newsome, J., et al.: The Sybil attack in sensor networks: analysis defenses. In: Third International Symposium on Information Processing in Sensor Networks 2004, IPSN 2004, pp. 259–268 (2004). https://doi.org/10.1109/IPSN.2004.1307346

3. Heilman, E., et al.: Eclipse attacks on bitcoin's peer-to-peer network. In: Proceedings of the 24th USENIX Conference on Security Symposium, SEC 2015, pp. 129–144. USENIX Association, Washington, D.C. (2015). ISBN 978-1-931971- 232. http://dl.acm.org/citation.cfm?id=2831143.2831152

4. Okupski, K.: Bitcoin Developer Reference. Technische Universiteit Eindhoven, Eindhoven (2014)

5. Eyal, I., Sirer, E.G.: Majority is not enough: bitcoin mining is vulnerable. arXiv e-prints, November 2013. arXiv:1311.0243 [cs.CR]

6. Nakamoto, S.: Bitcoin: A peer-to-peer electronic cash system (2008). http://bitcoin.org/bitcoin.pdf

7. BitFury Group: Proof of Stake versus Proof of Work. Whitepaper, BitFury Group (2015)

8. Li, W., Andreina, S., Bohli, J.-M., Karame, G.: Securing proof-of-stake blockchain protocols. In: Garcia-Alfaro, J., Navarro-Arribas, G., Hartenstein, H., Herrera-Joancomartí, J. (eds.) ESORICS/DPM/CBT -2017. LNCS, vol. 10436, pp. 297–315. Springer, Cham (2017). https://doi.org/10.1007/978-3-319-67816-0_17. ISBN 978-3-319-67816-0

9. Fanti, G., Viswanath, P.: Deanonymization in the bitcoin P2P network. In: Guyon, I., et al. (eds.) Advances in Neural Information Processing Systems 30, pp. 1364–1373. Curran Associates Inc (2017). http://papers.nips.cc/paper/6735-deanonymization-in-the-bitcoin-p2p-network.pdf

10. Reid, F., Harrigan, M.: An analysis of anonymity in the bitcoin system. arXiv e-prints, July 2011. arXiv:1107.4524 [physics.soc-ph]

11. Quesnelle, J.: On the linkability of Zcash transactions. CoRR abs/1712.01210 (2017). arXiv:1712.01210

12. Bhupatiraju, V., Kuszmaul, J., Vale, V.: On the viability of distributed consensus by proof of space (2017)

13. Kędziora, M., Kozłowski, P., Szczepanik, M., Jóźwiak, P.: Analysis of blockchain selfish mining attacks. In: Borzemski, L., Świątek, J., Wilimowska, Z. (eds.) ISAT 2019. AISC, vol. 1050, pp. 231–240. Springer, Cham (2020). https://doi.org/10.1007/978-3-030-30440-9_22. ISSN 2194-5357

14. Bitcoin.org. Bitcoin developer guide. https://bitcoin.org/en/developer-guide

15. Dziembowski, S., et al.: Proofs of Space. Cryptology ePrint Archive, Report 2013/796 (2013). https://eprint.iacr.org/2013/796

16. Fischer, M.J., Lynch, N.A., Paterson, M.S.: Impossibility of distributed consensus with one faulty process. J. ACM 32(2), 374–382 (1985). https://doi.org/10.1145/3149.214121. ISSN 0004-5411. http://doi.acm.org/10.1145/3149.214121

17. Nayak, K., et al.: Stubborn mining: generalizing selfish mining and combining with an eclipse attack. In: 2016 IEEE European Symposium on Security and Privacy (EuroSP), pp. 305–320 (2016). https://doi.org/10.1109/EuroSP.2016.32

18. Olfati-Saber, R., Fax, J.A., Murray, R.M.: Consensus and cooperation in networked multi-agent systems. Proc. IEEE 95(1), 215–233 (2007). https://doi.org/10.1109/JPROC.2006.887293. ISSN 0018-9219

19. Zheng, Z., et al.: Blockchain challenges and opportunities: a survey. Int. J. Web Grid Serv. 14(4), 352 (2017)

20. Sapirshtein, A., Sompolinsky, Y., Zohar, A.: Optimal selfish mining strategies in bitcoin. CoRR abs/1507.06183 (2015). arXiv:1507.06183

Reinforcement Learning Based Real-Time Pricing in Open Cloud Markets

Pankaj Mishra$^{(\boxtimes)}$, Ahmed Moustafa, and Takayuki Ito

Nagoya Institute of Technology, Nagoya, Japan
pankaj.mishra@itolab.nitech.ac.jp, {ahmed,ito}@nitech.ac.jp

Abstract. Recently, with the rise in demand for reliable and economical cloud services, there is a rise in the number of cloud providers competing among each other. In such a competitive open market of multiple cloud providers, providers aim to model the selling prices of their requested resources in real-time to maximise their revenue. In this regard, there is a pressing need for an efficient real-time pricing mechanism, that effectively considers a change in the supply and demand of the resources in a certain open cloud market. In this research, we propose a reinforcement learning-based real-time pricing mechanism for dynamically modelling the prices of the requested resources. In specific, the proposed real-time pricing mechanism in a reverse-auction based resource allocation paradigm, which utilises the supply/demand of the resources and undisclosed preferences of the cloud users. Further, we compare the proposed approach with two state-of-the-art resource allocation approaches and the proposed approach outperforms the other two resource allocation approaches.

Keywords: Open cloud markets · Resource allocation · Real-time pricing

1 Introduction

In recent years, with a tremendous rise in several cloud consumers and cloud providers, open cloud markets has become a very competitive and complex business market. In such a competitive and complex market, each providers aim at maximising their revenue. In this regard, providers aim to serve their resources to the highest possible number of consumers and also selling their resources with the maximum possible prices. On the other hand, consumers aim to satisfy their resource requests with minimum possible price and also within the desired deadline. Thus, there is a pressing need to design an efficient pricing mechanism, that efficiently handles the conflicting objectives of the participants. In this regard, to handle the above-mentioned conflicting objectives of the participants, we propose a learning-based dynamic pricing mechanism. In specific, we propose an efficient learning-based real-time pricing mechanism, that models the selling prices in real-time with least possible delay. Moreover, a dynamic pricing

© Springer Nature Switzerland AG 2020
H. Fujita et al. (Eds.): IEA/AIE 2020, LNAI 12144, pp. 419–430, 2020.
https://doi.org/10.1007/978-3-030-55789-8_37

mechanism, which not only considers the supply/demand of the resources but also the preferences of the consumers, while modelling the selling price. In this regard, the proposed pricing mechanism consists of two main steps, which are, (1) tracking the supply-demand resources in the market (i.e. tracking the availability and demand of resources) and (2) mechanism to analyse the undisclosed preferences of the dynamically arriving consumers. Besides, an optimal pricing mechanism must maintain the equilibrium and social welfare [8] in any auction paradigm.

In the literature, Samimi et al. [7] and Zaman et al. [9] proposed efficient combinatorial auction-based resource allocation approaches in cloud computing environments and geo-distributed data centres, respectively. However, these two approaches are consumer-centric, which maximise the utilities of consumer alone, i.e., by selecting a provider with the minimum selling price. Meanwhile, Kong et al. [3] and Li et al. [4] proposed two auction-based pricing mechanism for cloud resource allocation and cloud workflow scheduling, respectively. However, both of these pricing mechanisms were based on static mathematical model, which could only adapt to static or gradually changing open cloud markets. Besides, there exist certain behavioural interdependencies [6] among vendors and consumers when modelling their bid values (i.e. resource prices). However, analysing and incorporating these interdependencies is a challenging problem that is not yet practically addressed. Therefore, there arises the need for real-time pricing approaches, that can incorporate the behavioural interdependencies among vendors and consumers. Secondly, buying and selling in such open markets takes place in an auction paradigm, therefore maintaining the equilibrium in an auction paradigm is also a challenging problem. In specific, maintaining the equilibrium in open cloud markets requires promoting two main characteristics, which are competitiveness [8] and fairness [5]. In specific, the optimal pricing policy needs to maintain competitiveness by dynamically modelling the resource selling prices based on supply/demand in the market. Meanwhile, a provider selection strategy needs to observe fairness and gives equal winning opportunities to all the bidding vendors in the market. However, to the best of our knowledge, none of the existing resource allocation approaches focus on addressing these three challenges simultaneously.

Therefore, it becomes clear that the existing resource allocation approaches exhibit two key limitations, such that 1) they are only suitable for static or gradually changing environments, whereas, they fail to adapt to dynamically changing environments; 2) they fail to consider the fairness in the open cloud market. In order to address the above-mentioned challenges and limitations, we propose a novel two-stage resource allocation approach that employs a reverse auction paradigm in open cloud markets. The proposed approach works in two stages as follows: 1) learning-based real-time pricing policy for consumers, and 2) fairness-based provider selection strategy for consumers. Accordingly, the contributions of this research are as follows:

– First, a new learning-based real-time pricing approach is proposed which optimises the true resource valuation of provider, based on the supply and demand in the open cloud market.
– Second, fairness mechanism for provider selection policy is proposed, to aid potential cloud users to select the best available cloud providers.
– Third, an open cloud market simulated environment is developed, that can simulate the dynamic arrival and departure of cloud users with different resource requests.

The rest of this paper is organised as follows. The problem formulation of learning the resource allocation in open cloud markets is introduced in Sect. 2. Section 3 presents the modelling of the cloud open market into a Markov process. In Sect. 4 and 5, the proposed real-time pricing algorithm and cloud provider selection strategy is discussed. Whereas, the experimental results are presented for evaluating the proposed approach in Sect. 6. Finally, the paper is concluded in Sect. 7.

2   Problem Formulation

This section presents the formulation of the proposed real-time reverse-auction (*RTRA*) pricing approach in an open-cloud market. In this context, the proposed approach is a two-stage process, as depicted in Fig. 1, which are, the real-time pricing algorithm and the cloud provider selection policy. Usually, in a reverse-auction paradigm for open cloud markets, there are three participants, which are, the cloud auctioneer, the cloud consumers and the cloud providers. The auctioneer conducts the auction by coordinating between the providers and the users. Specifically, it accepts the cloud resource requests from all the dynamically arriving users at one hand and their respective selling prices from all providers on the other hand. Then, based on these values, auctioneer determines the resource allocation rule and the payment rule. Further, this allocation and payment rule is broadcast to all the participants in the reverse-auction. Besides, it is the responsibility of the auctioneer to maintain the stability in the auction paradigm by observing confidentiality and non-partisan in the open cloud markets.

In this regard, we consider an open cloud market with one auctioneer and a set of n providers denoted as $p_1, \ldots, p_n$. Furthermore, each provider is represented by an autonomous agent. In addition, each provider has g types of resources, denoted as $pr_1, \ldots, pr_g$; where pr_k represents the quantity of resource type $k \in g$. These resources are requested by m dynamically arriving consumers denoted as $c_1, \ldots, c_m$, each with resource request req_j, where $j \in [1, m]$. Further, each resource request is denoted as $req \equiv (\{rb_k\}, dl)$, where rb_k represents the quantity of the requested resource of type $k \in g$, whereas, dl denotes a strict deadline for serving each consumer's request. By following a reverse-auction paradigm, various cloud resources are dynamically allocated using the proposed *RTRA* approach as follows. Firstly, upon receiving a set of resource requests, a cloud auctioneer broadcasts these requests to all the available providers in the market. Then, each available provider sets its optimal selling price (i.e. bid) using a real-time pricing policy and submits it to the auctioneer. Finally, the auctioneer elicits a suitable provider to serve the lodged request by each consumer, based on a provider selection strategy. Besides, it should be noted that the proposed $RTRA$ approach sustains with the assumption that multiple independent provider can presumably participate in the proposed auction platform to sell their resources through an unbiased auctioneer. As a result, each provider is expected to agree to share its information to maximise its revenues in the long term. Towards this end, we model the open cloud market into multi-agent Markov decision process $(MMDP)$. In this context, we structure the state space of the proposed $MMDP$ model based on the feature vector of the requesting consumers stored in the transaction database. In the further section, we explain the proposed preference labelling scheme along with three other key aspects of the proposed $MMDP$ model, i.e., states, actions, and rewards.

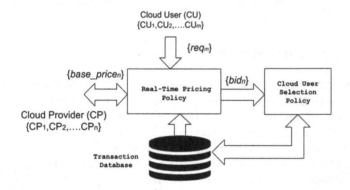

Fig. 1. The proposed architecture of real-time resource allocation

3   Modelling Cloud Open Markets

In this section, we model the resource allocation problem in a multi-agent Markov decision process $(MMDP)$. In this regard, the open cloud market is the Markov environment, wherein m dynamically arriving consumers change the state of the environment with their resource requests req. In turn, n autonomous agents on behalf of each provider optimises the selling price for each resource request req from consumers. In this context, the action space is defined as $A \equiv \{A_1, \ldots A_n\}$, wherein, A_i represents the action space for p_i, $i \in n$. Further, based on the pricing policy $\pi_i : s_i \mapsto A_i$ of p_i, the action $act_i \in A_i$ is determined. After executing the action act_i (after resource allocation), the state s is transferred to the next state s_{i+1}; which is based on the transition function $\tau * s \times A_1 \times \cdots \times A_n \mapsto \Omega(s)$, where $\Omega(s)$ represents the set of probability distributions and s represents the shared state. Finally, at the end of each episode, all the agents, which took part in the auction, receive rewards based on the outcome of the auction (win or loss), such that, reward $r_i : s_i \times A_i \times \ldots \times A_n \mapsto R_i$, where R is the resultant reward. In this regard, each agent (i.e. provider) models their selling price based on the demand-supply of the resources in the market. In further subsections, we explain the three key aspects of the proposed *MMDP* model; i.e., states, actions, and rewards.

3.1 State

In the proposed *MMDP* model, each state s represents the status of all the agents (i.e. providers) in the market. In this context, the existence of multiple consumers within each episode leads to multiple resource allocation transactions within each single episode. Therefore, the state s is represented by concatenating all the transactional states within an episode, that is denoted as $s \equiv [H, F]$. In this state representation, H represents the vector of concatenated transactional states, and F represents the feature vector of each consumer. In this regard, each transactional state is represented by a vector η which is represented based on the categories of providers and consumers. Specifically, η is represented as $\eta_{p_i} \equiv [rev_{p_i}, rb_{c_j}, avg\_win_{p_i}]$, where, rev_{p_i}, rb_{c_i}, and $avg\_win_{p_i}$ denote the revenue, requested resources, and average winning rate of the provider p_i, respectively. In this regard, at the end of each episode, all the transactional state vectors η are concatenated together, which is represented as concatenated transactional state vector H. Whereas, feature vector represents the information of the requesting consumers stored in the transaction database.

3.2 Action

Initially, each provider independently sets a base price for each bundle of the requested resources. This base price is denoted as $base\_price_j^i$ by the provider p_i for the requested bundle of resources from consumer c_j, where $i \in [1, n]$ and $j \in [1, m]$. In this context, different providers optimise their base prices based on a set of exclusive adjustment multipliers for each resource request. In this regard, these adjustment multipliers represent the optimised action values act, which are obtained based on a proposed RL-based algorithm. Specifically, based on the learnt adjustment multipliers (i.e. action values), the optimal selling price for each provider sp is computed using Eq. 1.

$$bid = sp(i) = base\_price_j^i \times (1 + act_i) \tag{1}$$

In this context, the action space for n different providers is denoted as $A \equiv \{A_1, \ldots A_n\}$, where A_i represents the action space for provider p_i, where $i \in n$. Further, based on provider p_i's pricing policy $\pi_i : s_i \mapsto A_i$, the action value $act_i \in A_i$ is determined. Finally after executing the chosen actions (allocating the resources), the proposed *MMDP* model transfers to the next state s_{i+1}. This state change occurs based on the transition function $\tau * s \times A_1 \times \cdots \times A_n \mapsto \Omega(s)$, where $\Omega(s)$ represents a set of probability distributions. Finally, at the end of each episode, all the bidding providers receive rewards based on their chosen actions, such that; $r_i : s_i \times A_i \times \cdots \times A_n \mapsto R$.

3.3 Reward

In open cloud markets, several providers are competing amongst each other to maximise their profits (rewards). In this context, the profit of each provider is maximised by winning the highest possible number of auctions, while simultaneously minimising its loss and non-participation rates. Therefore, in this work, the reward function is formulated to represent the win, the loss and the non-participation of each provider. In this regard, we chose to model the reward function r for each provider p_i as $r_i \equiv (\alpha \times win_i, \beta \times loss_i, \gamma \times non\_participation_i)$, wherein, win_i, $loss_i$ and $non\_participation_i$ denote the number of times provider p_i won, lost and did not participate in the auctions within a single episode. In this context, α, β and γ represent the impact of each of these outcomes, which are set independently by each provider. Finally, the episodic reward of each provider is computed via cumulatively adding all the rewards that were earned within each episode. In this regard, to reduce the complexity of updating the reward values after each auction; these reward values are updated only at the end of each episode. Finally, the proposed *MMDP* model transfers to the next state.

4   Real-Time Pricing

In this section, we discuss the proposed reinforcement learning-based real-time pricing algorithm. As mentioned before the proposed algorithm aims at learning the supply/demand of the resources as well as the consumers' preferences in the open cloud markets. In any competitive open markets, providers have a limited volume of resources, and they make profit only serving the resources to the consumers. Therefore, to maximise their profit, providers would have to offer their resources wisely at the right time, and importantly at the right price. In this regard, the provider is expected to carefully decide whether to bid for any particular request from the consumer by evaluating their undisclosed preferences. Considering this dynamism of open cloud markets, there is a need for a learning-based real-time pricing approach. In addition, given the presumed real-time scenario, with no available training data sets, supervised learning becomes infeasible. Therefore, the proposed real-time pricing algorithm adopts a reinforcement learning scheme [2] in order to handle the presumed real-time scenario. As a result, the proposed real-time pricing algorithm enables the bidding providers to dynamically optimise their bid values. Specifically, the proposed algorithm optimises the bid values based on the dynamically changing supply/demand in an open market, and the characteristics of the requesting consumers. In doing so, the proposed algorithm takes three inputs, which are (i) the characteristics of consumers in the form of feature vector (F), (ii) the concatenated categorical state H, and (iii) the cumulative rewards of all providers. Then, the proposed algorithm provides the adjustment multipliers for all the providers as output. Meanwhile, the initial state (s_0) is determined by a predefined distribution. Further, as depicted in Algorithm 1, each provider aims to select and leverage a certain adjustment multiplier (action), to maximise its total expected future revenues. These future revenues are discounted by the factor γ each time-step. In this regard, the future reward at time-step t for provider i is denoted as $R_i = \sum_{t=0}^{T_e} \gamma^t r_i^t$, where T_e is the time-step at which the bidding process ends. In addition, the Q function for provider i is denoted by Eq. 2.

$$Q_i^\pi(s, a) = \mathcal{E}_{\pi, \tau}[\sum_{t=0}^{T} \gamma^t r_i^t | s_0 = s, a] \tag{2}$$

where $\pi = \{\pi_1, \ldots, \pi_n\}$ is the set of joint-policies of all the $CB - agent$ and $a = [a_1, \ldots, a_n]$ denotes the list of bid-multipliers (actions) of all the $CB-agent$. Further, the next state s' and the next action a' are computed using Bellman equation as shown in Eq. 3:

$$Q_i^\pi(s, a) = \mathcal{E}_{r,s'}[r(s, a) + \gamma \mathcal{E}_{a' \sim \pi}[Q_i^\pi(s', s')]]. \tag{3}$$

On the other hand, the mapping function μ_i maps each shared state $s[H, F]$ of provider p_i to the selected action act_i, based on Eq. 4. This mapping function μ is the actor in the actor-critic architecture.

$$a_i = \mu(s) = \mu_i([H, F]) \tag{4}$$

Further, we get Eq. 5 from Eqs. 3 and 4. As shown in Eq. 5, $\mu = \{\mu_1, \ldots, \mu_n\}$ are the joint deterministic policies of all the providers.

$$Q_i^\mu(s, a_1, \ldots, a_n) = \mathcal{E}_{r,s'}[r(s, a_1, \ldots, a_n) + \gamma Q_i^\mu(s', \mu_1(s', \ldots, \mu_n(s')))] \tag{5}$$

In this regard, the goal of the proposed algorithm becomes to learn an optimal policy for each provider to attain the Nash equilibrium [1]. In addition, in such stochastic environments, each provider learns to behave optimally by learning an optimal policy (μ_i), which is also based on the optimal policies of the other co-existing providers. Further, in the proposed algorithm, the equilibrium of all providers are achieved by gradually reducing the loss function $LOSS(\theta_i^Q)$ of the critic Q_i^μ with the parameter θ_i^Q as denoted in Eqs. 6 and 7. In specific, in Eqs. 6 and 7, $\mu' = \{\mu_1', \ldots, \mu_n'\}$ represents the set of target actors; each of these actors has a delayed parameter $\theta_i^{\mu'}$. Meanwhile, $Q_i^{\mu'}$ represents the target critic, which also has a set of delayed parameters $\theta_i^{Q'}$ for each actor, and $(s, act_i, \ldots, act_n, r_i, s')$ represents the transition tuple that is pushed into the replay memory D. In this regard, each provider's policy μ_i, with parameters θ_i^μ, is learned based on Eq. 8, as demonstrated in Algorithm 1.

$$LOSS(\theta_i^Q) = (y - \gamma Q_i(s, act_1, \ldots, act_n))^2 \tag{6}$$

$$y = r_i + \gamma Q_i'(s', act_1', \ldots, act_n')|_{act_0'=[\mu_0'([Z', F_1']), \ldots, \mu_0'([Z', F_u'])]} \tag{7}$$

$$\nabla_{\theta_i^\mu} J \approx \sum_w \nabla_{\theta_i^\mu} \mu_i([Z, F_w]) \nabla_{act_{iw}^q} Q_i(s, act_1, \ldots, act_n) \tag{8}$$

Finally, the optimal bid values are utilised by the proposed provider selection algorithm, which is discussed in the next section.

Algorithm 1. Real-Time Pricing Algorithm

1: **Initialise:** $Q_i(s, act_1, \ldots, act_n | \theta_i^Q)$, replay memory D
2: **Initialise:** actor μ_i, target actor μ_i'
3: **Initialise:** target network Q' with $\theta_i^{Q'} \leftarrow \theta_i^Q$,
4: $\theta_i^{\mu'} \leftarrow \theta_i^\mu$ for each provider i.
5: **for** episode $= 1$ to E **do**
6:     **Initialise:** s_0
7:     **for** $t = 1$ to T **do**
8:         **for** arriving CU within T **do**
9:             Get a_i using Equation 4
10:            compute $\Upsilon \in [0, 1]$
11:            Get r_i^t, where $i \in N$ and F
12:        $r_i = \sum_{i=1}^n \sum_{t=1}^T r_i^t$ rewards within T
13:        Push (s, act_i, r_i, s') into D // s' is the next state
14:        $s' \leftarrow S$
15:        **for** provider $i = 1$ to n **do**
16:            Sample mini batch $(s, act_i, \ldots, act_n, r_i, s')$ from D
17:            Update critic and actor using Equation 6 and 8
18:            Update target network: $\theta' \leftarrow \tau\theta + (1 - \tau)\theta$

5   Cloud Provider Selection Strategy

In this subsection, we discuss the proposed provider selection strategy for every resource requests from consumers. The proposed strategy handles the primitive drop in bidder drop problem in the auction [5] by employing a priority-based fairness mechanism. Firstly, in order to remove the bidder drop problem in the proposed resource approach, we include a priority-based fairness mechanism. In this regard, on receiving the bids from all the participating providers in the auction, auctioneer attaches the priority label (pr) to every bid from the providers. In specific, on receiving the request $req_j = (rb_j, dl_j)$ from consumer c_j and all the corresponding bids $bid_i^j = (rb_j, sp_i)$ from all the provider, where $i \in n$ and $j \in n$, priority label pr_i is computed using Eq. 9 as follows.

$$pr = 1 - (loss/O) \tag{9}$$

wherein, $loss$ is the number of times consumers lost in the last O auctions, such that provider with $pr \in [0,1]$, $pr = 1$ has highest priority and $pr = 0$ has the lowest priority. Secondly, auctioneer computes the normalised values of priority label pr and optimised selling price sp (bid values) based on the based on the simple additive weighting (SAW) technique [10] using Eq. 10 as follows:

$$S = \begin{cases} \frac{C^{max} - C}{C^{max} - C^{min}}, & \text{if } C^{max} - C^{min} \neq 0. \\ 1, & \text{if } C^{max} - C^{min} = 0. \end{cases} \tag{10}$$

wherein, where C denotes value of either sp or pr. Then, finally, bid score ($bid-score$) is calculated using Eq. 11, and provider's bid with minimum $bid - score$ is the winner.

$$bid - score = S_{pr} \times W_{pr} + S_{sp} \times W_{sp} \tag{11}$$

wherein, S_{pr} and S_{sp} denote the scaled values of priority label and selling price respectively. Whereas, W_{sp} and W_{sp} denote the preference weight of priority label and selling price respectively. These weight preferences are decided independently by each consumer. In this regard, the winning provider serves the requesting consumer. In this regard, auctioneer attempts to select one provider for each resource request req with maximum $bid - score$ and consumer pay the corresponding sp.

6   Experimental Setup and Results

This section presents the results of the performed experiments, that are conducted in order to evaluate the proposed real-time reverse-auction ($RTRA$) pricing approach in an open cloud market. The open cloud environment is implemented in $SimPy$ python library, wherein, each cloud provider is initialised with quantities of four types of resources; namely, computer processing speed, memory, storage, and bandwidth (BW). We compare the proposed $RTRA$ approach with two other notable bidding-based resource allocation approaches, which are as follows: (1) the combinatorial double auction resource allocation approach ($CDARA$) [7]; and (2) the indicator-based combinatorial auction-based approach ($ICAA$) [3]. The $CDARA$ has fixed selling price strategy and no fairness based provider selection strategy, whereas, $ICAA$ the indicator-based combinatorial has a static mathematical model to model the price and also with no fairness baaed provider selection strategy. Finally, We run the three resource allocation approaches for 120 episodes, each episode of length 1500 s, wherein 25 consumers arrive dynamically in each episode and 12 cloud providers.

The performance of the provider is evaluated based on two parameters, which are: (1) $average\ revenue$; (2) $average\ unavailability$ of resources with the provider; and (3) active participation of the provider computed as a ratio of never-won provider to total-provider in the auction denoted as $fairness$, as depicted in Table 1. Firstly, from Table 1, it is clear that the average revenue $RTRA$ is approximately 55% higher as compared to the other two approaches. Moreover, in the $RTRA$, unavailability of the provider is decreased by $1/3^{rd}$ of that observed in $CDARA$. In addition, $RTRA$ increases the participation of the provider through the proposed priority-based allocation which is reflected through the fairness values in Table 1. Therefore, the proposed $RTRA$ approach is capable of maximising the performance of the provider in the open cloud markets.

Similarly, the performance of consumers is evaluated based on three parameters, which are, (1) the ratio of resource price paid and maximum offered price by all the providers (PMB), (2) waiting time (avg_wait_time) of the

Table 1. Performance of cloud providers

Performance	CDARA	ICAA	RTRA*
Average revenue	25000	23000	**40000**
Average unavailability	64%	42%	**28%**
Fairness	48%	53%	**65%**

served consumers, and (3) scalability (σ) of the serving providers. Firstly from Table 2, the consumer in the cloud market with $RTRA$ approach pays marginally lesser prices as compared to the maximum offered prices among all the other bids. Secondly, the cloud market with $RTRA$ approach has lesser waiting time as compared to the other two approaches. Finally, $RTRA$ allocates the consumers request to more scalable providers, where, scalability($sigma$) is defined as $\sigma = 1 - (requested\ resource/available\ resource)$. Therefore, the proposed $RTRA$ approach is capable of maximising the performance of the CU in the open cloud markets.

Table 2. Performance cloud user

Utilities	CDARA	ICAA	RTRA*
PMB	0.86	0.92	**0.76**
avg_wait_time	24	22	**12**
σ	0.4	0.6	**1.2**

Finally, the overall performance of the open cloud market is evaluated based on three parameters, which are; (1) the participation rate of providers in the auction (availability); (2) the non-participation rate of providers in auction (unavailability); and (3) the lost rate of provider in the auction. As shown in Fig. 2, providers in the proposed $RTRA$ approach has a higher participation rate, lower non-participation rate and also lower lost rate, as compared to the other two approaches. This proves the capability of the $RTRA$ approach to serve higher numbers of consumer as compared to the other two approaches. In conclusion, the proposed $RTRA$ approach outperforms the other two approaches by serving the highest number of consumer while enabling higher participation in the auctions.

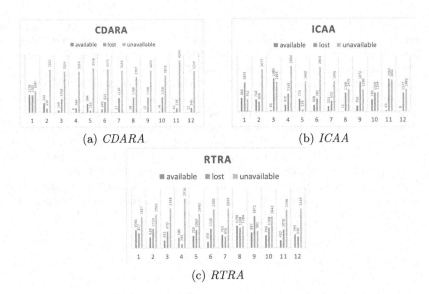

(a) *CDARA*   (b) *ICAA*

(c) *RTRA*

Fig. 2. The availability, unavailability and loss of providers in auction, based on the number of buyers handled

7 Conclusion

This paper proposes a learning-based real-time pricing approach for resource allocation in dynamic and complex open cloud markets. The proposed approach models the selling price of the requested resources in real-time, based on the proposed pricing approach, which considers the supply/demand of the resources in the market and the characteristics of the cloud users. As a result, the proposed approach enables both the cloud providers and cloud consumers to maximise their performances at the same time. The experimental results demonstrate the efficiency and fairness of the proposed resource allocation approach and its ability to maximise the overall performance of all the participants in an open cloud market. The future work is set to develop a mechanism that enables multiple preferences of the consumers in the provider selection strategy.

References

1. Hu, J., Wellman, M.P., et al.: Multiagent reinforcement learning: theoretical framework and an algorithm. In: ICML, vol. 98, pp. 242–250. Citeseer (1998)
2. Konda, V.R., Tsitsiklis, J.N.: Actor-critic algorithms. In: Advances in Neural Information Processing Systems, pp. 1008–1014 (2000)
3. Kong, Y., Zhang, M., Ye, D.: An auction-based approach for group task allocation in an open network environment. Comput. J. **59**(3), 403–422 (2015)
4. Li, X., Ding, R., Liu, X., Liu, X., Zhu, E., Zhong, Y.: A dynamic pricing reverse auction-based resource allocation mechanism in cloud workflow systems. Sci. Program. **2016**, 17 (2016)

5. Murillo, J., Muñoz, V., López, B., Busquets, D.: A fair mechanism for recurrent multi-unit auctions. In: Bergmann, R., Lindemann, G., Kirn, S., Pěchouček, M. (eds.) MATES 2008. LNCS (LNAI), vol. 5244, pp. 147–158. Springer, Heidelberg (2008). https://doi.org/10.1007/978-3-540-87805-6_14

6. Myerson, R.B.: Optimal auction design. Math. Oper. Res. 6(1), 58–73 (1981)

7. Samimi, P., Teimouri, Y., Mukhtar, M.: A combinatorial double auction resource allocation model in cloud computing. Inf. Sci. 357, 201–216 (2016)

8. Toosi, A.N., Vanmechelen, K., Khodadadi, F., Buyya, R.: An auction mechanism for cloud spot markets. ACM Trans. Auton. Adapt. Syst. (TAAS) 11(1), 2 (2016)

9. Zaman, S., Grosu, D.: Combinatorial auction-based allocation of virtual machine instances in clouds. J. Parallel Distrib. Comput. 73(4), 495–508 (2013)

10. Zeng, L., Benatallah, B., Dumas, M., Kalagnanam, J., Sheng, Q.Z.: Quality driven web services composition. In: Proceedings of the 12th International Conference on World Wide Web, pp. 411–421. ACM (2003)

Medical and Health-Related Applications

A New Integer Linear Programming Formulation to the Inverse QSAR/QSPR for Acyclic Chemical Compounds Using Skeleton Trees

Fan Zhang[1], Jianshen Zhu[1], Rachaya Chiewvanichakorn[1],
Aleksandar Shurbevski[1(✉)], Hiroshi Nagamochi[1], and Tatsuya Akutsu[2]

[1] Department of Applied Mathematics and Physics, Kyoto University, Kyoto, Japan
{fanzhang,zhujs,ch.rachaya,shurbevski,nag}@amp.i.kyoto-u.ac.jp
[2] Bioinformatics Center, Institute for Chemical Research, Kyoto University,
Uji, Japan
takutsu@kuicr.kyoto-u.ac.jp

Abstract. Computer-aided drug design is one of important application areas of intelligent systems. Recently a novel method has been proposed for inverse QSAR/QSPR using both artificial neural networks (ANN) and mixed integer linear programming (MILP), where inverse QSAR/QSPR is a major approach for drug design. This method consists of two phases: In the first phase, a feature function f is defined so that each chemical compound G is converted into a vector $f(G)$ of several descriptors of G, and a prediction function ψ is constructed with an ANN so that $\psi(f(G))$ takes a value nearly equal to a given chemical property π for many chemical compounds G in a data set. In the second phase, given a target value y^* of the chemical property π, a chemical structure G^* is inferred in the following way. An MILP $\mathcal{M}$ is formulated so that $\mathcal{M}$ admits a feasible solution (x^*, y^*) if and only if there exist vectors x^*, y^* and a chemical compound G^* such that $\psi(x^*) = y^*$ and $f(G^*) = x^*$. The method has been implemented for inferring acyclic chemical compounds. In this paper, we propose a new MILP for inferring acyclic chemical compounds by introducing a novel concept, skeleton tree, and conducted computational experiments. The results suggest that the proposed method outperforms the existing method when the diameter of graphs is up to around 6 to 8. For an instance for inferring acyclic chemical compounds with 38 non-hydrogen atoms from C, O and S and diameter 6, our method was 5×10^4 times faster.

1 Introduction

Recently, artificial intelligence techniques have been applied to various areas including pharmaceutical and medical sciences. Drug design is one of major tentative topics in such applications. Indeed, many computational methods have been developed for computer-aided drug design. In particular, extensive studies have been done on inverse QSAR/QSPR (quantitative structure-activity

© Springer Nature Switzerland AG 2020
H. Fujita et al. (Eds.): IEA/AIE 2020, LNAI 12144, pp. 433–444, 2020.
https://doi.org/10.1007/978-3-030-55789-8_38

and structure-property relationships) [12,18]. In this approach, desired chemical activities and/or properties are specified along with some constraints and then chemical compounds satisfying these conditions are inferred, where chemical compounds are usually represented as undirected graphs. Inverse QSAR/QSPR is often formulated as an optimization problem to find a chemical graph maximizing (or minimizing) an objective function under various constraints. In this formalization, objective functions reflect certain chemical activities or properties, and are often derived from a set of training data consisting of known molecules and their activities/properties using statistical machine learning methods.

Since it is difficult to directly handle chemical graphs in traditional statistical learning methods, chemical compounds are often represented as a vector of real or integer numbers, which is called a set of descriptors or a set of features. Using these descriptors, various heuristic and statistical methods have been developed for finding optimal or nearly optimal graph structures under given objective functions [8,12,16]. In such an approach, inference or enumeration of graph structures from a given feature vector is often required as a subtask. In order to solve this subtask, various methods have been developed [5,9,11,14] and some studies have been done on its computational complexity [1,13].

According to the recent rapid progress of Artificial Neural Network (ANN) and deep learning technologies, novel approaches have been proposed for inverse QSAR/QSPR, which include applications of variational autoencoders [6], recurrent neural networks [17,19], and grammar variational autoencoders [10]. In these approaches, neural networks are trained using known compound/activity data and then novel chemical graphs are generated by solving a kind of inverse problems on neural networks. Although these inverse problems are usually solved using various statistical methods, the optimality of the solution is not necessarily guaranteed. In order to guarantee the optimality mathematically, novel mixed integer linear programming (MILP)-based methods have been proposed [2] for ANNs with ReLU functions and sigmoid functions, in which activation functions on neurons are efficiently encoded as piece-wise linear functions so as to represent ReLU functions exactly and sigmoid functions approximately.

Chiewvanichakorn et al. [4] and Azam et al. [3] recently combined the MILP-based formulation of the inverse problem on ANNs [2] with efficient enumeration of tree-like graphs [5]. The combined framework for inverse QSAR/QSPR mainly consists of two phases. The first phase solves (I) PREDICTION PROBLEM, where each chemical compound G is transformed into a feature vector $f(G)$ and an ANN $\mathcal{N}$ is trained from existing chemical compounds and their values $a(G)$ on a chemical property π to obtain a prediction function $\psi_{\mathcal{N}}$ so that $a(G)$ is predicted as $\psi_{\mathcal{N}}(f(G))$. The second phase solves (II) INVERSE PROBLEM, where (II-a) given a target value y^* of the chemical property π, a feature vector x^* is inferred from the trained ANN $\mathcal{N}$ so that $\psi_{\mathcal{N}}(x^*)$ is close to y^* and (II-b) then a set of chemical structures G^* such that $f(G^*) = x^*$ is enumerated. In (II-a) of the above-mentioned methods [3,4], an MILP is formulated. In particular, Azam et al. [3] formulated an MILP for acyclic chemical compounds so that the following is guaranteed: either (i) every feature vector x^* inferred from a trained ANN

$\mathcal{N}$ in (II-a) admits a corresponding chemical structure G^* or (ii) no chemical structure exists for a given target value when no feature vector is inferred from the ANN $\mathcal{N}$. In this paper, we propose a new MILP for inferring acyclic chemical compounds with bounded degree, and conducted computational experiments on several chemical properties. The results suggest that the proposed method runs much faster than the previous method [3] does when the number of chemical elements and diameter are relatively small.

2   Preliminary

Let $\mathbb{R}$ and $\mathbb{Z}$ denote the sets of reals and non-negative integers, respectively. For two integers a and b, let $[a, b]$ denote the set of integers i with $a \leq i \leq b$.

Graphs. Let $H = (V, E)$ be a graph with a set V of vertices and a set E of edges. For a vertex $v \in V$, the set of neighbors of v in H is denoted by $N_H(v)$, and the *degree* $\deg_H(v)$ of v is defined to be $|N_H(v)|$. The length of a path is defined to be the number of edges in the path. The *distance* $\mathrm{dist}_H(u, v)$ between two vertices $u, v \in V$ is defined to be the minimum length of a path connecting u and v in H. The *diameter* $\mathrm{dia}(H)$ of H is defined to be the maximum distance between two vertices in H. The *sum-distance* $\mathrm{smdt}(H)$ of H is defined to be the sum of distances over all vertex pairs.

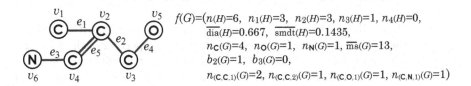

$f(G){=}(n{(H)}{=}6,\ n_1{(H)}{=}3,\ n_2{(H)}{=}3,\ n_3{(H)}{=}1,\ n_4{(H)}{=}0,$
$\overline{\mathrm{dia}}{(H)}{=}0.667,\ \overline{\mathrm{smdt}}{(H)}{=}0.1435,$
$n_{\mathsf{C}}{(G)}{=}4,\ n_{\mathsf{O}}{(G)}{=}1,\ n_{\mathsf{N}}{(G)}{=}1,\ \overline{\mathrm{ms}}{(G)}{=}13,$
$b_2{(G)}{=}1,\ b_3{(G)}{=}0,$
$n_{(\mathsf{C},\mathsf{C},1)}{(G)}{=}2,\ n_{(\mathsf{C},\mathsf{C},2)}{(G)}{=}1,\ n_{(\mathsf{C},\mathsf{O},1)}{(G)}{=}1,\ n_{(\mathsf{C},\mathsf{N},1)}{(G)}{=}1)$

Fig. 1. A chemical graph $G = (H, \alpha, \beta)$ and its feature vector $f(G)$.

Chemical Graphs. We represent the graph structure of a chemical compound as a graph with labels on vertices and multiplicity on edges in a hydrogen-suppressed model. Let Λ be a set of labels each of which represents a chemical element such as C (carbon), O (oxygen), N (nitrogen) and so on, where we assume that Λ does not contain H (hydrogen). Let mass(a) and val(a) denote the mass and valence of a chemical element $\mathsf{a} \in \Lambda$, respectively. In our model, we use integers $\mathrm{mass}^*(\mathsf{a}) = \lfloor 10 \cdot \mathrm{mass}(\mathsf{a}) \rfloor$, $\mathsf{a} \in \Lambda$. We introduce a total order $<$ over the elements in Λ according to their mass values; i.e., we write $\mathsf{a} < \mathsf{b}$ for chemical elements $\mathsf{a}, \mathsf{b} \in \Lambda$ with $\mathrm{mass}(\mathsf{a}) < \mathrm{mass}(\mathsf{b})$. Choose a set $\Gamma_<$ of tuples $\gamma = (\mathsf{a}, \mathsf{b}, k) \in \Lambda \times \Lambda \times [1, 3]$ such that $\mathsf{a} < \mathsf{b}$. For a tuple $\gamma = (\mathsf{a}, \mathsf{b}, k) \in \Lambda \times \Lambda \times [1, 3]$, let $\overline{\gamma}$ denote the tuple $(\mathsf{b}, \mathsf{a}, k)$. Set $\Gamma_> = \{\overline{\gamma} \mid \gamma \in \Gamma_<\}$, $\Gamma_= = \{(\mathsf{a}, \mathsf{a}, k) \mid \mathsf{a} \in \Lambda, k \in [1, 3]\}$ and $\Gamma = \Gamma_< \cup \Gamma_=$. A pair of two atoms a and b joined with a bond of multiplicity k is denoted by a tuple $\gamma = (\mathsf{a}, \mathsf{b}, k) \in \Gamma$.

A *chemical graph* in a hydrogen-suppressed model is defined to be a tuple $G = (H, \alpha, \beta)$ of a graph $H = (V, E)$, a function $\alpha : V \to \Lambda$ and a function $\beta : E \to [1, 3]$ such that (i) H is connected; and (ii) $\sum_{e=uv \in E} \beta(e) \leq \mathrm{val}(\alpha(u))$ for each vertex $u \in V$. Nearly 55% of the acyclic chemical graphs with at most 200 non-hydrogen atoms registered in the chemical database PubChem have maximum degree at most 3 in the hydrogen-suppressed model. Figure 1 illustrates an example of a chemical graph $G = (H, \alpha, \beta)$.

Descriptors. In our method, we use only graph-theoretical descriptors for defining a feature vector, which facilitates our designing an algorithm for constructing graphs. Given a chemical graph $G = (H = (V, E), \alpha, \beta)$, we define a *feature vector* $f(G)$ that consists of the following eight kinds of descriptors:
$n(H)$: the number of vertices in H;
$n_d(H)$ ($d \in [1, 4]$): the number of vertices of degree d in H;
$\overline{\mathrm{dia}}(H)$: the diameter of H divided by $|V|$;
$\overline{\mathrm{smdt}}(H)$: the sum of distances of H divided by $|V|^3$;
$n_{\mathsf{a}}(G)$ ($\mathsf{a} \in \Lambda$): the number of vertices with label $\mathsf{a} \in \Lambda$;
$\overline{\mathrm{ms}}(G)$: the average of mass* of atoms in G;
$b_i(G)$ ($i = 2, 3$): the number of double and triple bonds;
$n_\gamma(G)$ ($\gamma = (\mathsf{a}, \mathsf{b}, k) \in \Gamma$): the number of label pairs $\{\mathsf{a}, \mathsf{b}\}$ with multiplicity k.
Figure 1 illustrates an example of a feature vector $f(G)$.

3   A Method for Inferring Chemical Graphs

We review how MILPs are used in the method for the inverse QSAR/QSPR [3], which is illustrated in Fig. 2. For a specified chemical property π such as boiling point, we denote by $a(G)$ the observed value of the property π for a given chemical compound G, which is represented by a chemical graph $G = (H, \alpha, \beta)$. As the first phase, we solve (I) PREDICTION PROBLEM for the inverse QSAR/QSPR with the following three steps.

1. Prepare a data set $D = \{(G_i, a(G_i)) \mid i = 1, 2, \ldots, m\}$ of pairs of a chemical graph G_i and the value $a(G_i)$ for a specified chemical property π. Set reals $\underline{a}, \overline{a} \in \mathbb{R}$ so that $\underline{a} \leq a(G_i) \leq \overline{a}$, $i = 1, 2, \ldots, m$.
2. Set a graph class $\mathcal{G}$ to be a set of chemical graphs such that $\mathcal{G} \supseteq \{G_i \mid i = 1, 2, \ldots, m\}$. Introduce a feature function $f : \mathcal{G} \to \mathbb{R}^k$ for a positive integer k. We call $f(G)$ the *feature vector* of $G \in \mathcal{G}$, and call each entry of a vector $f(G)$ a *descriptor* of G.
3. Construct a prediction function $\psi_{\mathcal{N}}$ with an ANN $\mathcal{N}$ that, given a vector in $x \in \mathbb{R}^k$, returns a real $\psi_{\mathcal{N}}(x)$ with $\underline{a} \leq \psi_{\mathcal{N}}(x) \leq \overline{a}$ so that $\psi_{\mathcal{N}}(f(G))$ takes a value nearly equal to $a(G)$ for many chemical graphs in D.

See Fig. 2 for an illustration of Steps 1 to 3. As the second phase, we solve (II) INVERSE PROBLEM for the inverse QSAR/QSPR by treating the following inference problems.

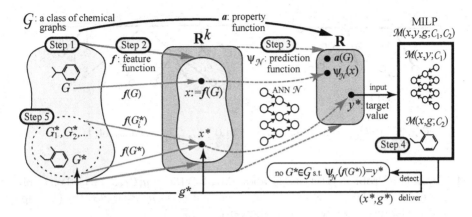

Fig. 2. An illustration of a property function a, a feature function f, a prediction function $\psi_{\mathcal{N}}$ and an MILP that either delivers a vector (x^*, g^*) that forms a chemical graph $G^* \in \mathcal{G}$ such that $\psi_{\mathcal{N}}(f(G^*)) = y^*$ (or $a(G^*) = y^*$) or detects that no such chemical graph G^* exists in $\mathcal{G}$.

(II-a) Inference of Vectors
Input: A real $y^* \in [\underline{a}, \overline{a}]$.
Output: Vectors $x^* \in \mathbb{R}^k$ and $g^* \in \mathbb{R}^h$ such that $\psi_{\mathcal{N}}(x^*) = y^*$ and g^* forms a chemical graph $G^* \in \mathcal{G}$ with $f(G^*) = x^*$.

(II-b) Inference of Graphs
Input: A vector $x^* \in \mathbb{R}^k$.
Output: All graphs $G^* \in \mathcal{G}$ such that $f(G^*) = x^*$.

To treat Problem (II-a), we rely on the next result.

Theorem 1 ([2]). *Let $\mathcal{N}$ be an ANN with a piecewise-linear activation function for an input vector $x \in \mathbb{R}^k$, n_A denote the number of nodes in the architecture and n_B denote the total number of break-points over all activation functions. Then there is an MILP $\mathcal{M}(x, y; \mathcal{C}_1)$ that consists of variable vectors $x \in \mathbb{R}^k$, $y \in \mathbb{R}$, and an auxiliary variable vector $z \in \mathbb{R}^p$ for some integer $p = O(n_A + n_B)$ and a set $\mathcal{C}_1$ of $O(n_A + n_B)$ constraints on these variables such that $\psi_{\mathcal{N}}(x^*) = y^*$ if and only if there is a vector (x^*, y^*) feasible to $\mathcal{M}(x, y; \mathcal{C}_1)$.*

We also introduce a variable vector $g \in \mathbb{R}^h$ for some integer h and a set $\mathcal{C}_2$ of constraints on x and g such that (x^*, g^*) is feasible to the MILP $\mathcal{M}(x, g; \mathcal{C}_2)$ if and only if g^* forms a chemical graph $G^* \in \mathcal{G}$ with $f(G^*) = x^*$ (see [3] for details). In MILPs, we can easily impose additional linear constraints or fix some variables to specified constants.

We design an algorithm to Problem (II-b) based on the branch-and-bound method (see [5] for enumerating acyclic chemical compounds).

The second phase consists of the next two steps.

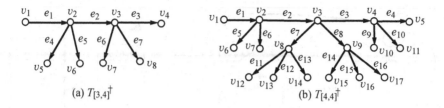

Fig. 3. (a) $T_{[3,4]}^{\dagger}$, where $n_{\max}(3,4) = 8$; (b) $T_{[4,4]}^{\dagger}$, where $n_{\max}(4,4) = 17$.

4. Formulate Problem (II-a) as the above MILP $\mathcal{M}(x, y, g; \mathcal{C}_1, \mathcal{C}_2)$ based on $\mathcal{G}$ and $\mathcal{N}$. Find a set F^* of vectors $x^* \in \mathbb{R}^k$ such that $(1-\varepsilon)y^* \le \psi_{\mathcal{N}}(x^*) \le (1+\varepsilon)y^*$ for a tolerance ε set to be a small positive real.
5. To solve Problem (II-b), enumerate all graphs $G^* \in \mathcal{G}$ such that $f(G^*) = x^*$ for each vector $x^* \in F^*$.

Figure 2 illustrates Steps 4 and 5.

As for an MILP $\mathcal{M}(x, g; \mathcal{C}_2)$, the previous formulation due to Azam et al. [3] is based on an idea that a required acyclic chemical graph G^* with n vertices will be constructed as a subset of $n - 1$ vertex pairs as edges over an $n \times n$ adjacency matrix (or a complete graph K_n with n vertices). In the next section, we introduce a new formulation for an MILP $\mathcal{M}(x, g; \mathcal{C}_2)$ so that G^* will be constructed as an induced subgraph of a larger acyclic graph, which we call "a skeleton tree".

4   MILPs for Representing Acyclic Chemical Graphs

In this section, we propose a new formulation to the MILP $\mathcal{M}(x, g; \mathcal{C}_2)$ in Step 4 of the method introduced in the previous section. We consider acyclic chemical graphs such that the maximum degree is at most 3 or 4.

For an integer D, let $\mathcal{T}_{[D,3]}$ (resp., $\mathcal{T}_{[D,4]}$) denote the set of trees H such that $\mathrm{dia}(H) = D$ and the maximum degree is at most 3 (resp., equal to 4), and define the *skeleton tree* $T_{[D,d]}^{\dagger}$, $d = 3, 4$ to be a tree in $\mathcal{T}_{[D,d]}$ with the maximum number of vertices, where $n_{\max}(D, d)$ denotes the number of vertices in $T_{[D,d]}^{\dagger}$. We assume that the vertices and edges in the skeleton tree $T_{[D,d]}^{\dagger} = (V^{\dagger} = \{v_1, v_2, \ldots, v_{n_{\max}(D,d)}\}, E^{\dagger} = \{e_1, e_2, \ldots, e_{n_{\max}(D,d)-1}\})$ are indexed with the following ordering σ: (i) $T_{[D,d]}^{\dagger}$ is rooted at vertex v_1, and $i < j$ holds for any vertex v_i and a child v_j of v_i; (ii) Each edge e_j joins two vertices v_{j+1} and v_k with $k \le j$, where we denote by $\mathrm{tail}(j)$ the index k of the parent v_k of vertex v_{j+1}; and (iii) For each $i = 1, 2, \ldots, D$, it holds that $v_i v_{i+1} \in E$; i.e., $(e_1, e_2, \ldots, e_D)$ is one of the longest paths in $T^{\dagger}_{[D,d]}$. Let $N_\sigma(i)$ for each $i = 1, 2, \ldots, n_{\max}(D, d)$ denote the set of indices j of edges e_j incident to vertex v_i, and $\mathrm{dist}_\sigma(i,j)$ denote the distance $\mathrm{dist}_T(v_i, v_j)$ in the tree $T = T^{\dagger}_{[D,d]}$. Figure 3 illustrates such an ordering σ for the skeleton trees $T^{\dagger}_{[3,4]}$ and $T^{\dagger}_{[4,4]}$.

Given integers $n^* \geq 3$, $\mathrm{dia}^* \geq 2$ and $d_{\max} \in \{3, 4\}$, consider an acyclic chemical graph $G = (H = (V, E), \alpha, \beta)$ such that $|V| = n^*$, $\mathrm{dia}(H) = \mathrm{dia}^*$ and the maximum degree in H is at most 3 for $d_{\max} = 3$ (or equal to 4 for $d_{\max} = 4$). We formulate the MILP $\mathcal{M}(x, g; \mathcal{C}_2)$ so that the underlying graph H is an induced subgraph of the skeleton tree $T^{\dagger}_{[\mathrm{dia}^*, d_{\max}]}$ such that $\{v_1, v_2, \ldots, v_{D+1}\} \subseteq V$.

To reduce the number of graph-isomorphic solutions in this MILP, we introduce a precedence constraint as follows. Let P_{prc} be a set of ordered index pairs (i, j) with $D + 2 \leq i < j \leq n_{\max}$. We call P_{prc} *proper* if the next conditions hold:

(a) For each pair of a vertex v_j and its parent v_i with $D + 2 \leq i < j \leq n_{\max}$ in $T^{\dagger}_{[\mathrm{dia}^*, d_{\max}]}$, there is a sequence $(i_1, i_2), (i_2, i_3), \ldots, (i_{k-1}, i_k)$ of index pairs in P_{prc} such that $i_1 = i$ and $i_k = j$; and

(b) Each subtree $H = (V, E)$ of $T^{\dagger}_{[\mathrm{dia}^*, d_{\max}]}$ with $\{v_1, v_2, \ldots, v_{D+1}\} \subseteq V$ is isomorphic to a subtree $H' = (V', E')$ with $\{v_1, v_2, \ldots, v_{D+1}\} \subseteq V'$ such that for each pair $(i, j) \in P_{\mathrm{prc}}$, if $v_j \in V'$ then $v_i \in V'$.

Note that a proper set P_{prc} is not necessarily unique. For example, we use $P_{\mathrm{prc}} = \{(5, 6), (7, 8), (6, 8)\}$ for the tree $T^{\dagger}_{[3,4]}$ in Fig. 3(a).

For a technical reason, we introduce a dummy chemical element ϵ, and denote by Γ_0 the set of dummy tuples (ϵ, ϵ, k), $(\epsilon, \mathrm{a}, k)$ and $(\mathrm{a}, \epsilon, k)$ ($\mathrm{a} \in \Lambda$, $k \in [0, 3]$). To represent elements $\mathrm{a} \in \Lambda \cup \{\epsilon\} \cup \Gamma_< \cup \Gamma_= \cup \Gamma_>$ in an MILP, we encode these elements a into some integers denoted by $[\mathrm{a}]$, where we assume that $[\epsilon] = 0$. For simplicity, we also denote n^* by n and $n_{\max}(\mathrm{dia}^*, d_{\max})$ by $n_{\max}$. Our new formulation is given as follows.

MILP $\mathcal{M}(x, g; \mathcal{C}_2)$
variables for descriptors in x:
$\mathrm{n}(d) \in [0, n]$ ($d \in [1, 4]$); $\mathrm{smdt} \in [0, n^3]$; $\mathrm{n}(\mathrm{a}) \in [0, n]$ ($\mathrm{a} \in \Lambda$);
$\mathrm{Mass} \in \mathbb{Z}$; $\mathrm{b}(k) \in [0, n - 1]$ ($k \in [1, 3]$); $\mathrm{n}(\gamma) \in [0, n - 1]$ ($\gamma \in \Gamma_< \cup \Gamma_=$)
variables for constructing H in g:
$v(i) \in \{0, 1\}$ ($i \in [1, n_{\max}]$); $\delta_{\mathrm{deg}}(i, d) \in \{0, 1\}$ ($i \in [1, n_{\max}]$, $d \in [0, 4]$);
$\mathrm{deg}(i) \in [0, 4]$ ($i \in [1, n_{\max}]$); $\mathrm{dist}(i, j) \in [0, \mathrm{dia}^*]$ ($1 \leq i < j \leq n_{\max}$);
$\widetilde{\alpha}(i) \in \{[\mathrm{a}] \mid \mathrm{a} \in \Lambda \cup \{\epsilon\}\}$ ($i \in [1, n_{\max}]$); $\widetilde{\beta}(j) \in [0, 3]$ ($j \in [1, n_{\max} - 1]$);
$\delta_\alpha(i, \mathrm{a}) \in \{0, 1\}$ ($i \in [1, n_{\max}]$, $\mathrm{a} \in \Lambda \cup \{\epsilon\}$);
$\delta_\beta(j, k) \in \{0, 1\}$ ($j \in [1, n_{\max} - 1]$, $k \in [0, 3]$);
$\delta_\tau(j, \gamma) \in \{0, 1\}$ ($j \in [1, n_{\max} - 1]$, $\gamma \in \Gamma \cup \Gamma_0$)

constraints in C_2:

$$\sum_{i\in[1,n_{\max}]} v(i) = n; \quad v(i) = 1 \ (i \in [1, \text{dia}^* + 1]);$$

$$v(i) \geq v(j) \ ((i,j) \in P_{\text{prc}}); \quad \sum_{i\in[1,n_{\max}]} \delta_{\deg}(i,d) = \text{n}(d) \ (d \in [0,4]);$$

$$\sum_{i\in[1,n_{\max}-1]} \delta_\beta(i,k) = \text{b}(k) \ (k \in [1,3]); \quad \sum_{1\leq i<j\leq n_{\max}} \text{dist}(i,j) = \text{smdt};$$

$$\sum_{i\in[1,n_{\max}]} \delta_\alpha(i,\mathsf{a}) = \text{n}(\mathsf{a}) \ (\mathsf{a} \in \varLambda); \quad \sum_{\mathsf{a}\in\varLambda} \text{mass}^*(\mathsf{a}) \cdot \text{n}(\mathsf{a}) = \text{Mass};$$

$$\sum_{i\in[1,n_{\max}-1]} (\delta_\tau(i,\gamma)+\delta_\tau(i,\overline{\gamma})) = \text{n}(\gamma) \ (\gamma\in\varGamma_<); \quad \sum_{i\in[1,n_{\max}-1]} \delta_\tau(i,\gamma) = \text{n}(\gamma) \ (\gamma\in\varGamma_=);$$

For each $i = 1, 2, \ldots, n_{\max}$,

$$\sum_{j\in N_\sigma(i)} v(j+1) = \deg(i), \quad \sum_{d\in[0,4]} \delta_{\deg}(i,d) = 1, \quad \sum_{\mathsf{a}\in\varLambda} \delta_\alpha(i,\mathsf{a}) = v(i),$$

$$\sum_{i\in[1,n_{\max}]} \delta_\alpha(i,\mathsf{a}) = \text{n}(\mathsf{a}), \quad \sum_{j\in N_\sigma(i)} \widetilde{\beta}(j) \leq \sum_{\mathsf{a}\in\varLambda} \text{val}(\mathsf{a}) \cdot \delta_\alpha(i,\mathsf{a});$$

For each pair (i,j) with $1 \leq i < j \leq n_{\max}$,

$$\text{dist}(i,j) \leq \text{dia}^* \cdot v(i), \quad \text{dist}(i,j) \geq \text{dist}_\sigma(i,j) - \text{dia}^* \cdot (2 - v(i) - v(j)),$$
$$\text{dist}(i,j) \leq \text{dia}^* \cdot v(j), \quad \text{dist}(i,j) \leq \text{dist}_\sigma(i,j) + \text{dia}^* \cdot (2 - v(i) - v(j));$$

For each $j = 1, 2, \ldots, n_{\max}-1$,

$$v(j+1) \leq \widetilde{\beta}(j) \leq 3v(j+1), \quad \sum_{k\in[1,3]} \delta_\beta(j,k) = v(j+1), \quad \sum_{k\in[1,3]} k\delta_\beta(j,k) = \widetilde{\beta}(j),$$

$$\sum_{\gamma\in\varGamma\cup\varGamma_0} \delta_\tau(j,\gamma) = 1, \quad \sum_{(\mathsf{a},\mathsf{b},k)\in\varGamma\cup\varGamma_0} [\mathsf{a}]\delta_\tau(j,(\mathsf{a},\mathsf{b},k)) = \widetilde{\alpha}(\text{tail}(j)),$$

$$\sum_{(\mathsf{a},\mathsf{b},k)\in\varGamma\cup\varGamma_0} [\mathsf{b}]\delta_\tau(j,(\mathsf{a},\mathsf{b},k)) = \widetilde{\alpha}(j+1), \quad \sum_{(\mathsf{a},\mathsf{b},k)\in\varGamma\cup\varGamma_0} k\delta_\tau(j,(\mathsf{a},\mathsf{b},k)) = \widetilde{\beta}(j).$$

5   Experimental Results

We implemented our new formulation for the MILP $\mathcal{M}(x,y,g;C_1,C_2)$ in Step 4 of the method for the inverse QSAR/QSPR [3], and conducted some experiments to compare the practical performance of the previous and new MILP formulations. We executed the experiments on a PC with Intel Core i5 1.6 GHz CPU and 8 GB of RAM running under the Mac OS operating system version 10.14.4. As a study case, we selected three chemical properties: heat of atomization (HA), octanol/water partition coefficient (Kow) and heat of combustion (HC).

Results on Phase 1. In Step 1, we collected a data set D of acyclic chemical graphs for HA from the article [15] and for Kow and HC from HSDB from Pub-Chem database. We set $\varLambda$ to be the set of all chemical elements of the chemical graphs in D; and Γ to be the set of all tuples $\gamma = (\mathsf{a}, \mathsf{b}, k) \in \Lambda \times \Lambda \times [1,3]$ of the chemical graphs in D. In Step 2, we set a graph class $\mathcal{G}$ to be the set of all possible acyclic chemical graphs over the sets Λ and Γ in Step 1. In Step 3, we used scikit-learn (version 0.21.3) to construct ANNs $\mathcal{N}$ where the tool and activation function are set to be MLPRegressor and Relu, respectively. To evaluate the performance of the resulting prediction function $\psi_{\mathcal{N}}$ with cross-validation, we partition a given data set D into five subsets D_i, $i \in [1,5]$ randomly, where $D \backslash D_i$ is used for a training set and D_i is used for a test set in five trials $i \in [1,5]$.

Table 1 shows the size and range of data sets that we prepared for each chemical property, and results on Phase 1, where we denote the following:

π: one of the chemical properties HA, Kow, and HC;

$|D|$: the size of data set D for a chemical property π;

Λ: the set of all chemical elements of the chemical graphs in D;

$[\underline{n}, \overline{n}]$: the minimum and maximum number of vertices in H over data set D;

$[\underline{a}, \overline{a}]$: the minimum and maximum values of $a(G)$ over data set D;

K: the number of descriptors in $f(G)$ for a chemical property π, where $K = |\Lambda| + |\Gamma| + 12$ for our feature vector $f(G)$;

Arch.: the size of hidden layers of ANNs, where $\langle 10 \rangle$ (resp., $\langle 30, 30 \rangle$) means an architecture $(K, 10, 1)$ with an input layer with K nodes, a middle layer with 10 nodes (resp., two hidden layers with 30 nodes) and an output layer with a single node;

L-time: the average time (sec.) to construct ANNs for each trial;

Test R^2: the coefficient of determination averaged over the five test sets.

Table 1. The results on Steps 1, 2 and 3

| π | Steps 1 and 2 | | | | | Step 3 | | | |
	$|D|$	Λ	$[\underline{n}, \overline{n}]$	$[\underline{a}, \overline{a}]$	K	Arch	L-time	Test R^2
HA	128	C,O,S	[2,11]	[450.3,3009.6]	19	$\langle 10 \rangle$	11.24	0.999
Kow	62	C,O,S	[2,16]	[−0.77,8.2]	19	$\langle 10, 10 \rangle$	0.373	0.969
Kow	430	C,Cl,O,N,S,Br,F	[1,36]	[−4.2,15.6]	42	$\langle 30 \rangle$	1.155	0.907
HC	204	C,O,N	[1,63]	[245.6,35099.6]	26	$\langle 10, 10 \rangle$	23.01	0.965
HC	222	C,O,N,F,S,Br	[1,63]	[245.6,35099.6]	34	$\langle 30, 30 \rangle$	17.376	0.961
HC	262	C,Cl,O,N,S,Br,F,Si,B,P	[1,63]	[245.6,35099.6]	47	$\langle 30, 30 \rangle$	5.194	0.965

Results on Phase 2. Let us call the previous MILP formulation based on adjacency matrix the AM method, and our new MILP formulation based on skeleton tree the ST method. We solved MILP instances in these methods using CPLEX (ILOG CPLEX version 12.9) [7]. We conducted an experiment for each property in {HA, Kow, HC} as follows. For several pairs $(d_{\max}, \text{dia}^*)$ of integers $d_{\max} \in \{3, 4\}$ and $\text{dia}^* \in [6, 13]$, choose each integer $n^* \in [14, n_{\max}(\text{dia}^*, d_{\max})]$ and six target values y_i^*, $i \in [1, 6]$. We tried to solve the six MILP instances with the AM and ST methods starting with $n^* = 14$ and increasing n^* up to $n_{\max}(\text{dia}^*, d_{\max})$. We terminated solving instances with each of the two methods

Table 2. The computation time of the methods AM and ST for HA

n^*	$d_{max} = 3$						$d_{max} = 4$					
	$dia^* = 8$		$dia^* = 12$		$dia^* = 13$		$dia^* = 6$		$dia^* = 8$		$dia^* = 9$	
	AM	ST	AM	ST	AM	ST	AM	ST	AM	ST	AM	ST
14	0.037	0.244	0.020	0.471	0.012	0.185	0.503	0.146	0.218	3.375	0.212	7.242
17	0.999	0.074	0.272	1.712	0.215	3.380	1.657	0.159	1.212	10.194	1.016	14.921
20	3.886	0.493	1.254	5.576	0.920	6.531	8.628	1.326	3.648	13.317	3.770	27.347
23	9.947	0.523	4.999	9.796	3.440	13.741	12.247	1.228	8.937	17.716	7.059	T.O.
26	50.422	0.539	10.713	17.758	8.613	32.541	73.451	1.522	22.141	24.742	20.998	–
29	T.O	0.601	24.904	13.587	T.O	94.952	T.O	1.206	66.298	40.580	T.O	–
32	–	0.554	T.O	25.978	–	121.37	–	1.572	T.O	33.445	–	–
36	–	0.252	–	46.940	–	77.627	$(2{\times}10^4)$	1.391	–	35.968	–	–
38	–	0.722	–	19.028	–	139.25	$(1{\times}10^5)$	2.133	–	42.081	–	–
41	–	0.445	–	28.145	–	90.231	–	1.141	–	30.332	–	–
44	–	0.152	–	60.771	–	T.O	–	1.258	–	41.772	–	–
47	n.a	n.a	–	71.990	–	–	–	0.799	–	34.532	–	–
50	n.a	n.a	–	72.139	–	–	–	0.445	–	T.O	–	–
53	n.a	n.a	–	T.O	–	–	–	0.050	–	–	–	–

whenever the running time of solving one of the six instances reaches a time limit of 300 s.

Table 2 and Fig. 4 show the computation time of the AM and ST methods in Step 4 for property HA, where we denote the following:

AM: the average time (sec.) to solve six instances with the method AM;

ST: the average time (sec.) to solve six instances with the method ST;

T.O.: the running time of one of the six instances exceeded 300 s.

We executed the method AM without a time limit for an instance with $n^* = 36$ (resp., $n^* = 38$ and 40) for HA, $dia^* = 6$ and $d_{max} = 4$, and the computation time was 21,962 (resp., 124,903 and 148,672) seconds. The method ST took 2.133 s for the same size of instances with $n^* = 38$. This means that the method ST was 58,557 times faster than the method AM for this size of instances.

For space limitation, we describe a summary of the experimental results on instances for Kow and HC: The method ST outperforms the method AM for the cases of $(\pi = \text{K}_{ow}, |\Lambda| = 3, d_{max} = 3, dia^* \le 11)$, $(\pi = \text{K}_{ow}, |\Lambda| = 3, d_{max} = 4, dia^* \le 7)$, $(\pi = \text{K}_{ow}, |\Lambda| = 7, d_{max} = 3, dia^* \le 8)$, $(\pi = \text{K}_{ow}, |\Lambda| = 7, d_{max} = 4, dia^* \le 5)$, $(\pi = \text{H}_{C}, |\Lambda| = 3, d_{max} = 3, dia^* \le 9)$, $(\pi = \text{H}_{C}, |\Lambda| = 3, d_{max} = 4, dia^* \le 6)$, $(\pi = \text{H}_{C}, |\Lambda| = 6, d_{max} = 3, dia^* \le 8)$, $(\pi = \text{H}_{C}, |\Lambda| = 6, d_{max} = 4, dia^* \le 7)$, $(\pi = \text{H}_{C}, |\Lambda| = 10, d_{max} = 3, dia^* \le 7)$ and $(\pi = \text{H}_{C}, |\Lambda| = 10, d_{max} = 4, dia^* \le 5)$.

From the experimental results, we observe that the ST method solves the problem of Step 4 faster than the AM method does when the diameter of graphs is up to around 11 for $d_{max} = 3$ and 8 for $d_{max} = 4$. Here we note that chemical graphs with diameter up to 11 for $d_{max} = 3$ and 8 for $d_{max} = 4$ account for about 35% and 18%, respectively, out of all acyclic chemical graphs with at most 200 non-hydrogen atoms registered in the PubChem chemical database, and about 63% and 40% out of the acyclic chemical graphs with at most 200 non-hydrogen atoms with $d_{max} = 3$ and $d_{max} = 4$, respectively.

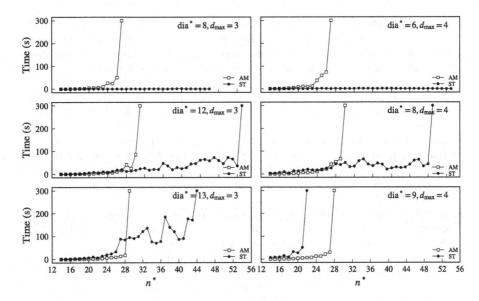

Fig. 4. The average computation time of the methods AM and ST for HA

6   Concluding Remarks

In this paper, we presented a new formulation of an MILP for inferring acyclic chemical graphs based on the method due to Azam et al. [3]. In the previous formulation, a tree H with n vertices was represented as a subset of vertex pairs over an $n \times n$ adjacency matrix. Contrary to this, our new formulation is based on a maximum tree of a specified diameter, called the skeleton tree, from which a required tree H is chosen as an induced subgraph. From the results on some computational experiments, we observe that the proposed method is more efficient than the previous method when the number of chemical elements and diameter are relatively small. Although this paper presented such an MILP for the class $\mathcal{G}$ of acyclic chemical graphs, we can apply a similar formulation to the acyclic part of any chemical graph.

References

1. Akutsu, T., Fukagawa, D., Jansson, J., Sadakane, K.: Inferring a graph from path frequency. Discrete Appl. Math. **160**(10–11), 1416–1428 (2012)
2. Akutsu, T., Nagamochi, H.: A mixed integer linear programming formulation to artificial neural networks. In: Proceedings of the 2nd International Conference on Information Science and Systems, pp. 215–220. ACM (2019)
3. Azam, N.A., Chiewvanichakorn, R., Zhang, F., Shurbevski, A., Nagamochi, H., Akutsu, T.: A method for the inverse QSAR/QSPR based on artificial neural networks and mixed integer linear programming. In: Proceedings of the 13th International Joint Conference on Biomedical Engineering Systems and Technologies - volume 3: BIOINFORMATICS, pp. 101–108 (2020)

4. Chiewvanichakorn, R., Wang, C., Zhang, Z., Shurbevski, A., Nagamochi, H., Akutsu, T.: A method for the inverse QSAR/QSPR based on artificial neural networks and mixed integer linear programming. In: Proceedings of the 10th International Conference on Bioscience, Biochemistry and Bioinformatics, pp. 40–46. ACM (2020). https://dl.acm.org/doi/10.1145/3386052.3386054

5. Fujiwara, H., Wang, J., Zhao, L., Nagamochi, H., Akutsu, T.: Enumerating treelike chemical graphs with given path frequency. J. Chem. Inf. Model. **48**(7), 1345–1357 (2008)

6. Gómez-Bombarelli, R., et al.: Automatic chemical design using a data-driven continuous representation of molecules. ACS Central Sci. **4**(2), 268–276 (2018)

7. IBM ILOG: CPLEX Optimization Studio 12.9. https://www.ibm.com/support/knowledgecenter/SSSA5P_12.9.0/ilog.odms.cplex.help/CPLEX/homepages/usrmancplex.html

8. Ikebata, H., Hongo, K., Isomura, T., Maezono, R., Yoshida, R.: Bayesian molecular design with a chemical language model. J. Comput. Aided Mol. Des. **31**(4), 379–391 (2017). https://doi.org/10.1007/s10822-016-0008-z

9. Kerber, A., Laue, R., Grüner, T., Meringer, M.: MOLGEN 4.0. Match Commun. Math. Comput. Chem. **37**, 205–208 (1998)

10. Kusner, M.J., Paige, B., Hernández-Lobato, J.M.: Grammar variational autoencoder. In: Proceedings of the 34th International Conference on Machine Learning, vol. 70, pp. 1945–1954 (2017)

11. Li, J., Nagamochi, H., Akutsu, T.: Enumerating substituted benzene isomers of tree-like chemical graphs. IEEE/ACM Trans. Comput. Biol. Bioinf. **15**(2), 633–646 (2016)

12. Miyao, T., Kaneko, H., Funatsu, K.: Inverse QSPR/QSAR analysis for chemical structure generation (from y to x). J. Chem. Inf. Model. **56**(2), 286–299 (2016)

13. Nagamochi, H.: A detachment algorithm for inferring a graph from path frequency. Algorithmica **53**(2), 207–224 (2009)

14. Reymond, J.L.: The chemical space project. Acc. Chem. Res. **48**(3), 722–730 (2015)

15. Roy, K., Saha, A.: Comparative QSPR studies with molecular connectivity, molecular negentropy and TAU indices. J. Mol. Model. **9**(4), 259–270 (2003)

16. Rupakheti, C., Virshup, A., Yang, W., Beratan, D.N.: Strategy to discover diverse optimal molecules in the small molecule universe. J. Chem. Inf. Model. **55**(3), 529–537 (2015)

17. Segler, M.H.S., Kogej, T., Tyrchan, C., Waller, M.P.: Generating focused molecule libraries for drug discovery with recurrent neural networks. ACS Central Sci. **4**(1), 120–131 (2017)

18. Skvortsova, M.I., Baskin, I.I., Slovokhotova, O.L., Palyulin, V.A., Zefirov, N.S.: Inverse problem in QSAR/QSPR studies for the case of topological indices characterizing molecular shape (Kier indices). J. Chem. Inf. Comput. Sci. **33**(4), 630–634 (1993)

19. Yang, X., Zhang, J., Yoshizoe, K., Terayama, K., Tsuda, K.: ChemTS: an efficient Python library for de novo molecular generation. Sci. Technol. Adv. Mater. **18**(1), 972–976 (2017)

Computing a Weighted Jaccard Index
of Electronic Medical Record
for Disease Prediction

Chia-Hui Huang[1], Yun-Te Liao[1], David Taniar[2], and Tun-Wen Pai[1,3(✉)]

[1] Department of Computer Science and Engineering, National Taiwan Ocean University,
Keelung, Taiwan
twp@csie.ntut.edu.tw
[2] Department of Software Systems and Cybersecurity, Monash University, Melbourne, Australia
[3] Department of Computer Science and Information Engineering, National Taipei
University of Technology, Taipei, Taiwan

Abstract. Amyotrophic lateral sclerosis (ALS) is a fatal neurodegenerative disease. Patient's spinal cord, brainstem, or motor cortex of the cerebral motor cortex are gradually degenerated and lead to systemic muscle atrophy and weakness. Medicine therapies used for treating ALS mainly focus on symptomatic treatment and delaying deterioration. At present, no precise treatment can effectively cure, halt, or reverse the progression of ALS disease. Therefore, it is extremely important to predict high-risk populations of ALS candidates and prevent from the early stages. This research analyzes the historical medical records of ALS patients from the Taiwan National Health Insurance Research Database (NHIRD). Through analyzing comorbidity within a specific time interval, we proposed a novel index, weighted Jaccard index, for effective prediction analyses. The weighted Jaccard index and logistic regression model were applied to build an ALS patient prediction system. Based on comparing electronic medical records testing subject and know ALS patient, we can effectively detect potential ALS patients at early stages. Early and accurate detection can provide medical doctors to conduct precise inspection and appropriate treatment with efficient and effective guidelines for decision making.

Keywords: Electronic medical record (EMR) · Secondary data · Disease prediction · ALS · Weighted Jaccard index

1 Background

Amyotrophic lateral sclerosis (ALS) is a fatal neurodegenerative disease [1]. The prevalence of ALS is uniformly distributed across most countries in the world, which is approximately 5 cases per 100,000 subjects per year. ALS lead to death within 2–3 years for bulbar symptom onset cases and 3–5 years for limb onset cases [2]. Up to now, the etiology of sporadic ALS is still unclear and there is no method to completely cure ALS. The medicines used to treat ALS are mainly applied to delay deterioration of the

© Springer Nature Switzerland AG 2020
H. Fujita et al. (Eds.): IEA/AIE 2020, LNAI 12144, pp. 445–456, 2020.
https://doi.org/10.1007/978-3-030-55789-8_39

disease, but not rooted out the symptoms [3, 4]. Therefore, if we could early predict the potential patients of ALS from his/her electronic medical records, early prevention could be achieved. In other words, doctors can persuade risk candidates to strengthen their personal immune system, to improve their own environmental factors for early prognosis. This would be the best therapy strategy to improve and slow down the progressive deterioration of ALS symptoms.

Medical big data play a big role in medical services such as clinical decision-making, medical quality monitoring, disease prediction models, clinical trial analysis, and personalized treatment. Taiwan possesses an internationally well-known National Health Insurance Research Database (NHIRD) [5], which maintains nationwide disease codes for hospitals, doctors, and patients. We can integrate, extract, and convert historical longitudinal medical records from the multi-aspects and multi-function databases into a required specification format according to patient- or disease-based records for further analytics.

ALS is difficult to be accurately predicted and is incurable. From the collected medical records, patients who have been hospitalized due to ALS, as many as 60% of patients were directly hospitalized at the first diagnosis of ALS. The remaining 40% of ALS patients also ended up with hospitalization with an average of two-year period after the first diagnosis of ALS. Therefore, analyzing and comparing the secondary data through extracting significant features and to establish a prediction model for ALS high-risk groups from experimental groups and control groups becomes an important strategy for at-risk ALS patients. Early prediction of ALS patients can help those high-risk patients to plan prognostic treatment, improve environmental factors and healthy diet arrangements. It may also increase the chance of effective disease control and avoid suffering from other comorbidities. To slow down progressive deterioration of ALS disease reflects an important treatment and healthcare policy, and extra important time period could reduce stress and anxiety for both patients and families.

In the past, some reports for ALS prediction were based on analyzing the correlation between living habits, environmental factors, and occupations respectively. The experimental data were obtained through questionnaires from patients [6]. Several studies focused on searching valuable biomarkers discovered from genome wide association study (GWAS). These approaches tried to find loci of genetic mutations through comparing sequenced genes from huge number of blood samples or biopsy from ALS and healthy subjects [7, 8]. Some other studies simulated survival rates from diagnosis to assist the judgment of prognosis treatment based on location of the disease, the age of onset of symptoms, gender, etc. [4].

The purpose of this research is to analyze comorbidities and trajectory patterns for identifying individuals who may possess high risk factors by comparing history medical records with ALS patients. It aims to identify high-risk populations at early stage by using a simple and fast detection mechanism, and recommendation for early precision inspection, treatment, and disease control could be suggested for a good decision making. The prediction index used in this paper is a modified version of Jaccard Index. The traditional Jaccard Index is a statistical value used to compare the similarity and diversity between two different sample sets. It measures the similarity condition between two limited sample groups. The larger the value, the more similar condition for the two groups [9]. However, in order to strengthen this traditional indicator and to enhance the importance of different populations of certain comorbidities within ALS patients, we propose a novel weighted Jaccard index indicator which could effectively reflect the comorbidity distribution model of medical records instead of using binary status of comorbidity, and the indices provide accurate prediction results compared to traditional approaches. Details and experimental results are described in the followings.

2 Materials and Methods

2.1 Jaccard Index

The traditional Jaccard index defines the similarity between two different sample sets of A and B. The index is shown in Eq. (1):

$$Jaccard(A, B) = \frac{|A \cap B|}{|A \cup B|} = \frac{|A \cap B|}{|A| + |B| - |A \cap B|} \tag{1}$$

where $A \cap B$ represents the number of overlapped items, and $A \cup B$ represents the number of union items. Each item is evaluated as "1" for existing status and "0" for nonexistent condition regardless of occurred frequencies of each item. In order to strengthen the differential effects of comorbidity distributions from historical records of known ALS patients, we propose an improved index called weighted Jaccard index (WJI) to evaluate similarities between two comorbidity sets. This index is based on the traditional Jaccard index [9] and add weighting coefficients proportionally to a enhanced set similarity. The calculation of WJI is explained as follows:

Table 1. Number of patients and corresponding weights for $AG*$ and $BG*$

$AG*$	$Disease_A$	d_{A1}	d_{A2}	d_{Ai}
	$Number_A$	N_{A1}	N_{A2}	N_{Ai}
	$Weight_A$	$\frac{N_{A1}}{\sum N_{Ai}}$	$\frac{N_{A2}}{\sum N_{Ai}}$	$\frac{N_{Ai}}{\sum N_{Ai}}$
$BG*$	$Disease_B$	d_{B1}	d_{B2}	d_{Bj}
	$Number_B$	N_{B1}	N_{B2}	N_{Bj}
	$Weight_B$	$\frac{N_{B1}}{\sum N_{Bi}}$	$\frac{N_{B2}}{\sum N_{Bi}}$	$\frac{N_{Bj}}{\sum N_{Bi}}$

Define AG as a set of comorbidity diseases of patient group A within an defined interval before a patient is diagnosed with a specific target disease. Then, we extracted the ordered top 20 diseases of the patient group by considering the same comorbidity diseases within a defined interval and the top 20 frequently co-occurred disease set is represented as $AG*$. Similarly, the comorbidity disease records for the other group B within the same interval is defined as BG, and the extracted set of the top 20 co-occurred diseases within this interval is defined as $BG *$. In the Table 1, $Disease_A$ represents the comorbidity diseases in patient group A, and d_{Ai} represents the ith comorbidity disease in group A. $Number_A$ represents the number of patients possessing a specific comorbidity disease in group A, and N_{Ai} represents the number of patients possessing the ith comorbidity disease in group A. $Weight_A$ represents the corresponding weights of a specific comorbidity diseases in group A, and $\frac{N_{Ai}}{\sum NA}$ represents the corresponding weight of the ith specific comorbidity disease in group A. Similarly, $Disease_B$ represents the comorbidity disease in patient group B, d_{Bj} represents the jth comorbidity disease in group B, N_{Bj} represents the number of patients possessing the jth comorbidity disease in group B, and $\frac{N_{B1}}{\sum NB}$ represents the corresponding weights of the jth specific comorbidity diseases in group B.

There are two cases for the weighting coefficient calculation. For the first case, when a specific disease does not appear in both $AG*$ and $BG*$ simueltaneously. When a certain disease only occurs in $AG*$, the corresponding weighting coefficient for such a specific disease is obtained by dividing the number of patients suffering the specific disease within $AG*$ by the total accumulated number of patients suffering all different diseases in $AG*$. In the same way, the algorithm can be applied to a certain disease only occurs in $BG*$. The second case occurs when the comorbidity disease appears both in $AG*$ and $BG*$ simultaneously. In this case, the corresponding weight of the specific disease in $AG*$ and $BG*$ is calculated by the same approach respectively and taking the averaged as the final corresponding weights. The formula is shown in Eq. (2):

$$Weight = \begin{cases} \frac{N}{\sum N}, & \text{if } d_{Ai} \neq d_{Bj} \\ \frac{\frac{N_{Ai}}{\sum N_{Ai}} + \frac{N_{Bj}}{\sum N_{Bj}}}{2}, & \text{if } d_{Ai} = d_{Bj} \end{cases} \tag{2}$$

The weighted Jaccard index (WJI) is obtained by summing up the weighting coefficient of $AG*$ and $BG*$ of overlapped diseases in both sets, and dividing by the sum of all weights within $AG*$ and $BG*$.

Here illustrate five examples to show the conception of our proposed index:

In Table 2, TTL_A represents the total number of patients suffering all different diseases in group A, TTL_B represents the total number of patients suffering all different diseases in group B.

Table 2. Weight and numbers of patients in groups A and B

$AG*$	$Disease_A$	d_{A1}	d_{A2}	d_{A3}	TTL_A
	$Number_A$	4	6	10	20
$BG*$	$Disease_B$	d_{B1}	d_{B2}	d_{B3}	TTL_B
	$Number_B$	3	2	5	10

Case 1: When the comorbidity diseases in both $AG*$ and $BG*$ are completely different from each other, in other words, disease contents meets the condition of $d_{Ai} \neq d_{Bj}$, and the weights are obtained according to the first calculation method. The corresponding

weights are calculated as follows:

$$Weight(d_{A1}) = \frac{4}{20} = 0.20, \quad Weight(d_{B1}) = \frac{3}{10} = 0.30,$$

$$Weight(d_{A2}) = \frac{6}{20} = 0.30, \quad Weight(d_{B2}) = \frac{2}{10} = 0.20,$$

$$Weight(d_{A3}) = \frac{10}{20} = 0.50, \quad Weight(d_{B3}) = \frac{5}{10} = 0.50$$

$$Jaccard(A, B) = \frac{0}{(3) + (3) - 0} = 0$$

$$Weighted\ Jaccard(A, B) = \frac{0}{(0.30 + 0.20 + 0.50) + (0.30 + 0.20 + 0.50) - 0} = 0$$

Case 2: When the contents of $AG*$ and $BG*$ completely hold identical condition as $d_{Ai} = d_{Bj}$, and the Weight is obtained through the second approach with a equation of $Weight = \frac{\frac{N_{Ai}}{\sum NA} + \frac{N_{Bj}}{\sum NB}}{2}$. The example of weights and corresponding WJI are calculated as follows:

$$Weight(d_{A1}) = \frac{\frac{4}{20} + \frac{3}{10}}{2} = 0.25, \quad Weight(d_{B1}) = \frac{\frac{4}{20} + \frac{3}{10}}{2} = 0.25,$$

$$Weight(d_{A2}) = \frac{\frac{6}{20} + \frac{2}{10}}{2} = 0.25, \quad Weight(d_{B2}) = \frac{\frac{6}{20} + \frac{2}{10}}{2} = 0.25,$$

$$Weight(d_{A3}) = \frac{\frac{10}{20} + \frac{5}{10}}{2} = 0.50, \quad Weight(d_{B3}) = \frac{\frac{10}{20} + \frac{5}{10}}{2} = 0.50$$

$$Jaccard(A, B) = \frac{3}{(3) + (3) - 3} = 1$$

$$Weighted\ Jaccard(A, B)$$
$$= \frac{0.25+0.25+0.50}{(0.25+0.25+0.50)+(0.25+0.25+0.50)-(0.25+0.25+0.50)} = 1$$

Case 3: When only d_{A1} and d_{B1} are identical in the two sets and all the other diseases are different. The weights for both d_{A1} and d_{B1} are calculated by the second approach, and the weights for all other diseases apply the first approach. Each weight and corresponding WJI are calculated as follows:

$$Weight(d_{A1}) = \frac{\frac{4}{20} + \frac{3}{10}}{2} = 0.25, \quad Weight(d_{B1}) = \frac{\frac{4}{20} + \frac{3}{10}}{2} = 0.25,$$

$$Weight(d_{A2}) = \frac{6}{20} = 0.30, \quad Weight(d_{B2}) = \frac{2}{10} = 0.20,$$

$$Weight(d_{A3}) = \frac{10}{20} = 0.50, \quad Weight(d_{B3}) = \frac{5}{10} = 0.50$$

$$Jaccard(A, B) = \frac{1}{(3) + (3) - 1} = \frac{1}{5}$$

$$Weighted\ Jaccard(A, B) = \frac{0.25}{(0.25+0.30+0.50)+(0.25+0.20+0.50)-(0.25)}$$
$$= \frac{1}{7}$$

Case 4: When only d_{A1} and d_{B2} identical diseases in the two sets, and all the other diseases are different. The weight of d_{A1} and d_{B2} are calculated by the second approach, and the weight for all other diseases apply the first approach. Each weight and corresponding WJI are calculated as follows:

$$Weight(d_{A1}) = \frac{\frac{4}{20} + \frac{2}{10}}{2} = 0.20, \quad Weight(d_{B1}) = \frac{3}{10} = 0.30,$$

$$Weight(d_{A2}) = \frac{6}{20} = 0.30, \quad Weight(d_{B2}) = \frac{\frac{4}{20} + \frac{2}{10}}{2} = 0.20,$$

$$Weight(d_{A3}) = \frac{10}{20} = 0.50, \quad Weight(d_{B3}) = \frac{5}{10} = 0.50$$

$$Jaccard(A, B) = \frac{1}{(3) + (3) - 1} = \frac{1}{5}$$

$$Weighted\ Jaccard(A, B) = \frac{0.20}{(0.20+0.30+0.50)+(0.30+0.20+0.50)-(0.20)}$$
$$= \frac{1}{9}$$

Case 5: When d_{A1} and d_{B2} are identical diseases in the two sets, d_{A2} and d_{B3} are identical as well, and all the other diseases are different. The weight of d_{A1}, d_{B2}, d_{A2}, and d_{B3} apply the second approach as before. However, the rest diseases apply the first approach. Each weight and corresponding WJI are calculated as follows:

$$Weight(d_{A1}) = \frac{\frac{4}{20} + \frac{2}{10}}{2} = 0.20, \quad Weight(d_{B1}) = \frac{3}{10} = 0.30$$

$$Weight(d_{A2}) = \frac{\frac{6}{20} + \frac{5}{10}}{2} = 0.40, \quad Weight(d_{B2}) = \frac{\frac{4}{20} + \frac{2}{10}}{2} = 0.20,$$

$$Weight(d_{A3}) = \frac{10}{20} = 0.50, \quad Weight(d_{B3}) = \frac{\frac{6}{20} + \frac{5}{10}}{2} = 0.40$$

$$Jaccard(A, B) = \frac{2}{(3) + (3) - 2} = \frac{1}{2}$$

$$Weighted\ Jaccard(A, B)$$
$$= \frac{0.20+0.40}{(0.20+0.40+0.50)+(0.30+0.20+0.40)-(0.20+0.40)} = \frac{3}{7}$$

2.2 Threshold

The threshold value is used to determine whether the subject possesses a potential status for ALS symptoms. First, we define the experimental group, control group, and feature group respectively. To collect the historical disease records within a period of time interval before the onset of ALS disease from clinical database, we identify the top 20 diseases from all know ALS patients and named as the experimental group. For control group, we focus on random selecting subject without ALS diagnosis records according to the same age and gender conditions, and five folds of the number of non-ALS subjects were selected as the control group. Similarly, their disease records within the same time interval were extracted and analyzed. Finally, we calculated the odds of each experimental group and control group regarding the comorbidity diseases, and the odds ratios were calculated for significant analysis of each comorbidity disease for both disease group compared to the healthy group. All the comorbidity diseases with odds ratio greater than 2 were collected as the feature group. Then, each person in the experimental group and the control group performs WJI similarity analysis for the feature group. After obtaining the similarity for each person, entering the data into the logistic regression model and sigmoid function for training a probability model and evaluating a threshold value. After that, a 5-fold cross validation was applied to obtain an average value as final threshold value to determine whether a new test subject possesses the risk potential of ALS disease.

2.3 Experimental Module

This research is mainly divided into two main modules and depicted in Fig. 1 and Fig. 2 respectively. Figure 1 shows the Data/Feature Extraction Analysis Module, and Fig. 2 shows the ALS Prediction Module. The following sections describe the function of each module:

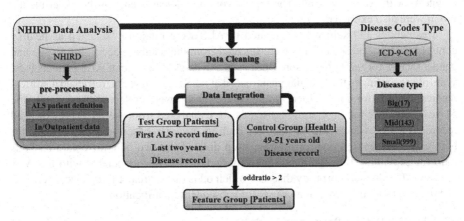

Fig. 1. Data/Feature extraction analysis module

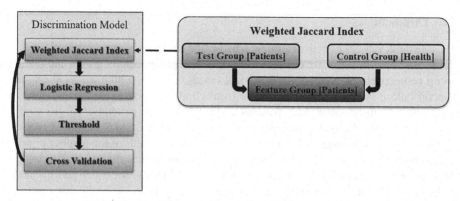

Fig. 2. ALS potential patient prediction module

Data/Feature Extraction Analysis Module. De-identified clinical data were obtained from the NHIRD medical database which contains electronic medical records of one million people from 1996 to 2013 (IRB: 104-543). After data cleaning and integration, the clustered medical records for representing experimental, control, and feature groups were extracted.

Extraction Analysis. A total of 162 patients with ALS diagnosis records were identified and retrieved from the medical database. In order to confirm that the initial candidates were indeed ALS patients, we only considered patients who were hospitalized due to ALS disease. Hence, only 71 subjects were selected for this study. Among them, 41 subjects were hospitalized directly at the first diagnosis and the remaining 30 subjects were hospitalized with an average of two-year period after the first diagnosis. In this study, the 71 hospitalized subjects due to ALS were considered as the experimental group. We further analyzed their two-year historical medical records before the first diagnosis with ALS symptom. The top 20 common diseases among the experimental group were also extracted.

Meanwhile, 399 subjects not related to ALS were randomly selected as healthy control subjects. These subjects were selected from the database according to age and gender factors for fair comparison. Because the average onset age of ALS patients is 51 years old according to the first diagnosis record of ALS. Therefore, the retrieved historical medical records for each subject in the control group were collected between 49 and 51 years old. The top 20 common diseases were analyzed from the previous extracted records. Finally, we calculated odds for each identified/exposed majority disease for both experimental and control groups. Then divided the two odds to obtain an odds ratio for explaining the significant factor of certain diseases associated with the ALS disease. After that, comorbidity diseases which odds ratios greater than 2 were selected and defined as an important feature set for subsequent identification.

ALS At-Risk Patient Prediction Module. Each subject either in the experimental group or the control group will be analyzed through weighted Jaccard index similarity analysis with respect to the feature group. After calculating WJI similarity for each subject, we applied logistic regression approach for training a probability model, and the model is used for predicting probability of suffering ALS disease within two years. The study adopted 5-fold cross validation to verify the accuracy and prediction stability of the trained model. Two different levels of disease definition were applied for comparison. The min-level analysis represents that each disease item is a single disease, while the mid-level analysis represents that each disease item is a disease group. In other words, there are 999 disease items in the min-level analysis, while these diseases were grouped into 143 disease groups for the mid-level analysis. The clustering rules were defined accord the definition by World Health Organization (WHO).

3   Results

The prediction model was constructed by employing logistic regression approach and the risk probability of suffering ALS is estimated and verified by a 5-fold cross validation for reliability and stability tests. According to Fig. 3 (a) and 4(a), it can be observed that the similarity value of WJIs between experimental group and feature group is generally higher than the WJIs obtained from the control and feature groups. In other words, it is observed from Fig. 3(b) and 4(b) that the WJIs between the control and the feature groups are generally lower than the experimental groups. If the training is divided into min-level classification and mid-level classification according to the disease classification, the verification results are shown in Fig. 5(a) and (b), Table 3 and Table 4. It can be observed that recall rate, specificity, and accuracy reached nearly 0.689, 0.716, and 0.702 in the min-level classification, while the AUC could reach 0.791. In comparison, the mid-level classification performs better than the min-level classification which recall rate, specificity, and accuracy could reach 0.731, 0.716, and 0.724 respectively, and an AUC of 0.845.

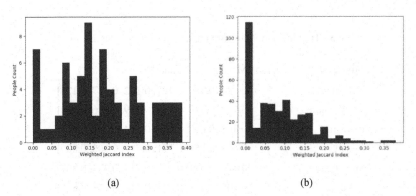

(a)   (b)

Fig. 3. WJI distributions for min-level classification. (a) WJIs between the experimental and the feature groups, (b) WJI distribution between the control group and the feature group.

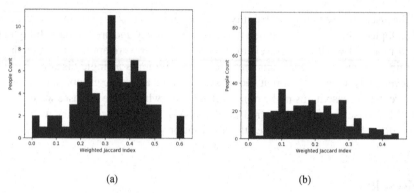

(a)   (b)

Fig. 4. WJI distributions for mid-level classification. (a) WJIs between the experimental and the feature groups, (b) WJI distribution between the control group and the feature group.

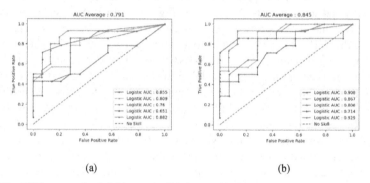

(a)   (b)

Fig. 5. The ROC curve of a 5-fold cross-validation verification. (a) Min-level classification; (b) Mid-level classification.

Table 3. Training and verification results [min-level classification]

Min-level classification	Mean [5 fold values]
Threshold	0.124 [0.119, 0.127, 0.122, 0.127, 0.124]
Model score	0.722 [0.719, 0.737, 0.719, 0.746, 0.688]
Precision	0.708 [0.667, 0.667, 0.714, 0.636, 0.857]
Recall	0.689 [0.857, 0.571, 0.714, 0.500, 0.800]
Specificity	0.716 [0.571, 0.714, 0.714, 0.714, 0.867]
Accuracy	0.702 [0.714, 0.643, 0.714, 0.607, 0.833]
F1 score	0.694 [0.750, 0.615, 0.714, 0.560, 0.828]
AUC	0.791 [0.855, 0.809, 0.760, 0.651, 0.882]

Table 4. Training and verification results [mid-level classification]

Mid-level classification	Mean [5 fold values]
Threshold	0.227 [0.220, 0.231, 0.226, 0.231, 0.227]
Model score	0.748 [0.754, 0.763, 0.728, 0.781, 0.714]
Precision	0.721 [0.667, 0.692, 0.750, 0.636, 0.857]
Recall	0.731 [0.857, 0.643, 0.857, 0.500, 0.800]
Specificity	0.716 [0.571, 0.714, 0.714, 0.714, 0.867]
Accuracy	0.724 [0.714, 0.679, 0.786, 0.607, 0.833]
F1 score	0.721 [0.750, 0.667, 0.800, 0.560, 0.828]
AUC	0.845 [0.908, 0.867, 0.806, 0.714, 0.929]

4 Discussion

This research applied historical medical records and the proposed weighted Jaccard index for a specific target disease prediction. Here we applied logistic regression technique to construct a model for predicting risk factor of suffering ALS disease as an example. The accuracy rates could reach as high as 70.2% for a min-level single-disease group condition and 72.4% for mid-level multiple-disease group classification, and an AUCs of 0.791 and 0.845 respectively. This is the first disease prediction model based on the historical medical records and the proposed novel weighted Jaccard indices. The ALS prediction module is considered as a time-saving and convenient non-invasive way to analyze at-risk potential, and it could be extended for any other specific disease. It can be detected and treated at early stages, prolong survival years and improve the quality of life. It is one of the *in silico* analytical tools to assist medical diagnosis. Through exploring huge amount of medical records to improve the preventive medical applications, we hope the proposed methods can facilitate doctors to make a good decision in terms of medical treatment and risk assessment for precision diagnosis in the future.

In this research, we tried to establish an ALS prediction module through the combination of disease trajectory and weighted Jaccard index. Due to the available small set of ALS patients and their corresponding medical records, several problems has occurred after performing disease trajectory analysis. There is no significant trajectory pattern could be found from such a small sample records. In fact, only short disease trajectory results could be observed and it is relatively difficult to adopt the patterns for accurate prediction. Hence, we proposed an alternative approach by considering the comorbidity diseases through a novel weighted Jaccard index to avoid the issue of overfitting. Nevertheless, the proposed novel index could be applied to other different target diseases, such as chronic diseases and cancers. There should be a large number of patients with such diseases, and the disease trajectories would be expected with long trajectories, and the prediction model would provide better prediction results and facilitate doctors to make right strategy for a specific disease.

References

1. Kiernan, M.C., et al.: Amyotrophic lateral sclerosis. Lancet **377**, 942–955 (2011)
2. Ingre, C., Roos, P.M., Piehl, F., Kamel, F., Fang, F.: Risk factors for amyotrophic lateral sclerosis. Clin. Epidemiol. **7**, 181–193 (2015)
3. Wijesekera, L.C., Leigh, P.N.: Amyotrophic lateral sclerosis. Orphanet J. Rare Dis. **4**, 3 (2009). https://doi.org/10.1186/1750-1172-4-3
4. Knibb, J.A., Keren, N., Kulka, A., et al.: A clinical tool for predicting survival in ALS. J. Neurol. Neurosurg. Psychiatry **87**, 1361–1367 (2016)
5. Wu, T.-Y., Majeed, A., Kuo, K.N.: An overview of the healthcare system in Taiwan. Lon. J Prim. Care **3**(2), 115–119 (2010)
6. Su, F.C., et al.: Association of environmental toxins with amyotrophic lateral sclerosis. JAMA Neurol. **73**(7), 803–811 (2016)
7. Van Rheenen, W., Diekstra, F.P., Harschnitz, O., et al.: Whole blood transcriptome analysis in amyotrophic lateral sclerosis: a biomarker study. PLoS one **13**(6), e0198874 (2018)
8. Chen, C.J., Chen, C.M., Pai, T.W., Chang, H.T., Hwang, C.S.: A genome-wide association study on amyotrophic lateral sclerosis in the Taiwanese Han population. Biomark. Med. **10**(6), 597–611 (2016)
9. Rahman, M., Hassan, M.R., Buyya, R.: Jaccard index based availability prediction in enterprise grids. Procedia Comput. Sci. **1**(1), 2707–2716 (2010)

The Differential Feature Detection and the Clustering Analysis to Breast Cancers

Juanying Xie[1(✉)], Zhaozhong Wu[1], Qin Xia[1], Lijuan Ding[1], and Hamido Fujita[2]

[1] School of Computer Sciecne, Shaanxi Normal University, Xi'an 710062, China
xiejuany@snnu.edu.cn
[2] Software and Information Science, Iwate Prefectural University, 152-52 Sugo, Takizawa, Iwate, Japan
HFujita-799@acm.org

Abstract. Breast cancer is one of the most common and fatal cancers in women today. To detect the differential features for correctly diagnosing breast cancer patients, a 2D feature selection algorithm is proposed in this paper. It is named as SDP for short because it adopts the Standard Deviation and Pearson correlation coefficient to define the feature discernibility and independence respectively. The feature importance is defined as the product of its discernibility and independence. A 2D space is constructed using discernibility as x-axis and independence as y-axis. The area of the rectangular enclosed by a feature coordinate lines and axes implies the importance of the feature. Those features in the upper right corner of the 2D space are with much higher importance than the rest ones. They comprise the differential feature subset. The spectral clustering algorithms SC_SD and SC_MD are adopted for clustering analysis, so the two algorithms SDP+SC_SD and SDP+SC_MD were developed for clustering analysis to the breast cancer data WBCD (Wisconsin Breast Cancer Database), WDBC (Wisconsin Diagnostic Breast Cancer), and WPBC (Wisconsin Prognostic Breast Cancer). The experimental results demonstrate that the proposed SDP can detect much better differential features of breast cancers than other compared feature selection algorithms, and the SDP+SC_SD and SDP+SC_MD algorithms outperform the algorithms without feature selection process embedded, such as SC_SD, SC_MD, DPC, SD_DPC, K-means and SD_K-medoids in terms of clustering accuracy, AMI (Adjusted Mutual Information), ARI (Adjusted Rand Index), sensitivity, specificity and precision.

Keywords: Breast cancer · Feature selection · Discernibility · Independence · Spectral clustering

Supported by the National Natural Science Foundation of China under Grant No. 61673251, and supported by the National Key Research and Development Program of China under Grant No. 2016YFC0901900, and the Scientific and Technological Achievements Transformation and Cultivation Funds of Shaanxi Normal University under Grant No. GK201806013, and the Innovation Funds of Graduate Programs at Shaanxi Normal University under Grant No. 2015CXS028, 2016CSY009, and 2018TS078.

H. Fujita et al. (Eds.): IEA/AIE 2020, LNAI 12144, pp. 457–469, 2020.
https://doi.org/10.1007/978-3-030-55789-8_40

1   Introduction

Breast cancer is a tumor that occurs in the epithelial tissue of breast. It is one of the cancers with the highest incidence and the fatal rate in women [14,22]. The morbidity and mortality of breast cancer have been going up, which leads to the serious harm to women's health [27]. To recognize the breast cancer is benign or malignant tumor rapidly and accurately is very important for taking timely and correct treatment to patients [4]. However, the traditional diagnosis for breast cancer is mainly based on experts' subjective observation and judgment of lesion location and pathological features, which requires experts with rich experience in the field of breast cancer diagnosis to achieve accurate diagnosis.

With the development of machine learning and data mining technology, there is a new way for diagnosing and identifying breast cancers. Watkins et al. proposed AIRS (Artificial Immune Recognition System) to classify the breast cancers, and got the accuracy of 97% on the WBCD [17]. Leung et al. proposed immune evolution based classifier SAIS (Simple Artificial Immune System) to get the accuracy of 96.6% on WBCD [10]. Jeleń et al. used the multiplane method to classify WBCD and got the accuracy of 93.5% (50% training, 50% test) and 95.9% (67% training, 33% test) [8]. The partial order structure graph combined lasso method got the sensitivity and specificity of 96.15% and 94.16% for WBCD (546 training samples and 137 test samples) by extracting features and reducing features and extracting rules [29]. The improved support vector machines obtained 98.59% recognition accuracy for WBCD [28]. Bayesian classifier and LVQ (Learning Vector Quantization) neural network obtained the recognition accuracy of 96.51% and 92.44% respectively on WDBC(70% training, 30% test) [25]. Tiwari et al. proposed a local search and global optimization based feature selection method to analysis the WDBC database and got the accuracy of 96.66% [15]. The LVQ neural network optimized by particle swarm obtained the 95.95% accuracy on WDBC [26]. The differential feature selection algorithm based on the minimum spanning tree and F statistic obtained 79.96% accuracy using 10% features [21]. The SVM classifier named GS.SVM got the 74.63% accuracy in WPBC database by using genetic algorithm to optimize subspaces [9]. The improved ST (Select and Test) algorithm proposed by Oyelade et al. can recognize the 26.60%, 56.17% and 54.05% breast cancer patients in WBCD, WDBC and WPBC datasets, respectively [12].

The aforementioned breast cancer studies used sample labels except for the reference in [21]. But it is usually difficult to get labels for samples. Furthermore, most of the aforementioned studies did not do feature selection, such that their study results could be affected by the redundant and irrelevant features in the data. Feature selection can eliminate those redundant and irrelevant features while preserving ones with strong classification information, so as to tell breast cancer patients from normal people. Spectral clustering is a kind of unsupervised learning theory, which does not require labels of samples. It transforms the clustering problem into the graph division one, and can find any arbitrary shape clusters and converge to the global optimal solution [2,16]. Therefore, we try to propose an unsupervised feature selection algorithm for breast cancer data in this paper. We named the feature selection algorithm as SDP (Standard Deviation and Pearson correlation coefficient based feature selection). SDP adopts features Standard Deviation to measure the feature discernibility, and Pearson correlation coefficient to value the feature independence. The product of a feature discernibility and its independence is used to evaluate its importance to classification. The first l important features are selected to constitute the feature subset. Then the spectral clustering analysis is carried out to the breast cancer samples with only those selected features using the true self-adaptive spectral clustering algorithms SC_SD (Spectral Clustering based on Standard Deviation) and SC_MD (Spectral Clustering based on Mean Distance) [18]. The corresponding algorithms are denoted as SDP+SC_SD, SDP+SC_MD for short.

2 SDP Algorithm

The SDP algorithm adopts feature standard deviation and Pearson correlation coefficient to define feature discernibility and feature independence, respectively, and the product of the feature discernibility and independence to measure the importance of the feature to classification. The top l important features are with good recognition capability for breast cancers.

2.1 Related Definitions

The Motivation of SDP algorithm is to discover the differential characteristics between breast cancer patients and normal people. Therefore the discernibility and independence and importance of a feature are defined as follows.

Definition 1. *Feature discernibility: A feature with strong discernibility must own similar values among samples within same category while different values for samples from other classes. Features with larger variance (or standard deviation) has a strong discernibility [7]. Therefore, we define the discernibility of feature i as its standard deviation in (1).*

$$dis_i = \sqrt{\frac{1}{n-1} \sum_{j=1}^{n} \left(f_{ji} - \frac{1}{n} \sum_{j=1}^{n} f_{ji} \right)^2} \, , \ i = 1, \cdots, d; j = 1, \cdots, n \quad (1)$$

where f_{ji} is the value of sample j on the feature i, and dis_i the discernibility of feature i as its standard deviation, d is the dimensions of samples, and n the number of samples.

Definition 2. *Feature independence: The Pearson correlation coefficient can measure the correlation between two features. The smaller the absolute value of the Pearson correlation coefficient between two features, the less correlation is between the two features. Therefore, the feature independence is defined in the Pearson correlation coefficient in (2).*

$$ind_i = \begin{cases} max_j \frac{1}{|r_{ij}|}, & dis_i = \max_{j=1,\cdots,d} \{dis_j\} \\ \min_{j:dis_j>dis_i} \frac{1}{|r_{ij}|}, & otherwise. \end{cases} \qquad (2)$$

Where r_{ij} is the Pearson correlation coefficient between features i and j. It can be seen from (2) that the independence of feature i with the highest discernibility is defined as the reciprocal of the absolute value of the Pearson correlation coefficient between feature i and its least relevant feature j. Otherwise, the independence of the feature i is defined as the reciprocal of the absolute value of the Pearson correlation coefficient between features i and j whose discernibility is just higher than that of feature i. Therefore the smaller the absolute value of Pearson correlation coefficient between features i and j, the stronger is the independence of feature i. This definition can guarantee that the feature with the highest capability of recognition will own the strongest independence, so that it will be selected in the feature selection process, and other selected features are those with both high discernibility and high independence as far as possible.

Definition 3. *Feature importance: The importance of feature i is defined as the product of its discernibility and its independence in (3). The $score_i$ is the numerical value of feature i's classification capability. The bigger the numerical value, the stronger is capability of feature i in classification. It is also a trade-off between feature i's discernibility and its independence.*

$$score_i = dis_i \times ind_i \qquad (3)$$

2.2   Algorithm Steps

Input: $\mathbf{D} \in \mathbb{R}^{n \times d}$, where n is the number of samples, and d the number of features.

Output: Feature subset S.

step1: Let $S = \varnothing$, and F comprises all features;
step2: Calculate the dis_i in (1) and ind_i in (2) for each feature $i(= 1, \cdots, n)$;
step3: Scatter all features in the 2-dimensional space with discernibility as x-axis and independence as y-axis;
step3: Calculate the $score_i$ for each feature $i(= 1, \cdots, n)$ in (3), and sort them in descending order;
step4: Select top l features whose $score$ values are much higher than that of the remaining ones, that is the features in the upper right corner of the 2-dimensional space and far away from the rest ones;
step5: These top l features comprise the feature subset S.

3   Spectral Clustering Analysis to Breast Cancer Data

It is reported that the neighborhood standard deviation based self-adaptive spectral clustering algorithm SC_SD and SC_MD algorithm [18] can effectively discover the true distribution of a dataset. Therefore, we adopt them to find the clustering of breast cancer data which has been done the feature selection by SDP algorithm. The clustering results are evaluated in terms of clustering accuracy, sensitivity, specificity, and precision.

The difference between SC_SD and SC_MD are their scale parameters σ_{SD_i} in (4) and σ_{MD_i} in (5), which are used to construct their affinity matrix in (6) and (7), respectively. The S_i in (4) and M_i in (5) are the number of samples in the neighborhood of sample i in SC_SD and SC_MD, respectively. The two neighborhood radius are, respectively, the standard deviation of sample i and the mean distance from sample i to the rest samples in training subset.

$$\sigma_{SD\_i} = \sqrt{\frac{1}{S_i - 1} \sum_{j=1}^{S_i} d^2(\mathbf{x}_j, \mathbf{x}_i)} \tag{4}$$

$$\sigma_{MD\_i} = \sqrt{\frac{1}{M_i - 1} \sum_{j=1}^{M_i} d^2(\mathbf{x}_j, \mathbf{x}_i)} \tag{5}$$

$$A_{ij} = \begin{cases} exp(-d^2(\mathbf{x}_i, \mathbf{x}_j)/\sigma_{SD\_i}\sigma_{SD\_j}), & i \neq j \\ 0, & i = j \end{cases} \tag{6}$$

$$A_{ij} = \begin{cases} exp(-d^2(\mathbf{x}_i, \mathbf{x}_j)/\sigma_{MD\_i}\sigma_{MD\_j}), & i \neq j \\ 0, & i = j \end{cases} \tag{7}$$

Here are the main steps of SC_SD and SC_MD for Breast cancer data.

Input: Breast cancer data $Data \in \mathbb{R}^{n \times l}$ and the number of clusters K, where n is the number of samples, l is the number of features selected by the proposed SDP algorithm.

Output: K clusters.

Step1: Construct affinity matrix $\mathbf{A} \in \mathbb{R}^{n \times n}$ in (6) and (7), respectively;

Step2: Calculate degree matrix $\mathbf{D}$ and Laplacian matrix $\mathbf{L} = \mathbf{D}^{-\frac{1}{2}}\mathbf{A}\mathbf{D}^{-\frac{1}{2}}$ of affinity matrix $\mathbf{A}$;

Step3: Let $\lambda_1 \geq \lambda_2 \geq \cdots \geq \lambda_K \geq 0$ be the first K maximum eigenvalues of L, and $\mathbf{v}^1, \mathbf{v}^2, \cdots, \mathbf{v}^K$ be their corresponding eigenvectors, then construct the matrix $\mathbf{V} = [\mathbf{v}^1, \mathbf{v}^2, \cdots, \mathbf{v}^K] \in \mathbb{R}^{n \times K}$;

Step4: Normalize matrix $\mathbf{V}$ in row to get matrix $\mathbf{U}$, that is $u_{ij} = v_{ij}/\left(\sum_k v_{ik}^2\right)^{\frac{1}{2}}$;

Step5: Take each row in $\mathbf{U}$ as a sample to cluster samples in $\mathbf{U}$ by SD_K-medoids algorithm in [19].

Step6: Assign sample $\mathbf{x}_i$ to cluster j if and only if row i of matrix $\mathbf{U}$ belonging to cluster j.

4   Time Complexity Analysis

The algorithms in this paper comprise the proposed SDP feature selection algorithm to reduce dimensions of breast cancer data, and the spectral clustering algorithms SC_SD or SC_MD to find the clustering of the reduced breast cancer samples. Assume that there are n samples with d features. The time complexity in SDP algorithm to calculate the discernibility and independence of features are $O(n^2)$ and $O(nd^2)$, respectively. The time complexity of calculating *Score* values of features and ranking them in descending order is $O(d \log d)$ in SDP. Because $d < n$ holds for breast cancer data, so the time complexity of proposed SDP is $O(n^2)$. The time complexity to construct the affinity matrix $\mathbf{A}$ and to calculate the Laplacian matrix $\mathbf{L}$ for spectral clustering to breast cancer data are both $O(n^2)$. The time complexity to construct the matrix $\mathbf{V}$ is $O(K^2)$, where K is the number of clusters. The time complexity to normalize the matrix $\mathbf{V}$ is $O(Kn)$. The time complexity of SD_K-medoids is $O(n^2 + Knt)$, where t is its iterations. The time complexity of assigning original samples to their proper clusters is $O(n)$. Therefore, the time complexity of spectral clustering analysis is $O(n^2)$. To summary the above analysis, it can be concluded that the totally time complexity of the algorithms in this paper is $O(n^2)$.

5   Experimental Results and Analyses

5.1   Datasets

The breast cancer datasets used in this paper are from UCI Machine Learning repository [5]. They are WBCD (Wisconsin Breast Cancer Database), WDBC (Wisconsin Diagnostic Breast Cancer), and WPBC (Wisconsin Prognostic Breast Cancer) datasets, respectively. There are small amount of missing data in WBCD and WPBC. We deleted those samples with missing data in experiments. The maximum and minimum normalization are adopted to preprocess data. The specific informations of these breast cancer data are shown in Table 1.

Table 1. Descriptions of breast cancer data

Datasets	#Samples (negative + positive)	#Attributes	#Clusters
WBCD	683 (444 + 239)	9	2
WDBC	569 (357 + 212)	30	2
WPBC	194 (148 + 46)	32	2

5.2 Feature Importance Definition Test

We test the correctness of feature importance in proposed SDP by WDBC. We calculate the discernibility and independence of features and scatter all features in the 2-dimensional space with discernibility and independence as x-axis and y-axis, respectively, in Fig. 1(a). Then we scatter all features in the 2-dimensional space with the importance and number of features as y-axis and x-axis, respectively, in Fig. 1(b). We calculate the importance of features and rank them in descending order. The proposed SDP is used to select those feature whose Score values are much higher than that of the remaining ones, which means the selected features by SDP are with much stronger capability in telling out breast cancer patients from normal people. The SVM classifier is constructed on the reduced data. The Libsvm [3] tool box is adopted to build the SVM classifier with linear kernel function and with the penalty factor $C = 20$. We compared the performance of the SVM classifiers corresponding to the feature subset by SDP and by the feature selection algorithms including EDPFS [23], RDPFS [23], MCFS [1], and Laplacian [6] in terms of the classification accuracy, F-measure, and sensitivity. The mean results of 10-fold cross validation experiments are shown in Table 2. The bold fonts mean the optimal results for the same feature subset.

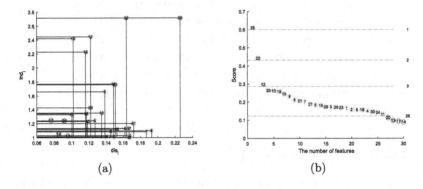

Fig. 1. Results of SDP on WDBC data

The results in Fig. 1 reveal that the features with both high discernibility and high independence are in the upper right corner of the 2-dimensional space with discernibility and independence as y-axis and x-axis, respectively. The scores of those features, that is the area of the rectangular enclosed by the axes and the coordinate lines, are much higher than that of the rest ones.

The results in Table 2 show that the proposed SDP can detect features with much better capability in classification than those compared feature selection algorithms. The sensitivities of SVM classifiers of the feature subset by SDP are absolutely superior to that of the other compared algorithms, which means that the proposed SDP can find features with the best recognition rate for malignant tumors compared to the other comparison feature algorithms. Therefore, we can conclude that the feature importance defined in SDP is very effective.

It should be noted that we had compared the performance of our SDP to that of the feature selection algorithms including EDPFS [23], RDPFS [23], MCFS [1], and Laplacian [6] in terms of accuracy, sensitivity, specificity, F-measure, precision and F2 [24] of the SVM and KNN classfiers of their selected feature subsets on the WDBC, WPBC and WBCD datasets. ALL the experimental results demonstrate that our SDP is superior to the compared feature selection algorithms. But because of the page limitation, we cannot display all the results.

Table 2. Performance comparison of SDP and other feature selection algorithms on WDBC

	SDP	EDPFS	RDPFS	MCFS	Laplacian	#Selected features
Accuracy	0.9122	0.8813	0.9125	0.8662	0.9125	1
	0.9198	0.8904	0.9115	0.8784	0.9177	2
	0.9191	0.9146	0.9269	0.8865	0.9287	3
	0.9754	0.9712	0.9729	0.9722	0.9747	26
F-measure	**0.8790**	0.8152	0.8654	0.7851	0.8654	1
	0.8884	0.8281	0.8651	0.8094	0.8764	2
	0.8872	0.8772	0.8911	0.8250	0.8978	3
	0.9659	0.9596	0.9621	0.9613	0.9647	26
Sensitivity	**0.8613**	0.7191	0.7756	0.6749	0.7756	1
	0.8632	0.7266	0.7822	0.7097	0.8009	2
	0.8603	0.8322	0.8246	0.7334	0.8558	3
	0.9445	0.9323	0.9359	0.9388	0.9416	26

5.3   Feature Subsets Detected by SDP

This subsection will show the specific features detected by SDP. Table 3 displays the feature subsets detected by SDP on WBCD, WDBC and WPBC.

Table 3. Feature subsets detected by SDP on three breast cancer datasets

Datasets	#Features	Selected features
WBCD	1	6
	5	1, 2, 4, 6, 8
	6	1, 2, 4, 6, 8, 9
WDBC	1	28
	2	22, 28
	3	12, 22, 28

(continued)

Table 3. (*continued*)

Datasets	#Features	Selected features
	26	1, 2, 3, 4, 5, 6, 7, 8, 9, 10, 11, 12, 15, 16, 18, 19, 21, 22, 23, 24, 25, 26, 27, 28, 29, 30
WPBC	1	32
	2	23, 32
	3	22, 23, 32
	4	22, 23, 31, 32
	7	6, 11, 19, 22, 23, 31, 32

The results in Table 3 show that the top 1 feature of WBCD selected by SDP is feature 6, that is the feature of Bare Nuclei. The top 5 features are features 1, 2, 4, 6, and 8, corresponding to physical features of Clump Thickness, Uniformity of Cell Size, Marginal Adhesion, Bare Nuclei and Normal Nucleoli. The top 6 features are features 1, 2, 4, 6, 8, and 9, where Mitoses physical feature was included besides the top 5 features.

The top 1 feature of WDBC by SDP is feature 28, that is the physical feature of Mean of the three largest values of concave points. The top 2 features are 22 and 28, where feature 22 is the Mean of the three largest values of texture. The top 3 features are 12, 22 and 28. The feature 12 is the Standard error of texture. When the top 26 features are selected, they are the features of feature 1 to 12, and features 15 and 16, 18 and 19, 21 to 30. These 26 features are related to cell radius, texture, perimeter, area, smoothness, compactness, concavity, concave points, symmetry, and fractal dimension.

For WPBC, the top 1 feature selected by SDP is feature 32, that is Lymph node status. The top 2 features are 23 and 32. They are physical features of Mean of the three largest values of perimeter, and Lymph node status. The top 3 features by SDP are features 22, 23, and 32, that is the Mean of the three largest values of texture is detected besides the top 2 features. The top 7 features detected by SDP are features 6, 11, 19, 22, 23, 31 and 32. These features are related to nuclear compactness, radius, texture, symmetry, perimeter, Tumor size, and Lymph node status.

5.4 Performance Comparison Between Algorithms

This subsection will compare the results of SDP+SC_SD and SDP+SC_MD proposed in this paper to that of other clustering algorithms without embedding feature selection, such as SC_SD, SC_MD, K-means [11], DPC [13], SD_DPC [20], SD_K-medoids [19] on WBCD, WDBC and WPBC datasets in terms of clustering accuracy, AMI, ARI, sensitivity, specificity and precision. The SDP+SC_SD and SDP+SC_MD mean that we adopt the proposed SDP to do feature selection, then use the SC_SD and SC_MD to do clustering analysis for reduced data, respectively. The other compared algorithms only do clustering analysis to original WBCD, WDBC and WPBC datasets. The parameters of both DPC and SD_DPC are 5, 9 and 5 for WBCD, WDBC, WPBC, respectively. The results of K-means are the mean results of its 20 runs. Table 4 display the results of the compared algorithms. The bold fonts mean the best results of algorithms on same dataset.

The experimental results of WBCD in Table 4 show that SDP+SC_SD and SDP+SC_MD algorithms get pretty good clustering results using about half number of features of the original ones in terms of clustering accuracy, AMI, ARI, sensitivity, specificity, and precision, especially the SDP+SC_MD algorithm gets the 100% sensitivity.

The results of WDBC in Table 4 show that both SDP+SC_SD and SDP+SC_MD algorithms can get a comparable good performances using only 2 features detected by SDP, especially SC_MD can get the best sensitivity of 92.92%. Furthermore, they can get the best clustering accuracy, AMI and ARI among the compared algorithms by using the first 26 important features detected by SDP algorithm. The results also show that the performances of SDP+SC_SD and SDP+SC_MD algorithms using top 26 features detected by SDP are better than that of using the top 2 features in terms of clustering accuracy, AMI, ARI, specificity and precision except for the sensitivity metric. Sensitivity means the ratio of recognizing as malignant tumor patients to true malignant tumor patients. Therefore the top 2 features detected by SDP are good enough to diagnosing malignant tumor patients in WDBC data.

Table 4. Results of different algorithms on WBCD, WDBC and WPBC

Datasets	Algorithms	#Features	Accuracy	AMI	ARI	Sensitivity	Specificity	Precision
WBCD	SDP+SC_SD	**5**	0.9619	0.7742	0.8526	0.9958	0.9437	0.9049
	SDP+SC_MD	**5**	0.9385	0.6992	0.7682	**1.0000**	0.9054	0.8505
	SC_SD	9	**0.9707**	**0.8120**	**0.8855**	0.9958	0.9572	0.9261
	SC_MD	9	0.9605	0.7683	0.8472	0.9958	0.9414	0.9015
	DPC	9	0.8272	0.3461	0.4055	0.5063	**1.0000**	**1.0000**
	SD_DPC	9	0.9224	0.6214	0.7557	0.8496	0.9865	0.9697
	K-means	9	0.9609	0.7448	0.8481	0.9259	0.9797	0.9609
	SD_K-medoids	9	0.9590	0.7354	0.8410	0.9205	0.9797	0.9607
WDBC	SDP+SC_SD	2	0.9033	0.5331	0.6495	0.9245	0.8908	0.8340
	SDP+SC_MD	2	0.9033	0.5351	0.6495	**0.9292**	0.8880	0.8312
	SDP+SC_SD	26	**0.9279**	**0.6110**	**0.7311**	0.9151	0.9356	0.8940
	SDP+SC_MD	26	**0.9279**	**0.6110**	**0.7311**	0.9151	0.9356	0.8940
	SC_SD	30	0.9262	0.6045	0.7251	0.9151	0.9328	0.8899
	SC_MD	30	**0.9279**	**0.6110**	**0.7311**	0.9151	0.9356	0.8940
	DPC	30	0.6151	0.0305	0.0506	0.5755	0.6387	0.4861
	SD_DPC	30	0.8576	0.4411	0.5016	0.6179	**1.0000**	**1.0000**
	K-means	30	0.9279	**0.6110**	0.7302	0.8491	0.9748	0.9524
	SD_K-medoids	30	0.9227	0.6011	0.7116	0.8160	0.9860	0.9719
WPBC	SDP+SC_SD	1	**0.7268**	0.0302	**0.1163**	0.3478	**0.8446**	0.4103
	SDP+SC_MD	1	**0.7268**	0.0302	**0.1163**	0.3478	**0.8446**	0.4103

(*continued*)

Table 4. (*continued*)

Datasets	Algorithms	#Features	Accuracy	AMI	ARI	Sensitivity	Specificity	Precision
	SDP+SC_SD	2	0.5928	**0.0314**	0.0308	**0.6957**	0.5608	**0.4762**
	SDP+SC_MD	2	0.5825	0.0279	0.0234	**0.6957**	0.5473	0.4688
	SC_SD	32	0.5412	0.005	0.0029	0.6087	0.5203	0.2828
	SC_MD	32	0.5412	0.0072	0.0026	0.6304	0.5135	0.2871
	DPC	32	0.6186	0.0123	0.0404	0.5217	0.6486	0.3158
	SD_DPC	32	0.6392	0.0285	0.0642	0.5870	0.6554	0.3462
	K-means	32	0.5925	0.0167	0.0289	0.6087	0.5875	0.3145
	SD_K-medoids	32	0.5979	0.0151	0.0313	0.5870	0.6014	0.3140

The results in Table 4 about WPBC data show that both SDP+SC_SD and SDP+SC_MD algorithms performed well only using the top 1 feature with best clustering accuracy, specificity and ARI among the compared algorithms. The best sensitivity among the compared algorithms can be got by SDP+SC_SD and SDP+SC_MD algorithms using the top 2 features detected by SDP, and SDP+SC_SD can also get the best AMI and precision. These facts reveal that the top 2 features detected by SDP are necessary to recognize the recurrence malignant tumor patients after recovering.

In addition, the experimental results in Table 4 also show that DPC and SD_DPC algorithms can get the optimal performance of 100% in terms of specificity and precision on WBCD and WDBC respectively, but the recognition rate of patients with malignant tumors is only 50.63% and 61.79% respectively.

The above analysis to the results in Table 4 demonstrate that the proposed SDP+SC_SD and SDP+SC_MD algorithms have good performance on WBCD, WDBC and WPBC datasets, which implies that the proposed SDP feature selection algorithm is powerful in finding the discriminative features to recognize the malignant tumor patients.

6 Conclusions

A feature selection algorithm named as SDP is proposed for detecting the differential features to recognize breast cancer patients. The feature discernibility, feature independence and feature importance are defined in SDP. The SDP is combined with the spectral clustering algorithms SC_SD and SC_MD, respectively, so as to get the SDP+SC_SD and SDP+SC_MD algorithms to analysis WBCD, WDBC and WPBC datasets.

The experimental results demonstrate that the proposed SDP algorithm can detect better differential features for breast cancers than compared feature selection algorithms, and SDP+SC_SD and SDP+SC_MD algorithms can get better performance on WBCD, WDBC and WPBC in recognizing the malignant tumor patients than other compared algorithms.

References

1. Cai, D., Zhang, C., He, X.: Unsupervised feature selection for multi-cluster data. In: Proceedings of the 16th ACM SIGKDD International Conference on Knowledge Discovery and Data Mining, pp. 333–342. ACM (2010)
2. Cai, X., Dai, G., Yang, L.: Survey on spectral clustering algorithms. Comput. Sci. **35**(7), 14–18 (2008)
3. Chang, C.C., Lin, C.J.: LIBSVM: a library for support vector machines. ACM Trans. Intell. Syst. Technol. (TIST) **2**(3), 27 (2011)
4. Deng, Z., Tan, G., Ye, J., Fan, B.: An immune classification algorithm for breast cancer diagnosis. J. Central South Univ. (Sci. Technol.) **41**(4), 1485–1490 (2010)
5. Dua, D., Graff, C.: UCI machine learning repository (2017). http://archive.ics.uci.edu/ml
6. He, X., Cai, D., Niyogi, P.: Laplacian score for feature selection. In: Advances in Neural Information Processing Systems, pp. 507–514 (2006)
7. Hu, M., Lin, Y., Yang, H., Zheng, L., Fu, W.: Spectral feature selection based on feature correlation. CAAI Trans. Intell. Syst. **12**(4), 519–525 (2017)
8. Jeleń, L., Fevens, T., Krzyżak, A.: Classification of breast cancer malignancy using cytological images of fine needle aspiration biopsies. Int. J. Appl. Math. Comput. Sci. **18**(1), 75–83 (2008)
9. Jiang, H., Yu, X.: Ga-based subspace classification algorithm for support vector machines. Comput. Sci. **40**(11), 255–260 (2013)
10. Leung, K., Cheong, F., Cheong, C.: Generating compact classifier systems using a simple artificial immune system. IEEE Trans. Syst. Man Cybern. Part B Cybern. **37**(5), 1344–1356 (2007)
11. MacQueen, J.B.: Some methods for the classification and analysis of multivariate observations. In: Proceedings of the 5th Berkeley Symposium on Mathematical Statistics and Probability, vol. 1:Statistics, pp. 281–297. University of California Press, Berkeley (1967)
12. Oyelade, O.N., Obiniyi, A.A., Junaidu, S.B., Adewuyi, A.S.: ST-ONCODIAG: a semantic rule-base approach to diagnosing breast cancer base on wisconsin datasets. Inform. Med. Unlocked **10**, 117–125 (2018)
13. Rodríguez, A., Laio, A.: Clustering by fast search and find of density peaks. Science **344**(6191), 1492–1496 (2014)
14. Shen, L., Margolies, L.R., Rothstein, J.H., Fluder, E., McBride, R.B., Sieh, W.: Deep learning to improve breast cancer early detection on screening mammography. Sci. rep. **9**, 12 (2019)
15. Tiwari, S., Singh, B., Kaur, M.: An approach for feature selection using local searching and global optimization techniques. Neural Comput. Appl. **28**(10), 2915–2930 (2017). https://doi.org/10.1007/s00521-017-2959-y
16. Ulrike, V.L.: A tutorial on spectral clustering. Stat. Comput. **17**(4), 395–416 (2007)
17. Watkins, A., Timmis, J., Boggess, L.: Artificial immune recognition system (AIRS): an immune-inspired supervised learning algorithm. Genet. Program Evolvable Mach. **5**(3), 291–317 (2004)
18. Xie, J., Ding, L.: The true self-adaptive spectral clustering algorithms. Acta Electronic Sinica **47**(05), 1000–1008 (2019)
19. Xie, J., Gao, R.: K-medoids clustering algorithms with optimized initial seeds by variance. J. Front. Comput. Sci. Technol. **9**(8), 973–984 (2015)
20. Xie, J., Jiang, W., Ding, L.: Clustering by searching density peaks via local standard deviation. In: Yin, H., et al. (eds.) IDEAL 2017. LNCS, vol. 10585, pp. 295–305. Springer, Cham (2017). https://doi.org/10.1007/978-3-319-68935-7_33

21. Xie, J., Li, Y., Zhou, Y., Wang, M.: Differential feature recognition of breast cancer patients based on minimum spanning tree clustering and f-statistics. In: Yin, X., Geller, J., Li, Y., Zhou, R., Wang, H., Zhang, Y. (eds.) HIS 2016. LNCS, vol. 10038, pp. 194–204. Springer, Cham (2016). https://doi.org/10.1007/978-3-319-48335-1_21

22. Xie, J., Liu, R., Luttrell, J., Zhang, C.: Deep learning based analysis of histopathological images of breast cancer. Front. Genetics 10, 80 (2019)

23. Xie, J., Qu, Y., Wang, M.: Unsupervised feature selection algorithms based on density peaks. J. Nanjing Univ. (Nat. Sci.) 52(4), 735–745 (2016)

24. Xie, J., Wang, M., Zhou, Y., Gao, H., Xu, S.: Differentially expressed gene selection algorithms for unbalanced gene datasets. Chin. J. Comput. 42(6), 1232–1251 (2019)

25. Ye, X., Wang, S.: Comparative study on the performances of Bayesian classification and LVQ neural network. Comput. Inform. Technol. 21(4), 14–17 (2013)

26. Zhang, C., Wei, S., Hu, X., et al.: Research and application of lvq neural network based on particle swarm optimization algorithm. J. Guizhou Univ. (Nat. Sci.) 30(5), 95–99 (2013)

27. Zhang, Y., Wu, C., Zhang, M.: The epidemic and characteristics of female breast cancer in china. China Oncol. 23(8), 561–569 (2013)

28. Zhang, Y., Shi, H., Shang, W., Xiaofeng, J.X.Z.: Improved method for computer-aided diagnosis of breast cancer based on support vector machines. Appl. Res. Comput. 30(8), 2373–2376 (2013)

29. Zheng, C., Hong, W., Wang, J.: Rule extraction method of breast cancer diagnosis based on partial orderd structure diagram. Comput. Eng. Des. 37(6), 1599–1603 (2016)

Left Ventricle Segmentation Using Scale-Independent Multi-Gate UNET in MRI Images

Mina Saber[1(✉)], Dina Abdelrauof[1(✉)], and Mustafa Elattar[2(✉)]

[1] Research and Development Division, Intixel Co. S.A.E., Cairo, Egypt
{mina.saber,dina.abdelrauof}@intixel.com
[2] Medical Imaging and Image Processing Group, Information Technology and Computer Science School, Nile University, Giza, Egypt
melattar@nu.edu.eg

Abstract. Left ventricle (LV) segmentation is crucial to assess left ventricle global function. U-Net; a Convolutional Neural Network (CNN); boosted the performance of many biomedical image segmentation tasks. In LV segmentation, U-Net suffered from accurately extracting small objects such as the apical short-axis slices. In this paper, we propose a fully automated left ventricle segmentation method for both short-axis and long-axis views. The proposed model utilizes U-Net architecture and Multi-Gate input block to enhance the performance by aggregating multi-scale features and adding different vision scopes providing a more robust model against scale variance of objects within the images. The proposed approach was validated against left ventricle segmentation challenge (LVSC) and Automated cardiac diagnosis challenge (ACDC). For LV myocardium segmentation, the proposed approach achieved mean dice index 82% on LVSC and 90% and 91% for end-diastole (ED) and end-systole (ES) time frames respectively on ACDC outperforming other published methods measurements at ED. For LV blood pool segmentation, mean dice index was 97% and 92.3% for the ED and ES time frames using ACDC, outperforming other methods' ED measurements.

Keywords: Left ventricle segmentation · U-Net · Inception · Magnetic resonance imaging · Segmentation

1 Introduction

Correct evaluation of the Left Ventricle (LV) function and mass is essential for diagnostic, therapeutic, and prognostic evaluation and risk assessment of patients with CAD [1]. Cardiac magnetic resonance (CMR) Steady-State Free Precision (SSFP) is regarded to be the gold standard modality for estimating global LV function because it is noninvasive and produces high temporal resolution [1,2]. Manual delineation of the LV is a time-consuming task due to the large number of images per patient. Therefore, fast, accurate, and fully automated LV segmentation is needed. Classical segmentation techniques have been

© Springer Nature Switzerland AG 2020
H. Fujita et al. (Eds.): IEA/AIE 2020, LNAI 12144, pp. 470–477, 2020.
https://doi.org/10.1007/978-3-030-55789-8_41

proposed such as multi-level Otsu thresholding [3], deformable models and level sets [4], graph cuts, and model-based approaches such as active and appearance shape models [5]. Although, those methods have achieved reasonable performance on some benchmark LV datasets, they weren't the preferred solution as they were not robust and had limited capacity to generalize over subjects with heart conditions outside the collected training set. Recently, fully convolutional encoder-decoder neural networks are gaining popularity in the medical image segmentation problems. U-Net; proposed by Ronneberger et al.; is an extension for encoder-decoder famous architecture [6]. The U-Net encoding stage is meant to obtain hierarchical features at different scales, however; in several studies; it was not capable of properly extracting the low-level features in the short-axis apical slices or handle the shape variability between short-axis and long-axis views. In this work, we introduce a new FCN-based encoder that incorporates low-level features together with multi-scale features extracted from our Multi-Gate block to refine the extracted features in the encoding path which in turn enhances the overall segmentation output. This method is capable of segmenting all short-axis view slices (apical, mid, and basal slices) as well as long-axis view images.

2 Related Work

A significant number of approaches have been proposed in the lines of automatic LV segmentation. Tran et al. have proposed fully convolutional neural network architecture for semantic segmentation in cardiac MRI however this approach was unable to perform well with segmenting the apical slices, as the dice index drops by 20% when moving from basal to apical slices [7]. Khened et al. proposed fully convolutional multi-scale residual dense nets; a variant of U-Net with dense and residual connections; to train the region of interest ROIs [8]. Those ROIs were extracted using Fourier analysis alongside Hough transform making use of the fact that a heart in the short-axis view is almost circular. Also, Li Kuo et al. detected the center of the heart then calculated discrete Fourier Transform of the temporal information along the cardiac cycle to guide segmentation through different time frames [9]. This approach was limited to CINE images where all time frames are available and will not perform properly on single or few time-frames such as ES and ED only. Isensee et al. have utilized the 3D U-Net by ensembling it with conventional 2D U-Net however, such a large number of parameters may lead to poor generalization [10]. Most of the mentioned techniques have common limitations such as being trained and validated only on the short-axis view, sub-optimal performance in segmenting apical slices and segmenting the papillary muscles as a part of the left ventricle. In this study, we introduce a light-weight fully automated left ventricle segmentation technique that handles long-axis and short-axis views. The proposed architecture can be considered as an extension for the U-Net architecture that utilizes the power of combining Multi-Gate Input block and dilated convolution in the encoding path to aggregate the most representative features from the input image without assuming any prior information.

3   Methods

3.1   Data

In this work, we utilized two datasets. The first one is the LVSC dataset [11] that has been released in 2011. LVSC consists of 100 annotated sets of SSFP cardiac MRI; including short-axis and long-axis views of the left ventricle; for patients with coronary artery disease and prior myocardial infarction. The second dataset is the ACDC dataset [12] which consists of 150 partially-annotated short-axis scans. The acquisition for the two datasets was done using two MR scanners of different magnetic strengths (1.5 T and 3.0 T). The two datasets feature non-isotropic image resolutions ranging from 192×256 to 512×512. In the context of this study, our model was trained on the ACDC training dataset (100 patients) and evaluated on the testing dataset (50 patients). To prove the generalization ability of the proposed approach, the LVSC training dataset was used for training the model with split-ratio of 60%, 20%, and 20% for the training, validation and testing respectively including both long-axis and short-axis samples. This split was done on patient level.

3.2   Network Architecture

Features scale has a significant impact on the overall performance of deep networks. The scale of the extracted features mainly depends on the convolution kernel size. Also increasing the kernel size increases the effective field of view and enhances the network performance [13]. However, usage of large receptive fields in the convolution layer will dramatically increase the number of parameters which in turn increases the probability of over-fitting on small datasets [14]. Dilated convolution is a special convolutional operation which is capable of enlarging receptive fields without increasing too many parameters. The proposed architecture; shown in Fig. 1; modifies the U-Net encoding path to include the multi-gate input block. The multi-gate block consists of two main components; multi-scale inception block as shown in Fig. 2 to employ the dilated convolution in extracting multi-scale features from the input image. The average pooling layer is to provide each level in the encoding path with a scaled version of the input capturing the full context of the image. The multi-scale inception block encapsulates four dilated convolution layers with different receptive fields and dilation factors in parallel paths. The resulting feature maps are then merged and propagated through 1×1 convolution to reduce the resulting number of maps.

In Fig. 1, The standard block consists of two successive 3×3 convolutions where each layer is followed by batch normalization and activation layer. Exponential Linear Unit was used as an activation function in the proposed architecture to speed up the convergence and stabilize the training profile [15]. The number of the standard blocks increases at each level of the encoding path to increase the context captured by increasing the receptive field of view for a given input.

3.3   Training

Adam algorithm [16] was used to optimized the cross entropy loss function
with initial learning rate = 0.001 that decay by a factor of 0.5 in case of the

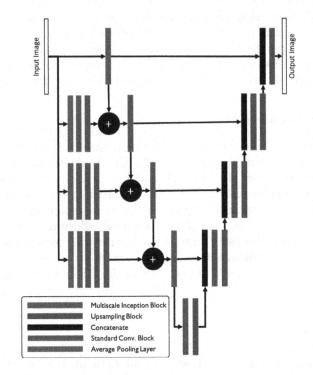

Fig. 1. A schematic diagram showing the proposed Multi-Gate U-Net architecture

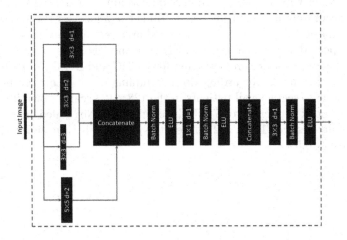

Fig. 2. A schematic diagram showing the Multi scale inception block architecture

validation loss didn't improve within five epochs. Training used a Minibatch size of 28 images. To reduce training and inference time, the input images were down sampled to 256×256 after zero-padding non-square images. Also, image normalization was applied using min max normalization. To enhance the overall performance and the generalization ability of the LV segmentation, we augmented the training dataset using different data augmentation options such as flipping left and right, random rotation, random brightness adjustment, and scaling the image. As a post processing step to eliminate any mis-segmented objects in the images, we take the average mask over the SA slices per patient, then we choose the closest object to that average mask excluding the remaining ones if any.

3.4   Evaluation Metrics

For LVSC we didn't have access to the testing data, so we evaluated the performance of our architecture against the native U-Net architecture. Regarding the ACDC dataset we report the Mean Dice and some clinical measurements that were obtained using the challenge website. In order to test our model against scale variation, especially small objects, we made an experiment where we take the input to our network and downsize it, then we padded the remaining area to 256×256 with zeros and feed this image to the network and calculate the same measurements. This experiment was repeated twice for our network and U-Net. The downsampling was 128×128 i.e. 25% of the original area and 192×192, i.e. 56%.

4   Results

Annotation quality of LVSC dataset was sub-optimal; however, the proposed architecture was able segment LV achieving Dice index of 82% outperforming the native U-Net with 1%. Also, the method was able to extract LV in the long-axis view accurately as shown in Fig. 4. Table 1 summarizes the results of our method on the ACDC test-set. The proposed architecture performs better than the other reported methods in terms of ED Dice and ED Mass bias. Visual result samples for success and failure cases and for ACDC and LVSC are reported and shown in Figs. 3 and 4. Regarding down sampling input size experiment, our model was able to segment all patients in the 128×128 scaled down images with only a dice index reduction of 7.5%. In contrast, U-Net failed in segmenting majority of images. For 192×192 images, the dice in this case slightly degraded with average of 0.5%.

Table 1. Measurements using ACDC dataset. The first four columns represent Dice measures. EF: Ejection Fraction, EDV: End Diastolic Volume, ED: End Diastole, corr: correlation, BP: Blood Pool, MYO: Myocardium

Methods	ES BP	ED BP	ES MYO	ED MYO	EDV corr	EDV bias	ED Mass corr	ED Mass bias
Proposed	0.91	0.97	0.91	0.90	0.996	−1.24	0.988	1.34
Isensee	0.933	0.965	0.919	0.34	0.997	1.59	0.986	4.05
Zotti	0.91	0.96	0.90	0.88	0.997	3.75	0.986	−1.83
Painchaud	0.91	0.96	0.89	0.88	0.997	3.82	0.987	−2.91
Khened	0.92	0.96	0.90	0.89	0.997	0.58	0.991	−2.87

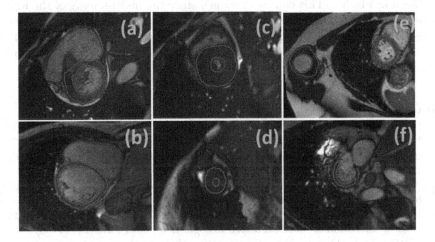

Fig. 3. Sample of segmentation results from ACDC dataset. (a–d) success test samples, (e, f) failure cases show the main limitations of the Multi-Gate U-Net

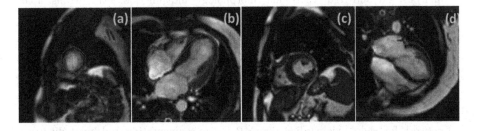

Fig. 4. Sample segmentation results from LVSC dataset. Red contours are the segmentation output and the green ones are the ground truth (Color figure online)

5   Discussion and Conclusion

In this paper, we introduce another component to the native U-Net architecture by integrating the Multi-Gate block in the encoding path. The introduced block mainly utilizes the power of multi-scale features aggregation and increase context awareness. Our model outperforms the U-Net in terms of dice index and other clinical measurements. From Figs. 3 and 4, we can conclude that the Multi-Gate Network is robust against small objects like apical slices in short-axis view and also able to compensate the modest quality of annotations. Table 1 shows reliability of the model in terms of clinical measurements while tested on ACDC test samples. Additionally, the method shows great robustness against input resolution reduction. These results support the applicability of the model for real-life clinical use. However, there are some limitations associated with the proposed architecture such as identifying more than one LV in a single image as shown in Fig. 3-e. This can be explained since the multi-scaling of the input; through average pooling; gives much larger context to the network, a blurry image, without fine details that makes the distinction slightly difficult. This problem appears only in one slice not all slices so it was successfully removed during post-processing.

References

1. Hazirolan, T., et al.: Comparison of short and long axis methods in cardiac MR imaging and echocardiography for left ventricular function. Diagn. Interv. Radiol. (Ankara, Turkey) **13**(1), 33–38 (2007)
2. Nabel, E.G., Braunwald, E.: A tale of coronary artery disease and myocardial infarction. N. Engl. J. Med. **366**(1), 54–63 (2012)
3. Huang, D.-Y., Wang, C.-H.: Optimal multi-level thresholding using a two-stage Otsu optimization approach. Pattern Recogn. Lett. **30**(3), 275–284 (2009)
4. Alattar, M.A., Osman, N.F., Fahmy, A.S.: Myocardial segmentation using constrained multi-seeded region growing. In: Campilho, A., Kamel, M. (eds.) ICIAR 2010. LNCS, vol. 6112, pp. 89–98. Springer, Heidelberg (2010). https://doi.org/10.1007/978-3-642-13775-4_10
5. Zhang, H., Wahle, A., Johnson, R., Scholz, T., Sonka, M.: 4-D cardiac MR image analysis: left and right ventricular morphology and function. IEEE Trans. Med. Imaging **29**(2), 350–364 (2010)
6. Ronneberger, O., Fischer, P., Brox, T.: U-Net: convolutional networks for biomedical image segmentation. In: Navab, N., Hornegger, J., Wells, W.M., Frangi, A.F. (eds.) MICCAI 2015. LNCS, vol. 9351, pp. 234–241. Springer, Cham (2015). https://doi.org/10.1007/978-3-319-24574-4_28
7. Tran, P.V.: A fully convolutional neural network for cardiac segmentation inshort-axis MRI. arXiv preprint arXiv:1604.00494 (2016)
8. Khened, M., Alex, V., Krishnamurthi, G.: Densely connected fully convolutional network for short-axis cardiac cine MR image segmentation and heart diagnosis using random forest. In: Pop, M., et al. (eds.) STACOM 2017. LNCS, vol. 10663, pp. 140–151. Springer, Cham (2018). https://doi.org/10.1007/978-3-319-75541-0_15

9. Tan, L.K., Liew, Y.M., Lim, E., McLaughlin, R.A.: Convolutional neural network regression for short-axis left ventricle segmentation in cardiac cine MR sequences. Med. Image Anal. **39**, 78–86 (2017)
10. Isensee, F., Jaeger, P.F., Full, P.M., Wolf, I., Engelhardt, S., Maier-Hein, K.H.: Automatic cardiac disease assessment on cine-MRI via time-series segmentation and domain specific features. In: Pop, M., et al. (eds.) STACOM 2017. LNCS, vol. 10663, pp. 120–129. Springer, Cham (2018). https://doi.org/10.1007/978-3-319-75541-0_13
11. Suinesiaputra, A., et al.: Left ventricular segmentation challenge from cardiac MRI: a collation study. In: Camara, O., Konukoglu, E., Pop, M., Rhode, K., Sermesant, M., Young, A. (eds.) STACOM 2011. LNCS, vol. 7085, pp. 88–97. Springer, Heidelberg (2012). https://doi.org/10.1007/978-3-642-28326-0_9
12. Bernard, O., Lalande, A., Zotti, C., Cervenansky, F., Humbert, O., Jodoin, P.-M.: Deep learning techniques for automatic MRI cardiac multi-structures segmentation and diagnosis: is the problem solved? IEEE Trans. Med. Imaging **37**(11), 2514–2525 (2018)
13. Peng, C., Zhang, X., Yu, G., Luo, G., Sun, J.: Large kernel matters–improve semantic segmentation by global convolutional network. In: Proceedings of the IEEE Conference on Computer Vision and Pattern Recognition, pp. 4353–4361 (2017)
14. Yu, F., Koltun, V.: Multi-scale context aggregation by dilated convolutions. arXiv preprint arXiv:1511.07122 (2015)
15. Clevert, D.-A., Unterthiner, T., Hochreiter, S.: Fast and accurate deep network learning by exponential linear units (ELUs). arXiv preprint arXiv:1511.07289 (2015)
16. Kingma, D.P., Ba, J.: Adam: a method for stochastic optimization. arXiv preprint arXiv:1412.6980 (2014)

Clustering-Based Data Reduction Approach to Speed up SVM in Classification and Regression Tasks

Adamo Santana[✉], Souta Inoue, Kenya Murakami, Tatsuya Iizaka, and Tetsuro Matsui

Fuji Electric, Co. Ltd., 1, Hino, Fujimachi, Tokyo 191-0064, Japan
{adamo-santana,inoue-souta,murakami-kenya,iizaka-tatsuya,
matsui-tetsuro}@fujielectric.com

Abstract. Support Vector Machine (SVM) is a popular machine learning algorithm, being able to tackle non-linear problem by use of appropriate kernels. However, the use of SVM can become unfeasible for many applications where relatively large datasets are used. Its application becomes particularly more prohibitive when considering environments where the hardware has strict memory and processing limitations. By appropriately reducing the size of the training set we can in succession speed up the learning and diagnosis of SVM. In this work we implement a data reduction approach using clustering to build a smaller and representative set. The approach is extended for both classification and regression problems. Results evaluated on both normal PC and low resource edge device showed a better performance with only a small loss in diagnosis accuracy for most cases. Still, on cases where a high loss was observed, the reduction approach allowed to regain the accuracy with a faster hyper-parameter optimization.

Keywords: Diagnosis · Data reduction · Clustering · SVM

1 Introduction

Support Vector Machine (SVM) is a popular machine learning (ML) algorithm in the area of diagnosis, being able to handle complex non-linear patterns, as found in real world applications, by using kernels (e.g. Gaussian). However, the use of SVM can become unfeasible for many applications where large datasets are used, since, among other things [1]:

- The optimization problem for training the model is quadratic with respect to the number of training samples;
- The decision (i.e. inference) cost grows linearly with the number of elected support vectors (SVs);
- The accuracy of the model, as it is for most ML algorithms, depends on the proper definition/adjustment of its hyper-parameters, particularly C and γ.

© Springer Nature Switzerland AG 2020
H. Fujita et al. (Eds.): IEA/AIE 2020, LNAI 12144, pp. 478–488, 2020.
https://doi.org/10.1007/978-3-030-55789-8_42

With the computational performance becoming impractical for very large datasets, and the quality of the results depending on fine-tuning hyper-parameters, which also has a very high computational cost, strategies to improve the general performance and applicability of SVM become necessary in scenarios where low computational resources are available, or when the diagnosis time is critical.

In this work we apply a clustering approach towards improving the training, diagnosis and hyper-parameter optimization time of SVM, with applicability for both regression and classification tasks. In addition to a performance evaluation on a standard desktop PC hardware, we intend to direct the focus of this application towards more constrained hardware scenarios, such as the case of low resource (i.e. memory and processing power) edge devices and/or programing controllers (PLC) that can be set in IoT environments; as such, the experiments are also applied to a low resource system with capabilities of a Fit SA Σ M4 IoT/M2M controller.

For SVM, the knowledge base that helps in better discriminating the patters is represented by its support vectors (SV), a set of samples estimated to be near the decision border of the hyperplane derived by the input feature set. While the support vector is crucial for the SVM's estimation, it is often the case that many of the points in the SV can be pruned, allowing to speed up the SVM inference while still retain its representation competence. Here, the clustering is used as a data reduction (DR) step, in order to reduce the number of training samples and, consequentially, the number of SV. The implemented approach uses the final centroids (i.e. matrix containing the centers of the clusters) as representative points, thus reducing the training process and removing points with or low contribution or considered unnecessary (due to a high similarity) to build the SV set.

The remainder of this paper is organized as follows: Sect. 2 presents related works in the objective of optimizing SVM training and/or inference. Section 3 details the implemented clustering-based approach for data reduction and speed up SVM. Experiments with different datasets for classification and regression problems are presented in Sect. 4, evaluating the results of the DR approach. Section 5 presents the final remarks of the paper.

2 Related Works

In the past years, various approaches have been proposed for improving the efficient of SVM for large datasets, as it can be impractical for SVM to handle them. The majority of which consider its offline optimization, while others yet still consider a similar setting, but for large scale online systems, where a stream of samples is continuously provided.

Looking into the more recent literature, three main groups of optimizations approaches can be seen for SVM:

- Optimize by decreasing the size of the training dataset (pre-processing);
- Optimize by decreasing the size of the parameter set (post-processing);
- Optimize the solver method for SVM (training/inference).

Practices presented for the first and second type concentrate on data reduction. Examples of such methods are: clustering [2–11], feature selection optimization [12, 13], sampling approaches (e.g. random, ranked, heuristic, nearest neighbor based) [5, 8, 14, 15].

For the third type, which focus on methods for faster SVM calculation by mathematical approximations, methods are also similarly varied; such as: stochastic gradient solvers [14, 16], Frank-Wolfe optimization [17], truncated Newton optimization [18], truncated Sequential Minimization Optimization [19], augmented Lagrange multipliers [20, 21], Mixed-Integer Linear Programming optimization [22].

In this work, we follow the practice from the first two groups, using clustering as methodology for data reduction. The use of clustering to improve the efficiency of SVM is an approach that we can also find being proposed by different works in recent literature. While they mainly focus on clustering application and decreasing samples with strong similarity to their respective cluster's centers, mixed approaches are often used; some of the algorithms that can be found are: k-means, which is the most employed clustering method [2, 3, 6, 8–11], fuzzy C-means [4], hierarchical clustering [5], random clustering [6], minimum enclosing ball [7].

With the initial goal of improving the performance for processing and memory use, we will also use k-means (KM) as the main algorithm for clustering. As our main application task is that of diagnosis, the goal is that of discriminating different behaviors in a system, whether it is by identifying a discrete class of fault (namely, as a classification problem), of estimating an output value that happens to exceed defined control limits (as a regression problem). For simplicity, we will take the example of the discrimination between normal and abnormal conditions when making future reference to the diagnosis task.

In our implementation, the clustering process is separately applied for each different class in the data (for regression, a distinction as inside or outside the control limits can also be used, if known a priori), which differentiates from the works of [7–9] that apply a joint clustering with all samples. Also, while independent clustering flows are executed for each class, instead of training local SVM models for each cluster [6, 10], only one global SVM estimator is trained with the joined output from all the clustering processes.

Also differentiating from the work of [11], we used only the reduced clustering results to train the SVM model, and extending the application to both classification and regression problems.

3   Support Vector Machine Speed up with Clustering-Based Data Reduction

Considering a standard procedure for ML/data mining application over a dataset, as shown in Fig. 1, the data reduction approach is to be applied to the Train data before building the SVM model (Step 3 – TrM); this way supporting the hyper-parameter optimization (POpt) step in tuning the parameters in a shorter time. For the SVM computation, the LIBSVM library will be used.

Since the new reduced set is composed solely by the centers of the final clusters, a reduction ratio r parameter is used to define how much the dataset should be reduced, which sets the number of clusters in the KM algorithm.

With the clustering application being responsible to tackle the large dataset and reduce it to a smaller number of samples (SV candidates), depending on the dataset's size, even a not too large number of clusters could require a considerable time. For this reason, when accounting for its application in an IoT environment, the clustering step is applied offline, in a more robust hardware. Following the offline (i.e. on a normal PC environment) DR application, two approaches can be implemented regarding the operations on the edge devices:

- Be responsible only for online diagnosis (or the test step – TeM – in a batch process), which benefits from the DR as the smaller number of SV is to be stored in memory, and would also speed up the diagnosis;
- Not only speed up diagnosis, but also train the SVM model, and update the SV set with new samples when necessary. Since the training time also drastically decrease with the reduced SV set, it is also made feasible to retrain the SVM in the border devices. For this case, samples identified (by the centralized center) with new patterns, that can significantly improve the estimation, can be added to the dataset in memory and used to retrain the SVM model.

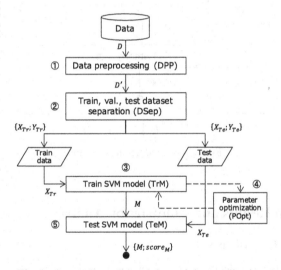

Fig. 1. Steps of a traditional data mining application.

The DR procedure is shown in Fig. 2. The procedure follows depending on the task T the dataset was designed for, whether regression or classification (Step 1). Step 2 branches depending on the task: for a classification task, the list of labels L ($L = \{L_1, L_2, \cdots, L_k\}$) is obtained from the output variable (Step C.1); the input set X^{in} is then divided into k sets, each set containing the samples of a different label (Step C.2). The clustering process is then applied separately on each of the subsets.

With a task of regression, if control limits or threshold values are given for the output variable, regression labels L are created (Step R.1), and the clustering is applied similarly to the classification approach in Step C.2; in this case the input and output variables are concatenated to form X^{in} ($X^{in} \leftarrow [X_{Tr}; Y_{Tr}]$), after the clustering, the reduced set is split, so that the reduced input set (X_{Tr}^R) and the respective regression output variable (Y_{Tr}^R) can be obtained. In case no control limits is known a priori, the clustering is applied over a single set (Step R.2).

After the clustering process is completed, as previously mentioned, the reduced set $\{X_{Tr}^R; Y_{Tr}^R\}$ is built from the centers of the final clusters (Step 4).

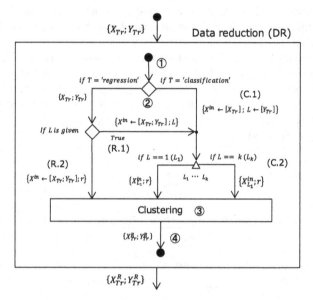

Fig. 2. Data reduction procedure.

4 Simulation Results and Comparison Analysis

4.1 Simulation Conditions

In the experiments, we selected some of the most used benchmark (BMark) datasets. We also apply the approach to different proprietary cases, in order to similarly evaluate the results on other real world application (RPvt) datasets for the task of diagnosis. All classification datasets have binary labels, and for the regression datasets we assume that no information on control limits is given to create new labels.

For a better comparison, we refrained from using datasets which are known to be unfeasible for standard SVM training (or have an extremely long training time), as the advantage of the DR in processing time would already be expected, and the quality of the diagnosis results would be difficult to compare. A total of 15 datasets were used (10 for classification and 5 for regression), details are shown in Table 1; BMark datasets can be obtained at [23, 24], while only assigned labels will be shown for RPvt ones, due to privacy reasons.

To evaluate the performance of the DR approach (DR-SVM) against the standard SVM application, we show results comparing the processing time for training the model, and also the diagnosis (testing). We run the experiments on two architectures, representing the specifications of a normal desktop computer (PC), and that of a PLC/edge device (Edge); the hardware and software details are as follows:

- PC – CPU: Intel 3.9 GHz (4 cores); Memory: 64 GB; OS: Windows 10;
- Edge – CPU: Arm Cortex-A7 1.0 GHz (2 cores); Memory: 1 GB; OS: Linux.

Besides the computational performance, the quality of the estimation is also evaluated, in order to compare the loss/gain from the data reduction. For classification, the F1-score is calculated; and for regression, the root mean square error (RMSE) is used as metric.

Table 1. Dataset details.

Dataset	Type	Task	Size (samples x variables)		
			Complete	Train set	Test set
Adult census (CBM01)	BMark	Classif.	48842×122	34189×122	14653×122
IJCNN (CBM02)	BMark	Classif.	126701×22	88690×22	38011×22
W8A (CBM03)	BMark	Classif.	64700×300	45290×300	19410×300
Satellite (CBM04)	BMark	Classif.	6435×36	4504×36	1931×36
Shuttle (CBM05)	BMark	Classif.	58000×9	40600×9	17400×9
CRP01	RPvt	Classif.	437703×29	306392×29	131311×29
CRP02	RPvt	Classif.	378714×21	265099×21	113615×21
CRP03	RPvt	Classif.	289440×12	202608×12	86832×12
CRP04	RPvt	Classif.	115200×21	80640×21	34560×21
CRP05	RPvt	Classif.	203040×11	142128×11	60912×11
Online video characteristics and transcoding time (RBM01)	BMark	Reg.	68784×20	48148×20	20636×20
Facebook comment volume (RBM02)	BMark	Reg.	50993×53	35695×53	15298×53
Combined cycle power plant (RBM03)	BMark	Reg.	9568×4	6697×4	2871×4
SGEMM GPU kernel performance (RBM04)	BMark	Reg.	241600×14	169120×14	72480×14
RRP01	RPvt	Reg.	4662×18	3263×18	1399×18

For the ML application, most conditions and parameters were set to standard values to use on all the experiments (Table 2). The value of the reduction ratio however, is set to vary depending on the dataset: r is set to 10% of the number of SV (n_{SV}) obtained when using the standard SVM (S-SVM) over the dataset. Gamma (γ) is calculated from (1), aiming to account for the scale variance across the input variables X.

We note that the value of r is set here to 10% of n_{SV} only to highlight the contrast in n_{SV} used by both approaches, and the representation capacity of the DR's support vectors. In practice, r can be defined empirically considering the desired tradeoff of expected accuracy and processing time, especially for datasets intractable for S- SVM.

$$\gamma = 1/(n_X \times var(X)) \tag{1}$$

Where n_X is the number of samples from the input feature set X, and $var(X)$ is the overall variance calculated from X.

Additionally, and analysis was made to estimate the performance, and the impact in the results, when also applying an optimization of the model's hyper-parameters (POpt) C and γ. The POpt was made using a grid search over 100 different combinations of the hyper-parameters, and cross-validated using 5 folds; the values for C and γ were logarithmically spaced in the ranges $\left[10^{-1}, 10^{3}\right]$ and $\left[10^{-5}, 10^{1}\right]$, respectively.

Table 2. Parameters used in the experiments.

Parameter	Values used
Train/Test set division ratio	70%/30%
Kernel (SVM)	RBF
C (SVM)	1.0
γ (SVM)	Equation (1)
r (DR)	10% n_{SV}

4.2 Experiments

Tables 3 and 5 show the performance results (in seconds) when running the methods on the PC and Edge based architectures, respectively. In cases where the analysis time took several hours and/or were impractical are represented here with the symbol "$\gg$". The diagnosis scores are presented in Table 4, comparing the F1-score (for classification) or RMSE (for regression) when using the S-SVM and the DR-SVM approaches.

Looking at the computational performance alone, it is clear that that DR step can take a considerable time for many datasets. For the SVM application to benefit from the DR approach, especially as its use is intended for devices with limited processing/storage capabilities, the clustering processes is done offline. Setting aside the DR time, we see the evident speed up obtained when looking at the online training and diagnosis time of the DR-SVM (both PC and Edge).

The speed up in training is a natural consequence of the reduced training set, which, in turn, would lead to a smaller number of SV and with it also bring a faster diagnosis. Likewise, being an iterative sequence of training and testing events, the POpt time is also much faster after the DR, whereas it was impractical for the S-SVM in almost all cases, as seen in Table 3.

Table 3. Computational performance results (sec.) from PC experiments.

Dataset	S-SVM			DR-SVM				POpt search time	
	Training time	Num. of SVs	Testing time	DR time	Training time	Num. of SVs	Testing time	S-SVM	DR-SVM
CBM01	124.865	13199	25.7131	59.0677	0.6597	1522	3.0743	≫	179.098
CBM02	41.8403	7086	8.1113	70.4681	0.0437	636	0.7127	≫	14.9530
CBM03	144.407	3536	22.5417	53.7642	0.0833	320	2.0496	≫	29.3932
CBM04	0.3853	1361	0.1095	0.3813	0.0035	176	0.0178	160.968	1.2746
CBM05	0.5744	367	0.1310	0.8388	0.0014	65	0.0342	358.752	0.5280
CRP01	19.2971	661	3.0058	46.0776	0.0182	130	0.6988	≫	0.6278
CRP02	20.9777	480	1.5293	30.2262	0.0038	95	0.3565	≫	0.5357
CRP03	21.7615	2896	4.8088	99.2477	0.0087	190	0.3409	≫	1.9987
CRP04	0.6737	129	0.1371	1.1339	0.0010	24	0.0431	≫	0.5204
CRP05	79.1667	2681	3.2034	57.9358	0.0114	337	0.4109	≫	1.8584
RBM01	100.772	42742	22.9535	102.022	0.7603	4127	2.1649	≫	515.374
RBM02	82.2185	28695	24.9093	120.365	0.5971	2833	2.9055	≫	187.374
RBM03	1.2851	6552	0.2642	1.7118	0.0215	645	0.0306	807.894	7.0752
RBM04	1167.25	168668	257.88	1802.07	3.8342	10107	15.5388	≫	1273.13
RRP01	0.3419	2129	0.0796	1.5988	0.0035	181	0.0089	195.250	1.1446

Table 4. Diagnosis results with F1-score (higher is better) and RMSE (lower is better).

Dataset	Score	S-SVM	S-SVM + POpt	DR-SVM	DR-SVM + POpt
CBM01	F1-score	0.77	–	0.74	0.68
CBM02		0.96	–	0.83	0.85
CBM03		0.88	–	0.85	0.58
CBM04		0.83	0.89	*0.68*	0.82
CBM05		0.99	0.99	*0.53*	0.99
CRP01		1.00	–	*0.49*	1.00
CRP02		1.00	–	0.91	1.00
CRP03		1.00	–	0.98	0.99
CRP04		1.00	–	*0.25*	1.00
CRP05		1.00	–	0.99	0.99
RBM01	RMSE	7.2421	–	9.6030	**3.3754**
RBM02		62.2705	–	**60.8163**	**54.6435**
RBM03		4.0978	3.7318	4.7575	4.2727
RBM04		261.8802	–	*334.4830*	**106.5831**
RRP01		2.3491	1.4896	*5.1460*	2.2785

From Table 4, we see that on many cases the score using DR was similar to the ones using the entire training dataset, even though only a fraction of the dataset (and number of SV) was used.

On cases where the DR-SVM score was noticeably worst (CBM04/05, CRP01/04, RBM04, RRP01, in italic), applying the POpt greatly improved the results, matching the original score, or even surpassing in some cases (RBM01/02/04, in bold). These occurrences could be attributed due to the final clusters' centers being positioned distant from samples on the border set by the initially fixed hyper-parameters; but could then be improved by applying POpt, or alternatively increasing the value of r.

Also, when looking at the results from the Edge experiments (Table 5), we find that in some instances the total learning time for the Edge application (i.e. DR offline and TrM/POpt in the device) can be faster than the TrM of the S-SVM in the PC (CBM01/03, CRP05, with underline). And when considering the TrM of the S-SVM in the Edge, the DR-SVM can remain faster even after the POpt is applied (CBM02/03/05, CRP01-05, with double underline).

Finally, from Table 5, we can ratify that the testing time using the DR approach is improved in all cases (in bold), showing the advantage over the S-SVM (even if all training had being done offline) in having a faster diagnosis per sample; with average speed up of 9x. In an online diagnosis, with the presented settings, the device with DR-SVM would then be able to process in average an additional 40K events per minute.

Table 5. Computational performance results (sec.) from Edge experiments.

Dataset	S-SVM		DR-SVM		POpt Search time		Total DR-SVM learning time	
	Training time	Testing time	Training time	Testing time	S-SVM	DR-SVM	DR + TrM	DR + TrM + POpt
CBM01	964.386	263.163	4.7326	**30.1954**	≫	2960.70	63.8003	3019.77
CBM02	521.879	124.051	0.4522	**9.6693**	≫	248.592	70.9203	319.060
CBM03	806.291	181.462	0.4342	**16.5759**	≫	419.036	54.1984	472.800
CBM04	4.6612	1.3633	0.0322	**0.1813**	2247.87	10.0960	0.4135	10.4774
CBM05	9.0797	1.8795	0.0065	**0.3574**	≫	3.5577	0.8452	4.3964
CRP01	252.353	39.5063	0.0295	**8.1053**	≫	10.4803	46.1071	56.5579
CRP02	288.676	21.5898	0.0087	**4.4591**	≫	4.0570	30.2349	34.2832
CRP03	404.872	81.2745	0.0636	**5.4669**	≫	18.5974	99.3113	117.845
CRP04	9.8558	1.8411	0.0047	**0.3694**	≫	3.1550	1.1385	4.2889
CRP05	1122.09	50.5072	0.0779	**6.5757**	≫	17.2324	58.0137	75.1681
RBM01	1461.19	378.156	10.3666	**35.0786**	≫	≫	112.388	6345.23
RBM02	847.685	301.503	5.9636	**29.2538**	≫	2513.22	126.329	2639.55
RBM03	20.8311	4.5892	0.1993	**0.4522**	≫	63.0560	1.9111	64.9671
RBM04	≫	≫	62.5067	**296.272**	≫	≫	1864.58	≫
RRP01	4.7701	1.1131	0.0268	**0.0969**	2773.94	9.8126	1.6256	11.4382

5 Conclusion

While SVM if generally known for its strengths as a ML algorithm, whether for novelty detection, classification or regression, its application can computationally demanding when using non-linear kernels on large datasets. Such cases can make training time alone unfeasible on standard computers, being even more challenging when considering its use on hardware with lower memory and processing capabilities, and online applications where inference has to be done fast.

With ordinarily being the case that the training on large datasets can result in a similarly large number of SV, not only the training, but also the testing (or online diagnosis) can become impractical, as the time complexity for inference also grows with the number of SV.

In this paper we used clustering for initial reduction of the training set, using its centroids as a representative subset of training points towards estimating the relevant SV. Results from experiments with classification and regression datasets showed that the reduced number of SV provided an evident improvement in the training, diagnosis and hyper-parameter optimization time of the model, without much compromise to the quality of the diagnosis.

Because the number of clusters is still a parameter that must be defined a priori, a procedure for its automatic adjustment is being realized, aiming to estimate a reduction ratio to the dataset that can provide a satisfactory approximation to the original data and minimize the difference in the diagnosis obtained between the complete and the reduced sets. This way, on cases where even a smaller number of clusters could be used, further gains in resource use and in the training/diagnosis time could be obtained.

Additionally, by making use of more specific computing engines, like the Apache Spark, we can improve the handling of the dataset in memory and reduce the clustering processing time for larger datasets.

References

1. Picard, D.: Very Fast Kernel SVM under Budget Constraints. arXiv, arXiv:1701.00167 (2016)
2. Yao, Y., et al.: K-SVM: an effective SVM algorithm based on k-means clustering. J. Comput. **8**(10), 2632–2639 (2013)
3. Wu, Y., Xu, L., Chen, Y., Zhang, X.: Research on voiceprint recognition based on weighted clustering recognition SVM algorithm. In: IEEE Chinese Automation Congress (CAC), pp. 1144–1148 (2017)
4. Wang, L., Li, Q., Sui, M., Xiao, H.: A new method of sample reduction for support vector classification. In: IEEE Asia-Pacific Services Computing Conference (APSCC), pp. 301–304 (2012)
5. Xiao, H., Sun, F., Liang, Y.: Support vector machine algorithm based on kernel hierarchical clustering for multiclass classification. In: IEEE International Conference on Electrical and Control Engineering (ICECE), pp. 2201–2204 (2010)
6. Zantedeschi, V., Emonet, R., Sebban, M.: L^3-SVMs: Landmarks-based Linear Local Support Vectors Machines. arXiv, arXiv:1703.00284 (2017)
7. Li, X., Yu, W.: Fast support vector machine classification for large data sets. Int. J. Comput. Intell. Syst. **7**(2), 197–212 (2014)

8. Wang, J., Wu, X., Zhang, C.: Support vector machines based on K-means clustering for real-time business intelligence systems. Int. J. Bus. Intell. Data Min. **1**(1), 54–64 (2005)

9. Wang, W., Xi, J.: A rapid pattern-recognition method for driving styles using clustering-based support vector machines. In: IEEE American Control Conference (ACC), pp. 5270–5275 (2016)

10. Cheng, H., Tan, P.N., Jin, R.: Efficient algorithm for localized support vector machine. IEEE Trans. Knowl. Data Eng. **22**(4), 537–549 (2010)

11. Chen, J., Zhang, C., Xue, X., Liu, C.L.: Fast instance selection for speeding up support vector machines. Knowl. Based Syst. **45**, 1–7 (2013)

12. Wani, M.A.: Hybrid method for fast SVM training in applications involving large volumes of data. In: IEEE International Conference on Machine Learning and Applications (ICMLA), vol. 2, pp. 491–494 (2013)

13. Rahimi, A., Recht, B.: Random features for large-scale kernel machines. In: Advances in neural information processing systems, pp. 1177–1184 (2008)

14. Melki, G., Kecman, V.: Speeding up online training of L1 support vector machines. In: IEEE SoutheastCon, pp. 1–6 (2016)

15. Qian, J., Root, J., Saligrama, V., Chen, Y.: A Rank-SVM Approach to Anomaly Detection. arXiv, arXiv:1405.0530 (2014)

16. Kecman, V., Melki, G.: Fast online algorithms for support vector machines. In: IEEE SoutheastCon, pp. 1–6 (2016)

17. Frandi, E., Nanculef, R., Gasparo, M.G., Lodi, S., Sartori, C.: Training support vector machines using Frank-Wolfe optimization methods. Int. J. Pattern Recognit. Artif. Intell. **27**(03), 1–40 (2013)

18. Pölsterl, S., Navab, N., Katouzian, A.: An Efficient Training Algorithm for Kernel Survival Support Vector Machines. arXiv, arXiv:1611.07054 (2016)

19. Xi, Y.T., Xu, H., Lee, R., Ramadge, P.J.: Online Kernel SVM for real-time fMRI brain state prediction. In: IEEE International Conference on Acoustics, Speech and Signal Processing (ICASSP), pp. 2040–2043 (2011)

20. Mao, X., Fu, Z., Wu, O., Hu, W.: Fast kernel SVM training via support vector identification. In: IEEE International Conference on Pattern Recognition (ICPR), pp. 1554–1559 (2016)

21. Nie, F., Huang, Y., Wang, X., Huang, H.: New primal SVM solver with linear computational cost for big data classifications. In: International Conference on International Conference on Machine Learning, vol. 32, pp. 505–513 (2014)

22. Fischetti, M.: Fast training of support vector machines with Gaussian kernel. Discrete Optim. **22**, 183–194 (2016)

23. UCI Machine Learning Repository (2020). https://archive.ics.uci.edu/ml/datasets/. Accessed 10 Jan 2020

24. LIBSVM Data (2020). https://www.csie.ntu.edu.tw/~cjlin/libsvmtools/datasets/binary.html. Accessed 10 Jan 2020

AI for Health – Knowledge-Based Generation of Tailor-Made Exercise Plans

Florian Grigoleit[✉], Peter Struss, and Florian Kreuzpointner

Technische Universität München, 85748 Garching bei München, Germany
{florian.grigoleit,kf}@tum.de, struss@in.tum.de

Abstract. There is an increasing awareness of the essential impact of physical exercising on health. Properly planning such activities requires knowledge about prerequisites and impact of different exercises and selecting those that reflect all relevant physical properties of the individual and contextual conditions. Otherwise, exercising may not only be ineffective, but even harmful to health. We developed an app for the generation of tailor-made exercise plans using techniques from knowledge-based configuration and consistency-based diagnosis. The paper focuses on presenting the systematic and formal representation of the domain knowledge and its exploitation by a generic configuration algorithm.

Keywords: Knowledge-based configuration · Health sports · Precision health · Knowledge representation · Constraint-satisfaction

1 Introduction

Physical exercising has become a global trend [10, 11]. Ambitious athletes populate fitness centers, others perform occasional jogging and other exercises to compensate for adverse working conditions, lack of physical activities and overweight, and physiotherapeutic practices help patients to recover from injuries and diseases or to prevent them. There is an increasing risk of cardiovascular diseases, but also of type 2 diabetes. Every year, approximately 600,000 people die because of inactivity, [11]. In all cases, it is essential that the selection of exercises reflects the goals of exercising, the physical abilities and properties of the individual, external conditions, such as available equipment, restrictions due to injuries etc. and finally, the amount of time dedicated to exercising is essential. Otherwise, inappropriate exercises may not only fail to contribute to the objectives, but even create a negative impact, such as overload, injuries, or discouragement. To counteract this problem, it is important to make the knowledge about daily activity, but especially about physical exercise, accessible to everyone.

The planning of exercising requires expert knowledge, which is usually not available to the exercising individual. Existing apps, such as [12, 13, 14] usually offer solely a selection from a set of pre-manufactured plans for certain standard profiles and, hence, do not offer sufficiently individualized solutions. This is important in particular for trainees without experience or health problems.

© Springer Nature Switzerland AG 2020
H. Fujita et al. (Eds.): IEA/AIE 2020, LNAI 12144, pp. 489–501, 2020.
https://doi.org/10.1007/978-3-030-55789-8_43

Therefore, we developed a knowledge-based solution to exercise planning. Especially for untrained people, such a personalized plan is essential to avoiding overextension and frustration or even harmful exercises.

A computer-based solution requires a systematic representation of domain knowledge and a general configuration algorithm, as described in [2, 3], for generating exercise plans based on this knowledge and the individual goals and conditions. This is what we present in this paper.

2 Health-Oriented Exercising

For persons intending to exercise, selecting a proper combination of exercises from a very large set is difficult to impossible, since too many aspects have to be taken into account, [15]. Often weight-loss is the goal. Another common motivation for exercising is to increase their general fitness. This includes the cardiovascular system as well as the musculoskeletal system. Different users pursue different sub-goals in this goal, such as the body region to be trained. While men tend to train the upper extremities, women tend to train the abdomen and buttocks. Selecting appropriate exercises requires consideration of the actual fitness of the person, because the training plan can only be of use if the stimuli lead to a positive adaptation. The trainee may be tempted to perform many exercises that seem to be related to his intentions, but as a result, running the risk of negative consequences, such as overloading of the musculoskeletal system, is too high. Finally, on top of all of this, particular conditions and preferences of the individual have to be taken into account, such as availability of equipment (weights, devices in a Fitness Studio), preference of outdoor activity, avoidance of overload as well as boring repetitions (which may not be an issue for ambitious athletes in the Fitness center, but for couch potatoes who want to earn some points from their health insurance and need to be motivated in order to continue.

Let us consider an introductory example in order to highlight the challenges of exercise planning. Imagine an elderly man with overweight who is mainly seated while working, who has never exercised, but now intends to do so in order to *lose weight*. The person may decide to exercise *Jogging* and *Pushups*. However, this is inadvisable and potentially harmful. First, it is unlikely that the Trainee would reach his goals with that selection. Second, both the fact that he is *untrained* and his working position *sitting* require easy exercises that strengthen the core (abdominal muscles and lower back). *Pushup* would be appropriate if the same person had a higher fitness level. *Jogging* is a very effective exercise for weight loss, but not advisable for an overweight beginner, because it is stressful for the knee joints.

Furthermore, the general goal of *weight loss* comprises several subgoals, for instance increasing the *strength of the lower back* muscles (even more so, for persons with working position *sitting*, which need to be covered by different exercises. This kind of knowledge is not easily available to an ordinary user. The composition of a set of exercises that reflects various subgoals is further complicated by the fact that the relationship between goals and exercises is not 1:1 but m:n:

- usually, achieving a goal requires several exercises which can vary in their amount of contribution (for instance, *Strength_LowerBack* needs two or three exercises out of {*back extension, dead lift, rowing with free weight, …*}).
- exercises typically have an impact on a several (sub)goals (for instance *rowing with free weights* contributes to *lower back*, *delta muscles* and to *triceps*. Hence, exercise planning becomes a combinatorial and optimization problem, given the usual limitations on time that can be dedicated to exercising and the fact that subgoals may vary in their importance under different circumstances.

3 The Exercising Knowledge Base

3.1 Structuring Knowledge About Exercise Planning

As illustrated by the example in Sect. 2, the creation of an exercise plan has to reflect three conceptual areas and their interrelationships:

- The characterization of the **exercising person** and the conditions for exercising
- The **objectives** of exercising
- The **exercises** themselves.

This can be seen as a specialization of the general organization of a knowledge base for the generic configuration tool, GECKO (see Sect. 4), that we exploit for exercise planning (see Fig. 1).

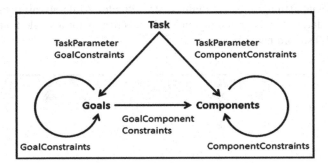

Fig. 1. Basic concepts in GECKO

The exercising knowledge base has to represent knowledge about

- the various **goals** of exercising and their relationships
- the set of possible **exercises** ("components") and their interdependencies
- the **relationship** between **goals** and suitable **exercises** (the GoalComponentConstraints), which is the core knowledge for creating exercise plans
- how to represent the **task** characteristics in terms of personal and contextual conditions that have an impact on the exercise plan
- the impact of the **task** on admissible **goals** (TaskParameterGoalConstraints)
- the influence of the task characterization on the selection of exercises.

In the following, we describe the content of these six partitions of the exercising knowledge base.

3.2 The Goal Structure

There are different categories of goals and different levels of refining them. Most of the goals are not known and not visible to the user; they reflect experts' knowledge, and their interdependencies are introduced automatically. A fitness goal represents an aspect of human physiology, often a body part to be trained with the objective of improving it in a distinct manner. For example, a muscle can be trained to either grow (muscle gain) or to become stronger (strength gain). If the represented body part is sufficiently trained by the selected exercises, the intended improvement will occur, and the goal is considered to be achieved. To free trainees from acquiring a deeper understanding of training sciences, we abstract the specific fitness goals, e.g. strength training biceps, into understandable high-level goals, trainee goals, which have an intuitive meaning for the user. The fitness goals that can be directly achieved by exercises and are structured in region, muscle group, and muscle goals.

Goals can be dependent on each other, as e.g. the trainee goal muscleGain requires the achievement of the training goal strength. This is encoded by requires (MuscleGain, Strength). In particular, the refinement of a higher-level goal into lower-level ones is realized by this constraint. For example, the goal muscle gain requires both strength and endurance training as indicated in Fig. 2. Goals can have different connections and priorities, depending on the parent goal. Hence, on each level, the Cartesian product of goals is created, e.g. instead of one Training Goal strength, there are seven training goals MuscleGain.strength, GeneralFitness.strength etc. While not all paths comprise all layers (endurance ends on the region level), this results for a single trainee goal on average in 900 subordinate fitness goals.

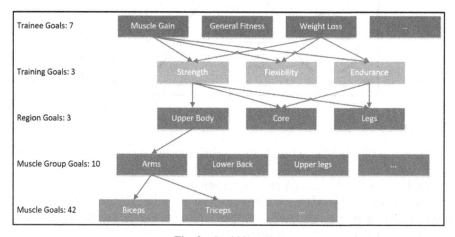

Fig. 2. Goal hierarchy

3.3 Exercises

An *exercise* is a physical activity that trains, i.e. *contributes* to, a set of *fitness goals*, as explained Sect. 3.4. An *exercise* has a set of parameters specifying the requirements for performing the *exercise*, e.g. the required *equipment*, as well as parameters describing the characteristics of the exercise, e.g. is it a strength or an endurance *exercise*. Here an exercise has the following attributes:

- Required fitness level
- Duration
- Required equipment.

An *exercise* has the predicate *active,* indicating whether an *exercise* is included in the exercise selection.

3.4 Exercise-Goal Relations

The knowledge about which exercises contribute to which goals is obviously a crucial part of the domain knowledge. Usually, *fitness goals* are not achieved by a *single exercise*, but require the combined contribution of a set of *exercises* selected from a set of suitable *exercises*. We represent these relationships by a representational scheme called *Choice*, which comprises a set of constraints. Such a set consists of all *exercises* contributing to the *achievement* of the same *fitness goal*.

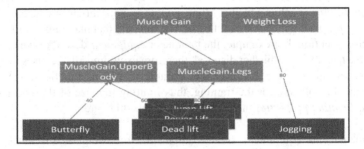

Fig. 3. Exercises contributing to multiple *fitness goals*

The *choice* associated with a fitness goal contains a set of *options*, i.e. all *exercises* contributing to the achievement of the respective *fitness goal*. Each link between an exercise and a goal has an associated *contribution* which is used to express different degrees in which the exercise helps to achieve a goal, if it is included in the exercise plan. An *exercise* may have a high *contribution* to one *fitness goal* and a low *contribution* to another *goal*. The domain of the contributions is: {0, 20, 40, 60, 80, 100}. If *active* holds for an exercise, the *ActContribution* (short for actual contribution) equals the potential contribution. Otherwise, it is zero. The choice has an *achievement threshold,* which represents the amount of the combined *ActContributions* considered necessary to *achieve* it. The *achievement level* is the combined value of all *ActContributions*. The *achievement level* is computed through a combine function, which in this study was a sum function.

Figure 3 contains a partial representation of a choice, indicating different contributions. For example, if *Butterfly* and *Dead lift* are active, *MuscleGain-UpperBody* will have an *achievementLevel* of 100.

3.5 Task

A *Task* is the triple of TraineeGoal (TaskGoal), and Trainee- and TrainingProperties (TaskParameters) and TaskRestrictions. *TaskGoals*, is the assignment of *goal.Achieved* = *T* to a set of user selected goals a solution to a configuration problem has to satisfy. *TaskRestriction* are user selected constraints on the activity of components, i.e. they define whether an exercise must be, or should not be, included in the selection. *TaskParameters* are domain-specific value assignments to parameters and are constants. They comprise a characterization of the exercising person in terms of *Trainee properties*. The ones considered in the study presented in Section 5 are

- Age: {below 19, 20–39, 40, 59, 60–75, >75}
- BMI: {underweight, normal, overweight, obese}
- Fitness level: {completely untrained, untrained, somewhat trained, trained, very trained}
- Working position: {sitting, standing, overhead}

Trainee properties form the basis for creating a customized exercise selection that enables the *trainee* to achieve his/her *fitness goal* in an appropriate and healthy manner.

Training properties are another set of *TaskParameters* that may imply restrictions on the exercise selection. For example, the parameter *equipment* describes which training material the *trainee* has at his/her disposal, therefore excluding all exercises that require equipment unavailable to the trainee. The parameter training duration represents the available amount of time for the training, thus limiting the size of the set of selected exercises. *Training properties* included in the present study are:

- Equipment: {no equipment, small equipment, free weights, machines, special equipment}
- Training Duration: {5, 10, 15,..., 90}

For equipment, we defined an ascending order. If the trainee states that he/she has small equipment, no equipment would also hold and so on. In this study, we assumed that all exercises require the same amount of time (5 min), thus the training duration allows 1 to 18 exercises.

3.6 TaskParameterGoalConstraints

In theory, some TaskParameters might prevent some goals from being realistic or reasonable, but the current knowledge base does not exploit this.

3.7 TaskParameterComponentConstraints

The parameters introduced so far help to characterize the training conditions, the trainee, the available exercises, and the goals of the training. The relations between parameters and exercise selection can roughly be divided in two groups: First, simple restrictions though binary relations (for example, exclusion of exercises requiring equipment that is not available), and second, complex relations. For the former, the following relations have been formalized so far:

Difficulty of exercises: for all exercises, it must hold that trainee.fitnesslevel is equal or slightly greater than exercise.requiredFitnesslevel, see (3).

$$\forall active(exercise_i) : exercise_i.fitnesslevel \geq trainee.fitnesslevel \land \qquad (1)$$

$$trainee.fitnesslevel - exercise_i.fitnesslevel \leq 2 \qquad (2)$$

Thus, a trainee must not be provided with exercises that are beyond his fitness level, but also not with such that are significantly easier.

Available Equipment: No exercise must be selected, for which no equipment is available, (3)

$$\forall active(exercise_i) : exercise_i.equipment = trainee.equipment \qquad (3)$$

4 Solution Algorithm

Exercise planning can be seen as a configuration problem [9] rather than a planning problem in the spirit of planning in AI: determining a set of components (i.e. a session with a number of exercises) that together fulfill a particular purpose under certain restrictions. We use our generic configuration system, GECKO (generic constraint based, Konfigurator, see [6], to generate exercise plans. The main objectives of the development of GECKO are

- a fairly **domain-independent** solution to configuration problems,
- based on a small set of **generic concepts** that support a clear structuring of the knowledge base,
- allowing its use **without** detailed **domain knowledge,**
- considering **optimality** criteria.

The creation of a knowledge base for a specific application system is done by providing

- domain specific **specializations** of the generic concepts and specification of variables associated with them and
- **constraints** of different types on these variables.

In Sect. 3, we showed how this can be done in a natural way for the exercising domain.

4.1   Consistency-Based Configuration

We formalize the configuration problem as identifying a subset of the components that satisfies the task specified by the user and is consistent with the configuration knowledge base **ConfigKB**. This can be seen as an assignment, AA, of activity to the components, which indicates the inclusion in or exclusion from the configuration.

Definition 1 (Activity Assignment)
An activity assignment for a set $COMPS_0 \subseteq COMPS$ is the conjunction $AA(COMPS_0)$
$=$

$$\left[\bigwedge_{comp \in COMPS_0} ACT(comp) \right]$$

$$\wedge \left[\bigwedge_{comp \in COMPS \setminus COMPS_0} \neg ACT(comp) \right]$$

$ACT(comp)$ is a literal which holds when a component $comp \in COMPS$ is part of a configuration.

Definition 2 (Configuration Task)
A configuration task is a pair (ConfigKB, Task) where:

- *ConfigKB, the knowledge base, containing the domain-specific objects and constraints,*
- *Task is a triple (TaskGoals, TaskParameters, TaskRestrictions) where:*

  - *TaskGoals, is the assignment of goal.Achieved $= T$ to a set of user selected goals a solution to a configuration problem has to satisfy*
  - *TaskParameters, domain-specific value assignments to parameters, are constants*
  - *TaskRestriction, user selected constraints on the activity of components*

Section 3.5 showed the specification if Task for the exercising domain.

To establish a solution to a configuration task, a set of active components has to be consistent with the task and the knowledge base.

Definition 3 (Configuration for a Task)
A configuration for a Task is an activity assignment $AA(\Gamma)$ such that $ConfigKB \cup TASK \cup \{AA(\Gamma)\}$ is satisfiable.

   $ConfigKB \cup TASK \cup \{AA(\Gamma)\} \not\models \bot$

A configuration is minimal iff for no proper subset Γ' of Γ is $AA(\Gamma')$ is a configuration.

Consistency seems to be a weak condition. After all, we want the configuration to satisfy the goals, not just be consistent with them. But this is ensured, as stated by the following proposition. Intuitively, if an activity assignment yields an AchievementLevel lower than the AchievementThreshold of a goal, it would be inconsistent with the goal.Achieved = T as required by the task.

Proposition 1
If AA(Γ) is a solution to a configuration task (ConfigKB, Task) then

$$AA(\Gamma) \cup ConfigKB \models$$
$$\forall goal \in TaskGoalsgoal.Achieved = T$$

This view on configuration was inspired by the formalization of consistency-based diagnosis [1, 7], where modes OK or ¬ OK are assigned to the components of a system, and a diagnosis is defined as a mode assignment MA(Δ) that is consistent with the model library, the structural description of the system and a set of observations (which are all sets of constraints, just like ConfigKB and Task):

ModelLib ∪ Structure ∪ Obs ∪ {MA(Δ)} ⊭ ⊥

In consequence, solutions to consistency-based diagnosis can also be exploited for generating configurations. This includes the introduction of a utility function and the application of best-first search to generate solution.

4.2 The Solution as a Configuration Algorithm

In consistency-based diagnosis, a utility function is often based on probabilities of component modes (assuming independent failures of components) ([7]) or, weaker, some order on the modes ([13]). In GECKO, we consider the contributions of components to the satisfaction of goals (possibly weighted by priorities of goals) and their cost.

Definition 4 (Utility Function)
A function h(AA(Γ), Task) is a utility function for a configuration problem iff it is admissible for A* search.

Definition 5 (Optimal configuration)
A configuration AA(Γ) is optimal regarding a utility function h(AA(Γ), Task) iff for no configuration AA(Γ') h(AA(Γ'), Task) is larger.

The utility of a configuration represents the fulfillment of the required goals and the cost of the configuration. It depends on its active components only. In the following, it is assumed that

- the contribution of a configuration is obtained solely as a combination of contributions of the active components included in the configuration and otherwise independent of the type of properties of the components,
- the cost of the contribution is given as the sum of the cost of the involved active components and will usually be numerical, and
- we can define a ratio "/" of contributions and cost.

The first defined function sums up the AchievementLevels (i.e. the combined contributions of all active components) multiplied with a weight dependent on the goal priority of all active goals and divides this by cost of all active components. (In the definition, we simplify the notation by writing Goalj.AchievementLevel instead of Goalj.Choicej. AchievementLevel etc.).

Definition 6 (GECKO Utility Function)

$$hl(AA(\Gamma, ActGoals)$$
$$:= \frac{\sum_{Goal_j \in ActGoals} weight(Goal_j.Priority) * Goal_j.AchievementLevel}{\sum_{Comp_i \in \Gamma} Comp_i.Cost}$$

This function ignores an important aspect: If the AchievementThreshold of some choice has already been reached, the utility of adding yet another component with a contribution to this choice is overestimated. The second utility function tries to capture this by disregarding any excesses above the AchievementThresholds.

Definition 7 (GECKO Utility Function with contribution limit)

$$hl(AA(\Gamma, ActGoals)$$
$$:= \frac{\sum_{Goal_j \in ActGoals} weight(Goal_j.Priority) * CurbedLevel(Goal_j)}{\sum_{Comp_i \in \Gamma} Comp_i.Cost}$$

Where CurbedLevel is defined by

$$CurbedLevel := \min(Goal_j.AchievementLevel, Goal_j.AchievementThreshold)$$

Based on this, we can exploit best-first search and solutions that have been developed in the context of consistency-based diagnosis.

5 Results and Evaluation

5.1 Evaluation Setup and Results

The test cases were conducted on an office laptop with an Intel Core i7-6820 2,70 GHz and 17 GB RAM. Average runtime for the test cases was 60 s.

The tested knowledge base contained 350 exercises, 5 trainee goals, 3 training goals, 3 region goals, 13 muscle group goals, and 42 muscle goals. Exercises contributed to one to six muscle goals.

Based on the formal representation from Sect. 3, we used the GECKO for generating exercise sessions, using haifacsp [14] as the constraint solver. Table 1 shows an exemplary test case and its result. Fifty training plans were generated for the evaluation.

Table 1. Exemplary test case

Parameter	Test Case	———Exercises———
Trainee Goal:	General Fitness	Leg extension (e)
Fitness Level:	Completely untrained_	Abdominal_Reverse_Crunch (e)
Equipment:	machines	Glutaeus Exercise
Working Position:	sitting	Running
Age:	below_19	Burpees
BMI:	underweight	Flexx_Chest (e)
Number Exercises:	6	

5.2 Evaluation Approach

As we explained in Sect. 2, exercise selection is a highly complex and often subjective task. Thus, it can be assumed that there is no such thing as a perfect exercise plan, at least not by formal standards. So, instead of merely testing, if the training plans are correct with regard to the formalization, we conducted an expert evaluation. To this end, we generated a set of fifty test cases and, subsequently, exercise sets.

Based on the implemented parameters and the purpose of the exercise selection, we defined the following criteria for single exercises:

- **Age** (whether a trainee is too old or young for an exercise)
- **BMI** (whether a trainee is too over- or underweight for an exercise)
- **Fitness Level** (is the difficulty of the level appropriate)
- **Working Position** (does the exercise support or harm the working position)
- **Equipment** (can the exercise be performed with the available equipment)
- **Not relevant** (is the exercise relevant to the trainee's goal)

And another set of criteria for assessing the combination of exercises, i.e. the plans.

- **Distribution of training** (is the time allocated to parts of the trained distributed appropriately)
- **Trained goal**(s) (does the training serve the intended purpose)
- **Missing exercises**
  - Dependency of exercises (do selected exercises require other exercises)
  - Superior to included exercises
- **Combination of exercises**
  - Required variation (is the selection to monotone)

Using these criteria, we created a questionnaire and asked domain experts, i.e. sport scientists and fitness coaches, to analyze the resulting plans. We differentiated between incorrect, i.e. exercise violates training principle or cannot be executed by trainee, and inappropriate, that is the exercise or session is not recommended by experts.

Three experts reviewed 50 test cases. The aim of this first study was both to find bugs in the knowledge base and to obtain an understanding of the quality and deficits of the generated plans.

5.3 Evaluation Results

Assessment of the evaluation: A plan met the quality criteria if the review revealed neither an incorrect nor an inappropriate judgement by the experts. In the first stage we ignored detected bugs in the knowledge base, because they don't contribute to the understanding of the quality of the solution. Therefore, the evaluation of the test runs that 72% of the training plans are appropriate for training. 28% of the plans where inappropriate because single exercises in a plan where not useful due to age, BMI or fitness level. Another point of criticism was that two test cases contained several almost identical exercises, e.g. *pushup* and *push up with narrow arms*.

6 Discussion

We presented a solution to automate the selection of exercises for physical exercise. To this end, we created a formal representation of the domain knowledge, based on knowledge-based configuration. As presented in 5, most test cases were considered appropriate. Consequently, the incorrect or inappropriate cases will be further analyzed.

For further improving the exercise selections, the next step in extending the knowledgebase is including the injuries and physical limitations, such as back pain or knee problems, as TaskParameters and establish TaskParameterComponentConstraints that exclude exercises based on them. Furthermore, training methods must be included. So far, the focus was on selecting exercises, but the training method, which describes how an exercise is performed, e.g. with how many repetitions and with how much weight, was omitted. Other topics for future work are the generation of multiple instances of an exercise, properties of combinations of exercises as well as the sequence of execution. Beyond the generation task, it is important to maintain the motivation of the trainee and provide appropriate incentives and reflect the user's feedback.

We established that most test cases were considered appropriate by domain experts and only a small minority was deemed incorrect. Before testing the solution with trainees, the next step will be a larger test series using an improved and extended knowledge base. Here, also physiotherapists, PE teachers, and fitness coaches will be asked to assess the training plan. Furthermore, the assessment will no longer focus on whether the results are correct or appropriate, but on an evaluation of their quality.

Acknowledgments. We would like to thank our project partners, especially the domain experts (Susan Halmheu and Mathias Siebenbürger) for the evaluation of the training plans.

References

1. de Kleer, J., Williams, B.C.: Diagnosing multiple faults. Artif. Intell. **32**(1), 97–130 (1987)

2. Friedrich, G., Stumptner, M.: Configuration. In: Faltings, B., Freuder, E.C., Friedrich, G., Felfernig, A. (eds.) Consistency-Based Configuration, pp. 1–6. AAAI Press. Papers from the AAAI workshop (99-05), Menlo Park, California (1999)
3. Felfernig, A., Hotz, L., Baglay, C., Tiihonen, L.: Knowledge-Based Configuration From Research To Business Cases. Morgan Kaufmann, Amsterdam (2014)
4. Struss, P.: Model-based problem solving. In: von Harmelen, F., Lifschitz, V., Porter, B. (eds.) Handbook of Knowledge Representation, pp. 395–465. Elsevier, Amsterdam (2008)
5. Grigoleit, F., Struss, P.: Configuration as diagnosis: generating configurations with conflict-directed A* - an application to training plan generation. In: DX, pp. 91–98 (2015)
6. Dressler, O., Struss, P.: Model-based diagnosis with the default-based diagnostic engine: effective control strategies that work in practice. In: ECAI-94 (1994)
7. HaifaCSP Technion, Haifa. https://strichman.net.technion.ac.il/haifacsp/
8. Junker, U.: Configuration. In: Rossi, F., Beek, P., Walsh, T. (eds.) Handbook of Constraint Programming. Elsevier, Boston (2006)
9. Dobbs, R., et al.: Overcoming obesity: An initial economic analysis, (2016). http://www.mckinsey.com/insights/economic_studies/how_the_world_could_better_fight_obesity Accessed 14 Dec 2019
10. Swinburn, B.A., et al.: The global obesity pandemic: shaped by global drivers and local environments. Lancet 378(9793), 804–814 (2011)
11. FREELATICS. https://www.freeletics.com
12. Bodbot. https://www.bodbot.com/
13. Fitbod. https://www.fitbod.me/
14. Kraemer, W.J., Ratamess, N.A.: Fundamentals of resistance training: progression and exercise prescription. Med. Sci. Sports Exerc. 36(4), 674–688 (2004)

Anomaly Detection and Automated Diagnosis

A Multi-phase Iterative Approach for Anomaly Detection and Its Agnostic Evaluation

Kévin Ducharlet[1(✉)], Louise Travé-Massuyès[2], Marie-Véronique Le Lann[2], and Youssef Miloudi[1]

[1] CARL Software - Berger-Levrault, Limonest, France
{kevin.ducharlet,youssef.miloudi}@carl.eu
[2] LAAS-CNRS, University of Toulouse, CNRS, INSA, Toulouse, France
{louise,mvlelann}@laas.fr

Abstract. Data generated by sets of sensors can be used to perform predictive maintenance on industrial systems. However, these sensors may suffer faults that corrupt the data. Because the knowledge of sensor faults is usually not available for training, it is necessary to develop an agnostic method to learn and detect these faults. According to these industrial requirements, the contribution of this paper is twofold: 1) an unsupervised method based on the successive application of specialized anomaly detection methods; 2) an agnostic evaluation method using a supervised model, where the data labels come from the unsupervised process. This approach is demonstrated on two public datasets and on a real industrial dataset.

Keywords: Anomaly detection · Machine learning · Agnostic evaluation · Industrial applications

1 Introduction

Nowadays, industrial systems are equipped with sets of sensors that perform different physical measures on the parts of the system. These measures are linked to the instant they are observed and form time series that can be used in various ways and give precious information about the behaviour of the system over time. An interesting use of these data is predictive maintenance that aims to anticipate breakdowns with the study of irregularities in data [1]. However, faulty sensors may generate irrelevant irregularities that disturb the analysis of data, leading to misinterpretations. For this reason, sensor faults must be detected so that the impact of faulty data on the predictive maintenance process is limited.

We chose to train anomaly detection models on selected training datasets. Yet, available datasets are usually not labelled, which means that we do not

This project is supported by ANITI through the French "Investing for the Future – PIA3" program under the Grant agreement noANR-19-PI3A-0004.

© Springer Nature Switzerland AG 2020
H. Fujita et al. (Eds.): IEA/AIE 2020, LNAI 12144, pp. 505–517, 2020.
https://doi.org/10.1007/978-3-030-55789-8_44

know if an observation is normal or generated by a faulty sensor. On top of that, for the same company, there are different systems with different operating contexts. Thus, we wish to build a solution that can be generalized to most industrial systems.

The contribution of this paper is twofold. We propose: 1) an unsupervised method, named SuMeRI (Successive Methods Run Iteratively), that performs anomaly detection on unlabelled datasets. It is based on applying successively different anomaly detection methods that detect different kinds of anomalies. Each method is applied iteratively on data where anomalies are removed at each iteration; 2) an agnostic evaluation method based on consistency checking, named CC-Eval (Consistency Checking Evaluation), that measures the similarity of the results of the unsupervised learned model with those of a supervised model learned with the training dataset labelled according to the unsupervised model (cf. Fig. 1).

The paper is organized as follows. Section 2 provides a state of the art of the field of anomaly detection and positions our contribution. Section 3 presents the two methods, SuMeRI and CC-Eval. In Sect. 4, some results obtained with public datasets and real case study datasets are provided to validate our approach. To conclude, Sect. 5 includes a discussion of the results and directions for future works.

2 Related Work on Anomaly Detection

Various communities have been interested in anomaly detection, starting with statisticians in the end of the 19th century with the works of Edgeworth [2], followed by control scientists interested in fault detection and isolation, i.e. the FDI community [3], and later by artificial intelligence scientists that proposed paradigms for diagnosis reasoning, i.e. the DX community [4]. Many of these latter works are based on the existence of a model built from physical knowledge [5–7]. However, the complexity of today's systems which makes it difficult to build models and the growing interest in data-based methods have led to the development of many machine learning methods. This paper is interested in these methods.

A commonly accepted definition of anomaly, or *outlier*, is the one from Hawkins that defines an anomaly as "an observation which deviates so much from the other observations as to arouse suspicions that it was generated by a different mechanism" [8]. A survey published by Chandola et al. in 2009 [9] acts as a reference in the domain and provides all the required elements to approach the field. This survey introduces three kinds of anomalies:

- *point anomalies* defined as pointwise observations that appear abnormal compared to the rest of the dataset,
- *contextual anomalies* defined as pointwise observations that appear abnormal with respect to the context in which they occur,
- *collective anomalies* defined as collections of observations that are solely abnormal if studied together.

While point anomalies are the most frequent and the most trivial to detect, collective anomalies are nonetheless common in time series and this makes their detection important in industrial context with sensor data.

The survey by Chandola [9] also divides the methods according to the availability of data labels during the training phase. Learning methods using entirely labelled datasets, for which labels indicate for each instance if it is normal or abnormal, are called *supervised* and usually refer to the classification problem. But the cases where the knowledge about the labels is available are quite uncommon, and this is why semi-supervised methods are preferred. In this later case, models are learned from training datasets in which only one specific class is present, usually normal data. The methods of the last type are known as unsupervised and they do not use any knowledge a priori. Because of this, unsupervised techniques are more generalizable than supervised or semi-supervised. However, unsupervised methods often make the assumption that abnormal data are very few in the training set and that it is possible to apply a sufficiently robust semi-supervised method without being disturbed by outliers.

There are two types of outputs for anomaly detection methods: a decision or a score. The decision is mostly used in a supervised context and only gives the predicted class of each instance. The score is defined as the degree of abnormality (or, in some cases, normality) of an instance, in a range that is defined by the method itself. Then, one usually defines a threshold on this score to make a decision.

As described in the introduction, the detection of faulty sensors needs to be done in an unsupervised context. Studies about unsupervised anomaly detection are common because of the wide applicability of such methods, especially those based on the famous nearest neighbor method [10–12]. Because of the large amount of methods that claim to be the best in response to specific needs and the lack of labelled data to evaluate the learned models, numerous comparative studies have been done to help selecting the one that fits the best a defined context [13–15].

However, authors of these studies hardly give the datasets that are used to get their results nor they indicate how they parameterize the algorithms that they compare. Even if they provide such information, their analyzes focus solely on a specific experimental context or dataset, and the results can not be generalized to other contexts, as proved by the differences in the results of the different works. Some comparative works try to be more generalizable using a wide variety of datasets and evaluation metrics [16], but it is still difficult to know if their results can be applied in a general context, in particular for the detection of sensors faults. Eventually, it is not possible to compare the proposed algorithms on our real case study datasets because the metrics computed to perform the comparison require the availability of validation data.

For this reason, we developed SuMeRI to apply different methods successively and CC-Eval to give a first sight on an evaluation of the model in a fully unsupervised way, bypassing the lack of validation data. Applying different methods offers the opportunity to detect different kinds of anomalies (point, contextual

and collective), the only requirement to use a method in SuMeRI being the availability of a score output. Interestingly, SuMeRI can be used in an univariate or
a multivariate setup, which is important to distinguish an anomaly caused by
a faulty sensor from an anomaly affecting the system components. To make it
more robust to anomalies in the training set, SuMeRI runs each method iteratively, so the training set is cleaned from the most outlying instances at each
iteration until a fixed stopping point. It returns an unsupervised model able to
compute an anomaly score on data and decide which instances are abnormal, this
decision being based on a learned threshold on the score. To evaluate the model
returned by SuMeRI, we propose CC-Eval that uses a supervised method on
the dataset labelled by the unsupervised SuMeRI model. CC-Eval measures the
consistency of the SuMeRI model by comparing its results to those of the supervised model learned on the training dataset labelled according to the SuMeRI
model. The rational under this idea is that the greater the consistency, the more
likely it is that SuMeRI's results follow a well-founded rule. The whole process
is represented in Fig. 1.

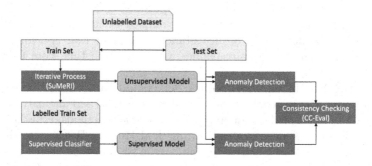

Fig. 1. Architecture of the proposed solution.

3   Unsupervised Anomaly Detection and Agnostic Evaluation

3.1   Successive Methods Run Iteratively (SuMeRI) Approach

Methods for anomaly detection that do not require labelled data usually need to
estimate the anomaly rate in the training set. Yet, this information is unknown
in most cases and fixing a too high or low rate leads to a high false positive or
high false negative rate respectively. On the other hand, anomalies may be of
different types, which may require different anomaly detection methods. That
is why we developed the SuMeRI approach that applies specialized anomaly
detection methods in successive phases and runs each method iteratively on
data that are increasingly free of anomalies. The complete SuMeRI algorithm is
given by Algorithm 1.

Iterative Process. Each iteration detects the current *most outlying instances* and removes them from the training set so that the next iteration trains a new model from a "cleaner" data set. This requires an anomaly score to decide which samples are the most outlying ones and a stopping condition to stop the iterations so that over fitting is avoided. The SuMeRI approach can hence use any anomaly detection method that meets the two following conditions:

- it computes an anomaly score s_i for each data instance i and discriminates anomalous instances from normal ones, hence defining a threshold l_1,
- it allows one to define a measure to evaluate the quality of the model over the iterations.

The "most outlying instances" are identified by a score above the threshold l_2 defined by:

$$l_2 = \operatorname*{argmax}_{s_i \in S_{out}} (\min_{s_j \neq s_i} (|s_i - s_j|)) \tag{1}$$

Let us define S_{in} (S_{out}) as the set of scores associated to normal (outlier) instances and assume that scores are scaled in $[-1, 1]$ at the end of each iteration.

The heuristic used to stop the iterations at each phase is based on the score distribution. Removing the most outlying instances at each iteration shifts the threshold l_1 towards greater score values. It also results in increasing the kurtosis, i.e. the *tailedness*, of the normal score distribution and decreasing the kurtosis of the outlier score distribution. The desired model is the one where the outliers detected by the model and the inliers have similar kurtosis. In practice, we compare the maximal distance between nearest neighbors score values for outliers and for inliers and stop when the condition of Eq. (2) is satisfied:

$$d_{max}(S_{out}) = \max_{s_i \in S_{out}} (\min_{s_j \neq s_i} (|s_i - s_j|)) \leq d_{max}(S_{in}) = \max_{s_i \in S_{in}} (\min_{s_j \neq s_i} (|s_i - s_j|)) \tag{2}$$

Algorithm 1. SuMeRI algorithm

Input: $data, methods$ ▷ *methods* contains One-Class SVM and Linear Regression
1: $data \leftarrow scale(data)$
2: $l \leftarrow length(methods)$
3: $models \leftarrow list[length \leftarrow l]$ ▷ One model learned per method
4: **for** $i \leftarrow 1, l$ **do**
5:     $method \leftarrow methods[i]$
6:     $stopping\_condition \leftarrow False$
7:     **while** $stopping\_condition \neq True$ **do**   ▷ Each loop on a method is a phase
8:        $models[i] \leftarrow learn\_model(method, data)$
9:        $score \leftarrow compute\_score(model, data)$
10:       $stopping\_condition \leftarrow stopping\_decision(score)$
11:       $data \leftarrow normal\_data(data, score)$
12:     **end while**
13: **end for**
14: **return** $models$

Successive Phases. The successive phases in SuMeRI apply different methods to detect different anomaly types; point, contextual, and collective. As mentioned in Sect. 2, many unsupervised methods make the assumption that there is a non-significant proportion of outliers in the dataset. These methods are thus close to semi-supervised methods, which assume that the training set is representative of normality, thanks to robustness to possible outliers. Because each phase of SuMeRI removes the detected outliers from the training set, SuMeRI allows the use of these methods for the last phases. Nevertheless, unsupervised methods able to learn without assumption on the anomaly rate in the training dataset should be used in the first phases.

3.2 Agnostic Evaluation Based on Consistency Checking (CC-Eval)

To evaluate the performances of anomaly detection models, one commonly uses metrics such as the receiver operating characteristic (ROC) and the precision-recall (PR) curves, using their area under the curve (AUC) and average precision (AP) to be compared between different methods [13]. However, these metrics require labelled data to be used which are not available in our case. An unsupervised solution has been suggested to evaluate methods without the use of labelled data [17], but the consistency of its results in comparison with the ROC and PR curves seems to depend on the datasets.

Thus, we propose a solution to overcome this issue and introduce CC-Eval. The purpose of CC-Eval is twofold: 1) check if there is a learnable logic behind the obtained results, and 2) check if the iterative learning has not gone too far in the search for anomalies, leading to overfitting and high false positive rates.

To achieve this, the datasets are divided into a training set and a testing set (cf. Fig. 1), representing 90% and 10% of the whole dataset respectively. To avoid temporal discontinuity in the data that could have an impact on learned models, the splitting is done by blocks from the tails of the dataset. SuMeRI is then applied on the learning set and the whole dataset is labelled by applying the learned model, considering one anomaly class for each SuMeRI phase and another class for normal instances. Using the learned labels as the ground truth, a supervised classifier is used to learn the classification of data. The learned classifier is then used to predict the classes for the test set.

Here, the consistency is defined as the logic of the discrimination learned between inliers and outliers. The assumption behind this method is that a supervised classifier is efficient if there is consistency in the classes to learn. Thus, the consistency is evaluated with the area under the ROC curve (ROC-AUC) which quantifies how well the model is able to distinguish the different classes. Because there are more than two classes, it is required to compute an aggregation on the different ROC curves. We propose to compute the ROC-AUC as the mean of four different approaches:

- macro average with one-vs-rest approach,
- weighted average with one-vs-rest approach,
- macro average with one-vs-one approach,
- weighted average with one-vs-one approach.

The macro average approaches do not take into account the imbalance in classes while the weighted average approaches do. The one-vs-rest approaches compute the mean of the ROC-AUC for each class against all the others (A vs B&C, B vs A&C, C vs A&B). The one-vs-one approaches compute the mean of the ROC-AUC for each binary combination of classes, excluding the data in the other classes (A vs B, A vs C, B vs C).

In this context, the main issue with the classification problem is that the classes are imbalanced. Usually, there are far more instances in the normal class than instances in the outlier classes, and the most populated class for outliers varies depending on the dataset. Nevertheless, studies have been made to resolve this issue, using for the most of them sampling strategies to reduce the gap between the size of the classes or cost-sensitive learning to reduce the effects of these gaps [18]. Four different classifying methods have been considered which use ensemble of classifiers trained on different samples generated by random under-sampling: 1) EasyEnsemble [19], 2) Bagging predictors [20], 3) Random Forest [21] and 4) RUSBoost [22].

To decide which method to use, we generate ten different models by applying SuMeRI with different parameters on a multivariate labelled public dataset. Then, we compute the ROC-AUC using the known labels on these ten models and establish a ground truth ranking among them based on the results. The best classifier is defined as the one that gives the closest ranking to the ground truth. This experience showed that the EasyEnsemble classifier with an under-sampling on the majority class gives the best results, closely followed by the RandomForest classifier with the same sampling strategy. Both of these methods will be used in the following section.

4   Experimental Results

4.1   SuMeRI and CC-Eval Settings

In our setting, we want to detect point anomalies and contextual anomalies. Hence our SuMeRI setting includes two phases:

- The first phase uses the One-Class SVM method. Leveraging the kernel trick, the One-Class SVM method learns a non linear separation between normal and anomalous classes. Then, the anomaly score is computed as the signed distance of each instance from the separation.
- The second phase is a prediction method. For each instance, we use its context as explanatory variables to predict its value. We define the context of an instance temporally with statistical metrics on the previous and following instances in a defined window. These metrics are the minimum, the first quartile, the median, the third quartile, the maximum, the mean, the standard deviation, and the mean squared error. The score is then computed based on the prediction error, as the Euclidean distance between the predicted value and the real value, with a standardization to be comparable with the score of the first phase. Two methods have been considered for this prediction: Linear

Regression and Neural Networks. The later gives slightly better results but at a high computational cost, that is why we have opted for the former.

For our experiments, we used the implementation of One-Class SVM and Linear Regression from scikit-learn, a Python library for Machine Learning, in SuMeRI. The ROC-AUC approaches are also implemented in this library. The classifiers used in CC-Eval are from the imbalanced-learn Python library with the names *EasyEnsembleClassifier* and *BalancedRandomForestClassifier*.

With the chosen configuration, SuMeRI only requires three parameters. The first one is the proportion of anomalies to compute the models, it should be low enough to drop only few anomalies at each step. It is used as the nu parameter for the scikit-learn implementation of One-Class SVM and also to fix the separation between anomalies and normal instances during the second phase. The second one is the kernel function used for the kernel trick in One-Class SVM. The last one is the window size on which are computed the contextual metrics.

4.2   Public Datasets

SuMeRI and CC-Eval, presented in Sect. 3, have been tested on two public datasets. The first one is from the Numenta Anomaly Benchmark [23] and it measures the temperature on a machine with known system failures in the dataset. The second contains the Http requests in the KDD Cup 1999 dataset, used for The Third International Knowledge Discovery and Data Mining Tools Competition, as it is presented in the ODDS library [24].

The machine temperature dataset is close to our industrial application domain. It is univariate and contains 22695 instances with three known anomalies: two being linked to a breakdown and a third one that appears as a hint for the occurrence of the second breakdown. The SuMeRI model is learned with the linear kernel, an anomaly rate fixed at 0.001 and a window of size 20, which

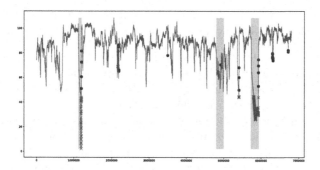

Fig. 2. Results of SuMeRI for the machine temperature dataset. Green areas represent the true anomalies, the left one is the first known anomaly, the middle one is the third and the right one is the second. Outliers detected by the first phase are represented with red crosses and those detected by the second phase with blue circles. (Color figure online)

means that we consider ten periods before and after each instance to compute the contextual metrics. The SuMeRI results are shown in Fig. 2 where the three green areas represent the actual anomaly positions.

The first iterative phase with One-Class SVM converges after three iterations and is able to detect the outliers in the left and right green areas that correspond to the two big failures. There is still one false positive due to the fact that the iterations stopped one step too late. The second phase converges after six iterations. Because of the use of statistical metrics in a window around each point for the prediction, anomalies are found where the signal has rapid variations. This phase is able to detect anomalies in the three green areas. Let us notice that only the HTM method makes the same detection [23], but it rises more false positives than true positives. The results of CC-Eval are given in Table 1.

Table 1. CC-Eval results for the model learned on the machine temperature dataset. The ROC-AUC are computed with different multi-class and average strategies for the chosen classification methods. The last column is the mean of the four previous ones.

	One-vs-one		One-vs-rest		Mean
	Macro	Weighted	Macro	Weighted	
EasyEnsemble	0.9915	0.9879	0.9881	0.9794	0.9867
BalancedRandomForest	0.9800	0.9733	0.9678	0.9865	0.9769

The Http dataset contains three variables and is entirely labelled. Among its 567498 instances, there are 2211 anomalies (0.4%) that have to be detected. However, the information on the cause of the anomalies is unknown and this example does not match our application domain, but proves the applicability of the presented solution for multivariate datasets. SuMeRI is applied with the linear kernel, a window of size 20 and an anomaly rate fixed at 0.0001. The results of both phases in comparison with the known labels are shown in Table 2. The first phase converges since the first iteration. The iterative process is not of great use in this case, but the stopping condition occurs at the right iteration according to the results. The second phase converges after three iterations and only detects 11 outliers in the remaining dataset. Its performances are poor in comparison with those from the first iteration with only one true positive. Because we know the labels for this dataset, it is possible to compute the ROC AUC of the model which is equal to 0.9989. We can also compute this score for each phase. The ROC AUC for the model learned during the first phase is equal to 0.9970 for the whole dataset while it is equal to 0.9959 for the model learned during the second phase on the data that are negatively labelled by the first phase (0.4354 when applied on the whole dataset, because the instances labelled positively by the first phase are not detected as outliers, which means the true positive rate is very low). The results of CC-Eval are given in Table 3 and are a bit lower than the true ones.

Table 2. Results of SuMeRI for the Http dataset.

	Phase 1	Phase 2	Total
True positives *(rate)*	2202 *(99.593%)*	1 *(11.111%)*	2203 *(99.638%)*
True negatives *(rate)*	565241 *(99.992%)*	565231 *(99.998%)*	565231 *(99.990%)*
False positives *(rate)*	46 *(0.008%)*	10 *(0.002%)*	56 *(0.010%)*
False negatives *(rate)*	9 *(0.407%)*	8 *(88.889%)*	8 *(0.362%)*

Table 3. CC-Eval results for the model learned on the Http dataset, computed as for Table 1.

	One-vs-one		One-vs-rest		Mean
	Macro	Weighted	Macro	Weighted	
EasyEnsemble	0.9331	0.8999	0.9611	0.9944	0.9471
BalancedRandomForest	0.9836	0.9754	0.9777	0.9986	0.9838

For these two public datasets, the results of CC-Eval are quite good and seem to be consistent with the model performances.

4.3 Real Case Study Datasets

SuMeRI and CC-Eval also have been applied on a real case study dataset where labels were not available but. where some anomalies were known, as in the machine temperature dataset used in the previous subsection. This dataset is also univariate and constituted of temperature measurements in a building that is equipped with an intelligent auto-regulating system. This dataset is one of the kind that is targeted by this solution, as a timeseries with changing in the operating mode which makes it difficult to model.

The dataset contains 446581 instances sampled every 15 s and there are two areas where we want to detect anomalies. The first one is a provoked growth in temperature with the heat of the sensor, the second one is a treatment issue on the signal with an added offset. The rbf kernel for the SVM method has been used in SuMeRI with an anomaly rate of 0.0001 and a window of size 20. The results are shown in Fig. 3. The offset is not detected by the first phase because it is considered as a second normal cluster which causes the score, computed based on the distance to the learned separation, to be the same as the more normal values for the real normal cluster. However, the instances where the changes happened are detected as anomalous by the second phase. The Table 4 displays the results of CC-Eval, which are quiet weak for the *EasyEnsembleClassifier* in comparison with the results for the public datasets. This is consistent with the fact that the learned model does not perform as well.

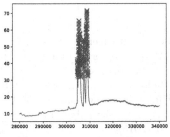

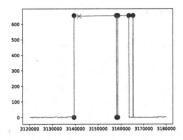

| (a) First anomaly to detect | (b) Second anomaly to detect |

Fig. 3. Results of SuMeRI for the building temperature dataset. The X-axis represents a number of seconds since the first observation. The first anomaly is well detected by the first phase (red crosses). The second anomaly is not detected by the One-Class SVM phase which consider these values as a normal cluster, but the offset is detected by the second phase (blue circles). (Color figure online)

Table 4. CC-Eval results for the model learned on the building temperature dataset, computed as for Table 1 and Table 3.

	One-vs-one		One-vs-rest		Mean
	Macro	Weighted	Macro	Weighted	
EasyEnsemble	0.8333	0.7500	0.6686	0.4927	0.6817
BalancedRandomForest	0.9997	0.9996	0.9998	0.9999	0.9998

5   Conclusions and Future Works

In this paper, an approach for anomaly detection in unlabelled industrial datasets has been proposed. The approach, called SuMeRI, is based on applying successive specialized anomaly detection methods in an iterative way. The methods admissible in SuMeRI should meet two requirements: the availability of an anomaly score and the ability to measure the quality of the model. The results presented in this paper are obtained with the application of two methods, one specialized in detecting point anomalies and the other in contextual anomalies. Because it is difficult to evaluate the results for unlabelled datasets, an agnostic consistency checking method named CC-Eval has been proposed.

Future works will focus on the improvement of SuMeRI in the following directions:

– Instantiate SuMeRI with more than two successive methods while still limiting the number of hyper-parameters to be tuned. This could be done with the use of ensemble methods.
– Include a specialized method for collective anomalies.
– Make explicit and explain automatically the rules underlying the consistency found by CC-Eval.
– Give a better evaluation of the whole solution and compare it to state-of-the-art methods.

References

1. Garcia, M.C., Sanz-Bobi, M.A., del Pico, J.: SIMAP: intelligent system for predictive maintenance: application to the health condition monitoring of a windturbine gearbox. Comput. Ind. **57**(6), 552–568 (2006). E-Maintenance Special Issue
2. Edgeworth, F.Y.: On discordant observations. Philos. Mag. **23**(5), 364–375 (1887)
3. Gao, Z., Cecati, C., Ding, S.X.: A survey of fault diagnosis and fault-tolerant techniques–part I: fault diagnosis with model-based and signal-based approaches. IEEE Trans. Industr. Electron. **62**(6), 3757–3767 (2015)
4. Peischl, B., Wotawa, F.: Model-based diagnosis or reasoning from first principles. IEEE Intell. Syst. **18**(3), 32–37 (2003)
5. Trave-Massuyes, L.: Bridging control and articial intelligence theories for diagnosis: a survey. Eng. Appl. Artif. Intell. **27**, 1–16 (2014)
6. Cordier, M.-O., Dague, P., Pencolé, Y., Travé-Massuyès, L.: Diagnosis and supervision: model-based approaches. In: Marquis, P., Papini, O., Prade, H. (eds.) A Guided Tour of Artificial Intelligence Research, pp. 673–706. Springer, Cham (2020). https://doi.org/10.1007/978-3-030-06164-7_21
7. Biswas, G., Cordier, M.O., Lunze, J., Staroswiecki, M., Trave-Massuyes, L.: Diagnosis of complex systems: bridging the methodologies of the FDI and DX communities. IEEE SMC Trans. - Part B **34**(5), 2159–2162 (2004). Special Issue
8. Hawkins, D.M.: Identification of Outliers. Monographs on Statistics and Applied Probability. Springer, Dordrecht (1980). https://doi.org/10.1007/978-94-015-3994-4
9. Chandola, V., Banerjee, A., Kumar, V.: Anomaly detection: a survey. ACM Comput. Surv. **41**(3), 1–58 (2009).
10. Breuning, M.M., Kriegel, H.-P., Ng, R.T., Sander, J.: LOF: identifying density-based local outliers. In: Proceedings of the ACM SIGMOD 2000 International Conference on Management of Data, Dallas, pp. 93–104. ACM (2000)
11. Kriegel, H.-P., Kröger, P., Schubert, E., Zimek, A.: LoOP: local outlier probabilities. In: Proceedings of the 18th ACM Conference on Information and Knowledge Management (CIKM 2009), Hong Kong, pp. 1649–1652. ACM (2009)
12. Padamitriou, S., Kitagawa, H., Gibbons, P.B., Faloustos, C.: LOCI: fast outlier detection using the local correlation integral. In: Proceedings of the 19th International Conference on Data Engineering, Bangalore, pp. 315–326. IEEE (2003)
13. Domingues, R., Filippone, M., Michiardi, P., Zouaoui, J.: A comparative evaluation of outlier detection algorithms: experiments and analyses. Pattern Recogn. **74**, 406–421 (2018)
14. Martin, R.A., Schwabacher, M., Oza, N.C., Srivastava, A.N.: Comparison of unsupervised anomaly detection methods for systems health management using space shuttle. In: Proceedings of the Joint Army Navy NASA Air Force Conference on Propulsion (2007)
15. Auslander, B., Gupta, K.M., Aha, D.W.: Comparative evaluation of anomaly detection algorithms for local maritime video surveillance. In: Proceedings of SPIE - The International Society for Optical Engineering, SPIE (2011)
16. Goldstein, M., Uchida, S.: A comparative evaluation of unsupervised anomaly detection algorithms for multivariate data. PLoS ONE **11**(4), e0152173 (2016)
17. Goix, N.: How to Evaluate the Quality of Unsupervised Anomaly Detection Algorithms? hal-01341809 (2016)
18. Ganganwar, V.: An overview of classification algorithms for imbalanced datasets. Int. J. Emerg. Technol. Adv. Eng. **2**(4), 42–47 (2012)

19. Liu, X.Y., Wu, J., Zhou, Z.H.: Exploratory undersampling for class-imbalance learning. IEEE Trans. Syst. Man Cybern. Part B (Cybern.) **39**(2), 539–550 (2008)
20. Breiman, L.: Bagging predictors. Mach. Learn. **24**(2), 123–140 (1996)
21. Chen, C., Liaw, A., Breiman, L.: Using random forest to learn imbalanced data, 110, pp. 1–12. University of California, Berkeley (2004)
22. Seiffert, C., Khoshgoftaar, T.M., Van Hulse, J., Napolitano, A.: RUSBoost: a hybrid approach to alleviating class imbalance. IEEE Trans. Syst. Man Cybern. Part A (Syst. Hum.) **40**(1), 185–197 (2009)
23. Lavin, A., Ahmad, S.: Evaluating real-time anomaly detection algorithms-the numenta anomaly benchmark. In: 14th International Conference on Machine Learning and Applications (ICMLA), Miami, pp. 38–44. IEEE (2015)
24. Rayana, S.: ODDS Library. Stony Brook University, Department of Computer Science (2016). http://odds.cs.stonybrook.edu

On the Use of Answer Set Programming for Model-Based Diagnosis

Franz Wotawa[✉][iD]

Institute for Software Technology, Graz University of Technology,
Inffeldgasse 16b/2, 8010 Graz, Austria
wotawa@ist.tugraz.at

Abstract. Model-based diagnosis has been an active area of AI for several decades leading to many applications ranging from automotive to space. The underlying idea is to utilize a model of a system to localize faults in the system directly. Model-based diagnosis usually is implemented using theorem provers or constraint solvers combined with specialized diagnosis algorithms. In this paper, we contribute to research in model-based diagnosis and present a way of using answer set programming for computing diagnoses. In particular, we discuss a specific coding of diagnosis problems as answer set programs, and answer the research question whether answer set programming can be used for diagnosis in practice. For this purpose, we come up with an experimental study based on Boolean circuits comparing diagnosis using answer set programming with diagnosis based on a specialized diagnosis algorithm. Although, the specialized algorithm provide diagnoses in shorter time on average, answer set programming offers additional features making it very much attractive to be used in practice.

Keywords: Model-based diagnosis · Answer set programming · Modeling for diagnosis · Experimental evaluation

1 Introduction

Automating activities like diagnosis, i.e., the detection and identification of faults, has been an active research area of AI having applications in medicine, e.g., the MYCIN [3] expert system, and technical systems, e.g., networks [13]. Although successful, classical expert systems require constructing knowledge bases that allow for deriving diagnoses from given symptoms, i.e., observed deviations from the expected behavior. These knowledge bases have to be very specifically developed for a certain task, can hardly be adapted to fit slightly different tasks, and require a lot of maintenance effort. In order to avoid these drawbacks, model-based diagnosis [6,11,20] was invented as an alternative to expert systems for diagnostic applications.

In model-based diagnosis (MBD) a model of the system is directly used to compute diagnoses. This model captures the structure and behavior of a system

© Springer Nature Switzerland AG 2020
H. Fujita et al. (Eds.): IEA/AIE 2020, LNAI 12144, pp. 518–529, 2020.
https://doi.org/10.1007/978-3-030-55789-8_45

and can be easily adapted in case of system changes or when being applied for modeling other systems reducing development and maintenance costs. MBD has been showed to be applicable in many domains including power transmission networks [2], automotive [14,17,21], space [16,18], programs [9,22], and even spreadsheets [1,10].

Models used in MBD represent at least the expected behavior of the system's components making assumptions about the health state of components explicit. Algorithms used for implementing MBD make use of these health states and search for a combination of health states that explains a detected misbehavior. There have been different MBD algorithms available. Some are based on hitting sets [20] of conflicting health states, the assumption-based truth maintenance system [11], or other means [8]. All these algorithms have in common to rely on a formal model where either a propositional or first-order logic theorem prover, or a constraint solver is used for providing the required reasoning capabilities.

In this paper, we focus on a more recent category of theorem provers relying on answer set programming (ASP) (see [7]). Informally speaking, an answer set is a model of a logical theory if every proposition in the model has an acyclic logical derivation from the given logical theory. ASP can be used for implementing non-monotonic logics like default logic [19]. For example, Chen et al. [4] introduced a system implementing default logic using ASP. Because of the fact that default logic can also be used for formalizing MBD problems (see [20]) it is obvious that ASP can be used for diagnosis as well. What remains, is to show that ASP solvers are fast enough to be of value for practical applications, which is exactly the objective behind this paper.

We discuss the formalization of MBD using ASP and further elaborate on the research questions, whether ASP solvers (and there in particular clingo[1]) are fast enough for being used in practice, where we compare the running time for computing diagnoses using clingo with a specialized diagnosis algorithm based on an optimized horn-clause SAT solver in combination with hitting set computation. In addition, we discuss the pros and cons regarding the use of ASP in the context of MBD.

We organize this paper as follows. First, we discuss the basic foundations behind MBD. Afterwards, we show how ASP can be used to represent MBD problems before presenting the results obtained using an experimental study, which is based on Boolean circuits of various size ranging from 160 to more than 3,500 gates. We further summarize the obtained results, and conclude the paper.

2   Model-Based Diagnosis

In this section, we briefly summarize the basic foundations of MBD following Reiter's definitions of diagnosis [20]. Besides introducing definitions, we make use of a running example for illustrating the underlying ideas. In Fig. 1 we see a small Boolean circuit comprising three inverter gates, an observed input value 0,

---

[1] See https://potassco.org/.

and an observed output value 0. Considering the stated behavior of an inverter, we immediately see that a value of 1 would be expected at the output instead of 0. Straightforward, we are interested in identifying the reasons behind this deviation. Such a situation perfectly characterizes diagnosis, where we have a system comprising interconnected components implementing a given behavior, and a set of observations that are in contradiction with the expectations. In case of MBD the expectations can be derived from the model of the system (i.e., the formalized behavior of the components).

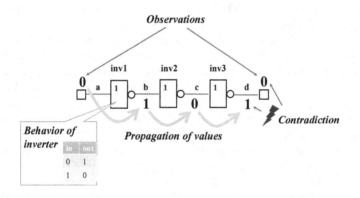

Fig. 1. A simple inverter chain circuit.

Using this discussed context, we are able to formalize a diagnosis problem as follows:

Definition 1 (Diagnosis problem). *Given a system* $(SD, COMP)$ *comprising a set of logical sentences SD representing the structure and behavior of a system, and COMP a set of components, and a set of observations OBS. A tuple* $(SD, COMP, OBS)$ *forms a diagnosis problem where we are interested in finding those components that are responsible for deviations between expected and observed values.*

From the definition of a diagnosis problem it is not given how the system description SD, i.e., the model of the system, should be formalized. For this purpose, Reiter [20] assumed to represent the correct behavior of components and the system structure in form of first-order logical sentences. To distinguish correct from incorrect behavior, Reiter furthermore introduced the special predicate ab. The predicate in negated form is used in SD for formally stating the component's correct behavior.

For example, for an inverter C we are able to specify its behavior using the following 4 rules:

$$\neg ab(C) \wedge in(C,0) \wedge inverter(C) \rightarrow out(C,1)$$
$$\neg ab(C) \wedge in(C,1) \wedge inverter(C) \rightarrow out(C,0)$$
$$\neg ab(C) \wedge out(C,0) \wedge inverter(C) \rightarrow in(C,1)$$
$$\neg ab(C) \wedge out(C,1) \wedge inverter(C) \rightarrow in(C,0)$$

In addition to these rules we add facts representing structural information to SD for the inverter chain example:

$$inverter(inv1) \wedge inverter(inv2) \wedge inverter(inv3)$$
$$\forall X \in \{0,1\} : out(inv1,X) \leftrightarrow in(inv2,X)$$
$$\forall X \in \{0,1\} : out(inv2,X) \leftrightarrow in(inv3,X)$$

Moreover, we have to state inconsistency of values. This can be handled using the following two rules where $\bot$ represent the contradiction (or false):

$$\forall X \in \{inv1,inv2,inv3\} : in(X,1) \wedge in(X,0) \rightarrow \bot$$
$$\forall X \in \{inv1,inv2,inv3\} : out(X,1) \wedge out(X,0) \rightarrow \bot$$

To complete formulating a diagnosis problem for the inverter chain, we define $COMP$ as set $\{inv1,inv2,inv3\}$, and OBS as $\{in(inv1,0),out(inv3,0)\}$. A diagnosis is an explanation for a deviation between expectations and observations. In MBD we are able to state whether a particular component is faulty or not using the ab predicate. Hence, in MBD an explanation for a deviation is an assignment of truth values to these health state predicates for each component in $COMP$.

Definition 2 (Diagnosis). *Given a diagnosis problem $(SD, COMP, OBS)$. A set $\Delta \subseteq COMP$ is a diagnosis if and only if $SD \cup OBS \cup \{\neg ab(C)|C \in COMP \setminus \Delta\} \cup \{ab(C)|C \in \Delta\}$ is satisfiable. A diagnosis Δ is called a minimal (or parsimonious) diagnosis, if none of its real subsets is itself a diagnosis.*

This definition considers a set of components to be a diagnosis that when assuming they are not working correctly and all other component are working as expected, there is no contradiction with the given observations. Obviously, a diagnosis exists if and only if $SD \cup COMP$ is not a contradiction. For our inverter chain, it is easy to see that we have three minimal diagnoses: $\{inv1\}$, $\{inv2\}$, and $\{inv3\}$, not being able to distinguish them using the available observations. Note that assuming all components to be faulty, is always a diagnosis. Hence, we are mainly interested in computing minimal diagnosis only. Moreover, in practice usually, we first search for single fault diagnoses, i.e., diagnoses considering only one component to be faulty, and larger diagnoses afterwards if needed.

Computing diagnoses can be easily done considering to search for subsets of the set of components $COMP$. Because of the exponential number of subsets there is a need for an efficient algorithm. Reiter [20] introduced such a diagnosis algorithm that is based on hitting set computation. For the experimental evaluation, we make use of an available Java implementation of this algorithm. For more information about diagnosis and its application, we refer the interested reader to [23].

3   Using Answer Set Programming for Diagnosis

ASP has gained a lot of attention over the past years because of its capabilities and the availability of fast tools. In particular, ASP tools allow to specify knowledge in a general form using predicates and variables and apply grounding, i.e., replacing variables with constants, before reasoning takes place. Hence, reasoning like checking for satisfiability can be done using efficient propositional SAT solvers. In the following, we explain how ASP can be used for computing diagnoses. We use the inverter chain from Fig. 1 for illustrating the use of ASP we rely on the input language of the ASP tool clingo, which is basically using a slightly extended version of Prolog [5].

As already mentioned an answer set is a satisfiable set of propositions that can be derived from the answer set program in an acyclic way. For example, the program

```
a :- b.
b :- a.
```

representing the two rules $b \rightarrow a$ and $a \rightarrow b$ has an empty answer set because they are satisfied in case a and b are both false and no proposition can be derived in an acyclic way. The situation is different in the following case:

```
a :- not b.
b :- not a.
```

In this example, we see that assuming a and b to be false would lead to an unsatisfiable set of rule. Only if either a or b is true, the rules can be satisfied. If a is false, then b must be true, because of the second rule. If b is false, then a must be true, because of the first rule. Hence, we obtain two answer sets $\{a\}$ and $\{b\}$ for this second example.

In order to use ASP for diagnosis, we have to come up with an ASP representation of our first-order model. Because of the fact that clingo and other ASP solver make use of Prolog syntax, this is simple. For the inverters, we introduce the following clingo representation:

```
out(X,1) :- inverter(X), in(X,0), nab(X).
out(X,0) :- inverter(X), in(X,1), nab(X).
in(X,1) :- inverter(X), out(X,0), nab(X).
in(X,0) :- inverter(X), out(X,1), nab(X).
```

The structure of the inverter chain can be represented as follows:

```
inverter(inv1).
inverter(inv2).
inverter(inv3).
```

```
in(inv2,X)  :- out(inv1,X).
out(inv1,X)  :- in(inv2,X).
in(inv3,X)  :- out(inv2,X).
out(inv2,X)  :- in(inv3,X).
```

Note that in the representation, we map the bijunction $\leftrightarrow$ to two horn clause rules. What is missing is the representation of the integrity constraints for values, and the way we handle diagnosis. We start with the integrity constraints:

```
:- in(X,1), in(X,0).
:- out(X,1), out(X,0).
```

It is worth noting that this representation is more or less a 1-to-1 representation of the logical model introduced in the previous section. clingo automatically grounds the used variables in the ASP programs given. Hence, there is no need to specify domains for variables in this case.

To make use of ASP for diagnosis, we have to be able representing the health assumptions $ab(C)$. Let us have a look at the second ASP example of this section. There we made use of negation to come up with a situation where the ASP solver can decide either a or b to be true. This is exactly what we need. Instead of propositions a and b, we consider $ab(C)$ and its negated representation $nab(C)$ for all components C. For our inverter example, we introduce the following rules for the predicates representing the health status of components:

```
comp(X)  :- inverter(X).

nab(X)  :- comp(X), not ab(X).
ab(X)  :- comp(X), not nab(X).
```

Note that this representation is very much general being able to handle different components. The first rule states that inverters are components. If we come up with new types of components, we only need to add similar rules. The last two rules state that either a component is 'not abnormal' (*nab*) in case it is not *ab*, or it is 'abnormal' but only if it is not *nab*.

What is missing is a possibility to state that we are only interested in single faults. This is the case where there is only one *ab* predicate true. clingo provides means to handle this case. We are able to formulate a rule counting the number of *ab*'s. Via further constraining the number of *ab*'s we are able to restrict the ASP solver searching for single faults only. The following two rules server this purpose when using clingo:

```
no_ab(N)  :- N = #count { C : ab(C) }.

:- not no_ab(1).
```

When running clingo using the described model together with two facts representing the observations, i.e., in(inv1,0). out(inv3,0), we obtain the following results:

```
clingo version 5.4.0
Reading from digital_circuit_diagnosis.pl
```

```
Solving...
Answer: 1
inverter(inv1) inverter(inv2) inverter(inv3) comp(inv1)
    comp(inv2) comp(inv3) in(inv1,0) out(inv3,0) nab(inv1)
    ab(inv2) nab(inv3) out(inv1,1) in(inv3,1) out(inv2,1)
    in(inv2,1) no_ab(1)
Answer: 2
inverter(inv1) inverter(inv2) inverter(inv3) comp(inv1)
    comp(inv2) comp(inv3) in(inv1,0) out(inv3,0) nab(inv1)
    nab(inv2) ab(inv3) out(inv1,1) in(inv2,1) out(inv2,0)
    in(inv3,0) no_ab(1)
Answer: 3
inverter(inv1) inverter(inv2) inverter(inv3) comp(inv1)
    comp(inv2) comp(inv3) in(inv1,0) out(inv3,0) ab(inv1)
    nab(inv2) nab(inv3) in(inv3,1) out(inv2,1) in(inv2,0)
    out(inv1,0) no_ab(1)
SATISFIABLE

Models      : 3
Calls       : 1
Time        : 0.002s (Solving: 0.00s 1st Model: 0.00s
    Unsat: 0.00s)
CPU Time    : 0.002s
```

The outcome contains the three single fault diagnosis and also all values of gate inputs and outputs that have to hold for a particular diagnosis.

In summary, making use of ASP solvers like clingo for diagnosis is rather straightforward. The logical models used for diagnosis can be more or less directly mapped to a form that corresponds with the syntax of the input of the solver. For all components $C \in COMP$, we have to introduce two additional rules allowing the ASP solver to select the truth value of the health states. In addition, we have to restrict the search space of the solver introducing rules stating that we are only interested in single fault diagnoses. It is worth noting that we also can make use of ASP for computing double, triple or even all diagnoses. In this case, we only have to change the corresponding cardinality constraint no_ab accordingly.

4   Experimental Evaluation

In order to answer the research question, whether ASP can be effectively used for implementing MBD in practice, we first carried out an experimental study that is based on Boolean circuits. In particular, the study is based on the well known ISCAS85 circuits[2]. In Table 1, we give an overview of the statistics of the different Boolean circuits including the number of inputs, outputs, and gates. These circuits have been often used for comparing diagnosis algorithms. Most recently, Nica et al. [15] compared 6 diagnosis algorithms making use of more than 300 different input/output combinations and considering single, double,

---

[2] See http://www.cbl.ncsu.edu:16080/benchmarks/ISCAS85/.

and triple faults to be identified and located. In our experimental evaluation, we rely on the same test cases but consider only those corresponding to single fault diagnoses. Hence, we only used 100 test cases, which correspond to single fault diagnoses, in our experimental evaluation.

Table 1. ISCAS85 circuit statistics.

Circuit	#Inputs	#Outputs	#Gates	Function
c432	36	7	160	27-ch. interrupt controller
c499	41	32	202	32-bit SEC circuit
c880	60	26	383	8-bit ALU
c1355	41	32	546	32-bit SEC circuit
c1908	33	25	880	16-bit SEC/DED circuit
c2670	233	140	1,193	12-bit ALU and controller
c3540	50	22	1,559	8-bit ALU
c5315	178	123	2,307	9-bit ALU
c6288	32	32	2,406	16-bit multiplier
c7752	207	108	3,512	32-bit adder/comparator

In the following we compare the ASP diagnosis implementation for the different ISCAS85 benchmarks handling single faults using `clingo` version 5.4.0, with an hitting set based implementation HSDIAG, which has been used previously in [15]. For carrying out the experiments, we made use of a MacBook Pro (15-in, 2016), 2.9 GHz Quad-Core Intel Core i7, with 16 GB 2133 MHz LPDDR3 main memory, running under macOS Catalina Version 10.15.2. In Table 2 we summarize the obtain running times for the ASP and the HSDIAG algorithms where we report the minimum, maximum, average, median, and standard deviation. In addition, we also give the number of atoms $\#A$ and rules $\#R$ considered by the ASP solver, and the number of literals $\#L$ and rules $\#R$ HSDIAG has to handle.

It is worth noting that in case of the ASP solver `clingo` optimization takes place before computing all answer sets. The original number of rules has to be higher than in the case of HSDIAG because we require two additional rules for each component in the circuit. For running ASP and HSDIAG, we set the number of diagnoses to be computed to 100. For both diagnosis algorithms, we made use of the standard setup provided and did not consider further optimizations.

From Fig. 2 we see that HSDIAG has a faster median running time for all ISCAS85 benchmark circuits. In case of the average running time, HSDIAG is faster than ASP in 7 from 10 cases. However, ASP provides a reasonable running time ranging from about 0.1 to 114 s for all benchmarks whereas the time required to compute 100 diagnosis in case of HSDIAG varies from about 0.01 to 474 s. In addition, ASP was able to provide diagnoses for all 100 test cases whereas in case of HSDIAG we were not able to compute diagnoses for

Table 2. Average diagnosis time (in milliseconds) required for computing all single fault diagnosis using ASP (a) and HSDIAG (b). For the evaluation we used 10 different diagnosis examples for each circuit

Circuit	#A	#R	Min	Max	Avg	Median	Std. Dev.
c432	3,324	1,667	107.0	128.0	119.0	118.5	5.7
c499	4,404	2,107	161.0	247.0	194.5	171.0	34.8
c880	7,574	3,995	268.0	423.0	366.9	394.0	50.6
c1355	10,700	5,547	484.0	675.0	537.3	520.0	52.7
c1908	16,779	8,861	1,098.0	3,520.0	2,023.3	1,931.0	701.4
c2670	23,014	12,480	1,490.0	3,842.0	2,838.0	2,935.5	763.2
c3540	31,979	17,222	4,682.0	10,802.0	7,680.3	7,387.5	2,002.7
c5315	45,227	23,907	6,046.0	31,400.0	18,491.5	18,336.0	8,422.7
c6288	47,775	26,316	6,393.0	11,431.0	7,577.2	6,942.5	1,582.8
c7552	67,516	36,060	14,503.0	114,238.0	53,229.1	43,902.5	31,450.0

(a) APS solver

Circuit	#L	#R	Min	Max	Avg	Median	Std. Dev.
c432	5,391	1,497	3.0	76.9	29.1	14.0	26.82
c499	8,350	2,262	4.0	27.0	12.9	7.0	9.9
c880	10,293	3,099	4.0	34.0	16.6	10.0	10.9
c1355	14,758	4,398	28.0	174.0	62.5	38.5	46.6
c1908	21,555	6,345	7.0	42,489.0	4,407.7	233.5	12,694.4
c2670	30,021	9,144	9.0	1,045.0	165.9	41.5	301.9
c3540	41,653	12,277	45.0	145,755.0	14,794.6	152.5	43,654.0
c5315	62,098	18,251	31.0	2,003.0	411.6	164.5	579.3
c6288[*]	65,008	19,328	200.0	474,400.0	123,695.3	2,796.0	195,766.4
c7552	86,791	25,977	150.0	9,082.0	1,899.7	391.0	3,045.8

(b) HSDIAG ([*] Failed to run all examples)

the c6288 in two cases. In these two cases, we stopped computing diagnosis using HSDIAG after 15 min. In summary, we conclude that HSDIAG provides often faster diagnosis computation, is not able to handle all cases, and has a larger running time variation compared to ASP. This is very much visible when considering Fig. 2 where we depict all running times for all executed tests.

Taking the obtained experimental results and other facts into account, we are able to answer our original research question. ASP can be used for practical applications, whenever there are no tight time constraints. Using ASP we are able to provide diagnosis solutions even in cases where specialized diagnosis algorithm fail. The running time of ASP seems to be more predictable and ASP offers enhanced functionalities like handling of numbers. Hence, for smaller

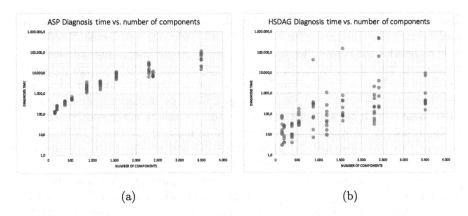

Fig. 2. Diagnosis time as a function of the number of components for ASP (a) and HSDIAG (b).

systems ASP seems to be a good bases for diagnostic applications. The pros speaking for using ASPs for diagnosis are:

- Time required for computing at least single fault diagnoses seems to be predictable considering the size of systems.
- ASP provides diagnoses even in cases where other methods cannot within a given time limit.
- The coding of diagnosis using ASP is rather straightforward.
- ASP solvers like `clingo` are able to handle other data types like numbers and not only Boolean values.
- The overall running time seems to be fast enough for a wider range of applications.
- ASP solvers might offer an interface to ordinary programming languages allowing an easy integration into existing applications.

The cons of ASP for diagnosis applications can be summarized as follows:

- Diagnosis using ASP cannot beat specialized diagnosis algorithms with respect to running time in most cases.
- Time required to compute diagnoses might be too high in cases where we need diagnoses to be computed within fractions of a second even for larger systems.

Hence, depending on the application domain using ASP solvers for diagnosis seems to be an appealing alternative to other theorem provers, SAT solvers, or constraint solvers.

5 Conclusion

In this paper, we have shown how to utilize current ASP solvers like `clingo` for computing diagnoses based on models. For this purpose, we introduced rules allowing the ASP solver setting the truth value of predicates representing the health state of components automatically. In addition, we showed how we are able to restrict the computation of diagnoses to single faults only. Furthermore, we answered the question whether today's ASP solvers are fast enough to be used for diagnosis in practice. We carried out an experimental evaluation comparing the diagnosis time required using `clingo` and a specialized implementation of MBD based on Reiter's algorithm. As a result we saw that the specialized diagnosis implementation was faster on average in most of the cases. However, the running time performance of `clingo` was still good enough for many practical applications not requiring immediate feedback.

Moreover, APS solvers offer additional functionality like handling numbers. There is no need to implement a specialized diagnosis algorithm, avoiding additional costs for verification and validation activities. `clingo` also allows a tighter integration with ordinary programming languages like Python. Hence, for a wider range of applications the use of ASP solvers for diagnosis seems to be an attractive and valuable alternative to specialized diagnosis implementation. Future research includes extending the evaluation to double and triple faults. We also plan to make use of ASP for diagnosing programs and spreadsheets. In addition, it would be of interest to do research on learning rules for diagnosis, e.g., considering association rule mining [12].

Acknowledgement. The research was supported by ECSEL JU under the project H2020 826060 AI4DI - Artificial Intelligence for Digitising Industry. AI4DI is funded by the Austrian Federal Ministry of Transport, Innovation and Technology (BMVIT) under the program "ICT of the Future" between May 2019 and April 2022. More information can be retrieved from https://iktderzukunft.at/en/ bmⓋⓕ.

References

1. Abreu, R., Hofer, B., Perez, A., Wotawa, F.: Using constraints to diagnose faulty spreadsheets. Softw. Qual. J., 1–26 (2014). https://doi.org/10.1007/s11219-014-9236-4
2. Beschta, A., Dressler, O., Freitag, H., Montag, M., Struss, P.: A model-based approach to fault localization in power transmission networks. Intell. Syst. Eng. (1992)
3. Buchanan, B.G., Shortliffe, E.H. (eds.): Rule-Based Expert Systems - The MYCIN Experiments of the Stanford Heuristic Programming Project. Addison-Wesley Publishing Company (1984)
4. Chen, Y., Wan, H., Zhang, Y., Zhou, Y.: dl2asp: implementing default logic via answer set programming. In: Janhunen, T., Niemelä, I. (eds.) JELIA 2010. LNCS (LNAI), vol. 6341, pp. 104–116. Springer, Heidelberg (2010). https://doi.org/10.1007/978-3-642-15675-5_11
5. Clocksin, W.F., Mellish, C.S.: Programming in Prolog, 5th edn. Springer, Heidelberg (2003). https://doi.org/10.1007/978-3-642-55481-0
6. Davis, R.: Diagnostic reasoning based on structure and behavior. Artif. Intell. **24**, 347–410 (1984)
7. Eiter, T., Ianni, G., Krennwallner, T.: Answer set programming: a primer. In: Tessaris, S., et al. (eds.) Reasoning Web 2009. LNCS, vol. 5689, pp. 40–110. Springer, Heidelberg (2009). https://doi.org/10.1007/978-3-642-03754-2_2

8. Felfernig, A., Schubert, M., Zehentner, C.: An efficient diagnosis algorithm for inconsistent constraint sets. AI EDAM **26**(1), 53–62 (2012). https://doi.org/10.1017/S0890060411000011

9. Friedrich, G., Stumptner, M., Wotawa, F.: Model-based diagnosis of hardware designs. Artif. Intell. **111**(2), 3–39 (1999)

10. Jannach, D., Schmitz, T.: Model-based diagnosis of spreadsheet programs: a constraint-based debugging approach. Autom. Softw. Eng., 1–40 (2014). https://doi.org/10.1007/s10515-014-0141-7

11. de Kleer, J., Williams, B.C.: Diagnosing multiple faults. Artif. Intell. **32**(1), 97–130 (1987)

12. Luna, J.M., Fournier-Viger, P., Ventura, S.: Frequent itemset mining: a 25 years review. Wiley Interdiscip. Rev. Data Mining Knowl. Discov. **9**(6) (2019). https://doi.org/10.1002/widm.1329

13. Mathonet, R., Cotthem, H.V., Vanryckeghem, L.: DANTES – an expert system for real-time network troubleshooting. In: Proceedings IJCAI, pp. 527–530. Morgan Kaufmann, Milano, August 1987

14. Milde, H., Guckenbiehl, T., Malik, A., Neumann, B., Struss, P.: Integrating model-based diagnosis techniques into current work processes - three case studies from the INDIA project. AI Commun. **13** (2000). Special Issue on Industrial Applications of Model-Based Reasoning

15. Nica, I., Pill, I., Quaritsch, T., Wotawa, F.: The route to success - a performance comparison of diagnosis algorithms. In: IJCAI, pp. 1039–1045. IJCAI/AAAI (2013)

16. Pell, B., et al.: A remote-agent prototype for spacecraft autonomy. In: Proceedings of the SPIE Conference on Optical Science, Engineering, and Instrumentation, Volume on Space Sciencecraft Control and Tracking in the New Millennium. Bellingham, Waschington, U.S.A., Society of Professional Image Engineers (1996)

17. Picardi, C., et al.: IDD: integrating diagnosis in the design of automotive systems. In: Proceedings of the European Conference on Artificial Intelligence (ECAI), pp. 628–632. IOS Press, Lyon (2002)

18. Rajan, K., et al.: Remote agent: an autonomous control system for the new millennium. In: Proceedings of the 14th European Conference on Artificial Intelligence (ECAI), Berlin, August 2000

19. Reiter, R.: A logic for default reasoning. Artif. Intell. **13**(1–2) (1980)

20. Reiter, R.: A theory of diagnosis from first principles. Artif. Intell. **32**(1), 57–95 (1987)

21. Sachenbacher, M., Struss, P., Carlén, C.M.: A prototype for model-based on-board diagnosis of automotive systems. AI Commun. **13** (2000). Special Issue on Industrial Applications of Model-Based Reasoning

22. Wotawa, F., Stumptner, M., Mayer, W.: Model-based debugging or how to diagnose programs automatically. In: Hendtlass, T., Ali, M. (eds.) IEA/AIE 2002. LNCS (LNAI), vol. 2358, pp. 746–757. Springer, Heidelberg (2002). https://doi.org/10.1007/3-540-48035-8_72

23. Wotawa, F.: Reasoning from first principles for self-adaptive and autonomous systems. In: Lughofer, E., Sayed-Mouchaweh, M. (eds.) Predictive Maintenance in Dynamic Systems, pp. 427–460. Springer, Cham (2019). https://doi.org/10.1007/978-3-030-05645-2_15

Decision-Support and Agent-Based Systems

Consensus-Based Protocol for Distributed Exploration and Mapping

Zilong Jiao[(✉)] and Jae Oh[(✉)]

Department of Electrical Engineering and Computer Science,
Syracuse University, Syracuse, NY 13244, USA
`{zijiao,jcoh}@syr.edu`

Abstract. Distributed exploration in multi-agent systems requires agents to retain a consistent view of the environment, even under limited communication ranges. We propose a consensus-based protocol for distributed exploration that enables agents to synchronize their local maps so that their collective global views are consistent. With the proposed protocol, agents can dynamically form communication networks and synchronize their local environment maps. In contrast to the existing consensus-based solutions, our work is computationally efficient and provides a convergence guarantee under communication loss. Through our extensive experiments, we show that the proposed protocol enables agents to build consistent environment maps collaboratively and efficiently. We also show that agents can significantly save their communication bandwidth and reach optimal solutions in the presence of communication loss.

Keywords: Multi-agent systems · Distributed exploration · Collaborative mapping · Consensus-based protocol

1 Introduction

In multi-agent distributed exploration tasks, each agent keeps its environmental view with a local environment map. The sum of the information from the local maps will give a collective view of the entire environment so that agents can navigate the environment and perform tasks collaboratively and effectively. With limited communication ranges, some local maps of the agents can have inconsistent views of the environment due to the difficulties of reconciling discrepancies among local maps. The goal of the exploration in this paper is that each location in the environment is visited by one agent only and only once so that the combined maps represent accurate global views. If the collective maps are not consistent, agents may visit the same location multiple times and therefore wasting resources.

Designing an effective consensus-based protocol is challenging. First, agents with limited resources cannot tolerate intensive communication. Second, the convergence guarantee for an optimal map building is difficult when a protocol has

© Springer Nature Switzerland AG 2020
H. Fujita et al. (Eds.): IEA/AIE 2020, LNAI 12144, pp. 533–544, 2020.
https://doi.org/10.1007/978-3-030-55789-8_46

complex rules. Third, it is hard to ensure the connectivity of the communication network formed by moving agents, particularly in the presence of communication loss [8]. Although the convergence of min, max, and average consensus solutions have been proven under different dynamics of communication networks [4,9,14–17], those existing solutions are communication heavy.

To address the challenges mentioned above, we propose a consensus-based protocol for agents to synchronize their grid-based local environment maps in distributed map building and exploration. Following the proposed protocol, agents can dynamically construct connected communication networks and update their local environment maps until convergence. The proposed protocol is computationally efficient because it allows each location of the grid to be visited only once. It allowed agents to keep track of their communication history and avoid exchanging redundant information. We experimentally show that the proposed protocol can preserve its convergence guarantee under communication loss. We evaluated the proposed protocol in the well-known distributed frontier-based exploration [3] problem. In extensive experiments, we analyzed the impact of the proposed protocol on agent exploration performance and demonstrated its efficiency in terms of its communication demand. We showed that the proposed protocol was robust and enabled agents to build consistent maps under communication loss.

Researchers actively study Consensus-based algorithms in distributed multi-agent systems. Aragues et al. [1] proposed an offline method where agents used a consensus algorithm to estimate their global frame of reference for merging local environment maps. Later, the authors [2] extended their work as an online method, which allowed agents to merge feature-based environment maps during exploration. In [2], agents estimated their average position through a consensus approach, so that they can correctly merge their local maps. Different from those existing methods, our work focused on grid-based environment maps. The proposed protocol enabled agents to avoid exchanging redundant information and was robust against communication loss.

We organized the rest of the paper as follows. In Sect. 2, we briefly discussed theoretical consensus methods and their applications in practice. In Sect. 3, we formulated multi-agent collaborative mapping as a consensus problem. We proposed consensus-based protocol in Sect. 4, followed by its evaluation in Sect. 5. Finally, Sect. 6 concluded the paper.

2   Related Work

In an exploration task, agents can navigate with and without environment maps [10–12]. In the map-based approach, agents must estimate a consistent view of an environment. To this end, the consensus in the context of multi-agent systems has been extensively studied in theory [4,15]. The convergence of max, min, and average consensus algorithms have been proven under different communication network dynamics [14,16,17]. However, the impact of communication loss on consensus-based algorithms is under-explored. In this paper, we prove that the proposed consensus-based protocol can preserve its convergence guarantee in the presence of problematically modeled communication loss.

There is a limited number of consensus-based algorithms for distributed exploration. In [1], agents use an offline consensus-based algorithm to estimate their global reference frame, to merge their independently built local environment maps after exploration. In [2], during exploration, agents periodically merge their local feature-based environment maps to a global map based on consensus. In general, solutions of consensus among agents have a high demand for communication and are sensitive to a topological structure of an underlying communication network. It is challenging to apply a consensus-based solution to a multi-agent system, where agents have constrained energy and dynamically changing communication networks. In contrast to those existing methods, we propose a consensus-based protocol that can be well scaled for a large number of agents to share a grid-based environment map in a distributed setting. The proposed protocol is computationally efficient. It saves communication bandwidth by preventing agents from sharing redundant information.

Besides consensus-based algorithms, alternative communication mechanisms were proposed for agents to exchange environmental information in distributed exploration. In [6], agents use a rendezvous technique to verify their relative locations before meeting each other to merge their environment maps. In [18], agents form auction and bid for their next places to visit during exploration. In recent literature, role-based communication mechanisms are proposed for agents with limited communication ranges. In [7], Beacon agents constitute a wireless mobile ad hoc network to provide gradient signals for Walker agents to navigate between targets and their home-base. In [5], Relay agents meet the other agents at way-points to collect and transport their gathered environmental information to their home-base. Different from the communication mechanisms mentioned above, the proposed consensus-based protocol avoids allocating dedicated agents for communication purposed and enables agents to better focus on exploring their environment.

3   Problem Statement

In this paper, we study a consensus-based communication protocol in the context of distributed exploration. The proposed protocol serves two purposes. It enables agents to form communication networks during exploration, and 2) allows agents within the same communication network to synchronize their local environment maps.

Let $\mathcal{N} = \{1, 2, \ldots, n\}$ be a set of agents. Each agent $i \in \mathcal{N}$ has an $n \times m$ grid M_i which represent the environment mapped by the agent. Each cell c_i in M_i corresponds to a square area in the environment where the agent i operates. c_i is associated with a label $L(c_i) \in \{open, occupied, unknown\}$. Let $M_i(t)$ be the environment map built by the agent i at time t; $x_i(t)$ be the state of the agent i at time t. We define $x_i(t)$ as $M_i(t)$, s.t.

$$x_i(t) = M_i(t) \tag{1}$$

Let $G(t) = (V, E)$ denote an unweighted undirected graph representing a communication network formed by some agents $V \subset \mathcal{N}$ at time t. E is a set of edges, where each $e_{ij} \in E$ indicates that the agent i and j can communicate with each other. For each agent $i \in V$, the agent i updates its state $x_i(t)$ according to a protocol $\mu_i(t)$, s.t.

$$\dot{x}_i(t) = \mu_i(t) \tag{2}$$

$\dot{x}_i(t)$ is the updated $x_i(t)$. Let $x_i(t)$ be a initial state of agent i at time t; $\boldsymbol{x}(t)$ be a vector of $x_i(t)$, $\forall i \in V$. We say agents in V reach χ-consensus, if there is a stable state $x^* = \chi(\boldsymbol{x})$, s.t. $\forall i \in V$, $x_i(t + \delta) = x^*$ with $\delta \to \infty$.

4   Consensus-Based Protocol

Agents following the proposed protocol can form connected communication networks and synchronize their local environment maps during exploration under possible communication loss. The consensus-based protocol allows agents to keeps track of their communication history and avoids communicating redundant information. We show, through experiments, that the proposed protocol can preserve its convergence guarantee in the presence of communication loss.

Let $G(t) = (V, E)$ be a connected graph representing a communication network constituted by a set of agents $V \subset \mathcal{N}$ at time t. For each $i \in V$, let $S_i^{kn}(t)$ be a set of cells known to agent i, s.t.

$$S_i^{kn}(t) = \{c_i \mid c_i \text{ is a cell of } M_i(t) \text{ and } L(c) \neq unknown\}$$

Let $N(i)$ be the neighbors of agent i; h be a function, s.t.

$$h(c_k, k) = \begin{cases} 1, & \text{if } c_k \in S_k^{kn}, \forall k \in N(i). \\ 0, & \text{otherwise.} \end{cases} \tag{3}$$

To avoid broadcasting redundant cells, at time t, an agent i will broadcast a set of cells $S_i^{st}(t + \delta)$, s.t.

$$S_i^{st}(t) = \{c_i \mid c_i \in S_i^{kn}(t) \text{ and } \exists k \in N(i), \ h(c_i, k) = 0\}$$

Intuitively, $S_i^{st}(t + \delta)$ contains all the cells in $M_i(t)$ that haven't been sent to all the neighbors of agent i. Let δ be the time that have passed after the initial time t. At time $t + \delta$, each $i \in V$ updates its state $x_i(t + \delta)$ according to the consensus protocol

$$\mu_i(t + \delta) = l(\bigcup_{k \in N(i)} S_k^{st}(t + \delta)) \tag{4}$$

l is a function which does error correction for combining received cell values into $M_i(t + \delta)$. Specifically, we represent each cell state with a unique value, s.t., $open = 0$, $occupied = 1$, and $unknown = 2$. When a cell have conflicting values, l set the value of the cell as the min of the conflicting values. With the predefined value setting, it will be guaranteed to set an unknown cell as a $open$ or $occupied$ cell, if there is a $open$ or $occupied$ among the conflicting values.

Theorem 1. *Let $G(t + \delta) = (V, E)$, $\forall \delta \in \mathbb{N}$, be a connected graph with a set of nodes, V and a set of edges, E, where t is the time when G is initially formed and δ is the passed time steps after t. Suppose $G(t + \delta), \forall \delta \in \mathbb{N}$, has a fixed topology, and there is no communication loss among agents. As $\delta \to |V|$, protocol 4 can have all $i \in V$ reach χ-consensus, where each $M_i(t)$ is merged with $M_j(t), \forall j \in V/\{i\}$.*

Proof. We first prove the convergence of the proposed protocol (i.e., Eq. 4) by contradiction, and then we derive the upper bound of δ for agents to reach consensus. $\mathcal{S} = \bigcup_{\forall i \in V} S_i^{kn}(t)$ is the all possible known occupancy probabilities that can be exchanged by all agents over the period of δ. According to the update function l, the number of known cells in agent i's environment map $M_i(t + \delta)$ is monotonically increasing with respect to δ. Here, we say a agent $i \in V$ has a cell $c \in \mathcal{S}$, if the cell c is known in $M_i(t + \delta)$.

Suppose $S_i^{kn}(t + \delta) \subset S^{kn}$, as $\delta \to \infty$. In this case, let j be an agent that can be reached by i, and j does not have a cell c. According to the definition of l, all the agents in $N(j)$ must not have c in their environment map. Besides, the neighbors of agents in $N(j)$ must not have c too. By induction, all the agents in G that can be reached from v_i must not contain c. Since $G(t + \delta)$ is connected and has a fixed topology for all $\delta \in \mathbb{N}$, there is always a path from i to each of the other agents. This contradicts the assumption of $S_i^{kn}(t + \delta) \subset S^{kn}$. Therefore, $S_i^{kn}(t + \delta) = S^{kn}$ for all $i \in V$, as $\delta \to \infty$.

To derive the upper bound of δ, we transform the application of protocol 4 to information cascade on $G(t)$. We say $c_j \in S_j^{kn}(t)$ is propagated from j to i, if i receives c_j from j. At each time step, each agent $i \in V$ propagate all $c_i \in S_i^{kn}(t)$ to all the other agents. For all agents in V, $S_i^{kn}(t)$ are propagated in parallel. Note that, at each time step, each cell c_i in $S_i^{kn}(t)$ can be propagated one hop between two agents, and the propagation stops when an agent has already had c_i. In this case, there is no cycle during the propagation. Because of the parallel propagation, the time for agents to reach consensus is determined by the longest simple path that c_i is propagated through. In $G(t)$, the longest simple path between any two agents has the length of $|V| - 1$. Therefore, the time step for agents in V to reach χ-consensus is bounded by $|V|$.

To synchronize local maps of agents, the protocol 4 requires $G(t + \delta)$ to remain connected for all $\delta \in \mathcal{N}$. To constitute a connected communication network during exploration, each agent in V try to connect to its nearby agents periodically. An agent i stops moving when it connects to another agent, and then it starts broadcasting $S_i(t)$. Let t be the time when agent i joins a communication network. Algorithm 1 presents the protocol for agents to constitute a connected communication network during exploration.

An agent considers agents in the same communication network reached a consensus if its local environment map has not been updated for a certain amount of time. Note that, according to protocol 4, two agents will broadcast empty sets of cells to each other, if they have already shared all the known occupancy probabilities in their own environment maps.

Algorithm 1. Forming connected communication networks

1: **function** CONSTITUTENETWORK()
2:     listen for messages from other agents
3:     **if** connected to agent k **then**
4:         **if** $k \notin N(i)$ **then**
5:             Stop moving
6:             $N(i) \leftarrow N(i) \cup \{k\}$
7:             use protocol 4 to update $M_i(t)$ based on $S_k^{st}(t + \delta)$
8:             broadcast $S_i^{st}(t + \delta)$
9:             $\delta \leftarrow \delta + 1$
10:     **if** $N(i) = \{\}$ or reached consensus **then**
11:         continue to explore the environment
12:         $\delta \leftarrow 0$

Theorem 2. *At any time t, the communication network $G(t + \delta), \forall \delta \in \mathbb{N}$, constituted by agents using Algorithm 1 is connected.*

Proof. Based on Algorithm 1, during exploration an communication network is initialized two agents. Given a communication network $G(t)$ consisting of n agents, $n \geq 2$. Suppose $G_0(t)$ is $G(t)$'s initial network consisting of two agents i and j. Let k be the third agent joining $G(t)$ at time $t + 1$. In this case, k must be able to communicate with either i or j. Since $G(t + 1)$ is undirected, there is a path between any two of i, j and k. Therefore, $G(t + 1)$ is connected.

To prove the network connectivity by induction, we suppose $G(t + \delta), \forall \delta \in \mathbb{N}$, is a connected undirected graph representing a communication network formed by a set of agents V at time $t + \delta$. Let x be an agent joining $G(t + \delta)$ at time $t + \delta + 1$. In this case, x must be able to communicate with at least one agent $i \in V$. Since $G(t + \delta)$ is connected, each $j \in V/\{i\}$ has at least one path to i, and vice versa. Since for each $i \in V$ there is at least one path between i and x, $G(t + \delta + 1)$ is a connected graph. By induction, $G(t + \delta), \forall \delta \in \mathbb{N}$ is connected. Therefore, a communication network constituted by agents using Algorithm 1 is always connected.

With Theorem 2 and 1, we prove that agents using Algorithm 1 can eventually form a connected communication network with fixed topology, and the proposed protocol can let agents eventually reach consensus of their environment maps. At last, Theorem 3 prove that the agents can preserve the convergence guarantee in presence of communication loss.

Theorem 3. *Let $G(t + \delta) = (V, E), \forall \delta \in \mathbb{N}$, be a connected communication network formed by a set of agents V at time t. Suppose $G(t + \delta)$ is a weighted graph, where an edge $e_{ij} \in E$ represents the probability of agent i receiving information from agent j through communication. As $\delta \to \infty$ all agents in V can reach χ-consensus, where each $M_i(t)$ is merged with $M_j(t), \forall j \in V/\{i\}$.*

Proof. Based on Theorem 1, we prove the convergence of the proposed consensus protocol in the weighted communication network, if we can guarantee that each agent in V can eventually receive all the known cells in all its neighbors' environment map. Note that the proposed protocol does not consider the cells whose occupancy probabilities are unknown to all agents. Therefore, those cells will not have any impact on convergence.

Let $N(v_i)$ be a set of neighbors of agent i. Suppose cell c_i unknown in $M_i(t)$ but known by at least one $k \in N(i)$. At each time step, the probability of agent i updating c_i to be known is $\Sigma_{v_k \in N(v_i)} \alpha e_{ik}$, where α is a decision variable. $\alpha = 1$ if c_i is known by k. Otherwise, $\alpha = 0$. According to Eq. 3 and 4, each $k \in N(i)$ will repeatedly broadcast c_i, if c_i known by agent k. Given δ time steps, the probability of agent i marking c_i to be known is $\delta \Sigma_{v_k \in N(v_i)} \alpha e_{ik}$. According to the equation, this probability will monotonically increase with respect to δ. Therefore, as $\delta \to \infty$, it is guaranteed that agent i will update c_i to be known based on the cells broadcast by its neighbors.

5 Experiment

5.1 Experiment Setup

In the experiments, we integrate the proposed protocol with the frontier-based distributed exploration [3]. The experiments are implemented in C++ using Gazebo [13], and they are executed on a server with 8 CPUs and 128 GB memory. The simulated agents have diameters of 0.5 m and communication ranges of 20 m. Each agent has a LiDar sensor with a maximum range of 10 m. The sensor scans an agent's surrounding environment with 360, evenly spaced lasers. The perceived range measures contain small Gaussian noises with the zero mean and the standard deviation of 0.1.

During exploration, an agent uses its LiDar sensor for both environment mapping and collision detection. It has a linear motion model [19] and moves at a constant speed. For collision detection, each agent senses the environment within the corn of $120°$ in its front at a constant rate. If the agent detects objects that are 0.8 m away, it stops moving immediately and then keeps turning to its right until no objects can be detected.

5.2 Impact on Exploration Efficiency

To understand the impact of the proposed protocol on environment exploration, we apply it to the frontier-based exploration, where agents always move to their closest frontiers.

Having a consistent view of the mapped environment can be easily achieved when agents have unlimited communication ranges (i.e., each agent can always communicate with all the other agents.). In contrast, the proposed protocol enables agents with limited communication ranges to have consistent local environment maps after reaching consensus. To evaluate its impact on exploration, we compare the performance of the frontier-based exploration algorithms with and without the proposed protocol. Without the proposed protocol, we assume

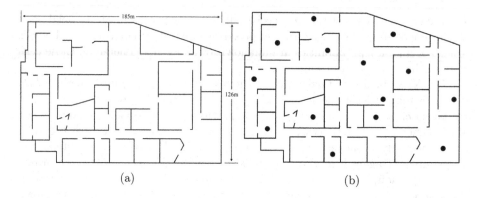

Fig. 1. An indoor environments with and without randomly distributed obstacles.

that agents have unlimited communication ranges. For each number of agents, the result is based on their average performance in 12 simulation runs, and the error bars are standard deviation.

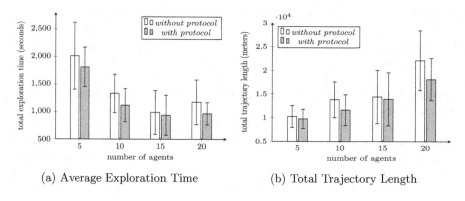

(a) Average Exploration Time   (b) Total Trajectory Length

Fig. 2. For the environment presented in Fig. 1.

As presented in Fig. 2a, the proposed protocol enabled agents with limited communication to conduct faster exploration. The results indicate that the time for agents to reach consensus is reasonable and does not introduce significant overhead. The resulting trajectory lengths show that the proposed protocol can enable agents to complete their exploration with shorter travel distances. However, according to exploration time, agents have decrease exploration performance, when their density exceeds a certain threshold (i.e., 15 agents). The reason is that, as the density of agents increases, their interference between each other (i.e., agents avoid colliding each other) becomes more significant.

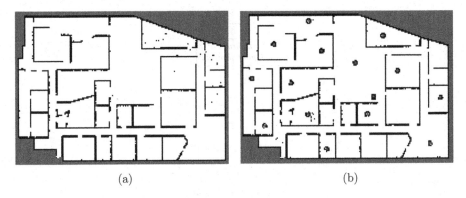

(a)   (b)

Fig. 3. The environment maps built by 10 agents following the proposed protocol.

5.3   Collaborative Mapping

The proposed protocol must enable agents to build correct environment map after the exploration. We let 10 agents map both environments presented in Fig. 1 using the proposed protocol. Figure 3 presents the resulting maps of both environments at the end of exploration. Each cell in an agent's local environment map corresponds to a 1 m × 1 m area in the environment. In general, agents following the proposed protocol can correctly merge their local environment maps and estimate the overall maps in both environments. The correctness of the map can be verified by comparing the structure of the mapped obstacles with ground-truth presented in Fig. 3. Besides the structures of mapped walls, one can clearly identify the randomly distributed obstacles based on their shapes and sizes in Fig. 3b. With a single LiDar sensor, an agent can not distinguish obstacles from the other agents. The small squares scattered in Fig. 3 are agents that are mistakenly detected as obstacles. Since sensor data of agents consists of small Gaussian noises, agents can map the same obstacles into slightly different cells in their local environment maps. That causes the mapped obstacles to have non-smooth edges.

5.4   Communication Demand

Following the proposed protocol, each agent keeps track of its communication history. This is done by having an agent maintaining a simple data structure to keep track of the cells that have been broadcast to each of the other agents. For the evaluation of communication demand, we compare the amount of information exchanged by 10 agents with and without following the proposed protocol during exploration. We approximate the amount of information exchange as the total number of occupancy probabilities broadcast by all the agents during exploration. Without following the proposed protocol, each agent in the same communication network broadcasts all the known occupancy probabilities in its local environment map until convergence. The experiment results are presented

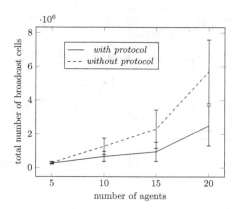

Fig. 4. The total amount of information exchanged by 10 agents with and without following the proposed protocol in their frontier-based exploration.

Table 1. The total number of times agents broadcast the cells of their occupancy matrices in the proposed approach and in the *closest frontier exploration*.

Number of agents	Total broadcast times	
	Proposed approach	*Closest frontier exploration*
5	251	335
10	690	1116
15	1428	1683
20	2261	3475

in Fig. 4, where the error bars represent the standard deviation. Table 1 summarizes the amount of exchanged information, given different number of agents. For each number of agents, the result is based on their average performance in 12 simulation runs.

With the proposed protocol, each agent only broadcasts the cells that haven't been sent to all its current neighbors. As presented in Fig. 4, the proposed protocol can significantly save the communication bandwidth for sharing information. In the naive approach, an agent has to broadcast all the known cells in its environment map. With the proposed protocol, the number of cells that agents have to broadcast grows mush slower as more agents are deployed for exploration.

5.5   Environment Mapping Under Communication Loss

We apply the proposed protocol to a group of 10 agents with communication loss. As we assumed that all the agents have the same capability, in this part of experiments, each agent has the same probability of losing the message broadcast by its neighbors. Specifically, we conducted two experiments where agents have communication loss probabilities of 0.2 and 0.4, respectively. Figure 5 shows the final environment mapped by the agents in both experiments.

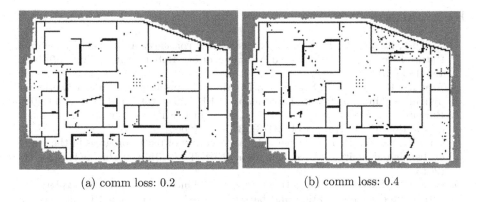

(a) comm loss: 0.2   (b) comm loss: 0.4

Fig. 5. The environments mapped by 10 agents with communication loss.

In both experiments, we set a sufficiently large threshold on the broadcasting iterations of agents, in order to ensure the agents can converge on their consensus. The results of both experiments suggest that the proposed protocol is robust against communication loss. As higher communication loss requires larger threshold for broadcasting iterations, agents with higher communication loss have to spend longer time on maintaining their communication network during exploration. In this case, each agent can have a higher probability of being falsely detected as an obstacle by other exploring agents. Therefore, in Fig. 5, the map built by agents with the higher communication loss probability (i.e., 0.4) is noisier.

6 Conclusion

This paper presented a consensus-based communication protocol that enabled agents to synchronize grid-based local environment maps in distributed exploration. We proved the convergence of the protocol under communication loss. In extensive experiments, we demonstrated that the proposed protocol could be well applied to distributed exploration. We showed that it could enable agents to build consistent environment maps under communication loss in a distributed setting. In practice, agents may operate under more complex communication dynamics, such as limited communication bandwidth and communication noises. The proposed protocol didn't address those issues, but we would consider them for future extension. As another direction of future work, we would enable agents to maintain the connectivity of their communication network while navigating in an environment. This will minimize the overhead of map synchronization and make the protocol be better applied to environment exploration.

References

1. Aragues, R., Cortes, J., Sagues, C.: Distributed consensus algorithms for merging feature-based maps with limited communication. Robot. Auton. Syst. **59**(3–4), 163–180 (2011)
2. Aragues, R., Cortes, J., Sagues, C.: Distributed consensus on robot networks for dynamically merging feature-based maps. IEEE Trans. Rob. **28**(4), 840–854 (2012)
3. Burgard, W., Moors, M., Stachniss, C., Schneider, F.E.: Coordinated multi-robot exploration. IEEE Trans. Rob. **21**(3), 376–386 (2005)
4. Cao, Y., Yu, W., Ren, W., Chen, G.: An overview of recent progress in the study of distributed multi-agent coordination. IEEE Trans. Industr. Inf. **9**(1), 427–438 (2013)
5. Cesare, K., Skeele, R., Yoo, S.H., Zhang, Y., Hollinger, G.: Multi-UAV exploration with limited communication and battery. In: IEEE International Conference on Robotics and Automation, pp. 2230–2235 (2015)
6. Fox, D., Ko, J., Konolige, K., Limketkai, B., Schulz, D., Stewart, B.: Distributed multirobot exploration and mapping. Proc. IEEE **94**(7), 1325–1339 (2006)
7. Hoff, N.R., Sagoff, A., Wood, R.J., Nagpal, R.: Two foraging algorithms for robot swarms using only local communication. In: 2010 IEEE International Conference on Robotics and Biomimetics (ROBIO), pp. 123–130. IEEE (2010)
8. Hsieh, M.A., Cowley, A., Kumar, V., Taylor, C.J.: Maintaining network connectivity and performance in robot teams. J. Field Robot. **25**(1–2), 111–131 (2008)
9. Iutzeler, F., Ciblat, P., Jakubowicz, J.: Analysis of max-consensus algorithms in wireless channels. IEEE Trans. Signal Process. **60**(11), 6103–6107 (2012)
10. Jiao, Z., Oh, J.: Simultaneous exploration and harvesting in multi-robot foraging. In: Mouhoub, M., Sadaoui, S., Ait Mohamed, O., Ali, M. (eds.) IEA/AIE 2018. LNCS (LNAI), vol. 10868, pp. 496–502. Springer, Cham (2018). https://doi.org/10.1007/978-3-319-92058-0_48
11. Jiao, Z., Oh, J.: Asynchronous multitask reinforcement learning with dropout for continuous control. In: 2019 18th IEEE International Conference on Machine Learning and Applications (ICMLA), pp. 529–534. IEEE (2019)
12. Jiao, Z., Oh, J.: End-to-end reinforcement learning for multi-agent continuous control. In: 2019 18th IEEE International Conference on Machine Learning and Applications (ICMLA), pp. 535–540. IEEE (2019)
13. Koenig, N., Howard, A.: Design and use paradigms for Gazebo, an open-source multi-robot simulator. In: Proceedings of the 2004 IEEE/RSJ International Conference on Intelligent Robots and Systems (IROS 2004), vol. 3, pp. 2149–2154. IEEE (2004)
14. Moreau, L.: Stability of multiagent systems with time-dependent communication links. IEEE Trans. Autom. Control **50**(2), 169–182 (2005)
15. Olfati-Saber, R., Fax, J.A., Murray, R.M.: Consensus and cooperation in networked multi-agent systems. Proc. IEEE **95**(1), 215–233 (2007)
16. Olfati-Saber, R., Murray, R.M.: Consensus problems in networks of agents with switching topology and time-delays. IEEE Trans. Autom. Control **49**(9), 1520–1533 (2004)
17. Ren, W., Beard, R.W.: Consensus seeking in multiagent systems under dynamically changing interaction topologies. IEEE Trans. Autom. Control **50**(5), 655–661 (2005)
18. Sheng, W., Yang, Q., Tan, J., Xi, N.: Distributed multi-robot coordination in area exploration. Robot. Auton. Syst. **54**(12), 945–955 (2006)
19. Thrun, S., Burgard, W., Fox, D.: Probabilistic Robotics. MIT Press, Cambridge (2005)

A Real-Time Actor-Critic Architecture for Continuous Control

Zilong Jiao[(✉)] and Jae Oh[(✉)]

Department of Electrical Engineering and Computer Science, Syracuse University,
Syracuse, NY 13244, USA
{zijiao,jcoh}@syr.edu

Abstract. Reinforcement learning achieved impressive results in various challenging artificial environments and demonstrated its practical potential. In a real-world environment, an agent operates in continuous time, and it is unavoidable for the agent to have control delay. In the context of reinforcement learning, we define control delays as the time delay before an agent actuates an action in a particular state. The high-variance of control delay can destabilize an agent's learning performance and make an environment Non-Markovian, creating a challenging situation for reinforcement learning algorithms. To address this issue, we present a scalable real-time architecture, RTAC, for reinforcement learning to continuous control. A reinforcement learning application usually consists of a policy training phase in simulation and the deployment phase of the learned policy to the real-world environment. We evaluated RTAC in a simulated environment close to its real-world setting, where agents operate in real-time and learn to map high-dimensional sensor data to continuous actions. In extensive experiments, RTAC was able to stabilize control delay and consistently learn optimal policies. Additionally, we demonstrated that RTAC was suitable for distributed learning even in the presence of control delay.

Keywords: Real-time architecture · Actor-critic methods ·
Continuous control · Distributed learning · Control delay

1 Introduction

Traditional reinforcement learning has been extensively studied in theory but had limited applications in practice. With recent advancement, deep reinforcement learning has achieved impressive results in various challenging domains [11,15,26] and demonstrated its potential on solving complex real-world problems. However, applying reinforcement learning to real-world environments is still challenging. It faces critical constrains [4], such as sample complexity, high-dimensional continuous state and action spaces, agent safety, partial observability, and a demand for real-time inference. Inspired by those impressive results achieved in artificial domains, off-policy model-free reinforcement learning methods have been actively studied and applied to various continuous control tasks, such as navigation [12], collision avoidance [3], and grasping [17].

© Springer Nature Switzerland AG 2020
H. Fujita et al. (Eds.): IEA/AIE 2020, LNAI 12144, pp. 545–556, 2020.
https://doi.org/10.1007/978-3-030-55789-8_47

In a typical reinforcement learning setting, an agent follows a Markov Decision Process (MDP) and plans its actions in discrete time. As reinforcement learning comes to a real-world environment, an agent must discretize continuous time into a certain resolution and takes action in real-time. In real-time systems, control delay is ubiquitous and is well-known for destabilizing system performance [16]. In the context of reinforcement learning, we defined control delay as the time delay before an agent actuates an action determined in a particular state. Figure 1 shows an example.

Because of the stochastic nature of the MDP, control delay can be adapted by a transition function. However, through theoretical analysis and empirical experiments, we found that control delay with high variance could make an environment Non-Markovian, making reinforcement learning difficult. Addressing this issue, we proposed a real-time actor-critic architecture (RTAC) for applying off-policy reinforcement learning methods for continuous real-time control. RTAC stabilized control delay by decoupling policy learning from environment interaction through multiple threads. Besides, it could be well scaled for distributed learning. We evaluated RTAC in a series of environments that were simulated close to real-world settings. Those environments were physics-enabled, where agents learned to map high-dimensional sensor data to continuous actions. Through extensive experiments, we demonstrated the effectiveness of RTAC in terms of stabilizing control delay, improving learning performance, and supporting distributed learning.

As optimal learning policy requires a large number of trials and errors, a practical application of reinforcement learning usually involves learning policy in simulation and deploying the learned policy to a real-world environment in the second phase. It is crucial to reduce the *reality* gap [24] and incorporate control delay in simulation to enable successful policy transfer. Schuitema et al. [20] extended the Q-learning and SARSA algorithms to incorporate delay time and archived improved performance in simulation. Hester et al. [5] proposed a real-time architecture for applying model-based reinforcement learning to autonomous vehicle control. Unlike the previous work, RTAC was based on the recent advancement in model-free off-policy reinforcement learning [11,23,25], and used experience replay [1] to improve the sample efficiency of a learning method. Besides, it well supported distributed learning even in the presence of control delay.

We structured the rest of the paper as follows. In Sect. 2, we briefly reviewed the reinforcement learning literature. In Sect. 3, we analyzed the effects of control delay on an MDP. Addressing the control delay issue, we presented the proposed architecture in Sect. 4, followed by its evaluation in Sect. 5. In Sect. 6, we discussed the evaluation in more detail and concluded the paper.

2   Related Work

Recently, reinforcement learning has achieved impressive results in various domains, including Go, Atari Games, Deepmind Lab, Statecraft2, and continuous control. Those methods are mainly off-policy and model-free, and they are historically considered unstable and sample inefficient. Utilizing deep neural networks [11,14], distributed learning methods [13] and advanced optimization techniques [21,22], those state-of-the-art methods are able to achieve super-human performance in the aforementioned domains. However, those impressive results were achieved in either artificial environments or in simulation. For deploying reinforcement learning policies to real-world environments, control delay is an important factor to be considered, but it has been rarely studied in the context of reinforcement learning.

Applying reinforcement learning to the real-world environment poses various challenges [4,7,8], including sample complexity, continuous high-dimensional state and action spaces, real-time inference, and systems delays. Hester et al. proposed a real-time architecture for applying model-based reinforcement learning to autonomous driving. To meet the real-time constraints, the authors utilize different threads to infer actions, plan roll-out trajectories, and optimize policies. Schuitema et al. studied the control delay issues in the context of reinforcement learning. The authors incorporate control delay into the Q-Learning and SARSA algorithms, and demonstrated their improved performance in a grid-world environment [6] and a two-link manipulator task in simulation.

Utilizing the recent advancement in deep reinforcement learning, we proposed RTAC, which applied an off-policy actor-critic method to continuous real-time control. RTAC utilized experience replay to improve the sample complexity of an off-policy method. Besides, RTAC has the scalability of distributed learning, which enabled it to better utilize the increasingly available computing resources for training state-of-the-art reinforcement learning methods for real-time control.

3   MDP Under Control Delay

Reinforcement learning models the interactions between an agent and an environment as a Markov Decision Process (MDP). An MDP is a tuple (S, A, T, R), where S and A are the state and action spaces of an agent. $T : S \times A \times S \rightarrow \mathbb{R}$ is a state transition function, and it defines a probability distribution over the states resulting from an action execution. $R : S \times A \times S \rightarrow \mathbb{R}$ is a reward function, and it determines a reward of an agent-based on the consequence of taking action in a specific state. In MDP, an agent takes actions in discrete-time, and incrementally constitutes a state-action trajectory. By definition, an environment is Markovian, if the probability distribution over the future states that an agent will visit only depends on the present state of the agent.

It is crucial to ensure the Markovian property of an environment to apply reinforcement learning to a control task. However, in practice, an agent operates in continuous time and always has time delay before actuating an action determined in an observed state. We defined the time delay of δ for an agent to take action as a control delay, and Figure 1 illustrates an example. In a control task, δ depends on an intrinsic property of an agent, and it can have high variance under different resource constraints.

Let s_t be the state observed by an agent at time t, and the agent determines its optimal action based on s_t. When δ has high variance, s_t can be arbitrarily

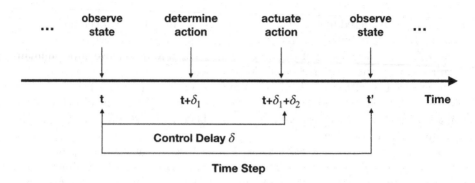

Fig. 1. An example of control delay.

different from the state $s_{t+\delta}$ where the agent actuate the action. In this case, the optimality of the determined action can not be guaranteed. To ensure the optimality of the determined action, the agent needs to predict $s_{t+\delta}$ based on s_t and determines its action based on the predicted state. In this case, we shall consider $s_{t+\delta}$ to be the present state, and determining an action involves the historical state s_t. This violates the Markovian property of an environment. Therefore, δ with high variance is prohibitive in reinforcement learning applications.

An MDP is able to tolerate control delay δ with low variance, as long as δ meets the real-time constraint of an agent (i.e., $\delta < t'-t$). Because of the stochastic nature of an MDP, the transition function T can incorporate the effect of the δ into the probability distribution over the transited states. Although the magnitude of δ does not affect Markovian property, it poses challenges for deploying reinforcement learning policy in practice, particularly for safety-critical tasks.

4 RTAC

This paper presents a Real-Time Actor-Critic architecture (RTAC) for continuous control. RTAC is designed for model-free off-policy learning, and it enables off-the-shelf actor-critic methods to be applied to a real-time system. Figure 2a shows the proposed architecture.

4.1 Stabilizing Control Delay

The high variance of control delay can destabilize reinforcement learning and can cause an environment being Non-Markovian. RTAC reduces the variance of control delay by decoupling environment exploration from policy learning. In a typical off-policy learning setting, an agent uses the behavior policy and the updated policy for environment exploration and policy optimization, respectively. In RTAC, an agent uses separate threads, the behavior thread, and the update thread to execute the behavior and update policy. The behavior thread reduces the variance of control delay δ by holding off actuation until δ is larger

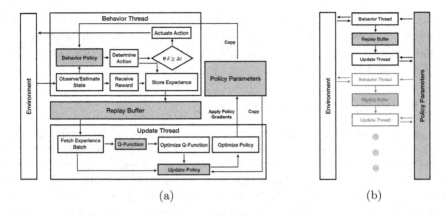

(a)   (b)

Fig. 2. (a): Real-time actor-critic architecture (RTAC); (b): Distributed RTAC

than a predefined threshold Δt. It is important to note that Δt must be reasonably small and must allow the agent to meet its minimum control frequency.

We assume that an agent represents its update policy and behavior policy as parameterized functions (e.g., neural networks). Also, the update policy and the behavior policy share the same set of parameter variables. RTAC stores the values of the parameter variables either in a shared memory or in a parameter server. During learning, the behavior thread periodically (i.e., every t time steps) updates the behavior policy by copying the shared parameter values. On the other hand, the update thread uses DDPG [11] algorithm to evaluate policy gradients and apply them to the shared parameter values, and evolves the update policy by copying the updated shared parameter values to its local memory. Those copying operations can be greatly accelerated by GPUs and can have very low latency.

In RTAC, each pair of update and behavior threads jointly maintains a replay buffer with a predefined maximum size. When the amount of experience exceeds the limit, the behavior thread will swap the oldest state-action transactions out of the replay buffer. Although the synchronization between those two threads can introduce some overhead, it can be accommodated by the predefined Δt. In an actor-critic method, an agent alternates between improving its Q-function and optimizing its policy. In RTAC, the Q-function is completely local to the update thread, and the update thread optimizes its Q-function and update policy in a sequence.

Without decoupling policy learning from environment exploration, an agent typically adopts a sequential architecture, where it improves its update policy every specific time steps or after certain trajectories are collected [15,22]. With the sequential architecture, policy optimization can hinder environment interaction, since an agent will not determine its action until policy optimization is complete. If the policy optimization happened before the end of an episode, it introduces additional time delay and increases the variance of control delay.

Although an agent could optimize the update policy at the end of an episode, we argue that the sequential architecture can slow down the learning process, particularly when the optimization is computationally demanding (e.g., computing returns of roll-out trajectories).

4.2   Scalability of Distributed Learning

A practical application of reinforcement learning typically involves training a policy in simulation and transferring the learned policy to a real-world environment in the following phase. As learning an effective policy requires a large number of trials and errors, a parallel training mechanism is desired and adopted by recent reinforcement learning methods [9,13]. To support distributed training, RTAC allows multiple worker agents to simultaneously learn in a different environment and synthesize the knowledge they have learned into a shared policy. Figure 2b presents the distributed version of RTAC.

In a distributed RTAC, each agent has its update thread, behavior thread and replay buffer. During learning, each agent is situated in its own environment and accumulate its experience in its own replay buffer through its behavior thread. All the worker agents share the same set of policy parameter variables, and the update thread of each agent applies its policy gradients to the shared policy parameter values either synchronously or asynchronously. Such a policy optimization mechanism enables various off-the-shelf distributed reinforcement learning methods [13] to be applicable to real-time systems.

It is important to note that the distributed RTAC requires worker agents to use the same Δt in their update threads. Otherwise, the experience collected by different agents can be inconsistent and hinders the convergence of the share policy parameters. In the distributed setting, RTAC uses separated replay buffers for worker agents. This reduces the overhead caused by merging the experience collected in different environments.

5   Experiments

5.1   Experiment Setting

Environment & Agent. We evaluate RTAC in navigation tasks in simulated maze-like environments. The proposed task environments are physic-enabled and implemented using ROS [18] and Gazebo [10]. Figure 3a to 3d illustrates those environments. In each environment, an agent needs to plan a collision-free path from a fixed initial location to a pre-defined goal in real-time. We simulate an agent as a mobile robot, i.e., a ROSbot. The robot has a simulated LiDar sensor, which has a range of 0.1 m to 2 m and can scan its surrounding environment with 180 evenly spaced lasers. During navigation, an agent moves at a constant speed in its heading direction, and the agent controls its moving direction through rotational velocities.

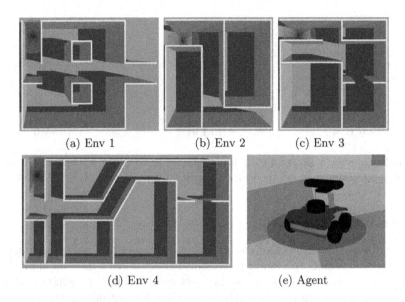

(a) Env 1   (b) Env 2   (c) Env 3

(d) Env 4   (e) Agent

Fig. 3. The agent and environments used in experiments

Learning Setup. For continuous control in real-time, an agent learns a deterministic policy using DDPG algorithm [11]. We define a state of an agent as a 3-tuple $(v, d, o_{t:t+l})$. v is the present velocity of the agent, d is a vector representing a normalized direction from the agent's current position to its goal. $o_{t:t+l}$ is a sequence of LiDar observation perceived by the agent from time t to $t+l$. With respect to the kinematic constraints of a ROSbot, we define an agent action as a target rotational velocity of $\omega \in [-\pi, \pi]$ in radian per second. During navigation, an agent receives a reward given by the following reward function.

$$r = \begin{cases} 0 & d' \leq d \\ \alpha(d' - d) & d' > d \\ -1 & \text{collided with an obstacle} \\ 1 & \text{reached the goal} \end{cases}$$

d' and d are the Euclidean distances from an agent's previous and present locations to its goal. ω is an agent's present rotational velocity. α is a coefficient, s.t. $0 < \alpha \leq 1$.

We represent the actor and critic functions of an agent as deep neural networks. Considering that the LiDar sensor data in $o_{t:t+l}$ is high-dimensional, and project $o_{t:t+l}$ into a lower-dimensional embedding through a LSTM. Specifically, an agent retrieves the last output of the LSTM as its embedding of $o_{t:t+l}$. In both actor and critic functions, the output of the LSTM (i.e., the embedding of $o_{t:t+l}$) is concatenated with other input and then feed into a multi-layer perceptron to determine the final output. Table 1 summarizes the hyper-parameters used for policy learning.

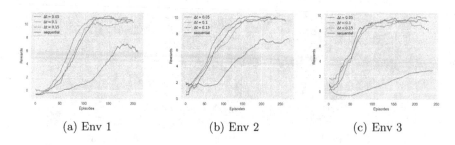

(a) Env 1   (b) Env 2   (c) Env 3

Fig. 4. The moving averages of episodic rewards achieved through RTAC under various control delay. Video demo of the learned policies: https://youtu.be/6dH7-0Miu7c

5.2   Learning with RTAC

RTAC reduces the variance of control delay by decoupling policy learning and environment interaction in order to stabilize an actor-critic method applied to a real-time system. In contrast, when an agent conducts policy learning and environment interaction in sequence, the time delay for optimizing the policy can increase the variance of control delay, particularly in systems with limited computing resources, such as robots. Table 2 compares the RTAC with the sequential architecture, in terms of their control delay statistics. The results presented in Table 2 is based on the average of mean and variance of control delay in 50 episodes. In the experiments, compare the control delay in RTAC based on different delay threshold of Δt. As we increase Δt, the variance of control delay does not change significantly across all environments. In contrast, the sequential architecture causes much higher variance.

We compare the performance of DDPG with the RTAC and the sequential architecture, in terms of a moving average of episodic rewards. Figure 4 presents the results collected in all environments. According to the results, the DDPG with RTAC significantly outperforms the DDPG with the sequential architecture. According Table 2, the sequential architecture causes much higher variance

Table 1. Hyper-parameters used in experiments for learning policies.

Hidden layer neurons	1024
Number of hidden layers	2
LSTM hidden state size	128
Hidden layer activation	ReLu
Actor learning rate	0.00001
Critic learning rate	0.001
Target network τ	0.01
Batch size	256
Relay buffer size	100000

Table 2. Control delay statistics for RTAC and the sequential architecture.

	$\Delta t = 0.05$		$\Delta t = 0.1$		$\Delta t = 0.15$		Sequential	
	Var	Mean	Var	Mean	Var	Mean	Var	Mean
Env 1	1.925e−05	0.051	2.431e−05	0.101	1.466e−05	0.151	0.004	0.127
Env 2	1.399e−05	0.051	3.739e−05	0.101	5.582e−06	0.151	0.003	0.123
Env 3	2.557e−05	0.051	1.68e−05	0.101	2.041e−05	0.151	0.004	0.129

of control delay, although the mean is in the reasonable range (i.e. $[0.05, 0.15]$ in our experiments). With different threshold, Δt, DDPG with RTAC achieved similar learning performance. According to the MDP formulation, reinforcement learning requires agents to take action in the same state as where the action is determined. This makes reinforcement learning is more vulnerable to control delay, comparing it to the conventional model-based control methods. Although the control delay variance of the sequential approach is within the acceptable range for a practical robotics application, the control delay can still significantly disturb the performance of a reinforcement learning agent. This proves our claim that the high-variance of control delay can make an environment Non-Markovian, but the magnitude of control delay does not affect the Markovian property, as long as it is in a reasonable range.

5.3   Distributed Learning

RTAC applies to a distributed setting. To learn a policy, distributed RTAC consists of multiple pairs of behavior and update threads, and each pair of threads corresponds to a worker agent. Those worker agents can optimize globally shared policy parameters in parallel. In this part of the experiment, we conduct the distributed learning using a single PC and implement the shared parameter variables as shared memory. In the distributed setting, each agent learns its own environment through DDPG and applies its policy gradients asynchronously to the shared policy parameters.

For the distributed learning, we apply RTAC to the environment presented in Fig. 3d, which consists of all the patterns presented in Env 1 to Env 3. Since the distributed RTAC has multiple worker agents learn the shared policy in parallel, one can expect the learned policy to represent more complex behavior, comparing to the policy learned by a single agent. Figure 5 shows the performance of distributed learning under various delay thresholds. Table 3 summarizes the control delay during the distributed learning.

As shown on the right side of Fig. 5b, we used 3 worker agents to learn the same environment, and each worker agent interacts with its copy of the environment. For evaluating the policy jointly optimized by those three worker agents, every 5 s, a separate evaluation agent makes a copy of the jointly optimized policy and evaluates its performance in terms of episodic rewards. The left side of Fig. 5b shows the evaluation agent.

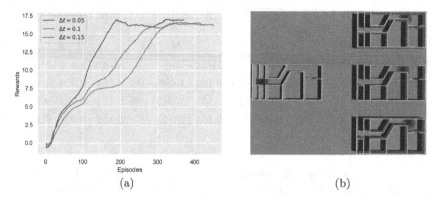

Fig. 5. Distributed learning with RTAC. (a): moving averages of episodic rewards during learning; (b): The environment setup for the distributed learning. Video demo of the learned policies: https://youtu.be/6dH7-0Miu7c

Table 3. Control delay statistics for distributed RTAC.

	$\Delta t = 0.05$		$\Delta t = 0.1$		$\Delta t = 0.15$	
	Var	Mean	Var	Mean	Var	Mean
Agent 1	1.559e−04	0.052	5.642e−05	0.102	2.584e−05	0.152
Agent 2	1.390e−04	0.052	3.148e−05	0.102	2.948e−05	0.152
Agent 3	1.591e−04	0.053	6.194e−05	0.102	4.441e−05	0.152

According to the results, RTAC enables the asynchronous DDPG to converge at optimal policies under all three delay thresholds (i.e., 0.05, 0.1, 0.15). For each experiment, we initialize the shared policy with random parameters, and the difference in terms of convergence time is caused by those initial parameter settings, instead of the delay time threshold Δt. According to Table 3, the distributed RTAC does not significantly increase the variance of control delay, and it can ensure each worker environment being Markovian throughout the learning. Besides, the results also prove that an MDP could tolerate stabilized control delay even in a distributed setting.

6   Discussion and Conclusion

This paper presented a reinforcement learning architecture RTAC, which applied an actor-critic method to continuous real-time control. We demonstrated that the control delay with high variance could greatly destabilize the learning performance of an agent. The RTAC architecture was able to reduce the variance and ensure an environment being Markovian effectively. We evaluated the RTAC in challenging navigation tasks in simulation, where agents operated in physics-enabled environments and learned to map high-dimensional sensor data to continuous actions in real-time. Through extensive experiments, we demonstrated that RTAC was salable and allowed multiple worker agents to asynchronously optimize a shared policy in the presence of various control delay.

In this paper, RTAC used DDPG to learn policies for continuous control. However, it is straight forward to incorporate alternative off-policy actor-critic methods into RTAC for learning either deterministic or stochastic policies. Besides, one can also combine RTAC with more advanced experience replay techniques [2,19]. Although we used a small number of agents to evaluate the distributed RTAC, the overall task to be learned was still challenging. Using a smaller number of worker agents allowed us to better focus on the effects of control delay on distributed learning in a close-to-real-world setting. Specifically, during the distributed learning, each worker agent learned to map high-dimensional sensor data (i.e., 720 dimensions) to continuous actions and asynchronously update over 1 million parameters of the shared policy.

The proposed RTAC is for the off-policy model-free reinforcement learning. In practice, many control tasks still favor model-based methods, particularly for robotic tasks. In future work, we would expand our study to model-based methods. It would be attractive to have a unified architecture for applying either model-based or model-free methods to real-time systems.

References

1. Adam, S., Busoniu, L., Babuska, R.: Experience replay for real-time reinforcement learning control. IEEE Trans. Syst. Man Cybern. Part C (Appl. Rev.) **42**(2), 201–212 (2011)
2. Andrychowicz, M., et al.: Hindsight experience replay. In: Advances in Neural Information Processing Systems, pp. 5048–5058 (2017)
3. Chen, Y.F., Liu, M., Everett, M., How, J.P.: Decentralized non-communicating multiagent collision avoidance with deep reinforcement learning. In: 2017 IEEE International Conference on Robotics and Automation (ICRA), pp. 285–292. IEEE (2017)
4. Dulac-Arnold, G., Mankowitz, D., Hester, T.: Challenges of real-world reinforcement learning (2019). arXiv preprint arXiv:1904.12901
5. Hester, T., Quinlan, M., Stone, P.: Rtmba: A real-time model-based reinforcement learning architecture for robot control. In: 2012 IEEE International Conference on Robotics and Automation, pp. 85–90. IEEE (2012)
6. Jiao, Z., Oh, J.: Simultaneous exploration and harvesting in multi-robot foraging. In: Mouhoub, M., Sadaoui, S., Ait Mohamed, O., Ali, M. (eds.) IEA/AIE 2018. LNCS (LNAI), vol. 10868, pp. 496–502. Springer, Cham (2018). https://doi.org/10.1007/978-3-319-92058-0_48
7. Jiao, Z., Oh, J.: Asynchronous multitask reinforcement learning with dropout for continuous control. In: 2019 18th IEEE International Conference On Machine Learning And Applications (ICMLA), pp. 529–534. IEEE (2019)
8. Jiao, Z., Oh, J.: End-to-end reinforcement learning for multi-agent continuous control. In: 2019 18th IEEE International Conference On Machine Learning And Applications (ICMLA), pp. 535–540. IEEE (2019)
9. Kapturowski, S., Ostrovski, G., Quan, J., Munos, R., Dabney, W.: Recurrent experience replay in distributed reinforcement learning (2018)

10. Koenig, N., Howard, A.: Design and use paradigms for gazebo, an open-source multi-robot simulator. In: Proceedings of the 2004 IEEE/RSJ International Conference on Intelligent Robots and Systems, (IROS 2004). vol. 3, pp. 2149–2154. IEEE (2004)
11. Lillicrap, T.P., et al.: Continuous control with deep reinforcement learning (2015). arXiv preprint arXiv:1509.02971
12. Long, P., Fan, T., Liao, X., Liu, W., Zhang, H., Pan, J.: Towards optimally decentralized multi-robot collision avoidance via deep reinforcement learning (2017). arXiv preprint arXiv:1709.10082
13. Mnih, V., et al.: Asynchronous methods for deep reinforcement learning. In: International Conference on Machine Learning, pp. 1928–1937 (2016)
14. Mnih, V., et al.: Playing atari with deep reinforcement learning (2013). arXiv preprint arXiv:1312.5602
15. Mnih, V., et al.: Human-level control through deep reinforcement learning. Nature **518**(7540), 529 (2015)
16. Nilsson, J., et al.: Real-time control systems with delays (1998)
17. Pinto, L., Gupta, A.: Supersizing self-supervision: learning to grasp from 50k tries and 700 robot hours. In: 2016 IEEE International Conference On Robotics and Automation (ICRA), pp. 3406–3413. IEEE (2016)
18. Quigley, M., et al.: Ros: an open-source robot operating system. In: ICRA Workshop on Open Source Software, vol. 3, p. 5. Kobe, Japan (2009)
19. Schaul, T., Quan, J., Antonoglou, I., Silver, D.: Prioritized experience replay (2015). arXiv preprint arXiv:1511.05952
20. Schuitema, E., Buşoniu, L., Babuška, R., Jonker, P.: Control delay in reinforcement learning for real-time dynamic systems: a memoryless approach. In: 2010 IEEE/RSJ International Conference on Intelligent Robots and Systems (IROS), pp. 3226–3231. IEEE (2010)
21. Schulman, J., Levine, S., Abbeel, P., Jordan, M., Moritz, P.: Trust region policy optimization. In: International Conference on Machine Learning, pp. 1889–1897 (2015)
22. Schulman, J., Wolski, F., Dhariwal, P., Radford, A., Klimov, O.: Proximal policy optimization algorithms (2017). arXiv preprint arXiv:1707.06347
23. Silver, D., Lever, G., Heess, N., Degris, T., Wierstra, D., Riedmiller, M.: Deterministic policy gradient algorithms. In: ICML (2014)
24. Tobin, J., Fong, R., Ray, A., Schneider, J., Zaremba, W., Abbeel, P.: Domain randomization for transferring deep neural networks from simulation to the real world. In: 2017 IEEE/RSJ International Conference on Intelligent Robots and Systems (IROS), pp. 23–30. IEEE (2017)
25. Van Hasselt, H., Guez, A., Silver, D.: Deep reinforcement learning with double q-learning. In: Thirtieth AAAI Conference on Artificial Intelligence (2016)
26. Vinyals, O., et al.: Starcraft ii: a new challenge for reinforcement learning (2017). arXiv preprint arXiv:1708.04782

Action-Based Programming with YAGI - An Update on Usability and Performance

Thomas Eckstein and Gerald Steinbauer$^{(\boxtimes)}$

Institute for Software Technology, Graz University of Technology, Graz, Austria
eckstein@student.tugraz.at, steinbauer@ist.tugraz.at

Abstract. YAGI is a declarative and procedural programming language based on Situation Calculus and Golog. The language allows the user to model a target domain as a set of actions and fluents, as well as an imperative control program or as a planning problem. YAGI was designed to broaden the use of action-based programming. The work presented in this paper starts with an existing realization of YAGI, which has major drawbacks such as low performance or a limited syntax. We redesigned the language syntax, added new data types and implemented a new, more efficient interpreter with an improved knowledge representation. Moreover, we evaluated our implementation using an example domain and showed that it performs several magnitudes better than the previous YAGI version regarding runtime and solution quality.

Keywords: Golog · Situation Calculus · Action-based programming · High-level control · Autonomous agents and robots · Usability

1 Introduction

In order to allow an intelligent agent or robot to perform a non-trivial task such as assistance in a home, smart manufacturing or transporting goods, the system needs to be equipped with some high-level decision making or deliberation mechanism. Usually these mechanisms combine planning, reasoning, action execution and monitoring to allow following a goal, integrating newly arrived information, executing actions and reacting to unforeseen situations [5]. Many different architectures and methods had been proposed over the years that allow developers to describe the task and its related information as well as the way the task is executed. Some of the systems follow the idea of knowledge-based systems with a separate planning and execution module where mainly the goal is specified while others follow a more rigid procedural/imperative approach where the solution to the task is specified in more detail.

An early but still used approach to formalize such problems is the action-based programming paradigm. Its representation is usually based on the Situation Calculus [9], a formal logic framework to reason about actions and their

© Springer Nature Switzerland AG 2020
H. Fujita et al. (Eds.): IEA/AIE 2020, LNAI 12144, pp. 557–569, 2020.
https://doi.org/10.1007/978-3-030-55789-8_48

effects. Golog, which was proposed in [6] extends the Situation Calculus towards an action programming language by providing control structures like while loops. Thus, Golog allows the user to combine planning and imperative programming respective to interleave action planning and action execution. During the years several extensions like IndiGolog [1] were developed, that incorporates concepts like sensing or concurrency which are important for robots and agents. Although, these concepts are very elegant and based on strong theoretical foundations [11] the interpreters for action-based languages are usually implemented in Prolog and lack of a sound language definition, a clear separation between the language and interpreter implementation and tool support. These factors together with the not commonly used logic-based programming approach limits the bread acceptance of the approach in the robotics community. Thus with YAGI (Yet Another Golog Interpreter) [4] a more accessible variant of the approach was proposed that kept the strong theoretical foundation. In [7] a first full realization of YAGI including a complete language definition, a sound interpreter and an interface to robot systems was presented. Despite that all design objectives were met, the implementation suffers from conceptional and implementation shortcoming such as a lack of support of data types and performance issues in planning. The persistent interest in making action-based programming is also shown by other lines of research. In [8] the authors work on a direct integration of the action-based paradigm into the C++ programming language.

The goal of the work presented in this paper is to improve and extend the YAGI ecosystem to allow a wider range of programmers to use it for realizing high-level control for robots or agents. The contribution of this papers is threefold. First the syntax of the language was reworked in order to be less ambiguous, more accessible to users without much knowledge in logic, and more compact. Second we introduced basic data types like enums and integers into the language while keeping compliance with the theoretical background provided by Situation Calculus and Golog. Moreover, an improved support for compile time type checking was introduced into the interpreter to better support programming and debugging. Third a improved reasoning mechanism and planning approach were integrated in order to improve the execution performance of YAGI control programs.

The remainder of the paper is organized as follows. In the next section we provide a brief introduction to the foundation of action-based programming and YAGI. In Sect. 3 we present the developed improvements in more details. Experimental results showing the effectiveness of the proposed improvements are presented in the next section. In Sect. 5 we draw some conclusions and give an outlook on future work.

2   Action-Based Programming Primer

2.1   Situation Calculus and the Golog Family

YAGI as a action-based robot and agent programming language has its roots in the Golog language family [6] and its formal foundations in the Situation Calculus [9]. The Situation Calculus is a first-order logic language with equality (and some limited second-order features), which allows for reasoning about actions and their effects [11]. The major concepts are situations, fluents, and primitive actions. Situations reflect the evolution of the world and are represented as sequence of executed actions rooted in the initial situation S_0. Properties of the world are represented by fluents, which are situation-dependent predicates. Actions have preconditions, which are logical formulas describing if an action is executable in a situation. Successor state axioms define whether a fluent holds after the execution of an action. The precondition and successor state axioms together with some domain-independent foundational axioms form the so-called basic action theory (BAT) describing a domain.

Golog [6] is an action-based agent programming language based on the Situation Calculus. Besides primitive actions it also provides complex actions that can be defined through programming constructs for control (e.g. while loops) and constructs for non-deterministic choice (e.g. non-deterministic choice between two actions). This allows to combine imperative and declarative programming easily. The program execution is based on the semantics of the Situation Calculus and querying the basic action theory whenever the Golog program requires the evaluation of fluents. Golog follows the so-called offline execution semantics where a complete execution trace is the result of a constructive proof.

The first Golog interpreter was implemented by [6] using the logic programming language Prolog. It is a set of clauses that interprets the operations discussed above. A major drawback is that there is no separation between BAT, the Golog program and the interpreter rules. Since Prolog uses a depth-first search algorithm, it is not guaranteed that programs terminate even if there is a finite execution trace.

IndiGolog [1] is an extension of the Golog programming languages with an improved interpreter written in Prolog. An online execution mode was introduced to allow quick reaction, during which the immediately next best action is derived and executed despite future needs. A search operator can be used to enforce conventional offline behavior, where the entire subprogram is searched for a valid execution trace. IndiGolog supports complex operations such as sensing, interrupts and concurrent execution.

2.2   Legacy YAGI

YAGI was the first attempt to establish a standalone Golog-based programming language that resolves the issues of former implementations [4]. The first YAGI interpreter, to which we refer as Legacy YAGI, was implemented in C++ [7]. The ecosystem provides a clear separation of back-end and front-end as well as a clearly defined syntax while maintaining a strong semantic relation to the strong theoretical foundations of the Situation Calculus and (Indi)Golog. In order to ease the use of the language for non-logic experts, fluents are represented as associative arrays and effects are modeled by queries and manipulations of those arrays. Moreover, the reasoning method was changed from Prolog-based regression to progression using an explicit model representation [2]. A drawback is that the language only supports strings as data types and lacks arithmetic operations along with enum types for organizing constants. This was a result of the strong formal coupling of YAGI's semantics with the foundations of the underlying Situation Calculus.

Besides limitations on the language the interpreter also suffers from short-comings in the implementation. Legacy YAGI programs are parsed and interpreted using a tree-walk interpreter. Semantical errors such as type mismatches are only detected during runtime. Situations are represented by SQLite databases, with each fluent being a table. A breadth-first search strategy is used, with one new thread per search branch. This means that even simple search problems are very resource demanding. As discussed in [7] there is no way to limit the scope of the search operator, e.g. cutting off irrelevant branches.

3   Concept and Implementation

In this chapter we present our contributions to improve YAGI. We proceed by giving an overview over the language specification and highlight our improvements. Moreover, we give an insight into implementation details and the revised knowledge representation and reasoning. Finally, we give a brief example program to better illustrate how the language can be used. Our implementation is open source and can be found online[1].

3.1   Concept

Starting from the existing YAGI implementation, we explore features that can be improved or added to the language. The syntax needs to become more compact and less verbose, but it should nevertheless be easy to learn (similarity to common programming languages) and connected to Golog. Support for fundamental data types such as booleans, integers, floats along with arithmetic operators needs to be added. The user should be able to define enum types for grouping related constants together. During compile time as many syntactical and semantic errors as possible should be detected and reported. Runtime and memory usage must be improved by using more efficient data structures for storing fluents and by letting the user have more control over the search process.

We decide to use the already existing ecosystem of the programming language Q3 as a basis for our implementation. Q3 is one of our side projects which we originally developed as an embedded scripting language for a Minecraft server[2].

---

[1] https://git.ist.tugraz.at/ais/yagi/tree/master/rewrite.

[2] Details to the Q3 framework can be found at https://git.rootlair.com/teckstein/q3/.

The YAGI program is compiled to Q3 Bytecode, which is executed by the Q3 Machine (Q3M). Since this ecosystem already provides support for concurrent tasks, event handling, an API, error handling, dynamic code loading, we decided to write a compiler in Java that compiles YAGI to Q3 Bytecode. This lets us implement Breadth-First-Search in a more efficient way. Instead of creating a new thread for each search branch, a task queue contained within Q3 is sufficient.

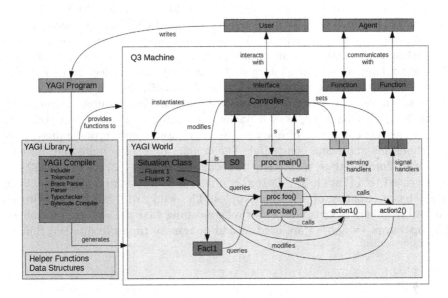

Fig. 1. The components of the revised YAGI ecosystem.

A YAGI program is compiled into a Q3 Class having all procedures, actions, facts (immutable fluents) and initial situations as members, as illustrated in Fig. 1. It also contains a nested class representing a situation with all fluents as members. Some helper functions are added by the compiler. A Q3 program instantiates and executes the generated class and attaches sensing and signal handlers, so the YAGI program can interact with the outside world. The YAGI Library interfaces with the Q3M and provides helper classes for sets, fluents and randomized iterators. It also contains the YAGI compiler, so YAGI programs can be compiled on the fly within the Q3M.

We give the syntax a general overhaul by shifting from keywords to operators and by adding some syntactic sugar. For example, `not(<$obj> in objectAt)` and `$x == $y` would become `!objectAt[$obj] && $x == $y`. Curly brackets are used for blocks. For example, `if(...) then ... end if` becomes `if(...) { ... }`. Trailing commas are allowed, which can be useful when formatting long lists.

3.2   Data Types

Our new YAGI language supports multiple data types, such as integers, floating point numbers, booleans, strings, sets, tuples and user defined enum types. The language is statically typed and all data types are passed by-value. The types of parameters (in actions, procedures, fluents) must be explicitly stated, while the types of local variables can be fully inferred.

Integer types can either be unrestricted (int) or ranged, meaning that the type int[1:10] represents all integers between 1 and 10. Floating point numbers (float) can only be unrestricted. Booleans are denoted by the constants true and false. Basic arithmetic and logic operators are also available and work like in familiar languages such as Java or C. An enum is a collection of logically related constants (enum Color { RED, GREEN, BLUE, }).

n-ary tuples are denoted by square brackets. For example, [true, 3, Color.RED] denotes a tuple of a boolean, an integer and an enum constant. Sets are always comprised of n-ary tuples of countable data types (ranged integers, booleans and enum constants). The domain of a set is finite and must be known at compile-time. {[1, _], [3, Color.GREEN], [2, Color.BLUE]} is typechecked to {[int[1:3], Color]}, unless explicitly stated otherwise. Such as set has a capacity of 9 distinct tuples. The wildcard _ will be substituted with all members of Color for this tuple, meaning that the set has 5 members. The operators += and -= are used to add or remove tuples from a set. Similar

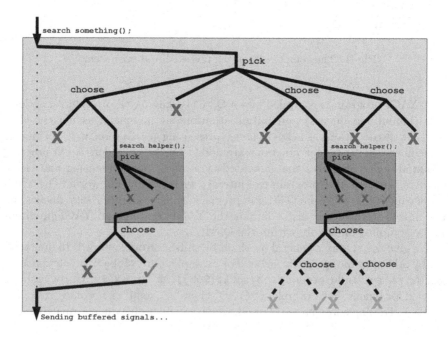

Fig. 2. Exemplary control flow during offline execution mode. Search branches are scheduled left-to-right, top-to-bottom.

to [7] tuples are related to the parameters of fluents and play a major role in representing their semantics. In contrast to Legacy YAGI, strings (`string`) are only used for communication and debugging.

The quantifier expressions `exists` and `for` take a tuple of free variables or values, an expression evaluating to a set and a boolean expression depending on the given variables. `exists [$obj, Location.KITCHEN] in objectAt such $obj != Object.COFFEE` tests if something other than coffee is in the kitchen. `for` has a similar syntax and is true if the given predicate holds for all tuples in the given set. Set membership can also be tested using the `[]` operator. For example, `objectAt[Object.COFFEE, Location.KITCHEN]` is true if the fluent holds true for the given tuple in the current situation.

3.3   Fluents and Actions

Fluents are declared in the global scope along with their data types. For example, `fluent objectAt[Object, Location]` holds tuples of the type `[Object, Location]`. Instead of being predicates depending on the current situation, the semantics of fluents in YAGI is represented by sets of tuples. Each action alters the members of those sets representing the action's effect on the fluents. In contrast to Legacy YAGI, where fluents are internally represented using SQLite tables, we use arrays of integers for fluents and for sets in general. Each tuple is encoded to an integer and appended to the array. In a previous version we used bitmaps to represent sets, with each tuple encoded to an index in the bitmap and a set bit indicates a present tuple. Due to sets being mostly sparsely populated in practice, we decided to use the former method. YAGI also supports facts, which have a syntax similar to fluents, but are immutable and have their values initialized at their declarations.

An action consists of a parameter list, a precondition denoted by the keyword `poss`, a `do` block stating the effects on the fluents and an optional `signal` call that triggers an event in the outside world. Additionally, sensing handlers can be called in the `do` block that read a value from the outside world and store it in a fluent. Calls to other actions or procedures are disallowed inside actions.

3.4   Procedures

Procedures define the behavior of the program and contain control structures responsible for planning, such as `pick`, `choose`, `if` or `while`. Other procedures and actions can be called as well. `if` and `while` share the same behavior with classical programming languages. A `choose` statement has two or more blocks and non-deterministically executes one of those. The `pick` statement works similar to `exists`, but the tuple for which the boolean expression holds true is accessible trough the variables afterwards. This can be used to query fluents.

There are two modes of execution, namely online and offline execution. The default mode is online execution, during which action signals are called instantly and `pick` and `choose` yield one random value regardless of whether successive preconditions will fail. If that is the case, the program terminates. If a procedure is called with a preceding `search` keyword, offline mode is entered. Each `pick` and `choose` forks the control flow and all possibilities are tried until one search path leaves the called procedure, as shown in Fig. 2. If no path succeeds, the program fails. Consequently, signals are buffered and executed once a result is found. Usually the path with the least amount of decisions gets chosen and all remaining paths are discarded. To restrict the scope of `search`, our new YAGI also supports `search` calls during offline mode. Calling a procedure during offline mode would normally return an arbitrary amount of paths, but a preceding `search` keyword limits this to at most one path. This makes sense when the result of an inner procedure doesn't affect the overall result. This concept is similar to cuts in Prolog.

3.5 Internal Representation of Situations and Formulas

Both YAGI and Legacy YAGI use progression [11] instead of regression. Queries are answered using a model-theoretic approach instead of resolution-based reasoning. Situations are represented as objects containing the current state of all fluents as members in YAGI. Each fluent explicitly stores the values for which it holds true. In the beginning a copy of the initial situation is passed to the main procedure. Each action updates the situation object. For each new search branch the current situation object is shallow-copied and all fluent objects are set to copy-on-write. This saves a considerable amount of memory, since most actions only change a handful of fluents. Each search branch is represented as a task on a scheduling queue. A task consists of the position in the program, the current situation and values of local variables. If a non-deterministic decision is taken during offline mode, all emerging tasks are pushed to the end of the queue, ensuring strict breadth-first search behavior. In [3] we argue that our implementation still conforms to the principles of Golog.

3.6 Example

The code shown below contains an excerpt from a YAGI program controlling a packet delivery robot. The robot can travel between various locations, which are sparsely interconnected. It can pick up or drop packets in these locations. The goal is to deliver all packets to their destination with as little steps as possible. The full code can be found in [3] or online.[3] `pickup` takes a packet as parameter and can be executed if the robot and the packet are at the same location. The fluents are updated and a signal is emitted. The procedure `findSolution` iterates until all requests have been fulfilled. Each iteration, the robot either chooses a random packet, goes there and picks it up or chooses a packet it holds, goes to its destination and drops it. If no matching packet is found or a precondition is violated, the current search branch fails and another one is continued. It would also be possible to serve requests serially, but that would yield a longer action sequence and would not utilize the robot's ability to hold multiple packets.

---

[3] https://pastebin.com/MaKhjyEZ.

```
enum Loc { A, B, C, D, E, F, G, H, I, }
enum Packet { P1, P2, P3, P4, }
fluent robotAt[Loc]
fluent packetAt[Packet, Loc]
fluent holding[Packet]
fluent requests[Packet, Loc]

action pickup(Packet $packet) {
  poss (exists [$loc] in robotAt such packetAt[$packet, $loc])
  do {
    packetAt -= [$packet, _];
    holding += [$packet]; }
  signal sigPickup($packet) }

proc main() {
  search findSolution(); }

proc findSolution() {
  while(requests[_, _]) {
    choose {
      pick [$packet, $loc] in packetAt such requests[$packet, _];
      search findPath($loc);
      pickup($packet); }
    or {
      pick [$packet, $loc] in requests such holding[$packet];
      search findPath($loc);
      drop($packet);
      removeRequest($packet); }}}

proc findPath(Loc $to) {
  while(!robotAt[$to]) {
    pick [$next] in [Loc];
    move($next); }}
```

4   Experimental Evaluation

4.1   Comparison with Legacy YAGI

In order to show the performance gain by our redesign of YAGI we performed an experimental evaluation. We modeled the same problem in both YAGI and Legacy YAGI and compared runtime and success rates for increasing problem sizes and execution strategies.

We chose the blocks world scenario from [10], where there are a total of b blocks available, which can be stacked on top of each other, as shown in Fig. 3. At most one block fits on top of another block. Only the uppermost block on a stack can be moved. A block x sitting on top of another block can be put on the table using the action $putOnTable(x)$, which creates a new stack. It is also allowed to move a block x on top of another block y with the action $stackBlocks(x, y)$. The initial situation starts with b blocks, which are randomly stacked to make s stacks. Our planning goal is to have block 0 stacked on top of block 1. The source code of our YAGI program can be found in [3] or online.[4]

---

[4] https://pastebin.com/4121c0fU.

In order to highlight the difference between offline and online execution the problem is encoded in three different ways, as already conducted by [7]:

1. **Non-deterministic, no Planning**: The non-deterministic control structures `pick` and `choose` are used in online execution mode, making an irreversible decision each time. A solution may only be found by chance.
2. **Conditional, no planning**: Also executed in online mode. `if`-statements are used to mitigate any violations of preconditions, so that the program can never fail, but termination is not guaranteed.
3. **Non-deterministic, planning**: The program is identical to the first one, except that it is executed in offline mode. All possible decisions are evaluated simultaneously. It doesn't matter if a single execution path fails.

The programs are then run n times, each time with a randomized initial situation. The execution time and success of each test is measured, with ≥ 10 seconds counting as timeout. These measurements are then evaluated for each scenario, yielding the mean execution time $t_\mu[ms]$, its standard deviation $t_\sigma[ms]$, the shortest and longest execution times $t_{min}[ms]$ and $t_{max}[ms]$. Only tests not running into a timeout are considered here. The success rate succ[%] tells how many of the total tests yield a valid solution. The timeout rate to[%] says how many test runs time out.

The same evaluation methodology is used for the domains implemented in Legacy YAGI and Golog, which is encoded as a Prolog program. Timeouts are set to 120 s for Legacy YAGI programs and to 5 s for Golog programs, because they finish very quickly in case they terminate.

The results depicted in Table 1 show that YAGI vastly outperforms Legacy YAGI regarding runtime. It is also more reliable than Golog regarding timeout rates. The advantage against Legacy YAGI might result from the use of more efficient data structures for fluents and from the single-threaded implementation of our search algorithm instead of starting one thread per search path. While Legacy YAGI is entirely string-based,

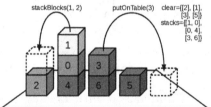

Fig. 3. An example scenario of the blocks world domain with 7 blocks and 4 stacks. The fluent values for this situation are given on the top right.

YAGI supports enum constants and integers, which also need less memory. The large number of runs n for the YAGI tests are necessary to mitigate the influence of Java Virtual Machine warmup.

Golog finds a solution very quickly in about half of the cases, but in the other half it doesn't terminate at all. Its depth-first search algorithm might repeatedly apply an action followed by an action reversing the previous action, so the searching scope is trapped between a few states. While YAGI (non-det., no plan.) only finds a solution by chance by applying purely random actions, YAGI (cond., no plan.) only applies random actions that are suitable, so action preconditions and test actions are never violated and thus the program never fails. The program can in theory take arbitrarily long for any given input to terminate, which results in the high $t_{max}[ms]$ values in our results. YAGI (non-det., plan.) yields the most suitable results, because it is expected to give a relatively short action sequence and is sufficiently fast. The large values of $t_{max}[ms]$ and $t_\sigma[ms]$ might be due to corner cases where the target block lies at the bottom of a large stack.

Table 1. Results for the blocks world test scenario $b = 10, s = 5$

Program	n	$t_\mu[ms]$	$t_\sigma[ms]$	$t_{min}[ms]$	$t_{max}[ms]$	succ [%]	to [%]
YAGI (non-det., no plan.)	1000	0.04	0.03	0.01	0.7	6	0
YAGI (cond., no plan.)	1000	6.3	176	0.01	5581	100	0
YAGI (non-det., plan.)	100	56	134	0.06	701	100	0
Legacy Y. (non-det., no plan.)	20	15	9.7	4	43	10	0
Legacy Y. (cond., no plan.)	20	200	305	12	1050	100	0
Legacy Y. (non-det., plan.)	20	29272	19790	9071	60253	30	70
Golog	20	0.09	0.04	0.03	0.1	55	45

4.2 Trial in Education

In order to evaluate the usability YAGI was used by students in the AI course Classical Topics in Computer Science at Graz University of Technology in winter term 2018/19. The 120 participants who worked in groups of 4 had to encode a planning problem in YAGI. Based on the given framework, the students had to implement a simplified item manufacturing process inspired by the video game Minecraft.

Out of the 28 groups that handed in a solution, 22 passed more than half of the private test cases. This means that they had definitely managed to implement basic planning to solve the problem in general. 7 groups could pass all test cases. Some implemented a purely imperative program using much domain knowledge. Others used planning to formulate the problem in a more abstract way. Groups with a deterministic approach had programs taking less than a second to execute, but the computed action sequences were usually longer than those of programs with planning.

Student feedback was widely positive, with some showing interest in the advanced concept of YAGI and its applications. Others noted that it was relatively easy but they stuck with the usual imperative programming. Common criticisms were the lack of debugging tools during search mode and the lack of functional fluents.

5 Conclusion

In this paper we presented a redesign of the action-based programming language YAGI, so that the language becomes faster and more usable. We examined the previous YAGI implementation (Legacy YAGI) and pointed out its drawbacks, such as its lack of proper data structures, limited syntax and poor performance.

We outlined a basic architecture of the new YAGI ecosystem, where YAGI code is compiled to a primitive opcode that is executed by a Java program. Furthermore we proposed a more compact syntax with more syntactic sugar and discussed the language specification. We added support for new data types such as ranged integers, user defined enums or booleans along with better control over the planning process. A strict typechecker was also one of the new core features. Moreover, we presented an alternative knowledge representation and reasoning approach.

We evaluated the performance compared to Legacy YAGI by measuring the runtime and success rate of different implementations of an example domain in both languages as well as in classic Golog. The results showed that YAGI clearly outperforms its predecessor. Finally, we discussed how YAGI had already been successfully used by students in an educational setting, showing that our implementation is more accessible to programmers for modelling control programs for agents.

Acknowledgment. This work is supported by the TU Graz LEAD project "Dependable Internet of Things in Adverse Environments".

References

1. De Giacomo, G., Lespérance, Y., Levesque, H.J., Sardina, S.: IndiGolog: a high-level programming language for embedded reasoning agents. In: El Fallah Seghrouchni, A., Dix, J., Dastani, M., Bordini, R.H. (eds.) Multi-Agent Programming, pp. 31–72. Springer, Boston, MA (2009). https://doi.org/10.1007/978-0-387-89299-3_2

2. De Giacomo, G., Mancini, T.: Scaling up reasoning about actions using relational database technology. In: Proceedings of the 19th National Conference on Artifical Intelligence AAAI 2004, pp. 245–250. AAAI Press (2004)

3. Eckstein, T.: Improvement and Extension of YAGI, a Golog Based Programming Language for Planning and Real-Time Interaction in Robotics. Bachelor's thesis, Graz University of Technology (2019). https://yagi.ist.tugraz.at/index.php/publications-2/

4. Ferrein, A., Steinbauer, G., Vassos, S.: Action-based imperative programming with yagi. In: Workshops at the Twenty-Sixth AAAI Conference on Artificial Intelligence (2012)

5. Ingrand, F., Ghallab, M.: Deliberation for autonomous robots: a survey. Artif. Intell. **247**, 10–44 (2017)

6. Levesque, H.J., Reiter, R., Lespérance, Y., Fangzhen, L., Scherl, R.B.: Golog: a logic programming language for dynamic domains. J. Logic Program. **19**(20), 1–679 (1994)

7. Maier, C.: YAGI - an easy and light-weighted action-programming language for education and research in artificial intelligence and robotics. Master's thesis, Graz University of Technology (2015)

8. Mataré, V., Schiffer, S., Ferrein, A.: golog++ : An integrative system design. In: Steinbauer, G., Ferrein, A. (eds.) Proceedings of the 11th Cognitive Robotics Workshop 2018, co-located with 16th International Conference on Principles of Knowledge Representation and Reasoning CogRob KR 2018, Tempe, AZ, USA, 27 October 2018, CEUR Workshop Proceedings, vol. 2325, pp. 29–36 (2018)
9. McCarthy, J.: Situations, actions, and causal laws. Stanford University, Technical report (1963)
10. Nilsson, N.J.: Principles of Artificial Intelligence. Springer, Heidelberg (1982)
11. Reiter, R.: Knowledge in Action: Logical Foundations for Specifying and Implementing Dynamical Systems. MIT press, Cambridge (2001)

A New Approach to Determine 2-Optimality Consensus for Collectives

Dai Tho Dang[1], Zygmunt Mazur[2], and Dosam Hwang[1]([⊠])

[1] Department of Computer Engineering, Yeungnam University,
Gyeongsan, Republic of Korea
daithodang@ynu.ac.kr, dosamhwang@gmail.com
[2] Faculty of Computer Science and Management,
Wroclaw University of Science and Technology, Wroclaw, Poland
zygmunt.mazur@pwr.edu.pl

Abstract. Systems and individuals commonly utilize information from multiple sources aim to make a decision. Because knowledge from various sources is often inconsistent, determining the 2-Optimality consensus of a collective, which is an NP-hard problem, is a complicated task, and heuristic algorithms are used to solve this task. The basic algorithm has been the most widely used for the 2-Optimality consensus determination in previous work. While this generates an acceptable consensus quality and reasonable time performance, a new approach having higher consensus quality with an acceptable running time is needed. This study proposes an approach by which the consensus is determined based on the optimal consensuses of its smaller parts. The experiment shows that the consensus quality of the proposed approach is 3.15% higher than that of the basic algorithm, and its time performance is acceptable.

Keywords: Collective · Collective knowledge · Consensus · 2-Optimality consensus

1 Introduction

A collective is generally considered as a set of intelligent units, such as agent systems, experts, and individuals, that make decisions independently [1]. Using knowledge of collectives for decision-making is of significant importance in our societies. One example is multi-agent systems, which enable decision-making based on a set of different sources from the knowledge bases of agents [2,24]. In a collective, members have individual knowledge about one problem, and the knowledge states are often conflicting [3]. Thus, integrating the knowledge states into a consistent state, called a consensus or collective knowledge, is a complex task. Consensus theory presents an efficient approach to this task [4].

The binary vector is helpful to express the knowledge state, and the knowledge of a collective containing such structure is necessary for e-commerce [5,6], distributed networks [7], robotic networks [8], and bioinformatics [9]. Several

© Springer Nature Switzerland AG 2020
H. Fujita et al. (Eds.): IEA/AIE 2020, LNAI 12144, pp. 570–581, 2020.
https://doi.org/10.1007/978-3-030-55789-8_49

postulates have been used to define consensus functions, of which 2-Optimality is one of the essential ones; the consensus meeting the postulate 2-Optimality is the collective's optimum representative, and distances from it to the collective members are uniform [4]. For a collective containing binary vectors, ascertaining consensus satisfying the postulate 2-Optimality is an NP-hard problem. While brute-force algorithm generates an optimal consensus, it is not a practical algorithm because its complexity is $O(n2^m)$. Thus, heuristic algorithms have been developed to solve this task.

In the existing literature, the basic algorithm is the most popular to ascertain the consensus of a collective containing binary vectors [10]. First, this algorithm randomly generates a candidate consensus, following which components of the candidate consensus are sequentially determined. The consensus of the basic algorithm depends on the random candidate consensus, and it can be a locally optimal solution. While the basic algorithm generates a reasonable consensus quality within an acceptable timeframe, a new approach that provides a higher consensus quality within an acceptable running time is needed. This work proposes a new approach for determining the consensus satisfying the postulate 2-Optimality of a collective containing binary vectors. This approach vertically divides the collective into several small parts; rather than using the algorithm brute-force for the primary collective, this algorithm is applied to ascertain the optimal consensuses of the parts. These optimal consensuses are applied to ascertain the consensus of the collective. The proposed approach reduces the time complexity of the brute-force approach, and the optimal consensuses of smaller parts can be applied to ascertain the consensus of the collective.

For the proposed approach, two critical problems require investigation. The first is calculating the time complexity of the proposed approach in general, and the second is measuring the quality of the consensus depending on the size of the smaller parts. In this study, we focus on the generalized calculation of the time complexity of the suggested approach and investigate the cases of a collective divided into two and three parts. The problem of measuring the consensus quality according to the size of the smaller parts will be investigated in the future.

The remainder of this paper is organized as follows. Sect. 2 presents the related work, followed by the discussion of some basic notions in Sect. 3. The suggested approach is introduced in Sect. 4. Experimental outcomes and evaluation are shown in Sect. 5, before finally presenting the overall conclusions in Sect. 6.

2 Related Work

In the field of computer science, the consensus problem has been studied for a long time [11]. It is the basis of distributed computing [11], and it has received increasing attention from computer scientists in recent years [9,12]. As the Internet of Things (IoT) develops at a fast pace, its applications can be seen in various areas of our life, for example, healthcare, smart cities, and smart homes. IoT has several consensus problems that require solutions, such as task allocation, resource allocation, and decision-making in service-oriented tasks [12].

Blockchain technology presents an alternative economic structure which could modify the economy in the future, and it is based on consensus [13,14]. In bioinformatics, the consensus problem has been considered for a long time, and several consensus problems require investigation, such as microRNA target ranking [15], gene prediction [16], and drug interactions [17]. In economics, various consensus problems are encountered; one example is the prediction market, used by several well-known companies such as IBM and Microsoft, in which the opinions of the crowd are aggregated [18].

There are three methods to solve the consensus problem [4,19]. In the axiomatic method, several postulates have been proposed for consensus choice. No consensus choice function exists that can concomitantly satisfy all these postulates. In all of them, the postulates 1-Optimality and 2-Optimality are the most well-known; if a consensus fulfils one of these two postulates, it will fulfil the bulk of postulates [4]. The constructive method handles consensus problems at microstructure and macrostructure levels. The microstructure is the structure of the members, and the macrostructure is the relation of the members [4]. The optimization method uses optimality rules to ascertain the consensus choice function. Condorcet's optimality, global optimality and maximal similarity are optimality rules that are often used for this task [4].

For a collective containing binary vectors, heuristic algorithms have been developed for consensus determination [10,20]. Initially, the basic algorithm, or algorithm H1, was proposed for this task. The time complexity of this algorithm is $O(n2^m)$. Based on the algorithm H1, algorithms H2 and H3 were developed. The time complexity of these two algorithms is also $O(n2^m)$, but the algorithm H1 is faster than the algorithms H2 and H3 by 3.8% and 3.71%, respectively. The difference between the consensus quality of these algorithms is not statistically significant [10]. An algorithm based on combining the basic algorithm and partition, with time complexity of $O(n2^m)$ is introduced in [21]. The difference between the consensus quality of this algorithm and the basic algorithm is not statistically significant.

Genetic algorithms have also been used to solve this task. The algorithms Gen1 and Gen2 use a roulette-wheel and a tournament selection approach, respectively [10]. The consensus quality of Gen1 is 1.73% lower than that of H1, and the consensus quality of Gen 2 is 0.01% higher than that of H1. Nonetheless, because the running time of these two algorithms is very high, they are not practical [10]. Thus far, the basic algorithm has remained the most popular to ascertain the consensus fulfilling the postulate 2-Optimality of collectives containing binary vectors.

3   Basic Notions

Let $\prod(U)$ be a finite set of objects describing all potential knowledge states for a given problem. Let $\prod_p(U)$ denote the set of all p-element subsets with repetitions of set U for $p \in N$ and let

$$\prod(U) = \bigcup_{p \in N} \prod_p(U)$$

Thus, $\prod(U)$ is the finite set of all nonempty subsets with repetitions of set U. A set $X \in \prod(U)$ is regarded as a collective, where each element $x \in X$ describes the knowledge state of a collective member [22, 23].

The members of set U have two structures. The microstructure represents the structure of the members of U, for example, linear order, n-tree, and binary vector. The macrostructure represents the relation between the members of U [4].

The macrostructure of the set U is a distance function $d: U \times U \longrightarrow [0, 1]$, and a consensus choice function in space (U, d) indicates a function

$$C : \prod(U) \longrightarrow 2^U$$

For collective X, the set $C(X)$ is termed the representation of X, and $x \in C(X)$ is termed a consensus of collective X [4].

Definition 1. *For a given collective, $X \in \prod(U)$, the consensus of collective X is determined by the 2-Optimality criterion iff*

$$d^2(c, X) = \min_{y \in U} d^2(y, X)$$

where c is the consensus of X and $d^2(c, X)$ is the sum of squared distances from the consensus c to members of X [4]. In case U is a finite set of binary vectors with length m, set U has 2^m elements.

Definition 2. *A collective containing binary vectors $X \in \prod(U)$ is defined as*

$$X = \{x_1, x_2, ..., x_n\}$$

where each $x_i(i = \overline{1, n})$ is a binary vector with length m. Each element x_i is represented as

$$x_i = (x_i^1, x_i^2, ..., x_i^m), x_i^k = \{0, 1\}, k = \overline{1, m}$$

The sum of squared distances from the consensus c to members of the collective X containing binary vectors is described as follows:

$$d^2(c, X) = \sum_{i=1}^{n} (\sum_{k=1}^{m} |c^k - x_i^k|)$$

4   Proposed Approach

The basic algorithm initially randomly generates a candidate consensus, following which the components of the candidate consensus are sequentially determined. The consensus of the basic algorithm depends on the random candidate consensus, and it can be a locally optimal solution.

Collective X includes n binary vectors, and the length of the vectors is m. To determine the optimal consensus, we perform the brute-force algorithm on the collective X. This approach is not practical because the time complexity of this algorithm is $O(n2^m)$.

It can be seen that if m reduces by 1, the brute-force approach's time complexity reduces two times. Thus, to decrease the time complexity for ascertaining the 2-Optimality consensus, instead of using the brute-force algorithm for the collective X, this algorithm is applied smaller to parts. The proposed approach is described in steps as follows:

- Step 1: Vertically dividing the collective X into k parts: $X_1, X_2, ..., X_k$
  • The difference between the length of any two parts is not greater than 1
- Step 2: Using the brute-force algorithm to determine the consensuses of these parts
- Step 3: Combining the consensuses of the parts to generate the consensus of the collective X

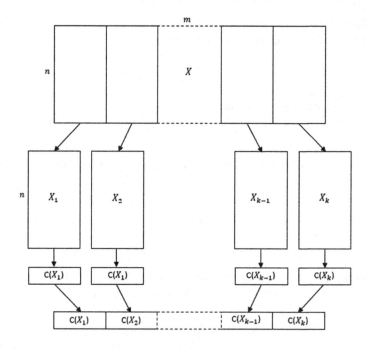

Fig. 1. Proposed approach.

This approach allows us to ascertain the consensus of X based on the optimal consensuses of its parts. The proposed approach is described in detail in Fig.1. Next, the complexity of the proposed approach is calculated. The computational complexity is computed in the following cases: the primary collective is divided into two parts, three parts, and the generalized case.

1. Collective X is divided into two parts:
   - If m is an even number, the suggested approach's time complexity is $O(n2^{\frac{m}{2}})$. It is $2^{\frac{m}{2}}$ times lower than that of the brute-force algorithm.
   - If m is an odd number, the suggested approach's time complexity is $O(n2^{\lfloor m/2 \rfloor+1})$. It is $O(2^{\lfloor m/2 \rfloor})$ times lower than that of the brute-force algorithm.
2. Collective X is divided into three parts:
   - If m is evenly divisible by 3, the suggested approach's time complexity is $O(n2^{\frac{m}{3}})$. It is 2 times lower than that of the brute-force algorithm.
   - If m is an aliquant part of 3, the suggested approach's time complexity is $O(n2^{\lfloor m/3 \rfloor+1})$. It is $O(2^{\lfloor 2m/3 \rfloor})$ times lower than that of the brute-force algorithm.
3. In general, collective X is divided into k parts:
   - If m is evenly divisible by k, the suggested approach's time complexity is $O(n2^{\frac{m}{k}})$. It is $O(2^{\lfloor (k-1)m/k \rfloor})$ times lower than that of the brute-force algorithm.
   - If m is an aliquant part of 3, the suggested approach's time complexity is $O(n2^{\lfloor m/k \rfloor+1})$. It is $O(2^{\lfloor (k-1)m/k \rfloor})$ times lower than that of the brute-force algorithm.

From the previous analysis, it is clear that the proposed approach is hugely efficient in decreasing time computation in consensus determination. Next, we present this approach in the case where the collective X is vertically divided into two and three parts.

1. Dividing collective X into two parts (called TW algorithm):
   - Step 1: Dividing the collective X into two parts: X_1 and X_2
     • X_1 has $\lfloor m/2 \rfloor$ columns, X_2 has $m - \lfloor m/2 \rfloor$ columns
   - Step 2: Determining the 2-Optimality consensus $C(X_1)$ of part X_1 and $C(X_2)$ of part X_2
   - Step 3: Determining the 2-Optimality consensus of X
2. Dividing X into three parts (called TH algorithm):
   - Step 1: Vertically dividing the collective X into three parts: X_1, X_2, X_3
     • The number of parts having $\lfloor m/3 \rfloor$ columns is r and the number of parts having $\lfloor m/3 \rfloor +1$ columns is g, where $m = 3 \times g + r$ $(0 \leq r < 3)$
   - Step 2: Determining the 2-Optimality consensus $C(X_1)$ of part X_1, $C(X_2)$ of part X_2, and $C(X_3)$ of part X_3
   - Step 3: Determining the 2-Optimality consensus of X

The next section will investigate the consensus quality and time performance of these two cases.

5   Experiments and Evaluation

This section evaluates the efficiency of the suggested approach through experiments. The running time and consensus quality of the TW and TH algorithms were investigated. Because the basic algorithm is the most popular for 2-Optimality determination, the consensus quality and running time of the suggested approach is compared to it, and the following two problems are investigated:

- Comparing the consensus quality of the TW, TH, and basic algorithms.
- Comparing the running time of the TW, TH, brute-force, and basic algorithms.

In this work, the significant value α was chosen as 0.05.

5.1   Consensus Quality of the Proposed Approach

The brute-force algorithm generates the optimal consensus. Let c^* be the optimal consensus of X, the following formula is applied to calculate the consensus quality of a heuristic algorithm:

$$Qu = 1 - \frac{\mid d^2(c, X) - d^2(c^*, X) \mid}{d^2(c^*, X)}$$

where c is the 2-Optimality consensus of the heuristic algorithm.

The experiment aims to compare the consensus quality of the suggested approach to that of the basic algorithm. In order to do this, the consensus quality of the TW, TH, and basic algorithms must be measured. Dataset, including 25 collectives, was randomly generated. Each collective contained 600 members, and the length of each member was 20. The three algorithms were run on the dataset. Finally, three samples, which were the quality of consensuses of the TW algorithm, TH algorithm, and basic algorithm, were obtained. These are shown in Table 1. The hypotheses are presented as follows:

H_0: there is no difference between the consensus quality of the three algorithms.

H_1: there is difference between the consensus quality of the three algorithms.

The Shapiro-Wilk test was applied to determine the distribution of samples. The results indicated that the samples do not come from a normal distribution; thus, the Kruskal-Wallis test was used to compare the samples. Because $p - value < \alpha$, the hypothesis H_0 is rejected. Therefore, medians of samples are compared in pairs. The consensus quality of the TW algorithm is 0.21% higher than that of the TH algorithm and 3.36% higher than that of the algorithm basic, and the consensus quality of the TH algorithm is 3.15% higher than that of the algorithm basic.

Table 1. Consensus quality of the TW, TH, and basic algorithms.

Collective	Consensus quality		
	TW algorithm	TH algorithm	Basic algorithm
1	1.0000	0.9933	0.9676
2	0.9991	0.9996	0.9717
3	0.9998	0.9959	0.9647
4	0.9979	0.9932	0.9641
5	0.9932	0.9941	0.9714
6	1.0000	0.9993	0.9618
7	1.0000	0.9975	0.9689
8	1.0000	0.9956	0.9670
9	0.9962	0.9936	0.9714
10	0.9979	0.9961	0.9654
11	0.9993	0.9993	0.9670
12	1.0000	0.9997	0.9772
13	0.9983	0.9983	0.9672
14	0.9988	0.9942	0.9641
15	1.0000	1.0000	0.9666
16	0.9983	0.9983	0.9621
17	1.0000	0.9987	0.9644
18	0.9985	0.9914	0.9734
19	1.0000	0.9972	0.9657
20	0.9987	1.0000	0.9654
21	0.9993	0.9975	0.9567
22	1.0000	0.9995	0.9549
23	0.9995	0.9964	0.9594
24	0.9952	0.9938	0.9735
25	1.0000	0.9908	0.9614

The outcomes are visually presented in Fig. 2. The red, green, and grey lines represent the consensus quality of the TH algorithm, TW algorithm, and basic algorithm, respectively.

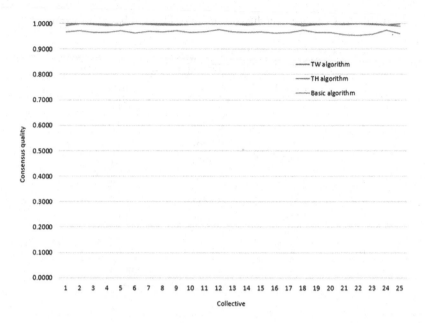

Fig. 2. Consensus quality of the TW, TH, and basic algorithms. (Color figure online)

Table 2. Running time of the brute-force, TW, TH, and basic algorithms.

Collective size	Running time (seconds)			
	Brute-force algorithm	TW algorithm	TH algorithm	Basic algorithm
300	85.770	0.0991	0.0260	0.0100
350	98.200	0.1140	0.0300	0.0115
400	112.600	0.1330	0.0325	0.0130
450	126.260	0.1470	0.0370	0.0130
500	140.445	0.1630	0.0410	0.0152
550	155.300	0.1793	0.0433	0.0176
600	168.370	0.1970	0.0478	0.0184
650	182.390	0.2090	0.0515	0.0195
700	198.830	0.2260	0.0559	0.0270
750	211.840	0.2420	0.0583	0.0216
800	226.950	0.2647	0.0610	0.0238
850	238.300	0.2770	0.0652	0.0245
900	257.330	0.2900	0.0685	0.0258
950	267.840	0.3070	0.0710	0.0262
1000	279.134	0.3220	0.0746	0.0289

5.2   Running Time of the Suggested Approach

This experiment aims to compare the time performance of the TW algorithm, TH algorithm, basic algorithm, and brute-force algorithm. The time performance of the algorithms was measured, and statistical tests were used to evaluate the experimental results. A dataset containing 15 collectives was randomly created. The length of the collective members was 20. The four algorithms were run on this dataset, and finally, running time samples of the algorithms were obtained.

Table 2 describes the running times of the algorithms, they were compared; the hypotheses are presented as follows:

H_0: there is no difference between the running time of the four algorithms.

H_1: there is a difference between the running time of the four algorithms.

The Shapiro-Wilk test is used to determine the distribution of samples. The $p-value > \alpha$, indicating that samples come from normal distribution. Therefore, the one-way ANOVA test was used, and it indicates that there is a difference between the running time of the four algorithms.

The variances are unequal; the Tamhane's T2 posthoc test was used to compare of running time of these algorithms. The result shows that the basic algorithm, TH algorithm, and TW algorithm, respectively, are 99.99%, 99.94%, and 99.89% faster than the brute-force algorithm. The basic algorithm is 90.66% faster than the TW algorithm and 61.24% faster than the TH algorithm. The TH and TW algorithms are practical because their running times are acceptable.

The running times are visually presented in Fig. 3. The green, red, yellow, and grey columns show the running time of the brute-force algorithm, basic algorithm, TH algorithm, and TH algorithm, respectively.

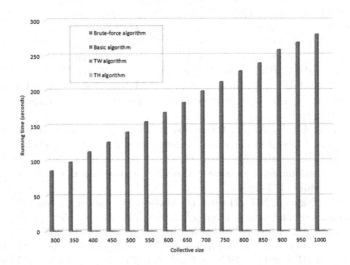

Fig. 3. Time performance of the brute-force, TW, TH, and basic algorithms.

6   Conclusion and Future Work

This study proposes a new approach to determine the 2-Optimality consensus of collectives containing binary vectors. It determines the consensus of the collective based on the optimal consensuses of its smaller parts. The running time of the suggested method is acceptable. The consensus quality of the proposed approach is 3.15% higher than that of the basic algorithm in the cases where the collective is vertically divided into two and three parts. Of the practical algorithms, the TW algorithm generates the optimum consensus quality, followed by the TH algorithm.

In future studies, we will utilize parallel processing to decrease the running time of the proposed approach. Additionally, we will investigate the second critical problem, that is, measuring the consensus quality according to the size of smaller parts.

Acknowledgement. This research was supported by the National Research Foundation of Korea (NRF) grant funded by the BK21PLUS Program (22A20130012009).

References

1. Maleszka, M., Nguyen, N.T.: Integration computing and collective intelligence. Expert Syst. Appl. **42**(1), 332–340 (2015)
2. Bosse, T., Jonker, C.M., Schut, M.C., Treur, J.: Collective representational content for shared extended mind. Cogn. Syst. Res. **7**(2–3), 151–174 (2006)
3. Huang, C., Yang, P., Hsieh, K.: Knowledge-based systems knowledge discovery of consensus and conflict interval-based temporal patterns: A novel group decision approach. Knowl. Syst. **140**, 201–213 (2018)
4. Nguyen, N.T.: Advanced Methods for Inconsistent Knowledge Management. AIKP. Springer, London (2008). https://doi.org/10.1007/978-1-84628-889-0
5. Aggarwal, C.C., Procopiuc, C., Yu, P.S.: Finding localized associations in market basket data. IEEE Trans. Knowl. Data Eng. **14**(1), 51–62 (2002)
6. Griva, A., Bardaki, C., Pramatari, K., Papakiriakopoulos, D.: Retail business analytics: customer visit segmentation using market basket data. Expert Syst. Appl. **100**, 1–16 (2018)
7. Fagiolini, A., Bicchi, A.: On the robust synthesis of logical consensus algorithms for distributed intrusion detection. Automatica **49**(8), 2339–2350 (2013)
8. Di Paola, D., Gasparri, A., Naso, D., Ulivi, G., Lewis, F.L.: Decentralized task sequencing and multiple mission control for heterogeneous robotic networks. In: ICRA, pp. 4467–4473 (2011)
9. Dang, D.T., Nguyen, N.T., Hwang, D.: Multi-step consensus: an effective approach for determining consensus in large collectives. Cybern. Syst. **50**(2), 208–229 (2019)
10. Kozierkiewicz, A., Sitarczyk, M.: Heuristic algorithms for 2-optimality consensus determination. In: Nguyen, N.T., Hoang, D.H., Hong, T.-P., Pham, H., Trawiński, B. (eds.) ACIIDS 2018. LNCS (LNAI), vol. 10751, pp. 48–58. Springer, Cham (2018). https://doi.org/10.1007/978-3-319-75417-8_5
11. Olfati-saber, B.R., Fax, J.A., Murray, R.M.: Consensus and cooperation in networked multi-agent systems. Proc. IEEE **95**(1), 215–233 (2007)

12. Lo, S.K., et al.: Analysis of blockchain solutions for IoT: a systematic literature review. IEEE Access **7**, 58822–58835 (2019)
13. Nawari, N.O., Ravindran, S.: Blockchain and the built environment: potentials and limitations. J. Buildi. Eng. **25**, 1–16 (2019)
14. Tuan, T., Dinh, A., Liu, R., Zhang, M., Chen, G.: Untangling blockchain: A data processing view of blockchain systems. IEEE Trans. Knowl. Data Eng. **30**(7), 1366–1385 (2018)
15. Sengupta, D., Pyne, A., Maulik, U., Bandyopadhyay, S.: Reformulated Kemeny optimal aggregation with application in consensus ranking of microRNA targets. IEEE/ACM Trans. Comput. Biol. Bioinform. **10**(3), 742–751 (2013)
16. Goel, N., Singh, S., Aseri, T.C.: A comparative analysis of soft computing techniques for gene prediction. Anal. Biochem. **438**(1), 14–21 (2013)
17. Ferdousi, R., Safdari, R., Omidi, Y.: Computational prediction of drug-drug interactions based on drugs functional similarities. J. Biomed. Inf. **70**, 54–64 (2017)
18. Bassamboo, A., Cui, R., Moreno, A.: The wisdom of crowds in operations: forecasting using prediction markets. Soc. Sci. Res. Netw. (2015). https://doi.org/10.2139/ssrn.2679663
19. Hudry, O.: Consensus theories: an oriented survey. Math. Soc. Sci. **47**(190), 139–167 (2010)
20. Kozierkiewicz-Hetmańska, A.: Comparison of one-level and two-level consensuses satisfying the 2-optimality criterion. In: Nguyen, N.-T., Hoang, K., Jędrzejowicz, P. (eds.) ICCCI 2012. LNCS (LNAI), vol. 7653, pp. 1–10. Springer, Heidelberg (2012). https://doi.org/10.1007/978-3-642-34630-9_1
21. Dang, D.T., Nguyen, N.T., Hwang, D.: A new heuristic algorithm for 2-Optimality consensus determination. In: 2019 IEEE International Conference on Systems, Man and Cybernetics, pp. 70–75 (2019)
22. Nguyen, N.T.: Inconsistency of knowledge and collective intelligence. Cybern. Syst. **39**(6), 542–562 (2008)
23. Nguyen, N.T.: Processing inconsistency of knowledge in determining knowledge of a collective. Cybern. Syst. **40**, 670–688 (2009)
24. Hernes, M., Nguyen, N.T.: Definition of a framework for acquiring and acquisition subprocesses in a collective knowledge processing in the integrated management information system. Vietnam J. Comput. Sci. **6**(3), 257–272 (2019)

A Decision Support System to Provide Criminal Pattern Based Suggestions to Travelers

Khin Nandar Win[1,2], Jianguo Chen[1,2(✉)], Mingxing Duan[1,2,3],
Guoqing Xiao[1,2], Kenli Li[1,2], Philippe Fournier-Viger[4], and Keqin Li[1,5]

[1] College of Computer Science and Electronic Engineering, Hunan University,
Changsha 410006, China
[2] National Supercomputing Center, Changsha, China
{knandarwin,jianguochen,xiaoguoqing,lkl}@hnu.edu.cn
[3] National University of Defense Technology, Changsha, China
duanmingxing16@nudt.edu.cn
[4] School of Humanities and Social Sciences, Harbin Institute of Technology
(Shenzhen), Shenzhen, China
philfv8@yahoo.com
[5] Department of Computer Science, State University of New York,
New Paltz, NY 12561, USA
lik@newpaltz.edu
http://www-en.hnu.edu.cn/

Abstract. Crimes of various types are occurring in different areas of each country, almost every day. Hence, observing, predicting and preventing crimes is a crucial issue to ensure a peaceful and safe environment. Although several systems have been designed for analyzing crime related data, to our best knowledge, there is no system designed for travelers. This paper addresses this issue by proposing a decision support system named CPD-DSST (Crime Pattern Discovery Decision Support System for Travelers) that allows users to learn about crime occurrences in specific areas and provide suggestions to ensure travel safety. To discover and locate crimes, the system applies an efficient algorithm named Crime Classifying Discovery and Location (CCDL) based on multinomial logistic regression, and a Crime Rate Evaluation (CRE) algorithm, on spatio-temporal crime data. Experiments show that the proposed system can perform accurate predictions. Moreover, preliminary feedback indicates that the system is appreciated by users.

Keywords: Decision support system · Law enforcement · Multinomial logistic regression · Spatio-temporal data · Travel recommendation

1 Introduction

Nowadays, a huge amount of crime related data is collected by law enforcement agencies. An important data type is crime reports, which contain critical and useful information for discovering and predicting crimes. They can be used for tasks

© Springer Nature Switzerland AG 2020
H. Fujita et al. (Eds.): IEA/AIE 2020, LNAI 12144, pp. 582–587, 2020.
https://doi.org/10.1007/978-3-030-55789-8_50

such as arresting criminals and recovering stolen property. For instance, the Los Angeles Regional Crime Stoppers, a program that encourages citizens to anonymously report crimes, has received 31,000 tips between 2012 and 2014, which has resulted in 1,404 arrests, seizing 121 weapons and recovering $584,129 [6]. Crimes are reported by various channels setup by law enforcement agencies and non-profit organizations (NGO) such as short message services, phone applications and online reporting systems (e.g. FBI's online tips submission system and Newark Police's crime stoppers).

Collecting crime data is desirable but to make sense of a large amount of crime data, specialized computer systems must be designed. Hence, various decision support systems have been developed. Many studies have focused on supporting different aspects of criminal investigation such as detecting serial crimes [1], discovering and preventing crimes [9], and supporting the work of law enforcement officers through decision support systems [1,6]. For instance, a decision support system having both text analysis and classification capabilities was proposed to automate crime report analysis and classification, for enhancing e-government's and law enforcement's services [6].

In another study, a decision support system for serial crime detection was deployed in China's public security bureau for more than a year [1]. And also, a decision support system was developed at the Institute for Canadian Urban Research Studies, for forensic science investigations and longitudinal statistical analysis of crime and law enforcement performance [4]. Taking advantage of social media data, Gerber and Matthew predicted crimes using Twitter data [3]. They used Twitter-driven predictive analytics and kernel density estimation for crime prevention. For spatio-temporal crime prediction, a public decision support system for low population density areas has been proposed [5] and one for predictive patrolling.

Decision support systems have not only been used for crime related tasks but also in many other domains [3,4] such as agriculture [7], disaster management, hotel selection, facility location analysis [2], and urban planning, etc. Though, the above studies have used crime data in many different ways and some decision support systems have been proposed, there is no crime data-based decision support system specifically for travelers.

This paper addresses this important research gap by proposing a Traveler Decision Support System named CPD-DSST (Crime Pattern Discovery Decision Support System for Travelers), which provides relevant information to a user about a selected destination city, including related crimes and correlation information. The main goal is to ensure the safety of travelers by suggesting safer destinations, and suggest travel precautions.

2 Proposed Method

This section firstly presents logistic regression and the CRE algorithms [9]. Then, the CCDL algorithm is described that is used in the proposed system to classify, discover and locate crimes from spatiotemporal data.

Logistic regression has been widely used in various fields such as machine learning, statistics, and other application areas such as medical, marketing and social science. In our work, data is multivariate categories and our aim is to classify data instances into multiple classes. Hence, multinomial logistic regression is used instead of normal logistic regression and also because of its robustness and flexibility in prediction technique. For multinomial logistic regression, the *Softmax Function* is applied. Hence, it is also called softmax regression. The function normalizes the data points by dividing by the sum of all the exponential values. The normalization for the sum of all the output vector $\sigma(\mathbf{z})$ elements should be 1. The equation for this softmax function is defined as:

$$\sigma(\mathbf{z})_i = \frac{e^{z_i}}{\sum_{j=1}^{K} e^{z_j}} \text{ for } i = 1, \ldots, K, \quad \mathbf{z} = (z_1, \ldots, z_K) \in \mathbb{R}^K$$

$$\sigma : \mathbb{R}^K \to \mathbb{R}^K$$

(1)

Crime Rate Evaluation (CRE) algorithm was applied for various crimes having different occurrence frequencies and also to assess the crime impact in terms of aspects such as crime categories and locations. The CCDL algorithm proposed in this paper applies softmax function for the classification of multiple crimes and the CRE algorithm for evaluating crimes in a selected time interval.

The Eq. (2) evaluates the crime rate for a place b_m, where $\mathcal{B}(x_i)$ is the place (country), and $T_p = (st_p, et_p)$ is the time period to be used for the analysis, where st_p is the starting time and et_p is the end time. $B = \{b_1, ..., b_m, ...\}$ is the set of distinct places (countries) in the dataset X. Each crime record $x_i (1 \leq i \leq N)$ of X is a vector $x_i \in \mathbf{R}^B$. The crime rate $CR_{b_m}(T_p)$ of the region/country b_m in the time period T_p is defined as:

$$CR_{b_m}(T_p) = \frac{\sum_{\mathcal{B}(x_i)=b_m} x_i}{\sum_{b_m \in B} \sum_{\mathcal{B}(x_i)=b_m} x_i}, \quad \forall x_i \in X,$$

(2)

Crime Classifying Discovering and Locating (CCDL) algorithm combines multinomial logistic regression (Eq. 1) for classification with the CRE algorithm (Eq. 2). The pseudo-code of the proposed CCDL algorithm is shown in Algorithm 2.1. It takes as input a set X of crime records (data points) and outputs a crime classification (occurrence probabilities for different types of crime classes). The CCDL algorithm is applied for a time period defined by a start time and end time given to the CRE algorithm. After applying CRE, the algorithm calculates probabilities using multinomial logistic regression.

3   Crime Pattern Discovery Decision Support System for Travelers (CPD-DSST)

This section presents the developed Crime Pattern Discovery Decision Support System for Travelers (CPD-DSST). First, we present the dataset that is used in

Algorithm 2.1. Crime Classifying Discovering and Locating (CCDL) algorithm

Input:
    X: The crime records (data points);
    $st_p(CRE)$: The Start Time from the CRE algorithm;
    $et_p(CRE)$: The End Time from the CRE algorithm;
Output:
    Crime classes
1: get time period from CRE algorithm
2: get the shape of input data points $z \leftarrow CCDL.shape[0]$;
3: $K \leftarrow$ The total number of data points;
4: $i, j \leftarrow K$;
5: **while** True **do**
6:    **if** $et_p(CRE) > st_p(CRE)$ **then**
7:       **for** each i in range(K) **do**
8:          **for** each j in range(K) **do**
9:             calculate the probability $\sigma(\mathbf{z})_i \leftarrow$ probability(CCDL);
10:             $\sigma(\mathbf{z})_i \leftarrow \frac{e^{z_i}}{\sum_{j=1}^{K} e^{z_j}}$;
11:          **end for**
12:       **end for**
         $st_p(CRE) \leftarrow st_p(CRE) + 1$
13:    **else**
14:       break;
15:    **end if**
16: **end while**
17: **return** Crime classes

our experiment briefly and then we discuss the implementation of the proposed system. Finally, we evaluate the proposed system by comparing different logistic regression algorithms and report user feedback.

An overview of the proposed system is presented in Fig. 1. The system mainly focuses on providing information to a traveler (user) about past crimes and predicted crimes for a given place. Information about past crimes is shown in Fig. 2(a). Moreover, the system allows a user to report crimes by providing a place, time, crime type and summary of the crime, as shown in Fig. 2(b). Additionally, a user can see useful information about a place from the system via the Information menu.

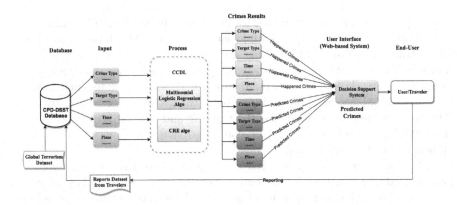

Fig. 1. An overview of the proposed Crime Pattern Discovery Decision Support System for Traveler (CPD-DSST)

The Global Terrorism DataBase (GTD) is an open-source database gathered from large-scale crime activities around the world from 1970 to 2018, covering approximately 228 countries and incidents that happened [8]. It was developed, and is manually maintained and updated, by the Maryland University. It includes more than 192,000 criminal records with more than 100 features.

For the demonstration of our web-based system, consider the scenario where we choose December 2019 as time and Changsha as city, as depicted in Fig. 2(a). As no crimes happened in Changsha in December 2019, the crime data is empty and we can assume that it is safe for a visit. In that scenario, we decided to analyze the data using a one month time period because most travelers visit a place for one month or less and may want to know what happened in the last month. If there are some crimes or accidents occurred in the selected place for the selected time period that have not been provided in the system, the system provides a reporting function Fig. 2(b) to the traveler/user that can be used to enter information into the system.

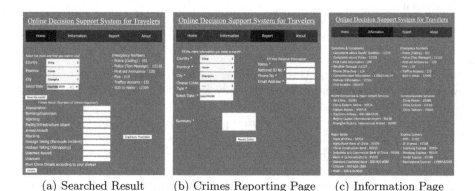

| (a) Searched Result | (b) Crimes Reporting Page | (c) Information Page |

Fig. 2. Traveler decision support system web pages

For **performance evaluation**, we evaluated the predictive function of the system by comparing the model with logistic regression. The dataset was divided into a training dataset and a testing dataset using a ratio of 7/3. Results are shown in Table 1 in terms of accuracy. The data used for this analysis contains target type, attack type, location and time where accidents happened from the GTD dataset [8]. For example, in India, more than 8000 accidents occurred from 2008 to 2018 and the accuracy for the training and testing datasets are 0.9435 and 0.93925, respectively. It can be concluded that the accuracy is not much influenced by the dataset size.

Additionally, we have also asked a group of persons to use our system to provide feedback. On overall the comments were very positive. They found the system to be useful and were looking forward for the system to be released publicly. They also provided a few minor comments to improve the user interface design that will be taken into account before the public release.

Table 1. Evaluation of the regression techniques in terms of accuracy

Country	Accidents (2008–2018)	Logistic regression		Multinomial logistic regression	
		Train dataset	Testing dataset	Train dataset	Testing dataset
China	100	0.56962	0.42857	0.94937	0.82857
Myanmar	300	0.48819	0.40909	0.99606	0.95454
India	8000	0.22801	0.22761	0.94350	0.93925

4   Conclusion

This paper presented a web-based traveler decision support system named CPD-DSST, which can highlight crime patterns using multinomial logistic regression and the CRE algorithm for a place that a traveler wants to visit. Moreover, the system can provide advices about touring each place. According to the evaluation of the system, the proposed CCDL algorithm and model can effectively detect crime patterns and gave useful suggestions to users. In future work, with the aspect of the traveler safety, we will further enhance the proposed system in supporting the work of government and law enforcement officers for specific critical regions.

References

1. Chi, H., Lin, Z., Jin, H., Xu, B., Qi, M.: A decision support system for detecting serial crimes. Knowl.-Based Syst. **123**, 88–101 (2017)
2. Erdoğan, G., Stylianou, N., Vasilakis, C.: An open source decision support system for facility location analysis. Decis. Support Syst. **125**, 113116 (2019)
3. Gerber, M.S.: Predicting crime using twitter and kernel density estimation. Decis. Support Syst. **61**, 115–125 (2014)
4. Ghaseminejad, A.H., Brantingham, P.: An executive decision support system for longitudinal statistical analysis of crime and law enforcement performance crime analysis system pacific region (CASPR). In: 2010 IEEE International Conference on Intelligence and Security Informatics, pp. 1–6. IEEE (2010)
5. Kadar, C., Maculan, R., Feuerriegel, S.: Public decision support for low population density areas: an imbalance-aware hyper-ensemble for spatio-temporal crime prediction. Decis. Support Syst. **119**, 107–117 (2019)
6. Ku, C.H., Leroy, G.: A decision support system: automated crime report analysis and classification for e-government. Gov. Inf. Q. **31**(4), 534–544 (2014)
7. Li, M., Sui, R., Meng, Y., Yan, H.: A real-time fuzzy decision support system for alfalfa irrigation. Comput. Electron. Agric. **163**, 104870 (2019)
8. of Maryland, U.: Global terrorism database (gtd). http://www.start.umd.edu/gtd
9. Win, K.N., Chen, J., Chen, Y., Fournier-Viger, P.: PCPD: a parallel crime pattern discovery system for large-scale spatiotemporal data based on fuzzy Clustering. Int. J. Fuzzy Syst. **21**(6), 1961–1974 (2019). https://doi.org/10.1007/s40815-019-00673-3

Model-Based Decision Support Systems - Conceptualization and General Architecture

Peter Struss[(✉)]

Computer Science Department, Technical University of Munich, Garching, Germany
`struss@in.tum.de`

Abstract. The paper presents an attempt to conceptualize decision support and various generic subtasks, to develop a general architecture of intelligent decision support systems, and to exploit previous work on process-oriented diagnosis within this architecture. The primary subtasks whose (intelligent) solution is heavily dependent on domain knowledge are situation assessment, i.e. inferring what is happening in a system from a set of observations, and therapy proposal, i.e. developing plans for interventions to achieve certain goals starting from the current situation. Both tasks can be solved by an extension of consistency-based diagnosis to process-oriented models.

Keywords: Decision support systems · Process-oriented modeling · Model-based systems

1 Introduction

When searching for definitions or characterizations of decision support systems (DSS) and for proposals for general architectures or (sub-) functions of DSS we obtained rather disappointing results. Many of the offered definitions boil down to "A DSS is a computer system (or set of tools) that supports making decisions". When architectures are proposed, they are often presented as a huge set of tools and computational steps embedded in a confusing web of interconnections, often heavily emphasizing data-driven techniques. The components are mainly characterized as various alternative or complementary techniques, rather than by the function they implement within the system. A survey of 100 papers on environmental decision systems (Sojda et al. 2012) shed a light on a variety of interesting and successful specialized DSS, but does not yield one or more principled ways of developing a DSS.

A systematic analysis and a conceptualization of DSS seem to be missing. Attempting this is not an academic question aiming at delivering a set of definitions. It is a practical necessity, if we are interested in the systematic re-use and integration of different tools and methods. And it is a prerequisite for establishing requirements on DSS and its components, especially when we want to build "intelligent DSS".

The conceptualization and architecture proposed in the following is based on our previous theory, implementation, and application of the process-oriented diagnosis system G+DE (Heller and Struss 2002). It does not claim to offer a general account for

© Springer Nature Switzerland AG 2020
H. Fujita et al. (Eds.): IEA/AIE 2020, LNAI 12144, pp. 588–600, 2020.
https://doi.org/10.1007/978-3-030-55789-8_51

all existing DSS. Its objective is to provide the foundation for the design and comparison of knowledge-based DSS and the identification of subtasks and modules and their principled interfaces. This leads to a second disclaimer: although we are fully aware that observations, knowledge, and inferences can be subject to a significant degree of uncertainty, we do not explicitly represent this at this stage of the formalization. The consideration behind this is that uncertainty (e.g. in terms of probabilities) can be added to the presented concepts, potentially "softening" results and introducing ambiguity and alternatives.

We start with the general concepts and tasks of decision making. Then we introduce the concepts and the algorithms underlying a model-based realization of a DSS und finally a number of challenges and open questions.

2   Conceptualization of a DSS

In the following, we use UML class diagrams in order to specify the relationships between different concepts.

2.1   Decision

A decision has to be taken only if there is a **choice**. It means choosing one from a set of alternative options. These options are different ways to **act**. If the possible ways to act are complex (rather than only a single action), we call them a plan and consider them as a set of actions (which simplifies the more general concept used in the AI field of planning). Hence, the concept decision has a set of plan options as an input and results in choosing one of them. The necessity to act and the criteria for the selection depend on two crucial concepts: the **goals** to be achieved (or maintained) and the current (or an assumed) **situation** the actions should be applied to in order to accomplish the goals. Hence, these basic concepts are related as depicted in Fig. 1.

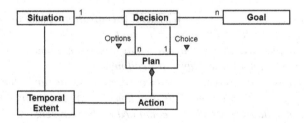

Fig. 1. Decision and related concepts

The activity of "selecting the best plan for achieving or pursuing goals in a given situation" characterizes decision making in the narrow sense only, since a prerequisite for this step is that

- an **understanding of the situation** and
- a set of **possibly effective plans**

have been generated beforehand. These are the main subtasks that require knowledge, domain expertise, and reasoning.

2.2 Example from Water Treatment

In the following we will use a simplification of a real problem encountered in a previous project on drinking water treatment in the city of Porto Alegre, Brazil (Roque *et al.* 2003) as an illustrative example, which is depicted in Fig. 2. Throughout the paper, we will use italic font when referring to this example in the following.

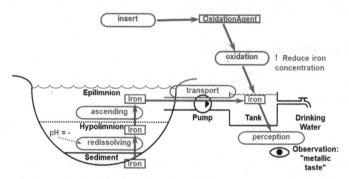

Fig. 2. The water treatment example including the initial information and observations (black), the causal explanation of the metallic taste (green), the added intermediate goal (orange), the proposed intervention and its impact (red) (Color figure online)

*In this scenario, customer complaints about an unpleasant **metallic taste of the drinking water** were the starting point. In order to determine appropriate countermeasures, experts from the municipal department of water and sewage in Porto Alegre, DMAE, analyzed the situation and finally identified the reasons behind the observation:*

- *The cause for the taste was an unusually **high concentration of iron** in drinking water, which was already present in the treatment plant.*
- *This had been **pumped in** from a particular reservoir, Lomba do Sabão, more specifically from its upper water layer, the epilimnion.*
- *The dissolved iron in this layer did **ascend** from the lower water layer, the hypolimnion.*
- *The increased iron concentration in the hypolimnion was caused by a higher rate of **re-dissolving of solid iron**, which was contained in the sediment.*
- *This increased re-dissolving was due to an unexpected **change in the pH** towards acidic conditions of the hypolimnion as a result of a phase of algal bloom that had occurred recently.*

Based on this analysis, reducing the iron concentration in the plant by creating an oxidation process through the introduction of oxidation agents such as chlorine or ozone was the appropriate intervention, since neither removal of the solid iron from the sediment nor the modification of the pH in the reservoir was a feasible solution.

2.3   High-Level Decomposition of a DSS

A DSS should support or automate the selection of a plan for achieving certain goals based on a set of observations. Doing this in a way that deserves the attribute "intelligent" certainly implies determining whether or not and to what extent carrying out a plan will effectively transform the given situation to one that satisfies the goals. This, in turn, requires the step to interpret the observations in order to derive a more complete picture of the current situation, which includes, in particular, a representation of the present causal interactions in the system, such as the causal story behind the metallic taste of the drinking water in the above example. We call this subtask, which certainly requires deep domain knowledge, **situation assessment**. Another subtask, which is necessary, unless the different options to choose from are given a priori, is the generation of plans that promise to achieve the goals. The inputs to this **therapy proposal** step are the situation, the goals (e.g. *iron concentration in drinking water below threshold*), and the potential actions that can be taken (such as *introduce ozone*). Together with the decision making step, i.e. the choice of one plan from the options, we obtain the high-level decomposition of the DSS shown in Fig. 3.

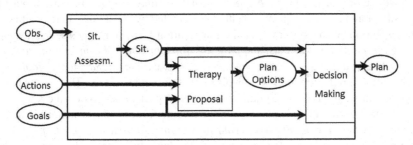

Fig. 3. Top-level components of DSS

2.4   Therapy Proposal

The initial step in therapy proposal (see Fig. 4) is to evaluate the situation with respect to the given goals, i.e. determining which goals are violated and, perhaps, in which way (e.g. a particular threshold exceeded or undershot, relevant objects missing or unwanted ones being present). *In our example, the goal iron concentration in drinking water below threshold is violated, and the actual iron concentration deviates positively).* This situation evaluation may not only be necessary for the current situation, but also for future situations with or without applying actions. This requires a prediction module, which produces the description of such forecast situations, e.g. through simulation. The discrepancies between situations and goals are an input to goal generation. This is needed, although we already have a set of explicit goals, because, usually, we need to determine some intermediate goals that guide the search for a plan.

If the iron concentration in drinking water is above the threshold, even after a remedial action, the respective goal will be violated for some time, while a feasible

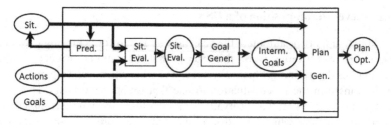

Fig. 4. Decomposition of therapy proposal

intermediate goal will be reducing the iron concentration. Together with the situation description and the available actions, they are an input to the plan generation module.

2.5 Decision Making

The last step leading to a decision is **plan evaluation**, by accumulating **situation evaluation**, i.e. checking whether and to what extent they establish situations that fulfill the (intermediate or general) goals. Again, this analysis may not only be carried out for a single situation, but for a sequence of situations created by executing a plan or situations in the evolution of the system after the intervention has been finished. *For example, shutting off the pump and, thus, stopping the transportation of iron into the tank may stop drinking water supply after a while and, hence, violate a fundamental goal.*

Therefore, prediction may be involved in this subtask, too. This is, usually, a knowledge-intensive task, since determining the impact of actions on the system and its evolution requires an understanding of the driving forces and interactions in the system. *One has to know about the relevant impact of oxidation in order to propose the action of introducing an oxidation agent into the tank.* In contrast, the step of picking one plan from a set of suitable ones based on the plan evaluation is a more or less formal task based on the plan evaluation, which is guided by weights and priorities of goals and cost of plan execution, factors that are external to the domain model.

So far, we characterized the inputs and outputs of the DSS and its modules by informal concepts. To lay the foundations for a model-based solution to DSS, we continue by providing more precise, but also specialized definitions of these concepts.

3 Process-Model-Based DSS - Concepts

The meanings of the main concepts we used in the description of the interfaces are:

- **Observation:** "What is the case" (information)
- **Situation:** "What is going on" (applied knowledge)
- **Goal:** "What should be the case" (objectives)
- **Action:** "What can be done" (potential)
- **Plan:** "What should be done" (intention)

The above concepts can all be stated as assertions about the world under different modalities, which means we need to fix a basic ontology for statements about the considered domain, a class of real or potential systems. We distinguish propositions about

- **Structure:** in terms of a set of existing objects and object relations between them
- **Behavior:** by constraining quantities or properties of the objects to a particular range

Figure 5 introduces the unifying concept of an assertion in a UML static structure diagram, depicting that (non-)existence is assigned to structural elements, which are objects and object relations. Objects have quantities, and assertions can associate a certain range to them. Observations and goals are collections of assertions. *The former ones are, for instance, DrinkingWater.Taste = metallic and the assumption, Hypolimnion.pH = neutral, if this denotes the neutral range.*

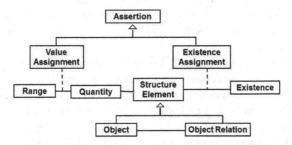

Fig. 5. Assertions w.r.t. quantities and objects and object relations, resp.

DrinkingWater.Iron.Concentration = [0, threshold) is an overall goal and Drinking-Water.Iron.DerivativeConcentration = − the intermediate goal replacing it. A situation or, more precisely, a representation of a situation is meant to capture not only an extended description of the observed system, but also reflect our understanding of the underlying causal interdependencies. Hence, we represent it as a set of assertions and a collection of processes (Fig. 6a).

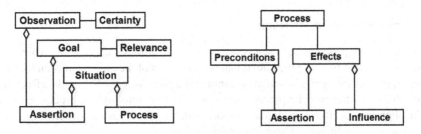

Fig. 6. a) Observation, goal, and situation, b) process

Processes (Fig. 6b) represent the mechanisms that are the driving forces of the system's dynamics (or, rather, our model thereof). Processes are characterized by a collection of **preconditions**, which are assertions, and **effects**, which are assertions or influences, i.e. their impact on the system dynamics, i.e. their contribution to changes of variables. *The preconditions of the process IronRedissolving contains existence assignments such as Exist(Sediment) = T and Exist(Contains (Sediment, SolidIron)) = T and the value assignment Hypolimnion.pH = − (meaning acidic), while the effects state Exist(Contains(Hypolimnion, DissolvedIron)) = T and Influence + (Hypolimnion, DissolvedIron.DerivativeConcentration), among others.*

4 Process-Model-Based DSS - Algorithm

The architecture and the interface and functional definitions of its modules are still very general. This is deliberate, because our goal is to provide a generic way of structuring DSS that allows us to combine different techniques and models that implement different modules or sub-modules. For instance, situation assessment might be realized by a case-based diagnosis system, while therapy proposal uses rules linking disturbances and corrective interventions.

In this section, we describe our model-based solution to implementing an intelligent DSS along the lines presented above. It will turn out to provide a very coherent way of implementing different modules of the architecture. Its core is the concept of a process, which was introduced in Sect. 3 and which is embedded in an inference scheme that opens ways, in particular, to implement (semi-)automatic situation assessment and therapy proposal. This implementation is based on a rigorous specification of the involved subtasks (see (Heller and Struss 2002)), which is grounded in a logical reconstruction of Qualitative Process theory (Forbus 1984) and exploits theories and techniques from model-based problem solving (Struss 2008).

In the following subsections, we summarize the representation of elementary processes, which are collected in a domain library, and then discuss the task of composing them to form a model of a complex system. Then we outline a solution to situation assessment, which is seen as the task of model composition. On this basis, situation evaluation and goal generation can be implemented. Therapy proposal turns out to be an instance of model composition, too.

4.1 Process-Oriented Modeling

In Sect. 3, a process was represented as a pair of conditions and effects, which both contained assertions, while the effects captured impacts of the process informally stated as **influences**. Turning the informal semantics of a process, namely that the effects will be established whenever the preconditions are satisfied, into logic, a process becomes an **implication**, i.e. if the conditions are satisfied, the effects are also true:

StructuralConditions ∧ QuantityConditions

⇒ StructuralEffects ∧ QuantityEffects,

where StructuralConditions and StructuralEffects are ExistenceAssignments, and QuantityConditions and QuantityEffects contain ValueAssignments as described in Sect. 3.

In addition, QuantityEffects contain influences. Influences capture the impact of a process on the dynamics of the systems, i.e. how quantities change, but, nevertheless, are beyond the expressiveness of differential equations. *The result of the process of iron re-dissolving cannot simply be stated as a positive derivative of the iron concentration in the hypolimnion. This is because, in the composed system model, there may be other, counteracting processes, which may compensate or override the impact of re-dissolving and lead to a negative derivative of the iron concentration. All the local model of re-dissolving may state is a positive* **contribution,** *a positive influence.* In an approximate way, it specifies a partial derivative of the iron concentration: *the iron concentration in the water can be a function of several variables, and the partial derivative w.r.t. the iron concentration in the sediment is positive. Whether it really increases or not, depends on other influencing variables, as well.*

In the step of model composition, the various influences on each variable will be combined, e.g. in a linear combination, like the different inflows and outflows to a container are added. An essential aspect of this step is that **all** relevant influences and, hence, processes are assumed to be known, which is an assumption that may have to be retracted in subsequent reasoning steps in order to find a plausible explanation of the observations. This step of influence resolution implies that, **if the model does not contain any influence on a variable, the variable does not change**. Similarly, objects and relations added to the model need to be introduced as part of the StructuralEffects of some process. As a consequence, a model that contains a change in a variable or an object or object relation that are not caused by (i.e. occurring in the effects of) at least one process is inconsistent.

4.2 Situation Assessment

Situation assessment was introduced as the task of constructing a causal account for what has been observed. This is done by creating situations, i.e. sets of assertions and processes that are consistent with the observations based on a library of processes.

In a process-oriented framework, situation assessment **is model composition**. The resulting model needs to be complete w.r.t. the effects of the included processes (and the processes triggered by these effects etc.). But the more important aspect in situation assessment is its completeness regarding causation – everything that has been observed or hypothesized requires a process in the situation causing it (with an important exception, as discussed below).

Candidates for such explanatory situations are constructed in a cycle of completing them in a (causally) forward, deductive direction (i.e. adding effects and their consequences) and in a backward, abductive direction (by hypothesizing preconditions and their potential causation). In contrast to the former step, the latter is usually not unambiguous (there may be different causal accounts for the observations). Hence, situation assessment may deliver alternative results, and the user has to pick one.

How can the extension of model in this causally upstream direction be enforced? The key to this lies in the fact stated above that a situation that is lacking a causal account for a change or a StructureElement is not simply incomplete, but inconsistent. For determining possible "repairs" of inconstant situation hypotheses, consistency-based diagnosis ((de Kleer-Williams 1984), (Struss 2008)) is applied.

The algorithm (depicted in Fig. 7) starts with the set of initial observations, i.e. Assertions, which may be treated as facts or assumptions, and enters the cycle.

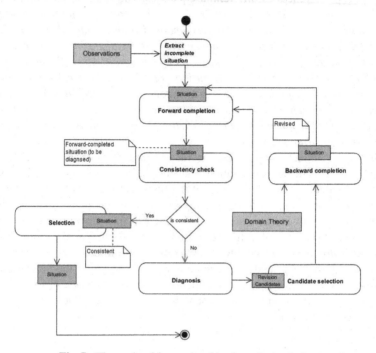

Fig. 7. The cycle of forward and backward completion

- **Forward completion:** iteratively instantiates processes (as implications) whose preconditions are satisfied
- **Consistency check:** if the situation is consistent, it is a candidate for a proper situation assessment
- **Diagnosis:** consistency-based diagnosis identifies assumptions that contribute to the inconsistency and, hence, are potential targets for revisions of the model that may (ultimately) lead to a consistent situation. Such assumptions can be initial hypothetical observations (*such as pH = 0*) or closed-world assumptions related to objects or object relations or the ones involved in influence resolution. *For instance, when the situation has been extended to hypothesize dissolved iron in the hypolimnion, the closed world assumption regarding this object is a candidate for revision (and in this case, the only one).*
- **Candidate selection:** usually, there are several possible revisions. Either they are pursued in parallel, or a heuristic or a user selects one option. If a closed-world assumption is involved, it is related to a variable or StructureElement.
- **Backward completion:** searches the library for processes that affect the variable or StructureElement associated with the retracted closed-world assumption, *in our case,*

the Redissolving process. Again, there is usually a choice to be made. The chosen process is added to the situation which is then entered to forward completion.

Re-entering the loop is necessary because the added process may have more effects than only the intended one, which could lead to the instantiation of new processes and also to inconsistencies.

In our example, backward completion in the loop produces a chain of processes back to the Redissolving process. Since its precondition Hypolimnion.pH = - contradicts the default Hypolimnion.pH = 0, the latter becomes a candidate and can be dropped. But the structural preconditions, i.e. the existence of solid iron in the sediment, remains unexplained. This illustrates that backward completion is inherently infinite; demanding for causal explanations would, due to the finiteness of the process library, always end in a missing explanation and, hence, in an inconsistency (unless there exists a process without preconditions or a cyclic causation, which would indicate a modeling error). In consequence, asking for causes of the preconditions has to be limited: *while existence of iron at high concentrations in the tank and the water layers demands for an explanation, the presence of iron in the sediment does not have to be justified by another process, because this would be beyond the scope of the model (and of the interest).* We call these preconditions that cut off the search for explanations **introducibles**. They characterize the boundaries of the model and the desired explanations.

What is treated as introducible may heavily depend on the specific task. *If the unexpected pH is taken for granted, the search may stop here, because it is possible to determine an appropriate intervention. In a different setting, one may be interested in tracing it back to algae bloom and its creation by an abundance of nutrients caused by agricultural runoff, in order to take preventive actions regarding this root cause.* Naturally, the search is confined by the scope of the domain library that is assumed to contain all the processes known to be potentially relevant to the class of problems to be solved. Each structural element and influence on a variable that does not appear as an effect of any process in the library has to be treated as an introducible. Do we have to consider this as a weakness of process-model-based situation assessment? No, because this is not different from the performance of a human expert who cannot and will not generate explanations beyond the causal accounts (s)he is aware of.

The result of this solution to situation assessment is a set of candidate situations, each of them being

- A **minimal** situation
- containing **all facts**
- otherwise only **introducibles** and
- **effects** of occurring processes
- **closed** w.r.t. **effects**,
- with a **maximal** set of **assumptions** holding.

This means, while the factual observations are preserved (*DinkingWater.Taste = metallic*), some assumptions may be dropped (*Hypolimnion.pH = 0*). In addition, there are introducibles (*Ex(Contains(Sediment, SolidIron))*) and their consequences.

4.3 Situation Evaluation and Goal Generation

If situation assessment, perhaps after filtering by the user, has committed to a particular situation (or a disjunction of situations), situation assessment has to check for violation of goals. This is a straightforward consistency check of the assertions established by the situation and the goals. An actual value (range) of a variable may be disjoint w.r.t. the goal range, more specifically, it may be above or below this range (*DrinkingWater.Iron.DeviationConcentration* =+). Or the (non-)existence of a structural object required by a goal may contradict the situation. If the existence of an object is required by a goal, but does not result from a process contained in the situation, this creates an inconsistency.

From the detected goal violations, intermediate goals can be generated in a mechanical way. The creation of a non-existent object remains a goal (*in the water treatment world, having (a limited amount of) chlorine contained in drinking water would be an example*). If a quantity is higher (lower) than the range specified by the goal, then a negative (positive) derivative becomes an intermediate goal, as for instance

$$DrinkingWater.Iron.DerivativeConcentration =$$
$$-DrinkingWater.Iron.DeviationConcentration = -.$$

4.4 Therapy Proposal

Therapy proposal operates similar to situation assessment, but aiming at consistency with the (intermediate) goals (rather than with observations). The causally upstream completion of the model is bounded by **action triggers** as the introducibles (i.e. every change applied to the situation is ultimately caused by actions), and the downstream completion helps to check whether the actions have "side-effects", which may violate some of the (intermediate) goals. Therapy proposal creates

- A **minimal** structure
- containing all **facts**
- otherwise only **action triggers** and
- **effects** of occurring processes
- **closed** w.r.t. **effects**
- with all **fixed goals** and a **maximal** set of **relaxable goals** holds.

As suggested by this characterization, situation assessment and therapy proposal can be performed by the same algorithm, by interchanging observations and goals, and natural introducibles and action triggers (see (Heller 2001)).

Although the basic outline of these inferences may suggest a straightforward solution, we indeed face a fundamental problem: which elements of the situation actually persist while the actions considered by therapy proposal are applied? After all, the actions aim at modifying the situation. The answer to this question is based on the consideration that absolute values change continuously (*DrinkingWater.Iron.DeviationConcentration* =+ *persists at least for a while*), while the derivatives may be discontinuously influenced by the action (*DrinkingWater.Iron.DerivativeConcentration becomes negative immediately unless other processes counteract the impact of oxidation*). As a consequence, we cannot simply preserve derivatives in the situation in which we apply the actions. However, we cannot drop them altogether, because only those affected by the actions and the processes triggered by them may change, while the others will persist. In consequence, we turn all derivatives into assumptions with the effect that they will be dropped if enforced by the therapy proposal and otherwise persist.

5 Discussion

The process-model-based solution described above has some inherent restrictions and practical problems that will be addressed by future work.

Snapshot-Based Analysis: right now, the analysis assumes that a situation, i.e. a causal explanation can be constructed within a (qualitative) temporal snapshot. More technically, the analysis does not go backward beyond integration steps. If they are included, there could be concurrent changes in the system, and different orders of their temporal occurrences would have to be considered, which complicates the analysis.

Intervention Plans: Similarly, the generated interventions are collections of actions, executed in parallel, rather than in a particular order or a certain point in time.

Observations: Uncertainty of observations is an issue in practical settings. Currently it is treated only in a black-grey-white fashion by allowing observations to be assumptions or defaults, that may be retracted.

Controlling the Search: in a rich domain, there will be many possible branches to pursue during the search for consistent situations. Lacking general heuristics for determining promising foci at this point, we envision an interactive solution, where the user selects the branches to be pursued. However, this creates the necessity of providing him with a clear and informative picture of the state of the search, which is non-trivial.

Determining Introducibles: It will usually be impossible to expect a user to define introducibles in a comprehensive way beforehand, because it would require anticipating the potential causal explanations generated by the DSS. Again, the only feasible solution appears to be an interactive one, where the user decides on the fly, whether or not something needs further causal analysis.

Integration of Components: DSS applications to environmental problems have to consider not only natural (or social) processes, but also man-made systems, artifacts that interact with them. For instance, in our current application domain water management and treatment (Struss et al. 2016) components, such as pipelines, pumps, dams, wells, and water treatment plants are subject to diagnosis and may implement interventions.

Acknowledgements. Many thanks to Ulrich Heller for contributing to the foundations of this work and to Hosain Ibna Bashar, Iliya Valchev, and Xin Ruan for discussing and implementing some of the ideas, our former Brazilian project partners, Elenara Lersch, Waldir Roque, Paulo Salles, as well as Radhika Selvamani and Udai Agarwal from VIT Chennai for helping to improve the paper.

References

de Kleer, J., Williams, B.C.: Diagnosing multiple faults. Artif. Intell. **31**(1), 97–130 (1987)

Forbus, K.: Qualitative process theory. Artif. Intell. **24**, 85–168 (1984)

Heller, U.: Process-oriented Consistency-based Diagnosis-Theory, Implementation and Applications. Dissertation, Fakultät für Informatik, TU München (2001)

Heller, U., Struss, P.: Consistency-based problem solving for environmental decision support. Comput. Aided Civ. Infrastruct. Eng. **17**, 79–92 (2002)

Roque, W., Struss, P., Salles, P., Heller, U.: Design de um sistema de suporte à decisão baseado em modelos para o tratamento de ÁGUA. In: I Workshop de tecnologia da informação apliacada ao meio ambiente - cbcomp2003. Itajaf, SC, Anais do III Congresso Brasileiro de Computação, pp. 1894–1906 (2003)

Sojda, R.S., et al.: Identifying the decision to be supported: a review of papers from environmental modelling and software. In: Proceedings of the International Environmental Modelling and Software Society (iEMSs) (2012)

Struss, P.: Model-based problem solving. In: Van Harmelen, F., Lifschitz, V., Porter, B. (eds.) Handbook of Knowledge Representation, pp. 395–465. Elsevier, Amsterdam (2008)

Struss, P., Steinbruch, F., Woiwode, C.: Structuring the domain knowledge for model-based decision support to water management in a Peri-urban region in India. In: 29th International Workshop on Qualitative Reasoning (QR16), New York, US (2016)

Multimedia Applications

Driver License Field Detection Using Real-Time Deep Networks

Chun-Ming Tsai[1(✉)], Jun-Wei Hsieh[2], Ming-Ching Chang[3], and Yu-Chen Lin[1]

[1] Department of Computer Science, University of Taipei, Taipei, Taiwan
cmtsai2009@gmail.com, sky81915@hotmail.com
[2] College of Artificial Intelligence and Green Energy, National Chiao Tung University,
Hsinchu, Taiwan
jwhsieh@nctu.edu.tw
[3] Department of Computer Science, University at Albany, State University of New York,
Albany, USA
mchang2@albany.edu

Abstract. We present an automatic system for real-time visual detection and recognition of multiple driver's license fields using an effective deep YOLOv3 detection network. Driver licenses are essential Photo IDs frequently checked by law enforcement and insurers. Automatic detection and recognition of multiple fields from the license can replace manual key-in and significantly improve workflow. In this paper, we developed an Intelligent Driving License Reading System (IDLRS) addressing the following challenging problems: (1) varying fields and contents from multiple types and versions of driver licenses, (2) varying capturing angles and illuminations from a mobile camera, (3) fast processing for real-world applications. To retain high detection accuracy and versatility, we propose to directly detect multiple field contents in a single shot by adopting and fine-tuning the recent YOLOv3-608 detector, which can detect 11 fields from the new Taiwan driver license with accuracy of 97.5%. Our approach does not rely on text detection or OCR and outperforms them when tested with large viewing angles. To further examine such capability, we perform evaluations in 4 large tilting view configurations (top, bottom, left, right), and achieve accuracies of 93.3%, 90.2%, 97.5%, 94.3%, respectively.

Keywords: Field detection · Driver license · OCR · Deep learning · YOLOv3

1 Introduction

Driver licenses are essential Photo IDs frequently checked by law enforcement, government, and insurers. Automatic detection and recognition of multiple fields from the license can replace manual key-in and significantly improve the workflow in the real-world usage scenario. Currently, Taiwan law enforcement uses the M-Police Mobile Computer System App to query license numbers and legality and routine checks. This process and workflow is hard to automate *e.g.* for integration with next-generation

© Springer Nature Switzerland AG 2020
H. Fujita et al. (Eds.): IEA/AIE 2020, LNAI 12144, pp. 603–613, 2020.
https://doi.org/10.1007/978-3-030-55789-8_52

smart city infrastructure. When reading from a mobile camera, field texts and numbers can appear to be very small. The lighting condition can be extremely dark or bright. Other circumstances such as dynamic vigilance and alerts need to be constantly maintained by police officers, thus such automatic Intelligent Driver License Reading System (IDLRS) must be effective and easy enough to use for practitioners. The use case of IDLAS includes photo ID check, age check for liquor purchase, package receiver check, insurance claim at a car accident, etc.

Problem Setup. Figure 1 shows an example of the new Taiwan driver license, which we aim to recognize and process. In Fig. 1(a), the front face consists of a table of fixed template containing 11 fields: "駕照號碼", "性別", "姓名", "出生日期", "駕照種類", "持照條件", "住址", "有效日期", "管轄編號", "發照日期", and "審驗日期." Figure 1(b) shows the English translation of the same information: *License No., Sex, Name, Date of Birth, Type, Condition, Address, Date of Expiry, Control No., Date of Issue, and Date of Inspection*. In this paper, we focus on detecting (identifying and localizing) all 11 fields, such that the results can be directly used for field content recognition, *e.g.* by OCR.

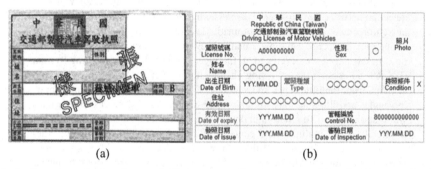

(a)   (b)

Fig. 1. Overview of driver license field detection and content recognition. In this example, the new Taiwan driver license consists of fixed templates of tables containing multiple fields in the front. (a) The front face of the new driving license in Taiwan. (b) The field names and field contents in the driving license table.

To ensure the real-world use for law enforcement, insurers, and practitioners, such automatic license reader much be able to detect and identify multiple license fields from a mobile camera in real time. Detailed functions of this system pipeline include: (i) capturing a driver license image, (ii) detecting field contents, (iii) classifying and recognizing field contents. Challenges in this pipeline include: (1) varying fields and contents from multiple types and versions of driver licenses, (2) varying capturing angles and illuminations from a mobile camera, (3) fast processing for real-world applications.

Recent Deep Learning (DL) methods have shown extraordinary performance in many computer vision tasks, including object detection [2–4]. The main reason behind this superior performance is that the deep network automatically learns discriminative features directly from the training data. To retain high detection accuracy and versatility, we propose to directly detect multiple field contents in a single shot by adopting and fine-tuning a widely-used deep object detection network, namely the You Only Look Once (YOLO) v3-608 detector [4]. YOLOv3 operates based on making prediction from performing regressions (rather than classifications of) anchor boxes in the network, which is both fast and accurate. Detection is performed at multiple scales in one shot (that can make predictions from 10x more number of bounding boxes compared to YOLO v2), thus it is good at detecting small objects in a large field-of-view.

To the best of our knowledge, our work is the first on adopting the cutting-edge DL visual object detector toward automatic license reader. The training of the proposed YOLOv3 pipeline is performed based on a much smaller driver license dataset, when compared to common use of YOLOv3 in standard datasets such as COCO or PASCAL VOC [5].

In this paper, we only focus on the detection of all 11 types of license fields as shown in Fig. 1. Since the table format of all fields in the driver license table are fixed (while the contents of the fields can vary case-by-case), our proposed approach of first detection all fixed fields is more effective than performing scene text recognition or OCR on a character-by-character level. We focus more on improving the accuracy and versatility in handling large viewing angles and luminance changes when designing and evaluation of our framework.

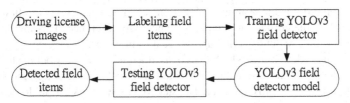

Fig. 2. Pipeline of works performed in this study.

Figure 2 shows a flowchart of works performed in this study, which includes the following steps: First, 11 field items are labeled from images of the front face of driving licenses. Second, YOLOv3 is used to train field detector from the collected images. Third, the YOLOv3 field detection results are tested. Finally, we confirm that the 11 fields in the front face driving license were accurately detected.

The rest of this paper is organized as follows. Section 2 describes the related works in visual document processing from images, particularly on table field detection and content understanding. Section 3 describes the design and training of the proposed driver license field detector based on YOLOv3. Experimental results and discussion are shown in Sect. 4, and conclusion is given in Sect. 5.

2   Related Works

As mobile devices with cameras are becoming ubiquitous, mobile digital cameras and smart edge devices have several advantages over flatbed scanners for document processing in terms of portability, fast response, and non-contact requirements [6]. Smart edge devices can nowadays perform real-time, online processing of images and data, as well as connect to cloud services more securely.

We focus the survey of the detection and recognition of the tabular formats in document processing, as tables are the essential element of documents for presenting structural information. Recognition and analysis of tables have been an active area of research. We organize the survey of prior works into three main categories: (1) table detection, (2) table layout recognition, and (3) table content understanding [10]. Table detection focuses on the identification and localization of table boundaries in a document image. Table recognition looks for internal table structures such as rows, columns and table cells. Table understanding is on extracting semantic meanings of the individual table contents. In Sect. 2.4 we survey Convolutional Neural Network (CNN) based methods, as these end-to-end deep learning methods are closely related in terms of similarity and design.

2.1 Table Detection in Documents

Conventional table detection methods perform line detection, corner detection, or character detection on the given document images, where the orientation of the tables is usually assumed to be known. Seo *et al.* [6] proposed a junction-based table detection in camera-captured document images. Junctions of line intersections are first detected in order to locate the corners of cells and connectivity inference. This method is only applicable for tables with horizontal and vertical ruling lines.

The work of [7] determines which lines in the document are likely belong to the tables of interest. The deep-learning method of Gilani *et al.* [8] consist of a Region Proposal Network followed by a fully connected Faster-RCNN network for table detection. It performs well on multiple document types (research papers and magazines) with varying layouts, and out-performs the popular Google Tesseract OCR on the publicly available UNLV dataset.

2.2 Table Layout Recognition

The recognition of table structure is essential for extracting its contents [10]. Table layout recognition methods focus on identifying the table structures and determining sub-structures of the tables. Hassan and Baumgartner [9] investigated table structures in three categories: (1) tables with both horizontal and vertical ruling lines, (2) tables with only horizontal lines, and (3) tables without ruling lines. This method can recognize spanning rows, spanning columns, and multi-line rows. Their method is effective in converting a wide variety of tabular structures into HTML for information extraction.

In contrast, OCR based approaches take an alternative route, by relying on detecting and recognize characters directly (without recognizing the actual table structure) for layout extraction. The deep learning network pipeline in [10] recognizes table contents in heterogeneous document images, where the textual contents are classified into table or non-table elements, followed by a contextual post processing. However, the exact table layout is still unknown after content extraction.

2.3   Table Content Understanding

Table understanding has been studied actively since the blooming of Big Data with huge volumes of tabular data in Web and PDF format [12].

Göbel *et al.* [11] evaluated table understanding methods in terms of (1) table detection, (2) table structure recognition, and (3) functional analysis for PDF understanding. Implementations and evaluations of each task are provided. The ICDAR 2013 Table Competition [12, 13] focused on table location and table structure recognition, which attracted both academic and industrial participations. Results show that the best performing systems can achieve average accuracies in the range of 84% to 87%.

2.4   CNN-Based Text Detection

Convolutional Neural Networks (CNNs) are widely used in scene text detection. The fully-convolutional regression network in [13] directly detects scene texts from natural images, where the network is trained using synthetic images containing texts in clutters overlaid with backgrounds. The Fully Convolutional Network (FCN) of Zhang *et al.* [14] detects texts from scene images, where text regions of individual characters are predicted and segmented out, and then the centroid of each character are estimated for the spatial understanding and recognition of the texts. This method can handle texts in multiple orientations, languages and fonts. Jaderberg *et al.* [15] localize and recognize scene texts based on an object-agnostic region proposal mechanism for detection and a CNN classification. Their model was trained solely on synthetic data without the need of manual labelling.

The Deep Matching Prior Network (DMPNet) [16] is CNN-based that can detect text with tighter quadrangle with F-score of 70.64%. A shared Monte-Carlo module computes the polygonal areas that localize texts with quadrangle fast and accurately. A *smooth Ln loss* is used to moderately adjust the predicted bounding box. The rotation-based framework in [17] is built upon a region-proposal-based architecture that can handle arbitrary-oriented texts in natural scenes effectively. The FCN model in [18] can detect and local a potentially large number of texts in the image view. Three strategies (based on the detection of boxes, corners, or left sides) are used to improve detection precision on a broad range of documents.

3   YOLOv3 Driving License Field Detector

We perform driver license understanding by directly detect the multiple types of fields using the YOLOv3 object detector. These detected fields can be used for field content recognition. In comparison to other document understanding approaches surveyed in Sect. 2, our approach bypasses the complex table layout recognition issues. Our approach is also more effective than OCR and scene-text-based methods, since we directly leverage the fixed template of the driver license table formats (which can vary between version but the change is rather rare).

3.1   YOLO Object Detector

The You Only Look Once (YOLO) [19] is a popular single-shot object detection network that can achieve real-time performance. The first version (YOLOv1) is developed in 2015, with the idea that object detection is re-framed as a regression problem. The detection of multiple objects can be performed in a single network pass, where the bounding boxes of all detected objects and classification probabilities are produced. The base network is originally GoogLeNet, and later VGG based Darknet implementation is also available. The input image is first divided into a grid of cells, such that object bounding boxes and class probabilities in each cell can be directly predicted. The pipeline includes a post-processing, merging, and non-maximal-suppression, and yields the final prediction. The YOLO sub-sequel, YOLOv2 [20] outperforms YOLOv1 in both accuracy and speed. YOLOv3 [4] is more accurate than YOLOv2 but not faster. There are also fast versions such as the tiny YOLOv3 that runs extremely fast, however with reduced accuracy.

3.2   Driving License Field Labeling

We use LabelImg [21], a publicly available tool, to label the driver license fields. LabelImg is a graphical image annotation tool that can quickly mark object bounding boxes in images. Annotations are saved as XML files in PASCAL VOC format, a format used by ImageNet challenge [22]. We convert the LabelImg.xml file into the Darknet [23] format in the following:

`<object-class > < center-x > < center-y > < width > < height>`

The $<$ object-class $>$ is an integer representing object class (various driver license fields), ranging from 0 to 10. The $<$ center-x $>$ and $<$ center-y $>$ are the bounding box center, normalized (divided) by the image width and height, respectively. The $<$ width $>$ and $<$ height $>$ are bounding box dimensions normalized image width and height, respectively. So these entries are values between 0 and 1.

3.3   Training and Testing YOLOv3 Field Detector

Our YOLOv3-608 field detectors can be trained following the standard steps in [23, 24], including the adjustment of the configuration file (driverLicense-yolov3-608.cfg which stores important training parameters), with slight tuning and modification. For the 416 × 416 input image, the cell size in the YOLOv3 network is 32 × 32. Detailed steps are described in the following.

1. We randomly select 70% to 90% from our driver license data as training set in several experiment trials. The remaining is used as the testing set.
2. Configure the settings for 11 field classes and 3 yolo-layers.
   # filters = (# classes + 5) x 3 = 48,
   since there are 3 convolutional layers before each yolo layer. The name file `field.names` should contain the field names of the 11 classes.

3. Data file `driverLicense.data` should contain:

```
classes = 11
train = data/driverLicenseTrain.txt
valid = data/driverLicenseTest.txt
backup = backup/
```

4. Start training:

```
./darknet detector train /path/to/driverLicense.data
/path/to/driverLicense-yolov3-608.cfg
./darknet53.conv.74
-map > /path/to/driverLicenseYolov3-608Train.log
```

5. Upon competition, resulting model `driverLicenseYolov3-608_final.weights` should be produced in the `backup/` directory.

6. We test the YOLOv3-608 field detector on the new Taiwan driver license test sets for performance evaluation.

4 Experimental Results and Discussion

All experiments were performed on a machine equipped with an Intel Xeon E3-1231 V3 @3.40 GHz CPU and having 8 GB of DDR5 memory on an NVIDIA Ge-Force 1070Ti GPU.

Dataset. The collected driver license data contains 617 licenses in total (Table 1). The number of the training samples is 556 and testing is 61. As aforementioned, we performed evaluation in five tilting viewing angles, including the no-tilt (up-right) angle. The number of data samples for the Positive (Pos) *i.e.* viewing from an up-right angle, Top-to-Bottom (TB), Bottom-to-Top (BT), Left-to-Right (LR), and Right-to-Left (RL) are 369, 20, 48, 79, 101, respectively.

Table 1. Driving license table with five capture angles.

CA	Pos	TB	BT	LR	RL
Number	369	20	48	79	101

We use the Intersection over Union (IoU) and mAP metrics [26] to evaluate the YOLOv3-608 performance. The IoU metric measures the accuracy of an object detector in terms of detected bounding box overlaps with ground-truth box labeling. Mean average precision (mAP) calculates as the mean value of average precisions for each class, where the average precision is area-under-curve (AUC) of Precision-Recall (PR) curve for each threshold for each class. Specifically, the True Positive (TP), False Positive (FP), False Negative (FN), and True Negative (TN) are all calculated for each class. Precision = $TP/(TP + FP)$ measures how accurate are the predictions. Recall = $TP/(TP + FN)$ measures how well all the positives found and is defined as. Accuracy is calculated as $(TP + TN)/(TP + TN + FP + FN)$. Finally, the F-score is calculated as 2 (precision x recall) /(precision + recall).

Table 2 shows the field detection results by using YOLOv3-608 field detector of our collected driving licenses. The accuracy (A), precision (P), recall (R), and F score (F) metrics for the positive (Pos) captured angle are 97.5, 99.7, 97.6, and 98.6, respectively. The accuracy, precision, recall, and F score metrics for from top to bottom (TB) captured angle are 93.3, 99.5, 93.4, and 96.4, respectively. The accuracy, precision, recall, and F score metrics for from bottom to top (BT) captured angle are 90.2, 100, 90.0, and 94.8, respectively. The accuracy, precision, recall, and F score metrics for from left to right (LR) captured angle are 97.5, 99.9, 97.6, and 98.7, respectively. The accuracy, precision, recall, and F score metrics for from right to left (RL) captured angle are 94.3, 99.8, 94.5, and 97.1, respectively.

Table 2. The field detection results of our collected driving licenses.

CA	TP	TN	FP	FN	A	P	R	F
Pos	4189	287	13	104	97.5	99.7	97.6	98.6
TB	211	11	1	15	93.3	99.5	93.4	96.4
BT	550	0	0	61	90.2	100	90.0	94.8
LR	990	21	1	23	97.5	99.9	97.6	98.7
RL	1184	16	3	69	94.3	99.8	94.5	97.1

Figure 3 shows the field detected results for the new Taiwan driver license for viewing from an up-right angle which is detected by our fine-tuned YOLOv3-608 field detector. The left and the right of the Fig. 3 are obtained from the Internet and the subject which simulates the police to see the driver license. As show in the left driver license of the Fig. 3, all fields were all correctly detected, except that the field "Condition" was incorrectly detected as "Control No." and "Condition". However, all fields in the right driver license were all correctly detected.

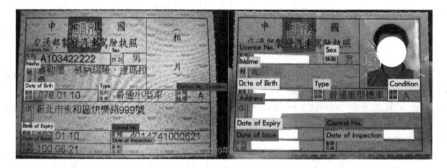

Fig. 3. Example of field detected result for viewing from an up-right angle.

Figures 4, 5, 6 and 7 show the field detected results for the Taiwan driver licenses which are viewing from top to bottom angle, from bottom to top angle, from left to right, and from right to left, respectively. From the left driver license of these examples, most of the 11 fields in each driver license can be detected by our YOLOv3-608 field detector. However, the field "Condition" was incorrectly detected as "Control No." in Figs. 4 and 7 and the field "Condition" cannot be detected in Figs. 5 and 6. However, the right real driver license of these examples, all fields were all correctly detected.

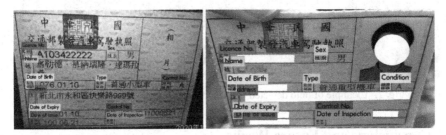

Fig. 4. Example of field detected result for viewing from top to bottom angle.

Fig. 5. Example of field detected result for viewing from bottom to top angle.

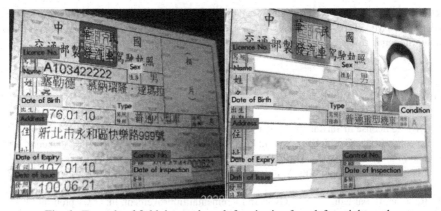

Fig. 6. Example of field detected result for viewing from left to right angle.

Fig. 7. Example of field detected result for viewing from right to left angle.

5   Conclusions and Future Works

This paper presented an automatic license field detection system based on the deep YOLOv3-608 networks, which are fine-tuned on a newly collected dataset of new Taiwan driver licenses. Detection accuracy of 97.5% is achieved in the test. Additional evaluations are performed on the YOLOv3-608 module with large viewing angles in 4 tilting configurations (top, bottom, left, right), and accuracies of 93.3, 90.2, 97.5, 94.3 are achieved, respectively.

Future works include the collection of a larger multi-national driver license dataset that can be used for model generalization. In addition, newer deep network backbones such as an improved YOLO variation can also be adapted. Finally, the developed field detection capability can be integrated with field content recognition for real-world usage evaluation and deployment.

Acknowledgements. This work is supported by the Ministry of Science and Technology, Taiwan, under Grants MOST 107-2221-E-845-005 – and MOST 108-2221-E-845 -003 -MY3. We thank Walter Slocombe for paper writing improvement.

References

1. Driving license in Taiwan. https://en.wikipedia.org/wiki/Driving_license_in_Taiwan
2. Ren, S., He, K., Girshick, R., Sun, J.: Faster R-CNN: Towards real-time object detection with region proposal networks. IEEE Trans. PAMI **39**(6), 1137–1149 (2017)
3. Liu, W. et al., : Single Shot MultiBox Detector. In: arXiv preprint (2016). arXiv:1512.023 25v5
4. Redmon, J., Farhadi, A.: YOLOv3: An Incremental Improvement. In: arXiv preprint (2018). arXiv:1804.02767
5. The PASCAL VOC project. http://host.robots.ox.ac.uk/pascal/VOC/#bestpractice
6. Seo, W., Koo, H.I., Cho, N.I.: Junction-based table detection in camera-captured document images. IJDAR **18**(1), 47–57 (2015)

7. e Silva, A.C., Jorge, A., Torgo, L.: Automatic selection of table areas in documents for information extraction. In: Pires, F.M., Abreu, S. (eds.) EPIA 2003. LNCS (LNAI), vol. 2902, pp. 460–465. Springer, Heidelberg (2003). https://doi.org/10.1007/978-3-540-24580-3_54

8. Gilani, A., Qasim, S.R., Malik, I., Shafait, F.: Table detection using deep learning. In: 14th IAPR International Conference on Document Analysis and Recognition (ICDAR), Kyoto, pp. 771–776 (2107)

9. Hassan, T., Baumgartner, R.: Table recognition and understanding from PDF files. In: Ninth International Conference on Document Analysis and Recognition (ICDAR 2007), Parana, pp. 1143–1147 (2007)

10. Rashid, S.F., Akmal, A., Adnan, M., Aslam, A.A., Dengel, A.: Table recognition in heterogeneous documents using machine learning. In: 14th IAPR International Conference on Document Analysis and Recognition (ICDAR), Kyoto, pp. 777–782 (2017)

11. Göbel, M., Hassan, T., Oro, E., Orsi, G.: A methodology for evaluating algorithms for table understanding in PDF documents. In: ACM Symposium on Document Engineering, pp. 45–48 (2012)

12. Göbel, M., Hassan, T., Oro, E., Orsi, G.: ICDAR 2013 table competition. In: 12th International Conference on Document Analysis and Recognition, Washington, DC, pp. 1449–1453 (2013)

13. Gupta, A., Vedaldi, A., Zisserman, A.: Synthetic data for text localization in natural images. In: IEEE Conference on Computer Vision and Pattern Recognition, Las Vegas (2016)

14. Zhang, Z., Zhang, C., Shen, W., Yao, C., Liu, W., Bai, X.: Multioriented text detection with fully convolutional networks. In: IEEE Conference on Computer Vision and Pattern Recognition, Las Vegas (2016)

15. Jaderberg, M., Simonyan, K., Vedaldi, A., Zisserman, A.: Reading text in the wild with convolutional neural networks. Int. J. Comput. Vis. **116**(1), 1–20 (2016)

16. Liu, Y., Jin, L.: Deep matching prior network: toward tighter multioriented text detection. In: Proceedings of the IEEE Conference on Computer Vision and Pattern Recognition, Honolulu, vol. 2, p. 8 (2017)

17. Ma, J., et al.: Arbitrary-oriented scene text detection via rotation proposals. IEEE Trans. Multimedia **20**(11), 3111–3122 (2018)

18. Moysset, B., Kermorvant, C., Wolf, C.: Learning to detect, localize and recognize many text objects in document images from few examples. IJDAR **21**(3), 161–175 (2018). https://doi.org/10.1007/s10032-018-0305-2

19. Redmon, J., Divvala, S., Girshick, R., Farhadi, A.: You Only Look Once: Unified, Real-Time Object Detection. In: arXiv preprint (2016). arXiv:1506.02640v5

20. Redmon, J., Farhadi, A.: YOLO9000: Better, Faster, Stronger. In: arXiv preprint (2016). arXiv:1612.08242v1

21. Tzutalin. LabelImg. Git code. https://github.com/tzutalin/labelImg

22. ImageNet. http://www.image-net.org

23. Darknet: Open Source Neural Networks in C. https://pjreddie.com/darknet/

24. AlexeyAB/darknet. https://github.com/AlexeyAB/darknet#how-to-train-to-detect-your-custom-objects

Calibration of a Microphone Array Based on a Probabilistic Model of Microphone Positions

Katsuhiro Dan[1(⊠)] ⓘ, Katsutoshi Itoyama[1] ⓘ, Kenji Nishida[1] ⓘ,
and Kazuhiro Nakadai[1,2] ⓘ

[1] School of Engineering, Tokyo Institute of Technology, 2-12-1, O-okayama,
Meguro, Tokyo 152-8552, Japan
{dan,itoyama,nishida}@ra.sc.e.titech.ac.jp
[2] Honda Research Insititute Japan Co., Ltd., 8-1 Honcho,
Wako, Saitama 351-0114, Japan
nakadai@jp.honda-ri.com

Abstract. This paper addresses a novel method for calibrating microphone positions included in a microphone array. The performance of microphone array processing deteriorates due to two factors: (1) differences between predetermined position and actual positions of the microphones. (2) sound source signal overlaps in frequency and time. To solve these problems, we propose a probabilistic generative model of the sound propagation process determined by microphone and sound source positions. The model is defined as the product of three probabilities: (1) prior probability of the microphone positions based on reference positions, (2) prior probability of the sound source spectrum, and (3) conditional probability of the recorded spectrum. Based on the model, an iterative algorithm to calibrate the microphone positions is derived as a solution of the maximum *a posteriori* estimation. Preliminary experiments through numerical simulation with an 8-ch microphone array revealed that the proposed method accurately estimated the microphone positions when using multiple sound sources. Preliminary experiments through numerical simulation with an 8-ch microphone array suggested the proposed method accurately estimated the microphone positions when using multiple sound sources.

Keywords: Array signal processing · Microphone array · Calibration

1 Introduction

In recent years, microphone arrays have been mounted on various devices including robots. Acoustic signal processing technologies such as sound source localization and sound source separation have been developed and investigated

This work was supported by JSPS KAKENHI Grant No. 16H02884, 17K00365, and 19K12017.

[3,5,6,16]. In particular, robot audition has gained the attention of researchers aiming to develop auditory functions for real-world-deployable systems including robots [4]. With the spread of microphone arrays, the method of calibration is very important. When including a microphone array in a robot, one challenge is a potential mismatch between the optimally pre-tuned parameter values of the array and optimal values, for reasons including device aging, measurement errors of microphone positions, and environmental change by robot motion. This means that inevitable errors occur in the transfer function that represents a sound propagation property between a sound source to the microphone array as shown in Fig. 1. Using a signal recorded by microphones located in their true positions, sound source localization can be performed accurately. However, If the microphones are not located in their true positions, accurate sound source localization cannot be performed. Therefore, correct calibration of the microphone positions is essential.

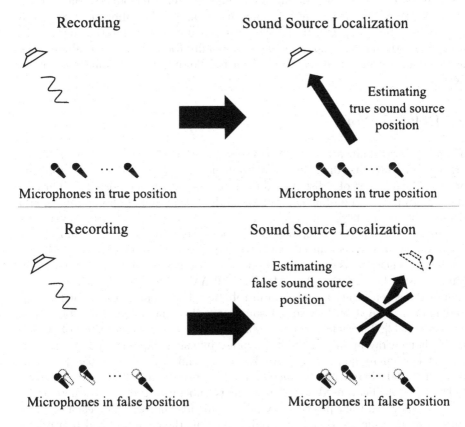

Fig. 1. Problem of sound source localization using false microphone positions

Easy-to-use calibration requires two main conditions. 1. Calibration using sound source signals recorded simultaneously, because multiple sound sources are often recorded among environmental sounds. 2. Calibration using any sound source signal, so calibration can be performed without the need to prepare a specific sound. However, most studies to date reported calibration methods necessitating a laborious calibration procedure before the advent of microphone array processing [2,8,9,11]. To realize the above conditions, there is a need for a calibration method which can calibrate using any sound source signals recorded at the same time, but it has not been realized yet.

To achieve the above, this paper proposes a novel method for calibration of microphone positions in the microphone array based on Maximum *A Posteriori* (MAP) estimation by assuming a prior distribution centered on the initial position of each microphone to avoid any mismatch problem in transfer functions. MAP estimation can calibrate using multiple simultaneous sound sources with white noise as an environmental sound, and also by using sound source signals that do not rise well, such as environmental sounds.

The rest of this paper is organized as follows: Sect. 2 introduces related work about signal processing of microphone arrays and their calibration. Section 3 shows the problem specification. Section 4 proposes our calibration method of a probabilistic model of microphone array calibration. Section 5 evaluates the accuracy of our proposed method, which is followed by concluding remarks in Sect. 6.

2   Related Work

There are two main streams for transfer function estimation of microphone arrays; asynchronous distributed microphone arrays and ad hoc microphone arrays. The first test of a distributed microphone array was reported by Thrun S [13]. He proposed an on-line calibration method using sound onset timings, and demonstrated its effectiveness using actual microphone devices. However, this was limited in that the sound source locations were predetermined, and microphones were fully synchronized. To overcome these limitations, Miura *et al.* proposed an asynchronous on-line microphone position estimation method based on Simultaneous Localization And Mapping (SLAM), called simultaneous estimation of robot self-location and map in robotics. They replaced a robot location and map in SLAM with a sound source position and microphone positions, respectively [2]. They successfully calibrated the microphone positions of an 8 ch microphone array in an incremental manner by using clapping while in motion. For ad hoc microphone arrays, Raykar *et al.* and Ono *et al.* separately proposed methods based on distance estimation using time difference of arrival (TDOA) [8,9]. They used sound recordings and playing devices, wherein the devices calibrated their positions by playing sounds and recording each other in turn using inter-device time synchronization. But their method had limitations in that Raykar *et al.* assumed the audio signals were sparse enough, and Ono *et al.* used a special acoustic signal called a time stretched pulse (TSP).

These studies [2,8,9,13] adopted hard assumptions that audio signals have sparseness or clear onsets as does a clapping sound. The present paper obviates these assumptions by proposing probabilistic calibration using prior distributions of the microphone positions.

3   Problem Specification

Let $x_m \in \mathbb{R}^d$ be the coordinates of the m-th microphone position constituting an M-channel microphone array in d-dimensional space (usually $d = 2$ or 3.) Although these microphones are arranged according to given reference positions $\bar{x}_m \in \mathbb{R}^d$, the actual positions may deviate from the reference position. The goal of the proposed method was to estimate the actual position of each microphone x_m using audio signals recorded by the microphone array. There are two assumptions:

1. The positions of the microphones and sound sources are time-invariant, and
2. The transfer function is time-invariant and given as a function of the microphone positions.

Let N be the number of sound sources and $s_{nft} \in \mathbb{C}$ be a complex spectrum obtained by the short-time Fourier transform (STFT) of the n-th source signal at the f-th frequency bin ($1 \leq f \leq F$) and the t-th time frame ($1 \leq t \leq T$). Let $z_{mft} \in \mathbb{C}$ be a complex spectrum of the mixture of sound sources recorded by the m-th microphone in the microphone array. By using a transfer function,

$$r_{nf} = (r_{n1f}, \ldots, r_{nMf})^T \tag{1}$$

from the n-th source to the m-th microphone, the relationship between the source and recorded spectra is represented as

$$z_{mft} = \sum_n r_{nmf} s_{nft} + \epsilon_{mft} \tag{2}$$

where ϵ_{mft} is an observed noise. On the basis of this relationship, the microphone positions

$$X = \{x_1, \ldots, x_M\} \tag{3}$$

are calibrated using the set of the recorded sound mixture $Z = \{z_{ft}\}_{1 \leq f \leq F, 1 \leq t \leq T}$.

4   Proposed Method

This section describes a method to calibrate microphone positions based on a probabilistic generative model that represents the physical model shown in Sect. 3. We constructed a probabilistic generative model to represent the physical model expressed in Eq. 2 with random and uncertain factors related to the diffusive noise in each microphone, the source spectrum, and the microphone positions. We derived an algorithm to calibrate the microphone positions by MAP estimation of the probabilistic generative model.

4.1 Probabilistic Generative Model

We formulated the microphone position calibration problem using a probabilistic framework in order to deal with multiple non-deterministic properties involved in generation and transmission processes of the acoustic signal. Let us consider the following three random variables:

- $X = \{x_m \in \mathbb{R}^d\}_{1 \le m \le M}$: unknown positions of microphones.
- $S = \{s_{nft} \in \mathbb{C}\}_{1 \le n \le N, 1 \le f \le F, 1 \le t \le T}$: unknown source spectrum.
- $Z = \{z_{mft} \in \mathbb{C}\}_{1 \le m \le M, 1 \le f \le F, 1 \le t \le T}$: known observed spectrum recorded by each microphone.

The calibration problem can be solved as the MAP estimation of the joint posterior probability $p(X, S|Z)$[1]. Although the source spectrum S does not have to be output in the proposed method, is secondarily obtained by MAP estimation. Therefore, the proposed method performs source separation at the same time as calibration.

Let $p(Z, X, S)$ be the joint probability distribution of all the random variables. Under the independence of the sound source and the microphone position, it is decomposed as

$$p(Z, X, S) = p(Z|X, S)p(X)p(S). \tag{4}$$

The first term $p(Z|X, S)$ represents the conditional distribution of the observed spectrum Z. Under the assumption that the observed spectrum has conditional independence between times, frequencies, and channels, we decompose the distribution $p(Z|X, S)$ as

$$P(Z|S, X) = \prod_m \prod_f \prod_t P(z_{mft}|s_{1ft}, \dots, s_{Nft}, X). \tag{5}$$

Let the observed noise ϵ_{mft} follow a circularly symmetric complex Gaussian distribution with zero-mean and variance of σ_z^2. Since

$$\epsilon_{mft} = z_{mft} - \sum_n r_{nmf} s_{nft} \tag{6}$$

from Eq. (2), the observed spectrum z_{mft} follows a complex Gaussian distribution with a mean of $\sum_n r_{nmf} s_{nft}$ and a variance of σ_z^2:

$$P(z_{mft}|s_{1ft}, \dots, s_{Nft}, X) \propto \tag{7}$$
$$\exp\left(-\frac{(z_{mft} - \sum_n r_{nmf} s_{nft})^\star (z_{mft} - \sum_n r_{nmf} s_{nft})}{\sigma_z^2}\right).$$

---

[1] Strictly, S should be marginalized and the MAP estimation of $p(X|Z)$ should be solved.

Here, $(\cdot)^{\star}$ represents the complex conjugate operator. To derive the transfer function r_{nmf}, the following methods are often used in numerical simulations calculated from microphone positions based on the propagation wave model [5,15].

The second term $p(\boldsymbol{X})$ represents the prior distribution of the microphone positions. The microphones are basically arranged at a predetermined reference positions, however, their actual positions include some deviation from the reference position due to manufacturing errors and uncertainty. Assuming this deviation is independent and isotropic for each microphone, $p(\boldsymbol{X})$ can be decomposed as

$$p(\boldsymbol{X}) = \prod_m p(\boldsymbol{x}_m). \tag{8}$$

Let the position of each microphone follow an isotropic Gaussian distribution with a mean reference position of $\bar{\boldsymbol{x}}_m$ and a variance of σ_x^2:

$$p(\boldsymbol{x}_m) \propto \exp\left(-\frac{(\boldsymbol{x}_m - \bar{\boldsymbol{x}}_m)^T(\boldsymbol{x}_m - \bar{\boldsymbol{x}}_m)}{2\sigma_x^2}\right). \tag{9}$$

The third term $p(\boldsymbol{S})$ represents the prior distribution of the source spectrum. Let the source spectrum be independent for each source. Regarding the representation of the prior distribution of the spectrum of a single sound source, various representations had been proposed, including a low-rank representation using non-negative matrix factorization (NMF) [1,10], and a non-linear representation using deep neural networks (DNN) [7,14]. Here we assume independence of the source spectrum in the time-frequency plane which is a simple expression and decomposes $p(\boldsymbol{S})$ as

$$p(\boldsymbol{S}) = \prod_n \prod_f \prod_t p(s_{nft}). \tag{10}$$

This assumes the source signal is time-invariant white noise. Assuming the source spectrum s_{nft} for each source, time, and frequency follow a circularly symmetric complex Gaussian distribution with zero-mean and a variance of σ_s^2, the distribution can be represented as

$$p(s_{nft}) \propto \exp\left(-\frac{s_{nft}^{\star}s_{nft}}{\sigma_s^2}\right). \tag{11}$$

4.2   Calibration of Microphone Array

We derived a calibration algorithm based on the probabilistic generative model of the microphone array described in the previous section. The optimal microphone positions for a given observed spectrum were obtained by maximizing the logarithmic posterior probability as

$$\hat{\boldsymbol{X}}, \hat{\boldsymbol{S}} = \arg\max_{\boldsymbol{X},\boldsymbol{S}} p(\boldsymbol{X}, \boldsymbol{S}|\boldsymbol{Z}) \tag{12}$$

$$= \arg\max_{\boldsymbol{X},\boldsymbol{S}} \log p(\boldsymbol{Z}, \boldsymbol{X}, \boldsymbol{S}).$$

Since the microphone position and the source spectrum share mutually dependent posterior probability, iterative maximization is performed to approximate simultaneous maximization. Maximization according to microphone positions X is achieved by a grid search using a predefined range and step size.

This is because the black-box nonlinear function r_{nmf} intervenes in the microphone positions X and the observed spectrum Z, and it is impossible to determine the derivative of the logarithmic posterior probability with respect to X. Maximization according to the source spectrum S is achieved by finding the maximal value of the posterior probability. Since the logarithmic posterior probability is quadratic with respect to S, the optimal source spectrum $\hat{s}_{ft} = \hat{s}_{1ft}, \ldots, \hat{s}_{Nft}]^T$ can be obtained from the partial derivative of the $\log p(Z, X, S)$. When the observed spectrum is $z_{ft} = z_{1ft}, \ldots, z_{Mft}$ and the transfer function is $r_f = r_{1f}, \ldots, r_{Nf}$, the estimation of the source spectrum is expressed as follows. (I represents an N-by-N identity matrix.)

$$\hat{s}_{ft} = \left\{ \left(r_f^T r_f^\star \right)^T + \frac{\sigma_z^2}{\sigma_s^2} I \right\}^{-1} \left(z_{ft}^T r_f^\star \right)^T. \tag{13}$$

The derived algorithm is described in Algorithm 1.

Algorithm 1. Iterative Estimation of X and S

    Initialize $X^{(0)}$ and set $t \leftarrow 0$
    repeat
        $S^{(t+1)} \leftarrow \arg \max_{S} \log P(X^{(t)}, S | Z)$ using Eq. (13)
        $X^{(t+1)} \leftarrow \arg \max_{X} \log P(X, S^{(t+1)} | Z)$ using grid search
        $t \leftarrow t + 1$
    until Until convergence

---

5   Evaluation

The proposed method was evaluated through numerical simulation. We measured errors among estimated positions of the microphones and their actual positions including displacement from reference positions. The sound source was placed at a sufficient distance that the signal was assumed to be a plane wave when it reached the array. In such cases, array signal processing techniques such as sound source localization can be correctly performed by correct relative microphone positions even if parallel movement is involved. Therefore, we fixed the position of the first microphone (CH1) to the reference position and estimated the positions of the other microphone.

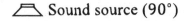
Sound source (90°)

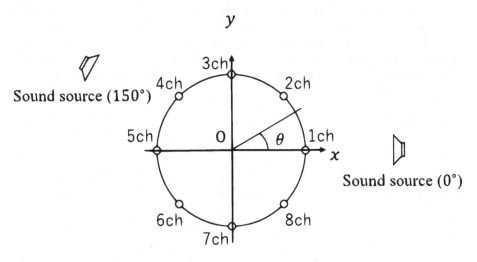

Fig. 2. Numerical simulation environment

5.1 Experimental Condition

A planer circular 8-channel microphone array with a radius of 12 cm was used as shown in Fig. 2. The sound sources were placed on the same plane as the microphone array, so the number of dimensions d was set to 2. The origin of the coordinate axes was set to coincide with the center of the reference microphone positions. The coordinates of the first microphone (CH1) were set to be on the positive direction of the X-axis, that is, angle $\theta = 0°$. The other microphones were numbered counterclockwise on the coordinate plane as CH2, CH3, ..., CH8.

Sound source at a sampling frequency of 24 kHz was emitted from each sound source with the source spectrum s_{nft} obtained by performing STFT with a window size of 1024 points and a shift size of 256 points.

The observed spectrum z_{mft} was obtained by multiplying the transfer function r_{nmf} calculated from the ground-truth microphone positions to the source spectrum s_{nft}. The parameters, σ_z^2 and σ_s^2 were set to 5×10^{-10} and 5×10^{-8}, respectively. σ_x^2 was set square of the scale of displacement. The grid size was set every one-tenth of the displacement size. Here, transfer function was calculated from microphone positions based on the propagation wave model [5, 15].

5.2 Evaluation 1: Calibration of Microphone Positions

In the first evaluation, we analyzed the properties of the proposed method while changing three factors: 1. Magnitude of microphone displacement; while the number of sound sources was fixed at 2 (the direction of the sources $\theta = 0°, 90°$), each microphone was given a displacement of 1, 2, or 3 cm, centered on the reference position. 2. Number of sound sources; while the displacement was fixed at 1 cm, the number of sources was changed to one ($\theta = 0°$), two ($\theta = 0°, 90°$), or three ($\theta = 0°, 90°, 150°$). 3. Type of sound source; experiment under the above conditions, using white noise and voice. This time we used the voice data that are in the JVS corpus [12]. For each condition, twenty trials were performed while changing the direction of displacement for each microphone chosen from a uniform distribution of $[0, 2\pi)$.

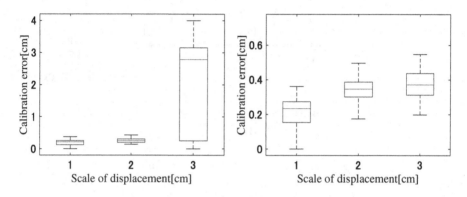

Fig. 3. Displacement error of each microphone (Left: White noise, right: Voice)

Results. Median errors for each case 1–3 are shown in Fig. 3. In case 1, the median value when using white noise was 0.21 cm, which reduced the initial error by 79%. Median value when using voice was 0.23 cm, which reduced the initial error by 77%. In case 2, former one was 0.25 cm, which reduced the initial error by 88%. Latter one was 0.35 cm, which reduced the initial error by 83%. In case 3, former one was 2.80 cm, which reduced the initial error by 6.7%. Latter one was 0.37 cm, which reduced the initial error by 88%.

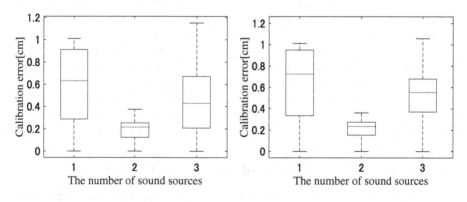

Fig. 4. Error of each microphone for the number of sound sources (Left: White noise, right: Voice)

Median errors for cases 1, 4–5 are shown in Fig. 4. In case 4, former one was 0.63 cm, which reduced the initial error by 37%. Latter one was 0.72 cm, which reduced the initial error by 28%. In case 5, former one was 0.43 cm, which reduced the initial error by 57%. Latter one was 0.55 cm, which reduced the initial error by 45%.

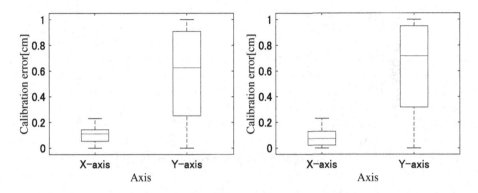

Fig. 5. X,Y-axis error of each microphone in case 4 (Left: White noise, right: Voice)

In addition, the X,Y-axis error of each microphone in case 4 is shown in Figure. 5. In case 4, median value of X axis when using white noise was 0.11 cm, which reduced the initial error by 89%. median value of X axis when using voice was 0.07 cm, which reduced the initial error by 93%. Median value of Y axis when using white noise was 0.63 cm, which reduced the initial error by 37%. Median value of X axis when using voice was 0.72 cm, which reduced the initial error by 28%.

5.3   Evaluation 2: Sound Source Localization

To improve sound source localization after calibration, we compared the sound source localization performance using a Robot-Audition open source software called HARK [4]. We assumed the sound sources were at y-axis infinity and obtained sound source localization results in the cases of (1)–(3) when using white noise.

Results. Average sound localization error are shown in Fig. 6. In case 1 the average error of sound source localization after calibration was 0°. In case 2, it was 1°. In case 3, it was 1.2°.

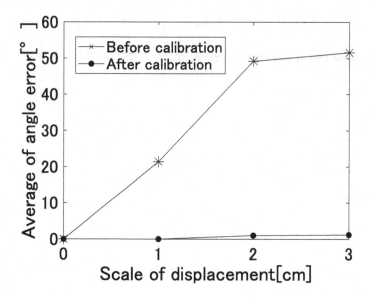

Fig. 6. Sound source localization error

5.4 Discussion

It was confirmed that if a reference position is displaced by 3 cm when using white noise, the estimated position will fall into a local optimum by iterative estimation. The following cause may be considered. If a white noise signal is recorded, the signal will be overlapped in all frequency bins and time indexes. Also, since the sampling frequency is 24 [kHz], the Nyquist frequency is 12 [kHz]. On top of that, the frequency of the wave with a wavelength of 3 cm was 11.3 [kHz], and as a result that the phase information in most frequency bands could not be used effectively, the calibration accuracy of the microphone position deteriorated. Sound source localization was significantly improved after calibration, possibly because calibration was performed well in the same direction as the sound source signal used for calibration. When the number of sound sources was 2 or more, the estimation accuracy of the microphone positions was improved. A reason why calibration results for three sound sources were slightly better than for two may be that estimation of the sound source spectrum was not success-ful. In addition, a reason for the poor estimation of one sound source may be that displacement in the Y-axis direction was not estimated from Fig. 5. This means if a sound source is distant, displacement only in its direction is estimated. Therefore, if one calibrates a microphone array using a distant sound source, it is necessary to use two or more sound sources.

6 Conclusion and Future Work

This paper presented a method to calibrate a microphone array based on a proba-bilistic model of microphone positions. We constructed a probabilistic generative model and proposed a calibration method through MAP estimation. To evaluate the accuracy of our proposed method, we used numerical simulation to calibrate a circular microphone array with a radius of 12 cm. The proposed method calibrated well under the condition that displacement is at least 1 cm and 2 cm and 2 sound sources are used. Limitations in accuracy of the estimation of sound sources occur with sound source signals recorded simultaneously and if input signals are white noise resulting in source signals overlapping in frequency and time. In future work, we aim to evaluate the proposed method under assumptions corresponding to an actual environment and conduct the experiment in practical form.

References

1. Févotte, C., Bertin, N., Durrieu, J.L.: Nonnegative matrix factorization with the Itakura-Saito divergence: with application to music analysis. Neural Comput. **21**(3), 793–830 (2009)
2. Miura, H., Yoshida, T., Nakamura, K., Nakadai, K.: SLAM-based online calibration of asynchronous microphone array for robot audition. In: IROS 2011, pp. 524–529 (2011)
3. Nakadai, K., Nakajima, H., Hasegawa, Y., Tsujino, H.: Sound source separation of moving speakers for robot audition. In: ICASSP 2009, pp. 3685–3688 (2009)
4. Nakadai, K., Okuno, H.G., Mizumoto, T.: Development, deployment and applications of robot audition open source software HARK. J. Robot. Mechatron. **29**(1), 16–25 (2017)
5. Nakamura, K., Nakadai, K., Asano, F., Hasegawa, Y., Tsujino, H.: Intelligent sound source localization for dynamic environments. In: IROS 2009, pp. 664–669 (2009)
6. Nishiura, T., Yamada, T., Nakamura, S., Shikano, K.: Localization of multiple sound sources based on a CSP analysis with a microphone array. In: ICASSP 2000, vol. 2, pp. 1053–1056 (2000)
7. Nugraha, A., Liutkus, A., Vincent, E.: Multichannel audio source separation with deep neural networks. IEEE/ACM Trans. Audio Speech Lang. Process. **24**(9), 1652–1664 (2015)
8. Ono, N., Shibata, K., Kameoka, H.: Self-localization and channel synchronization of smartphone arrays using sound emissions. In: APSIPA ASC 2016, pp. 1–5 (2016)
9. Raykar, V.C., Kozintsev, I.V., Lienhart, R.: Position calibration of microphones and loudspeakers in distributed computing platforms. IEEE Trans. Speech Audio Process. **13**(1), 70–83 (2005)
10. Smaragdis, P., Févotte, C., Mysore, G.J., Mohammadiha, N., Hoffman, M.: Static and dynamic source separation using nonnegative factorizations: a unified view. IEEE Signal Process. Mag. **31**(3), 66–75 (2014)
11. Su, D., Vidal-Calleja, T., Miro, J.V.: Simultaneous asynchronous microphone array calibration and sound source localisation. In: IROS 2015, pp. 5561–5567 (2015)
12. Takamichi, S., Mitsui, K., Saito, Y., Koriyama, T., Tanji, N., Saruwatari, H.: JVS corpus: free Japanese multi-speaker voice corpus (2019). arXiv:1908.06248 [cs.SD]
13. Thrun, S.: Affine structure from sound. In: NIPS 2005, pp. 1353–1360 (2005)
14. Uhlich, S., Giron, F., Mitsufuji, Y.: Deep neural network based instrument extraction from music. In: ICASSP 2015, pp. 2135–2139 (2015)
15. Valin, J.M., Rouat, J., Michaud, F.: Enhanced robot audition based on microphone array source separation with post-filter. In: IROS 2004, vol. 3, pp. 2123–2128 (2004)
16. Zhang, C., Florencio, D., Ba, D.E., Zhang, Z.: Maximum likelihood sound source localization and beamforming for directional microphone arrays in distributed meetings. IEEE Trans. Multimedia **10**(3), 538–548 (2008)

How to Handle Head Collisions in VR

Marek Kopel$^{(\boxtimes)}$ and Bartłomiej Stanasiuk

Faculty of Computer Science and Management,
Wroclaw University of Science and Technology, wybrzeze Wyspiańskiego 27,
50-370 Wroclaw, Poland
marek.kopel@pwr.edu.pl

Abstract. In this paper a comparison of selected algorithms used to handle VR head collisions was presented. The following four methods were chosen and implemented: screen fade, delayed push-back, instant push-back, and teleportation. This paper examined what effects these methods have on VR sickness, the sense of presence, and the usability level.

Keywords: VR · Navigation · Video game · Head collision · Screen fade · Teleportation · Push back

1 Introduction

Virtual reality (VR) has become increasingly popular over the past few years. Several tech companies, such as Google, Facebook, Sony, Samsung, or HTC, released many affordable consumer-grade VR systems. Room-scale and seated VR experiences are now easily accessible to anyone interested in this trending technology. However, while VR technology has seen rapid development and growth in recent years, it still has much to improve upon. Modern VR systems lack the ability to simulate realistic haptic sensations. The current challenge is to increase the sense of presence in virtual environments using the available technology.

1.1 Background

Head tracking systems play a big part in VR movement. They are used to estimate the position and rotation of VR headset. The users can now control the virtual camera with their physical head movements. There is no longer a need for a mouse or another input device to look around in the virtual world. Head tracking greatly increases the immersion experienced by the users and can be successfully applied to many locomotion techniques. It also drastically reduces VR sickness symptoms that are caused due to the mismatch between vestibular and visual senses [6]. Unfortunately, it introduces a significant problem with head collisions that is not normally present in video games. The problem occurs when the user's head should collide with some virtual object, but the user keeps

© Springer Nature Switzerland AG 2020
H. Fujita et al. (Eds.): IEA/AIE 2020, LNAI 12144, pp. 626–637, 2020.
https://doi.org/10.1007/978-3-030-55789-8_54

Fig. 1. In the game Fallout 4 VR, players can see through walls when they lean against them. (image source: [11]).

moving his head forward in the real world without any obstructions. The physical collision with the object is currently impossible to simulate due to the lack of advanced haptic systems. Instead, the user sees the insides of the virtual object and unexpected clipping artifacts (see Fig. 1), which can break the gameplay and immersion of the game.

The ability to see through the game objects can also have undesirable effects on game mechanics. In games with teleportation locomotion, the players can cheat by teleporting through walls to skip harder game areas [1]. In shooter games, the players can see and shoot through walls, giving them an unfair advantage over enemies. With the recent popularity of VR games, there is a need for solution that could at least restrict the forbidden view. For these reasons, many VR developers try various techniques for handling head collisions in their games. The most commonly used methods are:

- Screen fade: When the head collision is detected, the screen slowly fades to black or any other solid color. The view remains blacked-out for as long as the player's head collides with the object. Because this technique is relatively simple to implement, it attracts many VR developers looking for a quick solution to the problem. However, the method has one weakness that makes it not work well with teleportation locomotion. For example, the player can unintentionally teleport to some head-object colliding position. He is then stuck in a black void, without knowing in what direction he should move his head to escape the darkness.
- Delayed push-back: If the player collides with some object for a couple of seconds, he is gradually pushed backwards until he leaves the object's boundaries. This method is often used in older open-world games that were nowadays ported to VR, such as Skyrim VR or Fallout 4 VR. In most implementations of this technique, the head tracking is completely disabled during the

duration of the push-back effect. Many players of these games reported that the camera push-back is frustrating to experience and makes them feel dizzy [9–11].

- Instant push-back: In this solution, the collisions are handled similarly to how they are handled in most non-VR games; the player's virtual body, including the head, is never able to get past virtual walls and other obstacles. When the user collides with some obstacle, a collision vector is computed and its projection on the horizontal plane is added to the player's position. This effectively moves the user away from the object, preventing him from ever looking inside it. If the player decides to keep moving his head forward, despite the fact that there is a virtual wall in front of him, he will feel like he is using his head to push himself backwards from the wall.
- Teleportation: If the collision is detected and maintained for a couple of seconds, the player is teleported to a nearby collision-free location. This method works similarly to delayed push-back, but instead of gradually moving the player away from the obstacle, he is instantly teleported to the last know valid position.

1.2 Related Works

Because head tracking is a relatively new technology and most recent studies on VR movement were focused on comparisons of various locomotion techniques [2], there is a general lack of research into the described problem of VR head collisions. In recent years, VR developers were forced to develop their own solution to the problem if they wanted to prevent the players from clipping through walls and other objects. This led to the creation of many techniques, each with their advantages and disadvantages, and each with their supporters and critics.

In 2018, a group of researchers experimented with the instant push-back technique, which they called in their paper [14] a "not there yet" approach. The researchers compared this solution to the screen fade method in terms of immersion and VR sickness. They found out that while the "not there yet" approach yielded better immersion results, it contributed more to VR sickness symptoms.

2 Method

There are several possible solutions to the problem of VR head collisions, and they can be implemented in many different ways. In this paper, the four most popular methods were chosen and examined: screen fade, delayed push-back, instant push-back, and teleportation. The quality of VR experience is affected by many factors. VR sickness, the sense of presence in virtual world, and the usability of the user interface are some of the main considerations in designing comfortable VR experience.

2.1   Questionnaires

Two questionnaires were prepared for the study. Before the experiment began, participants were asked to fill in a demographic questionnaire. Each time after testing one of the solutions to the problem of VR head collisions, the participants filled in a post-test questionnaire. This questionnaire is composed of three sections. First, the participants described what VR sickness symptoms they felt during the experiment. Next, they answered to six questions about the sense of presence in the virtual environment. Finally, the participants rated the tested method with eight usability factors on the scale from 1 to 5.

The VR sickness section of post-test questionnaire was prepared using the Simulator Sickness Questionnaire (SSQ) [7], a widely acknowledged standard for studying simulator sickness. The SSQ was developed in 1993 with the purpose of enhancing training efficiency in flight simulators. Due to the similarity of simulator sickness with VR sickness, in the recent years the SSQ was also used by many researchers studying comfortable VR locomotion [4,5,8]. The SSQ consists of 16 most commonly occurring simulator sickness symptoms. These symptoms fall into three different categories: nausea-related, oculomotor-related, and disorientation-related. Each questionnaire item is scored on a four point scale (0–3): none, slight, moderate, and severe.

The Slayer-Usoh-Steed (SUS) [13] presence questionnaire was used to prepare the second section of post-test questionnaire. It was chosen for this study because it is the second most cited presence questionnaire applicable for VR [12], and it has a relatively short list of six questions in comparison to other, longer questionnaires. All SUS questions are based on one of the three themes:

- the extent to which the virtual environment becomes the dominant reality,
- the sense of being in the virtual environment,
- the extent to which the virtual environment is remembered as a "place".

Each question is answered on a scale from 1 to 7, where the higher score indicates the greater sense of presence. The final presence score is a number of answers that have a score of 6 or 7.

In the usability section, the participants rated eight different factors:

- difficulty in understanding the method,
- difficulty in operating the method,
- feeling of being in control while using the method,
- required effort to use the method,
- feeling of tiredness while using the method,
- feeling of enjoyment while using the method,
- feeling of being overwhelmed while using the method,
- feeling of frustration while using the method.

This exact approach was also used in two recent VR locomotion studies [3,5]. Each question was answered on a 5 point Likert scale, where 1 meant "not at all" and 5 meant "very much". The total usability score was calculated as follows: for each factor, except for the "feeling of being in control" and "feeling of enjoyment", the score was firstly subtracted from 5, and then it was added to the total score. In the two mentioned cases, the factor score was firstly subtracted by 1, and then it was added to the total score.

2.2   Virtual Environment

The virtual environment used in the experiment is implemented using the Unity game engine. The environment consists of a large, blue platform with 10 red objects (see Fig. 2). At the start of the application, the user of VR headset is placed at the center of the platform. User can move within virtual environment by using the "Point & Teleport" locomotion technique. When user presses and holds trackpad on the controller, a blue teleportation curve appears. He can then teleport by pointing the curve at the destination area and releasing the trackpad. User can only teleport to areas on the platform that are not occupied by the 10 objects. When pointed area is not viable, the curve's color is changed from blue to red.

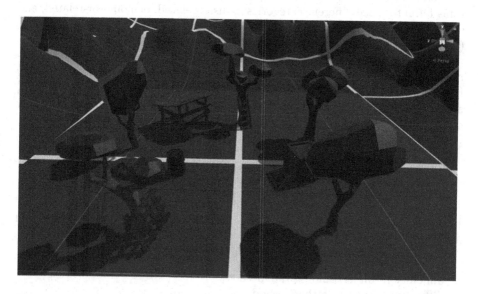

Fig. 2. The virtual environment used in the experiment: a large platform with 10 various objects on top. (Color figure online)

The goal for the user is to teleport closely to one of the objects in the virtual environment. Next, he has to lean his head towards the object with the purpose of colliding with it. Once the collision is detected, one of the implemented solutions to the problem of VR head collisions is activated. Finally, when user's head leaves the object's boundaries due to the workings of the method, the object's color is changed to green. Green color indicated that the interaction with an object was completed successfully. User has to repeat the process with each of the 10 objects until every obstacle is turned green. Because each of the four methods had to be tested separately, four different virtual environments were prepared for the study. All virtual environments are visually identical, with the same placement of the 10 virtual obstacles. The only difference is the method of handling head collisions.

The methods for handling head collisions are implemented by replicating the mechanisms seen in popular VR applications. In the screen fade method, the whole screen gradually fades to black in the span of a second when the collision is detected. In the delayed push-back method, last collision-free headset position is saved while user is not colliding with any object at the moment. After the collision is detected and maintained for the duration of a second, a collision vector between the last collision-free headset position and the current headset position is calculated. User is then pushed back in the direction of the normalized collision vector with the speed of 1 m/s. The push-back effect is working until user leaves the object's boundaries.

The implementation of the instant push-back method is similar to the delayed push-back with a couple of small differences. The collision vector is not normalized, and there is no delay before changing user position. In every frame of application (or every iteration of the game loop), user is instantly pushed back to the last collision-free position as long as the collision is still being detected. In the teleportation technique, the collision vector also is not normalized, and this time there is a delay before the method starts working. After the collision is detected and maintained for the duration of a second, screen fades to black for a moment and user is teleported to the last collision-free position.

2.3 Experiment

20 participants (3 females, 17 males) aged between 13 and 26 ($M = 19.65$, $SD = 5.29$) were recruited for the study. 12 of them were students of Computer Science master's degree (3 females, 9 males) aged between 23 and 26 ($M = 23.92$, $SD = 1.08$). The rest consisted of children from primary schools (8 males) aged between 13 and 14 ($M = 13.25$, $SD = 0.46$). The study employed a within-subject design. The independent variable was the solution to the VR head collisions problem, and it had four levels: screen fade, object fade, camera collider, camera push-back. For each participant, the order of tested solutions was assigned randomly with Latin Square counterbalancing, where each combination could only be used once. The experiment took place in teaching laboratories with play spaces of approximately 3 m × 3 m. The participants were equipped with the HTC Vive head-mounted display and its 6DoF hand controller. Additionally, two Base Stations of the HTC Vive Lighthouse System were used for tracking the head's position and rotation in a 3D environment.

Upon arriving in the teaching laboratory where the experiment took place, the participants were given a broad overview of the research problem. They were introduced to the main goal of the study and instructed about the task they had to do while present in the virtual environment. Once a participant was acquainted with the instructions, he was equipped with the headset and started testing one of the solutions to the problem of VR head collisions. His task was to collide with every obstacle in the presented virtual environment. The participant was not told which solution he was testing, and he had to discover on his own how the method was working. Once he successfully collided with an object and observed effects of the method, the obstacle's color was changed from red to green. The participant tested the method until every obstacle in the virtual environment was turned green. On average, it took participants 2.37 ($SD =$ 0.62) minutes to complete this task. Once done, the participant was asked to take off the headset and fill in a post-test questionnaire. In this questionnaire he described what VR sickness symptoms he felt during the experiment, answered to six questions about the sense of presence, and rated the tested method with eight usability factors. Once the post-test questionnaire was finished, the participant moved on to test next method. He repeated this process until each of the four methods were tested in this manner. Each participant tested the methods in a unique order.

3   Results

This study employed a repeated measures design, where each of 20 participants tested in succession all four solutions to the problem of VR head collisions. For each solution, 20 results were gathered from the questionnaires. If the assumptions of normality and sphericity were not violated, the difference between means were analyzed at 0.05 level of significance using a repeated measures ANOVA; otherwise, a Friedman test was used. The distribution of the data was checked for normality using Shapiro-Wilk test, and Mauchly's test was used to assess sphericity. If the results indicated statistical significance, additional post-hoc tests with Bonferroni correction were used to determine which means differed from each other.

3.1   VR Sickness

The total SSQ scores were calculated and were analyzed using a Friedman test due to non-normality. The test revealed that the total SSQ scores were significantly influenced by the method of handling collisions ($\chi^2(3) = 9.881, p = 0.02$). Post-hoc Wilcoxon signed-rank tests revealed that the screen fade method led to significantly lower scores than the delayed push-back method ($p < 0.001$). See Fig. 3 for an overview of the total SSQ scores.

3.2   Sense of Presence

Shapiro-Wilk tests indicated that the assumption of normality was not violated; and Mauchly's test indicated that the assumption of sphericity was met ($\chi^2(3) = 9.758, p = 0.082$). A repeated measures ANOVA revealed that the presence scores were significantly influenced by the method of handling collisions ($F(3, 76) = 11, p < 0.00001$). Post-hoc pairwise comparisons revealed higher presence scores

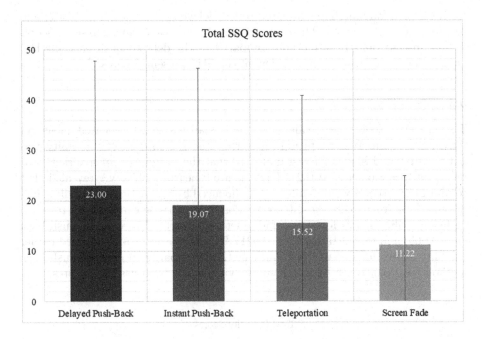

Fig. 3. The total SSQ scores.

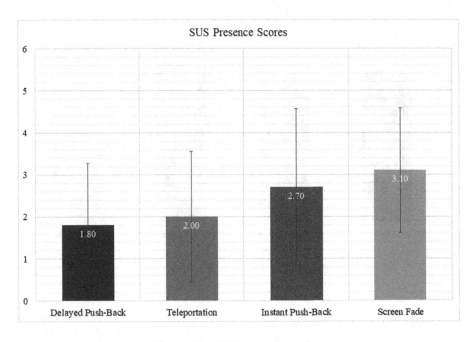

Fig. 4. The SUS presence scores.

for the screen fade method compared to the delayed push-back ($p < 0.001$) and teleportation ($p = 0.002$). The comparisons also revealed higher presence scores for the instant push-back method compared to the delayed push-back method ($p = 0.01$). See Fig. 4 for an overview of the SUS presence scores.

3.3   Usability

The usability scores were calculated and analyzed using repeated measures ANOVA. Shapiro-Wilk tests indicated that the assumption of normality was not violated; and Mauchly's test indicated that the assumption of sphericity was met ($\chi^2(3) = 8.08$, $p = 0.152$). The repeated measures ANOVA revealed that total usability scores were significantly influenced by the method of handling collisions ($F(3, 76) = 8.887$, $p < 0.0001$). Post-hoc pairwise comparisons revealed higher usability scores for screen fade method compared to delayed push-back ($p = 0.005$) and teleportation ($p = 0.021$). The comparisons also revealed higher usability scores for instant push-back method compared to delayed push-back method ($p = 0.03$). See Fig. 5 for an overview of the total usability scores.

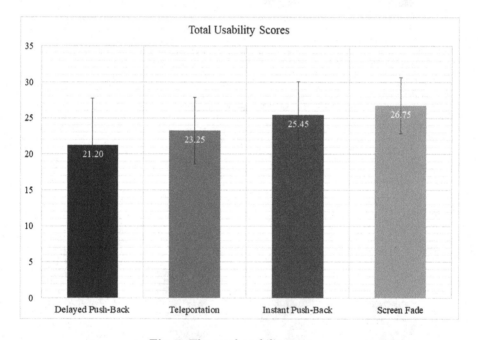

Fig. 5. The total usability scores.

4   Discussion

4.1   VR Sickness

The first research question in this study sought to determine how the proposed solutions to the problem of VR head collisions affect virtual reality sickness. The scores were significantly lower for the screen fade method than the delayed push-back method. The high scores of the delayed push-back method explain the complaints that can be found in many internet forums. This method was chosen by some VR developers and players reported that it had great impact on virtual reality sickness. The most likely reason for the high scores is that this method for a brief moment takes control of camera movements away from the player. Any unexpected accelerations and movements of the camera not initiated by the player's physical movement are some of the main causes of VR sickness.

4.2   Sense of Presence

The second research question sought to find how the studied methods affect the sense of presence. The screen fade method leads to a significantly higher sense of presence than the teleportation and the delayed push-back methods. Moreover, the instant push-back method provides a significantly higher sense of presence than the delayed push-back method. However, no significant difference was found between the screen fade method and the instant push-back method. There are several possible explanations for these results. During testing the screen fade method, participants often commented that fading screen to black feels most natural to them. Some participants explained that this effect resembles the darkness due to closing the eyes when their heads collide with objects in the real world. During testing the instant push-back method, some participants commented that they felt like they are pushing themselves away from the object by using their heads, which again in some way resembles how their feet would be pushed backwards on a slippery floor in the real world. Moreover, the screen fade and the instant push-back methods are the only techniques where players are never able to see the insides of objects and unexpected clipping artifacts. In the delayed push-back and the teleportation methods, players are able to see the insides of objects for a brief moment before the methods start to work and relocate them to a new position.

4.3   Usability

The third and final research question in this study sought to identify how usable are the proposed solutions. The exact same situation as in the sense of presence case repeated for total usability scores: the screen fade method leads to a significantly higher level of usability than the teleportation and the delayed push-back methods, and the instant push-back method provides a significantly higher level of usability than the delayed push-back method. Once again, no significant difference was found between the screen fade method and the instant push-back method. These results can be explained by many factors. Users lose some control of the movement for a longer while in the delayed push-back method. In the instant push-back method, users manually initiate the effect by pushing their heads forward in the direction of the collision and can stop doing it at any time. During testing the screen fade method, some participants liked the visual effect of gradual fading. They repeated moving the head in and out of the object, even if its color was already changed to green. Although participants mostly enjoyed discovering how the teleportation and the delayed push-back methods work, after experiencing them for the first time, some participants found it frustrating to wait a full second to trigger the effect with the remaining objects.

5   Conclusions

All methods presented in this work can be successfully used in video games. Overall, the screen fade method turned out to be the most efficient one of the four solutions and should be the first choice for VR developers. It is characterized by a simple implementation that can be easily integrated into any VR application. It has more positive effect than the delayed push-back method in terms of VR sickness, sense of presence, tiredness, enjoyment, frustration, and general usability. It also achieves better scores than the teleportation method in terms of usability, enjoyment, and sense of presence. The instant push-back method had mostly neutral results. It only has more positive effect than the delayed push-back method in terms of usability, feelings of control, and sense of presence. No significant differences were found between the delayed push-back and the screen fade method. It is recommended to use this solution in place of the screen fade method if the developer does not want to obstruct the vision of the players at any time. The teleportation method also had neutral scores for VR sickness. However, it achieved substandard scores for the sense of presence and usability, and it generally should not be the preferred solution. The delayed push-back method ranked the worst in all categories and should be avoided by VR developers.

5.1   Future Works

The screen fade method turned out to be the best in terms of many factors, however, this solution still has one minor issue that sometimes occurs in teleportation locomotion techniques. The users can unintentionally teleport to some head-object colliding position, and then be stuck with obscured screen without knowing in what direction they should move to escape the darkness. Further research in this field might explore the ways how to prevent this situation. For example, some visual cues that guide the users might help them move out of the collisions. In the case of other promising solutions to the problem of VR head collisions, a future study investigating the object fade method is strongly recommended. Object fade is a technique in which, rather than fading the whole screen to black, only parts of the colliding objects fade out as the camera gets closer to them. Due to the similarity to the screen fade method and the possibility to add some interesting fading visual effects, this solution seems now the most promising.

References

1. Bethesda Forums. Skyrim PlayStation VR Bug: Teleport through walls & doors. https://bethesda.net/community/category/216/skyrim-playstation-vr. Accessed 07 Jan 2019
2. Boletsis, C.: The new era of virtual reality locomotion: a systematic literature review of techniques and a proposed typology. Multimodal Technol. Interact. **1**, 24 (2017). https://doi.org/10.3390/mti1040024
3. Bozgeyikli, E., Raij, A., Katkoori, S., Dubey, R.: Point & teleport locomotion technique for virtual reality. In: Proceedings of the 2016 Annual Symposium on Computer-Human Interaction in Play, pp. 205–216 (2016)
4. Frommel, J., Sonntag, S., Weber, M.: Effects of controller-based locomotion on player experience in a virtual reality exploration game. In: Proceedings of the 12th International Conference on the Foundations of Digital Games (2017)
5. Habgoodand, J., Moore, D., Wilson, D., Alapont, S.: Rapid, continuous movement between nodes as an accessible virtual reality locomotion technique. In: 2018 IEEE Conference on Virtual Reality and 3D User Interfaces (VR), pp. 371–378 (2018)
6. Kemeny, A., George, P., Merienne, F., Colombet, F.: New vr navigation techniques to reduce cybersickness. Eng. Reality Virtual Reality **2017**, 48–53 (2017)
7. Kennedy, R.S., Lane, N.E., Berbaum, K., Lilienthal, M.G.: Simulator sickness questionnaire: an enhanced method for quantifying simulator sickness. Int. J. Aviat. Psychol. **3**, 203–220 (1993)
8. Langbehn, E., Lubos, P., Steinicke, F.: Evaluation of locomotion techniques for room-scale vr: Joystick, teleportation, and redirected walking. In: Proceedings of the Virtual Reality International Conference (2018)
9. Reddit. R/oculus - Rant: Player pushback is the absolute worst method to tackle "VR cheating" (head inside an object). https://www.reddit.com/r/oculus/comments/975rv1/rant_player_pushback_is_the_absolute_worst_method/. Accessed 07 Jan 2019
10. Reddit. R/skyrimvr - Is it possible to avoid the push-back from players/walls without changing bAllowVRCheating? https://www.reddit.com/r/skyrimvr/comments/8n533f/is_it_possible_to_avoid_the_pushback_from/. Accessed 07 Jan 2019
11. Reddit. R/Vive - Fallout 4 VR Thoughts + Locomotion issues. https://www.reddit.com/r/Vive/comments/7j9tgx/fallout_4_vr_thoughts_locomotion_issues/. Accessed 07 Jan 2019
12. Schwind, V., Knierim, P., Haas, N., Henze, N.: Using presence questionnaires in virtual reality. In: Proceedings of the 2019 CHI Conference on Human Factors in Computing Systems (2019)
13. Usoh, M., Catena, E., Arman, S., Slater, M.: Using presence questionnaires in reality. Presence Teleoperators Virtual Environ. **9**, 497–503 (2000)
14. Ziegler, P., Roth, D., Knots, A., Kreuzer, M., Mammen, S.: Simulator sick but still immersed: a comparison of head-object collision handling and their impact on fun, immersion, and simulator sickness. In: Proceedings of the 2018 IEEE Conference on Virtual Reality and 3D User Interfaces (VR), pp. 743–744 (2018)

Generation of Musical Scores from Chord Sequences Using Neurodynamic Model

Koji Masago[✉], Mizuho Amo, Shun Nishide, Xin Kang, and Fuji Ren

Faculty of Science and Technology, Tokushima University,
2-1 Minami-Josanjima-cho, Tokushima, Tokushima 770-8502, Japan
koji.masago@gmail.com

Abstract. Music generation is one of the important and challenging tasks in the field of music information processing. In this paper, we propose a method to generate music note sequences from a chord sequence based on the dynamics relation learned using a recurrent neural network model. For the dynamics learning model, Multiple Timescale Recurrent Neural Network (MTRNN) is used. The model comprises a hierarchical structure to learn different levels of information. The note sequence and chord sequence dynamics are self-organized into the model based on the similarity of the dynamics. The proposed method inputs a chord sequence into the trained MTRNN to calculate the latent parameter that represents the sequence. The parameter is used to generate the music note sequence using the forward calculation of MTRNN. Experiments were conducted using three music pieces arranged into three variations. The results of the experiment generating music note sequence from chord sequence for trained pieces showed the effectiveness of the method.

Keywords: Recurrent neural network · Music generation · Hierarchical learning

1 Introduction

With the spread of music sharing sites such as YouTube and popularization of smartphones, music has become an essential part of our daily lives. In addition to listening to music, enjoyment of music have started to infiltrate into arranging or creation of music. Creating music, which is far difficult from arranging music, requires understanding of music theory, in addition to talent and performance skills. To ease the difficulty of creating music, softwares to support music creation have been developed such as synthesizers which can test combinations of various music instruments without the need the actually play them. Although many researches have been conducted for music creation support, automatic music creation by computer is still a difficult task.

In this paper, we propose a method to use a neurodynamical model to generate music scores from a given chord sequence. The model comprises a hierarchical structure, learning the dynamics relation between the chord and musical note sequences of various music scores, self-organizing the dynamics of various scores within the model. Given a chord sequence, the model generates a music score

© Springer Nature Switzerland AG 2020
H. Fujita et al. (Eds.): IEA/AIE 2020, LNAI 12144, pp. 638–648, 2020.
https://doi.org/10.1007/978-3-030-55789-8_55

that best fits the chord sequence based on self-organized representation within the model. Basic experiments, training the model with a few music chords are done for evaluation of the method.

The rest of the paper is organized as follows. In Sect. 2, we present researches related to our work. In Sect. 3, the overview of the proposed method is presented. In Sect. 4, the experimental setup is described. In Sect. 5, the result of the experiment is presented. In Sect. 6, discussions based on the experimental results are presented. Conclusions and future work are presented in Sect. 7.

2 Related Research

Researches on music can be mainly divided into two categories: music recognition and music generation. In the next subsection, some works that have been conducted in the two categories are presented.

2.1 Music Recognition

With the advent of a wide variety of music, there is an increasing demand for a system that recommends and searches for a user's favorite music. In order to accomplish this goal, a system is required to analyze and recognize music audio signals from CDs or media. Various studies have been conducted in the field of music and chord recognition.

Sumi et al. recognized chord sequences based on the relationship between bass and chords in popular songs [1]. The work proposed a method that focuses on the interrelationships between musical elements that have not been considered in previous studies. Maruo et al. proposed a chord recognition method that eliminates singing voices and percussion instrument sounds to improve the performance of chord recognition [2]. Hidden Markov Models are also one of the traditional models for recognition of chords using chroma vectors [3]. Music recommendation systems have also been developed [4].

These works present the briskness of the field of music recognition.

2.2 Music Generation

Compared to music recognition, not much work have been done on music generation due to difficulties that lie in evaluation of the generated music. Regardless, music generation is an important topic, automatic music composition has been one of the hot topics for many years. The evolution of deep learning techniques has brought back researcher's attention to automatic music generation. Our research shares the same goal, to create an automatic music generation system.

Genetic Algorithms (GA) were originally used for music composition. Unehara et al. created a music composition system based on interaction between human and computer based on interactive GA [5]. Compared to completely automatic music composition, the system can reflect the user's preference into the composition. Göksu et al. proposed a combination of GA with multilayer perceptrons for music composition.

Recently, deep learning techniques have been used in music generation systems. The MiniBach System generates an accompaniment melody in Bach style by inputting a string of music notes [7]. DeepHear is a system that uses an autoencoder and feedforward network to learn jazz music [8]. By inputting a random seed into the middle layer of the autoencoder, the system could generate new music.

In our research, we utilize a deep learning model composed of hierarchical layers of neurons to learn sequences of music notes and chords hierarchically. The objective of the work is to generate music note sequences that match a given chord sequence based on the trained model, rather than completely random generation.

3   Proposed Method

For the learning model, we utilize Multiple Timescale Recurrent Neural Network (MTRNN) [9], which comprises a hierarchical structure to learn different levels of information. In this section, we present the overview of MTRNN with the modification of the model for music generation.

3.1   MTRNN

MTRNN is a predictor that takes the current state $IO(t)$ as input and outputs the next state $IO(t + 1)$. An overview of MTRNN is shown in Fig. 1. Unlike conventional RNN, MTRNN self-organizes the dynamics of multiple sequences into the model. Therefore, MTRNN can represent multiple sequences using a single model.

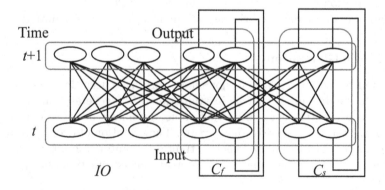

Fig. 1. MTRNN configuration

MTRNN is composed of a hierarchical structure with three neuron groups, IO neuron group, Fast Context neuron group (C_f), and Slow Context (C_s) neuron group. The connection of neurons between the input layer (neurons in time t) and output layer (neurons in time $t+1$) are fully connected except for the IO neuron group and C_s neuron group. The hierarchical structure of MTRNN is constructed by a parameter called time constant, which controls the firing rate of each neuron. Neuron groups with a large time constant have a low firing rate, while those with a small time constant have a high firing rate. The time constant is set in increasing order from IO, C_f, and C_s neuron groups. This model can be easily extended by adding a context layer with a larger time constant. Multiple sequences are learned in the model by self-organizing the dynamics into the initial value of C_s neurons ($C_s(0)$). The $C_s(0)$ values act as self-organized latent variables that are used to encode sequences. The three functions of MTRNN, learning, recognition, and generation are described in the following subsections.

Learning. Learning of MTRNN consists of forward and inverse calculations. By forward calculation, the output of each state is calculated. The output error is back propagated to update the weights of the connection using Back Propagation Through Time (BPTT) algorithm. In MTRNN, the $C_s(0)$ values, which are independently set for each sequence, are also updated during the process.

Recognition. Recognition is a process to calculate the $C_s(0)$ values from a given sequence. A similar calculation process is conducted as the learning, except the weights are fixed during the calculation. In recognition, only the $C_s(0)$ values are updated by BPTT.

Generation. Generation is a process to calculate the sequence from given $C_s(0)$ values. The calculation is similar to the forward calculation in learning. The calculation of the sequence is done iteratively, from the initial time 0 to the final.

3.2   Music Dynamics Learning Model.

In the proposed method, music scores, referring to note and chord information in this paper, are learned using MTRNN. Figure 2 shows the structure of the proposed method. Note that Fig. 2 is an equivalent of the model in Fig. 1 spanned along the temporal axis. Calling into account that notes changes frequently than chords, the construction of MTRNN is set so that notes are learned in IO group and chords are learned in Fast Context group. C_f neurons in the Fast Context group learn the contextual information. Neurons in Slow Context group are also divided into those that learn the context information (C_s) and those that self-organize the dynamics of the music scores (Control C_s). The difference between C_f neurons and C_s neurons is the difference in level of contextual information, where C_s neurons learn high level information compared to C_f neurons.

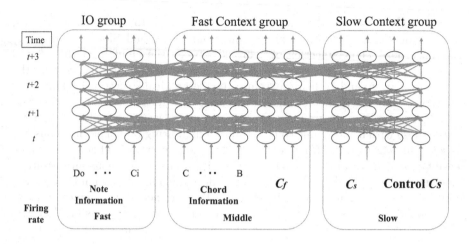

Fig. 2. Configuration of proposed model

The construction of the model is based on the work by Hinoshita et al., where the hierarchical structure of sentences were learned through self-organization [10]. In the work, sequences of alphabets were used instead of notes in Fig. 2. By learning the model, information of words were self-organized in C_f neurons, and gramatical information was self-organized in C_s neurons. The Control C_s neurons were used to encode and decode sentences. The model was capable of recognizing and generating sentences, implying the potential of learning music scores with hierarchical structure as in this paper.

The flow of this research is shown in Fig. 3. Learning of MTRNN is done using music scores containing both note sequences and chord sequences, self-organizing the Control $C_s(0)$ values of each music score. In recognition, only the chord sequence is input into MTRNN to calculate the Control $C_s(0)$ values representing the chord sequence. The calculated Control $C_s(0)$ is input into MTRNN through the generation function. The generation calculation derives the note sequence and chord sequence. In this manner, the Control $C_s(0)$ acts as a latent parameter that represents the music score. The proposed method estimates the latent parameter from the chord sequence, and uses the parameter to generate note sequence.

4   Experimental Setup

In this paper, we evaluate the model's capability to generate learned sequences. The learning process was done using three songs: Denderaryu, Twinkle Twinkle Little Star, and Bee March. Each song was modulated to C, E, G chords for a total of nine songs for learning. The assumptions of the experiment in this paper is shown in Fig. 4.

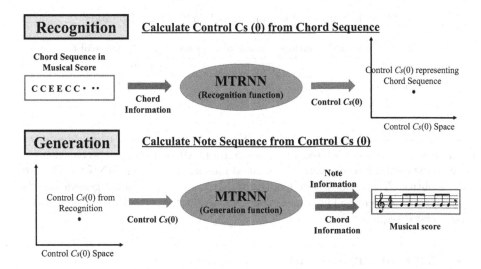

Fig. 3. Schematic flow of the proposed method

Note	1. Single tone only 2. Range of three octaves 3. Minimum note: Eighth note
Chord	1. Twelve root notes (C, C#, D, ... , A#, B) 2. Three types of chords (major, minor, seventh)

Fig. 4. Assumptions of the Experiment

For note/chord information, a neuron in MTRNN corresponds to each note/chord. An octave consists of 12 notes, and we consider 3 octaves with a rest, for a total of 37 neurons for those corresponding to notes. In the experiment, we assume that only one note exist in one time step corresponds to eighth note. For scalability to multiple notes, the sigmoid function is used as the activation function for neurons corresponding to notes. For chord neurons, we consider 12 root notes, with 3 types of chords for each root node, for a total of 36 neurons. Unlike notes, only a single chord exists at every time step, and thus the softmax function is used for the activation function.

The number of notes in each song was, 64 for Denderaryu, 96 for Twinkle Twinkle Little Star, and 128 for Bee March. Two types of MTRNN configuration, changing the number of C_f neurons, were used for training the model, shown in Table 1.

Table 1. Number of neurons in each MTRNN configuration

Model	Note information	Chord information	C_f	C_s	Control C_s
Model 1	37	36	100	30	6
Model 2	37	36	200	30	6

For each model shown in Table 1, the model was trained using nine songs (three songs each modulated in three chords). After training, the chord information is used for recognizing the trained nine songs with MTRNN to calculate the Control $C_s(0)$ values. The Control $C_s(0)$ values are used for generating the note/chord sequence. The experiment was evaluated based on the generated note/chord sequence.

5   Experimental Result

Evaluation of the experiment was done based on two criteria.

1. Accuracy of the model to reconstruct the chord sequence.
2. Similarity of the generated note sequence to the original.

For each time step, the generated and original sequences were matched to count the number of matched data, relative to the total number of steps. The matching rate was calculated as

$$M = \frac{N - l}{N} \times 100, \tag{1}$$

where M is the matching rate (in percent), N is the total number of notes/chords, and l is the number of notes/chords the did not match. The matching rate for the two models are shown in Table 2. From Table 2, it can be seen that chord data is relatively well reconstructed for both models, but a higher reconstruction rate could be achieved for a larger number of C_f neurons. The generation of note sequence is quite different between the two models, where Model 2 with a larger number of C_f neurons generates note sequence similar to the trained music score.

To visualize the generated note sequence, we colored each of the 36 notes (and a rest) with different colors for the original note sequence and generated note sequence for Model 2. As examples of the visualized results, the generated note sequence with the original note sequence of Bee March modulated to E chord and C chord are shown in Fig. 5 and Fig. 6, respectively. The left figures in Fig. 5 and Fig. 6 show the original note sequence, and the right figures show the generated sequence. As can be seen, a similar note sequence has been generated for both music scores.

Table 2. Matching rate for two models

Song	Modulation	Model 1		Model 2	
		Note	Chord	Note	Chord
Denderaryu (64 steps)	C	95.3	100.0	98.4	100.0
	E	89.1	95.3	95.3	100.0
	G	96.9	100.0	95.3	100.0
	Average	93.8	98.4	96.3	100.0
Twinkle Twinkle Little Star (96 steps)	C	97.9	87.5	100.0	99.0
	E	100.0	100.0	94.8	85.4
	G	55.2	68.8	92.7	92.7
	Average	84.4	85.4	95.8	92.4
Bee March (128 steps)	C	69.3	92.2	92.7	99.0
	E	90.1	97.9	95.3	97.4
	G	76.6	95.3	89.1	99.5
	Average	78.7	95.1	92.4	98.6

6   Discussion

From the results of the experiment, the model generated a similar note sequence as the original trained sequence from the recognized chord sequence. In this section, we discuss the results based on two aspects.

1. Model configuration and music generation
2. Generation of music note sequence using untrained chord sequence

6.1   Model Configuration and Music Generation

In this paper, we confirmed the effectiveness of the model using two types of models: one with 100 C_f neurons and the other with 200 C_f neurons. The learning capability of the MTRNN depends on the number of neurons, while the training cost increases as the number of neurons increase. Overfitting is also an issue that arises with a large number of neurons. Determination of the appropriate number of neurons, based on the training data is an issue left to be dealt as future work.

6.2   Generation of Music Note Sequence Using Untrained Music Scores

In this paper, experiments on music generation were done using chord sequences of trained music scores. The results of the experiment showed that the model is capable of reconstructing note sequences from chord sequence for trained music

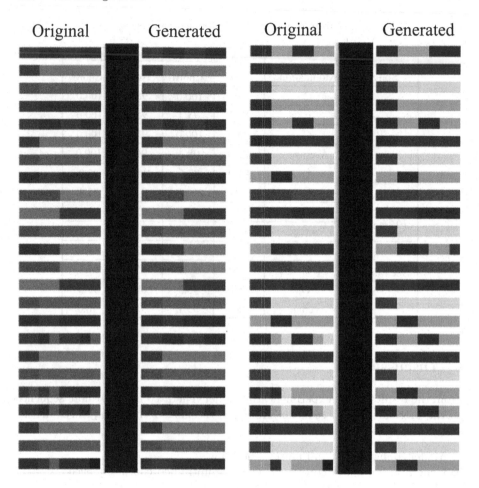

Fig. 5. Bee march E modulation result (Left figure: Original note sequence, Right figure: Generated note sequence)

Fig. 6. Bee march C modulation result (Left figure: Original note sequence, Right figure: Generated note sequence)

scores. The generation capability of the model with untrained chord sequences is an issue that still requires confirmation.

When generating a note sequence from an untrained chord sequence, neural networks tend to generate sequences that are similar to trained sequences based on the generalization capability. The ultimate goal of our work is to create a model that generates new musics. Therefore, the configuration of MTRNN (number of neurons) is a critical issue to prevent overfitting while maintaining enough neurons to achieve good training results.

7   Conclusion

In this paper, we presented a method to generate music scores from chord sequence using a neurodynamical model, namely MTRNN. The model comprises a hierarchical structure, and the note sequences and chord sequences are learned in different levels within the model. The proposed method first trains MTRNN using training music scores with both music note and chord sequences. During training of the model, the weights are updated and the Control $C_s(0)$ values are obtained for each training sequence. For music note generation, the chord sequence is input into MTRNN to calculate the Control $C_s(0)$ values that represent the chord sequence. The Control $C_S(0)$ is input into MTRNN to generate the note sequence.

Experiments were conducted using training data consisting of three songs modulated into three chords. The training data were used to train two MTRNNs with different number of C_f neurons. The results of the experiment showed that the method could reconstruct music note sequences from chord sequences for trained music scores.

Future work consists of analyzing the capability of the model relative to the configuration of MTRNN, and evaluation of the model using untrained chord sequences. For analysis of the model, we plan to conduct experiments by changing the number of neurons, and evaluate the generation capability. Furthermore, generation of music note sequence will be evaluated using untrained chord sequences. We believe that the model will contribute to the field of automatic music generation.

Acknowledgments. This research was supported by JSPS KAKENHI Grand Number 16H05877.

References

1. Sumi, K., Itoyama, K., Yoshii, K., Komatani, K., Ogata, T., Okuno, G.H.: Automatic chord sequence recognition based on integration of chord and bass pitch features. IPSJ J. **52**(4), 1803–1812 (2011)
2. Maruo, S., Yoshii, K., Itoyama, K., Mauch, M., Goto, M.: A feedback framework for improved chord recognition based on nmf-based approximate note transcription. In: Proceedings of the IEEE International Conference on Acoustics, Speech and Signal Processing, pp. 196–200 (2015)
3. Ueda, Y., Uchiyama, Y., Nishimoto, T., Ono, N., Sagayama, S.: HMM-based approach for automatic chord detection using refined acoustic features. In: Proceedings of the IEEE International Conference on Acoustics, Speech and Signal Processing, pp. 5518–5521 (2010)
4. Saito, Y., Itoh, T.: MusiCube : a visual music recommendation system featuring interactive evolutionary computing. In: Proceedings of the International Symposium on Visual Information Communication, Art. no. 5 (2011)
5. Unehara, M., Onisawa, T.: Music composition by interaction between human and computer. New Gener. Comput. **23**(2), 181–191 (2005)

6. Göksu, H., Pigg, P., Dixit, V.: Music composition using genetic algorithms (GA) and multilayer perceptrons (MLP). In: Wang, L., Chen, K., Ong, Y.S. (eds.) ICNC 2005. LNCS, vol. 3612, pp. 1242–1250. Springer, Heidelberg (2005). https://doi.org/10.1007/11539902_158

7. Hadjeres, G., Pachet, F., Nielsen, F.: DeepBach: a steerable model for bach chorales generation. In: Proceedings of the International Conference on Machine Learning, pp. 1362–1371 (2017)

8. Sun, F.: DeepHear - Composing and harmonizing music with neural networks. https://fephsun.github.io/2015/09/01/neural-music.html

9. Yamashita, Y., Tani, J.: Emergence of functional hierarchy in a multiple timescale recurrent neural network model: a humanoid Robot Experiment. PLoS Comput. Biol. 4(11), e1000220 (2008)

10. Hinoshita, W., Arie, H., Tani, J., Okuno, H.G., Ogata, T.: Emergence of hierarchical structure mirroring linguistic composition in a recurrent neural network. Neural Networks 24(4), 311–320 (2011). Elsevier

11. Berk, L.E., Winsler, A.: Scaffolding Children's Learning: Vygotsky and Early Childhood Education. National Association for Education of Young Children, 1995

Improving Variational Mode Decomposition-Based Signal Enhancement with the Use of Total Variation Denoising

Krzysztof Brzostowski$^{(\boxtimes)}$ and Jerzy Świątek

Faculty of Computer Science and Management, Wrocław University of Science and Technology, Wyb. Wyspiańskiego 27, 50-370 Wrocław, Poland
{krzysztof.brzostowski,jerzy.swiatek}@pwr.edu.pl

Abstract. In the paper, a novel approach to noise removing bases on Variational Mode Decomposition and Total Variation Denoising is presented. Variational Mode Decomposition is a state-of-the-art adaptive and non-recursive signal analysis method. The method is distinguished by the high accuracy of signal separation and noise robustness. In turn, Total Variation Denoising, which is defined in terms of a convex optimization problem, is a widely used regularizer in sparse signal denoising. In the paper, we proposed an approach in which Total Variation Denoising is applied to improve the Variational Mode Decomposition-based denoising method. The proposed method is an alternative to methods that have been already proposed, i.e., the methods combine hard and soft thresholding with Variational Mode Decomposition.

In our research, we found that the proposed approach based on Variational Mode Decomposition and Total Variation Denoising has the ability to improve the accuracy of the reference method. The approach was tested for two different synthetic signals. The used in our studies synthetic signals were corrupted by noise with different short and long dependencies.

The presented in work results show that the proposed novel approach gives a great improvement in signal enhancement and it is a promising direction of future research.

Keywords: Non-linear signals processing · Mode decomposition · Robotics

1 Introduction

Wearable robotics have been developed by both scholars and practitioners over the last few decades and they are used in a variety of military, medical, and industrial applications. However, existing wearable robots face numerous challenges with regard to both hardware and software. For example, one of the major challenges in the field of software is the effective processing of acquired data from sensors. Unfortunately, the available sensors suffer from some physical limitations. These limitations cause that gathered signals become prone to

© Springer Nature Switzerland AG 2020
H. Fujita et al. (Eds.): IEA/AIE 2020, LNAI 12144, pp. 649–660, 2020.
https://doi.org/10.1007/978-3-030-55789-8_56

disturbances during acquisition. We understand the noise as a signal distortion hindering the process of data acquisition. Hence, the enhancement of the gathered signal is required in order to suppress the noise to the desired extent.

The traditional approach to signal denoising based on filters. However, the filter-based methods have some limitations: a) inability to separate signal and noise in the same frequency band, and b) poor performance for non-stationary and short-term transient signals. An alternative approach to signal enhancement based on wavelet transform. Denoising the effects of wavelet-based algorithms are better in comparison to the filter-based approach. Unfortunately, the method is sensitive to periodic noise and it is difficult to select the exact value of threshold [1]. Moreover, the wavelet-based denoising algorithm relying on the predefined basis functions.

The other approach is Empirical Mode Decomposition (EMD) [7]. Unlike the wavelet-based approach, EMD has the advantage of not relying on the predefined basis functions. In the EMD method analyzed signal is decomposed into the set of oscillatory components (Intrinsic Mode Functions, IMF). Empirical Mode Decomposition method is widely used for analyzing non-stationary signals. However, there are many problems to use EMD. One of them is the problem of mode mixing, i.e. the existence of different time scales in a single IMF. Another is that Empirical Mode Decomposition still has not to sound mathematical theory.

Variational Mode Decomposition (VMD) is an adaptive and non-recursive signal analysis method [4]. Compared to EMD, the VMD approach has a strong mathematical background. This helps to improve the accuracy of signal separation and noise robustness [6]. Moreover, Variational Mode Decomposition effectively eliminates the effect of mode mixing during the decomposition process [16]. The set of determined, based on VMD algorithm, components are also named IMF.

VMD algorithm has been used in many fields. For example, in [9] the authors reported the application of Variational Mode Decomposition in forecasting day-ahead energy prices. While the authors of [17] proposed VMD in the analysis of seismic signals. The discussed method is also used for the enhancement of gyroscopic signals [15].

In the paper, a sparse signal processing is applied to design a new denoising algorithm based on Variational Mode Decomposition. The general idea of the proposed approach is to apply the Total Variation Denoising (TVD) algorithm to IMFs extracted from the signal with the use of the VMD algorithm. TVD is a widely used regularizer in signal processing which is defined in terms of a convex optimization problem containing a data fidelity and regularization terms. In the further part of the work, we prove that the TVD-based denoising algorithm proposed in the paper is an extension of the methods based on hard and soft thresholding.

2 Basic Principles

2.1 Variational Mode Decomposition

VMD is a multi-resolution analysis method which is used to extract from non-stationary signal a number of Intrinsic Mode Functions and their central frequencies. The extraction procedure bases on iterative scheme to find the optimal solution of the variational model of each finite bandwidth in which the components are adaptively separated according to their respective centre frequencies [4]. The objective function and the constrained variational problem is as follows:

$$\{u_k^*\}, \{\omega_k^*\} = \min_{\{u_k\},\{\omega_k\}} \left\{ \sum_{k=1}^{K} \| \frac{\partial}{\partial t} \left[\left(\delta(t) + \frac{j}{\pi t} \right) * u_k(t) \right] \exp^{-j\omega_k t} \|_2^2 \right\},$$

$$s.t. \sum_{k=1}^{K} u_k(t) = y(t)$$

(1)

where t is the time index, $\delta(t)$ is the Dirac distribution, u_k is shorthand notations for k-th IMF from the set of all IMFs, i.e. $u_K = \{u_1, u_2, \ldots, u_K\}$ and ω_k stands for k-th centre frequencies of k-th IMF, i.e. $\omega_K = \{\omega_1, \omega_2, \ldots, \omega_K\}$. The symbol $*$ denotes convolution. $y(t)$ stands for the signal to be decomposed.

By introducing the Lagrangian multiplier λ and the second penalty factor α (data-fidelity constraint parameter), the constrained variational problem (1) can be transformed into an unconstrained variational problem:

$$L(\{u_k\}, \{\omega_k\}, \lambda) = \alpha \sum_{k=1}^{K} \left\{ \sum_{k=1}^{K} \| \frac{\partial}{\partial t} \left[\left(\delta(t) + \frac{j}{\pi t} \right) * u_k(t) \right] \exp^{-j\omega_k t} \|_2^2 \right\}$$

$$+ \| y(t) - \sum_{k=1}^{K} u_k(t) \|_2^2 + \left\langle \lambda(t), y(t) - \sum_{k=1}^{K} u_k(t) \right\rangle$$

(2)

In order to solve the unconstrained optimization problem (2) it can be split into two different sub-problems, i.e, minimization of $u_k(t)$ and $\omega_k(t)$. Such separation of the formulated problem (2) is presented as:

$$u_k^{(n+1)} \leftarrow \arg\min_{u_k} L(u_{i<k}^{(n+1)}, u_{i\geqslant k}^{(n)}, \omega_i^{(n)}, \lambda^{(n)}) \tag{3a}$$

$$\omega_k^{(n+1)} \leftarrow \arg\min_{\omega_k} L(u_i^{(n+1)}, \omega_{i<k}^{(n+1)}, \omega_{i\geqslant k}^{(n)}, \lambda^{(n)}) \tag{3b}$$

where $k = 1, 2, \ldots, K$.

Based on continuous iteration updated, all IMFs and their corresponded centre frequencies are determined from [4]:

$$\hat{u}_k^{(n+1)}(\omega) = \frac{\hat{y}(\omega) - \sum_{i<k} \hat{u}_i^{(n+1)}(\omega) + \frac{\hat{\lambda}^{(n)}(\omega)}{2}}{1 + 2\alpha(\omega - \omega_k^{(n)})^2} \tag{4}$$

2.2 Sparsity-Based Signal Denoising

Let us assume the following discrete model of measured signal $y(n)$:

$$y(n) = s(n) + \xi(n), \quad n = 1, 2, \ldots, N \tag{5}$$

where $s(n)$ is the noise-free signal and $\xi(n)$ denotes the additive white Gaussian noise.

The components of introduced model (5) we consider as vectors in $\mathbb{R}^N$, i.e., $\mathbf{y} \in \mathbb{R}^N$, $\mathbf{s} \in \mathbb{R}^N$, and $\boldsymbol{\xi} \in \mathbb{R}^N$. In turn, based on (5), we formulate the general form of the unconstrained optimization problem:

$$\tilde{\mathbf{s}} = \arg\min_{\mathbf{s}}\{F(\mathbf{s}) = \frac{1}{2}||\mathbf{y} - \mathbf{s}||_2^2 + \eta\phi(\mathbf{s})\}, \tag{6}$$

where $\phi(\mathbf{s}) : \mathbb{R}^N \to \mathbb{R}$ is the regularization term, and $\eta > 0$ is the regularization parameter.

Generally, the optimization problem (6) can be formulated as convex or nonconvex. The convex approaches are based on sparsity-promoting convex penalty functions (e.g., ℓ_2 norm), while non-convex approaches are based on non-convex regularizers (e.g., ℓ_p pseudo norm with $p < 1$).

The typical form of regularizer $\phi(\mathbf{s})$, which can be used in the unconstrained optimization problems (6), is ℓ_2 norm of forward difference

$$\phi(\mathbf{s}) = ||\mathbf{D}\mathbf{s}||_2^2, \tag{7}$$

where

$$\mathbf{D} = \begin{bmatrix} -1 & 1 & & \\ & -1 & 1 & \\ & & \ddots & \ddots \\ & & & -1 & 1 \end{bmatrix}. \tag{8}$$

is the first-order difference matrix.

In turn, the norm ℓ_2, in the optimization task (6), can be replaced by ℓ_0 [19]. For this norm, the optimization task (6) is transformed into ℓ_0-norm based optimization task

$$\tilde{\mathbf{s}} = \arg\min_{\mathbf{s}}\{F(\mathbf{s}) = \frac{1}{2}||\mathbf{y} - \mathbf{s}||_2^2 + \eta||\mathbf{s}||_0\}. \tag{9}$$

To find the approximation of solution for the introduced optimization problem (9), the following operator can be applied [10]

$$\tilde{\mathbf{s}} = \text{hard}(\mathbf{s}, \eta) = \begin{cases} \mathbf{s} & |\mathbf{s}| > \eta \\ 0 & |\mathbf{s}| \leq \eta. \end{cases} \tag{10}$$

Equation (10) is called the *hard thresholding operator*. We illustrated the operator in Fig. 1.

Instead of ℓ_0 norm it is possible to use, for example, ℓ_1 norm. This implies the following penalty function $\phi(\mathbf{s}) = ||\mathbf{s}||_1$. Taking into account new penalty term, the optimization problem (6) can be rewritten as

$$\tilde{\mathbf{s}} = \arg\min_{\mathbf{s}}\{F(\mathbf{s}) = \frac{1}{2}||\mathbf{y} - \mathbf{s}||_2^2 + \eta||\mathbf{s}||_1\}. \tag{11}$$

The solution of the problem (11) can be determined by [10]

$$\tilde{\mathbf{s}} = \text{soft}(\mathbf{s},\eta) = \begin{cases} \text{sgn}\,(\mathbf{s})\,(|\mathbf{s}| - \eta) & |\mathbf{s}| > \eta \\ 0 & |\mathbf{s}| \le \eta, \end{cases} \tag{12}$$

where $|| \cdot ||_1$ represents ℓ_0 norm of $\mathbf{s}$.

The function $\text{soft}(\mathbf{s},\eta)$ is called the *soft thresholding operator*. The operator is illustrated in Fig. 2.

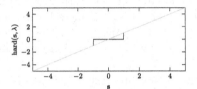

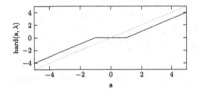

Fig. 1. The hard threshold operator   **Fig. 2.** The soft threshold operator

2.3 VMD-based Signal Denoising with Thresholding Operator

Discussed in the previous section hard and soft thresholding methods can be used to solve the problem of signal denoising based on Variational Mode Decomposition [6]. In the work 3 authors proposed an approach uses hard and soft threshold operators to modified VMD-related IMFs:

$$\tilde{u}_k = \text{hard}(\hat{u}_k, \eta_k) = \begin{cases} \hat{u}_k & |\hat{u}_k| > \eta_k \\ 0 & |\hat{u}_k| \le \eta_k. \end{cases} \tag{13}$$

and

$$\tilde{u}_k = \text{soft}(\hat{u}_k, \eta_k) = \begin{cases} \text{sgn}\,(\hat{u}_k)\,(|\hat{u}_k| - \eta_k) & |\hat{u}_k| > \eta_k \\ 0 & |\hat{u}_k| \le \eta_k, \end{cases} \tag{14}$$

where $k = 1, 2, \ldots, K$ is the k-th IMF. η_k denotes the threshold of the k-th IMF.

The value of threshold η_k for each IMF can be used with the use of universal formula [18]:

$$\eta_k = \sqrt{2E_k \ln N} \tag{15}$$

where N is the number of data points, while E_k is determined with

$$E_1 = \frac{\text{median}(|\hat{u}_1(n)|)}{0.6745} \tag{16a}$$

$$E_k = \frac{E_1^2}{\gamma}\rho^{-k}, \quad k = 2, 3, \ldots, K \tag{16b}$$

where values of ρ and γ have to be estimated. In [5] the authors proposed $\gamma = 0.719$ and $\rho = 2.01$.

In hard thresholding, each value of k-th IMF is compared against the threshold value and lower value is replaced by zero. However, the hard threshold leads to abrupt changes in processed signals and it generates artefacts in the estimated signal.

On the other hand, in soft thresholding values of the IMFs larger than the threshold value are modified by subtraction with the threshold value. The soft threshold operator-based signal denoising tends to smooth the signal.

In the next section, we propose a framework to design denoising algorithms based on penalty functions that promote sparsity more strongly.

3   The Proposed Approach

Both hard and soft thresholding operators introduce to the estimated signal some artifacts that deteriorate estimation outcomes. Hence some scholars propose modified forms of threshold method [8,11]. In the paper, we propose a new approach to VMD-based signal denoising by applying Total Variation Denoising (TVD).

TVD is a widely used regularizer in sparse signal processing and denoising. Total Variation Denoising is defined in terms of a convex optimization problem involving a quadratic data fidelity and a convex regularization term. It is worth mentioning that for 1-D TVD, the exact solution can be found using a fast direct method [3].

To estimate VMD related IMF we form the following unconstrained optimization problem:

$$\tilde{u}_k = \arg\min_{\hat{u}_k}\{F(\hat{u}_k) = \frac{1}{2}||\mathcal{T}(y) - \hat{u}_k||_2^2 + \eta_0||\mathbf{D}\mathcal{T}^{-1}(y)||_1\}. \tag{17}$$

where $k = 1, 2, \ldots, K$, η_0 is the regularization parameter, $\mathbf{D}$ denotes the first-order difference matrix (8), $\mathcal{T}$ stands for the Variation Mode Decomposition algorithm (see Sect. 2.1) and $\mathcal{T}^{-1}$ is the method of signal estimation based on IMFs determined with the use of VMD algorithm:

$$\tilde{y} = \sum_{k=1}^{K} \tilde{u}_k \tag{18}$$

The solution of the optimization problem (18) has the form

$$\tilde{y} = \text{tvd}(\hat{u}_k, \eta_k), \quad k = 1, 2, \ldots, K \tag{19}$$

where tvd is Total Variation Denoising algorithm.

The flow chart of the proposed methods is presented in Fig. 3. In the first step the processed signal $y(t)$ is decomposed into the set of K IMFs (denoted as $u_1, u_2, \ldots, u_K$) with the use of VMD-based algorithm described in Sect. 2.1. Subsequently, each IMFs is modified by TVD method (see Eq. 16), where thresholds are determined by formulas (15–16a,b). The last stage is reconstruction phase, ie. modified IMFs are used to form denoised signal with the use of formula (18).

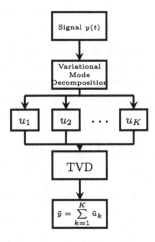

Fig. 3. The flow chart of the proposed algorithm.

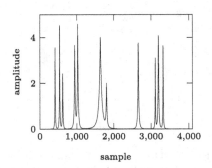

Fig. 4. The clear Bumps signal

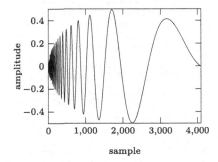

Fig. 5. The clear Doppler signal

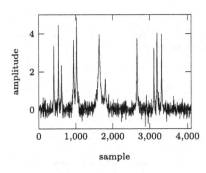

Fig. 6. The Bumps signal, SNR = 10 [dB], H = 0.8

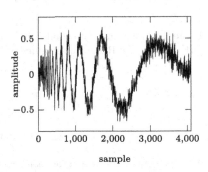

Fig. 7. The Doppler signal, SNR = 10 [dB], H = 0.8

4   Experiments

In order to quantitatively evaluate the denoising performance of the proposed method, some typical signals are generated. In order to simulate real-world signals, we added to the synthetic data noise, then we applied the proposed approach and some reference methods for comparison purpose. The performance of the proposed method, in the presented studies, is evaluated based on the signal-to-noise-ratio

$$SNR = 10\log_{10}\left(\frac{\sum_{n=1}^{N} s(n)}{\sum_{n=1}^{N}\left(s(n) - \tilde{y}(n)\right)^2}\right) \qquad (20)$$

where N is the number of data points, $s(n)$ is the noise-free (original) signal and $\tilde{y}(n)$ stands for denoised signal.

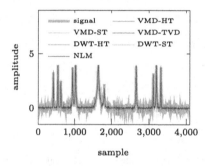

Fig. 8. The results of Bumps signal denoising (SNR = 10 [dB], H = 0.8)

Fig. 9. The results of Doppler signal denoising (SNR = 10 [dB], H = 0.8)

Table 1. The results of signal denoising for Bumps (left) and Doppler (right) signals with different SNR level and values of Hurst exponent H = 0.2

		SNR							SNR			
		0	5	10	15				0	5	10	15
VMD-HT	mean	3,49	7,89	12,36	16,57		VMD-HT	mean	0,20	5,39	10,94	15,22
	std	0,14	0,14	0,11	0,07			std	0,02	0,04	0,08	0,01
VMD-ST	mean	3,63	8,24	12,93	17,35		VMD-ST	mean	0,82	6,16	11,44	16,29
	std	0,05	0,05	0,05	0,05			std	0,01	0,01	0,03	0,02
VMD-TVD	mean	**17,44**	**20,80**	**23,97**	**25,83**		VMD-TVD	mean	**19,57**	**22,75**	**25,98**	**28,98**
	std	0,33	0,23	0,19	0,22			std	0,33	0,22	0,28	0,25
DWT-HT	mean	14,42	17,64	21,22	24,91		DWT-HT	mean	15,83	20,51	24,73	28,14
	std	0,27	0,33	0,30	0,41			std	0,37	0,32	0,26	0,36
DWT-ST	mean	14,47	17,45	19,90	22,51		DWT-ST	mean	15,89	20,63	24,78	27,99
	std	0,19	0,17	0,15	0,14			std	0,28	0,27	0,23	0,16
NLM	mean	9,51	14,22	18,58	23,16		NLM	mean	9,51	14,39	20,30	25,68
	std	0,11	0,10	0,15	0,19			std	0,11	0,10	0,10	0,13

4.1 Synthetic Data Analysis

In this study we test the performance of the proposed method based on simulated signals (i.e. noise-corrupted synthetic signals). We consider Bumps (see Fig. 4) and Doppler (Fig. 5) signals of length 4096 samples. The signals are generated by the *Wavelab* function *MakeSignal* [14].

As we mentioned, all generated signals are corrupted by noise. The signals have different SNR (0, 5, 10, and 15 [dB]) and various short and long dependencies. In order to simulate different short and long dependencies, we added to the synthetic signals fractional Gaussian noise (fGn) with Hurst exponent equal to 0.2, 0.5, and 0.8. In Fig. 6 we display the Bumps signal with SNR = 10 [dB]. In this case, the fractional Gaussian noise has the value of Hurst exponent equals 0.8. On the other hand, Fig. 7 presents the Doppler signal with SNR = 10 [dB]. The value of Hurst exponent is also equaled 0.8.

The proposed approach is compared with two types of VMD-based denoising methods, i.e. hard and soft thresholding. Variational Mode Decomposition-based signal denoising with the use of hard thresholding we named VMD-HT, while the algorithm based on soft thresholding is represented by VMD-ST. In order to perform in-depth analysis, additional reference methods were used. In the presented studies we applied two wavelet-based denoising methods and Non-Local Mean algorithm. In the further part of the works Non-Local Means algorithm, we denoted as NLM. While wavelet-based approaches based on hard and soft thresholding. An approach based on hard thresholding is denoted as DWT-HT, while an algorithm based on soft thresholding is represented by DWT-ST. The proposed approach which is based on VMD and TVD is named VMD-TVD.

In our studies, wavelet-based approaches (DWT-HT and DWT-ST) utilized Matlab's function *wden*. While Non-Local Means algorithm implementation based on [13]. NLM algorithm's parameters, i.e., the *bandwidth*, the *patch half-width*, and the *neighbourhood half-width* are determined empirically [13].

To verify the effectiveness of the proposed method, we carried out 30 trials for noisy synthetic signals (Bumps and Doppler).

4.2 Simulation Experiments

The results of conducted experiments on synthetic data are presented in Tables 1, 2 and 3. We tested Bumps and Doppler signals with SNR varying from 0 to 15 [dB]. While the value of the Hurst exponent was 0.2, 0.5 and 0.8.

As it can be seen, the proposed method (VMD-TVD) is always better than other VMD-related methods (VMD-HT and VMD-ST) and other reference methods (DWT-HT, DWT-ST, and NLM). It is true for both synthetic signals as well as for different SNR and the value of Hurst exponent.

The other successful methods are DWT-HT and DWT-ST denoising algorithms. However, these methods give better results depends on the type of synthetic signal and the value of SNR. For example, results obtained based on DWT-HT algorithm are mainly better for higher SNR, while DWT-ST is more precise for the lower value of SNR.

NLM denoising algorithm gives weaker results than the proposed algorithm (VMD-TVD) and DWT-based approaches (i.e. DWT-HT and DWT-ST). However, these results are better than the results obtained with the use of VMD-HT and VMD-ST algorithms.

The example results for test signals with SNR = 10 [dB] and value of Hurst Exponent equals 0.8 were shown in Fig. 8 (The Bumps signal) and Fig. 9 (The Doppler signal).

Comparing VMD-HT, VMD-ST and VMD-TVD we can see that in almost all cases VMD-ST is better than VMD-ST and VMD-TVD predominate over them. Weak results obtained by the VMD-HT algorithm is caused by discontinuous at the point of the threshold, which leads to additional oscillations. It affects on low SNR of the estimated signal. On the other hand, the soft thresholding avoids additional oscillations however there is a constant deviation between the estimated and true signal.

As we can see that the proposed approach (VMD-TVD) gives always a better result. The algorithm based on the Total Variation Denoising approach has the advantage of recovering the discontinuities in the estimated signal [2]. Moreover, this algorithm can smooth the high oscillation regions in the signal. The basic properties of the TVD method explain the high quality of the estimated signal. It must be added that the Total Variation Denoising has undesirable property. The property is named the staircase effect and it is related to the large constant zones generated by TVD in the signal [12]. This observation allows us to conclude that removing the staircase from the estimated signal can lead us to its better estimation.

Table 2. The results of signal denoising for Bumps (left) and Doppler (right) signals with different SNR level and values of Hurst exponent H = 0.5

		SNR							SNR			
		0	5	10	15				0	5	10	15
VMD-HT	mean	3,79	8,68	13,23	18,09		VMD-HT	mean	0,20	5,48	10,39	16,57
	std	0,13	0,22	0,17	0,17			std	0,00	0,06	0,03	0,14
VMD-ST	mean	3,73	8,98	13,62	18,37		VMD-ST	mean	2,02	6,12	11,23	16,21
	std	0,08	0,08	0,10	0,10			std	0,08	0,03	0,03	0,09
VMD-TVD	mean	**11,98**	**15,59**	**19,69**	**23,50**		VMD-TVD	mean	**13,83**	**17,34**	**20,85**	**24,88**
	std	0,20	0,28	0,19	0,19			std	0,33	0,29	0,26	0,25
DWT-HT	mean	8,68	13,17	17,66	22,32		DWT-HT	mean	9,00	13,93	18,72	23,45
	std	0,26	0,20	0,27	0,26			std	0,22	0,31	0,20	0,27
DWT-ST	mean	8,72	13,06	17,14	21,05		DWT-ST	mean	9,02	13,93	18,73	23,30
	std	0,27	0,16	0,21	0,21			std	0,23	0,31	0,21	0,23
NLM	mean	10,11	14,41	18,81	23,24		NLM	mean	9,04	13,74	18,96	23,62
	std	0,29	0,30	0,27	0,23			std	0,20	0,35	0,18	0,31

Table 3. The results of signal denoising for Bumps (left) and Doppler (right) signals with different SNR level and values of Hurst exponent H = 0.8

		SNR							SNR			
		0	5	10	15				0	5	10	15
VMD-HT	mean	4,02	9,08	13,58	18,10		VMD-HT	mean	0,15	5,49	11,00	16,67
	std	0,32	0,23	0,18	0,15			std	0,06	0,09	0,12	0,17
VMD-ST	mean	5,15	9,30	13,91	18,33		VMD-ST	mean	0,89	6,11	11,33	16,21
	std	0,20	0,16	0,18	0,17			std	0,02	0,04	0,10	0,11
VMD-TVD	mean	**9,60**	**13,83**	**17,77**	**21,38**		VMD-TVD	mean	**10,06**	**15,27**	**18,96**	**22,97**
	std	0,29	0,32	0,31	0,19			std	0,37	0,31	0,30	0,27
DWT-HT	mean	4,08	8,99	13,92	18,78		DWT-HT	mean	4,09	9,10	14,09	19,14
	std	0,26	0,23	0,21	0,27			std	0,27	0,31	0,26	0,28
DWT-ST	mean	5,81	10,59	15,33	19,83		DWT-ST	mean	5,88	10,90	15,92	20,72
	std	0,23	0,19	0,20	0,21			std	0,22	0,18	0,26	0,26
NLM	mean	0,87	13,40	17,80	22,34		NLM	mean	7,38	12,13	17,01	21,57
	std	0,31	0,34	0,34	0,27			std	0,26	0,24	0,28	0,24

5 Conclusions

In the paper, we proposed the novel approach to signal denoising based on Variational Mode Decomposition and Total Variation Denoising. The method is an alternative to existing approaches based on hard and soft thresholding. As the results of our experiments shown, the presented algorithm gives a great improvement in comparison to both VMD-HT and VMD-ST as well as widely used approaches like wavelet-based (DWT-HT and DWT-ST) and Non-Local Means.

Total Variation Denoising approach is an example of the application of sparse signal processing. TVD is defined in terms of a convex optimization problem involving a quadratic data fidelity and a convex regularization term. Since it is possible to formulate different optimization problems with various regularization terms it is possible to obtain other algorithms of signal enhancement. It means that the presented in the paper preliminary results show the promising direction of future research in which sparsity-based methods can improve the performance of VMD-based denoising methods.

References

1. Alfaouri, M., Daqrouq, K.: ECG signal denoising by wavelet transform thresholding. Am. J. Appl. Sci. **5**(3), 276–281 (2008)
2. Caselles, V., Chambolle, A., Novaga, M.: The discontinuity set of solutions of the TV denoising problem and some extensions. Multiscale Model. Simul. **6**, 879–894 (2007)
3. Condat, L.: A direct algorithm for 1-D total variation denoising. IEEE Signal Process. Lett. **20**(11), 1054–1057 (2013)
4. Dragomiretskiy, K., Zosso, D.: Variational mode decomposition. IEEE Trans. Signal Process. **62**(3), 531–544 (2013)

5. Flandrin, P., Gonçalves, P., Rilling, G.: EMD equivalent filter banks, from interpretation to applications. In: Hilbert-Huang transform and its applications, pp. 57–74. World Scientific (2005)
6. Hu, H., Zhang, L., Yan, H., Bai, Y., Wang, P.: Denoising and baseline drift removal method of MEMS hydrophone signal based on VMD and wavelet threshold processing. IEEE Access **7**, 59913–59922 (2019)
7. Huang, N.-E., et al.: The empirical mode decomposition and the hilbert spectrum for nonlinear and non-stationary time series analysis. Proc. Roy. Soc. London. Series A: Math. Phys. Eng. Sci. **454**(1971), 903–995 (1998)
8. Jiang, W.-W., et al.: Research on spectrum signal denoising based on improved threshold with lifting wavelet. J. Electron. Meas. Instrum. **28**(12), 1363–1368 (2014)
9. Lahmiri, S.: Comparing variational and empirical mode decomposition in forecasting day-ahead energy prices. IEEE Syst. J. **11**(3), 1907–1910 (2015)
10. Rish, I., Grabarnik, G.: Sparse Modeling: Theory, Algorithms, and Applications. CRC Press, Boca Raton (2014)
11. Ruizhen, Z., Guoxiang, S., Hong, W.: Better threshold estimation of wavelet coefficients for improving denoising. J. Northwest. Polytechnical Univ. **19**(4), 628–632 (2001)
12. Selesnick, I. W., Chen, P. Y.: Total variation denoising with overlapping group sparsity. In: 2013 IEEE International Conference on Acoustics, Speech and Signal Processing, pp. 5696–5700 (2013)
13. Tracey, B.-H., Miller, E.-L.: Nonlocal means denoising of ECG signals. IEEE Trans. Biomed. Eng. **59**(9), 2383–2386 (2012)
14. WaveLab850. https://statweb.stanford.edu/~wavelab/
15. Wu, Y., Shen, Ch., Cao, H., Che, X.: Improved morphological filter based on variational mode decomposition for MEMS gyroscope de-noising. Micromachines **9**(5), 246 (2018)
16. Xiao, Q., Li, J., Sun, J., Feng, H., Jin, S.: Natural-gas pipeline leak location using variational mode decomposition analysis and cross-time-frequency spectrum. Measurement **124**, 163–172 (2018)
17. Xue, Y.-J., Cao, J.-X., Wang, D.-X., Du, H.-K., Yao, Y.: Application of the variational-mode decomposition for seismic time-frequency analysis. IEEE J. Sel. Top. Appl. Earth Observations Remote Sens. **9**(8), 3821–3831 (2016)
18. Yang, G., Liu, Y., Wang, Y., Zhu, Z.: EMD interval thresholding denoising based on similarity measure to select relevant modes. Signal Process. **109**, 95–109 (2015)
19. Yang, Z., Ling, B.W.-K., Bingham, Ch.: Joint empirical mode decomposition and sparse binary programming for underlying trend extraction. IEEE Trans. Instrum. Measur. **62**(10), 2673–2682 (2013)

Machine Learning

Colored Petri Net Modeling for Prediction Processes in Machine Learning

Ibuki Kawamitsu and Morikazu Nakamura[(⊠)] [iD]

University of the Ryukyus, 1 Senbaru, Nishihara, Okinawa 903-0213, Japan
{e155745,morikazu}@ie.u-ryukyu.ac.jp

Abstract. This paper presents a modeling approach to visualize the prediction process of machine learning by colored Petri nets. The learning phase constructs a Petri net structure by feature selection, and extract arc functions specifying transition firing. Our method allows us to understand the relation between the causes and the prediction results quantitatively. As an application example, we apply our approach to the prediction of the number of daily customers and their orders in a Japanese restaurant. The experimental evaluation with real data shows the usefulness to understand the effect level of each feature for the prediction.

Keywords: Petri net · Colored Petri net · Machine learning · Regression · Classification

1 Introduction

With the advancement of IoT technologies, machine learning became more important in our smart life. In machine learning applications, understanding the actual process, which features contributed to the prediction, or how they did generate results, is crucially essential. To intuitively understand the steps of the machine learning processes, this paper investigates modeling the prediction process with Petri nets, a mathematical modeling language, where they can represent the relation between causes and results.

The authors of [1,2] emphasized the availability of a Petri net-based tool to visualize the complicated processes in automatic generation of the task schedule and the process mining. The research papers [3] and [4] reported modeling of the database transactions with their DB-net, an extension of the colored Petri net. They modeled the query processing with the firing of transitions to visualize complicated processes step by step.

The objective of our research is to visualize the machine learning process by modeling feature data, preprocessing, training the preprocessed data, and the evaluation with colored Petri nets. In our framework, during the feature selection phase, we acquire the net structure and extract the arc functions to calculate the colored token distribution. In [5], a class of high level Petri nets are proposed, with which they model probability and evidence propagation for

© Springer Nature Switzerland AG 2020
H. Fujita et al. (Eds.): IEA/AIE 2020, LNAI 12144, pp. 663–674, 2020.
https://doi.org/10.1007/978-3-030-55789-8_57

the Horn abduction. They compared their model with Bayesian network models and showed the usefulness of the Petri net models.

This paper presents colored Petri net models for classification and regression in machine learning applications. For the former, we treat the iris classification problem, well-known as a tutorial example problem, and visualize the prediction process with Petri nets and explain it by T-invariant, an essential characteristic of Petri net models. For the latter, we employ the Boston house price prediction problem; it is also well-known as an example problem for the regression. Our Petri net model expresses the prediction phases for the regression.

As a case study, we apply our method to the customer prediction of a restaurant located in Mie, Japan. With 50 features composed of real data such as weather conditions, the number of customers in the previous day, the number of tourists, we generate colored Petri net models for prediction processes of the number of customers and orders.

2   Petri Nets

This section explains some basic definition of the ordinary Petri net and a colored Petri net.

2.1   Petri Nets

Petri net $PN = (P, T, F, M_0)$ is composed of a directed bipartite graph $N = (P, T, F)$ and the initial marking M_0. The bipartite graph is represented by two vertex sets, P: a set of places and T: a set of transitions, and arc function F connecting these both vertices.

For mathematical convenience, matrix representation for the connection is often used. *Pre* and *Post* represent the incident matrix of size $P \times T$ to specify connection from places to transitions and from transition to places, respectively. That is, $Pre(p, t) = F((p, t))$ and $Post(p, t) = F((t, p))$, respectively. Tokens are located on each place $p \in P$, where the token distribution on P shows a status of the modeled system and is denoted by vector M of size $|P|$. The initial marking M_0 is the vector for the initial status of the system. Therefore, Petri net $PN = (N, M_0)$ models the network structure and the initial state of the modeled system.

Transition $t \in T$ is *enabled* when $M(p) \geq Pre(t, p)$ (or $F(p, t)$), $\forall p \in {}^\bullet t$ and t can fire only when it is enabled, where ${}^\bullet t$ and $t^\bullet$ denote the set of input places and the set of output places of t, respectively. An occurrence of firing corresponds to an event on the system. The firing of transition t removes $Pre(p, t)$ number of tokens from each $p \in {}^\bullet t$ and adds $Post(p, t)$ tokens to $p \in t^\bullet$. Therefore, the firing makes the token distribution changed, and it means the status change of the system by the event.

The status change can be represented by the following mathematical forms.

$$M' = M + Pre \cdot X + Post \cdot X \tag{1}$$
$$= M + A \cdot X, \tag{2}$$

where X is a vector of size T showing how many times each t fires during the status change and it is called *the firing count vector*.

Moreover, X is called as *T-semiflow* when the following condition holds

$$A \cdot X = \mathbf{0}. \tag{3}$$

Note that considering (2), the firing occurrence of specified times in *T-semiflow* for each transition returns to the original state M.

Figure 1 shows an example of the status change with a simple Petri net. The Petri net in Fig. 1 includes four places shown by circles and two transitions by boxes.

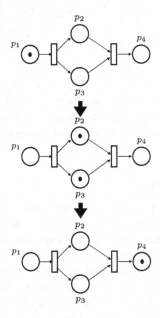

Fig. 1. Example of Petri net and state transition

The Petri net can be represented mathematically as follows:

$$PN = (N, M_0) \tag{4}$$
$$N = (P, T, Pre, Post) \tag{5}$$
$$P = \{p_1, p_2, p_3, p_4\} \tag{6}$$
$$T = \{t_1, t_2\} \tag{7}$$
$$Pre^\top = \begin{bmatrix} 1 & 0 & 0 & 0 \\ 0 & 1 & 1 & 0 \end{bmatrix} \tag{8}$$
$$Post^\top = \begin{bmatrix} 0 & 1 & 1 & 0 \\ 0 & 0 & 0 & 1 \end{bmatrix} \tag{9}$$

$$M_0^\top = (1,0,0,0) \tag{10}$$

where $\top$ expresses the transpose of a matrix.

In Fig. 1, the Petri net at the top is the initial status and then changes into the middle one by t_1's firing. Finally, the bottom one has obtained by t_2's firing at the middle status. Each status, marking, can be calculated by (2) as follows:

$$M_1 = M_0 + \begin{bmatrix} -1 & 0 \\ 1 & -1 \\ 1 & -1 \\ 0 & 1 \end{bmatrix} \cdot \begin{bmatrix} 1 \\ 0 \end{bmatrix} = \begin{bmatrix} 0 \\ 1 \\ 1 \\ 0 \end{bmatrix} \tag{11}$$

$$M_2 = M_1 + \begin{bmatrix} -1 & 0 \\ 1 & -1 \\ 1 & -1 \\ 0 & 1 \end{bmatrix} \cdot \begin{bmatrix} 0 \\ 1 \end{bmatrix} = \begin{bmatrix} 0 \\ 0 \\ 0 \\ 1 \end{bmatrix} \tag{12}$$

As shown above, Petri nets are useful not only for the visualization of the system but also for mathematical formulation. However, it is not enough when it comes to machine learning process modeling since we need to treat feature values of several types and to express the prediction process by the firing of transitions. Thus, we introduce a colored Petri net, an extension of the ordinary Petri net for the modeling.

2.2   Colored Petri Nets

Colored Petri nets are an extension of Petri nets, where we can assign values (colors) to tokens, guard functions to transitions for their firing conditions, arc functions to output arcs from transitions to calculate output colored tokens. The extension makes modeling power of Petri nets enriched.

The extended Petri net is denoted by $CPN = (\Sigma, P, T, F, V, C, G, E, M_0)$, where P, T, F represent a set of places, transitions, arcs, respectively. Σ shows a set of colors, $C : P \to \Sigma$ is a color function for places, that is, only tokens with the colors specified by $C(p)$ are allowed in place p. V is a set of arc variables, where $\mathrm{Type}(v) \subseteq \Sigma$ for $v \in V$. E denotes an arc function where $E((t,p))$ includes arc variables in V and should return a multiple set on $C(p)$ for each output place p of t when we set some value to each of the variables. G represents a guard function where G includes arc variables in V and should return a boolean value. A marking M in colored Petri nets is a mapping of multiple sets on $C(p)$ to each place, and M_0 is the initial marking.

Transition t is enabled at marking M with binding b when for each input place p of t,

$$M(p) \geq E((p,t))(b) \tag{13}$$

where $E((p,t))(b)$ denotes the result of the arc function for arc (p,t) when we bind $b(v)$ for each variable v in $E((p,t))$. At marking M, marking M' is generated by firing t with binding b:

$$M'(p) = M(p) - E((p,t))(b) + E((t,p))(b), \forall p \in P. \tag{14}$$

Colored Petri nets are very powerful in a sense that we can easily model complicated systems. Note that colored Petri net models can be converted into a corresponding ordinary Petri net.

3 Petri Net Model for Machine Learning

This section presents the classification and regression examples of the machine learning application. We use a well-known example, the iris classification for the classification, and the house price prediction in Boston city for the regression.

3.1 Classification

We show the process of the classification based on our colored Petri net modeling with a well-known classification example, iris classification, provided in the machine learning library for the Python language, called scikit-lean [6]. Places in our model are used as input buffers for feature data in which the input transitions generate feature data by their firing. Their output transitions correspond to some preprocessing or predicting. The output places of the predicting transition will have the answers.

In this example, we assume the feature selection process determined four feature data, *sepal length*, *sepal width*, *petal length*, and *petal width*, and for each we prepare an input place. These places connect to the predicting transitions, from which the token with the answer are generated and added into the output places. The answer labels for the classification problem are *setosa*, *versicolor*, and *virginica*.

Figure 2 shows the Petri net model generated by the SNAKES library [7]. The transitions, *sepal width input*, *petal length input*, *sepal length input*, and *petal width input* are always enabled and then fire to add tokens with feature data. The training process calculates the weight parameters for the prediction. We use here the naive Bayes algorithm and generates the function for the posterior probability.

The followings are the formal description of the colored Petri net, $CPN = (\Sigma, P, T, F, V, C, G, E, M_0)$, but we omit F since it is evident from Fig. 2.

$$\Sigma = \mathcal{R}^+, \mathcal{R}^+ \text{is the set of real positive numbers} \tag{15}$$

$$P = \{\text{sepal width, petal length, sepal length, petal width,}$$
$$\text{versicolor, virginica, setosa}\} \tag{16}$$

$$T = \{\text{sepal width input, petal length input, sepal length input,}$$
$$\text{petal width input, versilolor predictor, virginica predictor,}$$
$$\text{setosa predictor, versicolor output, verginica output, setosa output}\} \tag{17}$$

$$V = \{x_{11}, x_{12}, x_{13}, x_{21}, x_{22}, x_{23}, x_{31}, x_{32}, x_{33}, x_{41}, x_{42}, x_{43}, \tag{18}$$
$$\text{versicolorproba, virginicaproba, setosaproba}\} \tag{19}$$

$$C : P \to \Sigma \tag{20}$$
$$M_0 = (\phi, \phi, ..., \phi) \tag{21}$$

Arc functions correspond to the posterior probability calculation, where we employ naive Bayes algorithm as shown below.

$$E(a_1) = Pr(Y = \text{setosa}|X_1) = \frac{Pr(Y = \text{setosa})Pr(X_1|Y = \text{setosa})}{P(X_1)} \tag{22}$$

$$E(a_2) = Pr(Y = \text{versicolor}|X_2) = \frac{Pr(Y = \text{versicolor})Pr(X_2|y = \text{versicolor})}{P(X_2)} \tag{23}$$

$$E(a_3) = Pr(Y = \text{virginica}|X_3) = \frac{Pr(Y = \text{virginica})Pr(X_3|Y = \text{virginica})}{P(X_3)} \tag{24}$$

$$Y \in \{\text{versicolor}, \text{virginica}, \text{setosa}\} \tag{25}$$
$$X_1 = (x_{11}, x_{21}, x_{31}, x_{41}) \tag{26}$$
$$X_2 = (x_{12}, x_{22}, x_{32}, x_{42}) \tag{27}$$
$$X_3 = (x_{13}, x_{23}, x_{33}, x_{43}) \tag{28}$$
$$a_1 = (\text{setosa predictor}, \text{setosa}) \tag{29}$$
$$a_2 = (\text{versicolor predictor}, \text{versicolor}) \tag{30}$$
$$a_3 = (\text{virginica predictor}, \text{virginica}) \tag{31}$$

There are conflicts between the predicting transitions, *versicolor predictor*, *virginica predictor*, and *setosa predictor* for the tokens in each of their input places. The conflicts are solved by comparing the probability calculated by the arc functions (22, 23, 24), that is, the transition with the highest probability can fire.

In this example, for some input data, the answer to the prediction is *setosa* since the probability 0.919 is larger than the others, 0.042 and 0.039. The prediction fires and then add the token to the output place. Finally, the output transition *setosa output* fires as the result of the prediction.

The set of the fired transitions from the input to the output corresponds to a T-semiflow of the underlying Petri net. The followings are all the T-semiflow of the Petri nets, where each element of the vectors corresponds from the left element to *sepal width input*, *petal length input*, *sepal length input*, *petal width input*, *versicolor predictor*, *virginica predictor*, *setosa predictor*, *versicolor output*, *virginica output*, and *setosa output*. Each of them shows the prediction process for a label.

$$X_1^\top = (1, 1, 1, 1, 1, 0, 0, 1, 0, 0)$$
$$X_2^\top = (1, 1, 1, 1, 0, 1, 0, 0, 1, 0)$$
$$X_3^\top = (1, 1, 1, 1, 0, 0, 1, 0, 0, 1)$$

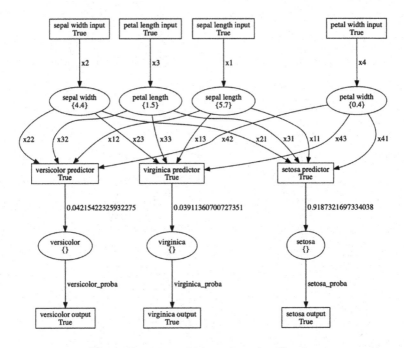

Fig. 2. Petri net of iris variety classification

3.2   Regression Problem

This section presents Petri net models for regression problems in machine learning. We use here an example, the house price prediction in Boston, provided in the *scikit-learn* [6]. Similar to the classification problem, the places for the feature data input, the predicting transitions, the answer place, and the answer output transition are generated.

Figure 3 denotes the Petri net model for the regression problem, where there are thirteen input transitions and the input buffer places, only one answer output place, and the output transition.

$$\Sigma = \mathcal{R}^+, \mathcal{R}^+ \text{is a set of real positive numbers} \tag{32}$$

$$P = \{\text{AGE}, \text{PTRATIO}, \text{RM}, \text{CHAS}, \text{RAD}, \text{TAX}, \text{B}, \text{CRIM}, \text{ZN},$$
$$\text{DIS}, \text{NOX}, \text{INDUS}, \text{LSTAT}, \text{PRICE}\} \tag{33}$$

$$T = \{\text{i-t1}, \text{i-t2}, \text{i-t3}, ..., \text{i-t13}, \text{t(predictor)}\} \tag{34}$$

$$V = \{\text{input1}, \text{input2}, ..., \text{input13}, x_1, x_2, x_3, ..., x_{13}\} \tag{35}$$

$$C : P \to \Sigma \tag{36}$$

$$M_0 = (\phi, \phi, ..., \phi) \tag{37}$$

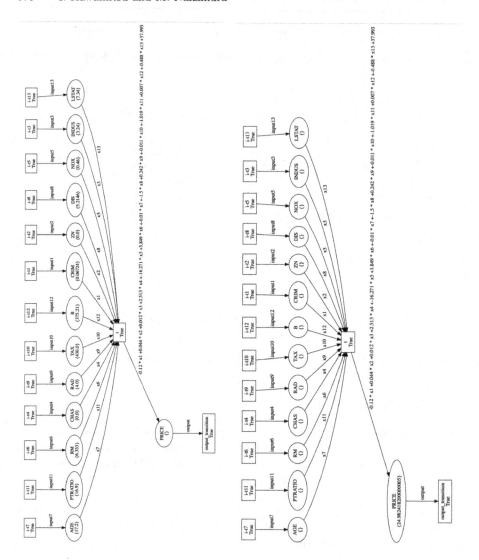

Fig. 3. Petri net showing the house price estimation process (Before the transition fires)

Fig. 4. Petri net showing the house price estimation process (After the transition fires)

Arc functions correspond to the prediction algorithm, where we employ the linear regression algorithm as shown below.

$$E((t(predictor), \text{PRICE})) = w^\top \cdot X \qquad (38)$$

$$X^\top = (x_1, x_2, x_3, ..., x_{13}) \qquad (39)$$

$$w^\top = (w_1, w_2, ..., w_{13}) \qquad (40)$$

The weight vector w is obtained by training for the linear regression.

The weighs on the arcs between the predicting transition t show the learned values by the training, and the equation on the arc between the predicting transition and the output place corresponds to the regression. We can calculate the predicted value by this equation. Figure 4 shows the process of the prediction on the Petri net model. The answer of the prediction is 25$.

4   Case Study: Customer Prediction at a Japanese Restaurant

In this section, we present a case study to apply our approach to a real regression problem. The problem is to predict the number of customers, the number of orders from feature data such as the past data, the weather conditions, the event schedule, the estimation of the tourists. For this purpose, the machine learning approach is quite suitable. In fact, a Japanese restaurant, Ebiya has good practice for machine learning-based customer prediction.

The following table shows a set of feature data used for our machine learning (Table 1).

Table 1. Available feature data

Dayofweek	Day of Week
Salestotal	Total sales per day (JPY)
Custnum	Total customers per day (people)
Tourists	Total tourists per day in Area (people)
Weather_today	Weather condition of the day
Rainfall_today	Rain amount per day (mm)
Rainfall_daytime	Rain amount for daytime (mm)

The categorical data such as *dayofweek, weather_today* are converted into binary data, therefore, the number of feature data becomes 33 in total.

4.1   Hillclimbing Based Feature Selection

In our Petri net-based approach, the feature selection leads to obtaining a Petri net structure. We employ a hillclimbing based algorithm for this purpose. We first start randomly selected feature data set, including 50% of the available feature data. Based on the accuracy value of the prediction evaluation, we perform hillclimbing based searching. The neighborhood is defined as one feature difference against the current set.

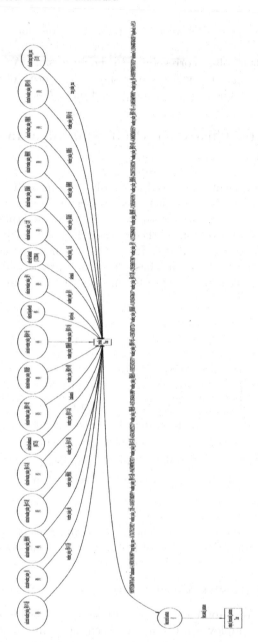

Fig. 5. Petri net model for customer prediction process

4.2   Petri Net Model for Customer Prediction

We generate a Petri net model by using the feature selection algorithm explained above with the Ridge Regression and showed it in Fig. 5. A Python program draws the Petri net model for the customer prediction process with a Petri net library SNAKES. The eighteen left side places correspond to the input feature data. The total number of available feature data is 33; therefore, the data was reduced by the feature selection algorithm.

The output arc from the predicting transition was labeled with the linear function including the generated coefficients by the machine learning. The firing of the predicting transition calculates the predicted number of customers and adds to the output place as a color token with the value. The transition of the Petri net model in Figure 5 is fired and then generates tokens with value.

The prediction for the number of customers is 553, while the answer was 561 for some day. Moreover, the experimental evaluation shows more than 95% accuracy of the prediction. By observing the Petri net model, we can understand intuitively which feature contributes to the number of customers.

5   Concluding Remarks

This paper proposed a Petri net approach to model the machine learning process. Colored Petri nets allow us to represent feature data as color tokens, and the trained model function as the arc function. By the feature selection algorithm with a machine learning algorithm obtained the Petri net structure and the function for the prediction. Thanks to Petri nets' graphical presentation, the machine learning process can be visualized. We can see which feature data contribute to the predicted value.

For a case study, we applied our method to a Japanese-style restaurant to predict the number of customers and orders. With real data provided from an authentic restaurant, the prediction accuracy was more than 95%, and we confirmed its capability for intuitive understanding. Our method is implemented and deployed on a cloud such as Microsoft Azure [8]. The prediction and generation of the Petri net model can be performed on Azure Function, Azure [8], FaaS (Function as a Service).

The examples shown in this paper are so simple, only tree structures with depth one of the relation between the causes and the results. However, more complicated prediction trees are also available and they should be effective to understand the machine learning processes for complicated cases.

Weka [9] is a well-known tool as a workbench for the machine learning, in which we can easily access to various technique in the machine learning. Our method can be led to useful tools for the machine learning process by introducing various algorithms for preprocessing for feature extracting and machine learning.

Acknowledge. To perform our research, Japanese Restaurant *Ebiya* located in Ise-city, Mie, Japan, provided real data. We sincerely thank them for their contribution.

References

1. Duque, D.A., Prietob, F.A., Hoyos, J.G.: Trajectory generation for robotic assembly operations using learning by demonstration. Robot. Comput. Integr. Manuf. **57**, 292–302 (2019). https://doi.org/10.1016/j.rcim.2018.12.007

2. Theis, J., Darabi, H.: behavioral petri net mining and automated analysis for human-computer interaction recommendations in multi-application environments. Proc. ACM Hum. Comput. Interact. **3**(EICS) (2019). https://doi.org/10.1145/3331155. Article No. 13

3. Montali, M., Rivkin, A.: DB-Nets: on the marriage of colored petri nets and relational databases. In: Koutny, M., Kleijn, J., Penczek, W. (eds.) Transactions on Petri Nets and Other Models of Concurrency XII. LNCS, vol. 10470, pp. 91–118. Springer, Heidelberg (2017). https://doi.org/10.1007/978-3-662-55862-1_5

4. Montali, M., Rivkin, A.: From DB-nets to coloured petri nets with priorities. In: Donatelli, S., Haar, S. (eds.) PETRI NETS 2019. LNCS, vol. 11522, pp. 449–469. Springer, Cham (2019). https://doi.org/10.1007/978-3-030-21571-2_24

5. Lautenbach, K., Philippi, S., Pinl, A.: Bayesian networks and petri nets. In Conference: 9. Fachtagung Entwurf komplexer Automatisierungssysteme At: Braunschweig, Germany (2006)

6. Pedregosa, F., et al.: Scikit-learn: machine learning in Python. J. Mach. Learn. Res. **12**, 2825–2830 (2011)

7. Pommereau, F.: SNAKES: a flexible high-level petri nets library (tool paper). In: Devillers, R., Valmari, A. (eds.) PETRI NETS 2015. LNCS, vol. 9115, pp. 254–265. Springer, Cham (2015). https://doi.org/10.1007/978-3-319-19488-2_13

8. Microsoft Azure. https://azure.microsoft.com/ja-jp/

9. Hall, M., Frank, E., Holmes, G., Pfahringer, B., Reutemann, P., Witten, I.H.: The WEKA data mining software: an update. ACM SIGKDD Explor. Newsl. **11**(1), 10–18 (2009). https://doi.org/10.1145/1656274.1656278

Enriching the Semantics of Temporal Relations for Temporal Pattern Mining

Ryosuke Matsuo[(✉)], Tomoyoshi Yamazaki, Muneo Kushima, and Kenji Araki

Faculty of Medicine, University of Miyazaki Hospital, Miyazaki, Japan
{ryosuke_matsuo,yama-cp,muneo_kushima,taichan}@med.miyazaki-u.ac.jp

Abstract. This paper proposes a method to enrich the semantics of before/after relations based on the closeness between two events and contexts surrounding two events identified by a key event in a period. The proposed method captures four types of before/after relations: continuous relation, discrete relation, same contextual relation and different contextual relation. We derive five embedding methods from the combination of the four relation types, and apply them to clinical data for the prediction of hospital length of stay where the events are treatments, the key event is surgery and the period is seven days after hospital admission. The experimental results showed that on the whole the embedding method employed all of the four relation types has higher scores of precision, recall, F1 score and accuracy than other embedding methods. This suggests that the potential for the elucidation of the candidates of medically meaningful temporal patterns increases by exploring the temporal patterns generated from the embedding method. A paired t-test indicated that significant differences are partially confirmed for discrete relation and same contextual relation but not confirmed for different contextual relation. We will apply the proposed method to a number of hospitals for performing the further analysis of the four relation types and elucidating medically meaningful temporal patterns.

Keywords: Temporal relations · Temporal pattern mining · Semantics

1 Introduction

In health care, Real-World Data (hereinafter called RWD) is derived from a diverse array of sources, such as electronic health records, medical claims, product and disease registries, laboratory test results and even cutting-edge technology paired with consumer mobile devices. RWD offer new opportunities to generate real-world evidence that can better inform regulatory decisions [3].

Temporal pattern discovery on electronic medical records is useful in understanding the clinical context around a drug and suspected adverse drug reaction combination and the probability of a causal relationship, especially as it relates to confounding by underlying disease [12]. In the assessment of comorbidities in population-based studies, time is a crucial parameter because it permits to identify more complex disease patterns apart from the pairwise disease associations

© Springer Nature Switzerland AG 2020
H. Fujita et al. (Eds.): IEA/AIE 2020, LNAI 12144, pp. 675–685, 2020.
https://doi.org/10.1007/978-3-030-55789-8_58

[4]. The identification of time-related disease associations is effective to predict the disease progression along time and potentially facilitate the early diagnosis of other comorbid diseases [4].

Temporal pattern mining can be classified into time point-based methods and time interval-based methods as well as univariate and multivariate methods [6]. Regarding the time-point based methods, there are three binary operators: before, equals and after for two points in time and temporal concept of order (sequential occurrence of time points) that corresponds to the three operators in univariate and multivariate data [6].

Temporal pattern mining has been applied to prediction and knowledge discovery in medicine [1,4,7,8]. [7] proposed a framework for open-ended pattern discovery in large patient records repositories that summarize and visualize the temporal association between the prescription of a drug and the occurrence of a medical event based on a graphical statistical approach. The graphical overview contrasts the observed and expected number of occurrences of the medical event in different time periods both before and after the prescription of interest. In order to screen for important temporal relationships, the authors introduced a new measure of temporal association, which contrasts the observed-to-expected ratio in a time period immediately after the prescription to the observed-to-expected ratio in a control period 2 years earlier. [1] proposed a pattern-based classification framework for complex multivariate time series data encountered in electronic health record systems. The method is based on temporal abstractions and temporal pattern mining to extract the classification features. The authors presented the Minimal Predictive Temporal Patterns framework to generate a small set of predictive and non-spurious patterns from a large number of temporal patterns. [8] used a temporal pattern mining algorithm to detect Temporal Association Rules (TARs) by identifying the most frequent temporal relationships among the derived basic temporal abstractions. TARs and their recurrence patterns are represented as features in a naive bayes classification model for coronary heart disease diagnosis. [4] employed disease-history vectors as time dimension which are represented by the disease history of individual patients as time sequences of ordered disease diagnoses to investigate the most common comorbidities and their underlying time-dependent characteristics.

The temporal relations of before/after are straightforward but essential between two points in time. The enrichment of the semantics of the before/after relations has been carried out by [2] and [10]. Fuzzy relations on the real line are used to grasp such usual notions as "much before", "closely after", etc. [2]. In TimeML which is a specification language for events and temporal expressions, "immediately before" and "immediately after" are included in the LINK tag called TLINK or a Temporal Link representing the temporal relationship holding between events or between an event and a time [10].

With an investigation of effect of a variety of before/after relations on prediction performance, the possibility to elucidate the candidates of medically effective temporal pattern for a certain purpose regarding prediction could potentially increase. To this end, we enrich the semantics of before/after relations by considering the closeness between two events and contexts surrounding two events identified by a key event in a period. We then capture four types of before/after relations: continuous relation, discrete relation, same contextual relation and different contextual relation. In order to analyze the effectiveness of the four relation types, we embed temporal patterns derived from the combination of the four types of before/after relations into independent variables, and apply five embedding methods to clinical data regarding treatments for the prediction of hospital length of stay.

2 Methods

2.1 Defining Four Types of Before/After Relations

Let $e1$ and $e2$ are two events in before/after relations in an established period such as hospital admission. In order to enrich the semantics of before/after relations, we consider the closeness between two events and contexts surrounding two events identified by a key event in the established period. Subsequently, we capture four types of before/after relations: continuous relation, discrete relation, same contextual relation, and different contextual relation as follows.

1. continuous relation: $e1$ continuously before $e2$, while $e2$ continuously after $e1$.
2. discrete relation: $e1$ discretely before $e2$, while $e2$ discretely after $e1$.
3. same contextual relation: $e1$ before $e2$ in $context1$, while $e2$ after $e1$ in $context1$.
4. different contextual relation: $e1$ before $e2$ in $context1$, while $e2$ after $e1$ in $context2$.

With regard to the closeness between two events, continuous relation is defined as an event $e1_t$ continuously appears before an event $e2_{t+C}$, while an event $e2_{t+C}$ continuously appears after an event $e1_t$ where C is a constant with same granularity of t such as a day, a week or a month. Furthermore, we consider discrete temporal patterns that are indecisive as meaningful patterns rather than continuous temporal patterns. Discrete relation is defined as an event $e1_t$ discretely appears before an event $e2_{t+C+\alpha}$, while an event $e2_{t+C+\alpha}$ discretely appears after an event $e1_t$ where α is a parameter to adjust the distance between $e1$ and $e2$ for identifying $e1$ and $e2$ as a temporal pattern.

We assume that a key event e^{key} in the established period such as surgery in hospital admission enriches the relations of before/after as it is possible to determine contexts surrounding two events $e1$ and $e2$. Therefore, we consider two relations: same contextual relation and different contextual relation by exploiting a key event e^{key}. Same contextual relation obey the following temporal order

$$e^{key} < e1 \cap e^{key} < e2 \tag{1}$$

or the following temporal order

$$e1 < e^{key} \cap e2 < e^{key} \tag{2}$$

In contrast, different contextual relation obey the following temporal order

$$e1 < e^{key} \cap e^{key} < e2 \tag{3}$$

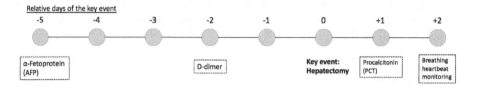

Fig. 1. The example of events and a key event in RWD in medicine.

Figure 1 shows the example of events and a key event in RWD in medicine where the events are treatments and the key event is surgery in seven days after hospital admission. The disease of the corresponding patient is classified into the category of malignant neoplasm of liver and intrahepatic bile ducts (including secondary). There are four treatments: "α-Fetoprotein (AFP)", "D-dimer", "Procalcitonin (PCT)" and "Breathing heartbeat monitoring" and "Hepatectomy" as surgery with the relative days of "Hepatectomy" in Fig. 1. The granularity of t is a day as well as the parameter C. "Procalcitonin (PCT)" and "Breathing heartbeat monitoring" are identified as continuous relation and same contextual relation after "Hepatectomy". "α-Fetoprotein (AFP)" and "D-dimer" are identified as discrete relation where α is 2 and same contextual relation before "Hepatectomy". In case of "D-dimer" and "Procalcitonin (PCT)", they are identified as different contextual relation based on "Hepatectomy".

We finally embed temporal patterns derived from the combination of the four types of before/after relations into independent variables for a prediction task.

2.2   Experimental Setting

The purpose of experiments is to analyze the effectiveness of the four types of before/after relations by comparing the prediction performance of embedding methods derived from the combination of the four types of before/after relations. We denote *cont* as continuous relation, *disc* as discrete relation, *c1* as same contextual relation and *c2* as different contextual relation. Five embedding methods: *cont*, *cont-c1*, *cont-disc*, *cont-disc-c1* and *cont-disc-c1-c2* are then derived from the combination of the four relation types. We use the term *nopattern* as the baseline, which considers not patterns but only an event as independent variables. Note that *nopattern* is initially embedded into the five embedding methods. *c1* and *c2* are considered for only data where a key event falls within the established period. *cont-c1-c2* is not included in the five embedding methods because continuous relation can not obey the temporal order of different contextual relation.

We use the Diagnosis Procedure Combination (hereinafter called DPC) data that include discharge abstract and administrative claims data [13]. The number of patients of DPC data is 44,257 in University of Miyazaki Hospital who discharged from Apr 1, 2014 to Mar 31, 2018. The codes of DPC classify patients based on the combination of diagnosis and procedures conducted within the hospitalization [5]. The DPC codes are structured with the following components: (i) 18 Major Diagnosis Categories and 2,927 diagnostic groups, as of 2014 (ii) type of admission, (iii) patient backgrounds (age, Japan Coma Scale, birth weight for neonates, etc.), (iv) surgical procedures, (v) adjuvant therapies, and (vi) comorbidities/complications [14]. We use the DPC code to identify a cluster of DPC data (hereinafter called DPC cluster). We select ten DPC clusters whose corresponding patients undergoing surgery frequently appeared in the DPC data.

The dependent variable is Hospital Length of Stay (hereinafter called HLOS). We detect the outliers by using the interquartile range (IQR), and removed patients whose HLOS is less than $Q1 - 1.5 \times IQR$ or greater than $Q3 + 1.5 \times IQR$. HLOS is then discretized by using its mean in each DPC cluster where the variable is 1 if HLOS is greater than its mean, otherwise 0. The independent variables are treatments at the admission of patients especially until seven days after the admission. We use only the treatments where the cost is greater than or equal to 1,000 yen. TFIDF [11] is employed to transform the independent variables into a computable form. We regard surgery as the key event for identifying same contextual relation and different contextual relation. We select the highest cost's surgery when the multiple surgery are recorded in the DPC data at the admission of a patient. If the multiple surgery's costs are the same, we select the surgery executed closely to the admission date of the patient.

In this experiments, five classifiers: logistic regression, naive bayes, neural network, random forest and SVM (linear) are employed after feature selection based on L2 regularization. The parameters of the classifiers and feature selection are default. We use 5-fold cross validation with four evaluation measures: precision, recall, F1 score and accuracy for evaluating prediction performance. The granularity of t regarding the parameter C is a day. The maximum of the parameter α is 6 in accordance with the range of the DPC data at the admission, while the minimum is 1. In this experiments, if the parameter α is greater than 1, temporal patterns fall within a range of 1 to α are identified as discrete relation. The same context of same contextual relation in this experiments is before or after surgery identified by the two types of temporal order. These experiments are executed by using Scikit-learn [9].

3   Experimental Results

Table 1 shows the statistics of patients who have surgery until seven days after the admission in each of the DPC clusters. The two types of before/after relations: same contextual relation $c1$ and different contextual relation $c2$ were considered for around 75 % or more of patients in each DPC cluster.

Table 2, 3, 4 and 5 give the summary of the four evaluation measures derived from the five embedding methods and their baseline for the five classifiers. In

Table 1. The statistics of patients having surgery.

DPC cluster	The number of patients	The number of patients having surgery	The ratio of the number of patients having surgery to the number of patients
Malignant neoplasm of liver and intrahepatic bile ducts (including secondary)	832	636	0.764
Femoral head avascular necrosis and coxarthropathy (including deformans)	571	530	0.928
Chronic suppurative tympanitis and middle ear cholesteatoma	565	530	0.938
Bladder tumor	545	480	0.881
Tachyarrhythmia	541	467	0.863
Malignant tumor of the stomach	535	421	0.787
Macular or posterior degeneration	490	467	0.953
Lung cancer	438	328	0.749
Head and neck cancer	392	314	0.801
External wounds of elbow or knee including sports injuries etc.	389	361	0.928

addition, Table 6 gives the average of the four evaluation measures by the embedding methods and their baseline for each classifier. Note that the indicated evaluation measures of embedding methods including discrete relation *disc* were the highest scores when varying the value of the parameter α.

Overall, *cont-disc-c1-c2* brought about high prediction performance measured by precision, recall, F1 score and accuracy for the five classifiers compared with other embedding methods and their baseline. In comparison with the precision and recall, the precision was better than the recall as shown in Table 6. This suggests that the embedding method could precisely predict long HLOS's patients but could not comprehensively predict long HLOS's patients. On the whole, *cont-disc-c1-c2* with NN as the classifier can have high evaluation measures in the experiments. Specifically, the accuracy was highest when using *cont-disc-c1-c2* with NN as the classifier among the four evaluation measures.

Table 2. The summary of precision.

Method	LR	NB	NN	RM	SVM
nopattern	0.618	0.695	0.762	0.678	0.675
cont	0.616	0.668	0.786	0.692	0.682
cont-c1	0.685	0.708	0.804	0.728	0.784
cont-disc	0.628	0.742	0.8	0.752	0.813
cont-disc-c1	0.663	0.789	0.817	0.757	0.826
cont-disc-c1-c2	0.669	0.774	0.82	0.754	0.832

Table 3. The summary of recall.

Method	LR	NB	NN	RM	SVM
nopattern	0.464	0.331	0.697	0.642	0.499
cont	0.533	0.443	0.761	0.639	0.603
cont-c1	0.617	0.534	0.775	0.647	0.676
cont-disc	0.554	0.482	0.781	0.683	0.646
cont-disc-c1	0.623	0.562	0.803	0.682	0.722
cont-disc-c1-c2	0.636	0.608	0.811	0.686	0.729

Table 4. The summary of F1 score.

Method	LR	NB	NN	RM	SVM
nopattern	0.487	0.392	0.702	0.642	0.518
cont	0.537	0.492	0.75	0.637	0.614
cont-c1	0.62	0.573	0.768	0.652	0.694
cont-disc	0.554	0.53	0.762	0.682	0.655
cont-disc-c1	0.621	0.595	0.785	0.68	0.732
cont-disc-c1-c2	0.625	0.632	0.792	0.682	0.739

Table 5. The summary of accuracy.

Method	LR	NB	NN	RM	SVM
nopattern	0.714	0.726	0.798	0.746	0.746
cont	0.721	0.745	0.819	0.751	0.78
cont-c1	0.744	0.769	0.832	0.772	0.807
cont-disc	0.724	0.753	0.826	0.781	0.807
cont-disc-c1	0.743	0.774	0.839	0.786	0.827
cont-disc-c1-c2	0.744	0.783	0.846	0.784	0.835

Table 6. The average of the four evaluation measures in the five classifiers.

Method	Precision	Recall	F1 score	Accuracy
nopattern	0.686	0.526	0.548	0.746
cont	0.689	0.596	0.606	0.763
*cont-c*1	0.742	0.65	0.661	0.785
cont-disc	0.747	0.63	0.636	0.778
*cont-disc-c*1	**0.77**	0.679	0.682	0.794
*cont-disc-c*1*-c*2	**0.77**	**0.694**	**0.694**	**0.798**

4 Discussion

A paired t-test is additionally carried out for analyzing the effectiveness of the four types of before/after relations by using following six pairs of the embedding methods: "*nopattern* and *cont*", "*nopattern* and *cont-disc*", "*cont* and *cont-disc*", "*cont* and *cont-c*1", "*cont-disc* and *cont-disc-c*1", and "*cont-disc-c*1 and *cont-disc-c*1*-c*2". We use the experimental results where the evaluation measure was accuracy for the paired t-test.

As shown in Table 7 and Table 8 which are the paired t-test results of "*nopattern* and *cont*" and "*nopattern* and *cont-disc*", considering not only continuous relation *cont* but also discrete relation *disc* was effective for HLOS prediction compared with *nopattern* that was considered only a treatment without its patterns as the independent variables. Moreover, as there were significant differences for three classifiers in case of the pair of "*cont* and *cont-disc*" represented in Table 9, the effectiveness of discrete relation was partially confirmed in the experimental setting. As for two contextual relations, Table 10 and Table 11 showed significant differences of same contextual relation *c*1 for three classifiers and four classifiers, respectively. Therefore, same contextual relation was also partially effective for the experiments. By contrast, as shown in Table 12, there was a significant difference of different contextual relation *c*2 for only one classifier in the experiments. For this reason, the effectiveness of different contextual relation *c*2 was not confirmed in the experimental setting.

Concerning the number of independent variables, basically, the higher the parameter α, the larger the number of independent variables for the embedding methods including discrete relation *disc*. Figure 2 shows the ratio of the values of the parameter α to all of the selected parameter α that gave highest scores

Table 7. The paired t-test results between *nopattern* and *cont*.

Method	LR	NB	NN	RM	SVM
nopattern	0.714	0.726	0.798	0.746	0.746
cont	0.721	0.745	**0.819 (p < 0.1)**	0.751	**0.78 (p < 0.1)**

Table 8. The paired t-test results between *nopattern* and *cont-disc*.

Method	LR	NB	NN	RM	SVM
nopattern	0.714	0.726	0.798	0.746	0.746
cont-disc	0.724	**0.753** (p $<$ 0.1)	**0.826** (p $<$ 0.1)	**0.781** (p $<$ 0.05)	**0.807** (p $<$ 0.01)

Table 9. The paired t-test results between *cont* and *cont-disc*.

Method	LR	NB	NN	RM	SVM
cont	0.721	0.745	0.819	0.751	0.78
cont-disc	0.724	**0.753** (p $<$ 0.1)	0.826	**0.781** (p $<$ 0.01)	**0.807** (p $<$ 0.01)

Table 10. The paired t-test results between *cont* and *cont-c1*.

Method	LR	NB	NN	RM	SVM
cont	0.721	0.745	0.819	0.751	0.78
cont-c1	**0.744** (p $<$ 0.1)	**0.769** (p $<$ 0.05)	0.832	0.772	**0.807** (p $<$ 0.05)

Table 11. The paired t-test results between *cont-disc* and *cont-disc-c1*.

Method	LR	NB	NN	RM	SVM
cont-disc	0.724	0.753	0.826	0.781	0.807
cont-disc-c1	**0.743** (p $<$ 0.1)	**0.774** (p $<$ 0.05)	**0.839** (p $<$ 0.1)	0.786	**0.827** (p $<$ 0.1)

Table 12. The paired t-test results between *cont-disc-c1* and *cont-disc-c1-c2*.

Method	LR	NB	NN	RM	SVM
cont-disc-c1	0.743	0.774	0.839	0.786	0.827
cont-disc-c1-c2	0.744	**0.783** (p $<$ 0.05)	0.846	0.784	0.835

for each evaluation measure when varying the value of the parameter α for each classifier and each DPC cluster. We found that the ratio of the high values of the parameter α such as 5 and 6 to all of the selected parameter α are low. Hence, the increase of the number of independent variables did not affect for prediction performance in the experiments. This suggests that adjusting the distance between two events to identify them as a temporal pattern by the parameter α is significant for enhancing prediction performance in the experimental setting.

In regard to the limitation of the proposed method, as only one key event is considered for determining same contextual relation and different contextual relation, multiple factors should be employed for the enrichment of the semantics of before/after relations according to the contexts surrounding two events.

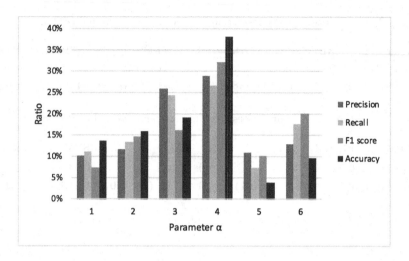

Fig. 2. The ratio of the values of the parameter α to all of the selected parameter α for *cont-disc-c1-c2*.

5 Conclusions

This paper considered four types of before/after relations by enriching the semantics of before/after relations based on the closeness between two events and the contexts surrounding two events identified by a key event in a period. In order to analyze the effectiveness of the four types of before/after relations, five embedding methods by combining the four types of before/after relations were applied to clinical data for HLOS prediction where the events were treatments, the key event was surgery and the period was seven days after hospital admission. The experimental results showed that on the whole the embedding method that employed all of the four types of before/after relations has high scores of the four evaluation measures compared with other embedding methods. This suggests that the potential to elucidate the candidates of medically meaningful temporal patterns increases by exploring the temporal patterns derived from the method embedded all of the four types of before/after relations. A paired t-test indicated that significant differences are partially confirmed for discrete relation and same contextual relation. However, the significant difference of different contextual relation was not revealed. As to the analysis of the number of independent variables regarding embedding methods including discrete relation, a large number of independent variables was not relative to the high scores of the four evaluation measures. We will apply the proposed method to a number of hospitals for both the further analysis of the four types of before/after relations and the elucidation of medically meaningful temporal patterns.

Acknowledgment. This work was supported by JSPS KAKENHI Grant Number JP18K09948.

References

1. Batal, I., Valizadegan, H., Cooper, G.F., Hauskrecht, M.: A temporal pattern mining approach for classifying electronic health record data. ACM Trans. Intell. Syst. Technol. **4**(4), 1–22 (2013)
2. Dubois, D., Prade, H.: Processing fuzzy temporal knowledge. IEEE Trans. Syst. Man Cybern. **19**(4), 729–744 (1989)
3. FDA: Statement from FDA Commissioner Scott Gottlieb, M.D., on FDA's new strategic framework to advance use of real-world evidence to support development of drugs and biologics (2018). https://www.fda.gov/news-events/press-announcements/statement-fda-commissioner-scott-gottlieb-md-fdas-new-strategic-framework-advance-use-real-world
4. Giannoula, A., Gutierrez-Sacristán, A., Bravo, Á., Sanz, F., Furlong, L.I.: Identifying temporal patterns in patient disease trajectories using dynamic time warping: a population-based study. Sci. Rep. **8**(1), 4216 (2018)
5. Matsuda, S., Fujimori, K., Kuwabara, K., Ishikawa, K.B., Fushimi, K.: Diagnosis procedure combination as an infrastructure for the clinical study. Asian Pac. J. Dis. Manag. **5**(4), 81–87 (2011)
6. Mörchen, F.: Unsupervised pattern mining from symbolic temporal data. ACM SIGKDD Explor. Newslett. **9**(1), 41–55 (2007)
7. Norén, G.N., Hopstadius, J., Bate, A., Star, K., Edwards, I.R.: Temporal pattern discovery in longitudinal electronic patient records. Data Min. Knowl. Disc. **20**(3), 361–387 (2010). https://doi.org/10.1007/s10618-009-0152-3
8. Orphanou, K., Dagliati, A., Sacchi, L., Stassopoulou, A., Keravnou, E., Bellazzi, R.: Incorporating repeating temporal association rules in Naïve Bayes classifiers for coronary heart disease diagnosis. J. Biomed. Inform. **81**, 74–82 (2018)
9. Pedregosa, F., et al.: Scikit-learn: machine learning in Python. J. Mach. Learn. Res. **12**, 2825–2830 (2011)
10. Pustejovsky, J., et al.: TimeML: robust specification of event and temporal expressions in text. New Dir. Quest. Answ. **3**, 28–34 (2003)
11. Salton, G., Buckley, C.: Term-weighting approaches in automatic text retrieval. Inf. Process. Manag. **24**(5), 513–523 (1988)
12. Star, K., Watson, S., Sandberg, L., Johansson, J., Edwards, I.R.: Longitudinal medical records as a complement to routine drug safety signal analysis. Pharmacoepidemiol. Drug Saf. **24**(5), 486–494 (2015)
13. Yasunaga, H., Matsui, H., Horiguchi, H., Fushimi, K., Matsuda, S.: Application of the diagnosis procedure combination (DPC) data to clinical studies. J. UOEH **36**(3), 191–197 (2014)
14. Yasunaga, H., Matsui, H., Horiguchi, H., Fushimi, K., Matsuda, S.: Clinical epidemiology and health services research using the diagnosis procedure combination database in Japan. Asian Pac. J. Dis. Manag. **7**(1–2), 19–24 (2015)

Integer Weighted Regression Tsetlin Machines

Kuruge Darshana Abeyrathna[(⊠)], Ole-Christoffer Granmo,
and Morten Goodwin

Centre for Artificial Intelligence Research, University of Agder, Grimstad, Norway
{darshana.abeyrathna,ole.granmo,morten.goodwin}@uia.no

Abstract. The Regression Tsetlin Machine (RTM) addresses the lack
of interpretability impeding state-of-the-art nonlinear regression models.
It does this by using conjunctive clauses in propositional logic to capture the underlying non-linear frequent patterns in the data. These, in
turn, are combined into a continuous output through summation, akin
to a linear regression function, however, with non-linear components and
binary weights. However, the resolution of the RTM output is proportional to the number of clauses employed. This means that computation
cost increases with resolution. To address this problem, we here introduce
integer weighted RTM clauses. Our integer weighted clause is a compact
representation of multiple clauses that capture the same sub-pattern—
w repeating clauses are turned into one, with an integer weight w. This
reduces computation cost w times, and increases interpretability through
a sparser representation. We introduce a novel learning scheme, based
on so-called stochastic searching on the line. We evaluate the potential
of the integer weighted RTM empirically using two artificial datasets.
The results show that the integer weighted RTM is able to acquire on
par or better accuracy using significantly less computational resources
compared to regular RTM and an RTM with real-valued weights.

Keywords: Tsetlin machines · Regression tsetlin machines · Weighted
tsetlin machines · Interpretable machine learning · Stochastic searching
on the line

1 Introduction

The recently introduced Regression Tsetlin Machine (RTM) [1,2] is a propositional logic based approach to interpretable non-linear regression, founded on
the Tsetlin Machine (TM) [3]. Being based on disjunctive normal form, like
Karnaugh maps, the TM can map an exponential number of input feature value
combinations to an appropriate output [4]. Recent research reports several distinct TM properties. The clauses that a TM produces have an interpretable
form (e.g., **if** X **satisfies** condition A **and not** condition B **then** Y = 1), similar to the branches in a decision tree [5]. For small-scale pattern recognition

© Springer Nature Switzerland AG 2020
H. Fujita et al. (Eds.): IEA/AIE 2020, LNAI 12144, pp. 686–694, 2020.
https://doi.org/10.1007/978-3-030-55789-8_59

problems, up to three orders of magnitude lower energy consumption and inference time has been reported, compared to neural networks alike [6]. Like neural networks, the TM can be used in convolution, providing competitive memory usage, computation speed, and accuracy on MNIST, F-MNIST and K-MNIST, in comparison with simple 4-layer CNNs, K-Nereast Neighbors, SVMs, Random Forests, Gradient Boosting, BinaryConnect, Logistic Circuits, and ResNet [7]. By introducing clause weights that allow one clause to represent multiple, it has been demonstrated that the number of clauses can be reduced up to 50×, without loss of accuracy, leading to more compact clause sets [4]. Finally, hyperparameter search can be simplified with multi-granular clauses, eliminating the pattern specificity parameter [8].

Paper Contributions: In the RTM, regression resolution is proportional to the number of conjunctive clauses employed. In other words, computation cost and memory usage grows proportionally with resolution. Building upon the Weighted TM (WTM) by Phoulady et al. [4], this paper introduces weights to the RTM scheme. However, while the WTM uses real-valued weights for classification, we instead propose a novel scheme based on *integer* weights, targeting *regression*. In brief, we use the stochastic searching on the line approach pioneered by Oommen in 1997 [9] to eliminate multiplication from the weight updating. In addition to the computational benefits this entails, we also argue that integer weighted clauses are more interpretable than real-valued ones because they can be seen as multiple copies of the same clause. Finally, our scheme does not introduce additional hyper-parameters, whereas the WTM relies on weight learning speed.

Paper Organization: The remainder of the paper is organized as follows. In Sect. 2, the basics of RTMs are provided. Then, in Sect. 3, the SPL problem and its solution are explained. The main contribution of this paper, the integer weighting scheme for the RTM, is presented in detail in Sect. 4 and evaluated empirically using two artificial datasets in Sect. 5. We conclude our work in Sect. 6.

2   The Regression Tsetlin Machine (RTM)

The RTM performs regression based on formulas in propositional logic. In all brevity, the input to an RTM is a vector $\mathbf{X}$ of o propositional variables x_k, $\mathbf{X} \in \{0, 1\}^o$. These are further augmented with their negated counterparts $\bar{x}_k = 1 - x_k$ to form a vector of literals: $\mathbf{L} = [x_1, \ldots, x_o, \bar{x}_1, \ldots, \bar{x}_o] = [l_1, \ldots, l_{2o}]$. In contrast to a regular TM, the output of an RTM is real-valued, normalized to the domain $y \in [0, 1]$.

Regression Function: The regression function of an RTM is simply a linear summation of products, where the products are built from the literals:

$$y = \frac{1}{T} \sum_{j=1}^{m} \prod_{k \in I_j} l_k. \tag{1}$$

Above, the index j refers to one particular product of literals, defined by the subset I_j of literal indexes. If we e.g. have two propositional variables x_1 and x_2, the literal index sets $I_1 = \{1, 4\}$ and $I_2 = \{2, 3\}$ define the function: $y = \frac{1}{T}(x_1 \bar{x}_2 + \bar{x}_1 x_2)$. The user set parameter T decides the resolution of the regression function. Notice that each product in the summation either evaluates to 0 or 1. This means that a larger T requires more literal products to reach a particular value y. Thus, increasing T makes the regression function increasingly fine-grained. In the following, we will formulate and refer to the products as *conjunctive clauses*, as is typical for the regular TM. The value c_j of each product is then a conjunction of literals:

$$c_j = \prod_{k \in I_j} l_k = \bigwedge_{k \in I_j} l_k. \tag{2}$$

Finally, note that the number of conjunctive clauses m in the regression function also is a user set parameter, which decides the expression power of the RTM.

Tsetlin Automata Teams: The composition of each clause is decided by a team of Tsetlin Automata (TAs) [10]. There are $2 \times o$ number of TAs per clause j. Each represents a particular literal k and decides whether to *include* or *exclude* that literal in the clause. The decision depends on the current state of the TA, denoted $a_{j,k} \in \{1, \ldots, 2N\}$. States from 1 to N produce an *exclude* action and states from $N + 1$ to $2N$ produce an *include* action. Accordingly, the set of indexes I_j can be defined as $I_j = \{k | a_{j,k} > N, 1 \le k \le 2o\}$. The states of all of the TAs are organized as an $m \times 2o$ matrix $\mathbf{A}$: $\mathbf{A} = (a_{j,k}) \in \{1, \ldots, 2N\}^{m \times 2o}$ where m is the number of clauses.

Learning Procedure: Learning in RTM is done through an online reinforcement scheme that updates the state matrix $\mathbf{A}$ by processing one training example $(\hat{X}_i, \hat{y}_i)$ at a time, as detailed below.

The RTM employs two kinds of feedback, Type I and Type II, further defined below. Type I feedback triggers TA state changes that eventually make a clause output 1 for the given training example $\hat{X}_i$. Conversely, Type II feedback triggers state changes that eventually make the clause output 0. Thus, overall, regression error can be systematically reduced by carefully distributing Type I and Type II feedback:

$$Feedback = \begin{cases} \text{Type I,} & \text{if } y < \hat{y}_i, \\ \text{Type II,} & \text{if } y > \hat{y}_i. \end{cases} \tag{3}$$

In effect, the number of clauses that evaluates to 1 is increased when the predicted output is less than the target output ($y < \hat{y}_i$) by providing Type I feedback. Type II feedback, on the other hand, is applied to decrease the number of clauses that evaluates to 1 when the predicted output is higher than the target output ($y > \hat{y}_i$).

Activation Probability: Feedback is handed out stochastically to regulate learning. The feedback probability p_j is proportional to the absolute error of the prediction, $|y - \hat{y}_i|$. Clauses activated for feedback are the stored in the matrix $\mathbf{P} = (p_j) \in \{0, 1\}^m$.

Type I Feedback: Type I feedback subdivides into Type Ia and Type Ib. Type Ia reinforces *include* actions of TAs whose corresponding literal value is 1, however, only when the clause output is 1. The probability of kth TA of the jth clause receives Type Ia feedback $r_{j,k}$ is $\frac{s-1}{s}$, where s ($s \geq 1$) is a user set parameter. Type Ib combats over-fitting by reinforcing *exclude* actions of TAs when the corresponding literal is 0 or when the clause output is 0. The probability of kth TA of the jth clause receives Type Ib feedback $q_{j,k}$ is $\frac{1}{s}$.

Using the complete set of conditions, the TAs selected for Type Ia feedback are singled out by the indexes $I^{\text{Ia}} = \{(j, k)|l_k = 1 \wedge c_j = 1 \wedge p_j = 1 \wedge r_{j,k} = 1\}$. Similarly, TAs selected for Type Ib are $I^{\text{Ib}} = \{(j, k)| (l_k = 0 \vee c_j = 0) \wedge p_j = 1 \wedge q_{j,k} = 1\}$.

Once the TAs have been targeted for Type Ia and Type Ib feedback, their states are updated. Available updating operators are $\oplus$ and $\ominus$, where $\oplus$ adds 1 to the current state while $\ominus$ subtracts 1. Thus, before a new learning iterations starts, the states in the matrix $\mathbf{A}$ are updated as follows: $\mathbf{A} \leftarrow (\mathbf{A} \oplus I^{\text{Ia}}) \ominus I^{\text{Ib}}$.

Type II Feedback: Type II feedback eventually changes the output of a clause from 1 to 0, for a specific input $\hat{X}_i$. This is achieved simply by including one or more of the literals that take the value 0 for $\hat{X}_i$. The indexes of TAs selected for Type II can thus be singled out as $I^{\text{II}} = \{(j, k)|l_k = 0 \wedge c_j = 1 \wedge p_j = 1\}$. Accordingly, the states of the TAs are updated as follows: $\mathbf{A} \leftarrow \mathbf{A} \oplus I^{\text{II}}$.

3   Stochastic Searching on the Line

Stochastic searching on the line, also referred to as stochastic point location (SPL) was pioneered by Oommen in 1997 [9]. SPL is a fundamental optimization problem where one tries to locate an unknown unique point within a given interval. The only available information for the Learning Mechanism (LM) is the possibly faulty feedback provided by the attached environment (E). According to the feedback, LM moves right or left from its current location in a discretized solution space.

The task at hand is to determine the optimal value λ^* of a variable λ, assuming that the environment is informative. That is, that it provides the correct direction of λ^* with probability $p > 0.5$. In SPL, λ is assume to be any number in the interval $[0, 1]$. The SPL scheme of Oommen discretizes the solution space by subdividing the unit interval into N steps, $\{0, 1/N, 2/N, ..., (N-1)/N, 1\}$. Hence, N defines the resolution of the learning scheme.

The current guess, $\lambda(n)$, is updated according to the feedback from the environment as follows:

$$\lambda(n+1) = \begin{cases} \lambda(n) + 1/N, & \text{if } E(n) = 1 \text{ and } 0 \leqslant \lambda(n) < 1, \\ \lambda(n) - 1/N, & \text{if } E(n) = 0 \text{ and } 0 < \lambda(n) \leqslant 1, \\ \lambda(n), & \text{Otherwise} . \end{cases} \quad (4)$$

The feedback $E(n) = 1$ is the environment suggestion to increase the value of λ and $E(n) = 0$ is the environment suggestion to decrease the value of λ. Asymptotically, the learning mechanics is able to find a value arbitrarily close to λ^* when $N \to \infty$ and $n \to \infty$.

4   Regression Tsetlin Machine with Weighted Clauses

We now introduce clauses with integer weights to provide a more compact representation of the regression function. The regression function for the integer weighted RTM attaches a weight w_j to each clause output c_j, $j = 1, ..., m$. Consequently, the regression output can be computed according to Eq. 5:

$$y = \frac{1}{T} \sum_{j=1}^{m} w_j \prod_{k \in I_j} l_k. \quad (5)$$

Weight Learning: Our approach to learning the weight of each clause is similar to SPL. However, the solution space of each weight is $[0, \infty]$, while the resolution of the learning scheme is $N = 1$. The weight attached to a clause is updated when the clause receives Type Ia feedback or Type II feedback. The weight updating procedure is summarized in Algorithm 1. Here, $w_j(n)$ is the weight of clause j at the nth training round.

Algorithm 1: Round n updating of clause weights
Initialization (round 0): $w_j(0) \leftarrow 0, j = 1, \ldots, m$
Initialization (round n): y is calculated according to Eq. 5.
for $j = 1, ..., m$ **do**
  if $y(n) < \hat{y}_i(n) \wedge c_j(n) = 1 \wedge p_j(n) = 1$ **then**
    $w_j(n+1) \leftarrow w_j(n) + N$
  else if $y(n) > \hat{y}_i(n) \wedge c_j(n) = 1 \wedge p_j(n) = 1 \wedge w_j(n) > 0$ **then**
    $w_j(n+1) \leftarrow w_j(n) - N$
  else
    $w_j(n+1) \leftarrow w_j(n)$
  end if
end for
Return $w_j(n+1), j = 1, \ldots, m$

Note that since weights in this study can take any value higher than or equal to 0, an unwanted clause can be turned off by setting its weight to 0. Further, sub-patterns that have a large impact on the calculation of y can be represented with a correspondingly larger weight.

5 Empirical Evaluation

In this section, we study the behavior of the RTM with integer weighting (RTM-IW) using two artificial datasets similar to the datasets presented in [1], in comparison with a standard RTM and a real-value weighted RTM (RTM-RW). We use Mean Absolute Error (MAE) to measure performance.

Artificial Datasets: Dataset I contains 3-bit feature input. The output, in turn, is 100 times larger than the decimal value of the binary input (e.g., the input $[0, 1, 0]$ produces the output 200). The training set consists of 8000 samples while the testing set consists of 2000 samples, both without noise. Dataset II contains the same data as Dataset I, except that the output of the training data is perturbed to introduce noise. Each input feature has been generated independently with equal probability of taking either the value 0 or 1, producing a uniform distribution of bit values.

Results and Discussion: The pattern distribution of the artificial data was analyzed in the original RTM study. As discussed, there are eight unique sub-patterns. The RTM is able to capture the complete set of sub-patterns utilizing no more than three types of clauses, i.e., $(1 * *)$, $(* 1 *)$, $(* * 1)$[1]. However, to produce the correct output, some clauses must be duplicated multiple times, depending on the input pattern. For instance, each dataset requires *seven* clauses to represent the three different patterns it contains, namely, $(4 \times (1 * *), 2 \times (* 1 *), 1 \times (* * 1))$[2]. So, with e.g. the input $[1, 0, 1]$, four clauses which represent the pattern $(1 * *)$ and one clause which represents the pattern $(* * 1)$ activate to correctly output 500 (after normalization).

Notably, it turns out that the RTM-IW requires even fewer clauses to capture the sub-patterns in the above data, as outlined in Table 1. Instead of having multiple clauses to represent one sub-pattern, RTM-IW utilizes merely one clause with the correct weight to do the same job. The advantage of the proposed integer weighting scheme is thus apparent. It learns the correct weight of each clause, so that it achieves an MAE of zero. Further, it is possible to ignore redundant clauses simply by giving them the weight zero. For the present dataset, for instance, decreasing m while keeping the same resolution, $T = 7$, does not impede accuracy. The RTM-RW on the other hand struggles to find the correct weights, and fails to minimize MAE. Here, the real valued weights were updated

---

[1] Here, $*$ means an input feature that can take an arbitrary value, either 0 or 1.

[2] In this expression, *"four* clauses to represent the pattern $(1 * *)$" is written as "4 $\times (1 * *)$".

Table 1. Behavior comparison of different RTM schemes on Dataset III.

	m	T	Pattern	I_j	$\bar{I}_j$	No. of clauses required	w_j	Training MAE	Testing MAE
RTM	7	7	$(1 * *)$	$\{1\}$	$\{\ \}$	4	–	0	0
			$(* 1 *)$	$\{2\}$	$\{\ \}$	2	–		
			$(* * 1)$	$\{3\}$	$\{\ \}$	1	–		
RTM-IW	3	7	$(1 * *)$	$\{1\}$	$\{\ \}$	1	4	0	0
			$(* 1 *)$	$\{2\}$	$\{\ \}$	1	2		
			$(* * 1)$	$\{3\}$	$\{\ \}$	1	1		
RTM-RW	3	7	$(1 * *)$	$\{1\}$	$\{\ \}$	1	3.987	1.857	1.799
			$(* 1 *)$	$\{2\}$	$\{\ \}$	1	2.027		
			$(* * 1)$	$\{3\}$	$\{\ \}$	1	0.971		

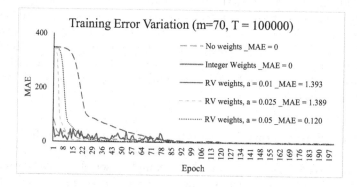

Fig. 1. The training error variation per training epoch for different RTM schemes.

with a learning rate of $\alpha = 0.01$, determined using a binary hyper-parameter search.

Figure 1 casts further light on learning behaviour by reporting training error per epoch for the three different RTM schemes with $m = 70$ and $T = 100000$. As seen, both RTM and RTM-IW obtain relatively low MAE after just one training epoch, eventually reaching MAE zero (training MAE at end of training are given in the legend of each graph). RTM-RW, on the other hand, starts off with a much higher MAE, which is drastically decreasing over a few epochs, however, fails to reach MAE 0 after becoming asymptotically stable.

We also studied the effect of T on performance with noise free data by varying T, while fixing the number of clauses m. For instance, RTM was able to reach a training MAE of 1.9 and a testing error of 2.1 with $m = T = 300$ [1]. For the same dataset, RTM-IW can reach a training error of 0.19 and a testing error of 1.87 with $m = 200$ and $T = 2000$. Further, for $m = 200$ and $T = 20\,000$, training error drops to 0.027 and testing error drops to 0.027. Finally, by increasing T to $200\,000$ training error falls to 0.0003 while testing error stabilises at 0.0002.

Table 2. Training and testing MAE after 200 training epochs by various methods with different m and T.

Model	RTM				RTM-RW				RTM-IW			
MAE	Training		Testing		Training		Testing		Training		Testing	
Dataset	I	II	I	II	I	II	I	II	I	II	I	II
m 7	0.000	7.400	0.000	5.000	2.230	7.702	2.217	5.955	1.172	8.019	1.171	6.236
20	14.600	13.800	14.200	14.500	1.023	7.863	1.036	6.007	0.487	9.844	0.493	8.499
70	0.000	6.600	0.000	4.200	0.292	7.365	0.295	5.735	0.189	7.602	0.189	5.532
300	1.900	5.800	2.100	3.300	0.104	5.800	0.106	2.226	0.078	5.685	0.078	2.234
700	1.000	5.900	1.000	3.400	0.013	5.551	0.013	1.968	0.044	5.532	0.044	2.149
2000	1.000	5.600	1.200	1.900	0.012	5.731	0.012	2.520	0.003	5.373	0.003	1.280
5000	0.900	5.500	1.000	2.700	0.010	5.635	0.010	2.252	0.001	5.412	0.001	1.501

To further compare the performance of RTM-IW with RTM and RTM-RW, each approach was evaluated using a wide rage of m and T settings. Representative training and testing MAE for both datasets are summarized in Table 2. Here, the number of clauses used with each dataset is also given. The T for the original RTM is equal to the number of clauses, while for the RTM with weights T is simply 100 times that number.

As seen, the training and testing MAE reach zero when the RTM operates with noise free data when $m = 7$. However, MAE approaches zero with RTM-IW and RTM-RW when increasing number of clauses m.

For noisy data, the minimum training MAE achieved by RTM is 5.500, obtained with $m = 5000$ clauses. The RTM-IW, on the other hand, obtains a lower MAE of 5.373 with less than half of the clauses ($m = 2000$). The accuracy of RTM-IW in comparison with RTM-RW is less clear, with quite similar MAE for noisy data. The average testing MAE across both the datasets, however, reveals that the average MAE of RTM-IW is lower than that of the RTM-RW (2.101 vs 2.168).

6   Conclusion

In this paper, we presented a new weighting scheme for the Regression Tsetlin Machine (RTM), RTM with Integer Weights (RTM-IW). The weights attached to the clauses helps the RTM represent sub-patterns in a more compact way. Since the weights are integer, interpretability is improved through a more compact representation of the clause set. We also presented a new weight learning scheme based on stochastic searching on the line, integrated with the Type I and Type II feedback of the RTM. The RTM-IW obtains on par or better accuracy with fewer number of clauses compared to RTM without weights. It also performs competitively in comparison with an alternative RTM with real-valued weights.

References

1. Abeyrathna, K.D., Granmo, O.-C., Jiao, L., Goodwin, M.: The Regression tsetlin machine: a tsetlin machine for continuous output problems. In: Moura Oliveira, P., Novais, P., Reis, L.P. (eds.) EPIA 2019. LNCS (LNAI), vol. 11805, pp. 268–280. Springer, Cham (2019). https://doi.org/10.1007/978-3-030-30244-3_23
2. Abeyrathna, K.D., Granmo, O.-C., Zhang, X., Jiao, L., Goodwin, M.: The regression Tsetlin machine: a novel approach to interpretable nonlinear regression. Philos. Trans. R. Soc. A **378**, 20190165 (2019)
3. Granmo, O.-C.: The tsetlin machine - a game theoretic bandit driven approach to optimal pattern recognition with propositional logic. arXiv:1804.01508
4. Phoulady, A., Granmo, O.-C., Gorji, S.R., Phoulady, H.A.: The weighted tsetlin machine: compressed representations with clause weighting. In: Ninth International Workshop on Statistical Relational AI (StarAI 2020) (2020)
5. Berge, G.T., Granmo, O.-C., Tveit, T.O., Goodwin, M., Jiao, L., Matheussen, B.V.: Using the Tsetlin Machine to learn human-interpretable rules for high-accuracy text categorization with medical applications. IEEE Access **7**, 115134–115146 (2019)
6. Wheeldon, A., Shafik, R., Yakovlev, A., Edwards, J., Haddadi, I., Granmo, O.-C.: Tsetlin machine: a new paradigm for pervasive AI. In: Proceedings of the SCONA Workshop at Design, Automation and Test in Europe (DATE) (2020)
7. Granmo, O.-C., Glimsdal, S., Jiao, L., Goodwin, M., Omlin, C. W., Berge, G.T.: The convolutional tsetlin machine. arXiv preprint:1905.09688 (2019)
8. Rahimi Gorji, S., Granmo, O.-C., Phoulady, A., Goodwin, M.: A tsetlin machine with multigranular clauses. In: Bramer, M., Petridis, M. (eds.) SGAI 2019. LNCS (LNAI), vol. 11927, pp. 146–151. Springer, Cham (2019). https://doi.org/10.1007/978-3-030-34885-4_11
9. Oommen, B.J.: Stochastic searching on the line and its applications to parameter learning in nonlinear optimization. IEEE Trans. Syst. Man Cybern. Part B (Cybern.) **27**(4), 733–739 (1997)
10. Tsetlin, M.L.: On behaviour of finite automata in random medium. Avtomat. i Telemekh **22**(10), 1345–1354 (1961)

Increasing the Inference and Learning Speed of Tsetlin Machines with Clause Indexing

Saeed Rahimi Gorji[1]([✉])[iD], Ole-Christoffer Granmo[1][iD], Sondre Glimsdal[1], Jonathan Edwards[2], and Morten Goodwin[1]

[1] Centre for Artificial Intelligence Research, University of Agder, Grimstad, Norway
{saeed.r.gorji,ole.granmo,sondre.glimsdal,morten.goodwin}@uia.no
[2] Temporal Computing, Newcastle, UK
jonny@temporalcomputing.com

Abstract. The Tsetlin Machine (TM) is a machine learning algorithm founded on the classical Tsetlin Automaton (TA) and game theory. It further leverages frequent pattern mining and resource allocation principles to extract common patterns in the data, rather than relying on minimizing output error, which is prone to overfitting. Unlike the intertwined nature of pattern representation in neural networks, a TM decomposes problems into self-contained patterns, represented as conjunctive clauses. The clause outputs, in turn, are combined into a classification decision through summation and thresholding, akin to a logistic regression function, however, with binary weights and a unit step output function. In this paper, we exploit this hierarchical structure by introducing a novel algorithm that avoids evaluating the clauses exhaustively. Instead we use a simple look-up table that indexes the clauses on the features that falsify them. In this manner, we can quickly evaluate a large number of clauses through falsification, simply by iterating through the features and using the look-up table to eliminate those clauses that are falsified. The look-up table is further structured so that it facilitates constant time updating, thus supporting use also during learning. We report up to 15 times faster classification and three times faster learning on MNIST and Fashion-MNIST image classification, and IMDb sentiment analysis.

Keywords: Tsetlin machines · Pattern recognition · Propositional logic · Frequent pattern mining · Learning automata · Pattern indexing

1 Introduction

The Tsetlin Machine (TM) is a recent machine learning approach that is based on the Tsetlin Automaton (TA) by M. L. Tsetlin [1], one of the pioneering solutions to the well-known multi-armed bandit problem [2,3] and the first Learning Automaton [4]. It further leverages frequent pattern mining [5] and resource allocation [6] to capture frequent patterns in the data, using a limited number of conjunctive clauses in propositional logic. The clause outputs, in turn,

H. Fujita et al. (Eds.): IEA/AIE 2020, LNAI 12144, pp. 695–708, 2020.
https://doi.org/10.1007/978-3-030-55789-8_60

are combined into a classification decision through summation and thresholding, akin to a logistic regression function, however, with binary weights and a unit step output function [7]. Being based on disjunctive normal form (DNF), like Karnaugh maps, the TM can map an exponential number of input feature value combinations to an appropriate output [8].

Recent Progress on TMs. Recent research reports several distinct TM properties. The clauses that a TM produces have an interpretable form (e.g., **if X satisfies** condition A **and not** condition B **then** Y = 1), similar to the branches in a decision tree [7]. For small-scale pattern recognition problems, up to three orders of magnitude lower energy consumption and inference time has been reported, compared to neural networks alike [9]. The TM structure captures regression problems, comparing favourably with Regression Trees, Random Forests and Support Vector Regression [10]. Like neural networks, the TM can be used in convolution, providing competitive memory usage, computation speed, and accuracy on MNIST, F-MNIST and K-MNIST, in comparison with simple 4-layer CNNs, K-Nereast Neighbors, SVMs, Random Forests, Gradient Boosting, BinaryConnect, Logistic Circuits, and ResNet [11]. By introducing clause weights that allow one clause to represent multiple, it has been demonstrated that the number of clauses can be reduced up to 50×, without loss of accuracy, leading to more compact clause sets [8]. Finally, hyper-parameter search can be simplified with multi-granular clauses, eliminating the pattern specificity parameter [12].

Paper Contributions and Organization. The expression power of a TM is governed by the number of clauses employed. However, adding clauses to increase accuracy comes at the cost of linearly increasing computation time. In this paper, we address this problem by exploiting the hierarchical structure of the TM. We first cover the basics of TMs in Sect. 2. Then, in Sect. 3, we introduce a novel indexing scheme that speeds up clause evaluation. The scheme is based on a simple look-up tables that indexes the clauses on the features that falsify them. We can then quickly evaluate a large number of clauses through falsification, simply by iterating through the features and using the look-up table to eliminate those clauses that are falsified. The look-up table is further structured so that it facilitates constant time updating, thus supporting indexing also during TM learning. In Sect. 4, we report up to 15 times faster classification and three times faster learning on three different datasets: IMDb sentiment analysis, and MNIST and Fashion-MNIST image classification. We conclude the paper in Sect. 5.

2   Tsetlin Machines

The Tsetlin Machine (TM) [13] is a novel machine learning approach that is particularly suited to low power and explainable applications [7,9]. The structure of a TM is similar to the majority of learning classification algorithms, with forward pass inference and backward pass parameter adjustment (Fig. 1). A key property of the approach is that the whole system is Boolean, requiring an encoder and decoder for any other form of data. Input data is defined as a set of feature vectors: $D = \{\mathbf{x}_1, \mathbf{x}_2, \ldots\}$, and the output Y is formulated as a function parameterized by the actions A of a team of TAs: $Y = f(x; A)$.

TA Team. The function f is parameterized by a team of TAs, each deciding the value of a binary parameter $a_k \in A$ (labelling the full collection of parameters A). Each TA has an integer state $t_k \in \{1, \ldots, 2N\}$, with $2N$ being the size of the state space. The action a_k of a TA is decided by its state: $a_k = 1$ if $t_k > N$ else 0.

Classification Function. The function f is formulated as an ensemble of n conjunctive clauses, each clause controlling the inclusion of a feature and (separately) its negation. A clause C_j is defined as:

$$C_j(\mathbf{x}) = \bigwedge_{k=1}^{o} (x_k \vee \neg a_k^j) \wedge (\neg x_k \vee \neg a_{o+k}^j). \tag{1}$$

Here, x_k is the kth feature of input vector $\mathbf{x} \in D_o$, o is the total number of features and a_k^j is the action of the TA that has been assigned to literal k for clause j. A variety of ensembling methods can be used – most commonly a unit step function, with half the clauses aimed at inhibition:

$$\hat{y} = 1 \quad \text{if} \quad \left(\sum_{j=1}^{n/2} C_j^+(\mathbf{x}) - \sum_{j=1}^{n/2} C_j^-(\mathbf{x}) \right) \geq 0 \quad \text{else} \quad 0. \tag{2}$$

When the problem is multi-classed (one-hot encoded [15]) thresholding is removed and the maximum vote is used to estimate the class:

$$\hat{y} = \mathbf{argmax}_i \left(\sum_{j=1}^{n/2} C_j^{i+}(\mathbf{x}) - \sum_{j=1}^{n/2} C_j^{i-}(\mathbf{x}) \right), \tag{3}$$

with i referring to the class that the clauses belong to.

Learning. Learning controls the inclusion of features or their negation. A TA k is trained using rewards and penalties. If $t_k > N$ (action 1), a reward increases the state, $t_k \mathrel{+}= 1$, up to $2N$, whilst a penalty decreases the state, $t_k \mathrel{-}= 1$. If $t_k \mathrel{<=} N$ (action 0), a reward decreases the state, $t_k \mathrel{-}= 1$, down to 1, whilst a penalty increases the state, $t_k \mathrel{+}= 1$. The learning rules are described in detail in [13], the basic premise being that individual TA can be controlled by examining their effect on the clause output relative to the desired output, the clause output, the polarity (an inhibitory or excitory clause), and the action. For example, if (i) a feature is 1, (ii) the TA state governing the inclusion of the feature is resolved to action 1, and (iii) the overall clause outputs 1 when a 1 output is desired, then the TA is rewarded, because it is trying to do the right action.

The learning approach has the following added randomness:

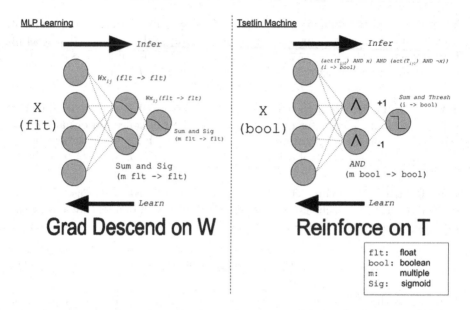

Fig. 1. The backwards pass requires no gradient methods (just manipulation of all TA states (t_k)). Due to the negation of features, clauses can singularly encode non-linearly separable boundaries. There are no sigmoid style functions, all functions are compatible with low level hardware implementations.

- An annealing style cooling parameter T regulates how many clauses are activated for updating. In general, increasing T together with the number of clauses leads to an increase in accuracy.
- A sensitivity s which controls the ratio of reward to penalty, based on a split of the total probability, with values $1/s$ and $1 - 1/s$. This parameter governs the fine-grainedness of the clauses, with a higher s producing clauses with more features included.

When comparing the TMs with a classical neural network [14] (Fig. 1), noteworthy points include: Boolean/integer rather than floating point internal representation; logic rather than linear algebra; and binary inputs/outputs.

3   Fast Clause Falsification Through Indexing

In this section we describe the details of our clause indexing scheme, including index based inference and learning.

Overall Clause Indexing Strategy. Recall that each TM clause is a conjunction of literals. The way that the TM determines whether to include a literal in a clause is by considering the action of the corresponding TA. Thus, to successfully evaluate a clause, the TM must scan through all the actions of the team of

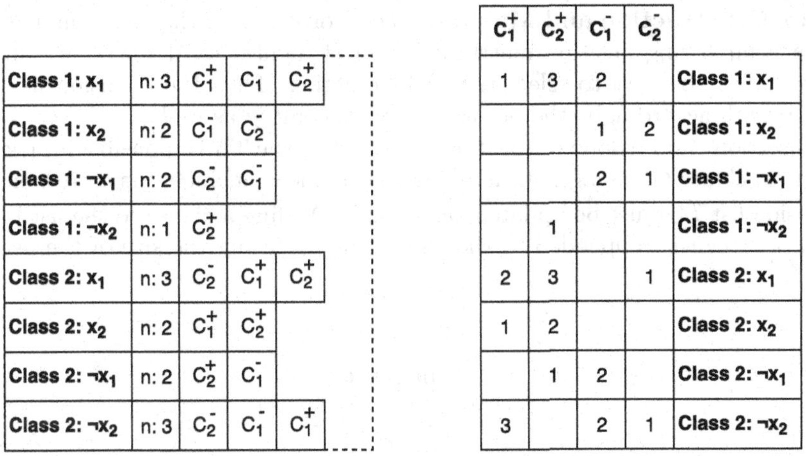

Class 1: x_1	n: 3	C_1^+	C_1^-	C_2^+
Class 1: x_2	n: 2	C_1^-	C_2^-	
Class 1: $\neg x_1$	n: 2	C_2^-	C_1^-	
Class 1: $\neg x_2$	n: 1	C_2^+		
Class 2: x_1	n: 3	C_2^-	C_1^+	C_2^+
Class 2: x_2	n: 2	C_1^+	C_2^+	
Class 2: $\neg x_1$	n: 2	C_2^+	C_1^-	
Class 2: $\neg x_2$	n: 3	C_2^-	C_1^-	C_1^+

C_1^+	C_2^+	C_1^-	C_2^-	
1	3	2		**Class 1: x_1**
		1	2	**Class 1: x_2**
		2	1	**Class 1: $\neg x_1$**
	1			**Class 1: $\neg x_2$**
2	3		1	**Class 2: x_1**
1	2			**Class 2: x_2**
	1	2		**Class 2: $\neg x_1$**
3		2	1	**Class 2: $\neg x_2$**

Fig. 2. Indexing structure.

TAs responsible for the clause, and perform the necessary logical operations to calculate its truth value. However, being a conjunction of literals, it is actually sufficient to merely identify a single literal that is false to determine that the clause itself is false. Conversely, if no false literals are present, we can conclude that the clause is true. Our approach is thus to maintain a list for each literal that contains the clauses that include that literal, and to use these lists to speed up inference and learning.

Index Based Inference. We first consider the clause lists themselves (Fig. 2, left), before we propose a position matrix that allows constant time updating of the clause lists (Fig. 2, right). For each class-literal pair (i, k), we maintain an *inclusion list* L_k^i that contains all the clauses in class i that include the literal l_k. Let the cardinality of L_k^i be $|L_k^i| = n_k^i$. Evaluating the complete set of class i clauses then proceeds as follows. Given a feature vector $\mathbf{x} = (x_1, x_2, \ldots, x_o)$, we define the index set L_F for the false literals as $L_F = \{k | l_k = 0\}$. Then $\mathcal{C}_F^i = \bigcup_{k \in L_F} L_k^i$ is the clauses of class i that have been falsified by $\mathbf{x}$. We partition $\mathcal{C}_F^i$ into clauses with positive polarity and clauses with negative polarity: $\mathcal{C}_F^i = \mathcal{C}_F^{i+} \cup \mathcal{C}_F^{i-}$. The predicted class can then simply be determined based on the cardinality of the sets of falsified clauses:

$$\hat{y} = \mathbf{argmax}_i \left(|\mathcal{C}_F^{i-}| - |\mathcal{C}_F^{i+}| \right), \tag{4}$$

Alongside the inclusion list, an $m \times n \times 2o$ matrix $\mathcal{M}$, with entries M_k^{ij}, keeps track of the position of each clause j in the inclusion list for literal k (Fig. 2, right). With the help of $\mathcal{M}$, searching and removing a clause from a list can take place in constant time.

Index Construction and Maintenance. Constructing the inclusion lists and the accompanying position matrix is rather straightforward since one can initialize all of the TAs to select the exclude action. Then the inclusion lists are empty, and, accordingly, the position matrix is empty as well.

Insertion can be done in constant time. When the TA responsible for e.g. literal l_k in clause C_j^i changes its action from exclude to include, the corresponding inclusion list L_k^i must be updated accordingly. Adding a clause to the list is fast because it can be appended to the end of the list using the size of the list n_k^i:

$$n_k^i \leftarrow n_k^i + 1$$

$$L_k^i[n_k^i] \leftarrow C_j^i$$

$$M_k^{ij} \leftarrow n_k^i.$$

Deletion, however, requires that we know the position of the clause that we intend to delete. This happens when one of the TAs changes from including to excluding its literal, say literal l_k in clause C_j^i. To avoid searching for C_j^i in L_k^i, we use M_k^{ij} for direct lookup:

$$L_k^i[M_k^{ij}] \leftarrow L_k^i[n_k^i]$$

$$n_k^i \leftarrow n_k^i - 1$$

$$M_k^{i,L_k^i[M_k^{ij}]} \leftarrow M_k^{ij}$$

$$M_k^{ij} \leftarrow NA.$$

As seen, the clause is deleted by overwriting it with the last clause in the list, and then updating $\mathcal{M}$ accordingly.

Memory Footprint. To get a better sense of the impact of indexing on memory usage, assume that we have a classification problem with m classes and feature vectors of size o, to be solved with a TM. Thus, the machine has m classes where each class contains n clauses. The memory usage for the whole machine is about $2 \times m \times n \times o$ bytes, considering a bitwise implementation and TAs with 8-bit memory. The indexing data structure then requires two tables with $m \times o$ rows and n columns. Consequently, assuming 2-byte entries, each of these tables require $2 \times m \times n \times o$ bytes of memory, which is roughly the same as the memory required by the TM itself. As a result, using the indexing data structure roughly triples memory usage.

Step-by-Step Example. Consider the indexing configuration in Fig. 2 and the feature vector $\mathbf{x} = (1, 0)$. To evaluate the clauses of class 1, i.e., C_1^+, C_1^-, C_2^+, and C_2^- (omitting the class index for simplicity), we initially assume that they all are true. This provides two votes for and two votes against the class, which gives a class score of 0. We then look up the literals $\neg x_1$ and x_2, which both are false, in the corresponding inclusion lists. We first obtain C_1^- and C_2^- from $\neg x_1$, which falsifies C_1^- and C_2^-. This changes the class score from 0 to 2 because two negative votes have been removed from the sum. We then obtain C_1^- and C_2^- from x_2, however, these have already been falsified, so are ignored. We have now derived a final class score of 2 simply by obtaining a total of four clause index values, rather than fully evaluating four clauses over four possible literals.

Let us now assume that the clause C_1^+ needs to be deleted from the inclusion list of x_1. The position matrix $\mathcal{M}$ is then used to update the inclusion list in constant time. By looking up x_1 and C_1^+ for class 1 in $\mathcal{M}$, we obtain the position of C_1^+ in the inclusion list L_1^1 of x_1, which is 1. Then the last element in the list, C_2^+, is moved to position 1, and the number of elements in the list is decremented. Finally, the entry 1 at row Class 1 : x_1 and column C_1^+ is erased, while the entry 3 at row Class 1 : x_1 and column C_2^+ is set to 1 to reflect the new position of C_2^+ in the list. Conversely, if C_1^+ needs to be added to e.g. the inclusion list of x_2, it is simply added to the end of the Class 1 : x_2 list, in position 3. Then $\mathcal{M}$ is updated by writing 3 into row Class 1 : x_2 and column C_1^+.

Remarks. As seen above, the maintenance of the inclusion lists and the position matrix takes constant time. The performance increase, however, is significant. Instead of going through all the clauses, evaluating each and every one of them, considering all of the literals, the TM only goes through the relevant inclusion lists. Thus, the amount of work becomes proportional to the size of the clauses. For instance, for the MNIST dataset, where the average length of clauses is about 58, the algorithm goes through the inclusion list for half of the literals (the falsifying ones), each containing about 740 clauses on average. Thus, instead of evaluating 20 000 clauses (the total number of clauses), by considering 1568 literals for each (in the worst case), the algorithm just considers 784 falsifying literals with an average of 740 clauses in their inclusion lists. In other words, the amount of work done to evaluate all the clauses with indexing is roughly 0.02 compared to the case without indexing. Similarly, for the IMDB dataset, the average clause length is about 116, and each inclusion list contains roughly 232 clauses on average. Without indexing, evaluating a clause requires assessing 10 000 literals, while indexing only requires assessing the inclusion lists for 5 000 falsifying literals. Using indexing reduces the amount of work to roughly 0.006 of the work performed without indexing. Thus, in both examples, using indexing reduces computation significantly. We now proceed to investigate how the indexing affect inference and learning speed overall.

4   Empirical Results

To measure the effect of indexing we use three well-known datasets, namely MNIST, Fashion-MNIST, and IMDb. We explore the effect of different number of clauses and features, to determine the effect indexing has on performance under varying conditions.

Our first set of experiments covers the MNIST dataset. MNIST consists of 28×28-pixel grayscale images of handwritten digits. We first binarize the images, converting each image into a 784-bit feature vector. We refer to this version of binarized MNIST as M1. Furthermore, to investigate the effect of increasing the number of features on indexing, we repeat the same experiment using 2, 3 and 4 threshold-based grey tone levels. This leads to feature vectors of size 1568 ($= 2 \times 784$), 2352 ($= 3 \times 784$) and 3136 ($= 4 \times 784$), respectively. We refer to these datasets as M2, M3 and M4.

Table 1 summarizes the resulting speedups. Different rows correspond to different number of clauses. The columns refer to different numbers of features, covering training and inference. As seen, increasing the number of clauses increases the effect of indexing. In particular, using more than a few thousands clauses results in about three times faster learning and seven times faster inference.

Table 1. Indexing speedup for different number of clauses and features on MNIST

Features	784		1568		2352		3136	
Clauses	Train	Test	Train	Test	Train	Test	Train	Test
1000	1.78	2.75	1.54	2.79	1.76	3.74	1.91	4.15
2000	1.67	2.83	2.63	5.74	2.46	5.95	2.87	6.06
5000	2.62	5.95	3.23	8.02	3.26	6.88	3.34	7.17
10000	2.82	6.31	3.49	8.31	3.47	7.60	3.43	6.78
20000	2.69	5.22	3.58	8.02	3.34	6.44	3.35	6.39

Figure 3 and Fig. 4 plots the average time of each epoch across our experiments, as a function of the number of clauses. From the figures, we see that after an initial phase of irregular behavior, both indexed and unindexed versions of the TM suggest a linear growth with roughly the same slopes. Thus, the general effect of using indexing on MNIST is a several-fold speedup, while the computational complexity of the TM algorithm remains unchanged, being linearly related to the number of clauses.

In our second set of experiments, we used the IMDb sentiment analysis dataset, consisting of 50,000 highly polar movie reviews, divided in positive and negative reviews. We varied the problem size by using binary feature vectors of size 5000, 10000, 15000 and 20000, which we respectively refer to as I1, I2, I3 and I4.

Table 2 summarizes the resulting speedup for IMDb, employing different number of clauses and features. Interestingly, in the training phase, indexing seems to slow down training by about 10%. On the flip side, when it comes to inference, indexing provided a speedup of up to 15 times.

Figure 5 and Fig. 6 compare the average time of each epoch across our experiments, as a function of the number of clauses. Again, we observe some irregularities for smaller clause numbers. However, as the number of clauses increases, the computation time of both the indexed and unindexed versions of the TM grows linearly with the number of clauses.

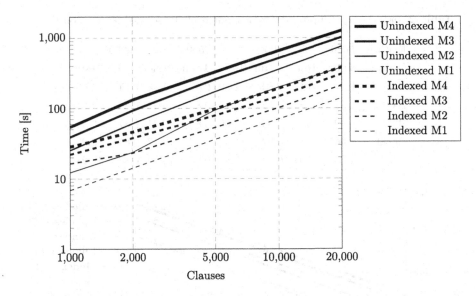

Fig. 3. Average training time on MNIST

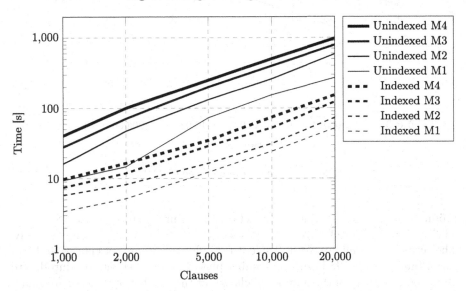

Fig. 4. Average inference time on MNIST

In our third and final set of experiments, we consider the Fashion-MNIST dataset [15]. Here too, each example is a 28 × 28-pixel grayscale image, associated with a label from 10 classes of different kinds of clothing items. Similar to MNIST, we binarized and flattened the data using different numbers of bits for each pixel (ranging from one to four bits by means of thresholding). This process

Table 2. Indexing speedup for different number of clauses and features on IMDb

Features	5000		10000		15000		20000	
Clauses	Train	Test	Train	Test	Train	Test	Train	Test
1000	0.76	1.95	0.84	1.59	0.81	1.78	0.81	2.03
2000	1.00	4.60	0.87	3.17	0.84	4.47	0.83	4.81
5000	0.99	8.58	0.93	8.66	0.85	9.41	0.87	9.49
10000	1.06	15.40	0.92	11.92	0.90	13.17	0.87	13.03
20000	1.05	15.19	0.93	15.87	0.86	13.84	0.88	13.28

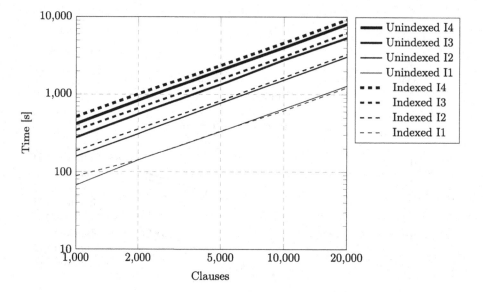

Fig. 5. Average training time on IMDb

produced four different binary datasets with feature vectors of size 784, 1568, 2352 and 3136. We refer to these datasets as F1, F2, F3 and F4, respectively. The results are summarized in Table 3 for different number of clauses and features. Similar to MNIST, in the training phase, we see a slight speedup due to indexing, except when the number of clauses or the number of features are small. In that case, indexing slows down training to some extent (by about 10%). On the other hand, for inference, we see a several-fold speedup (up to five-fold) due to indexing, especially with more clauses.

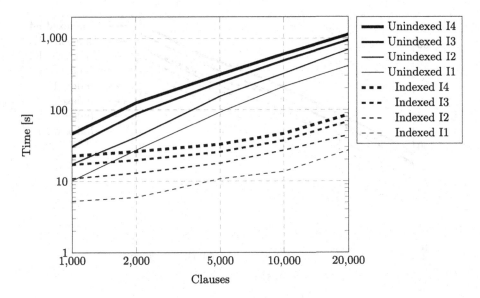

Fig. 6. Average inference time on IMDb

Table 3. Indexing speedup for different number of clauses and features on Fashion-MNIST

Features	784		1568		2352		3136	
Clauses	Train	Test	Train	Test	Train	Test	Train	Test
1000	0.86	0.82	0.97	1.33	1.04	1.68	1.04	1.77
2000	0.92	1.17	1.23	2.55	1.24	2.93	1.25	3.13
5000	1.23	2.21	1.32	2.85	1.33	2.84	1.33	2.95
10000	1.55	3.81	1.58	4.36	1.67	4.92	1.56	3.95
20000	1.55	3.54	1.65	4.54	1.70	4.74	1.67	5.05

Figure 7 and Fig. 8 compare the average time of each epoch as a function of the number of clauses. Again, we observe approximately linearly increasing computation time with respect to the number of clauses.

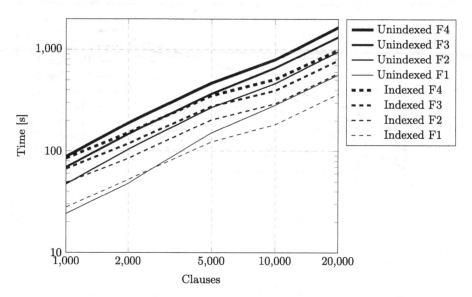

Fig. 7. Average training time on Fashion-MNIST

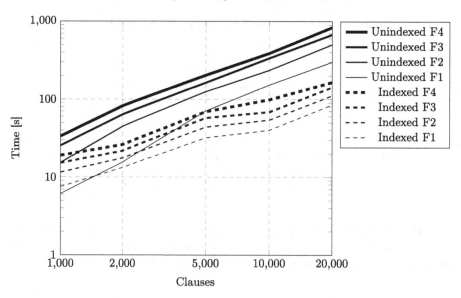

Fig. 8. Average inference time on Fashion-MNIST

5   Conclusion and Further Work

In this work, we introduced indexing-based evaluation of Tsetlin Machine clauses, which demonstrated a promising increase in computation speed. Using simple look-up tables and lists, we were able to leverage the typical sparsity of clauses. We evaluated the indexing scheme on MNIST, IMDb and Fashion-MNIST. For training, on average, the indexed TM produced a three-fold speedup on MNIST, a roughly 50% speedup on Fashion-MNIST, while it slowed down IMDb training. Thus, the effect on training speed seems to be highly dependent on the data. The overhead from maintaining the indexing structure could counteract any speed gains.

During inference, on the other hand, we saw a more consistent increase in performance. We achieved a three- to eight-fold speedup on MNIST and Fashion-MNIST, and a 13-fold speedup on IMDb, which is reasonable considering there is no index maintenance overhead involved in classification.

In our further work, we intend to investigate how clause indexing can speed up Monte Carlo tree search for board games, by exploiting the incremental changes of the board position from parent to child node.

References

1. Tsetlin, M.L.: On behaviour of finite automata in random medium. Avtomat. i Telemekh **22**(10), 1345–1354 (1961)
2. Robbins, H.: Some aspects of the sequential design of experiments. Bull. Am. Math. Soc. (1952)
3. Gittins, J.C.: Bandit processes and dynamic allocation indices. J. Roy. Stat. Soc. Ser. B (Methodological) **41**(2), 148–177 (1979)
4. Narendra, K.S., Thathachar, M.A.L.: Learning Automata: An Introduction. Prentice-Hall Inc. (1989)
5. Haugland, V., et al.: A two-armed bandit collective for hierarchical examplar based mining of frequent itemsets with applications to intrusion detection. TCCI XIV **8615**, 1–19 (2014)
6. Granmo, O.-C., et al.: Learning automata-based solutions to the nonlinear fractional knapsack problem with applications to optimal resource allocation. IEEE Trans. Syst. Man Cybern. Part B **37**(1), 166–175 (2007)
7. Berge, G.T., Granmo, O.-C., Tveit, T.O., Goodwin, M., Jiao, L., Matheussen, B.V.: Using the Tsetlin machine to learn human-interpretable rules for high-accuracy text categorization with medical applications. IEEE Access **7**, 115134–115146 (2019)
8. Phoulady, A., Granmo, O.-C., Gorji, S.R., Phoulady, H.A.: The weighted Tsetlin machine: compressed representations with clause weighting. In: Ninth International Workshop on Statistical Relational AI (StarAI 2020) (2020)
9. Wheeldon, A., Shafik, R., Yakovlev, A., Edwards, J., Haddadi, I., Granmo, O.-C.: Tsetlin machine: a new paradigm for pervasive AI. In: Proceedings of the SCONA Workshop at Design, Automation and Test in Europe (DATE) (2020)
10. Darshana Abeyrathna, K., Granmo, O.-C., Zhang, X., Jiao, L., Goodwin, M.: The Regression Tsetlin machine - a novel approach to interpretable non-linear regression. Philos. Trans. Roy. Soc. A **378** (2019)
11. Granmo, O.-C., Glimsdal, S., Jiao, L., Goodwin, M., Omlin, C.W., Berge, G.T.: The convolutional Tsetlin machine. arXiv preprint. arXiv:1905.09688 (2019)
12. Rahimi Gorji, S., Granmo, O.-C., Phoulady, A., Goodwin, M.: A tsetlin machine with multigranular clauses. In: Bramer, M., Petridis, M. (eds.) SGAI 2019. LNCS (LNAI), vol. 11927, pp. 146–151. Springer, Cham (2019). https://doi.org/10.1007/978-3-030-34885-4_11

13. Granmo, O.-C.: The Tsetlin machine - a game theoretic bandit driven approach to optimal pattern recognition with propositional logic. arXiv preprint arXiv:1804.01508 (2018)
14. Rumelhart, D.E., Hinton, G.E., Williams, R.J.: Learning representations by back-propagating errors. Nature **323**, 533–536 (1986)
15. Xiao, H., Rasul, K., Vollgraf, R.: Fashion-MNIST: a novel image dataset for benchmarking machine learning algorithms (2017)

Constrained Evolutionary Piecemeal Training to Design Convolutional Neural Networks

Dolly Sapra$^{(\boxtimes)}$ and Andy D. Pimentel

University of Amsterdam, Amsterdam, Netherlands
{d.sapra,a.d.pimentel}@uva.nl

Abstract. Neural Architecture Search (NAS), which automates the discovery of efficient neural networks, has demonstrated substantial potential in achieving state of the art performance in a variety of domains such as image classification and language understanding. In most NAS techniques, training of a neural network is considered a separate task or a performance estimation strategy to perform the architecture search. We demonstrate that network architecture and its coefficients can be learned together by unifying concepts of evolutionary search within a population based traditional training process. The consolidation is realised by cleaving the training process into pieces and then put back together in combination with evolution based architecture search operators. We show the competence and versatility of this concept by using datasets from two different domains, CIFAR-10 for image classification and PAMAP2 for human activity recognition. The search is constrained using minimum and maximum bounds on architecture parameters to restrict the size of neural network from becoming too large. Beginning the search from random untrained models, it achieves a fully trained model with a competent architecture, reaching an accuracy of 92.5% and 94.36% on CIFAR-10 and PAMAP2 respectively.

Keywords: Neural networks · Neural architecture search · AutoML · Constraint optimization

1 Introduction

Recent work to discover efficient neural networks automatically has proven to be a highly efficient methodology, where the discovered neural network architectures are outperforming the hand-crafted ones. Popular approaches for Neural Architecture Search (NAS) use reinforcement learning [26,37] and evolutionary algorithms [23,27]. However they take tens to thousands of GPU days to prepare, search and converge. Most of these methods rely on resource heavy training to guide the search process. In this paper, we look at NAS from a different perspective by exploring the possibility of finding optimal architectures during the training process itself as opposed to accuracy prediction or training as a separate performance estimation strategy.

© Springer Nature Switzerland AG 2020
H. Fujita et al. (Eds.): IEA/AIE 2020, LNAI 12144, pp. 709–721, 2020.
https://doi.org/10.1007/978-3-030-55789-8_61

Moreover, most of recent work in NAS techniques is focused on image classification tasks such as for CIFAR-10 [18] and Imagenet [11] leading to innovative research on complex search spaces that are highly suited to vision tasks. These search spaces are derived from hand-crafted previous state-of-the-art such as residual connections [16], cell based designs such as inception [32], dense net [17] or generated in a graph-like fashion [31]. While these search spaces have proven to be extremely efficient, these are not always easy to understand, train and implement. Most medium complexity tasks and domains such as human activity recognition [33], earth sciences [20,25] and astronomical studies [9] deploy plain convolutional networks as they are considered sufficient as well as easy to understand by scientists from non-AI background. A tool to find an efficient Convolutional Neural Network (CNN) in a few GPU days and to be usable in any domain by non-AI experts is also a step forward in simplifying and democratizing AI. In this paper we look at the Human Activity Recognition (HAR) domain with the PAMAP2 [29] dataset where inputs from multiple body worn sensors are used to predict the activity being undertaken by the wearer. We also demonstrate versatility of our approach using a very different CIFAR-10 dataset for image classification.

The main contribution of this paper is a novel NAS algorithm that searches for an efficient CNN architecture for a given task and converges in a few GPU hours. Our work leverages a population based computing technique which allows a group of CNNs to train in parallel. During this training, evolutionary operators are applied to some random CNNs at regular intervals which leads to architecture modification and hence exploration of the search space. A new architecture derived like this is always partially trained already as it was modified from another architecture undergoing training. In subsequent iterations derived architectures continue to train. Towards the end of this algorithm, the best candidates are selected from the population, which can then be post processed or trained further, if needed. All the CNNs generated and modified during search are bound by minimum and maximum values for each architecture parameter. These constraints are in place to make sure that architectures do not become too big and limits the resource consumption of the final neural network. This is an important factor to consider for tasks intended to be used on embedded systems like wearables in HAR tasks.

The rest of the paper is organized as follows. We discuss related published works for NAS in Sect. 2. We then present methodologies and detail our algorithm in Sect. 3 and describe experimental setup and results of our evaluations and validations in Sect. 4. Finally, we conclude the paper in Sect. 5.

2   Related Work

There are many published works in the field of NAS, roughly divided into three groups - reinforcement learning based, evolutionary and one-shot architecture search. In reinforcement learning (RL) and evolutionary search methods, training the neural network completely with each iteration is mandatory to evaluate its performance. In RL methods [4,26,37], the reward of the RL agent is dependent on the validation performance of the trained architecture. By continuously rewarding the agent the search is guided towards better performing neural networks. RL based approaches require the construction of an appropriate agent, which often itself is another neural network and its optimization also requires considerable effort in designing and fine tuning.

Evolutionary approaches [8,22,27,28,35] use genetic algorithms to optimize the neural architecture. These are population based algorithms where every iteration trains and evaluates all the architectures in the population. The subsequent generation of neural networks is chosen based on their prediction accuracy and average performance of the whole population increases with time leading to discovery of a highly efficient neural network architecture at convergence.

Unfortunately, a lot of these approaches require large computational resources as they need to train and validate hundreds to thousands of architectures. RL method [37] trained over 10,000 of neural architectures, requiring thousands of GPU days and another very efficient evolutionary search [13] still takes 56 GPU days to converge. Some works use proxy tasks and helpers such as hyper-networks [5], predictors [4,10] and controllers [26] to speed up the search, but even so they need significant preparation and construction time prior to the actual search. Our approach converges in a few GPU hours to a couple of GPU days without using any additional helper or proxy task.

One-Shot Architecture Search is another promising approach where NAS is modeled as a single training process of a super-network that comprises of all possible sub-networks. DropPath [38] drops out each path with some fixed probability and use the pre-trained super-network to evaluate architectures, which are sub-networks created by randomly zeroing out paths. DARTS [21] introduces an architecture parameter in addition for each path and jointly train weight parameters and architecture parameters using standard gradient descent. Other approaches might need to utilize proxy tasks to be viable such as [6] employs a memory-efficient scheme where only few paths are updated during search. The approaches which require the entire super-network to reside in GPU memory during NAS are restricted to relatively small architecture size, usually a cell that can be stacked to form multiple times to form the whole network. They usually also face a meta-architecture design problem after the search regarding number of cells to be used and how to connect them to build the actual model. Besides, a repeated single cell type may not be optimal for every application.

3   Methodology

We call our proposed approach Evolutionary Piecemeal Training, where piecemeal-training refers to training a neural network with a small 'data-piece' of size δ_k. A traditional continuous training is interceded by an evolutionary operator at regular intervals and the interval of intervention is dictated by the value of δ_k. An evolutionary operator modifies the architecture and the training further continues. This process is done for multiple neural networks forming a population and the candidates which are not able to achieve high accuracy during training keep dropping out from the population. This can also be seen as early training termination of candidates that are performing poorly. In this section, we give details of the key concepts and then outline the complete Algorithm.

3.1   Search Space

The search space for our algorithm is focused on plain Convolutional Neural Networks (CNNs), which consist of convolutional, fully-connected and pooling layers without residual connections, branching etc. Batch normalization and non-linear activations are configurable and can be added to the network. CNN architectures are defined by a group of blocks, each block is of a specific layer type and is bounded by minimum and maximum number of layers it can have. Additionally, each layer has upper and lower bounds on each of its parameters. For example, a convolutional layer will have bounds on the number of units, kernel sizes and stride sizes possible. Section 4.2 defines the architecture search spaces for CIFAR-10 and PAMAP2 respectively. The search space specifications along with its bounds are encoded as a collection of genes, also called a genotype. All possible combinations of parameters together form the gene pool from which individual neural networks are created and trained.

3.2   Population Based Training

We employ a population based training process where an initial population of neural networks is randomly created from the defined gene pool. In each iteration, all candidates of the population are piecemeal-trained and then evaluated using the validation set. Depending upon the available resources, all candidates can be trained in any combination of parallel and sequential manner. The size of the population is kept constant throughout the algorithm, though the candidates of the population keep changing as they are altered through the evolutionary operators applied in each iteration. The number of candidates in the population needs to be large enough to maintain enough diversity of CNN architectures in the population, while still satisfying the constraints applied to it.

3.3   Evolutionary Operators

Evolutionary algorithms are iterative population based algorithms where we evolve a better population over subsequent iterations. During each evolutionary step, some candidates from the population of CNNs are altered once using evolutionary operators called recombination and mutation operators.

Recombination works with two neural networks and swaps all layers in a gene-block of the same type. In this replacement, the layers being swapped are roughly in the same phase of feature extraction. The input and output feature map sizes of layer block being swapped are also identical in both of the selected networks. Figure 1 illustrates the recombination operator for swapping convolutional layers from two networks. Recombination is not a function preserving operator, but in the experiments they were found to be important to introduce diversity in the population by changing the total number of layers in a candidate through swapping. To reduce the negative effect of loss incurred due to recombination, we used a cooling-down approach to the recombination rate. In earlier iterations, where the training loss is already high, there are more swaps happening than in the later ones, where training loss is very low.

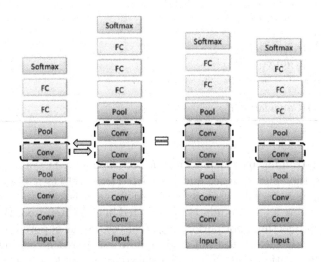

Fig. 1. Recombination operator on two neural networks swapping convolution layers

Mutation changes a layer's parameters such as number of kernels or kernel size and are designed to be function preserving. We apply the Net2Wider operator from [7] to increase the number of units and pruning [19] to reduce the number of filters in the layer. Filters are centrally cropped or zero-padded when their size is changed. Every mutation disrupts the ongoing training of the mutated candidates and some additional loss is incurred in the training-in-process. However, these specific operators were chosen because these are either completely or partially function preserving in nature, which means that the loss incurred from these operators is as small as possible and recoverable in later piecemeal-training.

3.4 Algorithm

Algorithm 1 outlines the complete algorithm, the goal is to find neural networks satisfying the architecture constraints with minimum prediction error.

$InitializePopulation()$ generates a neural network population of size N_p using a factory pattern class for the genotype and initializes them by training them for 1 epoch. Afterwards, this iterative algorithm runs for N_g generations. $PiecemealTrain()$ trains all individuals with randomly selected data of size δ_k

---

Algorithm 1: Evolutionary Piecemeal Training

    Evolutionary Inputs: N_g, N_p, P_r, P_m, α
    Training Inputs   : τ_{params}, δ_k

1 $\wp_o \leftarrow InitializePopulation(N_p)$
2 **for** $i \leftarrow 0 \ldots N_g$ **do**
3     $\wp_i \leftarrow PiecemealTrain(\wp_i, \tau_{params}, \delta_k)$
4     $E_v \leftarrow ValidationPopulation(\wp_i)$
5     $\wp_{best} \leftarrow BestSelection(\alpha, \wp_i, E_v)$
6     $\wp_r \leftarrow random((1-\alpha) * \wp_i)$
7   update $\wp_i \leftarrow \wp_{best} + \wp_r$
8     $\wp_{rc} \leftarrow Recombine(\wp_i, P_r)$
9     $\wp_{mu} \leftarrow Mutate(\wp_i, P_m)$
10   update $\wp_i \leftarrow \wp_{mu} + \wp_{rc} + \wp_{remaining}$
11 **end**
12 **return** $BestCandidatesOf(\wp_{N_g})$

---

using τ_{params} training parameters. $ValidationPopulation()$ evaluates the population using the validation set and $BestSelection()$ selects α best individuals using the accuracy on validation set achieved so far. α is kept at very high ratio $> 0.95 * N_p$ so only a few candidates are rejected in every iteration. By doing this our focus is on rejecting poor performing architectures, as opposed to promoting an architecture that learns fast but might not be able to reach very high accuracy in the end. Population size, N_p, is kept constant by randomly selecting $1 - \alpha$ networks from survivors and added back to the pool. Random selection ensures that children generations are not overwhelmed by only one type of architecture which might have reached high accuracy by luck. $Recombine()$ and $Mutate()$ are evolutionary operators, they select individuals from the population with selection probability of P_m and P_r respectively. P_r is linearly cooled to ≈ 0 from its initial value. The algorithm returns the best candidates of CNNs from the final population. Best candidates can be further processed, modified or trained as needed outside this algorithm.

4 Experiments

In this section, we evaluate our algorithm using two datasets and outline the experimental setup. We present the datasets and data augmentation techniques used, their respective constrained search spaces and finally the results obtained. We have used the Java based Jenetics library [1] for evolutionary computation and the Python based Caffe2 [3] library for training and testing. We used the ONNX [2] format to represent and transfer the neural networks across different modules. Our experiments run on only one GPU (GeForce RTX 2080).

4.1 DataSets

For experiments, we use the CIFAR-10 dataset for image classification and the PAMAP2 dataset of human activity recognition. CIFAR-10 consists of 60,000 labeled images of dimensions $32 \times 32 \times 3$, comprising of 50,000 training and 10,000 testing images. The images are divided into 10 classes. We use 5,000 images from the training set as a validation set and we use only this validation set to guide the search process. The test set is used only in the end to check final accuracy of the network, which is what we report in this paper. We use standard data augmentation as described in [14,30] with small translations, cropping, rotations and horizontal flips. For piecemeal-training, δ_k random images are picked from the training set from which approximately 50% are augmented images.

The PAMAP2 dataset provides recordings from three Inertial Measurement Units (IMU) and a heart rate monitor. Together, the input data is in the form of time-series from 40 channels. In this dataset, nine subjects performed twelve different household, sports and daily living activities. We do not consider the optional activities performed by some subjects. Following [15], recordings from participants 5 and 6 are used as validation and testing sets respectively. IMUs' recordings are downsampled to 30 Hz and a sliding window approach with a window size of 3s (100 samples) and step size of 660ms (22 samples) is used to segment the sequences. To augment the data, a sliding window is moved by different step sizes while keeping the window size the same at 3s.

4.2 Search Space

Table 1 and Table 2 describe the search space specifications. Total design points for CIFAR-10 and PAMAP2 are to the order of 10^8 and 10^5 respectively. Number of units per layer can be multiples of 16 for CIFAR-10 and 8 for PAMAP2.

Table 1. Cifar-10 architecture search space

Layer type	Layers		Units/Layer		Kernel-size		Stride
	L_{min}	L_{max}	U_{min}	U_{max}	K_{min}	K_{max}	S_t
Convolution	2	5	48	96	3×3	7×7	1
MaxPool	1	1	1	1	2×2	2×2	2
Convolution	2	7	80	320	3×3	7×7	1
MaxPool	1	1	1	1	2×2	2×2	2
Convolution	2	7	256	640	3×3	7×7	1
MaxPool	1	1	1	1	2×2	2×2	2
Fully Connected	2	3	128	1024	–	–	–
SoftMax	1	1	1	1	–	–	–

Table 2. PAMAP2 architecture search space

Layer type	Layers		Units/Layer		Kernel-size		Stride
	L_{min}	L_{max}	U_{min}	U_{max}	K_{min}	K_{max}	S_t
Convolution	2	4	64	128	3×1	7×1	1
MaxPool	1	1	1	1	2×1	2×1	2
Convolution	2	5	96	256	3×1	7×1	1
GlobalMaxPool	1	1	1	1	2×1	2×1	2
Fully Connected	1	3	128	512	–	–	–
SoftMax	1	1	1	1	–	–	–

4.3 Training Setup

We trained the CIFAR-10 dataset with 80 generations and population size of 80. δ_k for piecemeal training is set to 4k random images. Each convolution and fully connected layer is followed by ReLu activation. Training was done using the Adam optimizer and batch size of 80 with initial learning rate of $5e^{-4}$ with step learning rate decay, after every 20 iterations learning rate was reduced by $1e^{-4}$. Both evolutionary selection probabilities P_m and P_r are initialized to 0.3. P_m stays constant, while P_r is linearly decayed to reach 0.01 at the last iteration.

PAMAP2 was trained for 30 generations and population size of 50. δ_k for piecemeal training is set to 20k random sensor samples. Each convolution and fully connected layer is followed by ReLu activation. Training was done using the Adam optimizer and batch size of 100 with initial learning rate of $1e^{-4}$ with constant learning rate. Evolutionary selection probabilities P_m and P_r are initialized to 0.3 and 0.3 respectively. As with the CIFAR-10 experiment, P_m stays constant and P_r is linearly decayed.

Neural networks for CIFAR-10 are larger than for PAMAP2, with a single GPU with 11 GB memory 4 parallel training threads for CIFAR-10 and 7 parallel training threads could run simultaneously. The limitation on the level of parallelism was defined by the available memory on the GPU. Also to fasten up the search there was no Batch Normalization used as it consumes more memory and reduces possible parallelism. After the search, the best candidate was modified and every convolutional layer was appended with a batch normalization layer and further trained for 100 epochs.

4.4 Results

In this section, we evaluate our algorithm and compare with other state-of-the-art. Figure 2 shows the training curves for experiments on CIFAR-10 and PAMAP2, with each iteration there is a general increase in average accuracy of the population as well as the best accuracy of an individual in the population. The best model might change from an iteration to the next. The best models

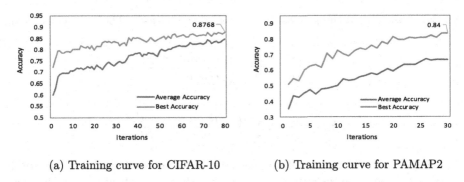

(a) Training curve for CIFAR-10   (b) Training curve for PAMAP2

Fig. 2. Training curves. Average Accuracy refers to the average performance of whole population at the given search iteration. Best accuracy refers to best found performance of an individual model from the population.

found at the end of all iterations were further trained to achieve final accuracy as reported in Fig. 2.

For CIFAR-10, the search took 2-GPU days and the best prediction accuracy was found to be 92.5% on the test set. Table 3 shows comparisons with other evolutionary approaches. We know that 92.5% is relatively low compared to other published works, but this is on a very simple and plain CNN without any architectural enhancements or advance data augmentation like mixup [36] or cutout [12]. Other approaches use a hybrid search space where different architecture blocks or cell modules as well as arbitrary skip connections are used. Instead of stacking conventional layers, these stack different blocks. The best model found in our experiments has 13 convolutional layers followed by 2 fully connected layers.

Table 3. CIFAR-10 accuracy comparisons with evolutionary approaches

Model	Search space	GPU-days	Accuracy (%)
CoDeepNeat [23]	Hybrid	–	92.7
GeneticCNN [35]	Hybrid	17	92.9
EANN-Net [8]	Hybrid	–	92.95
AmoebaNet [27]	Cell	3150	**96.6**
NSGANet [22]	Hybrid	8	96.15
Evolution [28]	Hybrid	1000+	94.6
EPT (ours)	Plain CNN	**2**	92.5

For the PAMAP2 dataset, we report the classification accuracy as well as the weighted F1-score ($F1_w$) and mean F1-score ($F1_m$), as this is an unbalanced set. These scores consider the correct classification of each class equally, using the precision, recall and the proportion of class in the dataset. These are calculated as below:

$$F1_w = \sum_i 2 \times \frac{n_i}{N} \times \frac{precision_i \times recall_i}{precision_i + recall_i}$$

$$F1_m = \frac{2}{N} \times \sum_i \frac{precision_i \times recall_i}{precision_i + recall_i}$$

where, n_i is the number of samples for each class $k \in K$ and N is the total number of samples in the dataset. Compared with the classification accuracy, the F1-scores are more convenient for evaluating the performance of the networks on highly unbalanced datasets. We calculate both F1-scores to compare our result with different published results.

Our approach shows very promising results on the PAMAP2 dataset. For PAMAP2, the search took only 10 GPU-hours and the best prediction accuracy was 94.36%. Table 3 compares our algorithm to other published works. In comparison, grid search [15] on CNN for PAMAP2 was able to achieve 93.7% and a hand-crafted CNN [24] was able to reach 93.21% clearly demonstrating that our approach is more efficient than naive search methods such as random or grid search, while also being resource efficient by evaluating far less design points. The best performance was found on a neural network that has 7 convolutional layers followed by 3 fully connected layers (Table 4).

Table 4. PAMAP2 accuracy comparisons

Model	Accuracy (%)	$F1_w$ (%)	$F1_m$ (%)
Hand designed [24]	93.13	93.21	–
Grid search (CNN) [15]	–	–	93.7
D^2C [34]	–	–	92.71
D^2CL [34]	–	–	93.2
EPT (ours)	**94.36**	**94.17**	**94.36**

5   Conclusion

In this paper, we presented a novel approach called Evolutionary Piecemeal Training which traverses the search space of plain CNNs to find an efficient architecture within reasonable constraints for a given task. We validated our algorithm on two different datasets which demonstrates the versatility of our method. We showed that for moderate complexity tasks such as the PAMAP2 dataset, our approach is better and more efficient than random or grid search methodologies.

We perceive that this process can easily be extended to perform multi-objective optimization, which allows the neural network to be optimized simultaneously for resource utilization along with its performance. As future work we aim to modify this search algorithm to incorporate hardware metrics and reduce resource usage of the neural network on the designated embedded system.

Acknowledgements. This project has received partial funding from the European Union's Horizon 2020 Research and Innovation programme under grant agreement No. 780788.

References

1. Jenetics library (2019). https://jenetics.io/
2. Onnx: Open neural network exchange formet (2019). https://onnx.ai/
3. Pytorch: An open source deep learning platform (2019). https://pytorch.org/
4. Baker, B., Gupta, O., Raskar, R., Naik, N.: Accelerating neural architecture search using performance prediction. arXiv preprint arXiv:1705.10823 (2017)
5. Brock, A., Lim, T., Ritchie, J.M., Weston, N.: Smash: one-shot model architecture search through hypernetworks. arXiv preprint arXiv:1708.05344 (2017)
6. Cai, H., Zhu, L., Han, S.: ProxylessNAS: direct neural architecture search on target task and hardware. In: International Conference on Learning Representations (2019)
7. Chen, T., Goodfellow, I., Shlens, J.: Net2net: accelerating learning via knowledge transfer. arXiv preprint arXiv:1511.05641 (2015)
8. Chen, Z., Zhou, Y., Huang, Z.: Auto-creation of effective neural network architecture by evolutionary algorithm and ResNet for image classification. In: 2019 IEEE International Conference on Systems, Man and Cybernetics (SMC), pp. 3895–3900. IEEE (2019)
9. Davies, A., Serjeant, S., Bromley, J.M.: Using convolutional neural networks to identify gravitational lenses in astronomical images. Mon. Not. Roy. Astron. Soc. (2019)
10. Deng, B., Yan, J., Lin, D.: Peephole: predicting network performance before training. arXiv preprint arXiv:1712.03351 (2017)
11. Deng, J., Dong, W., Socher, R., Li, L.J., Li, K., Fei-Fei, L.: Imagenet: a large-scale hierarchical image database. In: 2009 IEEE Conference on Computer Vision and Pattern Recognition, pp. 248–255. IEEE (2009)
12. DeVries, T., Taylor, G.W.: Improved regularization of convolutional neural networks with cutout. arXiv preprint arXiv:1708.04552 (2017)
13. Elsken, T., Metzen, J.H., Hutter, F.: Efficient multi-objective neural architecture search via Lamarckian evolution. In: International Conference on Learning Representations (2019)
14. Goodfellow, I.J., Warde-Farley, D., Mirza, M., Courville, A., Bengio, Y.: Maxout networks. arXiv preprint arXiv:1302.4389 (2013)
15. Hammerla, N.Y., Halloran, S., Plötz, T.: Deep, convolutional, and recurrent models for human activity recognition using wearables. arXiv preprint arXiv:1604.08880 (2016)
16. He, K., Zhang, X., Ren, S., Sun, J.: Deep residual learning for image recognition. In: Proceedings of the IEEE Conference on Computer Vision and Pattern Recognition, pp. 770–778 (2016)

17. Huang, G., Liu, Z., Van Der Maaten, L., Weinberger, K.Q.: Densely connected convolutional networks. In: Proceedings of the IEEE Conference on Computer Vision and Pattern Recognition, pp. 4700–4708 (2017)
18. Krizhevsky, A., Hinton, G., et al.: Learning multiple layers of features from tiny images. Technical report, Citeseer (2009)
19. Li, H., Kadav, A., Durdanovic, I., Samet, H., Graf, H.P.: Pruning filters for efficient convnets. arXiv preprint arXiv:1608.08710 (2016)
20. Ling, F., et al.: Measuring river wetted width from remotely sensed imagery at the sub-pixel scale with a deep convolutional neural network. Water Resources Research (2019)
21. Liu, H., Simonyan, K., Yang, Y.: Darts: Differentiable architecture search. arXiv preprint arXiv:1806.09055 (2018)
22. Lu, Z., et al.: NSGA-Net: a multi-objective genetic algorithm for neural architecture search. arXiv preprint arXiv:1810.03522 (2018)
23. Miikkulainen, R., et al.: Evolving deep neural networks. In: Artificial Intelligence in the Age of Neural Networks and Brain Computing, pp. 293–312. Elsevier (2019)
24. Moya Rueda, F., Grzeszick, R., Fink, G., Feldhorst, S., ten Hompel, M.: Convolutional neural networks for human activity recognition using body-worn sensors. Informatics **5**, 26 (2018)
25. Pan, B., Hsu, K., AghaKouchak, A., Sorooshian, S.: Improving precipitation estimation using convolutional neural network. Water Resour. Res. **55**(3), 2301–2321 (2019)
26. Pham, H., Guan, M., Zoph, B., Le, Q., Dean, J.: Efficient neural architecture search via parameter sharing. In: International Conference on Machine Learning, pp. 4092–4101 (2018)
27. Real, E., Aggarwal, A., Huang, Y., Le, Q.V.: Regularized evolution for image classifier architecture search. In: Proceedings of the AAAI Conference on Artificial Intelligence, vol. 33, pp. 4780–4789 (2019)
28. Real, E., et al.: Large-scale evolution of image classifiers. In: Proceedings of the 34th International Conference on Machine Learning-Volume 70, pp. 2902–2911 (2017). JMLR.org
29. Reiss, A., Stricker, D.: Introducing a new benchmarked dataset for activity monitoring. In: 2012 16th International Symposium on Wearable Computers, pp. 108–109. IEEE (2012)
30. Springenberg, J.T., Dosovitskiy, A., Brox, T., Riedmiller, M.: Striving for simplicity: the all convolutional net. arXiv preprint arXiv:1412.6806 (2014)
31. Suganuma, M., Shirakawa, S., Nagao, T.: A genetic programming approach to designing convolutional neural network architectures. In: Proceedings of the Genetic and Evolutionary Computation Conference, pp. 497–504. ACM (2017)
32. Szegedy, C., et al.: Going deeper with convolutions. In: Proceedings of the IEEE Conference on Computer Vision and Pattern Recognition, pp. 1–9 (2015)
33. Wang, J., Chen, Y., Hao, S., Peng, X., Hu, L.: Deep learning for sensor-based activity recognition: a survey. Pattern Recogn. Lett. **119**, 3–11 (2019)
34. Xi, R., Hou, M., Fu, M., Qu, H., Liu, D.: Deep dilated convolution on multimodality time series for human activity recognition. In: 2018 International Joint Conference on Neural Networks (IJCNN), pp. 1–8. IEEE (2018)
35. Xie, L., Yuille, A.: Genetic CNN. In: Proceedings of the IEEE International Conference on Computer Vision, pp. 1379–1388 (2017)
36. Zhang, H., Cisse, M., Dauphin, Y.N., Lopez-Paz, D.: mixup: beyond empirical risk minimization. arXiv preprint arXiv:1710.09412 (2017)

37. Zoph, B., Le, Q.V.: Neural architecture search with reinforcement learning. arXiv preprint arXiv:1611.01578 (2016)
38. Zoph, B., Vasudevan, V., Shlens, J., Le, Q.V.: Learning transferable architectures for scalable image recognition. In: Proceedings of the IEEE Conference on Computer Vision and Pattern Recognition, pp. 8697–8710 (2018)

Hierarchical Learning of Primitives Using Neurodynamic Model

Fusei Nomoto$^{(\boxtimes)}$, Tadayoshi Yasuda, Shun Nishide, Xin Kang, and Fuji Ren

Faculty of Science and Technology, Tokushima University,
2-1 Minami-Josanjima-cho, Tokushima, 770-8502 Tokushima, Japan
`fusei@docomo.ne.jp`

Abstract. Primitives are essential features in human's recognition capability. In this paper, we propose a method to form representations of motion primitives within a neural network model to learn sequences that consist of a combination of primitives. The training model is based on human development phenomena, namely motionese, where human infants learn simple motions first, and tend to learn complex motions as their skills develop. A neurodynamical model, Multiple Timescale Recurrent Neural Network (MTRNN), is used for the dynamics learning model. Neurons in MTRNN are composed hierarchically to learn different levels of information. The proposed method first trains the model using basic simple motions with neurons learning low level data. After training converges, neurons learning high level data are attached to the model, and complex motions are used to train the model. Experiments and analysis with drawing data show the models capability to form representations of primitives within the model.

Keywords: Recurrent neural network · Motion primitives · Dynamics learning

1 Introduction

In recent years, research on humanoid robots has been actively conducted. NAO and Pepper are examples of humanoid robots with high performance that have been commercialized for coexistence of robots and humans capable of performing various operations. A main issue required to be solved for such robots is the ability to adapt to general environment. One solution involves implementation of humans' environmental adaptation capability, which is developed through experience, into the robot's cognitive mechanism. Works in such field are known as cognitive developmental robotics field [1].

There are two main goals in the field of cognitive developmental robotics. The first goal is to understand human development mechanism through constructive approach using robots. The second goal is to create developmental robots based on human development. Our work is based on the latter goal aiming to create a developmental learning model for robots.

© Springer Nature Switzerland AG 2020
H. Fujita et al. (Eds.): IEA/AIE 2020, LNAI 12144, pp. 722–731, 2020.
https://doi.org/10.1007/978-3-030-55789-8_62

In this work, we specifically focus on "primitives" which are basic units in human cognition. For example, lines and arcs can be considered as the most basic primitives in recognition of shapes (which are composed of a combination of several lines and arcs). Through human growth, different levels of primitives (such as squares or circles) evolve to recognize complex shapes. We propose a training method to learn complex sequences as a combination of primitives within the model using a neurodynamical model, namely Multiple Timescale Recurrent Neural Network (MTRNN). The model is first trained using primitive sequences to form the representation of the primitives within the model. Then, the model is modified and trained using complex sequences composed of a combination of primitives. Experiments are conducted using drawing motions on a canvas to show the representation of learned sequences within the model, showing the capability of the method to learn complex sequences as a combination of primitives.

The rest of the paper is organized as follows. In Sect. 2, related work are presented. In Sect. 3, MTRNN model is described with the overview of the proposed training method. In Sect. 4, the experimental setup is presented. In Sect. 5, experiments with the analysis of the model is presented. In Sect. 6, discussions based on the experimental results are presented. Conclusions are presented in Sect. 7.

2   Related Work

In this section, we present studies related to this work.

2.1   Hierarchical Training of Primitives

In human infants, "scaffolding" is said to be an important feature for learning skills [2]. "Scaffolding" is a process where a caregiver adjusts the difficulty of a given task based on the developmental stage of the infant. At an early stage, the caregiver would provide simple tasks to learn primitive skills, and gradually raise the difficulty of the task as the infant develops. Such phenomenon have been shown to be effective in robotic applications. Triesch succeeded in promoting the training of creating joint attention through interaction between caregivers and learners [3]. Application of "scaffolding" to robot motion or behavior learning has been defined as motionese and have shown great potential [4]. Akgun proposed a system that extracts motion primitives based on a human's utterance to detect the key frame of a task [5]. In this paper, we focus more on self-organization of representation of primitives within the model using a hierarchically structured neural network model.

2.2   Drawing Development and Robots

In this study, we create a learning model using drawing data, aiming for application to developmental drawing robots. Drawing is one of the human skills that primitives take an important role. Based on Luquet, human infants are said to develop drawing skills based on five stages. Through the five stages, human infants learn to draw lines, circles, squares, and so on, which are primitives to learn further complex shapes. In this way, primitives evolve through the five stages through the infant's drawing experience.

Consideration of primitives has also been conducted in creating drawing robots. Mohan focused on Critical Points (CP), which are define primitive segments, and succeeded in drawing basic figures by combining predefined motion primitives [6]. Mochizuki focused more on the developmental aspect, incorporating the five stage development by Luquet into the robot's developmental model [8]. Insertion of pause in the caregiver's motion has made learning of primitives more clearer [9]. An issue left unsolved in these works is development of the learning model based on learning sequences. In this paper, we focus on modifying the learning model to acquire representation of primitives within the model.

3   Proposed Method

In this section, we present the proposed method with an overview of the recurrent neural network used in this work.

3.1   Multiple Timescale Recurrent Neural Network (MTRNN)

In this study, Multiple Timescale Recurrent Neural Network (MTRNN) [10] is used as the learning model. MTRNN is a two-layered recurrent neural network with a hierarchical structure of context neurons. The current state at time t is input into MTRNN to calculate the next state at time $t + 1$. It consists of three layers of neurons: IO (Input/Output) neurons, C_f (Fast-Context) neurons, and C_s (Slow-Context) neurons. The neurons in the input layer are fully connected to the neurons in the output layers except for the IO neurons and C_s neurons. A time constant value is set to each neuron, increasing in the order or IO, C_f, C_s, to determine the firing rate. The difference of time constants enables MTRNN to learn different levels of information within the model. The structure of MTRNN is shown in Fig. 1.

There are three functions in MTRNN: learning, recognition, and generation. Learning: The weight between each neuron is updated based on Back Propagation Through Time (BPTT). The initial values of C_f and C_s neurons (C_f and $C_s(0)$) which correspond to each data are obtained. Recognition: The $C_f(0)$ and $C_s(0)$ values that represent a given sequence is calculated using the trained MTRNN. Generation: Calculate a sequence for given $C_f(0)$ and $C_s(0)$ values through forward calculation.

Learning. Learning of MTRNN is performed by calculating the output value of each time series by forward calculation and updating the weights by BPTT. The data at each step in a training sequence is input into MTRNN, while the data of the next step is used as teacher data for the calculated output. The

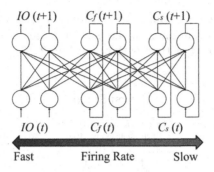

$IO(t+1)$ $C_f(t+1)$ $C_s(t+1)$

$IO(t)$ $C_f(t)$ $C_s(t)$

Fast   Firing Rate   Slow

Fig. 1. Structure of MTRNN

output error, calculated as the squared difference between the teacher data and calculated output, is back propagated through the time series to update the weight values and $C_f(0)$ and $C_s(0)$ values. The $C_f(0)$ and $C_s(0)$ values are calculated independently for each training sequence, representing the self-organized dynamic property of the sequence.

Recognition. Recognition is calculated through a similar process as learning. Given a sequence for recognition, the output values for each step are calculated through forward calculation. The output error is back propagated to update only the $C_f(0)$ and $C_s(0)$ values. The weights of MTRNN are fixed during the calculation process. The converged $C_f(0)$ and $C_s(0)$ values represent the given sequence.

Generation. Generation is done through forward calculation to obtain a sequence for given $C_f(0)$ and $C_s(0)$ values.

3.2 Proposed Method

In this paper, we propose a method to self-organize primitive motions in MTRNN to recognize complex motions as a combination of primitives. The proposed method is divided into two stages.

1. Train MTRNN with primitive motions.
2. Train MTRNN with complex motions.

In the first stage, primitive motions are used to train MTRNN. In this stage, the C_s neurons are not attached to MTRNN. Thus, training of MTRNN would self-organize the dynamics of primitive motions into the C_f neurons.

In the second stage, complex motions that consist of a combination of primitive motions are used to training MTRNN. The weights of MTRNN trained in the first stage are used, and the C_s neurons are attached to the model. Training of the model would self-organize dynamics of complex motions using the primitive motions in the C_f neurons.

In this research, we use single stroke drawing data for training MTRNN. Therefore, the (x, y) coordinates of the pen are input into the IO neurons. An overview of the proposed method is shown in Fig. 2.

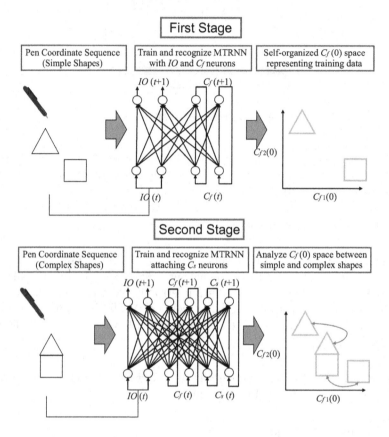

Fig. 2. Overview of proposed method

4   Experimental Setup

For acquisition of drawing data, we use a Liquid Crystal Display (LCD) tablet, Cintiq13HD (DTK-1301/KO), developed by Wacom Co., Ltd. (Fig. 3). During drawing with the LCD tablet, the coordinate values were obtained at 10 frames/second. For primitive motions, a human subject was asked to draw squares and triangles. Several types of data were acquired by drawing from different vertices and directions. A combination of the square and triangle, drawing a house, was used for complex motions. Each data consists of coordinate sequences obtained while drawing the figures three times recursively. An example of drawing a house is shown in Fig. 4.

Fig. 3. LCD tablet for data acquisition   **Fig. 4.** Drawing example of house

After data acquisition, MTRNN was trained using the proposed method. In the first stage, the data sequences of drawing squares and triangles were used to traing MTRNN without C_s neurons. Several setups of MTRNN were trained in the first stage. Evaluating the training error of each setup, the setup with the best performance was used for additional training in the second stage. In the second stage, C_s neurons were attached to the model trained in the first stage, and data sequences of drawing house were used to additionally train MTRNN. The setup of MTRNN used in the experiment, shown in Table 1, was decided experimentally.

Table 1. Setup of MTRNN

Neuron type	Number of neurons	Time constant
IO	2	2
C_f	60	5
C_s	2	70

5   Experimental Result and Analysis

Generation results of the model of drawing sequences of a house is shown in Figs. 5 and 6. In both figures, the purple line represents the actual drawing of the human subject, while the green line represents the generated drawing by the model. Figure 5 shows a success example of the drawing sequence generation, while Fig. 6 shows a failure example. In Fig. 6, the model failed to learn the primitives, and thus, generation of drawing sequence was mainly done in the vicinity of the square.

To analyze the success result shown in Fig. 5, we traced the sequence of C_f neurons. While most of the C_f neurons were inactive throughout the sequence,

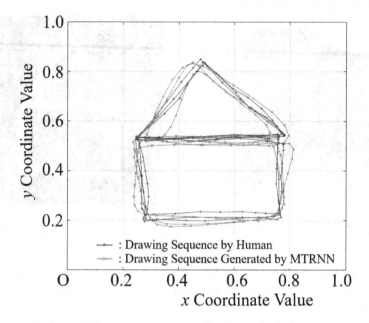

Fig. 5. Success example of drawing sequence generation of house

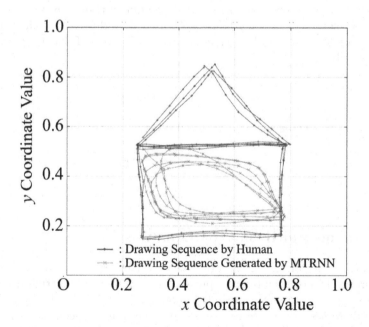

Fig. 6. Failure example of drawing sequence generation of house

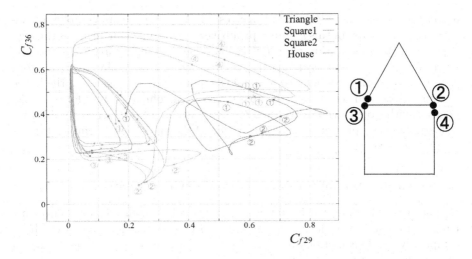

Fig. 7. Transition of Two C_f neuron values when drawing house

there were several C_f neurons that changed greatly along time. For analysis, we selected the two most active C_f neurons (C_{f29} and C_{f36}) which are expected to possess more information of the sequence than other inactive neurons. The transition of the values of the C_{f29} and C_{f36} neurons are shown in Fig. 7. The actual drawing of the house was done as shown in the right half of Fig. 7, starting from vertex ①, drawing a triangle to ②, then drawing the circle from vertex ③ to vertex ④. Drawing of the house was done three times recursively.

In Fig. 7, the yellow line represents the C_f value sequence of the house, the green and blue lines represent the C_f value sequences of two squares, and the purple line represents the C_f value sequence of a triangle. It is notable that the C_f value of the house starts in the vicinity of the triangle (① to ②), then moves to the C_f of the square (③) and then moves back to the triangle (④). The sequence is traced three times. The result of the analysis implies that the model learned that drawing of the house is a combination of drawing the triangle and the square.

6   Discussions

From the experimental results, it was confirmed that the proposed method is capable of learning sequences as a combination of primitive sequences. In this section, we discuss the results based on two aspects.

6.1   Self-organization of Primitives

In this paper, we regarded squares and triangles as primitives, to learn a house as complex shape. A more primitive figure could be considered as lines or arcs, while a house might be considered as primitives when drawing more complex scenes with several houses. In this manner, primitives are not fixed and evolves through experience. Representation of primitives is done by the hierarchical structure, where C_f neurons were used to learn squares and circles, while the C_s neurons were used to learn the whole sequence of the house. When considering the house as a primitive, and scene with multiple houses as a complex target for example, neurons with slower firing rates than C_s neurons could be attached to MTRNN to learn more complex sequences. Evaluation of the model using complex targets is left as future work.

6.2   Incorporation of the Model into Drawing Robot

An issue in the work by Mochizuki [8] is that, the robot is only capable of drawing basic figures, such as squares, triangles, and circles. While the robot starts from completely random motions to learn to draw basic shapes, the complexity of various shapes induces difficulty in learning and generating complex shapes. The proposed model lays out a potential for MTRNN to developmentally learn various shapes through evolution of primitives. In such robotics application, the association of drawing coordinate sequence and robot motion is still left as an issue for future work.

7   Conclusion

In this paper, we proposed a method to self-organize primitives within the model through two stages of training. MTRNN was used as the training model since it possesses several layers of neurons constructed hierarchically to learn different levels of information. In the first stage, primitive motions are used to train MTRNN without C_s neurons. In this stage, the primitives are represented within the C_f neurons of MTRNN. In the second stage, complex motions consisting of a combination of primitive motions are used to train MTRNN by attaching C_s neurons to the MTRNN trained in the first stage. Training in the second stage learns the complex motions based on the primitive motions self-organized in C_f neurons.

The method was evaluated using drawing sequences, squares and triangles as primitives and house as the complex motion. The results of the experiment showed that the model was capable of generating the drawing sequence of the house, while drawing failures did exist. Analysis of the C_f neurons for the succeeded drawing of the house has shown that the model learned the dynamics of the drawing the house as a combination of the dynamics of square and triangle (primitives).

As future work, we plan to evaluate the method with more complex targets. Further on, the model will be incorporated into a developmental drawing robot system. We expect that our work would contribute to the advancement of the field of cognitive developmental robotics.

Acknowledgments. This research was supported by JSPS KAKENHI Grand Number 16H05877.

References

1. Asada, M., MacDorman, K., Ishiguro, H., Kuniyoshi, Y.: Cognitive developmental robotics as a new paradigm for the design of humanoid robots. Rob. Auton. Syst. **37**, 185–193 (2001)
2. Berk L.E., Winsler A.: Scaffolding Children's Learning: Vygotsky and Early Childhood Education. National Association for Education (1995)
3. Triesch J., Teuscher C., Deák G.O., Carlson, E.: Gaze following: why (not) learn it? Dev. Sci. **9**(2), 125–147 (2006)
4. Brand, R.J., Baldwin, D.A., Ashburn, L.A.: Evidence for 'motionese': modifications in motions' infant-directed action. Dev. Sci. **5**(1), 72–83 (2002)
5. Akgun, B., Cakmak, M., Yoo, J.W., Thoma, A.L.: Augmenting kinesthetic teaching with keyframes. In: Proceedings of International Conference on Machine Learning, Workshop on New Developments in Imitation Learning (2011)
6. Mohan, V., Morasso, P., Metta, G., Kasderidis, S.: Teaching a humanoid robot to draw 'shapes'. Auton. Rob. **31**(1), 21–53 (2011)
7. Luquet, G.H.: Le Dessin Enfantin (1927)
8. Mochizuki, K., Nishide, S., Okuno, H.G., Ogata, T.: Developmental human-robot imitation learning of drawing with a neuro dynamical system. In: Proceedings of IEEE International Conference on Systems, Man, and Cybernetics, pp. 2336–2341 (2013)
9. Nishide, S., Mochizuki, K., Okuno, H.G., Ogata, T.: Insertion of pause in drawing from babbling for robot's developmental imitation learning. In: Proceedings of IEEE International Conference on Robotics and Automation, pp. 4785–4791 (2014)
10. Yamashita, Y., Tani, J.: Emergence of functional hierarchy in a multiple timescale recurrent neural network model: a humanoid Robot Experiment. PLoS Comput. Biol. **4**(11), e1000220 (2008)

Compressing and Interpreting SOM-Based Convolutional Neural Networks

Ryotaro Kamimura[1,2(✉)]

[1] Kumamoto Drone Technology and Development Foundation,
Techno Research Park, Techno Lab 203, 1155-12 Tabaru Shimomashiki-Gun,
Kumamoto 861-2202, Japan
ryotarokami@gmail.com
[2] IT Education Center, Tokai Univerisity,
4-1-1 Kitakaname, Hiratsuka, Kanagawa 259-1292, Japan

Abstract. The present paper aims to propose a new type of interpretation method by model compression to deal with complex multi-layered neural networks. For illustration, we use here a new type of learning method with self-organizing maps (SOM) accompanied by convolutional neural networks (CNN) and controlled selectivity to choose important connection weights. Naturally, we face much difficulty in interpreting the SOM with CNN due to complicated feature extraction procedures. A new type of model compression is introduced, where the convolutional parts are black-boxed to make the model compression easier. We applied the new method to the wireless localization data set. The results showed that improved generalization performance was obtained by the extraction of features of four classes by the SOM. In addition, the model compression method made it possible to extract linear relations between inputs and outputs.

Keywords: Compression · Linear interpretation · SOM · CNN · Generalization · Selectivity

1 Introduction

The present paper aims to propose a new model compression method for interpreting complex multi-layered neural networks. The new method is based on the model compression by partially black-boxing some parts of multi-layered neural networks. The model is applied to a new learning method where the self-organization maps (SOM) and convolutional neural networks (CNN) are combined, accompanied by a method to control the selectivity of connection weights or neurons.

Machine learning and, in particular, neural networks have been extensively used in many application fields recently, producing an ambivalent attitude either

© Springer Nature Switzerland AG 2020
H. Fujita et al. (Eds.): IEA/AIE 2020, LNAI 12144, pp. 732–744, 2020.
https://doi.org/10.1007/978-3-030-55789-8_63

toward or against them. One of the main problems lies in difficulty in explaining the inference mechanism; namely, we have had much difficulty in showing how and why decisions are made [9,31]. Thus, there have had many attempts to explain the inference mechanism in machine learning. Basically, we have two types of interpretation methods, namely, internal and external methods. The external methods consider the neural networks as black-box and try to infer the mechanism by seeing the responses to the inputs. In the convolutional neural networks, dealt with in this paper, the majority of interpretation methods should be considered external. For example, one of the most popular interpretation methods is the activation maximization and selectivity or perturbation analysis, which lies in detecting the input patterns maximizing the corresponding outputs [8,26] or the most sensitive features [18,22], including the layer-wise relevance propagation (LRP) [4]. In those methods, the interpretation of the internal mechanism is replaced by the examination of input patterns or the corresponding features, and they are heavily dependent on the intuitively interpreted properties of images.

The internal methods have been said to interpret the inference mechanism itself [16,30]. However, and actually, those methods tried to interpret the mechanism by using external methods such as symbolic rules. Moreover, they tried to include methods to force the learning to produce more interpretable results. It has been said that the interpretable models should incorporate additional procedures exclusively for interpretation [16]. On the contrary, we propose a method called "neral compressor" to interpret internally the main inference mechanism. This method aims to extract the simplest prototype model by compressing original multi-layered neural networks. Model compression has been used to transform complex networks into simpler ones [2,6,10], to deal with more and more complicated network architectures. In model compression, the complicated networks are imitated by the corresponding simpler ones. This is realized mainly by using the outputs from the complex and original networks by black-boxing the inner processes. Our proposed method aims to interpret the main inference mechanism internally by direct model compression. The original and complex networks can be compressed to interpret the main inference mechanism by examining the compressed information. In particular, the compression method can produce linear interpretability, which has received due attention in interpreting convolutional networks [1,5,17], though they are typical external methods. However, in complex multi-layered neural networks, there are several parts against the direct compression, where the information flow between layers becomes weaker or cut off, preventing us from compressing the networks. In this context, we use a hybrid method for interpretation, meaning that we try to interpret the internal inference mechanism for itself as much as possible, while some complicated parts are temporally black-boxed.

We show here one example of application of compression in which the SOM and the CNN are combined. In neural networks, one of the most important feature-extraction methods is the SOM. The SOM has been one of the well-known techniques in neural networks and used to compress multi-dimensional data into a lower-dimensional space [19,20]. In particular, SOM have been modified extensively for visualization and classification [7,25], [32,33] and applied to a variety of application fields. Naturally, rich information by SOM has produced many methods to make full use of it for supervised learning [12,13,21,28,29]. However, we have not yet developed methods to make full use of rich information produced by the SOM. This is because the main information produced by SOM tends to be represented over two-dimensional or three-dimensional feature spaces. The majority of those hybrid methods have not considered those higher-dimensional features spaces themselves. We think that this limitation is one of the most serious problems in SOM-based supervised learning. For this problem, we introduce here the convolutional neural network, which has been mainly developed for the two- or three-dimensional image data sets. By introducing the CNN, we can expect to develop a method to take into account higher-dimensional information created by the SOM.

Naturally, we must face difficulty in the interpretation of inference mechanism by introducing the CNN. In the CNN, many different kinds of layers, such as convolutional, max-pooling, must be introduced. For example, in the process of max-pooling, the inputs to the layer are reduced abruptly, which cuts off, in a certain sense, the smooth information flow from inputs to outputs. Our proposed method aims to interpret the main inference mechanism internally by direct model compression. The original and complex networks must be compressed as much as possible, and we can expect to interpret the main inference mechanism by examining the compressed information. However, at the present stage of research, the convolutional or max-pooling layers cannot be compressed by this method. Thus, these layers should be black-boxed and estimated by some other learning methods.

Finally, we should point out another problem with CNN, namely, the complexity problem. CNN have a naturally complicated network architecture, and usually combines many different types of layers such as convolutional, max-pooling, and so on, as was mentioned, and the number of parameters to be tuned tends to increase rapidly. Though many different types of regularization methods have been developed, such as the dropout, noise-injection, weight decay [3,11,24,27], those methods take a passive attitude for detecting important connection weights as well as neurons, meaning that important features can be obtained by randomly suppressing or choosing connection weights or neurons. We think that this passive attitude has been against the interpretation of the final results, because the interpretation is very contrary to the random feature extraction. Contrary to the passive attitude, we use the positive attitude for extracting important features, and choose those considered important by predetermined criteria. This extraction of important components is called "selective control," because we first define the selectivity of components and try to control the selectivity, making it possible to control the number of parameters inside. This selective control method can contribute to improved generalization as well as interpretation.

2   Theory and Computational Methods

2.1   SOM-Based Convolutional Neural Networks

We propose here a model that combines the SOM with the CNN. For extracting information from input patterns, the SOM have been one of the well-known methods in neural networks, because the method lies in imitating and representing input patterns as faithfully as possible. In particular, the SOM has an ability to visualize input information over multi-dimensional spaces. However, this rich information has not necessarily been utilized for training supervised learning. This is mainly because the SOM can produce multi-dimensional feature spaces, and the conventional supervised neural networks cannot extract important information represented over multi-dimensional spaces.

In this context, we here propose a method to combine the SOM with the CNN, because the CNN has been developed to deal with two- or three-dimensional data of images. We have roughly seven layers, including an input (1) and output (7) layer, as shown in Fig. 1(a). This network architecture was used actually in the experiments discussed below. In the convolutional components, there are three layers, accompanied by the max-pooling layer to reduce

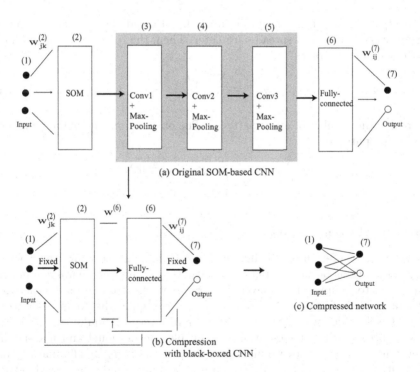

(a) Original SOM-based CNN

(b) Compression with black-boxed CNN

(c) Compressed network

Fig. 1. Original network architecture (a), black-boxed CNN (b) and final compressed network (c).

the number of neurons as much as possible to make the computation shorter and easier.

2.2 SOM

We briefly explain here the learning algorithm for SOM. Let $\mathbf{x}^s$ and $\mathbf{w}_j$ denote input and weight column vectors, then distance between input patterns and connection weights is computed by

$$\|{}^s\mathbf{x} - \mathbf{w}_j^{(2)}\|^2 = \sum_{k=1}^{n_1} \left({}^s x_k - w_{kj}^{(2)}\right)^2. \tag{1}$$

The ${}^s c$th winning neuron is obtained by

$${}^s c = \operatorname{argmin}_j \|{}^s\mathbf{x} - \mathbf{w}_j^{(2)}\| \tag{2}$$

We consider the following neighborhood function usually used in self-organizing maps

$$h_{j^s c} = \exp\left(-\frac{\|\mathbf{r}_j - \mathbf{r}_{c^s}\|^2}{2\sigma_\gamma^2}\right) \tag{3}$$

where $\mathbf{r}_j$ and $\mathbf{r}_{s c}$ denote the position of the jth and the ${}^s c$th neuron on the output space and σ_γ is a spread parameter. Then, the re-estimation equation in the batch mode becomes

$$\mathbf{w}_j^{(2)} = \frac{\sum_{s=1}^{q} h_{j^s c}{}^s\mathbf{x}}{\sum_{s=1}^{q} h_{j^s c}}. \tag{4}$$

where q is the number of input patterns. The output from the SOM component can be computed by

$${}^s v_j^{(2)} = \exp\left(-\frac{\|{}^s\mathbf{x} - \mathbf{w}_j^{(2)}\|^2}{\sigma}\right) \tag{5}$$

2.3 Model Compression

We try to interpret the main inference mechanism of the neural networks in terms of relations between inputs and outputs by using the neural compressor, which aims to compress multi-layered neural networks into the simplest networks without hidden neurons [14]. However, when we apply the compressor to the SOM+CNN model, we have a problem of hard interpretation of the CNN component, because the interpretation of multiple convolutional and max-pooling layer is not so easy. Thus, we black-box those layers here. By eliminating three convolutional layers in Fig. 1(a), we can have a simpler and compressed network with only four layers in Fig. 1(b), which can be further compressed into the simplest one in Fig. 1(c). The learning of this four-layered neural network is done by fixing weights for the SOM as well as the fully connected layer. Note that, for making the adjustment easier, the bias neurons are trained in this four-layered network.

Now, suppose that the learning of four-layered neural networks is successfully done, then we can compress the network into the simplest one. First, two connection weights of the last two layers, namely, $w_{ij}^{(7)}$ and $w_{jj'}^{(6)}$, are combined into

$$w_{ij'}^{(7*6)} = \sum_{j=1}^{n_6} w_{ij}^{(7)} w_{jj'}^{(6)} \tag{6}$$

Then, the final compressed weights are

$$w_{ik}^{(7*2)} = \sum_{j=1}^{n_2} w_{ij}^{(7*6)} w_{jk}^{(2)} \tag{7}$$

The final and compressed weights represent relations between inputs and outputs directly for easy interpretation. For stable interpretation, we should average the results by compression over many different initial conditions. However, for now we could obtain very stable results, presented below, almost independently of initial conditions.

2.4 Selectivity

For making the interpretation easier and having improved generalization, we propose a selectivity of neurons actually measured by the strength of connection weights [15]. The selectivity becomes higher when the number of connection weights, considered important, becomes smaller. Thus, the method is similar to the conventional regularization method. However, one of the main differences is that we try to pre-define the importance, and we try to extract important components as much as possible. The basic hypothesis is that in neural networks the selectivity of neurons or connection weights should be appropriately controlled for improved generalization.

We define here the selectivity of connection weights for three convolutional layers as well as the fully connected layer. The computational procedures for the selectivity for all those layers are completely the same. For simplicity, we introduce here the selectivity for the connection weights of the fully connected layer. First, we suppose that the importance of connection weights can be measured by the absolute strength of connection weights as the first approximation. For the connection weights to the output layer $w_{ij}^{(7)}$ the absolute values of the weights are computed by

$$u_{ij}^{(7)} = \mid w_{ij}^{(7)} \mid \tag{8}$$

Then, normalizing those absolute weights by maximum absolute weights, we have

$$\alpha z_{ij}^{(7)} = \left(\frac{u_{ij}^{(7)}}{u_{\max}^{(7)}} \right)^{\alpha} \tag{9}$$

where α is a parameter to control the strength of selectivity, and the maximum operation is applied to all connection weights in the corresponding layer. The selectivity can be defined by

$$\alpha_{\phi^{(7)}} = \frac{n_7 n_6 - \sum_{i=1}^{n_7} \sum_{j=1}^{n_6} {}^{\alpha}z_{ij}^{(7)}}{n_7 n_6 - 1} \tag{10}$$

where the number of neurons in the layers should be larger than one, because the selectivity can be defined only for multiple neurons. When this selectivity becomes one, only one connection weight become larger, while all the other weights become smaller. On the contrary, when the selectivity becomes smaller, all connection weights become as equal as possible, and the selectivity becomes zero. Naturally, when the parameter α becomes larger, the selectivity becomes larger, which makes it possible to control the selectivity

The selectivity is incorporated into learning by multiplying connection weights by the corresponding selectivity. For example, for the nth learning step, we have

$$w_{ij}^{(7)}(n+1) = {}^{\alpha}z_{ij}^{(7)} w_{ij}^{(7)}(n) \tag{11}$$

This multiplication is applied only once in a learning step, followed by several learning epochs within a learning step to assimilate the selectivity.

3   Results and Discussion

3.1   Wireless Indoor Localization Data Set

Experimental Outline. We present here one experimental result to show that the present method could extract the very linear relations between inputs and outputs by compressing the original SOM+CNN networks. The data set was used to determine indoor locations by the WiFi signal strengths [23]. The number of inputs was seven, representing the strength of WiFi signals at four different locations, corresponding to four outputs. The number of input patterns was 2,000. The network architecture was the same as that described in Fig. 1 except for the number of neurons. The number of neurons in the SOM component was 400 (20 by 20), which could be determined so as to increase generalization performance. The three convolutional components with max pooling were used, where 50 filters with 5 by 5 size were used, and 2 by 2 max pooling operations were adopted for all layers to reduce the number of neurons. The data set was first divided into the training (70%) and testing (30%) data sets. The selectivity parameter α was set to 0.1 for the convolutional layers, while it was set to 0.05 for the fully connected layer. All these parameter values were used to increase generalization performance by the preliminary experiments. Then, with at least ten different initial conditions, we trained the same networks, and the results with the best generalization performance were reported. We should report that little differences were observed, depending on different initial conditions.

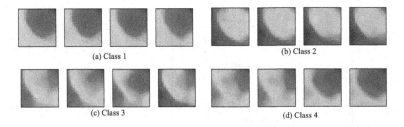

(a) Class 1   (b) Class 2

(c) Class 3   (d) Class 4

Fig. 2. The outputs from the SOM component for four classes.

Outputs from the SOM Component and Generalization. Figure 2 shows the outputs from the SOM component. As can be seen in the figures, all outputs were similar to each other, and inputs seem to be classified by the neurons on the diagonal position. However, four classes are explicitly distinguished by the strength of outputs. For example, class No. 1 in Fig. 2(a) divided the upper and lower parts of the feature maps the most clearly, and gradually only neurons on the diagonal position remained strong. This means that the SOM could clearly distinguish between different classes on the feature maps, and the information on this division surely can be used for the supervised learning. Thus, we tried to confirm whether those differences could contribute to the overall improvement, in particular, for generalization performance.

Generalization and Selectivity. Figure 3 shows generalization errors by five methods. All the errors were the best ones of those obtained by ten different initial conditions. The generalization errors by the present method of selectivity control decreased rapidly when the number of learning steps increased, and reached the lowest error of 0.0067 with 13 learning steps. Then, the errors gradually increased when the number of steps increased, meaning that over-training occurred. The second best error of 0.0117 was obtained by the four-layered neural networks with only a SOM component and without weights from the CNN component. The simple networks with three layers and without a SOM and CNN component produced the third best error of 0.015. The compressed networks from the networks with SOM and CNN components produced the fourth best error of 0.02. This means that the weights from the CNN and SOM could not contribute to the improvement of generalization of compressed networks. Finally, the worst error of 0.0283 was by the random forest method. Thus, the selectivity control method with the SOM and CNN component surely increased the generalization performance, but the knowledge could not necessarily contribute to the improvement of compressed networks. This can be explained by the fact that the SOM+CNN networks could produce the linear relations between inputs and outputs, which could not be extracted by the simple correlation coefficients between inputs and targets discussed below and shown in Fig. 5(a).

Figure 3(b) shows the values of selectivity of all layers in the CNN component. The selectivity of the fully connected and the first convolutional layer were

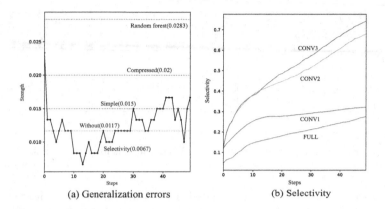

(a) Generalization errors   (b) Selectivity

Fig. 3. Generalization errors by five methods (a) and selectivity for four layers (b) for the wireless localization data set.

smaller. As mentioned above, the selectivity parameter for the fully connected layers was set to be smaller than those for the convolutional layer, which explains well the lower selectivity. However, the parameters for all three convolutional layers were the same (0.1). Thus, the lower selectivity of the first convolutional layer should be explained.

Figure 4 shows the input patterns that maximize the activations of the neurons or filters of the second and the third convolutional layer [8]. The filters of the first convolutional layer could not produce meaningful activation patterns, and thus the figures were omitted. Compared with the activations of the filters of the first convolutional layer, the filters of the second and the third convolutional layers showed explicit features. Those features seem to represent those extracted by the SOM, as shown in Fig. 2. The results suggest that, as the selectivity becomes higher, more explicit features can be obtained.

Interpretation. Finally, we tried to examine the relations between inputs and the corresponding targets by compressing the original networks with the SOM and CNN component.

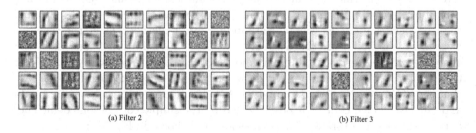

(a) Filter 2   (b) Filter 3

Fig. 4. Patterns with maximum activations from the filters of the second and third convolutional layer for the wireless localization data set.

First, we computed the correlation coefficients between inputs and the first target of four for the original and full data set for simplicity's sake. For all the other targets, similar results could be obtained. Figure 5(a) shows the correlation coefficients between inputs and targets of the original data set. As can be seen in the figure, the first, the fourth, the sixth, and the seventh input should play major roles to identify the first location. Figure 5(b) shows the compressed weights from the simple three-layered neural networks (one hidden layer) without the SOM and CNN component. As can be seen in the figure, these weights were similar to the correlation coefficients in Fig. 5(a). The only difference was that the sixth and seventh inputs showed less importance in terms of their strength. Compressed networks from the SOM and CNN networks showed the same tendency, that the compressed weights were similar to the correlation coefficients in Fig. 5(c). However, the compressed networks emphasized the negative parts of correlation coefficients. For example, the second, the third, and fifth input had negative correlation coefficients between inputs and targets. Those negative coefficients were strengthened in the compressed weights in Fig. 5(c). As explained in Fig. 3(a), the compressed networks could not improve generalization performance, compared with the simple networks without compression. These results show that the features not detected by the corresponding simple networks prevented the compressed networks from having better generalization performance. Finally, the random forest could not produce importance values similar to the correlation coefficient, partially because the random forest cannot deal with the negative effects. However, when we examine closely the importance values in Fig. 5(d), the random forest extracted features different from the correlations

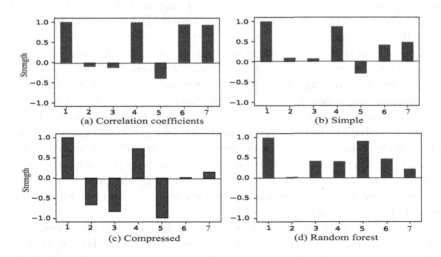

Fig. 5. Original correlation coefficients between inputs and targets for the full data set(a), weights by the simple network without knowledge from SOM and CNN (b), compressed weights from SOM and CNN networks with selectivity control (c), and the importance by the random forest (d) for the wireless localization data set. All the values were normalized by the maximum absolute ones for easy comparison.

coefficients. For example, input variable No.4 showed smaller importance, while it had the higher correlation coefficient in the original correlation coefficients in Fig. 5(a). As shown in Fig. 3, the random forest produced the worst generalization errors, which may be explained by this fact.

4 Conclusion

The present paper aimed to apply the compression method to more complicated network architecture by black-boxing some complicated parts. For the application, we used a new type of learning method for training multi-layered neural networks for interpretation. The new method tried to make full use of rich information created by SOM and the ability to process higher-dimensional data sets by the CNN component. Because we had difficulty in interpreting the CNN component, we introduced a model compression method in which the inference mechanism of the CNN component is estimated by using networks without a CNN component. In the CNN component, we need a number of weights, degrading the generalization and interpretation. For this, we introduced a method to control the selectivity of neurons as well as connection weights. When the selectivity becomes higher, a smaller number of neurons or connections tend to be used.

The method was applied to the wireless localization data set in which we must localize four positions by seeing the wireless strength. The results confirmed that the SOMs could clearly produce distinguished class features on a two-dimensional feature space. This detection of class features made it possible to improve generalization performance by the ability of CNN to process the two-dimensional feature maps. In addition, by examining the filters in the CNN component, we had a possibility that the selectivity increase could be related to the feature detection by the filters.

Then, we tried to estimate the relations between inputs and outputs by model compression. The results showed that the correlation coefficients between inputs and targets were extracted by simple neural networks. The present method seems to emphasize some specific correlations coefficients, which could not be detected by simply computing the correlation coefficients of the original data set. These results show that the model compression from SOM and CNN components show a possibility that the complex behaviors of neural networks can be mainly interpreted by the linear relations between inputs and outputs (linear interpretation), and then some differences between the correlations by the present method and correlation coefficients of the original data set are related to the improved generalization of neural networks.

One of the major problems of this method lies in the long computational time. This is because we try to assimilate the effect of selectivity gradually so as to smoothly compromise between improved generalization and interpretation. Thus, for more practical purposes, we need to shorten the computational time as much as possible.

The present results were very preliminary, and a lot of works remains to be done. For example, we can point out at least three future directions. First, we need to interpret more exactly which features created by the SOM are related actually to the improved generalization performance. Second, a more exact approach to the interpretation of the CNN component is necessary, and we should open up the black-box. Third, we should elaborate a more exact approach to interpret the relations between inputs and outputs.

References

1. Alain, G., Bengio, Y.: Understanding intermediate layers using linear classifier probes. arXiv preprint arXiv:1610.01644 (2016)
2. Ba, J., Caruana, R.: Do deep nets really need to be deep? In: Advances in Neural Information Processing Systems, pp. 2654–2662 (2014)
3. Bach, F., Jenatton, R., Mairal, J., Obozinski, G.: Optimization with sparsity-inducing penalties. Found. Trends® Mach. Learn. **4**(1), 1–106 (2012)
4. Bach, S., Binder, A., Montavon, G., Klauschen, F., Müller, K.R., Samek, W.: On pixel-wise explanations for non-linear classifier decisions by layer-wise relevance propagation. PLoS ONE **10**(7), e0130140 (2015)
5. Bau, D., Zhou, B., Khosla, A., Oliva, A., Torralba, A.: Network dissection: quantifying interpretability of deep visual representations. In: Proceedings of the IEEE Conference on Computer Vision and Pattern Recognition, pp. 6541–6549 (2017)
6. Buciluă, C., Caruana, R., Niculescu-Mizil, A.: Model compression. In: Proceedings of the 12th ACM SIGKDD International Conference on Knowledge Discovery and Data Mining, pp. 535–541. ACM (2006)
7. De Runz, C., Desjardin, E., Herbin, M.: Unsupervised visual data mining using self-organizing maps and a data-driven color mapping. In: 2012 16th International Conference on Information Visualisation (IV), pp. 241–245. IEEE (2012)
8. Erhan, D., Bengio, Y., Courville, A., Vincent, P.: Visualizing higher-layer features of a deep network. Univ. Montreal **1341**(3), 1 (2009)
9. Goodman, B., Flaxman, S.: European union regulations on algorithmic decision-making and a right to explanation. arXiv preprint arXiv:1606.08813 (2016)
10. Hinton, G., Vinyals, O., Dean, J.: Distilling the knowledge in a neural network. arXiv preprint arXiv:1503.02531 (2015)
11. Hinton, G.E., Srivastava, N., Krizhevsky, A., Sutskever, I., Salakhutdinov, R.R.: Improving neural networks by preventing co-adaptation of feature detectors. arXiv preprint arXiv:1207.0580 (2012)
12. Ismail, S., Shabri, A., Samsudin, R.: A hybrid model of self-organizing maps (SOM) and least square support vector machine (LSSVM) for time-series forecasting. Expert Syst. Appl. **38**(8), 10574–10578 (2011)
13. Jin, F., Qin, L., Jiang, L., Zhu, B., Tao, Y.: Novel separation method of black walnut meat from shell using invariant features and a supervised self-organizing map. J. Food Eng. **88**(1), 75–85 (2008)
14. Kamimura, R.: Neural self-compressor: collective interpretation by compressing multi-layered neural networks into non-layered networks. Neurocomputing **323**, 12–36 (2019)
15. Kamimura, R., Takeuchi, H.: Sparse semi-autoencoders to solve the vanishing information problem in multi-layered neural networks. Appl. Intell. **49**(7), 2522–2545 (2019). https://doi.org/10.1007/s10489-018-1393-x
16. Kim, B., Shah, J.A., Doshi-Velez, F.: Mind the gap: a generative approach to interpretable feature selection and extraction. In: Advances in Neural Information Processing Systems, pp. 2260–2268 (2015)

17. Kim, B., et al.: Interpretability beyond feature attribution: quantitative testing with concept activation vectors (TCAV). arXiv preprint arXiv:1711.11279 (2017)
18. Koh, P.W., Liang, P.: Understanding black-box predictions via influence functions. arXiv preprint arXiv:1703.04730 (2017)
19. Kohonen, T.: Self-Organization and Associative Memory. Springer, New York (1988). https://doi.org/10.1007/978-3-642-88163-3
20. Kohonen, T.: Self-Organizing Maps. Springer, Heidelberg (1995). https://doi.org/10.1007/978-3-642-56927-2
21. Ohno, S., Kidera, S., Kirimoto, T.: Efficient automatic target recognition method for aircraft SAR image using supervised SOM clustering. In: 2013 Asia-Pacific Conference on Synthetic Aperture Radar (APSAR), pp. 601–604. IEEE (2013)
22. Ribeiro, M.T., Singh, S., Guestrin, C.: Why should i trust you?: Explaining the predictions of any classifier. In: Proceedings of the 22nd ACM SIGKDD International Conference on Knowledge Discovery and Data Mining, pp. 1135–1144. ACM (2016)
23. Rohra, J.G., Perumal, B., Narayanan, S.J., Thakur, P., Bhatt, R.B.: User localization in an indoor environment using fuzzy hybrid of particle swarm optimization & gravitational search algorithm with neural networks. In: Deep, K., et al. (eds.) Proceedings of Sixth International Conference on Soft Computing for Problem Solving. AISC, vol. 546, pp. 286–295. Springer, Singapore (2017). https://doi.org/10.1007/978-981-10-3322-3_27
24. Rudy, J., Ding, W., Im, D.J., Taylor, G.W.: Neural network regularization via robust weight factorization. arXiv preprint arXiv:1412.6630 (2014)
25. Shieh, S.L., Liao, I.E.: A new approach for data clustering and visualization using self-organizing maps. Expert Syst. Appl. 39(15), 11924–11933 (2012)
26. Simonyan, K., Vedaldi, A., Zisserman, A.: Deep inside convolutional networks: visualising image classification models and saliency maps. arXiv preprint arXiv:1312.6034 (2013)
27. Srivastava, N., Hinton, G., Krizhevsky, A., Sutskever, I., Salakhutdinov, R.: Dropout: a simple way to prevent neural networks from overfitting. J. Mach. Learn. Res. 15(1), 1929–1958 (2014)
28. Titapiccolo, J.I., et al.: A supervised SOM approach to stratify cardiovascular risk in dialysis patients. In: Roa Romero, L. (ed.) XIII Mediterranean Conference on Medical and Biological Engineering and Computing 2013, pp. 1233–1236. Springer, Cham (2014). https://doi.org/10.1007/978-3-319-00846-2_305
29. Tsai, C.F., Lu, Y.H.: Customer churn prediction by hybrid neural networks. Expert Syst. Appl. 36(10), 12547–12553 (2009)
30. Ustun, B., Traca, S., Rudin, C.: Supersparse linear integer models for interpretable classification. arXiv preprint arXiv:1306.6677 (2013)
31. Varshney, K.R., Alemzadeh, H.: On the safety of machine learning: cyber-physical systems, decision sciences, and data products. Big Data 5(3), 246–255 (2017)
32. Xu, L., Chow, T.: Multivariate data classification using PolSOM. In: Prognostics and System Health Management Conference (PHM-Shenzhen), pp. 1–4. IEEE (2011)
33. Xu, Y., Xu, L., Chow, T.W.: PPoSOM: a new variant of polsom by using probabilistic assignment for multidimensional data visualization. Neurocomputing 74(11), 2018–2027 (2011)

Data Management and Data Clustering

A Quality Assessment Tool for Koblenz Datasets Using Metrics-Driven Approach

Szymon Pucher[1] and Dariusz Król[2]([⊠]) [iD]

[1] Faculty of Computer Science and Management, Wrocław University of Science
and Technology, Wrocław, Poland
[2] Department of Applied Informatics, Wrocław University of Science and Technology,
Wrocław, Poland
Dariusz.Krol@pwr.edu.pl

Abstract. This work presents metrics-based approach to data quality assessment for real-world open data from the Web – the Koblenz Network Collection. The entire dataset is evaluated from the point of consistency and completeness incorporating semantic check, integrity of references, value completeness and uniqueness. Additionally, statistical measures are calculated to increase insight into quality of the data and be able to detect outliers and anomalies, understand relations and correlations between variables, extract and control limits or determine settings for quality dimensions. The assessment finishes with generation of report containing results displayed using the graphs and the charts of data quality shortcomings. 226+ consolidated quality reports are generated. And three, for dblp, citation network of U.S. patents and twitter, are provided in detail. These reports demonstrate the overall level of good and very good quality for Koblenz repository. The correctness of the implementation by passing a quality control test is confirmed.

Keywords: Big data · Data analysis · Data management · Data visualization · Decision-making

1 Introduction

Data quality has gained vital importance and a decisive role since more and more companies started using data warehouse systems and other data driven systems [10–12]. The quality model defined in the standard ISO/IEC 25012:2008 is composed of 15 characteristics. In Table 1, we list all quality components defined in the ISO standard. The definitions do not provide single measure, therefore one or more metrics are to be considered. However, the benefit of data depends most often on its quality in the components such as *completeness, correctness (accuracy), timeliness* and *consistency* [1].

The main purpose of this research is to develop an effective tool, an application, to assess quality of data by transforming information into valuable knowledge with a resulting impact on efficiency and the validity of decisions based on

© Springer Nature Switzerland AG 2020
H. Fujita et al. (Eds.): IEA/AIE 2020, LNAI 12144, pp. 747–758, 2020.
https://doi.org/10.1007/978-3-030-55789-8_64

Table 1. Data quality characteristics in ISO/IEC 25012:2008.

Characteristic	Definition
accessibility	The degree to which data can be accessed
accuracy	The degree to which data has attributes that correctly represent the true value
availability	The degree to which data has attributes that enable it to be retrieved
completeness	The degree to which subject data associated with an entity has values
compliance	The degree to which data has attributes that adhere to standards, conventions or regulations
confidentiality	The degree to which data has attributes that ensure that it is only accessible and interpretable by authorized users
consistency	The degree to which data has attributes that are free from contradiction and are coherent with other data
credibility	The degree to which data has attributes that are regarded as true and believable
currentness	The degree to which data has attributes that are of the right age
efficiency	The degree to which data has attributes that can be processed and provide the expected levels of performance
portability	The degree to which data has attributes that enable it to be installed, replaced or moved from one system to another preserving the existing quality
precision	The degree to which data has attributes that are exact or that provide discrimination
recoverability	The degree to which data has attributes that enable it to maintain and preserve a specified level of operations and quality
traceability	The degree to which data has attributes that provide an audit trail of access to the data and of any changes made to the data
understandability	The degree to which data has attributes that enable it to be read and interpreted by users

those data. The assessment of the quality is carried out using metrics-oriented approach focusing on completeness and accuracy of the real-world data along with calculation of specific and statistical measures that have influence on data quality. The relying application should be capable of automatic data quality analysis of all Koblenz collection[1] in the areas of web science, network science and related fields, attracted by the Institute of Web Science and Technologies at the University of Koblenz–Landau, running upon single execution of the program and should generate full assessment report.

The purpose of this study is achieved through:

---

[1] 261+ datasets available at http://konect.uni-koblenz.de/.

- exploring the issues and challenges related to data quality requirements and find their impact on key quality components,
- investigating the best practices using metrics-driven approach to provide support for improving the quality of data,
- and delivering an effective tool to assess data quality for selected datasets.

In the case of quality assessment for Koblenz datasets considered here, no other quality evaluations have been reported and so a valid comparison with other work is not possible. To achieve their objectives, the research asks the following questions:

1. What are the concepts of quality to be measured? What are the important requirements, specific measurements and statistics for quality measures?
2. How to properly aggregate metrics into completeness, accuracy and total quality score? How can you be confident that these measurements are accurate? How are these metrics tied to the performance?
3. To what extent the existing set of metrics challenges data quality and has impact on decision-making process?

The remainder of the paper is structured to correspond with our research questions. In the next section, we provide a description of the metrics-driven approach for data quality assessment. Section 3 comprises the findings for the research questions through the demonstration the applicability and discusses the reliability and validity of the results. The last section concludes the paper. It analyses the results and offers suggestions for future research.

2 The Metrics-Driven Approach

The quality assessment using the metrics-driven approach is developed in three main steps:

1. **Data analysis** - study datasets, associated meta and type of report to be generated.
2. **Measures processing** - calculate data quality metrics and statistical measures; use lazy loading techniques for processing big data; this guarantees the application to process data files of any size.
3. **Report generation** - aggregate results assigning total quality score and suggestions on how to improve quality of the data; the latter is still a work in progress.

The assessment of data quality is performed using metrics-based approach. It provides data quality measures to calculate total quality score that user is able to perceive, reference and quantify. Generally, two perspectives on the measurement of data quality are defined. Those perspectives are: (1) **quality of design**, which focuses on the degree of correspondence between the user's requirements and the specification of the information system, and (2) **quality of conformance** which focuses on the correspondence between specification of the system and the existing implementation in information system. For the data from

Table 2. Requirements for quality measures.

Requirement	Description
adaptivity	Requires that it is able to consider adequately the context of a particular application
aggregation	States that it must be applicable to aggregate values with interpretation consistency on different levels
feasibility	Claims that it should be determinable, not very cost-intensive and possible to calculate the values in an automated way
interpretability	Means that the values have to be interpretable with a consistent interpretation
normalization	Assures that the values of the metrics are comparable and normalized as a percentage with 100% for the perfect score or have to take values within a bounded range, e.g., [0; 1]
scale	States that the difference between any two metric values can be determined and is meaningful, for example, the difference between 45% and 50% should be the same as for 95% and 100%, so that comparison of the scores is carrying out correctly

Koblenz collection **quality of conformance** perspective is used, since no information about user's requirements is attached [6]. In the literature a few intuitive approaches for quantifying data quality are present. They differ in the dimensions and then consecutively differ in the metrics and the way of measurement of those metrics [3,5]. Metrics-based approach enables multiple measurements and they should be defined with conformity to some general requirements [4,9]. Table 2 provides an overview of these requirements, a checklist associated with all implemented measures.

The metrics-oriented approach for our work is based on the following specific measurements: (a) **semantic consistency** – consistency check provides information about unambiguity that the assessed dataset is free of internal contradictions and thus has a clear interpretation, (b) **referential integrity** – also known as inclusion dependency, which imposes that some set of values is contained in another set of values due to lack of possibility of direct comparison between them, (c) **value completeness** – to capture the presence of null values, (d) **uniqueness** – refers to the singularity and is the inverse of the level of unwanted duplication of identified items within the dataset.

We propose an analysis of a range of statistics and then comparison between datasets, since those measures can point out some anomalies in the data that would be not noted otherwise [7,8]. Table 3 summarizes the findings related to these statistics.

Specific and statistical measures are aggregated into two dimensions *completeness* and *accuracy*, which describe two most important attributes of data regarding data quality. If data is not complete, this may bias the results. Such data may contain only a specific subset of data it describes. If data is not accu-

Table 3. Statistics for quality measures.

Statistic	Description
size	Means the cardinality of its edge set and is used for calculation of average degree and fill of the dataset
volume	Equals the number of edges within the dataset and has indirect influence on data quality. It is necessary for average degree and fill calculation which have impact on data quality assessment score
average degree	Denotes the average number of edges attached to one node and often indicates that the data may be of lower quality
fill	Represents the probability that an edge is present between two randomly chosen nodes
Gini coefficient	Measures the inequality of distribution. It takes values between zero and one, with zero denoting total equality between degrees, and one denoting the dominance of a single node

rate enough, then results may have many errors and may not be precise enough. The metric for *completeness* is calculated based on *value completeness, referential integrity, average degree, fill* and *Gini coefficient*. The purpose of this metric is to show, to what degree the data reflects the real world. On the other hand, *accuracy* describes the degree to which the result of a measurement conforms to the correct value or a standard based on *semantic consistency* and *uniqueness*. The purpose of this is to show to which degree the data describes events in precise and unambiguous way.

Total quality score is derived by a set of metrics, which can be combined to give a Global Quality Index [2]. When aggregating the data, we must consider following two assumptions: (1) aggregated quality metrics should be independent, (2) metrics should be weighted according to influence of the measured property. These two assumptions are challenging to address in practice. In reality, there may be some dependencies among the quality metrics. Moreover, assigning the correct values of weights to different metrics is significantly application-dependent as well as subjective. However, these analyses go beyond our present work. Therefore, we base purposively on our knowledge and experience. For this implementation, the following values were determined for the final score:

- PERFECT – all data quality metrics are at 100% and no statistical anomalies within the dataset were found.
- GOOD – no data quality metric is below certain threshold. This indicates that data is clean enough to use it for analysis or operations and no cleaning of the data is required, but some inconsistencies are present.
- SUFFICIENT – data set denoted as sufficient should be cleaned as soon as possible, however, it has high enough quality to use it for analysis and operations, still data cleaning is advised.

– INSUFFICIENT – data is not clean enough to use and cleaning process is required. The results of the quality assessment does not meet requirements imposed by any previous score.

3   A Quality Assessment Tool

3.1   Architecture and Implementation Platform

The architecture is developed on the concept of three objects provided by Python packages: (1) *File searcher object*, (2) *Data Quality Assessor object*, and (3) *Report Generator object*. *File searcher object* main task is to search for data files in the directory, categorize them and find metadata file for each of those files. When path to a file is found new object is created and the following attributes are assigned to: *File name*, *File path*, *File size*, *Group* and *Metadata file path*. A list of created *Data Quality Assessor objects* is returned by File Searcher as the last task to complete. Next, *Data Quality Assessor objects* load the data and metadata, perform calculations of all required quality metrics. Data is either lazy loaded or loaded entirely depending on the size of the data. If lazy loading is applied, an application splits data into multiple files (chunks). Each file is loaded on demand. After each load, the calculations on data are performed and graph data is cleared to make space for next chunk of data. When all calculations are performed, *Report Generator object* processes list of all *Data Quality Assessor objects* and creates a report containing relevant information about quality assessment. There are two types of reports: detailed, where all properties are described in detail and mass report, where only aggregated information is printed into table. When the report is generated and saved, the execution of the program is stopped. As concerns the implementation platform, besides a Python interpreter, several Python libraries and modules are required to run the code, for example, *Pandas*, *NumPy*, *Math*, *Os*, *Inflect*, *Matplotlib*, *ReportLab* and *Unittest*.

3.2   Results

For this research, we demonstrate the applicability of our application by evaluating Koblenz collection. In particular, we consider and process 226+ datasets. These results are not reported here due to space limit. The full report numbers 226 pages and can be accessed upon request. Thus, we focus our attention and deliver detailed outcomes for *dblp*, *patent citation* and *twitter* datasets. All computation was performed on an Intel Core i7 processor, 16 GB RAM machine under Windows 10.

The results are presented in a form of graphical reports. These reports are generated based on the original data and metadata files downloaded from Koblenz repository. Generally, we use *ReportLab* package and the process is handled by *Report Generator objects*. To print report, only data quality assessment results stored inside *Data Quality Assessor objects* are required. Charts are created by *Report Generator objects*. The detailed report consists of: (1) general information including *File name*, *File size*, *Category of data*, *Group of data*, *Metadata file path* associated, (2) specific attributes and statistical measures including *size*, *volume*, *average degree*, *fill*, *Gini coefficient*, bar chart representing edge frequency distribution, and (3) results of data quality in a form of a bar chart, data quality inconsistencies distribution – a pie chart, information on statistical anomalies – a bar chart, values of completeness and accuracy and total quality score. For each selected dataset, Table 4 lists the statistics.

Table 4. Statistics of the selected datasets.

File name	Size	Degree	Volume	Fill	Gini
dblp	317,080	6.62	1,049,866	1.04e-05	0.47
dblp-cite	12,591	7.9	49,743	3.14e-04	0.58
dblp-author	5,425,963	3.19	8,649,016	2.94e-07	0.37
dblp-coauthor	1,282,468	28.45	18,241,510	1.11e-05	0.67
patentcite	3,774,768	8.75	16,518,947	1.16e-06	0.46
ego-twitter	23,370	2.83	33,101	6.06e-05	0.45
twitter	41,652,230	70.51	1,468,365,182	8.46e-07	0.82
twitter-mpi	52,579,682	74.68	1,963,263,821	7.10e-07	0.83

Similar total quality scores from the same group can be observed. It may be due to referential integrity problems or differences in statistics, leading to anomalies and decrease of data quality score. Table 5 reports the final assessment results.

Table 5. Quality assessment results for selected datasets.

File name	Group	Completeness [%]	Accuracy [%]	Total score
dblp	dblp	97.38	100.0	SUFFICIENT
dblp-cite	dblp	98.51	100.0	GOOD
dblp-author	dblp	99.4	100.0	GOOD
dblp-coauthor	dblp	99.42	95.67	SUFFICIENT
patentcite	patent	100.0	100.0	PERFECT
ego-twitter	twitter	98.19	95.67	GOOD
twitter	twitter	100.0	100.0	PERFECT
twitter-mpi	twitter	98.19	100.0	GOOD

Figure 1 illustrates a sample detailed report generated from quality analysis of four dblp datasets. Total quality scores for them are SUFFICIENT or GOOD, but individual quality metrics like consistency and value completeness are at 99.99% or 100%. Main reason of decrease in quality are poor values of referential integrity and uniqueness. For example, low value of uniqueness for dblp co-author is due to redundancy of the edges, since each edge represents one common publication between two authors and two authors may have many common publications. Significant differences in average degree between two associated datasets for statistical anomalies calculation are due to redundancy of edges for one of the datasets. On the other hand, low values of referential integrity indicate, that there is a possibility of wrong assignment of references between datasets. To sum up, more research into finding correct references and influence of referential integrity on data quality score is necessary.

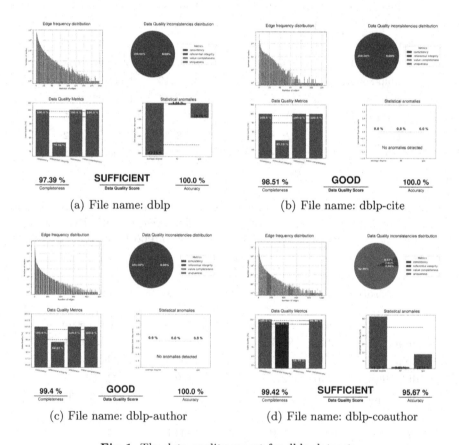

(a) File name: dblp

(b) File name: dblp-cite

(c) File name: dblp-author

(d) File name: dblp-coauthor

Fig. 1. The data quality report for dblp datasets.

While the calculations are running, loading of the data is fast, around 50 MB/s. The engine of the *Data Frame loading function*, from standard *Pandas* library we used, is written in C programming language which is much faster than Python programming language. However, computational time of *size* calculation according to its complexity is a problem. The optimization of this part should be a priority, since it has the significant influence on data quality assessment duration. The results on computational time in seconds varied significantly as shown in Table 6. The two first numerical columns display running time for 29,628,167KB *twitter* file and for 497,949KB *dblp* file, respectively. The default chunk size extracts 1 million rows.

Table 6. Computational time [s] for *twitter* and *dblp* datasets.

Process	*twitter* chunk	*dblp* chunk	*twitter*	*dblp*
Loading	0.30	0.33	592.49	6.23
Size	20.43	0.22	40132.95	4.26
Volume	0.01	0.01	0.06	0.01
Consistency	0.05	0.05	97.52	0.86
Value completeness	0.05	0.05	99.86	0.89
Uniqueness	0.23	0.24	451.90	4.60
Total time	21.07	16*	41374.88	303.92*

* Total time for *dblp* additionally includes referential integrity checking.

3.3   Control Test

In the context of software tools, correctness of an algorithm is an important means to ensuring the collectedness of the results. However, an in-depth discussion of it is beyond the scope of this paper. To conserve space, the correctness of the tool is verified by reducing the quality of dblp co-authorship network. To evaluate new data the following modifications are performed:

- 1,250,000 empty values added to first and second column of nodes,
- 1,000,000 string values added to first column and 1,250,000 string values to second column of nodes,
- 5,000,000 rows of redundant data added to the dataset,
- 2,000,000 values without reference added to the dataset.

Necessarily, calculating formulas used as metrics for the new dataset is the proper step towards quality reduction. This update results in a decrease of consistency, referential integrity, value completeness, uniqueness, completeness, accuracy and the total quality score. The addition of 2,250,000 inconsistent values should result in decrease of consistency. The difference between expected and actual value of referential integrity is due to inconsistencies in redundant data created during data modification process for reduction of uniqueness. This does not make the relevant metrics dependent upon uniqueness, but on the process of data modification. Increase in referential integrity after addition of 2,000,000 values without reference is due to the addition of 9,750,000 rows of inconsistent, empty and redundant data. The independence between the calculation of consistency and completeness is very important, since metrics are aggregated and therefore should be independent. This is the premise of metrics-driven approach. The calculation of completeness is done after addition of 2.5 million empty values to the dataset expanding it by the same number of rows. To express the impact on quality before and after modifications, an overall comparison of six metrics is outlined in Table 7. The detailed report for modified dblp co-author dataset is depicted in Fig. 2.

Table 7. Control test results for dblp co-authorship dataset.

Quality metrics	Before modification	After modification
Consistency [%]	100	↓ 96.11
Referential integrity [%]	96.74	↑ 97.95*
Value completeness [%]	99.99	↓ 95.68
Uniqueness [%]	56.79	↓ 55.55
Completeness [%]	99.42	↓ 95.14
Accuracy [%]	95.67	↓ 92.05
TOTAL SCORE	*SUFFICIENT*	↓ *INSUFFICIENT*

* The increase of referential integrity is due to inconsistencies in redundant data created during data modification. In the remaining cases, the decrease of quality metrics proves that the assessment produces consistent and correct results.

4   Final Remarks

The research described in this work ensures the Koblenz collection is reliable data which fulfils high quality standards in designing and conducting scientific experiments. We demonstrate that the tool described above is able to correctly assess the quality. It is very impressive that the results meet the requirements and expectations for datasets containing over 1 billion of entries, where some key metrics get perfect scores. An extensive set of calculations confirms the correctness through the quality control test and efficiency with throughput of up to 50MB/s and shows, that data errors and defects detected here are eminently skippable.

Based on the findings of this study, we plan to refine our set of metrics by gaining a better understanding of their impact on data quality score and consequently on the decision-making. The individual impact of the average degree of the vertex, graph fill and Gini coefficient is set in the range of 0 to 1.5%. However, more extensive research into the impact of these statistical measures on the quality of a particular data should be carried out. In addition, the classification of data based on the metadata provided with them is not accurate enough, only

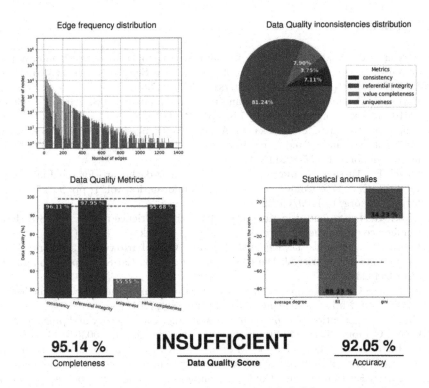

Fig. 2. Data Quality Assessment report for modified dblp co-author dataset.

small number of classes is given, to accurately assess the quality. It would be necessary to extend the class set in order to achieve more accurate results.

Regarding current solution, the degree of influence and correlation of other measures including spatial, spectral, and statistical properties [6] and which measures should be considered during data quality assessment would be beneficial. It also is possible to take advantage how the degradation of the data quality influences the quality of the results based on the data. Finally, further development of the application for measurement of data quality for broader set of datasets, including Stanford SNAP[2], Kaggle[3], Data.gov[4], UCI Machine Learning Repository[5], going beyond Koblenz data would expand possibilities of use of this application.

Acknowledgments. This research received financial support from the statutory funds at the Wrocław University of Science and Technology.

---

[2] 107+ networks available at http://snap.stanford.edu/data.
[3] 19,000+ datasets available at https://www.kaggle.com/.
[4] 200,000+ datasets available at https://www.data.gov/.
[5] 488+ datasets available at https://archive.ics.uci.edu/ml/index.php.

References

1. Azeroual, O., Saake, G., Wastl, J.: Data measurement in research information systems: metrics for the evaluation of data quality. Scientometrics **115**(3), 1271–1290 (2018). https://doi.org/10.1007/s11192-018-2735-5
2. Batini, C., Scannapieco, M.: Data and Information Quality. DSA. Springer, Cham (2016). https://doi.org/10.1007/978-3-319-24106-7
3. Chung, Y., Krishnan, S., Kraska, T.: A data quality metric (dqm): how to estimate the number of undetected errors in data sets. PVLDB **10**(11), 1094–1105 (2017). https://doi.org/10.14778/3115404.3115414
4. Gitzel, R., Turring, S., Maczey, S.: A data quality dashboard for reliability data. In: 2015 IEEE 17th Conference on Business Informatics, vol. 1, pp. 90–97 (2015). https://doi.org/10.1109/CBI.2015.24
5. Gitzel, R.: Data quality in time series data: an experience report. In: 18th IEEE Conference on Business Informatics - Industrial Track, pp. 41–49 (2016)
6. Heinrich, B., Hristova, D., Klier, M., Schiller, A., Szubartowicz, M.: Requirements for data quality metrics. J. Data Inf. Qual. **9**(2), 1–32 (2018). https://doi.org/10.1145/3148238
7. Johnson, S.G.: A data quality framework for the secondary use of electronic health information. Ph.D. thesis, University of Minnesota (2016)
8. Król, D.: Propagation phenomenon in complex networks: theory and practice. New Gener. Comput. **3**, 187–192 (2014). https://doi.org/10.1007/s00354-014-0400-y
9. Król, D., Skowroński, J., Zaręba, M., Bartecki, K.: Development of a decision support tool for intelligent manufacturing using classification and correlation analysis. In: 2019 IEEE International Conference on Systems, Man and Cybernetics (SMC), pp. 88–94 (Oct 2019). https://doi.org/10.1109/SMC.2019.8914222
10. Sadiq, S.: Handbook of Data Quality. Springer, Heidelberg (2013). https://doi.org/10.1007/978-3-642-36257-6
11. Stvilia, B., Gasser, L., Twidale, M.B., Smith, L.C.: A framework for information quality assessment. J. Assoc. Inf. Sci. Technol. **58**(12), 1720–1733 (2007). https://doi.org/10.1002/asi.20652
12. Veiga, A.K., et al.: A conceptual framework for quality assessment and management of biodiversity data. PLOS ONE **12**(6), 1–20 (2017). https://doi.org/10.1371/journal.pone.0178731

Applying Cluster-Based Zero-Shot Classifier to Data Imbalance Problems

Toshitaka Hayashi$^{(\boxtimes)}$ ⬚, Kotaro Ambai ⬚, and Hamido Fujita ⬚

Faculty of Software and Information Science, Iwate Prefectural University, Iwate, Japan
G236r002@s.iwate-pu.ac.jp, kotaro.ambai@gmail.com,
HFujita-799@acm.org

Abstract. Imbalanced data classification is an important issue in machine learning. To solve the data imbalance problem, rebalance algorithms are utilized. However, the rebalance algorithm has a lot of problems. Hence, the specific classification algorithm without a rebalancing algorithm is required. In this paper, we apply the zero-shot classifier to imbalance problem. The zero-shot classifier works despite the number of minority class is zero. Hence, the zero-shot classifier should also work in weaker data imbalance problems. We utilize the cluster-based zero-shot learning algorithm to do imbalanced data classification. The proposed method is evaluated using an imbalanced-learn dataset and compared with Decision Tree, K-Nearest Neighbor, and Over-sampling methods.

Keywords: Imbalance classification · Zero-shot learning · Binary classification

1 Introduction

Imbalanced data classification is an important issue in machine learning and data mining. Many real-world application has imbalanced problems such as listing prediction [1, 18], email filtering [2], and medical diagnosis [3]. Traditional classification algorithms are difficult to classify imbalanced data. Generally, the degree of imbalance in binary classification is defined by Imbalance Ratio (IR) [4]. IR is the value of the instance of the majority class divided by the instance of the minority class. Equation (1)

$$IR = \frac{Number\ of\ majority\ class}{Number\ of\ minority\ class} \tag{1}$$

In order to solve the imbalance problem, a rebalance algorithm is utilized. As a rebalance algorithm, over-sampling [4–7] and under-sampling [8–11] are utilized. However, in over-sampling, increasing data is unreliable because the fed data is fake ones. Also, in under-sampling, decreasing data is wasted. Hence, imbalanced data classification algorithms without any sampling methods are required.

We think a zero-shot learning algorithm is suitable to solve the imbalance problem. The zero-shot problem is one of the strongest imbalance problems. In the zero-shot problem, the number of minority class is zero. In such a problem, the sampling method is

© Springer Nature Switzerland AG 2020
H. Fujita et al. (Eds.): IEA/AIE 2020, LNAI 12144, pp. 759–769, 2020.
https://doi.org/10.1007/978-3-030-55789-8_65

not available. However, the zero-shot learning algorithm works even though the number of minority class is zero. Hence, the zero-shot classifier should also work in weaker imbalance problems.

In this paper, we proposed the imbalanced data classification method. The proposed method does not use the rebalance algorithm. We have applied the cluster-based zero-shot classification algorithm [12] to imbalanced data classification. Proposed method is evaluated using imbalanced-learn dataset [13]. As the experiment results, we have confirmed zero-shot classifier is suitable for imbalanced data classification.

2 Related Work

2.1 Imbalanced Data Classification

Imbalanced data classification is an important issue in machine learning. In order to solve the imbalance problem, many methods are proposed but roughly classified into two levels: data level and algorithm level. In the data-level approach, rebalance the class distribution by applying to resample to the training data. As a rebalance algorithm, over-sampling [4–7, 18] or under-sampling [8–11], and the hybrid methods are utilized. The over-sampling method increases the minority class and the under-sampling method decreases the majority class. The algorithm level approaches [14, 15, 18] apply a specific classification algorithm. In this approach, introduce costs, probabilistic estimation, thresholds, etc. and classify minority class samples more accurately.

2.2 Zero-Shot Learning

The goal of zero-shot learning is to predict class that is not in training data. Therefore, zero-shot learning predicts minority class without training data. In Zero-shot learning, the number of minority class is zero. Hence, the imbalance ratio is infinity. In such a case, the rebalance algorithm is not suitable. Hence, a specific classification algorithm is utilized. In zero-shot learning, there are two types of class, seen class and unseen class. Seen class is in the training data. Unseen class is not in the training data. The zero-shot learning algorithms train only seen class and predict unseen class.

Zero-shot learning algorithms are roughly classified into two groups, using the transfer learning method [16] or using the outlier detection method [12, 17]. In the transfer learning method, the mapping function from feature space to label space is trained from data for seen class. Then the mapping function model is applied to data for unseen class. Transfer learning method requires semantic label information [12]. In the outlier detection method, unseen data is treated as an outlier. The outlier detection model is trained from seen data. If the model predicts data as an outlier, the data is considered as unseen data.

We think zero-shot learning algorithms could be applied to imbalanced data classification because it works in a strong imbalance problem. In this paper, we apply outlier detection method. This is because imbalance-learn dataset does not have semantic label information [13].

3 Proposed Method

In this section, we propose how to apply zero-shot classifier to imbalance classification. In this paper, we consider binary data imbalance problem.

3.1 Zero-Shot Classifier

Zero-shot classifier framework is shown in Fig. 1. The zero-shot classifier trains only seen class and predict seen class and unseen class. To apply the zero-shot classifier to imbalance problems, only one class (seen class) is utilized for training. Hence, selecting one-class for training is necessary. There are two options, training minority class, or training majority class. Training and Prediction is done as Algorithm 1.

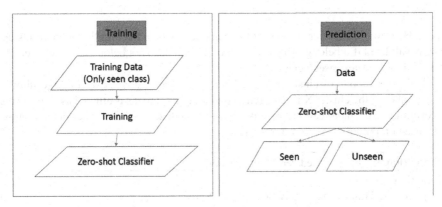

Fig. 1. Zero-shot learning framework

Algorithm 1. Zero-shot Learning framework

Function: Training
    Input: Training data (only Seen class)
    Output: The behavior (Distribution or Rules) of Seen class
    Estimate behavior of Seen data
Function: Prediction
    Input: Data
    Output: Class (Seen or Unseen)
    If Data fills the behavior of seen class:
        Output is **Seen**
    Else:
        Output is **Unseen**

To build a zero-shot learning framework, the estimation on the behavior of a seen class is an important problem. In this paper, we use cluster-based approach [12]. The framework of cluster-based zero-shot learning is shown in Fig. 2. In the training step, K-clusters are created from training data. Training data consists of only data for the seen

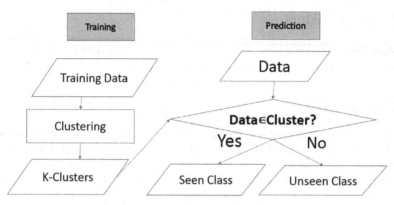

Fig. 2. Cluster-based zero-shot learning

class. Therefore, K-clusters represent the behavior of seen data. In the prediction step, data is validated if it belong to a cluster or not. If data does not belong to any cluster, the data is treated as an unseen class.

The pseudo code of clustering based zero-shot learning is shown in Algorithm 2. The number of the cluster K is required as parameter. K-means is utilized as a clustering algorithm. Also, Euclidean distance is utilized as a distance. For further details on the distance computation, please refer to [12].

Algorithm 2. Cluster-based zero-shot learning

Function: Training
    Input: Training data (only Seen class), K
    Output: K-Clusters and the thresholds
    Do clustering for training data
    For each data:
        Assign cluster
    For each cluster:
        Threshold = 0
        Centroid = average (data in cluster)
        For data in cluster:
            Distance = distance (centroid, data)
            If Distance > Threshold:
                Threshold = Distance
Function: Prediction
    Input: Data
    Output: Class (Seen or Unseen)
    Find the nearest cluster
    Distance = distance (data, centroid of nearest cluster)
    If Distance<Threshold of the nearest cluster:
        Output is **Seen**
    Else:
        Output is **Unseen**

---

4   Experiment

The proposed method has been validated against data listed in Sect. 4.1. The measurement of evaluation is shown in Sect. 4.2. The experiment results are shown in Sect. 4.3.

Table 1. Imbalanced dataset for experiment

Data	Dimension	Positive	Negative	IR
ecoli	7	35	301	8.60
optical_digits	64	554	5066	9.14
satimage	36	626	5809	9.28
pen_digits	16	1055	9937	9.42
abalone	10	391	3786	9.68
sick_euthyroid	42	293	2870	9.80
spectrometer	93	45	486	10.80
car_eval_34	21	134	1594	11.90
isolet	617	600	7197	12.00
us_crime	100	150	1844	12.29
yeast_ml8	103	178	2239	12.58
scene	294	177	2230	12.60
libras_move	90	24	336	14.00
thyroid_sick	52	231	3541	15.33
coil_2000	85	586	9236	15.76
arrhythmia	278	25	427	17.08
solar_flare_m0	32	68	1321	19.43
oil	49	41	896	21.85
car_eval_4	21	65	1663	25.58
wine_quality	11	183	4715	25.77
letter_img	16	734	19266	26.25
yeast_me2	8	51	1433	28.10
webpage	300	981	33799	34.45
ozone_level	72	73	2463	33.74
mammography	6	260	10923	42.01
protein_homo	74	1296	144455	111.46
abalone_19	10	32	4145	129.53

4.1 The Data

In this paper, we use imbalanced-learn datasets [13] for evaluation. Imbalanced-learn dataset consists of 27 datasets and is provided by imbalanced-learn package [13] on Python. For each dataset, number of dimension, number of instance for each class and IR are shown in Table 1. Positive class is minority class and negative class is majority class. All datasets are normalized based on z-score.

4.2 Measurement of the Evaluation

Evaluation is done using Area under the ROC Curve (AUC). AUC is calculated using TPR and FPR as given in Eq. (2)–Eq. (4). The confusion matrix is shown in Table 2.

$$\text{TPR} = \frac{TP}{TP + FN} \tag{2}$$

$$\text{TNR} = \frac{TN}{FP + TN} \tag{3}$$

$$\text{AUC} = \frac{TPR + TNR}{2} \tag{4}$$

Table 2. Confusion matrix

		Predicted	
		Positive	Negative
Actual	Positive	TP	FN
	Negative	FP	TN

4.3 Experiment Result

In the experiment, each dataset is separated into training and testing. 60% of the dataset is for training. 40% of the dataset is for testing. Training data consists of positive data and negative data. To evaluate the proposal, two different experiments (ZSL_POS, ZSL_NEG) are done. ZSL_POS uses only positive data for training, also, ZSL_NEG uses only negative data for training.

The result of ZSL_POS and ZSL_NEG depends on K-values. In this experiment, we show the best scores where $1 \leq K \leq 50$.

In the experiment, the proposed method is compared to existing classification algorithms. Decision Tree and K-Nearest Neighbor is utilized as classification algorithm.

First, comparison result to Decision Tree (DT) and K-Nearest Neighbor (KNN) without any sampling algorithms are as shown in Table 3. For each data, a bold score is the best result. In 12 datasets, ZSL_POS outperforms DT and KNN. Therefore, the

Table 3. Comparison result against DT and KNN

Data	DT	KNN	Proposed method	
			ZSL_NEG	ZSL_POS
ecoli	0.797	0.832	0.735	**0.835**
optical_digits	0.863	**0.979**	0.720	0.933
satimage	0.752	**0.818**	0.505	0.798
pen_digits	0.962	**0.994**	0.791	0.979
abalone	0.611	0.600	0.500	**0.759**
sick_euthyroid	**0.911**	0.577	0.500	0.681
spectrometer	0.775	0.803	**0.909**	0.781
car_eval_34	0.922	0.773	0.760	**0.928**
isolet	0.772	**0.913**	0.503	0.901
us_crime	0.667	0.629	0.579	**0.738**
yeast_ml8	0.497	0.534	0.536	**0.555**
scene	0.560	0.520	0.533	**0.602**
libras_move	0.700	0.800	**0.898**	0.735
thyroid_sick	**0.915**	0.567	0.504	0.679
coil_2000	0.550	0.519	0.500	**0.561**
arrhythmia	**0.844**	0.547	0.541	0.595
solar_flare_m0	0.520	0.529	0.542	**0.591**
oil	**0.736**	0.560	0.570	0.605
car_eval_4	0.959	0.746	0.886	**0.982**
wine_quality	**0.610**	0.543	0.566	0.571
letter_img	0.953	**0.991**	0.625	0.953
yeast_me2	0.590	0.547	0.522	**0.751**
webpage	**0.868**	0.777	0.526	0.598
ozone_level	0.537	0.498	0.519	**0.664**
mammography	0.755	**0.791**	0.620	0.556
protein_homo	**0.864**	0.702	0.516	0.639
abalone_19	0.534	0.500	0.536	**0.630**

zero-shot classifier is suitable for imbalanced data classification. Also, training only positive samples provides better AUC score.

Also, the proposed method is compared to over-sampling methods with DT and KNN. SMOTE [5], ADASYN [6], BorderlineSMOTE-1 [7], BorderlineSMOTE-2 [7] are utilized as over-sampling methods.

A comparison result against over-sampling methods with DT is shown in Table 4. The bold score is the best result. In 16 datasets, the proposed method has better result than combination of over-sampling methods and DT.

Table 4. Comparison result against over-sampling methods with DT

Data	SMOTE	ADASYN	BORDER1	BORDER2	ZSL_NEG	ZSL_POS
ecoli	0.713	0.625	0.729	0.713	0.735	**0.835**
optical_digits	0.898	0.864	0.893	0.869	0.720	**0.933**
satimage	0.769	0.761	0.770	0.787	0.505	**0.798**
pen_digits	0.966	0.956	0.959	0.973	0.791	**0.979**
abalone	0.600	0.577	0.636	0.603	0.500	**0.759**
sick_euthyroid	0.923	0.908	0.906	**0.939**	0.500	0.681
spectrometer	0.828	0.848	0.795	0.823	**0.909**	0.781
car_eval_34	0.958	0.968	**0.995**	**0.995**	0.760	0.928
isolet	0.841	0.794	0.774	0.797	0.503	**0.901**
us_crime	0.703	0.611	0.628	0.676	0.579	**0.738**
yeast_ml8	0.505	0.522	0.545	0.514	0.536	**0.555**
scene	0.561	0.572	0.599	0.579	0.533	**0.602**
libras_move	0.743	0.743	0.743	0.739	**0.898**	0.735
thyroid_sick	0.905	**0.923**	0.890	0.915	0.504	0.679
coil_2000	0.543	0.553	0.556	0.543	0.500	**0.561**
arrhythmia	**0.844**	**0.844**	**0.844**	0.612	0.541	0.595
solar_flare_m0	0.558	0.501	0.523	0.539	0.542	**0.591**
oil	0.688	**0.692**	0.637	0.660	0.570	0.605
car_eval_4	0.961	0.961	0.961	0.961	0.886	**0.982**
wine_quality	**0.649**	0.633	0.605	0.588	0.566	0.571
letter_img	0.947	**0.958**	0.948	0.951	0.625	0.953
yeast_me2	0.647	0.696	0.613	0.593	0.522	**0.751**
webpage	0.773	0.726	0.737	**0.794**	0.526	0.598
ozone_level	0.605	0.660	0.636	0.544	0.519	**0.664**
mammography	0.787	**0.840**	0.796	0.796	0.620	0.556
protein_homo	**0.881**	0.883	0.881	0.877	0.516	0.639
abalone_19	0.594	**0.669**	0.533	0.491	0.536	0.630

Comparison result against over-sampling methods with KNN is shown in Table 5. In 16 datasets, the proposed method has better result than combination of over-sampling methods and KNN.

Table 5. Comparison result against over-sampling methods with KNN

Data	SMOTE	ADASYN	BORDER1	BORDER2	ZSL_NEG	ZSL_POS
ecoli	0.803	0.803	0.803	0.779	0.735	**0.835**
optical_digits	0.986	0.986	**0.987**	0.983	0.720	0.933
satimage	0.901	**0.906**	0.898	0.883	0.505	0.798
pen_digits	**0.999**	0.997	0.996	0.996	0.791	0.979
abalone	0.722	0.726	0.743	0.725	0.500	**0.759**
sick_euthyroid	0.669	0.678	0.668	**0.699**	0.500	0.681
spectrometer	0.879	0.853	0.851	0.871	**0.909**	0.781
car_eval_34	0.883	0.858	0.856	0.896	0.760	**0.928**
isolet	**0.945**	0.945	0.944	0.929	0.503	0.901
us_crime	0.771	**0.794**	0.762	0.783	0.579	0.738
yeast_ml8	0.572	0.589	0.598	**0.598**	0.536	0.555
scene	**0.717**	0.694	0.675	0.684	0.533	0.602
libras_move	**0.946**	**0.946**	**0.946**	0.943	0.898	0.735
thyroid_sick	0.666	0.672	0.698	**0.704**	0.504	0.679
coil_2000	0.566	0.565	**0.572**	0.569	0.500	0.561
arrhythmia	**0.768**	0.718	0.615	0.513	0.541	0.595
solar_flare_m0	0.540	0.540	0.549	0.579	0.542	**0.591**
oil	0.733	**0.778**	0.709	0.743	0.570	0.605
car_eval_4	0.946	0.948	0.946	0.920	0.886	**0.982**
wine_quality	0.590	0.586	0.588	**0.618**	0.566	0.571
letter_img	0.996	**0.996**	0.997	0.994	0.625	0.953
yeast_me2	**0.841**	0.811	0.784	0.774	0.522	0.751
webpage	**0.898**	0.898	0.899	0.882	0.526	0.598
ozone_level	0.611	0.644	0.541	0.546	0.519	**0.664**
mammography	0.852	0.853	0.859	**0.862**	0.620	0.556
protein_homo	0.809	0.808	0.792	**0.817**	0.516	0.639
abalone_19	0.615	**0.694**	0.532	0.570	0.536	0.630

5 Discussion

In this section, we discuss about experiment results and highlighted issues in the experiment.

In Table 3, cluster-based zero-shot classifier outperform DT and KNN for 14 datasets. Therefore, zero-shot classifier is better than exist classification algorithms for imbalanced data classification. Hence, zero-shot classifier is suitable for imbalance data classification. In Table 4 and Table 5, proposed method can compete with oversampling

+ DT/KNN. Therefore, zero-shot classifier could be treated as one good alternatives solution for imbalanced data classification.

Where $1 <= K <= 50$, ZSL_POS has better result than ZSL_NEG. We think ZSL_NEG needs larger K-value. This is because negative class has larger number of data than positive data, in order to represent large number of instances, large number of cluster is required. Hence, AUC of ZSL_NEG should be increased with larger K-value.

Proposed methods train only one class. Therefore, training data for other class is not utilized. For example, ZSL_POS does not train negative data and ZSL_NEG does not train positive data. In such case, unseen data is wasted. Hence, algorithm should use training data effectively. We think other ways to use the data effectively are to validate the zero-shot classifier to optimize the K value, or combine the ZSL_POS classifier with the ZSL_NEG classifier.

In this paper, cluster-based zero-shot learning is based on K-means and Euclidean distance. However, other clustering algorithms or distance could improve classification scores. Also, zero-shot learning approach using other method than clustering could be utilized.

6   Conclusion and Future Work

In this paper, we apply cluster-based zero-shot classifier for imbalance data classification. We confirm cluster-based zero-shot classifier is suitable for imbalance data classification. Therefore, zero-shot classifier could be treated as one of the good solution for imbalanced data classification.

The proposed method train only data for seen class. Data for unseen class is ignored while training. We think that the proposed method wastes those ignored data, using data for unseen class effectively could improve classification score. In order to improve classification score, we consider future work as follows:

- In training, validate zero-shot classifier to optimize the K-value.
- Combine the ZSL_POS classifier and the ZSL_NEG classifier.
- Consider other clustering algorithms and distances.
- Consider other methods than clustering to estimate seen class.

References

1. Sun, J., Lang, J., Fujita, H., Li, H.: Imbalanced enterprise credit evaluation with DTE-SBD: decision tree ensemble based on SMOTE and bagging with differentiated sampling rates. Inf. Sci. **425**, 76–91 (2018)
2. Sanz, E.P., Gómez Hidalgo, J.M., Cortizo Pérez, J.C.: Chapter 3 Email spam filtering. In: Advances in Computers, vol. 74, pp. 45–114 (2008)
3. Li, D.-C., Liu, C.-W., Hu, S.C.: A learning method for the class imbalance problem with medical data sets. Comput. Biol. Med. **40**, 509–518 (2010)
4. Ambai, K., Fujita, H.: MNDO: Multivariate Normal Distribution Based Over-Sampling for Binary Classification. Frontiers in Artificial Intelligence and Applications, Volume 303: New Trends in Intelligent Software Methodologies, Tools and Techniques, pp. 425–438 (2018)

5. Chawla, N.V., Bowyer, K.W., Hall, L.O., Philip Kegelmeyer, W.: SMOTE: synthetic minority over-sampling technique. J. Artif. Intell. Res. **16**, 321–357 (2002)
6. He, H., Bai, Y., Garcia, E.A., Li, S.: ADASYN: adaptive synthetic sampling approach for imbalanced learning. In: Proceedings of the 2008 IEEE International Joint Conference on Neural Networks (IJCNN 2008), pp. 1322–1328 (2008)
7. Han, H., Wang, W.-Y., Mao, B.-H.: Borderline-SMOTE: a new over-sampling method in imbalanced data sets learning. In: Huang, D.-S., Zhang, X.-P., Huang, G.-B. (eds.) ICIC 2005. LNCS, vol. 3644, pp. 878–887. Springer, Heidelberg (2005). https://doi.org/10.1007/11538059_91
8. Wilson, D.L.: Asymptotic properties of nearest neighbor rules using edited data. IEEE Trans. Syst. Man Commun. **2**(3), 408–421 (1972)
9. Tomek, I.: Two modifications of CNN. IEEE Trans. Syst. Man Cybern. **6**, 769–772 (2010)
10. Smith, M.R., Martinez, T., Giraud-Carrier, C.: An instance level analysis of data complexity. Mach. Learn. **95**(2), 225–256 (2013). https://doi.org/10.1007/s10994-013-5422-z
11. Kubat, M., Matwin, S.: Addressing the curse of imbalanced training sets: one-sided selection. In: ICML, vol. 97, pp. 179–186 (1997)
12. Hayashi, T., Fujita, H.: Cluster-based zero-shot learning for multivariate data. J. Ambient Intell. Humanized Comput. (2020). https://doi.org/10.1007/s12652-020-02268-5
13. Lemaître, G., Nogueira, F., Aridas, C.K.: Imbalanced-learn: a Python toolbox to tackle the curse of imbalanced datasets in machine learning. J. Mach. Learn. Res. **18**, 1–5 (2017)
14. Zhang, C., et al.: Multi-imbalance: an open-source software for multi-class imbalance learning. Knowl.-Based Syst. **174**, 137–143 (2019)
15. Bia, J., Zhang, C.: An empirical comparison on state-of-the-art multi-class imbalance learning algorithms and a new diversified ensemble learning scheme. Knowl.-Based Syst. **158**, 81–93 (2018)
16. Wang, W., Zheng, V.W., Yu, H., Miao, C.: A survey of zero-shot learning: settings, methods, and applications. ACM Trans. Intell. Syst. Technol. (TIST) **10**(2) (2019). Article 13. https://doi.org/10.1145/3293318
17. Socher, R., Ganjoo, M., Manning, C.D., Ng, A.Y.: Zero-shot learning through cross-modal transfer. In: NIPS 2013 Proceedings of the 26th International Conference on Neural Information Processing Systems, vol. 1, pp. 935–943 (2013)
18. Sun, J., Li, H., Fujita, H., Binbin, F., Ai, W.: Class-imbalanced dynamic financial distress prediction based on Adaboost-SVM ensemble combined with SMOTE and time weighting. Inform. Fusion **54**, 128–144 (2020)

Determining Sufficient Volume of Data for Analysis with Statistical Framework

Tanvi Barot[1], Gautam Srivastava[1,2(✉)], and Vijay Mago[1]

[1] Department of Computer Science, Lakehead University,
Thunder Bay, ON P7B 5E1, Canada
{tbarot,gsrivast,vmago}@lakeheadu.ca
[2] Department of Mathematics and Computer Science, Brandon University,
270 18th Street, Brandon R7A 6A9, Canada
srivastavag@brandonu.ca

Abstract. In an era of Big Data where bigger is considered better, there lies an inherent question that pertains to this ideology. To what extent does this big data contribute to improved/enhanced predictive models? Current research on this topic suggests that the commonplace theory of the bigger, the better, is inaccurate and the bigger, does not always indicate better. In this paper, we provide a methodology that can correctly determine a sufficient volume of data ideally required for accurate data analysis. This paper focuses on data gathered from a popular social media site called Twitter and analyzes it to a point of desired statistical significance. In essence, relevant themes are extracted from the social media data, and are used as a basis to perform a meaningful statistical analysis. A typical approach to perform this requires gathering a sample of relevant tweets to identify the main themes, and declare them as significant results. This standard methodology rarely measures any statistical significance. As constructs such as the p-value are not applicable, researchers often find themselves under the assumption that their sample size was sufficient. In contrast to the typical approach, this paper proposes the use of Modified Confidence Intervals (CI) methods to determine a sample size necessary to statistically support a conclusion. The end goal of this study is to provide aid to researchers and big data practitioners on the determination of a sufficient sample size required for accurate data analysis.

Keywords: Big data · Modified confidence interval · Data analysis · Social network sites · Topic modelling · Volume of data

1 Introduction

The widespread presence of big data use in day to day lives has a deep impact on many fields, including economics and public health [22]. Almost all research areas and the field, including business, medicine, and computer science are currently majorly involved in this spreading computational culture of big data because of

© Springer Nature Switzerland AG 2020
H. Fujita et al. (Eds.): IEA/AIE 2020, LNAI 12144, pp. 770–781, 2020.
https://doi.org/10.1007/978-3-030-55789-8_66

its vast reach of influence and potential within so many disciplines. Moreover, the recent advancements and deployments of computing resources by public [10] and private [1] organizations have allowed analysis of big data possible in near real-time environments.

As a highly interdisciplinary subject, social media and related fields are also largely affected by the new technical wave of big data. There are frameworks [24] and platforms that allow collection and visualization of social media data in real time if deployed on High Performance Computing (HPC) resources [23]. For the past several years we have seen popular applications of big data. Some examples include:

- inferring people's daily social media behavior and their interaction with their peers [6]
- analysing health related behaviours or mining political discourses [21]

We can expect that this wave of big data on social media will eventually be extended to other domains and city wide applications such as real-time population census and monitoring the travel habits of people. Such attempts may be an easy target for spreading rumors or creating spam on social media platforms [20].

The key question is not only technological but also organizational. However, big data is the new wave for research. However, it does not matter how "big" big data is, it is just data. It is only if it has actual knowledge that can be inferred from the data does it actually matter. Big data also brings big problems in data usage and quality, which compromise the practicality and usability of the data. Research based on data with ambiguity does not meet the requirements of quality research studies in terms of originality and accuracy. This type of research will likely result in biased or erroneous conclusions if we do not have a deep and clear understanding of the quality issues of big data and related problems of it.

This paper focuses on the volume and quality problems of big data. The volume of big data is very subjective from application to application. However, there should be thresholds from where you have enough data to start your research or application. This study proposes the methodology to determine if provided volume of data is the threshold to start the analysis or the application need more data before it begins.

The rest of the paper is organized as follows. In Sect. 2, a systematic literature survey of related works is given. This is followed by our proposed methodology for determining sufficient data amounts for data analysis in Sect. 3. Our results are summarized in Sect. 4. Finally, conclusions and indication of future work are given in Sect. 6 and Sect. 5 respectively.

2   Literature Review

Big data is generally related information that have large data volumes and complex data structures [13], such as social media data [7], mobile phone call records, commercial website data (e.g., Alibaba, Amazon), blogs data, search engine data, smart car data, and smart home data [27].

The most popular and common description of big data is the "3V" model, where "3V" stands for volume, variety, and velocity [14]. The **volume** of big data refers to big data having a large volume to it. For example, there are 2 billion users on Facebook [5], YouTube videos can even have larger data volume in terms of data storage. The **variety** refers to the fact that big data should have diverse sources of data with different data structures, and potential applications. The **velocity** refers to real time data generation and updating. For instance, weather data are often updated several times each day. In addition to the "3V" model, "4V" and "5V" models are emerging as different researchers seek to redefine big data. IBM recommends to accept the **veracity** to describe the bias problems brought by big data and understand that the "4V" model can accurately describe big data. Some researchers argue that big data also have the features of **value**, **variability**, and **visualization**.

One of the reasons why big data is popular in most fields and disciplines is that to a great extent it improves the data availability and accessibility of various research subjects which lets researchers study distinct topics that were tough to study because of poor data availability. Big data provides the capacity to collect and analyze data with an unprecedented breadth and depth and scale [15].

Big data research is mainly data driven in every field and discipline. Therefore, big data research focuses on methodological transformation or computes the application of big data on various topics in urban research. The scope of big data studies are complex to summarize because big data have different types and each type has distinct applications. Methodological big data research is basically computation intensive. For instance, a few researchers have proposed innovative computational system for data mining on big data [25]. There are some studies that include urban computing from big data where researchers studies urban air quality with the help of big data [28] and the implementation of latest machine learning and deep learning techniques such as neural network for big data analysis [18,19].

Most of the current research that uses social media data are following the assumption that social media data has a really large number of users, which indicates that the data has a large and enough sample coverage. Therefore, social media data can indicate the entire population or feasible population sampling bias can be dismissed [16]. However, this is not the case and this assumption is wrong. The large size and volume of big data does not mean that the data is random and represents the entire population [2], and a larger quantity of data does not increase the quality [4].

Big Data quality changes from one type to another and from one technology to another. Generally, data quality issues scale with the increase of data collection. So, if data quality problems exist in a current, traditionally small sized data set, the same data quality problems will be larger, directly proportional to the increase in scale of data collection.

There are issues of using biased samples when researchers rely on social media for data in studies of social behaviour. Big data collected from social media has the potential to shed light on many questions about social behavior. The bigger is not always better and size is not all that matters sayings also hold true when it comes to social media datasets. Thinking carefully about research design is one way to address potential biases [9].

There are several strategies for determining the sample size when analyzing social media data. Perhaps the most often asked question concerning sampling is, "What size sample do I need?". The answer is not simple and the answer to this question is influenced by a number of factors, including the purpose of the study, population size, the risk of selecting a "bad" sample, and the allowable sampling error [11]. This papers attempts to address the issue of "What size sample do I need?". There tends to be a lot of challenges in social media data and general big data. Therefore, researchers spend much more time in determining the sample size to start their study. This paper provides a statistical framework and methods that researchers will find useful to answer this question. However, there are lot of factors researchers should consider before determining the sample size such as risks with their study and feasibility for trail and error in their study. The given methods will require the trail and error approach as the sample size is highly subjective from case to case.

3   Proposed Work

As discussed earlier, the most important question is the sample size. In this section, we would like to propose our methodology to determine sample size for text analytic purposes. This methodology is connected to social media data, as our methods have been focused and studied on Twitter data. There are mainly 3 main parts in this methodology as listed below. Each part of this method is crucial to obtaining a good sample size for commencing analysis. The overall process of the determining the sample size is depicted in Fig. 1.

1. Data Cleaning
2. Topic Modelling
3. Modified Confidence Interval

3.1   Data Cleaning

Data cleaning, also known as data cleansing, is the process of detecting and correcting (or removing) corrupt or inaccurate records from a record set, table, or database and refers to identifying incomplete, incorrect, inaccurate or irrelevant parts of the data and then replacing, modifying, or deleting the dirty or coarse data [26]. The accuracy of data is very crucial as incorrect/inconsistent data can lead to false results. False results can impact both public/private sector businesses due to misdirected investments. For example, the government may

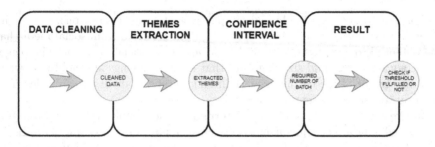

Fig. 1. Overall process flow chart

decide which regions require further investment on community infrastructure and services by analyzing population census figures. In this case, it is critical to have access to accurate and consistent data to avoid erroneous fiscal decisions. In the business sector, inaccurate data can be expensive as businesses run on profit. Many private companies use customer information databases that store data such as customer's contact info, mailing address and their preferences for a given company's products. As an example, if the preferences are not accurate, a given company may be affected by the cost of re-sending preferred items or even losing that customer who received incorrect items. There are many disciplines that use data cleaning to help create more accurate data. The forensic accounting and fraud investigating uses data cleaning in preparing its data and is typically done before data is sent to a data warehouse for further investigation [17]. There are some application programming interfaces (APIs) available for data cleaning purposes.

With the emergence of big data, data cleaning has become a more essential part of data analysis. Finance, healthcare, retail, entertainment, and education industries are now faced with larger amounts of raw data. As the data pool is getting bigger, the number of things that are going wrong also is increasing. Each error/fault becomes more difficult to resolve when you are unable to view all the data all at once.

In this paper, we have used several techniques to make sure the data we are using is cleaned and free from errors, including

- removing extra spaces
- removing duplicate tweets
- removing stop words from English Dictionary
- removing keywords that are used to collect the dataset
- removing special characters
- removing all the text to Lower Case
- spell checking
- deleting all formatting

Once the above measures are implemented on the tweets, these cleaned tweets are stored in a file that we can use for our next section.

3.2 Topic Modelling

With the use of topic modelling we can achieve our goal for **Theme Extraction**. Topic modelling is an unsupervised learning method as well as an easy way to analyze large volumes of raw (unlabeled) text which is known as corpus in Natural Language Processing terms. A "topic" contains a cluster of words that frequently appear together in the corpus. Topic models can connect words that are related to each other or on somewhat level with similar meaning and differentiate between uses of same words with other meanings.

The Latent Dirichlet Allocation (LDA) algorithm is a topic modelling technique that is widely used and has shown to be very consistent. When we provide the corpus to LDA, it extracts the relevant topics and its weight from the corpus. The weight of the topic gives the probability of the occurrence in the corpus. By sorting the weights, we can see the dominant topics in the corpus. LDA offers a fast and inexpensive way to get an overview of a collected data so that general questions can be answered instantly, specially it can get us answers about the comprehensive patterns in the corpus. It is very useful for performing initial analysis before investing time and money on more traditional automatic/manual content analysis [12].

In this paper, we run LDA to get k number of themes from the given corpus. The variable factor here is clearly k. The value for k can vary from case to case. The k value depends on the corpus size and how relevant topics your study needs. A simple way to obtain k is to run LDA with your corpus with minimum 10 topics, in other words setting $k = 10$. Next, check the probability of the occurrence for each topic and if you want to find highly related topics, drop the topics where probability of the occurrence is less than 50%. For example, if the top 5 themes give above 50% of the probability of occurrence, we can drop the rest of them by setting $k = 5$. Once LDA is run on the text, the results are shown in Fig. 2. Each topic has words and words have their probability of occurrence. Thus, we can determine what topics are in the text data and related words for each topic.

Now that we have our top-k themes as well as probability of occurrence, the values should be stored in a comma separated values (csv) file. As we are working with social media big data, we need to divide our data into batches before it is fed into data cleaning and LDA. To discover the optimal number of batches we use Grid search. Grid Search will give us the number of batches. The process of data cleaning and Topic Modelling will be performed for each batch. For instance, let us say we have 150 batches. First clean the batch, perform LDA on it and store the results of LDA. Repeat it until all batches are covered.

3.3 Modified Confidence Interval

Confidence Interval (CI) is a range of values defined such that there is a specified probability that the value of a parameter lies within it. The mathematical equation to find Confidence Interval is given in Eq. 1.

```
Topic: 0
Words: 0.035*"govern" + 0.024*"open" + 0.018*"coast" + 0.017*"tasmanian" + 0.017*"gold" + 0.014*"australia" + 0.013*"beat" + 0.
010*"win" + 0.010*"ahead" + 0.009*"shark"
Topic: 1
Words: 0.023*"world" + 0.014*"final" + 0.013*"record" + 0.012*"break" + 0.011*"lose" + 0.011*"australian" + 0.011*"leagu" + 0.0
11*"test" + 0.010*"australia" + 0.010*"hill"
Topic: 2
Words: 0.018*"rural" + 0.018*"council" + 0.015*"fund" + 0.014*"plan" + 0.013*"health" + 0.012*"chang" + 0.011*"nation" + 0.010
*"price" + 0.010*"servic" + 0.009*"say"
Topic: 3
Words: 0.025*"elect" + 0.022*"adelaid" + 0.012*"perth" + 0.011*"take" + 0.011*"say" + 0.010*"labor" + 0.010*"turnbul" + 0.009
*"vote" + 0.009*"royal" + 0.009*"time"
Topic: 4
Words: 0.032*"court" + 0.022*"face" + 0.020*"charg" + 0.020*"home" + 0.018*"tasmania" + 0.017*"murder" + 0.015*"trial" + 0.012
*"accus" + 0.012*"abus" + 0.012*"child"
Topic: 5
Words: 0.024*"countri" + 0.021*"hour" + 0.020*"australian" + 0.019*"warn" + 0.016*"live" + 0.013*"indigen" + 0.011*"call" + 0.0
09*"victorian" + 0.009*"campaign" + 0.008*"show"
Topic: 6
Words: 0.027*"south" + 0.024*"year" + 0.020*"interview" + 0.020*"north" + 0.019*"jail" + 0.018*"west" + 0.014*"island" + 0.013
*"australia" + 0.013*"victoria" + 0.010*"china"
Topic: 7
Words: 0.031*"queensland" + 0.029*"melbourn" + 0.018*"water" + 0.017*"claim" + 0.013*"hunter" + 0.012*"green" + 0.012*"resid" +
0.011*"darwin" + 0.010*"young" + 0.009*"plead"
Topic: 8
Words: 0.017*"attack" + 0.016*"kill" + 0.012*"victim" + 0.012*"violenc" + 0.010*"hobart" + 0.010*"rugbi" + 0.010*"secur" + 0.01
0*"say" + 0.009*"state" + 0.008*"domest"
Topic: 9
Words: 0.052*"polic" + 0.020*"crash" + 0.019*"death" + 0.017*"sydney" + 0.016*"miss" + 0.016*"woman" + 0.015*"die" + 0.015*"cha
```

Fig. 2. LDA results

$$CI = \bar{X} \pm t_{n-1,\alpha/2}\frac{S}{\sqrt{n}} \tag{1}$$

We can define the components of Eq. 1 as follows:

- X = mean of the output data from the replications

- S = standard deviation of the output data from replications

- n = number of replications

- $t_{n-1,\alpha/2}\frac{S}{\sqrt{n}}$ = value from Student t-distribution with $n-1$ degree of freedom and a significance level $\frac{\alpha}{2}$, $\alpha = 0.05$

By rearranging the Confidence Interval formula we can get n [8] which is the required number of batches to perform the analysis:

$$n = (\frac{100St_{n-1,\alpha/2}}{d\bar{X}})^2 \tag{2}$$

where d is the percentage deviation of the CI about the mean.

Using Eq. 2, we can obtain the threshold number which determines a good sample size. Here, we strongly recommend to use a 95% Modified Confidence Interval. For example, 95% Modified Confidence Interval means that if we measure the heights of 40 randomly chosen adult males, and get a mean height of 175 cm, our Modified Confidence Interval will be 175 cm ± 6.2 cm. This states that the true mean of all adult males is likely to be between 168.8 cm and 181.2 cm. The "95%" says that 95% of experiments similar to ours will include the true mean, but 5% will not [3]. This method will be applied on the extracted themes of all the batches and we will get the threshold number for the sample size. All numbers will be stores in a csv file.

3.4   Checking Results

After completing all the methodology above, we can now verify the results. The final numbers are shown in Table 1.

Table 1. Result of modified confidence interval

Themes	n
Theme 1	120
Theme 2	36
Theme 3	134
Theme n	45

First, get the maximum of all the themes given as $M = max(b_1, b_2, b_3, \ldots, b_n)$. Next, check the result value of R, as shown in Eq. 3

$$R = (M - number\_of batches\_that\_we\_divided\_in\_the\_first\_step\_n) \qquad (3)$$

If R is 0 then your sample size is good enough to start the data analysis project. Otherwise, R number of batches are required to perform the data analysis. Therefore, collect R number of data and try again.

4   Results

Our methodology was tested extensively on Twitter data. The tweets were collected for diabetes patients. The purpose of this dataset was to examine the diabetes related participation on Twitter by the frequency and timing of diabetes-related tweets. Before starting the actual analysis of diabetes related tweets, we applied the proposed methodology to determine if the collected data set possessed an adequate amount of diabetes related tweets.

We used 640,000 diabetes related tweets which were filtered based on diabetes related terms and hashtags using an in-house platform[1]. First, the dataset was cleaned by filtering it through various parameters. Each tweet text is filtered through all the given criteria as stated earlier. Next, all cleaned tweets were combined in a file, and the Latent Dirichlet Allocation (LDA) algorithm was applied on the cleaned tweet csv file. The 640,000 cleaned tweet dataset was divided into 160 equal sized batches. Each batch had approximately 4000 tweets. Next, LDA algorithm was applied on each batch to get the top-k topics. The results of each batch were saved in the same file which will combine the results of all 160 batches. As seen in Fig. 3, each topic consists of the number of words and probability of occurrence of each word in a topic.

---

[1] http://twitter.datalab.science/.

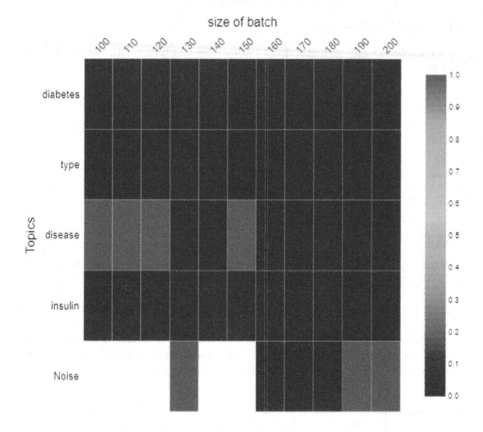

Fig. 3. Heatmap of the probability distribution

Finally, the modified confidence interval method was applied to the LDA results of all 160 batches to get number of batches needed for topics. As shown in Table 2, Column A represents topic and Column B represents number of batches needed for a given topic. Taking the maximum number from Column B which is 28, tweets were divided into 4000 tweets per batch. Next, we calculate $28 \times 4000 = 112000$ tweets. The data set originally had $640,000$ tweets. Therefore, we can conclude that this data set is good for diabetes related analysis.

As described in the example of this methodology, to summarize, we divided our data set into 160 equal parts. However, dividing the data-set into 160 equal parts is not random. We manually tested first 100 equal parts of dataset to 200 equal parts and checked the results and noise then decided to go with 160 equal parts. The results of 100 to 200 equal parts are shown in the Fig. 3. Figure 3 is a heat-map of the topics and number of batches. There are distinguishable topics on the Y-axis and number of batches on the X-axis. The red coloured tiles represents insufficient data for that particular topic in that particular size of batch. For example, 100 batches (in equal parts) of the dataset does not have enough data for topic 'disease'. As mentioned before, topic modeling is an unsupervised

Table 2. Result of modified confidence interval

Topic	Batch number
Diabetes	4
Type	2
Disease	28
Free	5
Insulin	6

learning method so we also experienced some noise in our results. Therefore, there are 160, 170, and 180 sized batched which have sufficient data that represents diabetes related data. Even though we experienced noise in those sized batches, there is sufficient data as prescribed by our methodology. Therefore, we chose 160 equal parts which is lowest of all size of batches which was shown to have sufficient volume of data for reliable results.

5   Future Work

This paper focused on determining a sufficient volume of data for reliable analysis. The methodology may be able to be automated. As mentioned earlier, batch size was tested manually and determined the final batch size as 160 equal parts. This process can be automated with sufficient computational resources where a system can determine the batch size automatically. An algoitthm can easily be developed where a range is inputted and the algorithm tests various batch sizes and presents final results as in the form of a suitable batch size for dividing the dataset into equal parts.

6   Conclusion

This paper began with presenting some real-world challenges with determining the sample size for reliable social media big data analysis. The sample size should be appropriate for the data analysis that is planned. Each study in text analytics is unique and there can be one or more attributes that need to be taken into consideration by researchers. Big data is helpful in all disciplines but it comes with the cost. It has so much potential in almost every field. As discussed earlier, big data cannot be presented in a simple spreadsheet and that makes it difficult to troubleshoot proper analysis techniques sometimes. No matter the size of the data, early feasibility studies have to focus on determining a relevant sample size. Our aim with this paper is to simplify the feasibility study for research related to unstructured text analysis. This paper shows a way to determine a sample size for social media data for text analysis purposes. There are some variable factors that big data analysis practitioners should consider as the parameters themselves may differ from analysis to analysis.

References

1. Baldwin, S.: Compute Canada: advancing computational research. J. Phys. Conf. Ser. **341**, 012001 (2012)
2. Boyd, D., Crawford, K.: Critical questions for big data: provocations for a cultural, technological, and scholarly phenomenon. Inf. Commun. Soc. **15**(5), 662–679 (2012)
3. Bruckman, A.: Can educational be fun. In: Game Developers Conference, vol. 99, pp. 75–79 (1999)
4. Cheshire, J., Batty, M.: Visualisation tools for understanding big data (2012)
5. Fatehkia, M., Kashyap, R., Weber, I.: Using facebook ad data to track the global digital gender gap. World Dev. **107**, 189–209 (2018)
6. Gavrila, V., Garrity, A., Hirschfeld, E., Edwards, B., Lee, J.M.: Peer support through a diabetes social media community. J. Diab. Sci. Technol. **13**(3), 493–497 (2019)
7. Ghani, N.A., Hamid, S., Hashem, I.A.T., Ahmed, E.: Social media big data analytics: a survey. Comput. Hum. Behav. **101**, 417–428 (2019)
8. Giabbanelli, P.J., Baniukiewicz, M.: Navigating complex systems for policymaking using simple software tools. In: Giabbanelli, P.J., Mago, V.K., Papageorgiou, E.I. (eds.) Advanced Data Analytics in Health. SIST, vol. 93, pp. 21–40. Springer, Cham (2018). https://doi.org/10.1007/978-3-319-77911-9_2
9. Hargittai, E.: Is bigger always better? Potential biases of big data derived from social network sites. Ann. Am. Acad. Polit. Soc. Sci. **659**(1), 63–76 (2015)
10. High, R.: The era of cognitive systems: an inside look at IBM Watson and how it works, pp. 1–16. IBM Corporation, Redbooks (2012)
11. Israel, G.D.: Determining sample size (1992)
12. Jacobi, C., Van Atteveldt, W., Welbers, K.: Quantitative analysis of large amounts of journalistic texts using topic modelling. Digit. J. **4**(1), 89–106 (2016)
13. Khoury, M.J., Ioannidis, J.P.: Big data meets public health. Science **346**(6213), 1054–1055 (2014)
14. Laney, D.: 3D data management: controlling data volume, velocity and variety. META Group Res. Note **6**(70), 1 (2001)
15. Lazer, D., et al.: Computational social science. Science **323**(5915), 721–723 (2009)
16. Mayer-Schönberger, V., Cukier, K.: Big Data: A Revolution that Will Transform How We Live, Work, and Think. Houghton Mifflin Harcourt (2013)
17. Nigrini, M.J.: Forensic analytics: methods and techniques for forensic accounting investigations, vol. 558. Wiley (2011)
18. O'Leary, D.E.: Artificial intelligence and big data. IEEE Intell. Syst. **28**(2), 96–99 (2013)
19. Pijanowski, B.C., Tayyebi, A., Doucette, J., Pekin, B.K., Braun, D., Plourde, J.: A big data urban growth simulation at a national scale: configuring the GIS and neural network based land transformation model to run in a high performance computing (HPC) environment. Environ. Model. Softw. **51**, 250–268 (2014)
20. Robinson, K., Mago, V.: Birds of prey: identifying lexical irregularities in spam on twitter. Wireless Netw. **11**, 1–8 (2018)
21. Sandhu, M., Vinson, C.D., Mago, V.K., Giabbanelli, P.J.: From associations to sarcasm: mining the shift of opinions regarding the supreme court on Twitter. Online Soc. Netw. Media **14**, 100054 (2019)
22. Shah, N., Srivastava, G., Savage, D.W., Mago, V.: Assessing Canadians health activity and nutritional habits through social media. Frontiers Public Health **7** (2019)

23. Shah, N., Willick, D., Mago, V.: A framework for social media data analytics using elasticsearch and kibana. Wireless Netw. **11**, 1–9 (2018)
24. Sharma, G., Srivastava, G., Mago, V.: A framework for automatic categorization of social data into medical domains. IEEE Trans. Comput. Soc. Syst. **7**(1), 129–140 (2019)
25. Gao, S., Li, L., Janowicz, K., Zhang, Y.: Constructing gazetteers from volunteered big geo-data based on hadoop. Comput. Environ. Urban Syst. **6**, 172–186 (2017)
26. Wu, S.: A review on coarse warranty data and analysis. Reliab. Eng. Syst. Saf. **114**, 1–11 (2013)
27. Yassine, A., Singh, S., Hossain, M.S., Muhammad, G.: IoT big data analytics for smart homes with fog and cloud computing. Future Gener. Comput. Syst. **91**, 563–573 (2019)
28. Zheng, Y., Liu, F., Hsieh, H.P.: U-air: when urban air quality inference meets big data. In: Proceedings of the 19th ACM SIGKDD International Conference on Knowledge Discovery and Data Mining, KDD 2013, pp. 1436–1444. ACM, New York (2013). http://doi.acm.org/10.1145/2487575.2488188

A Fuzzy Crow Search Algorithm for Solving Data Clustering Problem

Ze-Xue Wu[1], Ko-Wei Huang[1(✉)] [iD], and Chu-Sing Yang[2]

[1] Department of Electrical Engineering,
National Kaohsiung University of Science and Technology, Kaohsiung City, Taiwan
{1103104102,elone.huang}@nkust.edu.tw
[2] Institute of Computer and Communication Engineering,
Department of Electrical Engineering, National Cheng Kung University,
Tainan, Taiwan
csyang@ee.ncku.edu.tw

Abstract. The crow is one of the most intelligent bird and infamous for observing other birds so that they can steal their food. The crow search algorithm (CSA), a nature-based optimizer, is inspired by the social behavior of crows. Scholars have applied the CSA to obtain efficient solutions to certain function and combinatorial optimization problems. Another popular and powerful method with several real-world applications (e.g., energy, finance, marketing, and medical imaging) is fuzzy clustering. The fuzzy c-means (FCM) algorithm is a critical fuzzy clustering approach given its efficiency and implementation easily. However, the FCM algorithm can be easily trapped in the local optima. To solve this data clustering problem, this study proposes a hybrid fuzzy clustering algorithm combines the CSA and a fireworks algorithm. The algorithm performance is evaluated using eight well-known UCI benchmarks. The experimental analysis concludes that adding the fireworks algorithm improved the CSA's performance and offered better solutions than those by other meta-heuristic algorithms.

Keywords: Fuzzy c-means · Data clustering · Crow search algorithm · Fireworks algorithm · Meta heuristic algorithm

1 Introduction

In several research domains, such as gene clustering [1,2], medical image segmentation [3,4], outlier detection [5], pattern recognition [6,7], sensor clustering [8,9], social network analysis [10], and vector quantization [11]. there has been an increasing interest in data clustering. The concept of this clustering process involves assigning unlabeled data objects based on similarity criteria. The aim of this process is to produce groups containing data objects that are similar to each other and also different from the data objects in other groups.

The clustering problem involves two main types of clustering algorithms: hierarchical algorithms and partitional clustering algorithms [12]. The partitional

© Springer Nature Switzerland AG 2020
H. Fujita et al. (Eds.): IEA/AIE 2020, LNAI 12144, pp. 782–791, 2020.
https://doi.org/10.1007/978-3-030-55789-8_67

approach has been one of the most popular research domains in data clustering community when the k-means algorithm been proposed [13] and been proven as the NP-hard problem.

The meta-heuristic algorithms can be used to solve partitional models, such as the ant colony optimization [14], particle swarm optimization [15] and whale optimization algorithm [16], have made significant progress toward solving these complex optimization problems.

In this paper, we proposed the hybrid CSA and fireworks algorithm (FCFA) to solve data clustering problems. The FCFA processes all the advantages of the CSA, such as a search strategy and minimal parameters, and adapts the fireworks algorithm (FA) [17] to process the local search. The performance evaluation of FCFA was compared with the FCM algorithm [18], FPSO [19]. The results demonstrated that the FCFA achieved an better fitness values, as compared to other algorithms.

The remains of the article is organized as follows. In Sect. 2, the background knowledges are discussed. The proposed FCFA algorithm is presented detailed in Sect. 3. The simulation result are presented in Sect. 4. Finally, the conclusions and future work are given in Sect. 5.

2   Background Knowledge

2.1   Fuzzy C-Means

Bezdek [18,20] proposed the well-known clustering algorithm, FCM. The fuzzy theory has been adopted in many research fields including computer vision and robot moving path. The algorithm is an iterative optimization approach that aims to minimize the objective function. To address the fuzzy data clustering problem, a fuzzy membership function is used to assign each data object to a group. Every set of data objects, O, O_i, has an i-th data object in the set O.

$$J_{FCM} = \sum_{i=1}^{c} \sum_{j=1}^{n} \mu_{ij}^{m} \times \|d_{ij}\| \tag{1}$$

where μ_{ij} denotes the membership matrix. m parameter used to control the fuzziness of the cluster result (ex, $m¿1$), and d_{ij} denotes the Euclidean distance measure between the data object O_i and the cluster center Z_j see Eqs. (2) and (3).

$$d(O_i, O_j) = \sqrt{\sum_{t=1}^{m}(O_{i,t} - O_{j,t})^2} \tag{2}$$

where the Euclidean distance between data objects i and j can be obtained as follows Eq. (2):

$$f(O, Z) = \sum_{i=1}^{n} \sum_{j=1}^{k} w_{ij}\|O_i - Z_j\|^2 \tag{3}$$

where n is the number of data objects and k is number of clusters, respectively; w_{ij} is the weight of data object O_i within cluster Z_j, which will be 0 or 1 (depending on whether O_i is assigned to cluster Z_j or not); and $\|O_i - Z_j\|$ denotes the Euclidean distance between the data object O_i and the group center Z_j.

The following equation demonstrates the update in cluster center Z_j.

$$Z_j = \frac{\sum_{i=1}^{n} \mu_{ij}^m O_i}{\sum_{i=1}^{n} \mu_{ij}^m} \tag{4}$$

In the FCM algorithm, membership function μ_{ij} is the key concept and the degree of membership is updated as follows:

$$\mu_{ij} = \left[\sum_{j=1}^{k} \left(\frac{d_{ij}}{d_{ik}} \right)^{\frac{2}{m-1}} \right]^{-1} \tag{5}$$

The first step in the FCM algorithm is setting parameter m and to initialize membership matrix function μ. Next, cluster center C is derived using Eq. (4), following which the μ is updated by using Eq. (5).

2.2 Crow Search Algorithm (CSA)

The CSA was first introduced by Askarzadeh in 2016 [21], is a newly natural inspired algorithm that is designed for solving optimization problems. It is based on the behavior of crows that would hide place for caching food and to protect location.

Assuming that the searching space of each crow is a d-dimensional space and that the number of crows is equal to N, at iteration t as follows formularized:

$$X_i^t = \{x_{i,1}^t, x_{i,2}^t, x_{i,j}^t, \ldots, x_{i,d}^t\} \text{ for } i = 1, 2, 3 \ldots, N. \tag{6}$$

where $x_{i,j}^t$ is the j-th position of the crow i.

Each crow can memorize its hiding position. For example, the position of crow i at iteration t can be denoted as m_i^t, which implies that this is the optimal position of crow i. At iteration t, if crow i follows crow j, two cases may occur:

Case 1: Crow j is unaware that crow i is following. The crow i will be closed to the hiding place of crow j.

Case 2: Crow j aware that crow i is following. To protect the food, crow j deceives crow i by leading crow i to another location.

The new position of crow i at iteration $t + 1$ will be calculated in these cases as following Eq. (7):

$$x_{i,j}^{t+1} = \begin{cases} x_{i,j}^t + rand_i \times fL \times (m_{c,j}^i - x_{i,j}^i) & \text{if } rand_i \geq AP, \\ \text{random position} & \text{if } rand_i < AP. \end{cases} \tag{7}$$

where fL denotes the flight length of crow i at iteration t, $rand_i$ is a random variable that falls within the interval $[0,1]$, and AP is the awareness probability value of crow j at iteration t.

2.3 Fireworks Algorithm

The Fireworks algorithm (FA) was inspired by the process of setting off fireworks [17,22]. The FA implements searching strategies, such as the number of sparks, amplitude of the explosion, and selection of locations. The FA is based on the fact that the searching space is random, and each location explodes a firework that generates many sparks. In addition, there are two specific behavior patterns that fireworks can exhibit, known as good and bad. Good fireworks are located in a promising area that is close to the optimal location, while bad fireworks are located far from the optimal location. Then, in the next time interval, only the good fireworks are selected. In the FA, mathematical formulas for the number of sparks, amplitude of the explosion, generating sparks and selection of locations can be defined as follows.

Assume that there are N fireworks, with each firework in iteration t referred to as firework i, each firework can be represented as follows.

$$Y_i^t = \{Y_{i,1}^t, Y_{i,2}^t, Y_{i,j}^t, \ldots, Y_{i,d}^t\} \quad i = 1, 2, 3 \ldots, N. \tag{8}$$

where d denotes the number of dimensions, and $Y_{(i,j)}^t$ is the location of firework i in the dimension j in time t.

Number of Sparks. The number of sparks generated by each firework in location $Y_{(i,j)}^t$ can be represented as follows

$$s_i = m \times \frac{worst - fit(i) + \lambda}{\sum_{i=1}^n (worst - fit(i)) + \lambda} \tag{9}$$

where m is the control parameter for controlling the total number of sparks generated by N fireworks, $worst$ denotes the worst search fireworks, $fit(i)$ is the fitness value of firework i, and λ is the small constant necessary to prevent errors due to dividing by zero. Finally, there are bounds that limit the number of sparks. This can be modeled as follows

$$\tilde{s} = \begin{cases} round(a \times m), & \text{if } s_i < (a \times m) \\ round(b \times m), & \text{if } s_i > (b \times m), a < b < 1 \\ round(s_i), & \text{otherwise} \end{cases} \tag{10}$$

where a and b are the constants use to set bounds on the number of sparks, and round function is the rounding function.

Amplitude of Explosion. In addition, the better the quality of the firework, the lower the explosion amplitude will be. Thus, in contrast to the spark number, better fireworks will have a smaller amplitude. For each firework $Y_{(i,j)}^t$, the amplitude of explosion A_i can be represented as follows

$$A_i = \tilde{A} \times \frac{fit(i) - best + \lambda}{\sum_{i=1}^n (fit(i) - best) + \lambda} \tag{11}$$

where $\widetilde{A}$ is constant that represents the maximum explosion amplitude, and *best* denotes the best search result among all N fireworks.

Generating Sparks. The location of a spark around each firework can be determined as follows

$$\widetilde{X}_i^t, j = 1, 2, 3, ..., s_i, i = 1, 2, 3, ..., N \tag{12}$$

The location of the sparks can then be updated based on the displacement of operator h as follows

$$\widetilde{X}_{j,d}^t = \widetilde{X}_{j,d}^t + h \tag{13}$$

$$h = A_i \times rand(-1, 1) \tag{14}$$

where d is the randomized selected dimension of the searching space.

If the location of the sparks is out of range or not within the searching space, the new position will be assigned as follows

$$\widetilde{X}_{j,d}^t = X_{min,d}^t + round(\widetilde{X}_{j,d}^t) \mod (X_{max,d}^t - X_{min,d}^t) \tag{15}$$

where $X_{max,d}^t$ and $X_{min,d}^t$ represent the boundaries of spark $\widetilde{X}_{i,d}^t$ and $\widetilde{X}_{j,d}^t$ is the location of the spark when it goes outside of the searching space.

In addition, in order to create balance and diversity within the sparks, there is another approach used to update the location of a specific spark $\hat{X}_j^t$ that uses the Gaussian explosion coefficient g as follows

$$\hat{X}_{j,d}^t = \hat{X}_{j,d}^t \times g \tag{16}$$

If the location of specific sparks is out of range or out of the searching space, a new position will be assigned as follows

$$\hat{X}_{j,d}^t = X_{min,d}^t + round(\hat{X}_{j,d}^t) \mod (X_{max,d}^t - X_{min,d}^t) \tag{17}$$

where $X_{max,d}^t$ and $X_{min,d}^t$ represent the boundaries of spark $\hat{X}_{i,d}^t$ and $\hat{X}_{j,d}^t$ is the location of the specific spark when it goes outside of the searching space

Selection Locations. In each iteration, the best explosion location will be selected, and this location will be used for the next FA iteration. The other fireworks will be selected based on their similarity using the route-wheel approach of the genetic algorithm. In the route-wheel method, the probability of each individual is generated based on the fitness value, but in the FA algorithm, the Euclidean distance is applied instead. Assuming that the NP includes the set of all current locations of both fireworks and sparks after executing the belows updated formulas. The selected probability $P_b(X_i^t)$ of individual i can be calculated as follows:

$$P_b(X_i^t) = \frac{X_i^t}{\sum_{j=1}^{NP}(Dist(X_i^t)}\tag{18}$$

$$Dist(X_i^t) = \sum_{S=1}^{NP} \| X_i^t - X_S^t \|\tag{19}$$

where $Dist(X_i^t)$ represents the Euclidean distance between the individual i and other individuals such S among all location dimensions.

3 Proposed Algorithm

This study combines the FCM algorithm and CSA to propose a new algorithm, FCFA, that can be used to determine the optimal membership function.

3.1 Objective Function

In this paper, the fitness function is used to determine the quality of each agent's search solution as follows.

$$F(X_i) = \frac{K}{J_{FCM}(X_i)}\tag{20}$$

where K represents a constant, $J_{FCM}(X_i$ is the agent $i's$ objective value, and X_i is the agent $i's$ search solution. A smaller $J_{FCM}(X_i$ obtains the smaller value represents a better clustering result for agent X_i.

3.2 Solution Encoding

The use of a meta-heuristic algorithm suitable to resolve the clustering problem warrants the encoding of individuals as cluster centers within the appropriate solution. The CSA was originally proposed as a population-based algorithm to address continuous problems. Thus, it is easy to encode the cluster center of each agent. For n agents, there will generate n possible solutions to the membership degree matrix. The membership degree matrix of agent i considering the fuzzy relationship between the set of data objects O and cluster center C is represented as follows:

$$X_i = \begin{bmatrix} \mu_{11} & \mu_{12} & \mu_{13} & \cdots & \mu_{1k} \\ \mu_{21} & \mu_{22} & \mu_{23} & \cdots & \mu_{2k} \\ \vdots & \vdots & \vdots & \ddots & \vdots \\ \mu_{n1} & \mu_{n2} & \mu_{n3} & \cdots & \mu_{nk} \end{bmatrix}\tag{21}$$

The degree of the membership function from the i-th data object to the j-th cluster center is represented by the elements of μ_{ij} of row i and column j.

3.3 Proposed FCFA Algorithm

This study aims to resolve the fuzzy clustering problem using FCFA, which is a combination of the FCM and CSA algorithms. FCFA enables the minimization of the fitness value of each agent. Figure 1 lists the steps of the proposed FCFA. The first step is to the randomly generate each agent. Next, the CSA is executed to obtain the new cluster center and membership matrix. The local search strategy will be triggered to fine-tune each agent within a specific number of iterations.

Step	Description
1.	Compute FCM algorithm and remap membership function as one of the agents.
2.	Initiate membership matrix function for each agent.
3.	Identify each agent's cluster center according to Eq. (4).
4.	Estimate the fitness value for each agent using Eq. (20).
5.	Update each crow's position matrix according to Eq. (7).
6.	Repeat steps 3–5 if the criterion to stop CSA procedure is not exceeded.
7.	Execute local search procedure according to FA algorithm using specific iterations Eq (9) to Eq (19) and obtain each new membership function μ according to Eq. (5).
8.	Return to step 3 if the criterion to stop has not been exceeded.

Fig. 1. Proposed FCFA algorithm

4   Experimental Results

All simulations work on a computer with a 3.30 GHz Intel CPU and 16 GB of main memory and the programming was implemented using Python.

4.1   Parameter Setting

The parameter of the FCFA algorithm are detailed in Table 1.

Table 1. Parameters setting of FCFA

Number of simulations	30
Number of iterations	1000
Number of individual (crows) N	20
Cluster fuzziness mf	2.6
Maximum explosion amplitude $\hat{A}$	40
Number of sparks controlling parameter m	50
Constants a and b	0.04, 0.8

Table 2. The testing UCI benchmarks

Dataset	Number of data objects	Number of dimensions	Number of groups
Iris	150	4	3
Breast cancer	683	9	2
Car evaluation	1728	6	4
CMC	1473	3	9
Statlog	6435	36	7
Glass	214	6	9
Wine	178	13	3
Yeast	1484	8	10

4.2 Local Search Setting

In this section, we set three different number of iterations ranging from 30 to 100 for processing using the FA algorithm. In addition, we set the three number of iterations to be from 10 to 50 for executing the local search. The result are summarized in Table 3, and the results show that each 30 iterations of the CSA algorithm and 10 time per local search will archive better result. Thus, executing 30 iteration of CSA triggers 10 times local search FA algorithm.

Table 3. Comparison of different local search settings of the Iris dataset

Number of CSAiterations	30	50	100
Number of FA iterations 10	68.8	69.1	69.3
30	69.1	69.3	69.2
50	69.1	69.0	69.4

4.3 Comparison Result

The proposed FCFA algorithm will compare with the FCM [18], FCFA without the FA algorithm and fuzzy particle swarm optimization (FPSO) [19] in this section. We use the eight UCI benchmarks. The detail benchmark as presented in Table 2.

The experimental results are presented in Table 4. It can be observed that the proposed FCFA obtains the better result, especially in the Statlog and Wine instances. In addition, a comparison between the FCFA with that without local search indicates that the use of the FA algorithm leads to a performance improvement.

Table 4. Performance evaluation with FCFA, FCM, FPSO, and FCFA with FA

Dataset	FCFA	FCM [18]	FCFA without FA	FPSO [19]
Iris	**69.0**	70.5	69.1	70.2
Breast cancer	**2281.1**	2308.8	2282.7	2291.4
Car evaluation	**1181.7**	1181.9	**1181.7**	1181.9
CMC	3744.0	3785.1	**3734.5**	3762.0
Glass	**1291.0**	1295.1	1291.0	1293.4
Statlog	**64858.3**	65006.9	64906.0	64901.0
Wine	**11724.9**	12076.0	11729.6	11860.2
Yeast	**38.2**	38.4	38.2	38.3
Average	**10648.5**	10720.3	10654.1	10674.8

5   Conclusion

This paper proposed a hybrid crow search optimization algorithm and fireworks algorithm (FCFA) for solving fuzzy data clustering problems. The FCFA combines the advantages of the CSA, including the search strategy and fewer parameters, with the fireworks algorithm, which was subsequently used to process the local search. The performance evaluation of the proposed algorithm using eight UCI benchmarks and results demonstrated that the proposed algorithm performed better performance. As a future study, we plan to further investigate the FCFA for image segmentation of KUB medical images.

Acknowledgement. This research was supported by the Ministry of Science and Technology of Taiwan, under grants MOST 108-2221-E-006-111-.

References

1. Liu, H., Zhao, R., Fang, H., Cheng, F., Yun, F., Liu, Y.-Y.: Entropy-based consensus clustering for patient stratification. Bioinformatics **33**(17), 2691–2698 (2017)
2. Nguyen, T.P.Q., Kuo, R.J.: Partition-and-merge based fuzzy genetic clustering algorithm for categorical data. Appl. Soft Comput. **75**, 254–264 (2019)
3. Mistry, K., Zhang, L., Neoh, S.C., Lim, C.P., Fielding, B.: A micro-GA embedded pso feature selection approach to intelligent facial emotion recognition. IEEE Trans. Cybern. **47**(6), 1496–1509 (2017)
4. Perone, C.S., Ballester, P., Barros, R.C., Cohen-Adad, J.: Unsupervised domain adaptation for medical imaging segmentation with self-ensembling. NeuroImage **194**, 1–11 (2019)
5. Liu, H., Li, J., Wu, Y., Fu, Y.: Clustering with outlier removal. IEEE Trans. Knowl. Data Eng. (2019)
6. Gan, W., Lin, J.C.W., Fournier-Viger, P., Chao, H.C., Philip, S.Y.: HUOPM: high-utility occupancy pattern mining. IEEE Trans. Cybern. **50**, 1195–1208 (2019)

7. Lin, J.C.W., Yang, L., Fournier-Viger, P., Hong, T.P.: Mining of skyline patterns by considering both frequent and utility constraints. Eng. Appl. Artif. Intell. **77**, 229–238 (2019)
8. Zhou, Y., Wang, N., Xiang, W.: Clustering hierarchy protocol in wireless sensor networks using an improved PSO algorithm. IEEE Access **5**, 2241–2253 (2017)
9. Wang, J., Gao, Y., Liu, W., Sangaiah, A.K., Kim, H.J.: An improved routing schema with special clustering using pso algorithm for heterogeneous wireless sensor network. Sensors **19**(3), 671 (2019)
10. Yang, Q., Yang, N., Browning, T.R., Jiang, B., Yao, T.: Clustering product development project organization from the perspective of social network analysis. IEEE Trans. Eng. Manag. (2019)
11. Chiranjeevi, K., Jena, U., Prasad, P.M.K.: Hybrid cuckoo search based evolutionary vector quantization for image compression. In: Lu, H., Li, Y. (eds.) Artificial Intelligence and Computer Vision. SCI, vol. 672, pp. 89–114. Springer, Cham (2017). https://doi.org/10.1007/978-3-319-46245-5_7
12. Jain, A.K., Duin, R.P.W., Mao, J.: Statistical pattern recognition: a review. IEEE Trans. Pattern Anal. Mach. Intell. **22**(1), 4–37 (2000)
13. MacQueen, J.: Some methods for classification and analysis of multivariate observations. In: The 5th Berkeley Symposium on Mathematical Statistics and Probability, pp. 281–297 (1967)
14. Krishnan, M., Yun, S., Jung, Y.M.: Enhanced clustering and ACO-based multiple mobile sinks for efficiency improvement of wireless sensor networks. Comput. Netw. **160**, 33–40 (2019)
15. Zhao, F., Chen, Y., Liu, H., Fan, J.: Alternate PSO-based adaptive interval type-2 intuitionistic fuzzy C-means clustering algorithm for color image segmentation. IEEE Access **7**, 64028–64039 (2019)
16. Wu, Z.X., Huang, K.W., Chen, J.L., Yang, C.S.: A memetic fuzzy whale optimization algorithm for data clustering. In: IEEE Congress on Evolutionary Computation, CEC 2019, Wellington, New Zealand, 10–13 June 2019, pp. 1446–1452. IEEE (2019)
17. Tan, Y., Zhu, Y.: Fireworks algorithm for optimization. In: Tan, Y., Shi, Y., Tan, K.C. (eds.) ICSI 2010. LNCS, vol. 6145, pp. 355–364. Springer, Heidelberg (2010). https://doi.org/10.1007/978-3-642-13495-1_44
18. Bezdek, J.C., Ehrlich, R., Full, W.: FCM: the fuzzy c-means clustering algorithm. Comput. Geosci. **10**(2–3), 191–203 (1984)
19. Izakian, H., Abraham, A.: Fuzzy C-means and fuzzy swarm for fuzzy clustering problem. Expert Syst. Appl. **38**(3), 1835–1838 (2011)
20. Bezdek, J.C.: Fuzzy mathematics in pattern classification. Ph. D. Dissertation, Applied Mathematics, Cornell University (1973)
21. Askarzadeh, A.: A novel metaheuristic method for solving constrained engineering optimization problems: crow search algorithm. Comput. Struct. **169**, 1–12 (2016)
22. Li, J., Zheng, S., Tan, Y.: Adaptive fireworks algorithm. In: 2014 IEEE Congress on Evolutionary Computation (CEC), pp. 3214–3221. IEEE (2014)

Distributed Density Peak Clustering of Trajectory Data on Spark

Yunhong Zheng[1], Xinzheng Niu[1(✉)], Philippe Fournier-Viger[2], Fan Li[3], and Lin Gao[4]

[1] School of Computer Science and Engineering, University of Electronic Science and Technology of China, Chengdu, Sichuan, China
2386100@qq.com
[2] School of Humanities and Social Sciences, Harbin Institute of Technology (Shenzhen), Shenzhen, Guangdong, China
[3] Key Laboratory of Fundamental Synthetic Vision Graphics and Image Science, Sichuan University, Chengdu, Sichuan, China
[4] Department of Computing Science and Technology, Southwest University of Science and Technology, Mianyang, Sichuan, China

Abstract. With the widespread use of mobile devices and GPS, trajectory data mining has become a very popular research field. However, for many applications, a huge amount of trajectory data is collected, which raises the problem of how to efficiently mine this data. To process large batches of trajectory data, this paper proposes a distributed trajectory clustering algorithm based on density peak clustering, named DTR-DPC. The proposed method partitions the trajectory data into dense and sparse areas during the trajectory partitioning and division stage, and then applies different trajectory division methods for different areas. Then, the algorithm replaces each dense area by a single abstract trajectory to fit the distribution of trajectory points in dense areas, which can reduce the amount of distance calculation. Finally, a novel density peak clustering-based method (E-DPC) for Spark is applied, which requires limited human intervention. Experimental results on several large trajectory datasets show that thanks to the proposed approach, runtime of trajectory clustering can be greatly decreased while obtaining a high accuracy.

Keywords: Trajectory data · Distributed clustering · Spark · DPC · Dense areas

1 Introduction

With the massive use of personal mobile devices and vehicle-mounted positioning devices, and the increasing demand for hidden information in trajectory data, trajectory data mining has emerged as an important branch of data mining. Due to several years of research, the trajectory mining process is now quite stable. It

© Springer Nature Switzerland AG 2020
H. Fujita et al. (Eds.): IEA/AIE 2020, LNAI 12144, pp. 792–804, 2020.
https://doi.org/10.1007/978-3-030-55789-8_68

consists of various sub-tasks such as classic trajectory acquisition and abnormal trajectory exploration.

Though trajectory data mining has many applications, most algorithms have been designed to deal with a small volume of data, and are thus unsuitable for applications where a large amount of data must be analyzed. In the data pre-processing stage, to protect the locality of each trajectory, a sub-trajectory segmentation strategy [1] and grid-based division [2] can be applied, but these techniques increase the amount of calculations. In the trajectory distance calculation stage, most studies directly calculate the distance between two sub-trajectories [3]. If the traditional trajectory division strategy was used in the previous stage, the amount of calculations will be further increased. Then, in the data mining stage, the most widely used model is clustering, using techniques such as k-means [4] and DBSCAN [5]. However, most clustering methods require human intervention. For example, k-means requires setting the number of clusters, while the *Eps* and *MinPts* parameters must be set for DBSCAN.

Considering these limitations of previous work, this paper proposes a novel approach for distributed trajectory clustering based on density peak clustering [6], named DTR-DPC (Distributed Trajectory-Density Peak Clustering). It is implemented on the Spark distributed computing platform to support the analysis of large trajectory data sets. The proposed method is optimized for fast calculations at all stages of the trajectory data mining process. In the trajectory partitioning and division stage, we propose to partition the data into dense and sparse areas, where sparse areas are divided into sub-trajectory segments, and entire trajectories are kept for dense areas. In the sub-trajectory similarity measurement stage, the proposed method replaces each dense area by a special trajectory called *ST* to reduce unnecessary calculations. For the clustering stage, a novel algorithm named E-DPC is applied, which requires limited human intervention and performs calculations in a distributed manner. An experimental evaluation was performed. Results show that thanks to the proposed approach, runtime of trajectory clustering can be greatly decreased while obtaining a high accuracy.

The rest of this paper is organized as follows. Section 2 surveys related work. Section 3 defines the trajectory data mining problem. Section 4 presents the propsed DTR-DPC approach. Then, Sect. 5 describes experimental results. Finally, a conclusion is drawn in Sect. 6.

2 Related Work

This section first reviews state-of-the-art techniques for trajectory data mining. Then it surveys related work on distributed trajectory clustering.

2.1 Trajectory Data Mining

Trajectory data mining models can be mainly grouped into three categories: pattern mining, classification, and clustering.

Pattern mining is a set of unsupervised techniques that are used for discovering interesting movement patterns in a set of trajectories [7]. Algorithms have been designed to extract several types of patterns such as gathering patterns [7], sequential patterns [8], and periodic patterns [9]. A gathering pattern represents the movements of groups of persons such as for celebrations, parades, protests and traffic jams [10]. A sequential pattern is a subsequence of locations that frequently appears in trajectories [8], while a periodic pattern is movements that periodically appear over time, found in trajectory data containing time information [9].

Classification consists of using supervised learning to predict the next location for a trajectory. But because trajectories generally do not have reliable labels, classification is rarely used. However, a classification based approach using Support Vector Machine (SVM) was applied for vehicle trajectory analysis and provided some good results [11].

Clustering is the most commonly used method for trajectory data mining. Most of the related clustering work fall in one of three categories: k-means-based trajectory clustering [12], fuzzy-based trajectory clustering [13], and density-based trajectory clustering [14,15]. Yeen and Lorita [12] proposed a k-means and fuzzy c-means clustering-based algorithm, which iteratively recalculates the centroids (means) of clusters in the trajectory dataset. Because k-means has the disadvantage that the number of clusters must be preset, researchers have gradually changed toward using fuzzy and density-based clustering. Nadeem et al. [13] first proposed an unsupervised fuzzy approach for motion trajectory clustering, which performs hierarchical reconstruction only in the region affected by a new instance, rather than updating all the hierarchies. Most recent trajectory clustering studies have focused on density-based clustering using algorithms such as DBSCAN [14] and density peak clustering (DPC) [15]. The biggest advantage of this type of algorithms is their high accuracy, while the biggest disadvantage is that they requires considerable human intervention.

2.2   Distributed Trajectory Clustering

An important limitation of most trajectory data mining algorithms is that they cannot handle large-scale trajectory data. To address this limitation, several parallel and distributed trajectory algorithms were proposed. Wang et al. [16] designed a Spark in-memory computing algorithm using data partitioning to implement parallelism for DPC and reduce local density calculations. Hu et al. [17] proposed a method for the fast calculation of trajectory similarity based on coarse-grained Dynamic Time Warping, implemented using the Hadoop MapReduce model to handle massive dynamic trajectory data with time information. Myamoto et al. [18] proposed a flight path generation method that uses a dynamic distributed genetic algorithm. This method differs from others in that the number of groups into which individuals are partitioned is not fixed, and hierarchical clustering is done using a distributed genetic algorithm.

Some of the most recent studies [19,20] are based on density-based clustering. Chen et al. [19] and Wang et al. [20] have used the DBSCAN and DPC algorithms, respectively. The former is implemented using Hadoop MapReduce for distributed computing, while the latter uses the Spark in-memory computing model. Compared with the former, the latter has higher computing efficiency because of the advantages of DPC's fast search strategy and Spark's memory-based operation. Though these algorithms are useful, efficiency can still be improved. Hence, this paper proposes optimizations for each stage of the trajectory data clustering process.

3   Problem Formulation

This section introduces preliminary definitions and then defines the studied problem of trajectory data clustering.

Definition 1. Trajectory (TR): *A trajectory TR_i is an ordered list of coordinate trajectory points collected at different time, denoted as $TR_i =< p_i^1, p_i^2, \ldots p_i^j, \ldots, p_i^{n_i} >$, where a superscript j is used to denote the j-th point of TR_i, and p_i^j is a pair containing a location l_i^j (longitude and latitude) and a time t_i^j.*

Definition 2. Sub-trajectory (STR): *A sub-trajectory STR of a trajectory TR_i is a list of consecutive and non-repeating points from that trajectory, denoted as $STR_i =< p_{STR_i}^{start}, \ldots, p_{STR_i}^j, p_{STR_i}^k, \ldots, p_{STR_i}^{end} >$, where $p_{STR_i}^{start}$ and $p_{STR_i}^{end}$ are the start and end point of STR_i, respectively.*

Definition 3. Trajectory Distance (TRD): *The trajectory distance (TRD) between two sub-trajectories STR_i and STR_j according to a distance metric, is denoted as $TRD_i^j = dist(STR_i, STR_j)$.*

Definition 4. Trajectory Clustering (TRC): *Given a trajectory distance matrix TRD, which specifies the distance between each sub-trajectory pair, TRC consists of dividing the sub-trajectories into clusters $C = \{C_1, C_2, \ldots, C_{N_C}\}$ based on the distance matrix TRD.*

This study uses the Davies-Bouldin Index (DBI) [21] to evaluate TRC accuracy. The DBI is defined as

$$DBI = \frac{1}{k} \sum_{i=1}^{k} \max_{j \neq i}(\frac{\overline{C_i} + \overline{C_j}}{dist(C_c^i, C_c^j)}) \tag{1}$$

where $\overline{C_i}$ and $\overline{C_j}$ are the average distance inside clusters C_i and C_j respectively, C_c^i and C_c^j are the cluster centers of C_i and C_j, and k is the number of clusters.

In general, a lower DBI indicates a more accurate clustering. We define the distributed trajectory data clustering problem as follows:

Definition 5. *(Problem Definition): Let there be a set of trajectories $TR = \{TR_1, TR_2, \ldots, TR_i, \ldots, TR_{n_{TR}}\}$, and assume that each TR_i is divided into a set of sub-trajectories $\{STR_1, \ldots, STR_i, \ldots STR_{n_{STR}}\}$. Moverover, consider the distance matrix TRD between each STR_i and STR_j. The problem is to efficiently perform TRC using TRD to generate a clustering C having a low DBI value.*

4   A Distributed Density Peak Clustering Approach

To solve the defined problem, this section proposes a distributed trajectory density peak clustering algorithm for Spark, named DTR-DPC, which consists of three steps, namely, trajectory partitioning and division, sub-trajectory similarity calculation, enhanced density peak clustering. The proposed approach first partitions trajectory points into several dense or sparse grid areas. Then, approximate calculations are done in dense areas, while a traditional trajectory division strategy is applied in sparse areas. This dramatically reduce the amount of calculations. Secondly, the similarity between sub-trajectories is measured using three different methods. Finally, an enhanced density peak clustering (E-DPC) algorithm is called to cluster sub-trajectories. The two key advantages of E-DPC is that it reduces the need for human intervention and that calculations are distributed (using the Spark model).

An overview of the proposed approach is presented in Fig. 1.

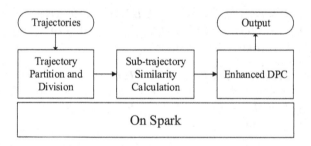

Fig. 1. Overview of the DTR-DPC approach

Phase 1 (trajectory partition and division): Data collection methods typically record a batch of trajectories TR containing many useless trajectory points which are continuously recorded in a small range and have no practical significance. To deal with this issue, the proposed method divides the area where trajectory points appear into several grid areas GA. The method assigns each point p_i^j to a different GA to form dense areas DA and sparse areas SA, for which different sub-trajectory partitioning strategies are applied. After that, trajectories are segmented into two different areas to form a sub-trajectory set STR.

Phase 2 (sub-trajectory similarity calculation): Then, using the trajectory division result of Phase 1, the similarity matrix (SM) of all sub-trajectory pairs is calculated using three different similarity measures, defined for various GA. The distance between two sub-trajectories in DA is denoted as $dist_{<da,da>}$, the distance between two sub-trajectories in SA is denoted as $dist_{<sa,sa>}$, while $dist_{<da,sa>}$ refers to the distance between a trajectory in a DA and one in a SA.

Phase 3 (enhanced density peak clustering): The SM matrix obtained in Phase 2 is used by an adaptive method to calculate the distance and density thresholds required by the DPC algorithm, to select some special points as center points. Then, each remaining point is assigned to its nearest center point to form a set of points called a cluster. Finally, the algorithm is optimized to run on Spark to reduce the runtime.

4.1   Trajectory Partition and Division

This first phase is performed in three steps. Firstly, a trajectory coordinate system (x, y) and grid area GA is built. Then, each point of the trajectory data is assigned to different grid areas to form dense areas and sparse areas based on a threshold Th_{NP} indicating a minimum number of points per area. Finally, different trajectory division strategies are applied for different area types to obtain the sub-trajectory set.

The trajectory coordinates system (x, y) is mainly used to map trajectory points to the two-dimensional coordinate axis. The X and Y axis values of a point denotes longitude and latitude, respectively. Furthermore, x-axis and y-axis scale values $|x_{scale}|$ and $|y_{scale}|$ are defined as:

$$|x_{scale}| = \frac{x_{max} - x_{min}}{NE} \tag{2}$$

$$|y_{scale}| = \frac{y_{max} - y_{min}}{NE} \tag{3}$$

where x_{max} and y_{max} represent the maximum longitude and latitude in all p_i^j, respectively, x_{min} and y_{min} represent the minimum longitude and latitude in all p_i^j, respectively, and NE is a number of equal parts, which needs to be set by the user.

Then, the method set grid lines at each x_{scale} and y_{scale} increment on the X and Y axis, respectively, to form GA, and each point p_i^j is assigned to different GA to form DA and SA, indicating that the number of points NP falling in an area is greater or smaller than the Th_{NP} threshold, respectively. An example of DA and SA partition diagram is shown in Fig. 2.

After partitioning trajectories, the algorithm divides trajectories for different areas to protect the locality of each trajectory and reduce the runtime. The algorithm scans all the trajectory segments of each entire trajectory TR_i, excluding all sub-trajectory segments that pass through DA, to save all the sub-trajectory segments that only pass through SA. The result is a set

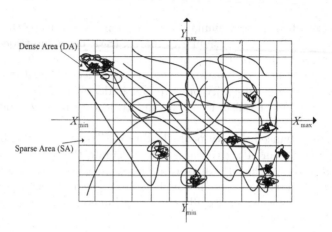

Fig. 2. A DA and SA partition diagram.

$SATR_i = \{SATR_i^1, \ldots, SATR_i^j, \ldots, SATR_i^n\}$, where $SATR_i^j$ represent the j-th sub-trajectory in TR_i, which only pass through SA.

All sub-trajectories that only pass through one dense area DA_i, are replaced by an abstract trajectory $DATR_i$, which is represented as:

$$DATR_i = < DA_i^{left_{lower}}, DA_i^{center}, DA_i^{right_{upper}} > \qquad (4)$$

where $DA_i^{left_{lower}}$, $DA_i^{right_{upper}}$, DA_i^{center} represents the left lower corner point, right upper corner point, and the center point of DA_i's center area, respectively.

For each sub-trajectory $SATR_i^j$, based on the MDA trajectory division strategy [1], the algorithm adds corner angles to obtain a better sub-trajectory division. A corner angle θ is defined as:

$$\theta = \begin{cases} \pi - \alpha, & \alpha \le \dfrac{\pi}{2}, \\ \alpha - \pi, & \alpha > \dfrac{\pi}{2}, \end{cases} \qquad (5)$$

A schematic diagram of this trajectory division approach is shown in Fig. 3, where α is calculated as:

$$\alpha = arccos((a^2 + b^2 - c^2)/2ab). \qquad (6)$$

Calculations made using the MDA strategy produce a trajectory division result $SASTR_k = < p_{SASTR_k}^{start}, \ldots, p_{SASTR_k}^j, \ldots, p_{SASTR_k}^{end} >$ of $SATR_i^j$. Hence the sub-trajectory set STR can be rewritten as:

$$\{DATR_1, \ldots, DATR_i, \ldots, DATR_{n_{DA}}, SASTR_1, \ldots, SASTR_j, \ldots, SASTR_{n_{SA}}\} \qquad (7)$$

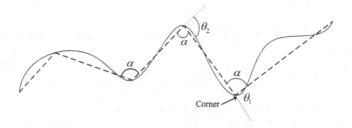

Fig. 3. The schematic diagram of trajectory division

where n_{DA} refers to the number of abstract sub-trajectories representing trajectories passing through DA and n_{SA} is the number of sub-trajectories obtained by dividing $SATR$.

4.2 Sub-trajectory Similarity Calculation

A sub-trajectory similarity measure is then applied to calculate the distance between each pair of trajectories. Because the proposed method handles two types of sub-trajectories ($DATR$ and $SASTR$), three types of distance are used: $dist_{<da,da>}^{<i,j>}$ between $DATR_i$ and $DATR_j$, $dist_{<da,sa>}^{i,j}$ between $DATR_i$ and $SASTR_j$, and $dist_{<sa,sa>}^{<i,j>}$ between $SASTR_i$ and $SASTR_j$.

For instance, two sub-trajectories $DATR_i$ and $DATR_j$ are shown in Fig. 4(a), and the distance $dist_{<da,da>}^{<i,j>}$ between two such sub-trajectories is calculated as:

$$dist_{<da,da>}^{<i,j>} = \beta dist_1 + \gamma dist2, \quad \beta + \gamma = 1, \tag{8}$$

where β and γ are distance weights, $dist_1$ is the distance between DA_i^{center} and DA_j^{center}, and $dist_2$ represents the average distance between the remaining two points $DA^{right_{upper}}$ and $DA^{left_{lower}}$.

For example, two sub-trajectories $SASTR_i$ and $SASTR_j$ are depicted in Fig. 4(b). The distance $dist_{<sa,sa>}^{<i,j>}$ between two such sub-trajectories is calculated using the MDA strategy [1], which is defined as :

$$dist_{<sa,sa>}^{<i,j>} = \omega_\perp \cdot dist_\perp(SASTR_i, SASTR_j) + \\ \omega_\parallel \cdot dist_\parallel(SASTR_i, SASTR_j) + \omega_\theta \cdot dist_\theta(SASTR_i, SASTR_j) \tag{9}$$

where $\omega_\perp$, $\omega_\parallel$ and ω_θ are the weights of three distance types. Those are $dist_\perp(SASTR_i, SASTR_j)$, $dist_\parallel(SASTR_i, SASTR_j)$ and $dist_\theta(SASTR_i, SASTR_j)$ of the MDA strategy [1], which are called vertical distance, parallel distance and angular distance.

Two sub-trajectories $DATR_i$ and $DASTR_j$ are shown in Fig. 4(c). The distance $dist_{<da,sa>}^{<i,j>}$ between two such sub-trajectories is also calculated using the MDA strategy [1], which treats $DA_i^{left_{lower}}$ as $p_{SASTR_j}^{start}$, and treats $DA_i^{right_{upper}}$ as $p_{SASTR_j}^{end}$.

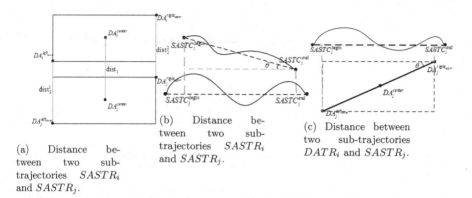

(a) Distance between two sub-trajectories $SASTR_i$ and $SASTR_j$.

(b) Distance between two sub-trajectories $SASTR_i$ and $SASTR_j$.

(c) Distance between two sub-trajectories $DATR_i$ and $SASTR_j$.

Fig. 4. The three distance types.

4.3   Enhanced Density Peak Clustering

To improve the efficiency of clustering trajectory data and reduce the impact of artificially set parameters on the DPC algorithm, this paper proposes an enhanced DPC algorithm, called E-DPC. It uses the Log-Normal distribution to adaptively select a density threshold $th_{density}$ and a distance threshold $th_{distance}$. The Log-Normal distribution is used instead of other distributions because its shape is narrow and upward, which means that we may choose more cluster centers to offset side-effects of data reduction in dense areas.

To apply DPC, the algorithm then selects some special trajectories as center points. Remaining sub-trajectories are then assigned to their nearest center points to form clusters. This assignment of a trajectory is done by considering the local density ρ_i and the minimum distances from any other trajectory with higher density δ_i. The greater the ρ_i and δ_i are, the more likely a sub-trajectory is to be a center point. However, a drawback of this traditional approach is that the thresholds of ρ_i and δ_i need to be set by hand.

To address this issue, the proposed aproach performs Log-Normal distribution fitting on all the calculated ρ and δ sets, and takes the upper quantile of the Log-Normal distribution as the local density threshold $th_{density}$ and distance threshold $th_{distance}$. The Log-Normal distribution is shown in Fig. 5, and ρ's Log-Normal distribution is:

$$f(\rho; \mu; \sigma) = \frac{1}{\sqrt{2\pi}\rho\sigma} e^{-\frac{(\ln \rho - \mu)}{2\sigma^2}} \tag{10}$$

5   Experimental Evaluation

This section reports results of experiments comparing the proposed TDR-DPC approach with two distributed trajectory clustering algorithms from the literature, namely DBSCAN [19] and DPC [16]. A Spark platform was used, having

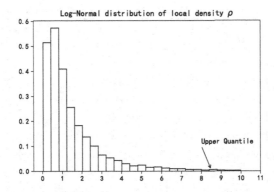

Fig. 5. The Log-Normal distribution of local density ρ

one master and five slaves, where each node is equipped with 8 GB of memory, a 500 GB hard disk and a Core i7@2.3 GHz CPU. The software settings are: Hadoop 2.7.7, Spark 2.4.4 and Python 2.7.11. The three compared algorithms were applied to two large-scale trajectory datasets. The first dataset (Gowalla) was obtained from a location-based social networking website where users share their locations by checking-in. It contains 6,442,890 check-ins with GPS information. The second dataset (GeoLife) consists of three years of trajectory data for 182 participants of the Geo-Life project. Both datasets are too big to be mined on a typical desktop computer.

An experiment was done to evaluate the influence of NE on the proposed method, and compare its performance to other methods. The parameter Th_{NP} was set to 1500, the distance weights β and γ were set to 0.6 and 0.4, respectively, and the distance weights $\omega_\perp$, $\omega_\parallel$ and ω_θ were set to 0.33. Then, the following algorithms were ran on each dataset: DBSCAN, DPC, DTR-STR($NE = 200$), DTR-STR($NE = 400$) and DTR-STR($NE = 800$). The DBI clustering accuracy and runtimes were recorded for each algorithm. This experimental design can not only verify the quality of the algorithm but also verify the impact of the partition size NE on the algorithm. Results in terms of clustering accuracy and runtimes are shown in Table 1 and Fig. 6.

Table 1. Clustering accuracy comparison (DBI)

Dataset	DBSCAN	DPC	DTR-DPC ($NE = 200$)	DTR-DPC ($NE = 400$)	DTR-DPC ($NE = 800$)
Gowalla	0.752	0.683	0.712	0.695	0.689
GeoLife	0.779	0.724	0.751	0.746	0.737

It can be observed in Table 1 that all DTR-DPC algorithms are slightly more accurate than the DPC algorithm, but both are less accurate than DBSCAN. These results are obtained because the DTR-DPC algorithm deletes

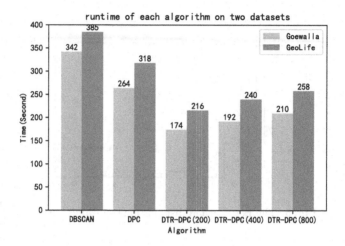

Fig. 6. Runtime comparison (s)

and abstracts numerous trajectory points. This has only a slight effect on clustering results because many points of a dense area are repeated stay points of the same trajectory, which has little effects on trajectory data mining. Moreover, it can also be observed that the larger the partition size NE of GA, the closer the accuracy is to that of the DPC algorithm. The reason is that the larger the NE is, the fewer points are deleted, and the greater the similarity is with the original data. In Fig. 6, the runtime of DTR-DPC is much less than those of the other two algorithms, and the larger the NE is, the longer the runtime, which was expected. In summary, NE is a value that needs to be reasonably set. If NE is set to a larger value, it may be faster but the accuracy may decrease. In future work, we will consider designing a methodology to automatically set NE. The experimental results reported in this paper have demonstrated that the proposed algorithm considerably reduces the runtime while preserving accuracy.

6   Conclusion

This paper proposed a distributed trajectory clustering approach based on density peak clustering on Spark, named DTR-DPC, which preprocesses data using a concept of dense and sparse areas, to reduce the number of invalid trajectories and calculations. Several different sub-trajectory distance measures are defined to compare sub-trajectories. Then, an enhanced E-DPC clustering algorithm is run on Spark to make large-scale trajectory data processing possible. Extensive experiments demonstrated that DTR-DPC guarantees almost constant clustering accuracy, while greatly reducing the runtime, and can perform large-scale trajectory data mining. In the future work, we will focus on finding the relationship between the number of equal parts NE and the size of the dataset to obtain maximum efficiency.

Acknowledgments. This research is sponsored by the Scientific Research Project of State Grid Sichuan Electric Power Company Information and Communication Company under Grant No. SGSCXT00XGJS1800219, and the Joint Funds of the Ministry of Education of China.

References

1. Lee, J.G., Han, J.: Trajectory clustering: a partition-and-group framework. In: Proceedings of ACM SIGMOD international conference on Management of data, pp. 593–604 (2007)
2. Wang, Y., Lei, P.: Using DTW to measure trajectory distance in grid space. In: IEEE International Conference on Information Science and Technology, pp. 152–155 (2014)
3. Lindahl, E., Hess, B., van der Spoel, D.: GROMACS 3.0: a package for molecular simulation and trajectory analysis. J. Mol. Model. **7**, 306–317 (2001). https://doi.org/10.1007/s008940100045
4. Huey-Ru, W., Yeh, M.-Y.: Profiling moving objects by dividing and clustering trajectories spatiotemporally. IEEE Trans. Knowl. Data Eng. **25**(11), 2615–2628 (2013)
5. Li, X., Ceikute, V.: Effective online group discovery in trajectory databases. IEEE Trans. Knowl. Data Eng. **25**(11), 2752–2766 (2013)
6. Rodriguez, A., Laio, A.: Clustering by fast search and find of density peaks. Science **344**(6191), 1492 (2014)
7. Zheng, K., Zheng, Y.: On discovery of gathering patterns from trajectories. In: IEEE International Conference on Data Engineering, pp. 242–253 (2013)
8. Pei, J., Han, J.: Mining sequential patterns by pattern-growth: the prefixSpan approach. IEEE Trans. Knowl. Data Eng. **16**(11), 1424–1440 (2004)
9. Lin, T.H., Yeh, J.S.: New data structure and algorithm for mining dynamic periodic patterns. In: IET International Conference on Frontier Computing. Theory, Technologies and Applications, pp. 55–59 (2010)
10. Zheng, K., Zheng, Y.: Online discovery of gathering patterns over trajectories. IEEE Trans. Knowl. Data Eng. **26**(8), 1974–1988 (2014)
11. Boubezoul, A., Koita, A.: Vehicle trajectories classification using support vectors machines for failure trajectory prediction. In: International Conference on Advances in Computational Tools for Engineering Applications, pp. 486–491 (2009)
12. Choong, M.Y., Angeline, L.: Modeling of vehicle trajectory using k-means and fuzzy c-means clustering. In: IEEE International Conference on Artificial Intelligence in Engineering and Technology, pp. 1–6 (2018)
13. Anjum, N., Cavallaro, A.: Unsuoervised fuzzy clustering for trjectory analysis. IEEE Int. Conf. Image Process. **3**, 213–216 (2007)
14. Zhao, L., Shi, G.: An adaptive hierarchical clustering method for ship trajectory data based on DBSCAN algorithm. In: IEEE International Conference on Big Data Analysis, pp. 329–336 (2017)
15. Ailin, H., Zhong, L.: Movement pattern extraction based on a non-parameter clustering algorithm. In: IEEE 4th International Conference on Big Data Analytics, pp. 5–9 (2019)
16. Wang, N., Gao, S.: Research on fast and parallel clustering method for trajectory data. In: IEEE International Conference on Parallel and Distributed Systems. pp. 252–258 (2018)

17. Hu, C., Kang, X.: Parallel clustering of big data of spatio-temporal trajectory. In: International Conference on Natural Computation, pp. 769–774 (2015)

18. Miyamoto, S., Matsumoto, T.: Dynamic distributed genetic algorithm using hierarchical clustering for flight trajectory optimization of winged rocket. In: International Conference on Machine Leraning and Applications, pp. 295–298 (2013)

19. Chen, Z., Guo, J.: DBSCAN algorithm clustering for massive AIS data based on the hadoop platform. In: International Conference on Industrial Informatics - Computing Technology, Intelligent Technology, Industrial Information Integration, pp. 25–28 (2017)

20. Wang, N., Gao, S.: Research on fast and paraller clustering method for trajectory data. In: IEEE International Conference on Parallel and Distributed Systems, pp. 252–258 (2018)

21. Davies, D.L., Bouldin, D.W.: A cluster seperation measure. IEEE Trans. Pattern Anal. Mach. Intell. 1(2), 224–227 (1979)

Pattern Mining

Parallel Mining of Partial Periodic Itemsets in Big Data

C. Saideep[1], R. Uday Kiran[2,3,4(✉)], Koji Zettsu[2], Cheng-Wei Wu[5],
P. Krishna Reddy[1], Masashi Toyoda[2], and Masaru Kitsuregawa[2,6]

[1] International Institute of Information Technology-Hyderabad, Telangana, India
[2] National Institute of Information and Communications Technology, Tokyo, Japan
[3] The University of Tokyo, Tokyo, Japan
[4] The University of Aizu, Aizu wakamatsu, Japan
udayrage@u-aizu.ac.jp
[5] National Ilan University, Yilan, Taiwan, ROC
[6] National Institute of Informatics, Tokyo, Japan

Abstract. Partial Periodic itemsets are an important class of regularities that exist in a temporal database. A Partial Periodic itemset is something persistent and predictable that appears in the data. Past studies on Partial Periodic itemsets have been primarily focused on centralized databases and are not scalable for Big Data environments. One cannot ignore the advantage of scalability by using more resources. This is because we deal with large databases in a real-time environment and using more resources can increase the performance. To address the issue we have proposed a parallel algorithm by including the step of **distributing transactional identifiers** among the machines and **mining the identical itemsets independently** over the different machines. Experiments on Apache Spark's distributed environment show that the proposed approach speeds up with the increase in a number of machines.

Keywords: Data mining · Periodic pattern mining · Time series · MapReduce · Big data analytics · Spark

1 Introduction

Partial periodic itemsets are an important class of regularities that exist in a temporal database. A partial periodic itemset represents something that is persistent and predictable with in the data. Finding partial periodic itemsets is thus useful to understand data. A classic application is a market-basket analytics. It involves extracting the useful information regarding the itemsets that are being purchased by the customers regularly in the market basket data. An example of a partial periodic itemset is as follows:

$$\{Bread,\ Butter\}[periodic\text{-}frequency = 1000]$$

© Springer Nature Switzerland AG 2020
H. Fujita et al. (Eds.): IEA/AIE 2020, LNAI 12144, pp. 807–819, 2020.
https://doi.org/10.1007/978-3-030-55789-8_69

The above itemset provides the information that 1000 customers have regularly (say, once in an hour) purchased the items 'Bread' and 'Butter.' This information may found to be useful in several applications, such as customer relation management, inventory management and recommendation systems.

Uday et al. [11] proposed a pattern-growth algorithm, called Partial Periodic Pattern-growth (3P-growth), to find all partial periodic itemsets in a temporal database. Briefly, This algorithm compresses the given database into a tree structure, called partial periodic pattern tree (3P-tree), and recursively mines 3P-tree by building conditional pattern bases. As 3P-growth is developed for running on a single machine, it suffers from the scalability and fault tolerant problems while dealing with big data.

Apache software foundation [1] proposed a parallel computing framework, called Hadoop [3], to efficiently process big data. Hadoop is primarily composed of storage section and processing section. The storage section of Hadoop is called HDFS (Hadoop Distributed File System). It is a scalable distributed file system that replicates the data across multiple nodes for reaching high reliability. The processing section of Hadoop refers to the Map-Reduce framework. The algorithms developed based on the Map-Reduce framework can take the nice property of key-value pairs to distribute the mining tasks among multiple computers. However, Hadoop is a disk-based architecture and it may spend a lot of time on I/O, which degrades the overall performance for processing.

In view of this, another parallel computing platform, called Spark [4], was proposed to handle large scale data. Spark employs a novel data structure, called RDD (Resilient Distributed Dataset), to manage the in-memory data distributed among the nodes in the cluster. Two important features of spark are: (i) fault-tolerant (i.e., if a portion of RDD is unrecoverable during the mining process, it can be easily reconstructed by using the previous RDD) and (ii) in-memory storage (i.e., data is kept in memory, leading to a large reduction of I/O costs.) These two advantages of Spark motivate us to design a new framework for parallel mining of partial periodic patterns on Spark platform.

To the best of our knowledge, the concept of partial periodic pattern mining has not yet been incorporated with Spark in-memory computing architecture. In view of this, this paper proposes a novel framework for parallel mining of partial periodic patterns in Spark in-memory computing architecture. This work is the first work that incorporates the partial periodic pattern mining with Spark in-memory computing architecture. We propose an efficient algorithm named 4P-growth (Parallel Partial Periodic Pattern-growth) for efficiently discovering the complete set of partial periodic patterns using Spark in-memory computing architecture. An efficient load balance technique has also been discussed to find the desired itemsets efficiently. The proposed parallel algorithm contains two Map-Reduce phases. In the first Map-reduce phase partial periodic item (length 1 items) are extracted. In the second map-reduce phase, multiple 4P-trees are built for mining partial periodic itemsets on multiple machines independently. Experimental results on three real-world datasets show that the proposed algorithm speeds up with the increase in the number of machines.

The rest of the paper is organized as follows. Section 2 discusses the background and related work on periodic itemset mining. The proposed in-memory 4P-growth algorithm to find all partial periodic itemsets in temporal databases is presented in Sect. 3. Performance evaluation is reported in Sect. 4. Finally, Sect. 5 concludes the paper with future research directions.

2   Background and Related Work

2.1   Periodic Itemset Mining

Periodic itemset mining is an important model in data mining. Most previous studies [10,13] focused on finding full periodic itemsets and did not take into account the partial periodic behavior of the items in the data. In the literature, some researchers have described models to find partial periodic itemsets in transactional databases [6–9,14]. Unfortunately, these models are impracticable on real-world very large databases, because the generated itemsets do not satisfy the *anti-monotonic property*. Recently, Uday et al. [11] proposed a model of partial periodic itemset that may exist in a temporal database. On contrary to the previous models, this model is practicable on real-world very large databases to because the generated itemsets satisfy the anti-monotonic property. A sequential algorithm, called Partial Periodic Pattern-Growth (3P-growth), was proposed to find the desired itemsets. In this paper, we propose a parallel algorithm to find partial periodic itemsets.

2.2   Model of Partial Periodic Itemsets

The model of partial periodic itemsets is as follows [11]: Let $I = \{i_1, i_2, \cdots, i_n\}$, $n \geq 1$, be the set of 'n' items appearing in a database. A set of items $X \subseteq I$ is called an **itemset** (or a **pattern**). An itemset containing k items is called a k-itemset. The length of this itemset is k. A transaction tr consists of *transaction identifier, timestamp* and an *itemset*. That is, $tr = (tid, ts, Y)$, where tid represents the transactional identifier, $ts \in \mathbb{R}^+$ represents the transaction time (or timestamp) and Y is an itemset. A temporal database TDB is an ordered collection of transactions, i.e. $TDB = \{tr_1, tr_2, \cdots, tr_m\}$, where $m = |TDB|$ represents the database size (or the total number of transactions). Let ts_{min} and ts_{max} denote the minimum and maximum timestamps of all transactions in TDB, respectively. For a transaction $tr = (tid, ts, Y)$, such that $X \subseteq Y$, it is said that X occurs in tr and such a timestamp is denoted as ts^X. Let $TS^X = \{ts_a^X, ts_b^X, \cdots, ts_c^X\}$, $ts_{min} \leq ts_a^X \leq ts_b^X \leq ts_c^X \leq ts_{max}$, be an ordered list of timestamps of transactions in which X appeared in TDB. The *frequency* of X, denoted as $freq(X) = |TS^X|$, where $|TS^X|$ represents the number of transactions containing X in TDB. Let $ts_j^X, ts_k^X \in TS^X$, $ts_{min} \leq ts_j^X \leq ts_k^X \leq ts_{max}$, denote any two consecutive timestamps in TS^X. The time difference between ts_k^X and ts_j^X is referred as an inter-arrival time of X, and denoted as iat_p^X, $p \geq 1$. That is, $iat_p^X = ts_k^X - ts_j^X$. Let $IAT^X = \{iat_1^X, iat_2^X, \cdots, iat_{freq(X)-1}^X\}$, be the list of all inter-arrival times of X in TDB. An inter-arrival time of X is said to be periodic (or interesting) if it is no more than the user-specified *period (per)*. That is, an $iat_i^X \in IAT^X$ is said to be periodic if $iat_i^X \leq per$. Let $\widehat{IAT^X}$ be the set of all inter-arrival times in IAT^X that are no more than per. That is, $\widehat{IAT^X} \subseteq IAT^X$ such that if $\exists iat_k^X \in IAT^X : iat_k^X \leq per$, then $iat_k^X \in IAT^X$. The periodic-frequency of X, denoted as $PF(X) = |\widehat{IAT^X}|$. An itemset X is a partial periodic itemset if $PF(X) \geq minPF$, where $minPF$ represents the user-specified *minimum periodic-frequency*.

Example 1. Consider the temporal database shown in Table 1. The set of items, $I = \{a, b, c, d, e, f, g\}$. This database contains 16 transactions. Therefore, the size of this database, i.e., $|TDB| = 16$. Each transaction in this database consists of a *transaction identifier* (denoted as, tid), a *timestamp* (denoted as, ts) and an itemset. The set of items, a and b, i.e., $\{a, b\}$ (in short, ab), is an itemset. This itemset contains two items. Therefore, it is a 2-itemset. The itemset ab appears in the first transaction. Therefore, $ts_1^{ab} = 1$. The set of all transactions containing ab in Table 1, i.e., $TS^{ab} = \{1, 3, 5, 9, 11, 12, 12\}$. An inter-arrival time of ab in TS^{ab}, i.e., $iat_1^{ab} = 3 - 1 = 2$. Similarly, other inter-arrival times of ab in TS^{ab} are: $iat_2^{ab} = 5 - 3 = 2$, $iat_3^{ab} = 9 - 5 = 4$, $iat_4^{ab} = 11 - 9 = 2$, $iat_5^{ab} = 12 - 11 = 1$, and $iat_6^{ab} = 12 - 12 = 0$. Thus, the set of all inter-arrival times of ab in TDB, i.e., $IAT^{ab} = \{2, 2, 4, 2, 1, 0\}$. If the user-specified $per = 2$, then $\widehat{IAT^{ab}} = \{2, 2, 2, 1, 0\}$. The *periodic-frequency* of ab in TDB, i.e., $PF(ab) = |\widehat{IAT^{ab}}| = |\{2, 2, 2, 1, 0\}| = 5$. In other words, the itemset ab has regularly appeared 5 times in the entire database. If the user-specified $minPF = 3$, then ab is a partial periodic itemset because $PF(ab) \geq minPF$. (The *periodic-frequency* can also be represented in the percentage of database size. However, we use the former definition for brevity.)

Table 1. Running example: temporal database

TID	ts	itemset	TID	ts	itemset	TID	ts	itemset	TID	ts	itemset
101	1	a, b, g	105	5	a, b, g	109	9	a, b, e, f	113	12	a, b, c, d
102	1	a, c, d	106	6	c, d	110	9	a, d, e	114	12	a, b, c, d
103	3	a, b	107	7	b, g	111	10	c, d, g	115	14	a, d, g
104	4	a, e, g	108	8	c, d, e, f	112	11	a, b, e, f	116	17	e, f

2.3 3P-Growth

Uday et al. [11] proposed a sequential pattern-growth algorithm, called 3P-growth, to find all desired patterns in a temporal database. Briefly, the algorithm involves the following two steps: (i) compress the database into a tree known as partial periodic pattern-tree (or 3P-tree) and (*ii*) finding all partial periodic patterns by recursively mining the 3P-tree. Unfortunately, 3P-growth being a sequential algorithm suffers from both scalability and fault tolerant problems. The proposed algorithm does not suffer from the scalability problem.

2.4 Parallel Algorithms for Finding User Interest-Based Patterns

Li et al. [12] proposed parallel FP-growth to extract frequent patterns using two Map-Reduce phases. Since then several parallel algorithms have been proposed to find user interest-based patterns [5]. Unfortunately, these algorithms cannot be directly used to mine partial periodic patterns. It is because these algorithms capture only the *frequency* and ignore the temporal occurrence information of a pattern in the data. We propose a variant of parallel FP-growth algorithm that captures both *frequency* and *temporal* occurrence information of a pattern in the data. The key difference between the proposed algorithm and current parallel frequent pattern mining algorithms lies in their *tree* structure, where an additional *ts* information is processed to compute the *periodic-frequency* of a pattern.

3   Proposed 4P-Growth Algorithm

The proposed 4P-growth algorithm to discover partial periodic patterns is presented in Algorithm 1, 2 and 3. Briefly the algorithm involves the following three steps:

1. Scan the database and construct (global) 4P-list.
2. Load balancing and construction of local 4P-trees in each worker machine by performing another scan on the database.
3. Recursively mine each local 4P-trees independently.

We now discuss each of the above steps in detail.

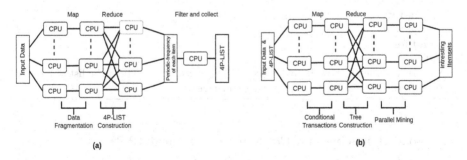

Fig. 1. Two phases of partial periodic itemset mining. (a) Generation of 4P-list (b) Construction of 4P-tree and parallel mining of patterns

3.1   Construction of 4P-List

Determine the periodic-frequency of each item by scanning the database using map-reduce computation. A global 4P-list is generated by pruning the uninteresting items (i.e., items which have periodic-frequency less than the user-specified $minPF$). The map-reduce computation is shown in Fig. 1(a), and described as follows:

- **Map:** output key-value pairs of each item in a transaction such that *key* is an item and *value* is the concatenation of tid and ts using a separator.
- **Reduce:** It groups all ts of each item in a ts-list. The ts-list is sorted and then its periodic-frequency is computed over the multiple machines (see Fig. 2(a) and 2(b)).

Next, all items in 4P-list are sorted in decreasing order of their supports and are assigned a rank (to simplify the distribution of transactions using a hash function). The most frequent item is assigned a rank of 0, the second most frequent item is assigned a rank of 1 and so on. For the example showed in Table 1 as the $minPF$ value is 4 items e, f and g can be pruned from TDB. The order of the items in 4P-List is as follows: a, b, d and c (in the decreasing order of their support).

Algorithm 1. 4P-List Construction (TDB)

Procedure Map(key=NUll,value=TDB_i)
  for each transaction $t_{CUR} \in TDB_i$ **do**
    for each item it in t_{cur} **do**
      Output (it,(ts,TID)) //ts is the current timestamp
    end for
  end for
Procedure Reduce(key=it,value=ts-list(list of timestamps))
  for each item $it \in I$ **do**
    sort ts-list of item it and initialise $id_l = 0, PF_{it} = 0$
    for each ts $\in$ ts-$list$ **do**
      if $ts - id_l \le per$ **then**
        $PF_{it} + = 1$
      end if
      $id_l = ts$
    end for
  end for
After this prune all those items whose $PF_{it} < minPF$ and return the list of left over items in the decreasing order of their frequencies

---

3.2   Load Balancing and Construction of Local 4P-Trees

Load Balancing. The items in global 4P-list are assigned by the worker machine by employing the following load balance formula:

$$item[machine\_id] = \frac{rank[item]}{\#workerMachines} \tag{1}$$

where, # workerMachines represents the total number of worker machines available for the computation.

Structure of 4P-Tree. In each worker machine, the given database is compressed into local trees, called 4P-trees. The structure of 4P-tree is as follows: A 4P-tree has two components: a 4P-list and a prefix-tree. The 4P-list consists of each distinct item (i), periodic-frequency (pf) and a pointer pointing to the first node in the prefix-tree carrying the item. Two types of nodes are maintained in prefix-tree: ordinary node and tail-node. The former is the type of node similar to that used in FP-tree, whereas the latter node represents the last item of any sorted transaction. In 4P-tree, we maintain the temporal occurrences of that branch's items only at the tail-node of every transaction. The tail-node structure maintains ts-list as shown in Fig. 2 (c)–(d).

Constructing 4P-Trees. In this step, local 4P-trees are constructed in each worker machine. The Map-Reduce steps are as follows.

Table 2. Running example: sub-patterns of machine-id 0

TID	ts	itemset	TID	ts	itemset	TID	ts	itemset	TID	ts	itemset
101	1	a	105	5	a	109	9	a	113	12	a, b, d
102	1	a, d	106	6	d	110	9	a, d	114	12	a, b, d
103	3	a	107	7	–	111	10	d	115	14	a, d
104	4	a	108	8	d	112	11	a	116	17	–

Table 3. Running example: sub-patterns of machine-id 1

TID	ts	itemset	TID	ts	itemset	TID	ts	itemset	TID	ts	itemset
101	1	a, b	105	5	a, b	109	9	a, b	113	12	a, b, d, c
102	1	a, d, c	106	6	d, c	110	9	–	114	12	a, b, d, c
103	3	a, b	107	7	b	111	10	d, c	115	14	–
104	4	–	108	8	d, c	112	11	a, b	116	17	–

– **Map:** For each transaction, the items which are not present in parallel 4P-list are filtered, translated into their ranks and are sorted in ascending order. Then all the sub-patterns (n sub-patterns will be generated) are extracted and are assigned to a machine based on the load balancing function shown in Eq. 1. This hash function gives a machine-id for which the pattern is responsible for further computation. Each sub-pattern is outputted as a key-value pair, with key as the machine-id and value as a tuple of sub-pattern and current ts, i.e., (machine-id, (sub-pattern, ts)). This generation of sub-patterns helps us to remove the communication cost by building independent 4P-trees shown in

Fig. 2(c) and 2(d). For the example, using the hash function shown in Eq. 1, the items a and d are allocated to machine-0, while the items b and c are allocated to machine-1. Based on this independent transactional databases are generated for both machines during the Map phase as shown in Algorithm 2. The sub-patterns which are used to build local 4P-trees for machine-id's 0 and 1 are shown in Table 2 and 3, respectively.

- **Reduce:** Independent local 4P-trees are constructed by inserting all the sub-patterns into the tree in the same order as the 4P-list with ts stored only in the tail-node of the branch. The process of construction local 4P-trees as stated in Algorithm 2 is the same as that of the 3P-tree construction in [11]. For the example, 4P-trees are constructed over machines using sub-patterns in Table 2 and 3. Two local 4P-Trees constructed on two different machines 0 and 1 are shown in Fig. 2 (c) and Fig. 2 (d), respectively.

Algorithm 2. 4P-Tree Construction (TDB,4P-LIST)

Procedure Map(key=NUll,value=TDB_i)
  for each transaction $t_{CUR} \in TDB_i$ **do**
    Prune and Sort the items in t_{cur} which are in 4P-List
    H=[ ]
    for $j = len(t_{cur}) - 1$ to 0 **do**
      machine-id=getmachine($t_{cur}[j]$)
      if machine-id not in H **then**
        H.append(machine-id)
        output(machine-id,($t_{cur}[0 : j]$,ts))
      end if
    end for
  end for
Procedure Reduce(key=machine-id,value=transactions)
  Initialise 4P-Tree
  for each transaction $t_{CUR} \in TDB_i$ **do**
    Set Cur_{node}=4P-Tree.Root
    for each item it in t_{cur} **do**
      if it not in children of Cur_{node} **then**
        create $NewNode$ and append to its parent Cur_{node}
      end if
      Update Cur_{node} to the corresponding child node
    end for
    add the current timestamp to the ts-list of the tail node
  end for

3.3   Recursive Mining of Local 4P-Trees

Note that for any suffix item, the complete conditional tree information is available in the corresponding machine. So, during conditional pattern building, communication is not required between the machines and partial periodic itemsets can be mined in parallel. Parallel mining of partial periodic itemsets as stated in

Algorithm 3, is similar to the mining process of 3P-growth but worker machine processes only for those suffix items for which it is responsible for computation. This is checked by using the hash function $\dfrac{rank[suffixItem]}{numberOfmachines}$. Mining for a suffix item (in increasing order of support) is done only if the output of the hash function is equal to the machine-id of that machine.

In the example, the local machines, machine-0 recursively mines patterns for items starting with suffix a or d and machine-1 recursively mines patterns for items starting with suffix b or c independently and parallel.

Mining of 4P-Trees: Note that for any suffix item, the complete conditional tree information is available in the corresponding machine. So, during conditional pattern building, communication is not required between the machines and partial periodic itemsets can be mined in parallel. Parallel mining of partial periodic itemsets as stated in Algorithm 3, is similar to the mining process of 3P-growth but worker machine processes only for those suffix items for which it is responsible for computation. This is checked by using the hash function $\dfrac{rank[SuffixItem]}{NoOfMachines}$. Here, it is the chosen suffix item. Mining for a suffix item (in increasing order of support) is done only if the output of the hash function is equal to the machine-id of that machine.

Algorithm 3. 4P-Growth (Machine-id,4P-Tree)

Procedure Map(key=machine-id,value=4P-*Tree*)
    for each item it in 4P-List **do**
        if current machine is responsible for item *it* **then**
            Generate Conditional Tree CT_{it} and mine recursively in CT_{it} for patterns with
            suffix *it*
        end if
    end for

In the example the local machines with machine-0 recursively mines patterns for items starting with suffix a or d and machine-1 recursively mines patterns for items starting with suffix b or c independently and parallelly

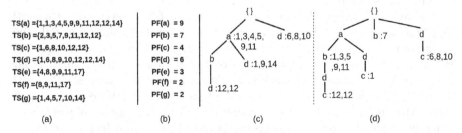

Fig. 2. Two phases of partial periodic itemset mining. **(a)** ts-list of length 1 items **(b)**periodic-frequency of length 1 items **(c)** 4P-tree on machine-id 0 and **(d)** 4P-tree on machine-id 1

4 Experimental Results

In this section, we compare the proposed 4P-growth against 3P-growth algorithm on various databases and demonstrate that the proposed algorithm is efficient when multiple machines were employed to generate desired patterns in very large real-world databases.

4.1 Experimental Setup

The proposed 4P-Growth algorithm was implemented using Apache Spark architecture and it is performed in a cluster of 25 machines with 2 GB memory each. The experiments on the reference 3P-Growth algorithm are performed on one machine of the same cluster for fairness purposes. Both 3P-growth and 4P-growth algorithms have been written in Python 3.5. Both synthetic (**T10I4D100K** and **T10I4D1000K**) and real-world (**Mushroom**) databases were used for evaluating the algorithms. The **T10I4D100K** is a sparse synthetic database, which is widely used for evaluating various pattern mining algorithms. This database contains 100,000 transactions and 870 distinct items. The **T10I4D1000K** is a very large sparse synthetic database that is widely used for evaluating various pattern mining algorithms. This database consists of 983,155 transactions and 30,387 distinct items. The **Mushroom database** is a real-world dense database containing 8,124 transactions and 119 distinct items. All of the above databases can be downloaded from [2]. Please note that all experiments have been conducted by varying the number of machines from 1 to 25.

4.2 Runtime Requirements of 3P-Growth and 4P-Growth with Varying Machine Count

In this experiment, we report the runtime results of 3P-growth and 4P-growth by varying the number of machines. The *per* and *minPF* values were set to a fixed value. The *per* values in T10I4D100K, T10I4D1000K, and Mushroom databases were set at 3%, 0.1% and 0.37%, respectively. The *minPF* values in T10I4D100K, T10I4D1000K, and Mushroom databases were set at 0.01%, 0.05% and 10%, respectively.

Figure 3(a)–(c) show the runtime requirements of 3P-growth and 4P-growth algorithm in various databases by varying the number of machines. The following observations can be drawn from these figures: (*i*) As 3P-growth is a sequential algorithm based on a single machine, its runtime (or execution time) is not influenced by the number of machines. Thus, the execution time of 3P-growth remains a straight line. (*ii*) The executive time of 4P-growth decreases significantly with the increase in machine count. (*iii*) When the number of machines is equal to 1, then 3P-growth performs better than 4P-growth, which is obvious, as parallel algorithms have an additional overhead of distributing the data across the machines. (*iv*) Overall, it can be observed that the proposed 4P-growth can efficiently handle very large and/or dense databases using multiple machines.

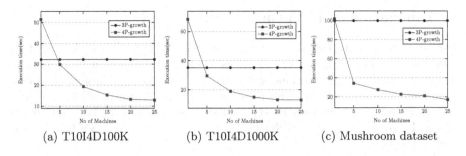

(a) T10I4D100K   (b) T10I4D1000K   (c) Mushroom dataset

Fig. 3. Evaluation of 3P-Growth and 4P-Growth algorithms on various databases

4.3 Runtime Requirements of 3P-Growth and 4P-Growth with Varying Minimum Periodic-Frequency

In the previous experiment, we have fixed the $minPF$ value and reported the runtime requirements of 3P-growth and 4P-growth by varying the number of machines. In this experiment, we fix the number of machines to 10 and report the runtime requirements of 3P-growth and 4P-growth by varying the $minPF$ values.

Figure 4(a)–(c) show the runtime requirements of 3P-growth and 4P-growth algorithm in various databases by varying $minPF$. The following observations can be drawn from these figures: (i) Execution time of 3P-growth algorithm decreases with an increase in $minPF$ value. This is due to the reduction of search space with an increase in $minPF$ value. (ii) The executive time of 4P-growth decreases significantly with the increase in $minPF$, similar to 3P-growth algorithm. However, the execution time of 4P-growth algorithm is nearly 50% less than that of the 3P-growth algorithm (iii) Overall, it can be observed that the proposed 4P-growth can efficiently handle very large and/or dense databases using multiple machines.

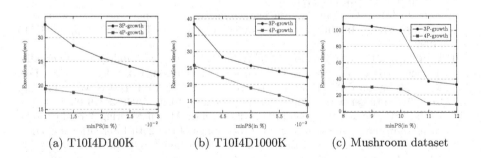

(a) T10I4D100K   (b) T10I4D1000K   (c) Mushroom dataset

Fig. 4. Evaluation of 3P-Growth and 4P-Growth algorithms

5   Conclusions and Future Work

Partial periodic patterns are an important class of regularities that exist in temporal databases. Most previous works focused on finding these patterns in databases using a single machine architecture. Consequently, these works suffer from the scalability problem. This paper proposes an efficient parallel algorithm, called 4P-growth, to discover partial periodic patterns in very large databases efficiently. By conducting experiments on very large databases, we have shown that our algorithm is efficient.

As a part of future work, we would like to investigate models and algorithms to find partial periodic patterns in data streams.

References

1. Apache Software Foundation. http://www.apache.org/
2. Datasets. http://www.tkl.iis.u-tokyo.ac.jp/udayrage/datasets.php
3. Hadoop. http://hadoop.apache.org
4. Spark. http://spark.apache.org/
5. Anirudh, A., Kiran, R.U., Reddy, P.K., Toyoda, M., Kitsuregawa, M.: An efficient map-reduce framework to mine periodic frequent patterns. In: Bellatreche, L., Chakravarthy, S. (eds.) Big Data Analytics and Knowledge Discovery, pp. 120–129. Springer, Cham (2017). https://doi.org/10.1007/978-3-319-64283-3_9
6. Aref, W.G., Elfeky, M.G., Elmagarmid, A.K.: Incremental, online, and merge mining of partial periodic patterns in time-series databases. IEEE TKDE 16(3), 332–342 (2004)
7. Berberidis, C., Vlahavas, I., Aref, W.G., Atallah, M., Elmagarmid, A.K.: On the discovery of weak periodicities in large time series. In: Elomaa, T., Mannila, H., Toivonen, H. (eds.) PKDD 2002. LNCS, vol. 2431, pp. 51–61. Springer, Heidelberg (2002). https://doi.org/10.1007/3-540-45681-3_5
8. Cao, H., Cheung, D.W., Mamoulis, N.: Discovering partial periodic patterns in discrete data sequences. In: Dai, H., Srikant, R., Zhang, C. (eds.) PAKDD 2004. LNCS (LNAI), vol. 3056, pp. 653–658. Springer, Heidelberg (2004). https://doi.org/10.1007/978-3-540-24775-3_77
9. Han, J., Dong, G., Yin, Y.: Efficient mining of partial periodic patterns in time series database. In: International Conference on Data Engineering, pp. 106–115 (1999)
10. Kiran, R.U., Kitsuregawa, M.: Novel techniques to reduce search space in periodic-frequent pattern mining. In: Bhowmick, S.S., Dyreson, C.E., Jensen, C.S., Lee, M.L., Muliantara, A., Thalheim, B. (eds.) DASFAA 2014. LNCS, vol. 8422, pp. 377–391. Springer, Cham (2014). https://doi.org/10.1007/978-3-319-05813-9_25
11. Kiran, R.U., Shang, H., Toyoda, M., Kitsuregawa, M.: Discovering partial periodic itemsets in temporal databases. In: Proceedings of the 29th International Conference on Scientific and Statistical Database Management. SSDBM 2017, New York, NY, USA. Association for Computing Machinery (2017). https://doi.org/10.1145/3085504.3085535
12. Li, H., Wang, Y., Zhang, D., Zhang, M., Chang, E.Y.: PFP: parallel FP-growth for query recommendation. In: Proceedings of the 2008 ACM Conference on Recommender Systems, RecSys 2008, New York, NY, USA, pp. 107–114. Association for Computing Machinery (2008). https://doi.org/10.1145/1454008.1454027

13. Tanbeer, S.K., Ahmed, C.F., Jeong, B.-S., Lee, Y.-K.: Discovering periodic-frequent patterns in transactional databases. In: Theeramunkong, T., Kijsirikul, B., Cercone, N., Ho, T.-B. (eds.) PAKDD 2009. LNCS (LNAI), vol. 5476, pp. 242–253. Springer, Heidelberg (2009). https://doi.org/10.1007/978-3-642-01307-2_24
14. Yang, R., Wang, W., Yu, P.: Infominer+: mining partial periodic patterns with gap penalties. In: ICDM, pp. 725–728 (2002)

A Fast Algorithm for Mining Closed Inter-transaction Patterns

Thanh-Ngo Nguyen[1], Loan T.T. Nguyen[2,3], Bay Vo[4(✉)], and Ngoc Thanh Nguyen[1]

[1] Department of Applied Informatics, Faculty of Computer Science and Management,
Wroclaw University of Science and Technology, Wroclaw, Poland
{thanh-ngo.nguyen,ngoc-thanh.nguyen}@pwr.edu.pl
[2] School of Computer Science and Engineering, International University,
Ho Chi Minh City, Vietnam
nttloan@hcmiu.edu.vn
[3] Vietnam National University, Ho Chi Minh City, Vietnam
[4] Faculty of Information Technology, University of Technology (HUTECH),
Ho Chi Minh City, Vietnam
vd.bay@hutech.edu.vn

Abstract. One of the important tasks of data mining is frequent pattern mining (FPM) in which frequently closed pattern mining (FCPM) is also a very interesting topic. So far, many algorithms have been proposed for mining frequent closed patterns (FCPs) of items occurring within transactions, Frequent Closed Intra-Transaction Patterns (FCITraPs), but only a few methods have been proposed for mining FCPs across transactions, Frequent Closed Inter-Transaction Patterns (FCITPs). In this study, we propose an N-list based algorithm, named NCITPs-Miner, for efficient mining FCITPs. The proposed method creates Frequent Inter-Transaction Patterns (FITPs) at 1-pattern level along with their N-lists, and then quickly generates all FCITPs by applying our proposed theorems, the closure properties, and subsumption check. Experiments show that NCITP-Miner outperforms state-of-the-art algorithms.

Keywords: Data mining · Inter-transaction pattern · Frequent closed inter-transaction pattern

1 Introduction

Frequent pattern mining (FPM) is one of the essential tasks of data mining. Since Agrawal et al. presented the first method, Apriori [1], for mining FPs, many algorithms have been proposed to mine frequent patterns (FPs) from large databases, such as Eclat [2], FP-Growth [3], etc. In addition, association rule mining (ARM) from a very large number of frequent patterns (FPs) may generate a lot of redundant association rules and thus makes it harder to predict and make decisions. To solve this problem, condensed representation solutions are used to not only reduce the overall size of the frequent pattern collection but also obtain non-redundant association rules. The two main types of condensed representation are maximal frequent patterns (FMPs) [4] and frequent closed patterns (FCPs)

© Springer Nature Switzerland AG 2020
H. Fujita et al. (Eds.): IEA/AIE 2020, LNAI 12144, pp. 820–831, 2020.
https://doi.org/10.1007/978-3-030-55789-8_70

[5]. Although a set of FCPs is a subset of FPs, it still has necessary properties for generating essential association rules that help significantly reduce search space, computation time, and memory usage. So far, many methods have been presented for mining FCPs, consisting of the Close algorithm [5], CLOSET [6], CLOSET+ [7] CHARM [8], and DCI_PLUS [9].

In recent years, Deng et al. proposed the PrePost algorithm and N-list structure for mining FPs [10]. The N-list is a compact structure allowing the determination of the support of FPs by performing N-list intersection operations in linear time. Applying the N-list structure to FPM helps reducing memory usage and mining time. Therefore, research groups have proposed several N-list-based methods to exploit different kinds of FPs, some of which are PrePost+ [11], negFIN [12], and NSFI [13] for mining FPs, NAFCP [14] for mining FCPs.

However, most of the previous studies only consider the task of FPM of items occurring within transactions of a database (Frequent Intra-Transaction Patterns, FIaTPs) and can only predict rules like *R1* given below. One of the major drawbacks of the FIaTP approaches is that it does not take into account the problem of mining frequent patterns of items occurring across several different transactions in the database (Frequent Inter-Transaction Patterns, FITPs), so it cannot predict rules like *R2*, which is generated from FITPs.

R1: *Whenever there is a visit to the website of BBC News and CNN News, there is also a visit to the websites of CNBC News and New York Times.*

R2: *If there is a visit to the BBC News and CNN News websites, then there will be a visit to the CNBC News website 4 min later, and a visit to the New York Times website 9 min later.*

So far, several methods have been proposed for mining FITPs and frequent closed inter-transaction patterns (FCITPs), such as FITI [15], ITP-Miner [16], PITP-Miner [17], ICMiner [18]. Nguyen et al. proposed an algorithm based on diffset concept and pruning strategies, DITP-Miner [19], to exploit FITPs efficiently. In addition, ICMiner, FCITP [20] algorithms were proposed for mining FCITPs, in which they used Tidsets, diffsets, DFS traversing, and closure properties to find FCITPs, but they still have many problems with FCITP mining that need to be improved to make them more efficient.

In this study, we propose an efficient algorithm for mining FCITPs, named NCITP-Miner. Our proposed algorithm has the following contributions. First of all, NCITP-Miner algorithm applies a proposed theorem to reduce the search space and find all FITPs at the 1-pattern level. Next, a tree structure named PPC-IT-Tree is constructed by inserting all FITPs of 1-patterns. Then, the approach traverses the PPC-IT-Tree to generate N-lists of all nodes from the tree, and thus we get all FITPs of 1-patterns with their N-lists. Next, we use DFS (Depth-First-Search) traversing and apply proposed theorems and subsumption check to find all FCITPs. Finally, experiments are conducted to prove the effectiveness of the NCITP-Miner algorithm.

The remainder of the paper is structured as follows. Section 2 outlines typical methods related to FITP and FCITP mining in recent years. Section 3 presents the basic concepts about FITPs, FCITPs, and N-list structure. The proposed algorithm NCITP-Miner is discussed and described in Sect. 4. Section 5 shows the experimental results of this study. Section 6 summarizes the results of this work and draws conclusions.

2 Related Works

Applications of ITP mining in predicting the movements of stock prices were presented by Lu et al. [21], and studying meteorological data was also proposed by Li et al. [22]. Several other algorithms based on Apriori have also been proposed to mine FITP, such as E/EH-Apriori [23] and FITI [15]. Expanding the scope of the association rules mined from traditional one-way internal transaction association rules to multi-dimensional inter-transaction association rules was also introduced by Li et al. [24]. Lee et al. recently presented two algorithms ITP-Miner [16] and ICMiner [18] to mine FITPs and FCITPs. To mine FITPs, the ITP-Miner algorithm is based on IT-Tree and DFS (Depth-First-Search) traversing. Based on the IT-Tree and CHARM properties, the ICMiner algorithm is proposed to mine all frequent closed inter-transaction patterns. Wang et al. proposed the PITP-Miner [17] algorithm, which relies on tree projection to mine the entire sets of FITPs in a database. In addition, FITP mining has been applied to mine profit rules from stock databases by Hsieh et al., with approaches such as PRMiner [25], JCMiner and ATMiner [26]. The authors also presented ITR-Miner and NRIT [27] to mine non-redundant inter-transaction association rules. Nguyen et al. recently proposed efficient methods, DITP-Miner [19] and FCITP [20], to effectively mine FITPs and FCITPs, respectively.

Among the existing algorithms for mining FITPs and FCITPs, some should be mentioned such as ITP-Miner, PITP-Miner, DITP-Miner, ICMiner, and FCITP algorithm, in which ITP-Miner used tidset to store the information of ITPs and DFS (Depth First Search) traversing to generate all FITPs. Despite significant improvements, there are still several limitations. Using intersections between tidsets to create new tidsets and calculate the support of new patterns can lead to very high computing costs in case of large databases with dense items in transactions. The PITP-Miner algorithm invokes multiple recursions to scan projected databases to determine the inter-transaction patterns' (ITPs) supports and form projected databases in the next levels resulting in consuming numerous resources regarding memory and runtime. The DITP-Miner algorithm uses diffset to store the ITPs' information to calculate the support of ITPs. DITP-Miner only works well in the case of dense databases. In addition, FCITP is a subset of FITP and the exploitation of FCITPs has brought great benefits such as eliminating a large number of redundant FITPs while the rules generated from FCITPs still ensure that they are the essential and non-redundant. The existing methods for mining FCITPs such as ICMiner and FCITP, both of which use the DFS strategy and the closure properties to remove non-closed sets to collect all FCITPs. On the one hand, ICMiner uses tidsets to store information to calculate the support of FCITPs. On the other hand, FCITP algorithm uses diffset to store information to determine the support of FCITPs. Therefore, FCITP algorithm is more efficient than ICMiner in the case of the sparse databases. In general, the main limitations of current methods of exploiting FCITPs and FITPs are multiple database scans, high computation cost for determining the support of ITPs, especially for databases with a large number of transactions and dense items. Therefore, existing methods consume too many resources regarding mining time and memory usage.

In recent years, after Deng et al. presented the PrePost [10] algorithm along with N-list structure [10] for efficiently mining FPs, several N-list based approaches have also been presented to mine FPs, such as PrePost+ [11], negFIN [12]. The N-list structure showed that it is a compact structure and applying it to FPM is very efficient regarding runtime and memory usage.

3 Basic Concepts

A transaction database (*TDB*) consists of *n* transactions, $T = \{T_1, T_2, ..., T_n\}$, in which each transaction is defined as a subset of a set of distinct items, $I = \{i_1, i_2, ..., i_m\}$. A *tidset* is a subset of a set of transaction identifies (*tid*) within a *TDB*, *tidset* $\subseteq$ *TID* = $\{tid_1, tid_2, ..., tid_n\}$. Therefore, a *TDB* can be expressed as a set of tuples $<tid, T_{tid}>$, where $T_{tid} \subseteq I$, and T_{tid} is a pattern occurring at *tid* transaction. The support of pattern *X*, denoted by *support(X)*, is the number of transactions in *TDB* containing the pattern *X*. Assuming that $\langle \alpha, T_\alpha \rangle$ and $\langle \beta, T_\beta \rangle$ are two transactions of *TDB*. The value of $(\alpha - \beta)$ is called the relative distance between α and β, where $\alpha > \beta$, and β is called the reference point. For β, an item i_k at α is called an extended item, denoted by $i_k(\alpha - \beta)$, where $(\alpha - \beta)$ is called the *Span* of the extended item. In the same way, with respect to the transaction at β, a transaction T_α at α is called an extended transaction and denoted as $T_\alpha(\alpha - \beta)$. Therefore, $T_\alpha(\alpha - \beta) = \{i_1(\alpha - \beta), ..., i_p(\alpha - \beta)\}$, where *p* is the number of items in T_α.

An example database *TDB* consists of 6 transactions (n = 6) and the set of items *I* = {A, B, C, D, T, W} is shown in Table 1. This database used throughout this article in illustrated examples. Therefore, with respect to the transaction at *tid* = 1 in the example database in Table 1, the extended transaction of the transaction at *tid* = 2 is {C(1), D(1), B(1), W(1)}.

Assuming that $x_i(\omega_i)$ and $x_j(\omega_j)$ are two extended items, $x_i(\omega_i) < x_j(\omega_j)$ if $(\omega_i < \omega_j)$ or $(\omega_i = \omega_j$ and $x_i < x_j)$. In addition, $x_i(\omega_i) = x_j(\omega_j)$ if $\omega_i = \omega_j$ and $x_i = x_j$. For instance, $C(0) < C(1), C(0) < D(0)$, and $C(1) = C(1)$. An inter-transaction pattern (ITP) is defined as a set of extended items, $\{x_1(\omega_1), x_2(\omega_2), \cdots, x_k(\omega_k)\}$, where $\omega_1 = 0$, $\omega_1 \leq maxSpan$, *maxSpan* is a maximum *Span* given by user, $x_i(\omega_i) < x_j(\omega_j)$, and $1 \leq i < j \leq l$. A pattern is called a *l*-pattern (or a pattern of length *l*), if it contains *l* extended items. For example, {A(0), T(0), W(1)} is a 3-pattern. Let *X* be an ITP, *X* is called an FITP if *support(X)* $\geq$ *minSup*, where *minSup* is a minimum support threshold given by user. An FITP *X* is called an FCITP if there does not exist any superset having the same support with it. For instance, using example *TDB* in Table 1 with *minSup* = 3 and *maxSpan* = 1, we have A(0)W(0) and A(0)C(0)W(0) are two FITPs because *support(A(0)W(0))* = *support(A(0)C(0)W(0))* = 4 > 3. A(0)W(0) is not an FCITP because A(0)C(0)W(0) is superset of A(0)W(0) and they have the same support value.

All of the frequent extended 1-patterns, denoted as I_1, are inserted into an PPC-IT-tree, which is a tree structure rooted at $\Re$, where each node in the tree consists of six fields, *Name, support, childnodes, Pre, Post,* and *Span,* in which *Name* is the name of 1-patterns, *support* is the frequency of extended 1-patterns, *childnodes* is the set of child nodes associated with their ancestor node, *Pre* and *Post* are the order numbers of the node when traversing PPC-IT-tree in Pre-order and Post-order ways, respectively, and

Span is the relative distance between the transaction containing 1-pattern of the node and the reference point, which are defined in the above section.

Table 1. Example *TDB* with *minSup* = 3 and *maxSpan* = 1 used to illustrate the NCITP-Miner algorithm.

Tid	Items
1	$\{A, W, T, C\}$
2	$\{C, D, W, B\}$
3	$\{A, W, C, T\}$
4	$\{A, W, C, D\}$
5	$\{A, C, W, D, T\}$
6	$\{C, D, B, T\}$

N-list associated with extended 1-patterns: The ITPP-code of each node nod_i in a PPC-IT-tree consists of a tuple of the form $K_i = <nod_i \cdot pre, nod_i \cdot post, nod_i \cdot support>$. The N-list associated with an extended pattern $B(Span_i)$, denoted by $N(B(Span_i))$, is the set of ITPP-codes associated with nodes in the PPC-IT-tree that they have the same *Name* of B and *Span* of $Span_i$. Thus $N(B(Span_i))$ is calculated as follows.

$$N(B(Span_i)) = \bigcup_{(nod_i \in \Re | (nod_i, \, Name=B) \wedge (nod_i, \, Span=Span_i))} K_i \tag{1}$$

Where K_i is the ITPP-code associated with nod_i. The support for $B(Span_i)$ is determined as follows.

$$support(B(Span_i)) = \sum_{K_i \in N(B(Span_i))} support(K_i) \tag{2}$$

N-list associated with 2-patterns: Let *PA* and *PB* be two extended 1-patterns with the same prefix *P*, where *A* precedes *B* in the order of increasing I_1's support. $N(PA)$ and $N(PB)$ are two N-lists linked with *PA* and *PB*, respectively. The N-list linked with *PAB* is determined as follows

- For each ITPP-code $K_i \in N(PA)$ and $K_j \in N(PB)$, if K_i is an ancestor of K_j and $K_i.Span = K_j.Span = 0$, the algorithm will add $<K_j.pre, K_j.post, K_i.support>$ to $N(PAB)$.
- For each ITPP-code $K_i \in N(PA)$ and $K_j \in N(PB)$, if K_i is an ancestor of K_j, and $K_j.Span > K_i.Span = 0$, the algorithm will add $<K_i.pre, K_i.post, K_j.support>$ to $N(PAB)$.
- The support of *PAB* is calculated by the following formula

$$support(PAB) = \sum_{K_i \in N(PAB)} support(K_i) \tag{3}$$

N-list associated with l-patterns ($l > 2$): Let PA and PB be two extended (l-1) patterns with the same prefix P, in which A is before B. $N(PA)$ and $N(PB)$ are two N-lists linked with PA and PB, respectively. The N-list linked with PAB is determined as follows:

For each ITPP-code $K_i \in N(PA)$ and $K_j \in N(PB)$, if K_i is an ancestor of K_j and one of the following conditions is satisfied:

- If $\exists\ K_z \in N(PAB)$ such that $pre(K_z) = pre(K_i)$ and $post(K_z) = post(K_i)$, then the algorithm updates the support of K_z, $support(K_z) = support(K_z) + support(K_i)$.
- Otherwise, the algorithm adds $<pre(K_i), post(K_i), support(K_j)>$ to $N(PAB)$. The support of PAB, $support(PAB)$, is also calculated by the formula (3).

4 The Proposed Algorithm NCITP-Miner for Mining FCITPs

Definition 1. Assuming that x is an item contained in the transactions with $tidset = \{t_1, ..., t_k\}$, where $tidset \subseteq T$. The $maxpoint$ of item x, denoted as $maxpoint(x)$, is defined as the order of the last element t_k of $tidset$ and $maxpoint(x)$ is also the $support(x)$.

For example, in Table 1, item A occurs in TDB at the transactions $<1,3,4,5>$, respectively. We have $maxpoint(A) = support(A) = 4$.

Theorem 1. [20] Assuming that $x(0)$ is an inter-transaction 1-pattern of item x at $Span = 0$ with $tidset(x(0)) = \{t_1, t_2, ..., t_u\}$, where $t_i \in T$ and $x(k)$ is an inter-transaction 1-pattern of item x at $Span = k$, where $0 \le Span \le maxSpan$ and $1 \le k \le maxSpan$. If $t_{u-ms+1} \le k$ or $maxpoint(x(0)) - minSup + 1 \le k$, where $minSup$ is a minimum support threshold given buy user and $maxpoint(x(0)) = support(x(0))$, then $x(k)$ cannot be a FITP.

Definition 2. Let PA and PB be two FITPs, $N(PA)$ and $N(PB)$ be two N-lists linked with PA and PB, respectively. A precedes B in the order of increasing I_1's support. $N(PB) \subseteq N(PA)$ if and only if $\forall\ K_i \in N(PB)$, $\exists\ K_j \in N(PA)$ such that K_j is an ancestor of K_i.

Theorem 2. Let PA and PB be two FITPs, $N(PA)$ and $N(PB)$ be two N-lists linked with PA and PB, respectively. If $N(PB) \subseteq N(PA)$, then PB is not a FCITP.

Proof. According to the way to determine the N-list linked with an inter-transaction l-pattern presented in the above section and $N(PB) \subseteq N(PA)$, we have:

$$N(PAB) = \bigcup pre(K_i), post(K_i), support(K_j)$$

Where $K_i \in N(PA)$, $K_j \in N(PB)$, and K_i is an ancestor of K_j.

Therefore,

$support(PAB) = \sum_{K_i \in N(XAB)} support(K_i) = \sum_{K_j \in N(PB)} support(K_j) = support(PB)$.

Based on the FCP definition mentioned in Sect. 3, PB is not FCITP. Therefore, Theorem 2 has been proven.

Algorithm: NCITP-Miner(*TDB, minSup, maxSpan*)
Input: A transaction *TDB, minSup, maxSpan.*
Output: all of *FCITPs.*
Method:
Step 1. Scan *TDB* to find all FIT *1*-patterns and their *support* $(0 \le Span \le maxSpan)$ by applying theorem 1.
Step 2. Construct the tree structure PPC-IT-Tree, traverse the PPC-IT-Tree to create N-lists of nodes of the tree, and combine N-lists associated with their appropriate FIT 1-patterns.
Step 3. Add FIT 1-patterns (*Span* = 0) with their N-lists into the set of FCITPs.
Step 4. Create FIT 2-patterns with the given *Span* thresholds $(1 \le Span \le maxSpan)$ by applying the properties presented in the N-lists associated with 2-patterns.
Step 6. Perform FCITP_DFS([*P*] of FIT 2-patterns, *minSup, maxSpan*) in the set of FCITPs recursively.
Step 7. Output *CITPs.*

Function: FCITP_DFS([*P*] of FIT 2-patterns, *minSup, maxSpan*)
Input: [*P*] of FIT 2-patterns (all 2-patterns having the same prefix *P* is being processed), *minSup, maxSpan.*
Output: the updated set for next level
Step 1. Sort the list of IT 2-patterns in the processing class in increasing order of their support.
Step 2. Apply Theorem 2 to eliminate non-closed patterns.
Step 3. Add *Xi* to *FCITPs* if it cannot be subsumed by any other patterns.
Step 4. FCITP_DFS([*Pi*], *CITPs, minSup, maxSpan*)

Fig. 1. The proposed algorithm NCITP-Miner for mining FCITPs.

5   Experimental Results

The algorithms used in the experiments were implemented in Microsoft Visual C# 2019 and tested on a computer equipped with the following configurations: 5th generation Intel® Core™ i5-5287U processor @ 2.9 Ghz, 8.00 GB RAM LPDDR3 and running macOS Catalina 10.15.2.

The databases which were used in our experiment evaluations can be downloaded at http://fimi.uantwerpen.be/data/. The features of these databases are described in Table 2.

Table 2. Databases used for experimentation

Dataset	#distinct items	#records	#Average length
Chess	76	3,196	37
T10I4D100K	870	100,000	10

The structure N-list is a compact one, so utilizing it in FCITP mining is a very useful task. Applying theorem 1 to eliminate infrequent patterns at the 1-pattern level, using the N-list structures to quickly calculate the support of patterns with linear complexity, and applying theorem 2 to determine whether a pattern is a closed pattern or not, all of which have contributed to the efficiency of the proposed algorithm, NCITP-Miner. The experimental results shown in Figs. 2, 3, 4, 5, 6, 7, 8 and 9 indicate that NCITP-Miner is more efficient than the ICM-Miner [18] and FCITP [20] in terms of runtime and memory usage.

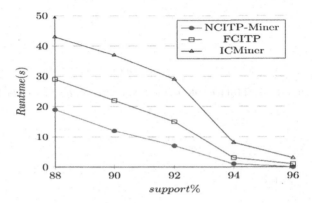

Fig. 2. Runtime on the CHESS database with various *Support* and *maxSpan* = 1.

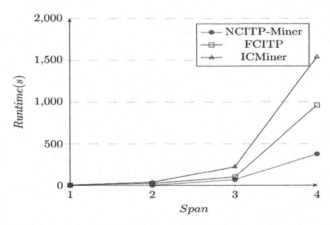

Fig. 3. Runtime on the CHESS database with various *Span* and *minSup* = 96%.

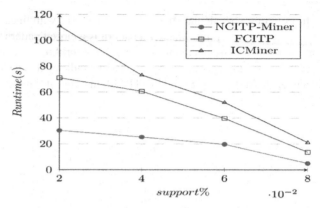

Fig. 4. Runtime on the T10I4D100K database with various *support* and *maxSpan* = 1.

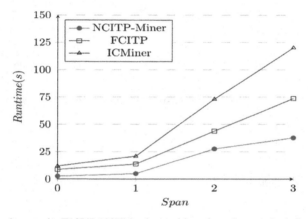

Fig. 5. Runtime on the T10I4D100K database with various *Span* and *minSup* = 0.08%.

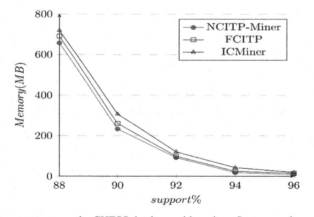

Fig. 6. Memory usage on the CHESS database with various *Support* and *maxSpan* = 1.

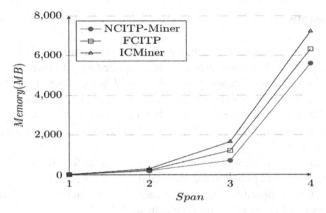

Fig. 7. Memory usage on the CHESS database with various *Span* and *minSup* = 96%.

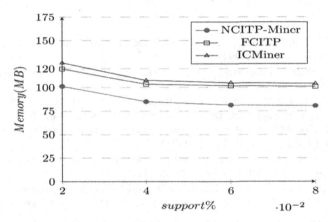

Fig. 8. Memory usage on the T10I4D100K database with various *support* and *maxSpan* = 1.

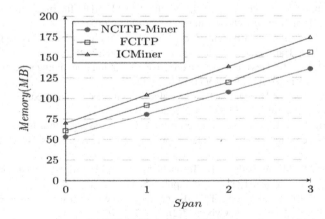

Fig. 9. Memory usage on the T10I4D100K database with various *Span* and *minSup* = 0.08%.

6 Conclusions

In this study, we have proposed an efficient algorithm for mining FCITPs, in which our proposed theorems and pruning strategies were applied to cut down the search space and quickly determine the frequency of ITPs as well as the closed ITPs. We have also performed experiments on real datasets that are widely used in the data mining research community. The proposed algorithm, NCITP-Miner, is more efficient than state-of-the-art algorithms for mining FCITPs, ICMiner and FCITP, in terms of runtime and memory usage in most cases.

In our future works, we will develop efficient methods for mining FITPs, FCITPs, and FMITPs, which are applied to incremental databases. Besides, we also will present more novel efficient algorithms and apply them to more massive databases with the parallel systems and the cloud computing platforms.

References

1. Agrawal, R., Ramakrishnan, S.: Fast algorithms for mining association rules. In: Proceedings -the VLDB, pp. 487–499 (1994)
2. Zaki, M.J.: Scalable algorithms for association mining. IEEE Trans. Knowl. Data Eng. 12(3), 372–390 (2000)
3. Han, J., Pei, J., Yin, Y., Mao, R.: Mining frequent patterns without candidate generation: a frequent-pattern tree approach. Data Min. Knowl. Discov. 8(1), 53–87 (2004)
4. Gouda, K., Zaki, M.J.: GenMax: an efficient algorithm for mining maximal frequent itemsets. Data Min. Knowl. Discov. 11(3), 223–242 (2005)
5. Pasquier, N., Bastide, Y., Taouil, R., Lakhal, L.: Efficient mining of association rules using closed itemset lattices. Inf. Syst. 24(1), 25–46 (1999)
6. Pei, J., Han, J., Mao, R.: CLOSET: an efficient algorithm for mining frequent closed itemsets. In: ACM SIGMOD Workshop on Research Issues in Data Mining and Knowledge Discovery, pp. 21–30 (2000)
7. Wang, J., Han, J., Pei, J.: {CLOSET+:} searching for the best strategies for mining frequent closed itemsets. In: Proceedings of the Ninth ACM SIGKDD International Conference on Knowledge Discovery and Data Mining, pp. 236–245 (2003)
8. Zaki, M.J., Hsiao, C.-J.: Efficient algorithms for mining closed itemsets and their lattice structure. IEEE Trans. Knowl. Data Eng. 17(4), 462–478 (2005)
9. Sahoo, J., Das, A.K., Goswami, A.: An effective association rule mining scheme using a new generic basis. Knowl. Inf. Syst. 43(1), 127–156 (2015)
10. Deng, Z.H., Wang, Z.H., Jiang, J.J.: A new algorithm for fast mining frequent itemsets using N-lists. Sci. China Inf. Sci. 55(9), 2008–2030 (2012)
11. Deng, Z.H., Lv, S.L.: PrePost+ : an efficient N-lists-based algorithm for mining frequent itemsets via children-parent equivalence pruning. Expert Syst. Appl. 42(13), 5424–5432 (2015)
12. Aryabarzan, N., Minaei-Bidgoli, B., Teshnehlab, M.: negFIN: An efficient algorithm for fast mining frequent itemsets. Expert Syst. Appl. 105, 129–143 (2018)
13. Vo, B., Le, T., Coenen, F., Hong, T.P.: Mining frequent itemsets using the N-list and subsume concepts. Int. J. Mach. Learn. Cybern. 7(2), 253–265 (2016)
14. Le, T., Vo, B.: An N-list-based algorithm for mining frequent closed patterns. Expert Syst. Appl. 42(19), 6648–6657 (2015)

15. Tung, A.K.H., Lu, H., Han, J., Feng, L.: Efficient mining of intertransaction association rules. IEEE Trans. Knowl. Data Eng. **15**(1), 43–56 (2003)
16. Lee, A.J.T., Wang, C.S.: An efficient algorithm for mining frequent inter-transaction patterns. Inf. Sci. (Ny) **177**, 3453–3476 (2007)
17. Wang, C.S., Chu, K.C.: Using a projection-based approach to mine frequent inter-transaction patterns. Expert Syst. Appl. **38**(9), 11024–11031 (2011)
18. Lee, A.J.T., Wang, C.S., Weng, W.Y., Chen, Y.A., Wu, H.W.: An efficient algorithm for mining closed inter-transaction itemsets. Data Knowl. Eng. **66**(1), 68–91 (2008)
19. Nguyen, T.N., Nguyen, L.T.T., Nguyen, N.T.: An improved algorithm for mining frequent inter-transaction patterns. In: Proceedings - IEEE International Conference on INnovations in Intelligent SysTems and Applications (INISTA), pp. 296–301 (2017)
20. Nguyen, T., Nguyen, L.T.T., Vo, B., Nguyen, N.: An efficient algorithm for mining frequent closed inter- transaction patterns. In: Proceedings - 2019 IEEE International Conference on Systems, Man and Cybernetics (SMC), pp. 2019–2024 (2019)
21. Lu, H., Feng, L., Han, J.: Beyond intratransaction association analysis: mining multidimensional intertransaction association rules. Proc. ACM Trans. Inf. Syst. **18**, 423–454 (2000)
22. Feng, L., Dillon, T., Liu, J.: Inter-transactional association rules for multi-dimensional contexts for prediction and their application to studying meteorological data. Data Knowl. Eng. **37**(1), 85–115 (2001)
23. Feng, L., Yu, J.X., Lu, H., Han, J.: A template model for multidimensional inter-transactional association rules. VLDB J. **11**(2), 153–175 (2002)
24. Li, Q., Feng, L., Wong, A.: From intra-transaction to generalized inter-transaction: landscaping multidimensional contexts in association rule mining. Inf. Sci. (Ny) **172**(3–4), 361–395 (2005)
25. Hsieh, Y.L., Yang, D.L., Wu, J.: Effective application of improved profit-mining algorithm for the interday trading model. Sci. World J. **2014**, 13 (2014). ID 874825
26. Hsieh, Y.L., Yang, D.L., Wu, J., Chen, Y.C.: Efficient mining of profit rules from closed inter-transaction itemsets. J. Inf. Sci. Eng. **32**(3), 575–595 (2016)
27. Wang, C.: Mining non-redundant inter-transaction rules. J. Inf. Sci. Eng. **31**(6), 1521–1536 (2015)

TKE: Mining Top-K Frequent Episodes

Philippe Fournier-Viger[1(✉)], Yanjun Yang[2], Peng Yang[2], Jerry Chun-Wei Lin[3], and Unil Yun[4]

[1] School of Natural Sciences and Humanities, Harbin Institute of Technology (Shenzhen), Shenzhen, China
philfv@hit.edu.cn
[2] School of Computer Sciences and Technology, Harbin Institute of Technology (Shenzhen), Shenzhen, China
juneyoung9724@gmail.com, pengyeung@163.com
[3] Department of Computing, Mathematics and Physics, Western Norway University of Applied Sciences (HVL), Bergen, Norway
jerrylin@ieee.org
[4] Department of Computer Engineering, Sejong University, Seoul, Republic of Korea
yunei@sejong.ac.kr

Abstract. Frequent episode mining is a popular data mining task for analyzing a sequence of events. It consists of identifying all subsequences of events that appear at least *minsup* times. Though traditional episode mining algorithms have many applications, a major problem is that setting the *minsup* parameter is not intuitive. If set too low, algorithms can have long execution times and find too many episodes, while if set too high, algorithms may find few patterns, and hence miss important information. Choosing *minsup* to find enough but not too many episodes is typically done by trial and error, which is time-consuming. As a solution, this paper redefines the task of frequent episode mining as top-k frequent episode mining, where the user can directly set the number of episodes *k* to be found. A fast algorithm named TKE is presented to find the top-k episodes in an event sequence. Experiments on benchmark datasets shows that TKE performs well and that it is a valuable alternative to traditional frequent episode mining algorithms.

Keywords: Pattern mining · Frequent episodes · Top-k episodes

1 Introduction

Sequences of symbols or events are a fundamental type of data, found in many domains. For example, a sequence can model daily purchases made by a customer, a set of locations visited by a tourist, a text (a sequence of words), a time-ordered list of alarms generated by a computer system, and moves in a game such as chess. Several data mining tasks have been proposed to find patterns in sequences of events (symbols). While some tasks such as sequential pattern mining were proposed to find patterns common to multiple sequences [11,27] or

© Springer Nature Switzerland AG 2020
H. Fujita et al. (Eds.): IEA/AIE 2020, LNAI 12144, pp. 832–845, 2020.
https://doi.org/10.1007/978-3-030-55789-8_71

across sequences [29], other tasks have been studied to find patterns in a single long sequence of events. For example, this is the case of tasks such as *periodic pattern mining* (finding patterns appearing with regularity in a sequence) [17] and *peak pattern mining* (finding patterns that have a high importance during a specific non-predefined time period) [13]. But the task that is arguably the most popular of this type is *frequent episode mining* (FEM) [15,23], which consists of identifying all episodes (subsequences of events) that appear at least *minsup* times in a sequence of events. FEM can be applied to two types of input sequences, which are *simple sequences* (sequences where each event has a unique timestamp) and *complex sequences* (where simultaneous events are allowed). From this data, three types of episodes are mainly extracted, which are (1) *parallel episodes* (sets of events appearing simultaneously), (2) *serial episodes* (sets of events that are totally ordered by time), and (3) *composite episodes* (where parallel and serial episodes may be combined) [23]. The first algorithms for FEM are WINEPI and MINEPI [23]. WINEPI mines all parallel and serial episodes by performing a breadth-first search and using a sliding window. It counts the support (occurrence frequency) of an episode as the number of windows where the episode appears. However, this *window-based frequency* has the problem that an occurrence may be counted more than once [16]. MINEPI adopts a breadth-first search approach but only look for minimal occurrences of episodes [23]. To avoid the problem of the window-based frequency, two algorithms named MINEPI+ and EMMA [15] adopt the *head frequency measure* [16]. EMMA utilizes a depth-first search and a memory anchor technique. It was shown that EMMA outperforms MINEPI [23] and MINEPI+ [15]. Episode mining is an active research area and many algorithms and extensions are developed every year such as to mine online episodes [3] and high utility episodes [21,30].

Though episode mining has many applications [15,23], it is a difficult task because the search space can be very large. Although efficient algorithms were designed, a major limitation of traditional frequent episode mining algorithms is that setting the *minsup* parameter is difficult. If *minsup* is set too high, few episodes are found, and important episodes may not be discovered. But if *minsup* is set too low, an algorithm may become extremely slow, find too many episodes, and may even run out of memory or storage space. Because users typically have limited time and storage space to analyze episodes, they are generally interested in finding enough but not too many episodes. But selecting an appropriate *minsup* value to find just enough episodes is not easy as it depends on database characteristics that are initially unknown to the user. Hence, a user will typically apply an episode mining algorithm many times with various *minsup* values until just enough episodes are found, which is time-consuming.

To address this issue, this paper redefines the problem of frequent episode mining as that of *top-k frequent episode mining*, where the goal is to find the *k* most frequent episodes. The advantage of this definition is that the user can directly set *k*, the number of patterns to be found rather than setting the *minsup* parameter. To efficiently identify the top-k episodes in an event sequence, this paper presents an algorithm named TKE (Top-K Episode mining). It uses an internal *minsup* threshold that is initially set to 1. Then, TKE starts to search for episodes and gradually increases that threshold as frequent episodes are found. To increase that threshold as quickly as possible to reduce the search space, TKE applies a concept of dynamic search, which consists of exploring the most promising patterns first. Experiments were done on various types of sequences to evaluate TKE's performance. Results have shown that TKE is efficient and is a valuable alternative to traditional episode mining algorithms.

The rest of this paper is organized as follows. Section 2 introduces the problem of frequent episode mining and presents the proposed problem definition. Section 3 describes the proposed algorithm. Then, Sect. 4 presents the experimental evaluation. Finally, Sect. 5 draws a conclusion and discuss research opportunities for extending this work.

2   Problem Definition

The traditional problem of frequent episode mining is defined as follows [15,23]. Let $E = \{i_1, i_2, \ldots, i_m\}$ be a finite set of events or symbols, also called items. An *event set* X is a subset of E, that is $X \subseteq E$. A *complex event sequence* $S = \langle (SE_{t_1}, t_1), (SE_{t_2}, t_2), \ldots, (SE_{t_n}, t_n) \rangle$ is a time-ordered list of tuples of the form (SE_{t_i}, t_i) where $SE_{t_i} \subseteq E$ is the set of events appearing at timestamp t_i, and $t_i < t_j$ for any integers $1 \leq i < j \leq n$. A set of events SE_{t_i} is called a *simultaneous event set* because they occurred at the same time. A complex event sequence where each event set contains a single event is called a *simple event sequence*. For the sake of brevity, in the following, timestamps having empty event sets are omitted when describing a complex event sequence.

For instance, Fig. 1 (left) illustrates the complex event sequence $S = \langle (\{a, c\}, t_1), (\{a\}, t_2), (\{a, b\}, t_3), (\{a\}, t_6), (\{a, b\}, t_7), (\{c\}, t_8), (\{b\}, t_9), (\{d\}, t_{11}) \rangle$. That sequence indicates that a appeared with c at time t_1, was followed by event a at t_2, then a and b at t_3, then a at t_6, then a and b at t_7, then c at t_8, then b at t_9, and finally d at t_{11}. That sequence will be used as running example. This type of sequence can model various data such as alarm sequences [23], cloud data [2], network data [18], stock data [21], malicious attacks [26], movements [14], and customer transactions [1,3,4,7].

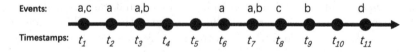

Fig. 1. A complex event sequence

Frequent episode mining aims at discovering all subsequences of events called episodes [15,23] that have a high support (many occurrences) in a sequence. Formally, an *episode* (also called *composite episode*) α is a non-empty totally ordered set of simultaneous events of the form $\langle X_1, X_2, \ldots, X_p \rangle$, where $X_i \subseteq E$

and X_i appears before X_j for any integers $1 \leq i < j \leq p$. An episode containing a single event set is called a *parallel episode*, while an episode where each event set contains a single event is called a *serial episode*. Thus, parallel episodes and serial episodes are special types of composite episodes.

There are different ways to calculate the support of an episode in a sequence. In this paper, we use the *head frequency* support, used by MINEPI+ and EMMA [16], which was argued to be more useful than prior measures [15]. The support is defined based on a concept of occurrence.

Definition 1 (Occurrence). *Let there be a complex event sequence* $S = \langle (SE_{t_1}, t_1), (SE_{t_2}, t_2), \ldots, (SE_{t_n}, t_n) \rangle$. *An occurrence of the episode* α *in* S *is a time interval* $[t_s, t_e]$ *such that there exist integers* $t_s = z_1 < z_2 < \ldots < z_w = t_e$ *such that* $X_1 \subseteq SE_{z_1}, X_2 \subseteq SE_{z_2}, \ldots, X_p \subseteq SE_{z_w}$. *The timestamps* t_s *and* t_e *are called the start and end points of* $[t_s, t_e]$, *respectively. The set of all occurrences of* α *in a complex event sequence is denoted as* $occSet(\alpha)$.

For instance, the set of all occurrences of the composite episode $\langle \{a\}, \{a, b\} \rangle$ is $occSet(\langle \{a\}, \{a, b\} \rangle) = \{[t_1, t_3], [t_1, t_7], [t_2, t_3], [t_2, t_7], [t_6, t_7]\}$.

Definition 2 (Support). *The support of an episode* α *is the number of distinct start points for its occurrences. It is denoted and defined as* $sup(\alpha) = |\{t_s | [t_s, t_e] \in occSet(\alpha)\}|$ [15].

For instance, if $winlen = 6$, the occurrences of the episode $\alpha = \langle \{a\}, \{a, b\} \rangle$ are $[t_1, t_3], [t_2, t_3], [t_3, t_7]$ and $[t_6, t_7]$. Because α has four distinct start points $(t_1, t_2, t_3,$ and $t_6)$ and $sup(\alpha) = 4$. The problem of FEM is defined as follows.

Definition 3 (Frequent Episode mining). *Let there be a user-defined window length* $winlen > 0$, *a threshold* $minsup > 0$ *and a complex event sequence* S. *The problem of frequent episode mining consists of finding all frequent episodes, that is episodes having a support that is no less than* $minsup$ [15].

For example, the frequent episodes found in the sequence of Fig. 1 for $minsup = 2$ and $winlen = 2$ are $\langle \{a, b\} \rangle$, $\langle \{a\}, \{b\} \rangle$, $\langle \{a\}, \{a, b\} \rangle$, $\langle \{a\}, \{a\} \rangle$, $\langle \{a\} \rangle$, $\langle \{b\} \rangle$, and $\langle \{c\} \rangle$. Their support values are 2, 2, 2, 3, 5, 3, and 2 respectively.

To solve the problem of FEM, previous papers have relied on the downward closure property of the support, which states that the support of an episode is less than or equal to that of its prefix episodes [15].

This paper redefines the problem of FEM as that of identifying the top-k frequent episodes, where $minsup$ is replaced by a parameter k.

Definition 4 (Top-k Frequent Episode mining). *Let there be a user-defined window* $winlen$, *an integer* $k > 0$ *and a complex event sequence* S. *The problem of top-k frequent episode mining consists of finding a set* T *of* k *episodes such that their support is greater or equal to that of any other episodes not in* T.

For example, the top-3 frequent subgraphs found in the complex event sequences of Fig. 1 are $\langle\{a\},\{a\}\rangle$, $\langle\{a\}\rangle$, and $\langle\{b\}\rangle$.

One should note that in the case where multiple episodes have the same support, the top-k frequent episode mining problem can have multiple solutions (different sets of episodes may form a set of top-k episodes). Moreover, for some datasets and *winlen* values, less than k episodes may be found.

The proposed problem is more difficult than traditional frequent episode mining since it is not known beforehand for which *minsup* value, the top-k episodes will be obtained. Hence, an algorithm may have to consider all episodes having a support greater than zero to identify the top-k episodes. Thus, the search space of top-k frequent episode mining is always equal or larger than that of traditional frequent episode mining with an optimal *minsup* value.

3   The TKE Algorithm

To efficiently solve the problem of top-k frequent episode mining, this paper proposes an algorithm named TKE (Top-k Episode Mining). It searches for frequent episodes while keeping a list of the current best episodes found until now. TKE relies on an internal *minsup* threshold initially set to 1, which is then gradually increased as more patterns are found. Increasing the internal threshold allows to reduce the search space. When the algorithm terminates the top-k episodes are found. TKE (Algorithm 1) has three input parameters: an input sequence, k and *winlen*. The output is a set of top-k frequent episodes. TKE consists of four phases, which are inspired by the EMMA algorithm [15] but adapted for efficient top-k episode mining. EMMA was chosen as basis for TKE because it is one of the most efficient episode mining algorithms [15].

Step 1. Finding the top-k events. TKE first sets $minsup = 1$. Then, the algorithm scans the input sequence S once to count the support of each event. Then, the *minsup* threshold is set to that of the k-th most frequent event. Increasing the support at this step is an optimization called *Single Episode Increase* (SEI). Then, TKE removes all events having a support less than *minsup* from S, or ignore them from further processing. Then, TKE scans the database again to create a vertical structure called *location list* for each remaining (frequent) event. The set of frequent events is denoted as E'. The location list structure is defined as follows.

Definition 5 (Location list). *Consider a sequence $S = \langle(SE_{t_1}, t_1), (SE_{t_2}, t_2),$ $\ldots, (SE_{t_n}, t_n)\rangle$. Without loss of generality, assume that each event set is sorted according to a total order on events $\prec$ (e.g. the lexicographical order $a \prec b \prec c \prec d$). Consider that an event e appears in an event set SE_{t_i} of S. It is then said that event e appears at the position $\sum_{w=1,\ldots,i-1} |SE_{t_w}| + |\{y|y \in SE_{t_i} \wedge y \prec e\}|$ of sequence S. Then, the location list of an event e is denoted as $locList(e)$ and defined as the list of all its positions in S. The support of an event can be derived from its location list as $sup(e) = |locList(e)|$.*

For instance, consider the input sequence $S = \langle(\{a,c\}, t_1), (\{a\}, t_2),$ $(\{a,b\}, t_3), (\{a\}, t_6), (\{a,b\}, t_7), (\{c\}, t_8), (\{b\}, t_9), (\{d\}, t_{11})\rangle$, that $k = 3$ and

$winlen = 2$. After the first scan, it is found that the support of events a, b, c and d are 5, 3, 2 and 1, respectively. Then, $minsup$ is set to the support of the k-th most frequent event, that is $minsup = 2$. After removing infrequent events, the sequence S becomes $S = \langle(\{a,c\},t_1), (\{a\},t_2), (\{a,b\},t_3), (\{a\},t_6),$ $(\{a,b\},t_7), (\{c\},t_8), (\{b\},t_9)\rangle$ Then, the location lists of a, b and c are built because their support is no less than 2. These lists are $locList(a) = \{0,2,3,5,6\}$, $locList(b) = \{4,7,9\}$ and $locList(c) = \{1,8\}$.

Step 2. Finding the top-k parallel episodes. The second step consists of finding the top-k parallel episodes by combining frequent events found in Step 1. This is implemented as follows. First, all the frequent events are inserted into a set $PEpisodes$. Then, TKE attempts to join each frequent episode $ep \in PEpisodes$ with each frequent event $e \in E' | e \notin ep \land \forall f \in ep, f \prec e$ to generate a larger parallel episode $newE = ep \cup \{e\}$, called a *parallel extension* of ep. The location list of $newE$ is created, which is defined as $locList(newE) = \{p | p \in locList(e) \land \exists q \in locList(ep) | t(p) = t(q)\}$, where $t(p)$ denotes the timestamp corresponding to position p in S. Then, if the support of $newE$ is no less than $minsup$ according to its location list, then $newE$ is inserted in $PEpisodes$, and then $minsup$ is set to the support of the k-th most frequent episode in $PEpisodes$. Finally, all episodes in $PEpisodes$ having a support less than $minsup$ are removed. Then, $PEpisodes$ contains the top-k frequent parallel episodes.

Consider the running example. The parallel extension of episode $\langle\{a\}\rangle$ with event b is done to obtain the episode $\langle\{a,b\}\rangle$ and its location list $locList(\langle\{a,b\}\rangle)$ $= \{4,7\}$. Thus, $sup(\langle\{a,b\}\rangle) = |locList(\langle\{a,b\}\rangle)| = 2$. This process is performed to generate other parallel extensions such as $\langle\{a,c\}\rangle$ and calculate their support values. After Step 2, the top-k parallel episodes are: $\langle\{a\}\rangle$, $\langle\{b\}\rangle$, $\langle\{c\}\rangle$, and $\langle\{a,b\}\rangle$, with support values of 5, 3, 2, 2, respectively, and $minsup = 2$.

Step 3. Re-encoding the input sequence using parallel episodes. The next step is to re-encode the input sequence S using the top-k parallel episodes to obtain a *re-encoded sequence S'*. This is done by replacing events in the input sequences by the top-k parallel episodes found in Step 2. For this purpose, a unique identifier is given to each top-k episode.

For instance, the IDs #1, #2, #3 and #4 are assigned to the top-k parallel episodes $\langle\{a\}\rangle$, $\langle\{b\}\rangle$, $\langle\{c\}\rangle$, and $\langle\{a,b\}\rangle$, respectively. Then, the input sequence is re-encoded as: $S = \langle(\{\#1\#3\},t_1), (\{\#1\},t_2), (\{\#1,\#2,\#4\},t_3), (\{\#1\},t_6),$ $(\{\#1,\#2,\#4\},t_7), (\{\#3\},t_8), (\{\#2\},t_9)\rangle$.

Step 4. Finding the top-k composite episodes. Then, the TKE algorithm attempts to find the top-k composite episodes. First, the top-k parallel episodes are inserted into a set $CEpisodes$. Then, TKE attempts to join each frequent composite episodes $ep \in CEpisodes$ with each frequent event $e \in PEpisodes$ to generate a larger episode called a serial extension of ep by e. Formally, the *serial extension* of an episode $ep = \langle SE_1, SE_2, \ldots, SE_x \rangle$ with a parallel episode e yields the episode $serialExtension(ep, e) = \langle SE_1, SE_2,$ $\ldots, SE_x, e \rangle$. The bound list of $newE$ is created, which is defined as follows:

Definition 6 (Bound list). *Consider a re-encoded sequence $S' = \langle(SE_{t_1}, t_1),$ $(SE_{t_2}, t_2), \ldots, (SE_{t_n}, t_n)\rangle$, a composite episode ep and a parallel episode e. The* *bound list of e is denoted and defined as* $boundList(e) = \{[t, t] | e \in SE_t \in S'\}$. *The bound list of the serial extension of the composite episode ep with e is defined as:* $boundList(serialExtension(ep, e)) = \{[u, w] | [u, v] \in boundList(ep) \land [w, w] \in boundList(e) \land w - u < winlen \land v < w\}$. *The support of a composite episode ep can be derived from its bound list as* $sup(ep) = |\{t_s | [ts, te] \in boundList(ep)\}|$.

If the support of $newE$ is no less than $minsup$ according to its bound list, then $newE$ is inserted in $CEpisodes$, and then $minsup$ is set to the support of the k-th most frequent episode in $CEpisodes$. Finally, all episodes in $CEpisodes$ having a support less than $minsup$ are removed. At the end of this step, $CEpisodes$ contains the top-k frequent composite episodes and the algorithm terminates.

For instance, the bound-list of $S = \langle\{a\}, \{a\}\rangle$ is $boundList(\langle\{a\}, \{a\}\rangle) = \{[t_1, t_2], [t_2, t_3], [t_6, t_7]\}$. Thus, $sup(\langle\{a\}, \{a\}\rangle) = |\{t_1, t_2, t_6\}| = 3$. Then, TKE considers other serial extensions and calculate their support values in the same way. After Step 4, $minsup = 3$ and the top-k composite episodes are: $\langle\{a\}\rangle$, $\langle\{b\}\rangle$ and $\langle\{a\}, \{a\}\rangle$, with a support of 5, 4, and 3 respectively (the final result).

Completeness. It is easy to see that the TKE algorithm only eliminates episodes that are not top-k episodes since it starts from $minsup = 1$ and gradually raises the $minsup$ threshold when k episodes are found. Thus, TKE can be considered complete. However, it is interesting to notice that due to the definition of support, mining the top-k episodes may result in a set of episodes that is slightly different from that obtained by running EMMA with an optimal minimum support. The reason is that extending a frequent episode with an infrequent episode can result in a frequent episode. For example, if $minsup = 3$ and $winlen = 6$, $sup(\langle\{a\}\rangle) = 5$, and can be extended with $\langle\{a, b\}\rangle$ having a support of 2 to generate the episode $\langle\{a\}, \{a, b\}\rangle$ having a support of $4 > 2$. Because EMMA has a fixed threshold, it eliminates $\langle\{a, b\}\rangle$ early and ignore this extension, while TKE may consider this extension since it starts from $minsup = 1$ and may not increase $minsup$ to a value greater than 2 before $\langle\{a\}, \{a, b\}\rangle$ is generated. Hence, TKE can find episodes not found by EMMA.

Implementation Details. To have an efficient top-k algorithm, data structures are important. An operation that is performed many times is to update the current list of top-k events/episodes and retrieve the support of the k most frequent one. To optimize this step, the list of top-k events, $PEpisodes$ and $CEpisodes$ are implemented as priority queues (heaps).

Dynamic Search Optimization. It can also be observed that the search order is important. If TKE finds episodes having a high support early, the $minsup$ threshold may be raised more quickly and a larger part of the search space may be pruned. To take advantage of this observation, an optimization called *dynamic search* is used. It consists of maintaining at any time a priority queue of episodes that can be extended to generate candidate episodes. Then, TKE is modified to always extend the episode from that queue that has the highest support before extending others. As it will be shown in the experiments, this optimization can greatly reduce TKE's runtime.

Algorithm 1. The TKE algorithm

---

input : S: an input sequence, k: a user-specified number of patterns, $winlen$: the window length

output: the top-k frequent episodes

1 $minsup \leftarrow 1$;
2 Scan S to calculate $sup(e)$ for each event $e \in E$;
3 $minsup \leftarrow$ support of the k-th most frequent event $e \in E$;
4 $E' \leftarrow \{e | e \in E \wedge sup(e) \geq minsup\}$;
5 Remove (or thereafter ignore) each event $e \notin E'$ from S;
6 Scan S to create the location list of each frequent event $e \in E'$;
7 $PEpisodes \leftarrow E'$;
8 **foreach** *parallel episode* $ep \in PEpisodes$ *such that* $sup(ep) \geq minup$ **do**
9 $\quad$ **foreach** *event* $e \in E'$ *such that* $sup(e) \geq minup$ **do**
10 $\quad\quad$ $newE \leftarrow parallelExtension(ep, e);$// and build $newE$'s location list
11 $\quad\quad$ **if** $sup(newE) \geq minsup$ **then**
12 $\quad\quad\quad$ $PEpisodes \leftarrow PEpisodes \cup \{newE\}$;
13 $\quad\quad\quad$ $minsup \leftarrow$ support of the k-th most frequent in $PEpisodes$;
14 $\quad\quad$ **end**
15 $\quad$ **end**
16 **end**
17 Re-encode the sequence S into a sequence S' using the parallel episodes;
18 $CEpisodes \leftarrow PEpisodes$;
19 **foreach** *composite episode* $ep \in CEpisodes$ *such that* $sup(ep) \geq minup$ **do**
20 $\quad$ **foreach** *event* $e \in PEpisodes$ *such that* $sup(e) \geq minup$ **do**
21 $\quad\quad$ $newE \leftarrow serialExtension(ep, e);$// and build $newE$'s bound list
22 $\quad\quad$ **if** $sup(newE) \geq minsup$ **then**
23 $\quad\quad\quad$ $CEpisodes \leftarrow CEpisodes \cup \{newE\}$;
24 $\quad\quad\quad$ $minsup \leftarrow$ support of the k-th most frequent episode in $CEpisodes$;
25 $\quad\quad$ **end**
26 $\quad$ **end**
27 **end**
28 Return $CEpisodes$;

---

4 Experimental Evaluation

Experiments have been done to evaluate the performance of the TKE algorithm. It has been implemented in Java and tested on a worsktation equipped with 32 GB of RAM and an Intel(R) Xeon(R) W-2123 processor. All memory measurements were done using the Java API. Three benchmark datasets have been used, named *e-commerce*, *retail* and *kosarak*. They are transaction databases having varied characteristics, which are commonly used for evaluating itemset [13, 22] and episode [3, 12, 30] mining algorithms. As in previous work, each item is considered as an event and each transaction as a simultaneous event set at a time point. *e-commerce* and *retail* are sparse customer transaction datasets, where each event set is a customer transaction and each event is a purchased item, while *kosarak* is a sequence of clicks on an Hungarian online portal website, which contains many long event sets. While *e-commerce* has real timestamps, *retail* and *kosarak* do not. Hence, for these datasets, the timestamp i was assigned to the i-th event set. The three datasets are described in Table 1 in terms of number of time points, number of distinct events, and average event set size.

Table 1. Dataset characteristics

Dataset	#Timestamps	#Events	Average event set size
e-commerce	14,975	3,468	11.71
retail	88,162	16,470	10.30
kosarak	990,002	41,270	8.10

4.1 Influence of k and Optimizations on TKE's Performance

A first experiment was done to (1) evaluate the influence of k on the runtime
and memory usage of TKE, and to (2) evaluate the influence of optimizations on
TKE's performance. The *winlen* parameter was set to a fixed value of 10 on the
three datasets, while the value of k was increased until a clear trend was observed,
the runtime was too long or algorithms ran out of memory. Three versions of
TKE were compared: (1) TKE (with all optimizations), (2) TKE without the
SEI strategy and (3) TKE without dynamic search (using the depth-first search
of EMMA). Results in terms of runtime and memory usage are reported in Fig. 2.

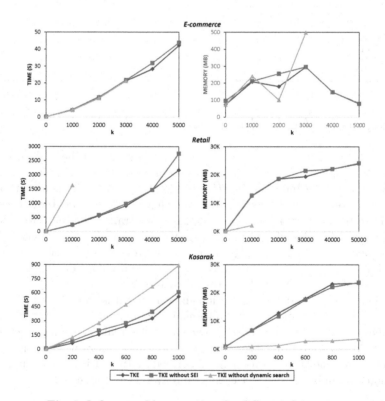

Fig. 2. Influence of k on runtime for different datasets.

It is first observed that as k is increased, runtime and memory usage increase. This is reasonable because more episodes must be found, and thus more episodes must be considered as potential top-k episodes before *minsup* can be raised.

Second, it is observed that using the dynamic search greatly decreases runtime. For example, on the *koarak* dataset, when $k = 800$, TKE with dynamic search can be twice faster than TKE with depth-first search, and on the *retail* dataset, when $k = 10000$, TKE with dynamic search is up to 7.5 times faster than TKE with depth-first search. These results are reasonable since the dynamic search explores episodes having the highest support first, and thus the internal *minsup* threshold may be raised more quickly than when using a depth-first search, which helps reducing the search space and thus the runtime. But on the *e-commerce* dataset, TKE with dynamic search is about as fast as TKE with depth-first search. The reason is that this dataset is relatively small and the time required for maintaining the candidate priority queue offset the benefits of using the dynamic search.

Third, it is observed that TKE with dynamic search generally consumes more memory than TKE without. This is because the former needs to maintain a priority queue to store potential frequent episodes and their bound lists. Note that on the *e-commerce* dataset, when $k \geq 4000$, TKE without dynamic search runs out of memory. Hence, results are not shown in Fig. 2.

Fourth, it is found that the SEI strategy generally slightly reduces the runtime on these three datasets. This is because TKE raises the internal *minsup* threshold early using that strategy after the first scan of the input sequence, which helps to reduce the search space. Moreover, TKE with the SEI strategy consumes a little more memory than TKE without the SEI strategy since the former keeps a priority queue of size k to store the support of each frequent event during Step 1 of the algorithm.

4.2 Performance Comparison with EMMA Set with an Optimal *minsup* Threshold

A second experiment was done, to compare the performance of TKE with that of EMMA. Because TKE and EMMA are not designed for the same task (mining the top-k episodes and mining all the frequent episodes), it is difficult to compare them. Nonetheless, to provide a comparison of TKE and EMMA's performance, we considered the scenario where the user would choose the optimal *minsup* value for EMMA to produce the same number of episodes as TKE. In this scenario, mining the top-k frequent episodes remains much more difficult than mining frequent episodes because for the former problem, the internal *minsup* threshold must be gradually increased from 1 during the search, while EMMA can directly use the optimal *minsup* threshold to reduce the search space. For this experiment, TKE without dynamic search was used. It was run with k values from 1 to 1000 and *winlen* = 10 on *kosarak*, while EMMA was run with *minsup* equal to the support of the least frequent episode found by TKE. It is to be noted that EMMA with this optimal *minsup* value can find less frequent episodes than TKE because EMMA can ignore some extensions with infrequent events, as mentioned in the previous section.

Results in terms of runtime, memory usage and number of patterns found by EMMA for an optimal *minsup* value are shown in Table 2. Results for *e-commerce* and *retail* are not shown but follow similar trends. It is observed that the runtime of TKE is more than that of EMMA, which is reasonable, given that top-k frequent episode mining is a more difficult than traditional frequent episode mining. In terms of memory, TKE requires more memory than EMMA, which is also reasonable since TKE needs to consider more potential patterns and to keep priority queues in memory to store potential top-k episodes and candidates.

It is important to notice that EMMA is run with an optimal *minsup* value. But in real life, the user doesn't know the optimal *minsup* value in advance. To find a desired number of episodes, a user may need to try and adjust the *minsup* threshold several times. For example, if the user wants between 400 to 600 frequent episodes, *minsup* must be set between 0.3293 and 0.3055. That is to say, if the user doesn't have any background knowledge about this dataset, there is only $0.3293 - 0.3055 = 2.38\%$ chance of setting *minsup* accurately, which is a very narrow range of options. If the user sets the parameter slightly higher, EMMA will generate too few episodes. And if the user sets *minsup* slightly lower, EMMA may generate too many episodes and have a long runtime. For example, for *minsup* = 0.1, EMMA will generate more than 62 times the number of desired episodes and be almost 4 times slower than TKE. This clearly shows the advantages of using TKE when the user doesn't have enough background knowledge about a dataset. Thus, to avoid using a trial-and-error approach to find a desired number of frequent episodes, this paper proposed the TKE algorithm, which directly let the user specify the number of episodes to be found.

Table 2. Comparison of TKE without dynamic search and EMMA with optimal *minsup* threshold on the *kosarak* dataset

k	minsup	#patterns	TKE runtime (s)	EMMA runtime (s)	TKE memory (MB)	EMMA memory (MB)
1	0.6075	1	3	3	644	239
200	0.3619	159	124	9	1066	788
400	0.3293	210	279	10	1321	2184
600	0.3055	511	473	24	2934	2077
800	0.2862	625	666	29	3038	2496
1000	0.2794	684	891	31	3702	1485

5   Conclusion

This paper has proposed to redefine the task of frequent episode mining as top-k frequent episode mining, and presented an efficient algorithm named TKE for this task. To apply the algorithm a user is only required to set k, the desired number of patterns. To increase its internal *minsup* threshold as quickly as possibleng the most promising patterns first. to reduce the search space, TKE applies a dynamic search, which consists of always explori A performance evaluation on real-life data has shown that TKE is efficient and is a valuable alternative to traditional episode mining algorithms. The source code of TKE as well as datasets can be downloaded from the SPMF data mining software http://www.philippe-fournier-viger.com/spmf/ [10]. For future work, extensions of TKE could be considered for variations such as mining episode using a hierarchy [5], weighted closed episode mining [19,31], high-utility episode mining [12,21,25] and episode mining in a stream [20,24]. Besides, we will consider designing algorithms for other pattern mining problems such as discovering significant trend sequences in dynamic attributed graphs [6], frequent subgraphs [8], high utility patterns [28] and sequential patterns with cost/utility values [9].

References

1. Achar, A., Laxman, S., Sastry, P.S.: A unified view of the apriori-based algorithms for frequent episode discovery. Knowl. Inf. Syst. **31**(2), 223–250 (2012). https://doi.org/10.1007/s10115-011-0408-2
2. Amiri, M., Mohammad-Khanli, L., Mirandola, R.: An online learning model based on episode mining for workload prediction in cloud. Future Gener. Comput. Syst. **87**, 83–101 (2018)
3. Ao, X., Luo, P., Li, C., Zhuang, F., He, Q.: Online frequent episode mining. In: Proceedings 31st IEEE International Conference on Data Engineering, pp. 891–902 (2015)
4. Ao, X., Luo, P., Wang, J., Zhuang, F., He, Q.: Mining precise-positioning episode rules from event sequences. IEEE Trans. Knowl. Data Eng. **30**(3), 530–543 (2018)
5. Ao, X., Shi, H., Wang, J., Zuo, L., Li, H., He, Q.: Large-scale frequent episode mining from complex event sequences with hierarchies. ACM Trans. Intell. Syst. Technol. (TIST) **10**(4), 1–26 (2019)
6. Cheng, Z., Flouvat, F., Selmaoui-Folcher, N.: Mining recurrent patterns in a dynamic attributed graph. In: Kim, J., Shim, K., Cao, L., Lee, J.-G., Lin, X., Moon, Y.-S. (eds.) PAKDD 2017. LNCS (LNAI), vol. 10235, pp. 631–643. Springer, Cham (2017). https://doi.org/10.1007/978-3-319-57529-2_49
7. Fahed, L., Brun, A., Boyer, A.: DEER: distant and essential episode rules for early prediction. Expert Syst. Appl. **93**, 283–298 (2018)
8. Fournier-Viger, P., Cheng, C., Lin, J.C.-W., Yun, U., Kiran, R.U.: TKG: efficient mining of top-K frequent subgraphs. In: Madria, S., Fournier-Viger, P., Chaudhary, S., Reddy, P.K. (eds.) BDA 2019. LNCS, vol. 11932, pp. 209–226. Springer, Cham (2019). https://doi.org/10.1007/978-3-030-37188-3_13
9. Fournier-Viger, P., Li, J., Lin, J.C.W., Chi, T.T., Kiran, R.U.: Mining cost-effective patterns in event logs. Knowl. Based Syst. **191**, 105241 (2020)
10. Fournier-Viger, P., et al.: The SPMF open-source data mining library version 2. In: Berendt, B., et al. (eds.) ECML PKDD 2016. LNCS (LNAI), vol. 9853, pp. 36–40. Springer, Cham (2016). https://doi.org/10.1007/978-3-319-46131-1_8
11. Fournier-Viger, P., Lin, J.C.W., Kiran, U.R., Koh, Y.S.: A survey of sequential pattern mining. Data Sci. Pattern Recogn. **1**(1), 54–77 (2017)

12. Fournier-Viger, P., Yang, P., Lin, J.C.-W., Yun, U.: HUE-span: fast high utility episode mining. In: Li, J., Wang, S., Qin, S., Li, X., Wang, S. (eds.) ADMA 2019. LNCS (LNAI), vol. 11888, pp. 169–184. Springer, Cham (2019). https://doi.org/10.1007/978-3-030-35231-8_12

13. Fournier-Viger, P., Zhang, Y., Lin, J.C.W., Fujita, H., Koh, Y.S.: Mining local and peak high utility itemsets. Inf. Sci. **481**, 344–367 (2019)

14. Helmi, S., Banaei-Kashani, F.: Mining frequent episodes from multivariate spatiotemporal event sequences. In: Proceedings 7th ACM SIGSPATIAL International Workshop on GeoStreaming, pp. 1–8 (2016)

15. Huang, K., Chang, C.: Efficient mining of frequent episodes from complex sequences. Inf. Syst. **33**(1), 96–114 (2008)

16. Iwanuma, K., Takano, Y., Nabeshima, H.: On anti-monotone frequency measures for extracting sequential patterns from a single very-long data sequence. In: Proceedings IEEE Conference on Cybernetics and Intelligent Systems, vol. 1, pp. 213–217 (2004)

17. Venkatesh, J.N., Uday Kiran, R., Krishna Reddy, P., Kitsuregawa, M.: Discovering periodic-correlated patterns in temporal databases. In: Hameurlain, A., Wagner, R., Hartmann, S., Ma, H. (eds.) Transactions on Large-Scale Data- and Knowledge-Centered Systems XXXVIII. LNCS, vol. 11250, pp. 146–172. Springer, Heidelberg (2018). https://doi.org/10.1007/978-3-662-58384-5_6

18. Li, L., Li, X., Lu, Z., Lloret, J., Song, H.: Sequential behavior pattern discovery with frequent episode mining and wireless sensor network. IEEE Commun. Mag. **55**(6), 205–211 (2017)

19. Liao, G., Yang, X., Xie, S., Yu, P.S., Wan, C.: Mining weighted frequent closed episodes over multiple sequences. Tehnički vjesnik **25**(2), 510–518 (2018)

20. Lin, S., Qiao, J., Wang, Y.: Frequent episode mining within the latest time windows over event streams. Appl. Intell. **40**(1), 13–28 (2013). https://doi.org/10.1007/s10489-013-0442-8

21. Lin, Y., Huang, C., Tseng, V.S.: A novel methodology for stock investment using high utility episode mining and genetic algorithm. Appl. Soft Comput. **59**, 303–315 (2017)

22. Luna, J.M., Fournier-Viger, P., Ventura, S.: Frequent itemset mining: a 25 years review. In: Lepping, J., (ed.) Wiley Interdisciplinary Reviews: Data Mining and Knowledge Discovery, vol. 9, no. 6, p. e1329. Wiley, Hoboken (2019)

23. Mannila, H., Toivonen, H., Verkamo, A.I.: Discovering frequent episodes in sequences. In: Proceedings 1st International Conference on Knowledge Discovery and Data Mining (1995)

24. Patnaik, D., Laxman, S., Chandramouli, B., Ramakrishnan, N.: Efficient episode mining of dynamic event streams. In: 2012 IEEE 12th International Conference on Data Mining, pp. 605–614 (2012)

25. Rathore, S., Dawar, S., Goyal, V., Patel, D.: Top-k high utility episode mining from a complex event sequence. In: Proceedings of the 21st International Conference on Management of Data, Computer Society of India (2016)

26. Su, M.Y.: Applying episode mining and pruning to identify malicious online attacks. Comput. Electr. Eng. **59**, 180–188 (2017)

27. Truong, T., Duong, H., Le, B., Fournier-Viger, P.: Fmaxclohusm: an efficient algorithm for mining frequent closed and maximal high utility sequences. Eng. Appl. Artif. Intell. **85**, 1–20 (2019)

28. Truong, T., Duong, H., Le, B., Fournier-Viger, P., Yun, U.: Efficient high average-utility itemset mining using novel vertical weak upper-bounds. Knowl. Based Syst. **183**, 104847 (2019)

29. Wenzhe, L., Qian, W., Luqun, Y., Jiadong, R., Davis, D.N., Changzhen, H.: Mining frequent intra-sequence and inter-sequence patterns using bitmap with a maximal span. In: Proceedings 14th Web Information System and Applications Conference, pp. 56–61. IEEE (2017)
30. Wu, C., Lin, Y., Yu, P.S., Tseng, V.S.: Mining high utility episodes in complex event sequences. In: Proceedings 19th ACM SIGKDD International Conference on Knowledge Discovery, pp. 536–544 (2013)
31. Zhou, W., Liu, H., Cheng, H.: Mining closed episodes from event sequences efficiently. In: Zaki, M.J., Yu, J.X., Ravindran, B., Pudi, V. (eds.) PAKDD 2010. LNCS (LNAI), vol. 6118, pp. 310–318. Springer, Heidelberg (2010). https://doi.org/10.1007/978-3-642-13657-3_34

TKU-CE: Cross-Entropy Method for Mining Top-K High Utility Itemsets

Wei Song$^{(\boxtimes)}$ (iD), Lu Liu, and Chaomin Huang

School of Information Science and Technology, North China University of Technology,
Beijing 100144, China
songwei@ncut.edu.cn

Abstract. Mining high utility itemsets (HUIs) is one of the most important research topics in data mining because HUIs consider non-binary frequency values of items in transactions and different profit values for each item. However, setting appropriate minimum utility thresholds by trial and error is a tedious process for users. Thus, mining the top-k high utility itemsets (top-k HUIs) without setting a utility threshold is becoming an alternative to determining all of the HUIs. In this paper, we propose a novel algorithm, named TKU-CE (Top-K high Utility mining based on Cross-Entropy method), for mining top-k HUIs. The TKU-CE algorithm follows the roadmap of cross entropy and tackles top-k HUI mining using combinatorial optimization. The main idea of TKU-CE is to generate the top-k HUIs by gradually updating the probabilities of itemsets with high utility values. Compared with the state-of-the-art algorithms, TKU-CE is not only easy to implement, but also saves computational costs incurred by additional data structures, threshold raising strategies, and pruning strategies. Extensive experimental results show that the TKU-CE algorithm is efficient, memory-saving, and can discover most actual top-k HUIs.

Keywords: Data mining · Heuristic method · Cross-entropy · Combinatorial optimization · Top-k high utility itemset

1 Introduction

High utility itemset (HUI) mining [7] is an extension of frequent itemset (FI) mining [1] used to discover high-profit itemsets by considering both the quantity and value of a single item. However, it is difficult for non-expert users to set an appropriate threshold. Consequently, top-k high utility itemset (top-k HUI) mining [13] is drawing researchers' attention. The k value is a more intuitive and direct parameter for users to set than the minimum threshold.

Top-k HUI mining uses the same concept of utility as HUI mining; that is, an item's utility is mainly reflected by the product of its profit and occurrence frequency. The k itemsets with the highest utility values comprise the target of top-k HUI mining. Wu et al. [13] first introduced the problem of top-k HUI mining, and proposed the TKU algorithm following the widely used two phase routine in HUI algorithms [7]. Later, they

H. Fujita et al. (Eds.): IEA/AIE 2020, LNAI 12144, pp. 846–857, 2020.
https://doi.org/10.1007/978-3-030-55789-8_72

proposed a more efficient algorithm TKO using the one phase routine [11]. Recently, several other top-k HUIs mining algorithms, such as kHMC [3] and TKEH [9], have also been proposed.

Whether top-k HUI mining is approached using a two-phase or one-phase algorithm, it is equivalent to HUI mining with the minimum utility threshold set to zero. Thus, the key techniques of these algorithms amount to various threshold raising strategies; that is, the minimum utility threshold increases gradually as the intermediate top-k results progress. Thus, in this paper, we aim to use a different strategy that does not require constant threshold-raising. We achieve this goal using the cross-entropy method.

The cross-entropy (CE) method is a combinatorial and multi-extremal optimization approach [2]. The CE method approaches the optimal values using an iterative procedure where each iteration is composed of two phases: generating a random data sample, and updating parameters to produce better samples in the next iteration. According to the number of samples N, better results are retained and worse results are abandoned in each iteration. After many cycles of iterations, the best N results are obtained. This approach is essentially consistent with the problem of top-k HUI mining. Thus, we use the CE method to formulate the novel top-k HUI mining algorithm proposed in this paper. The major contributions of this work are summarized as follows:

First, TKU-CE directly uses the utility value as the fitness function, and models the problem of top-k HUI mining through the perspective of combinatorial optimization. Second, TKU-CE updates a probability vector gradually. In this probability vector, there is a higher likelihood of updating the probability corresponding to itemsets with higher utility values. The experimental results show that TKU-CE is not only efficient, but also requires less memory resources. Furthermore, TKU-CE can discover more than 90% of the actual top-k HUIs in most cases.

2 Preliminaries

2.1 Top-K HUI Mining Problem

Let $I = \{i_1, i_2, ..., i_m\}$ be a finite set of items, and $X \subseteq I$ is called an *itemset*. Let $D = \{T_1, T_2, ..., T_n\}$ be a transaction database. Each transaction $T_i \in D$, with unique identifier *tid*, is a subset of I.

The *internal utility* $q(i_p, T_d)$ represents the quantity of item i_p in transaction T_d. The *external utility* $p(i_p)$ is the unit profit value of item i_p. The *utility* of item i_p in transaction T_d is defined as $u(i_p, T_d) = p(i_p) \times q(i_p, T_d)$. The utility of itemset X in transaction T_d is defined as $u(X, T_d) = \sum_{i_p \in X \wedge X \subseteq T_d} u(i_p, T_d)$. The utility of itemset X in D is defined as $u(X) = \sum_{X \subseteq T_d \wedge T_d \in D} u(X, T_d)$. The transaction utility of transaction T_d is defined as $TU(T_d) = u(T_d, T_d)$. The *minimum utility threshold* δ is given as a percentage of the total transaction utility values of the database, while the *minimum utility value* is defined as $min\_util = \delta \times \sum_{T_d \in D} TU(T_d)$. An itemset X is called a *high utility itemset* if $u(X) \geq min\_util$. The set of all HUIs in D w.r.t. min_util is denoted by $f_H(D, min\_util)$.

An itemset X is called a top-k HUI in a database D if there are less than k itemsets whose utilities are larger than $u(X)$ in $f_H(D, 0)$. Top-k HUI mining aims to discover the k itemsets with the highest utilities, where k is a parameter set by the user.

Table 1. Example database

TID	Transactions	TU
T_1	$(A, 1) (B, 1) (C, 1) (F, 2)$	19
T_2	$(B, 1) (D, 1) (E, 1)$	9
T_3	$(A, 1) (B, 1) (C, 1) (F, 1)$	15
T_4	$(C, 3) (D, 2) (F, 1)$	17
T_5	$(A, 1) (C, 2)$	13

Table 2. Profit table

Item	A	B	C	D	E	F
Profit	7	1	3	2	6	4

Example 1. Consider the database in Table 1 and the profit table in Table 2. For convenience, we write an itemset $\{B, C\}$ as BC. In the example database, the utility of item C in transaction T_1 is: $u(C, T_1) = 3 \times 1 = 3$, the utility of itemset BC in transaction T_1 is: $u(BC, T_1) = u(B, T_1) + u(C, T_1) = 4$, and the utility of itemset BC in the transaction database is $u(BC) = u(BC, T_1) + u(BC, T_3) = 8$. The transaction utility of T_5 is: $TU(T_5) = u(AC, T_5) = 13$. The utilities of the other transactions are given in the third column of Table 1. In this example, the set of top-3 HUIs is $\{ABCF: 34, AC: 33, ACF: 32\}$, where the number beside each itemset indicates its utility.

2.2 Cross-Entropy Method

The CE method can be used either for estimating probabilities of rare events in complex stochastic networks, or for solving difficult combinatorial optimization problems (COP). In this paper, we determine the top-k HUIs following the COP methodology.

The classical CE for COPs involving binary vectors is formalized as follows. Let $y = (y_1, y_2, ..., y_n)$ be an n-dimensional binary vector, that is, the value of y_i $(1 \leq i \leq n)$ is either zero or one. The goal of the CE method is to reconstruct the unknown vector y by maximizing the function $S(x)$ using a random search algorithm.

$$S(x) = n - \sum_{j=1}^{n} |x_j - y_j| \tag{1}$$

A naive way to find y is to repeatedly generate binary vectors $x = (x_1, x_2, ..., x_n)$ until a solution is equal to y, leading to $S(x) = n$. Elements of the trial binary vector x, namely $x_1, x_2, ..., x_n$ are independent Bernoulli random variables with success probabilities $p_1, p_2, ..., p_n$, and these probabilities can comprise a probability vector $p' = (p_1, p_2, ..., p_n)$. The CE method for COP consists of creating a sequence of probability vectors $p'_0, p'_1, ...$ and levels $\gamma_1, \gamma_2, ...$, such that the sequence $p'_0, p'_1, ...$ converges to the optimal probability vector, and the sequence $\gamma_1, \gamma_2, ...$ converges to the optimal performance.

Initially, $p_0' = (1/2, 1/2, \ldots, 1/2)$. For a sample $x_1, x_2, \ldots, x_N$ of Bernoulli vectors, calculate $S(x_i)$ for all i, and order the elements according to descending $S(x_i)$. Let γ_t be a ρ sample quantile of the performances. That is:

$$\gamma_t = S_{(\lceil \rho \times N \rceil)} \tag{2}$$

Then each element of the probability vector is updated by:

$$p_{t,j}' = \sum_{i=1}^{N} I_{\{S(x_i) \geq \gamma_t\}} I_{\{x_{ij}=1\}} \Big/ \sum_{i=1}^{N} I_{\{S(x_i) \geq \gamma_t\}} \tag{3}$$

where $j = 1, 2, \ldots, n$, $x_i = (x_{i1}, x_{i2}, \ldots, x_{in})$, t is the iteration number, and $I(\cdot)$ is the indicator function defined as:

$$I_E = \begin{cases} 1, & \text{if } E \text{ is true} \\ 0, & \text{otherwise} \end{cases} \tag{4}$$

where E is an event.

Equation 3 is used iteratively to update the probability vector until the stopping criterion is met. There are two possible stopping criteria: γ_t does not change for a number of subsequent iterations or the probability vector has converged to a binary vector.

3 Existing Algorithms

3.1 Top-K HUI Mining

The basic concepts of top-k HUI mining were given by Wu et al. [13]. Since the anti-monotonicity-based pruning strategies of top-k FI mining [12] cannot be used directly for top-k HUI mining, Wu et al. introduced the concept of the optimal minimum utility threshold, and used a threshold raising method to improve the mining efficiency. REPT [8] is another top-k HUI mining algorithm that follows the two phase methodology. The algorithm constructs a global tree structure to generate candidate top-k HUIs using three threshold raising strategies, and exploits exact and pre-evaluated utilities of itemsets with a length of one or two to reduce the number of candidates.

Recent algorithms focus on mining top-k HUIs directly without generating candidates. The TKO algorithm [11] utilizes a utility list data structure to maintain itemset information during the mining process. Furthermore, TKO also uses three pruning strategies to facilitate the mining process. kHMC [3] also mines the top-k HUIs in one phase. Besides a utility list, kHMC proposes the concept of coverage to raise the intermediate thresholds.

For existing top-k HUI mining algorithms, the major challenge differentiating top-k HUI mining and traditional HUI mining is the threshold raising strategies. Gradually raising the minimum utility threshold constricts the search space during the mining process. Thus, new methods that achieve suitable performance without using threshold raising strategies are pertinent for top-k HUI mining.

3.2 HUI Mining Using Heuristic Methods

Inspired by biological and physical phenomena, heuristic methods are effective for solving combinatorial problems, and have been used to traverse immense candidate itemset spaces within an acceptable time for mining FIs and HUIs.

Two HUI mining algorithms, $HUPE_{UMU}$-GARM and $HUPE_{WUMU}$-GARM, based on the genetic algorithm (GA) are proposed in [5]. Premature convergence is the main problem of these two algorithms; that is, the two algorithms easily fall into local optima. Particle swarm optimization (PSO) is another heuristic method used for mining HUIs. The PSO-based algorithm discovers HUIs comprehensively using local optimization and global optimization [6].

Unlike GAs and PSOs, ant colony optimization (ACO) produces a feasible solution in a constructive way. Wu et al. proposed an ACO-based algorithm called HUIM-ACS for mining HUIs [14]. This algorithm generates a routing graph before all of the ants start their tours. Furthermore, positive pruning and recursive pruning are used to improve the algorithm's efficiency.

Song and Huang studied the problem of HUI mining from the perspective of the artificial bee colony (ABC) algorithm. The proposed HUIM-ABC discovers HUIs by modeling the itemsets as nectar sources [10]. For each nectar source, three types of bees are used: employed bee, onlooker bee, and scout bee, for sequential optimization within one iteration. During one iteration, the algorithm outputs an itemset when it is verified as an HUI. This process is executed iteratively until the maximal cycle number is reached.

To the best of our knowledge, the heuristic method of cross-entropy has neither been used in HUI mining nor in top-k HUI mining.

4 The Proposed TKU-CE Algorithm

4.1 Bitmap Item Information Representation

The first component of the proposed TKU-CE algorithm is the representation of items. We use bitmap, an effective representation of item information in FI mining and HUI mining algorithms, in TKU-CE to identify transactions containing the target itemsets. We can calculate the utility values of the target itemsets efficiently using bit-wise operations.

Specifically, TKU-CE uses a *bitmap cover* representation for itemsets. In a bitmap cover, there is one bit for each transaction in the database. If item i appears in transaction T_j, then bit j of the bitmap cover for item i is set to one; otherwise, the bit is set to zero. This naturally extends to itemsets. Let X be an itemset, $Bit(X)$ corresponds to the bitmap cover that represents the transaction set for the itemset X. Let X and Y be two itemsets, $Bit(X \cup Y)$ can be computed as $Bit(X) \cap Bit(Y)$, i.e., the bitwise-AND of $Bit(X)$ and $Bit(Y)$.

4.2 Modeling Top-K HUI Mining Based on the CE Method

After transforming the database into a bitmap, it is natural to encode each solution in a binary vector. To discover the top-k HUIs from the transaction database, we use the utility of the itemset to replace Eq. 1 directly. That is, for an itemset X:

$$S(X) = u(X) \tag{5}$$

In each iteration t, we sort a sample $X_1, X_2, \ldots, X_N$ in descending order of $S(x_i)$ ($1 \leq i \leq N$), and update the sample quantile γ_t and the probability vector p'_t accordingly.

4.3 The Proposed Algorithm

Algorithm 1 describes our top-k HUI mining algorithm TKU-CE.

Algorithm 1	TKU-CE
Input	Transaction database D, the number of desired HUIs k, sample numbers N, the quantile parameter ρ, the maximum number of iterations max_iter
Output	Top-k high utility itemsets
1	Initialization( );
2	**while** $t \leq max\_iter$ **and** p'_t is not a binary vector **do**
3	Calculate p'_t using Eq. 3;
4	**for** i=1 to N **do**
5	**for** j=1 to $|I|$ **do**
6	Generate X_{ij} with the probability of $p'_{t,j}$;
7	**end for**
8	**end for**
9	Sort the N itemsets by descending order of utility;
10	Update the set of top-k HUIs KH using the new sample;
11	Calculate γ_t using Eq. 2;
12	t++;
13	**end while**
14	Output top-k HUIs.

In Algorithm 1, the procedure Initialization(), described in Algorithm 2, is first called in Step 1. The main loop from Step 2 to Step 13 calculates the top-k HUIs iteratively. Besides the maximal number of iterations, the probability vector becoming a binary vector is also a stopping criterion. With a binary probability vector, all of the N itemsets are the same in one iteration, because each item is definitely included or not included in each itemset. For example, there are five items A, B, C, D, and E. After many iterations, the probability vector converges to $(1, 1, 1, 0, 0)$, then the itemsets within the sample are all ABC because the probabilities of the bits corresponding to D and E are zero. Step 3 calculates the probability vector of the new iteration. The loop from Step 4 to Step 8 generates N new itemsets bit by bit. Here, $|I|$ is the number of items in I. Specifically, for itemset X_i and its jth bit, we randomly generate a probability $p_{i,j}$, then determine the value X_{ij} by:

$$X_{ij} = \begin{cases} 1, & \text{if } p_{i,j} \leq p'_{t,j} \\ 0, & \text{if } p_{i,j} > p'_{t,j} \end{cases} \qquad (6)$$

Step 9 arranges the itemsets in descending order of utility. We update KH, the set of the top-k HUIs, according to the latest sample in Step 10. The new ρ sample quantile is calculated in Step 11. Step 12 increments the iteration number by one. Finally, Step 14 outputs all of the discovered top-k HUIs.

It should be noted that the number of resulting itemsets of top-k HUI mining may be either less than k or more than k. As in TKU and TKO [11], we output the actual results regardless of whether the number of results is more or less than k.

Algorithm 2	Initialization( )
1	Represent the database using a bitmap;
2	p'_0=(1/2, 1/2, ..., 1/2);
3	**for** i=1 to N **do**
4	**for** j=1 to $|I|$ **do**
5	Generate X_{ij} with a probability of $p'_{0,j}$;
6	**end for**
7	**end for**
8	Sort the N itemsets by descending order of utility, and denote them as $S_1, S_2, ... S_N$;
9	Initialize the set of top-k HUIs KH with $S_1, S_2, ..., S_k$;
10	Calculate γ using Eq. 2;
11	$t = 1$;

In Algorithm 2, we first construct the bitmap representation of the database in Step 1. Step 2 initializes all of the probabilities in the probability vector to 1/2. That is the probability of being one or zero is 0.5. The loop (Steps 3–7) initializes the N itemsets of the first iteration. Step 8 reorders the itemsets by descending order of utility. Step 9 initializes KH according to the sample of the first iteration. Step 10 then calculates the ρ sample quantile. Finally, Step 11 sets the iteration number to one.

5   Performance Evaluation

In this section, we evaluate the performance of our TKU-CE algorithm and compare it with the TKU [13] and TKO [11] algorithms. We downloaded the source code of the two comparison algorithms from the SPMF data mining library [4].

5.1   Test Environment and Datasets

We conducted the experiments on a computer with a 4-Core 3.40 GHz CPU and 8 GB memory running 64-bit Microsoft Windows 10. We wrote our programs in Java. We used both synthetic and real datasets to evaluate the performance of the algorithms. We generated two synthetic datasets from the IBM data generator [1]: T25I100D5k and T35I100D7k. The parameters are as follows: T is the average size of the transactions, I is the total number of different items, and D is the total number of transactions. We downloaded real datasets from the SPMF data mining library [4]. Both Chess and Connect datasets originate from game steps. Table 3 gives the characteristics of the datasets used in the experiments.

The two real datasets already contain the utility information, while the two synthetic datasets do not contain the utility value or quantity of each item in each transaction. As

Table 3. Characteristics of datasets used for experiment evaluations

Datasets	Avg. trans. length	No. of items	No. of trans
T25I100D5k	25	100	5,000
T35I100D7k	35	100	7,000
Chess	37	76	3,196
Connect	43	130	67,557

in TKU [13] and TKO [11], we generate the unit profits for items between 1 and 1,000 using a log-normal distribution and generate the item quantities randomly between 1 and 5.

For all experiments, we set the sample number to 2,000, ρ to 0.2, and the maximum number of iterations to 2,000.

5.2 Runtime

We demonstrate the efficiency of our algorithm and the comparison algorithms with a varying number of desired itemsets, namely k, for each dataset.

Figure 1(a) shows the execution time comparison between the algorithms on the T25I100D5k dataset. As k increases from 20 to 100, TKU-CE is 8.70 times faster than TKU, and one order of magnitude faster than TKO, on average. In this set of experiments, TKU is faster than TKO. This is because the utility list and the pre-evaluation matrix structure of TKO are not effective on this sparse dataset, and these operations add computational complexity with respect to the other two algorithms.

Figure 1(b) shows the execution time comparison between the algorithms on the T35I100D7k dataset as k increases from 3 to 11. For this dataset, TKU did not return any results even after four days, thus, we did not plot its results. TKU-CE's superior efficiency over TKO is more obvious on T35I100D7k. TKU-CE is consistently three orders of magnitude faster than TKO. For 11 itemsets, the runtime comparison between TKU-CE and TKO is 35.696 s to 223,016.616 s, respectively. Because TKO incurs a very long runtime for this dataset, we set a low number of desired itemsets to obtain the output within a reasonable time.

For the Chess dataset, when the range of k is between 20 and 100, TKU runs out of memory. Thus, we again omit the results of TKU in Fig. 1(c). This time, TKU-CE is faster than TKO by 46.09% on average.

As with the results on Chess, we did not plot the TKU results in Fig. 1(d) because it again runs out of memory for the Connect dataset. On average, TKU-CE is 4.43 times faster than TKO.

As we can see from Fig. 1, the proposed heuristic algorithm TKU-CE always demonstrates superior efficiency. With respect to the other two algorithms, the one-phase TKO algorithm outperforms the two-phase TKU algorithm in most cases. The runtime of TKU-CE does not increase as k increases because when the sample number N is greater than k, k only affects the initialization and update of the set of top-k HUIs. This renders TKU-CE not only efficient but also easy to implement.

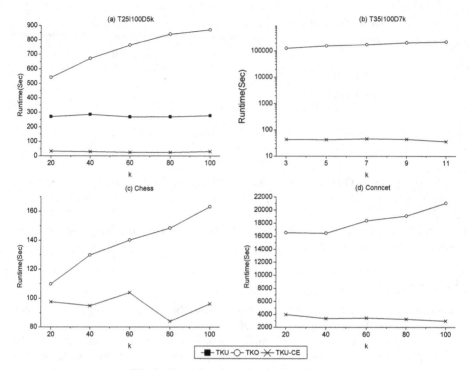

Fig. 1. Execution times for the four datasets

5.3 Memory Consumption

We also compare the memory usage of the three algorithms for the four datasets. The results are shown in Fig. 2. For the same reasons stated in Sect. 5.2, we only plot the results of TKU for T25I100D5k.

Figure 2(a) shows the memory usage for the T25I100D5k dataset. On average TKU-CE consumes 2.69 times less memory than TKU, and 73.45% less memory than TKO. Although Fig. 1(a) shows that TKU is more efficient than TKO, TKO outperforms TKU in terms of storage consumption.

Figures 2(b) and 1(b) show that the proposed TKU-CE algorithm is not only remarkably more efficient than TKO, but also uses very little memory—less than 10 MB on T35I100D7k.

Figures 2(c) and 2(d) show the memory consumption for the two real datasets Chess and Connect. TKU-CE still outperforms TKO with respect to memory usage. On average TKU-CE consumes 2.75 times and 4.92 times less memory than TKO on Chess and Connect, respectively.

From Fig. 2, we can see that TKU-CE consumes less memory than TKU and TKO on these four datasets. Furthermore, the memory usage of TKU-CE is nearly constant. This is because TKU-CE does not consider additional tree or list structures for storing and transforming the original information and does not include any threshold raising and pruning strategies. Thus, it is more suitable for the problem of top-k HUI mining without a user-specified minimum utility threshold.

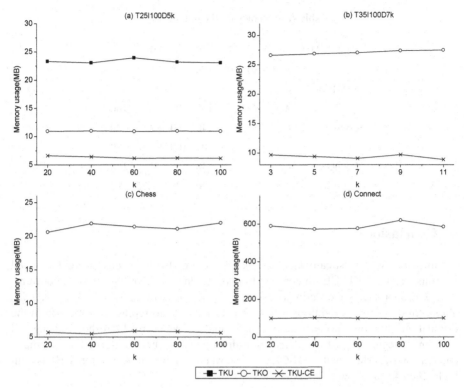

Fig. 2. Memory usage for the four datasets

5.4 Accuracy

A heuristic top-k HUI mining algorithm cannot ensure the discovery of all of the correct itemsets within a certain number of cycles; that is, some itemsets discovered by TKU-CE may not correspond to the actual top-k itemsets of the entire dataset. In this section, we compared the percentage of discovered actual top-k HUIs by k. We use the TKO algorithm to discover the actual complete list of top-k HUIs from the four datasets. We use the following equation to calculate the accuracy of top-k HUIs discovered by TKU-CE.

$$Acc_k = CE_k/k \times 100\% \tag{7}$$

where CE_k is the number of actual top-k HUIs discovered by TKU-CE.

Table 4 shows that TKU-CE can discover more than 90% of the actual top-k HUIs within 2000 iterations, except when k is set to 100 for the T25I100D5k dataset. Furthermore, TKU-CE outputs 100% of true top-k HUIs with 12 times for the four datasets. As the number of experiments for each dataset is 5, the probability of TKU-CE returning the exact top-k HUIs is 60%. Thus, we see that TKU-CE outputs most of the actual top-k HUIs within less time and consuming less memory.

Table 4. Accuracy for the four datasets

T25I100D5k	k	20	40	60	80	100
	Acc_k (%)	100	100	90	91.25	75
T35I100D7k	k	3	5	7	9	11
	Acc_k (%)	100	100	100	100	100
Chess	k	20	40	60	80	100
	Acc_k (%)	100	100	100	97.50	99
Connect	k	20	40	60	80	100
	Acc_k (%)	95	100	100	96.25	97

6   Conclusions

In this paper, we heuristically tackle top-k HUI mining by proposing a novel CE-based algorithm called TKU-CE. In contrast to existing one-phase or two-phase algorithms, TKU-CE approaches the optimal results using a stochastic approach. The TKU-CE algorithm does not use additional tree or list structures to represent or transform the original information. Furthermore, we also avoid the widespread threshold raising and pruning strategies of existing related algorithms. Experimental results on both synthetic and real datasets show that TKU-CE can discover a suitable number of top-k HUIs with high efficiency and low memory usage.

Acknowledgments. This work was partially supported by the National Natural Science Foundation of China (61977001), the Great Wall Scholar Program (CIT&TCD20190305), and Beijing Urban Governance Research Center.

References

1. Agrawal, R., Srikant, R.: Fast algorithms for mining association rules in large databases. In: Proceedings of the 20th International Conference on Very Large Data Bases, pp. 487–499 (1994)
2. de Boer, P.-T., Kroese, D.P., Mannor, S., Rubinstein, R.Y.: A tutorial on the cross-entropy method. Ann. Oper. Res. **134**(1), 19–67 (2005). https://doi.org/10.1007/s10479-005-5724-z
3. Duong, Q.H., Liao, B., Fournier-Viger, P., Dam, T.-L.: An efficient algorithm for mining the top-k high utility itemsets, using novel threshold raising and pruning strategies. Knowl. Based Syst. **104**, 106–122 (2016)
4. Fournier-Viger, P., et al.: The SPMF open-source data mining library version 2. In: Berendt, B., et al. (eds.) ECML PKDD 2016. LNCS (LNAI), vol. 9853, pp. 36–40. Springer, Cham (2016). https://doi.org/10.1007/978-3-319-46131-1_8
5. Kannimuthu, S., Premalatha, K.: Discovery of high utility itemsets using genetic algorithm with ranked mutation. Appl. Artif. Intell. **28**(4), 337–359 (2014)
6. Lin, J.C.-W., Yang, L., Fournier-Viger, P., Hong, T.-P., Voznak, M.: A binary PSO approach to mine high-utility itemsets. Soft. Comput. **21**(17), 5103–5121 (2016). https://doi.org/10.1007/s00500-016-2106-1

7. Liu, Y., Liao, W.-K., Choudhary, A.N.: A two-phase algorithm for fast discovery of high utility itemsets. In: Ho, T.B., Cheung, D., Liu, H. (eds.) PAKDD 2005. LNCS (LNAI), vol. 3518, pp. 689–695. Springer, Heidelberg (2005). https://doi.org/10.1007/11430919_79

8. Ryang, H., Yun, U.: Top-k high utility pattern mining with effective threshold raising strategies. Knowl.-Based Syst. **76**, 109–126 (2015)

9. Singh, K., Singh, S.S., Kumar, A., Biswas, B.: TKEH: an efficient algorithm for mining top-k high utility itemsets. Appl. Intell. **49**(3), 1078–1097 (2018). https://doi.org/10.1007/s10489-018-1316-x

10. Song, W., Huang, C.: Discovering high utility itemsets based on the artificial bee colony algorithm. In: Phung, D., Tseng, V.S., Webb, G.I., Ho, B., Ganji, M., Rashidi, L. (eds.) PAKDD 2018. LNCS (LNAI), vol. 10939, pp. 3–14. Springer, Cham (2018). https://doi.org/10.1007/978-3-319-93040-4_1

11. Tseng, V.S., Wu, C.-W., Fournier-Viger, P., Yu, P.S.: Efficient algorithms for mining top-k high utility itemsets. IEEE Trans. Knowl. Data Eng. **28**(1), 54–67 (2016)

12. Wang, J., Han, J., Lu, Y., Tzvetkov, P.: TFP: an efficient algorithm for mining top-k frequent closed itemsets. IEEE Trans. Knowl. Data Eng. **17**(5), 652–664 (2005)

13. Wu, C.-W., Shie, B.-E., Tseng, V.S., Yu, P.S.: Mining top-k high utility itemsets. In: Proceedings of the 18th ACM SIGKDD International Conference on Knowledge Discovery and Data Mining, pp. 78–86 (2012)

14. Wu, J.M.T., Zhan, J., Lin, J.C.W.: An ACO-based approach to mine high-utility itemsets. Knowl.-Based Syst. **116**, 102–113 (2017)

Mining Cross-Level High Utility Itemsets

Philippe Fournier-Viger[1](✉), Ying Wang[2], Jerry Chun-Wei Lin[3],
Jose Maria Luna[4], and Sebastian Ventura[4]

[1] School of Natural Sciences and Humanities, Harbin Institute of Technology
(Shenzhen), Shenzhen, China
philfv@hit.edu.cn
[2] School of Computer Sciences and Technology, Harbin Institute of Technology
(Shenzhen), Shenzhen, China
iwangying_919@163.com
[3] Department of Computing, Mathematics and Physics, Western Norway University
of Applied Sciences (HVL), Bergen, Norway
jerrylin@ieee.org
[4] Department of Computer Sciences, University of Cordoba, Cordoba, Spain
{jmluna,sventura}@uco.es

Abstract. Many algorithms have been proposed to find high utility
itemsets (sets of items that yield a high profit) in customer transactions.
Though, it is useful to analyze customer behavior, it ignores information
about item categories. To consider a product taxonomy and find high
utility itemsets describing relationships between items and categories,
the ML-HUI Miner was recently proposed. But it cannot find cross-level
itemsets (itemsets mixing items from different taxonomy levels), and it
is inefficient as it does not use relationships between categories to reduce
the search space. This paper addresses these issues by proposing a novel
problem called cross-level high utility itemset mining, and an algorithm
named CLH-Miner. It relies on novel upper bounds to efficiently search
for high utility itemsets when considering a taxonomy. An experimental
evaluation with real retail data shows that the algorithm is efficient and
can discover insightful patterns describing customer purchases.

Keywords: High utility mining · Cross-level itemset · Taxonomy

1 Introduction

Pattern mining is a sub-field of data mining that aims at discovering interesting
patterns in data to better understand the data or support decision-making. One
of the most popular pattern mining tasks is frequent itemset mining (FIM),
which consists of finding all frequently co-occurring itemsets (sets of values) in
a customer transaction database [6,15]. Although FIM is useful, it does not
consider purchase quantities of items and their relative importance. To address
this issue, a more general problem was proposed called high utility itemset mining
(HUIM) [5,7,13,19], where items in transactions have purchase quantities and

© Springer Nature Switzerland AG 2020
H. Fujita et al. (Eds.): IEA/AIE 2020, LNAI 12144, pp. 858–871, 2020.
https://doi.org/10.1007/978-3-030-55789-8_73

items have weights indicating their relative importance. The goal of HUIM is to find itemset that have a high utility (importance) such as those that yield a high profit in a customer transaction database. As HUIM is more general than FIM, it is also more difficult. The reason is that the utility measure is not anti-monotonic (the utility of an itemset may be larger, smaller thann or equal to that of its subsets). Thus, FIM techniques cannot be directly used in HUIM.

Although HUIM has many applications, it ignores that items are often organized in a taxonomy. For example, two items *dark chocolate* and *milk chocolate* are both specializations of the abstract item *chocolate*, which can in turn be a specialization of an item *snack and treats*. Traditional HUIM can only find patterns involving items at the lowest abstraction level. This is a major problem because although some items like *milk chocolate* and *dark chocolate* may not appear in HUIs, the item *chocolate* could be part of some HUIs. Thus, important information may be missed by traditional HUIM algorithms.

To mine frequent itemsets that contain items of different abstraction levels, many algorithms were developed [1, 10, 12, 14, 17, 21, 22]. But these algorithms all rely on the fact that the support (frequency) measure is anti-monotonic to reduce the search space. But this property does not hold for the utility. Hence, it is an important but difficult challenge to mine high utility itemsets containing items of different abstraction levels. Recently, a first work has been done in this direction by Cagliero et al. [2]. They proposed an algorithm named ML-HUI Miner to mine HUIs where items are of different abstraction levels. However, the algorithm has two important limitations. First, it can only find HUIs where all items are of the same abstraction level. This assumption makes the problem easier to solve but results in missing all patterns where items are of different levels, which are often interesting. Second, the algorithm utilizes simple properties to reduce the search space. It mines the different abstraction levels independently and does not use the relationships between abstraction levels to reduce the search space.

To address these limitations, this paper defines the more general problem of mining all cross-level high utility itemsets and studies its properties, and an algorithm named CLH-Miner is proposed to find all these itemsets efficiently. The algorithm relies on novel upper bounds and pruning properties to reduce the search space using relationships between items of different abstraction levels. Experiments on transaction data collected from two chains of retail stores show that interesting patterns are discovered, and that the algorithm's optimizations considerably improve its performance.

The rest of this paper is organized as follows. Section 2 reviews related work. Section 3 defines the problem of cross-level HUIM. Section 4 presents CLH-Miner. Section 5 describes experiments. Lastly, Sect. 6 draws a conclusion.

2   Related Work

Several algorithms were designed to discover frequent itemsets in transaction databases [6]. FIM was then generalized as HUIM to find itemsets of high importance (e.g. profit) in datasets where items are annotated with purchase quantities and weights (e.g. unit profit) [5]. Some representative HUIM algorithms are Two-Phase [13], HUI-Miner/HUI-Miner* [19] and FHM [7]. Two-Phase [13] adopts a breadth-first search to explore the search space of itemsets and an upper bound on the utility measure called TWU that is anti-monotonic to reduce the search space. But Two-Phase has two important drawbacks: (1) the TWU upper bound is loose, and (2) Two-Phase can perform numerous database scans. HUI-Miner [19] addressed these limitations by introducing the tighter remaining utility upper bound and relying on a novel vertical structure called utility-list. HUI-Miner avoids performing numerous database scans by directly joining utility-lists to calculate the utility and upper bound values of itemsets. An improved version of HUI-Miner called FHM was then proposed [7] and developing efficient HUIM algorithms is an active research area [5]. Although traditional HUI mining algorithms are useful to identify important (e.g. profitable) patterns, most of these algorithms ignore item taxonomies. Thus, they are unable to reveal insightful relationships between items of different taxonomy levels (e.g. product categories).

In FIM, several studies have aimed at extracting generalized patterns using a taxonomy. Srikant and Agrawal [21] first proposed using a taxonomy of items linked by *is-a* relationships to extract frequent itemsets and association rules containing items from different abstraction levels. The Cumulate algorithm [21] proposed in that study requires that the user sets a minimum support threshold. Then, Cumulate performs a breadth-first search starting from single items to generate larger itemsets. Cumulate can find *cross-level patterns*, that is patterns containing items from different abstraction levels, with the constraint that an item and its taxonomy ancestor may not appear together in a pattern. Then, Hipp et al. proposed the Prutax algorithm [12] to more efficiently find cross-level frequent itemsets using a depth-first search and a vertical database format. Prutax uses two search space pruning strategies: eliminating an itemset from the search space if one of its subsets is infrequent or if it has a taxonomy ancestor that is infrequent. Another depth-first search algorithm called SET was proposed [22] for cross-level FIM, which was claimed to outperform Prutax. Han and Fu [10] proposed a variation of the above problem called *multi-level pattern mining*, where a different minimum support threshold can be set for each taxonomy level but items in a frequent itemset must be from the same taxonomy level. They designed breadth-first search algorithms that start from abstract itemsets and recursively specialize frequent itemsets. Lui and Chung [14] then proposed a variation to mine cross-level itemsets using multiple thresholds. Their algorithm performs a breadth-first search and finds generalized itemsets by recursively generalizing infrequent itemsets based on a concept of item taxonomy distance. Ong et al. [17] then proposed a FP-tree based pattern-growth algorithm for multi-level itemset mining using a concept of taxonomy-based quantities. Another FP-tree based algorithm was proposed by Pramudiono for cross-level itemset mining [18], and an algorithm based on the AFOPT-tree structure [16] was designed. Rajkumar et al. [20] proposed using different thresholds for different taxonomy levels and itemset lengths in multi-level itemset mining. Other variations of the above problems have also been studied such as to find misleading multi-level itemsets [1], concise representations of generalized itemsets [11], and mining multi-level association rules using evolutionary algorithms [23].

To our knowledge only one algorithm named ML-HUI Miner was proposed to consider taxonomies in high utility itemset mining [2]. This algorithm extends HUI-Miner but has several important drawbacks: (1) it is unable to find cross-level patterns (containing items from multiple taxonomy levels), (2) it mines each taxonomy levels independently, and (3) it does not use relationships between taxonomy levels for search space pruning with the utility measure. This paper addresses these issues by defining a more general problem of cross-level HUIM and an efficient algorithm named CLH-Miner to find the desired patterns.

3   Preliminaries and Problem Definition

This section introduces key definitions and then defines the proposed problem.

Definition 1 (Transaction database). *Let $I = \{i_1, i_2, \ldots, i_m\}$ be a set of items. A transaction database is a multiset of transactions $D = \{T_1, T_2, \ldots, T_n\}$ such that for each transaction T_c, $T_c \in I$, and T_c has a unique identifier c called its Tid. Each item $i \in I$ is associated with a positive number $p(i)$, called its external utility (e.g. unit profit). For each item $i \in T_c$, a positive number $q(i, T_c)$ is called the internal utility of i (e.g. purchase quantity of i in T_c).*

Definition 2 (Taxonomy). *A **taxonomy** τ is a directed acyclic graph (a tree) defined for a database D. It contains a leaf for each item $i \in I$. An inner node represents an abstract category that aggregates all descendant leaf nodes (items) or descendant categories into a higher-level category. A child-parent edge between two (generalized) items i, j in τ represents an is-a relationship. Inner nodes are called **generalized items** or **abstract items**. The set of all generalized items is denoted as GI, and the set of all generalized or non-generalized items is denoted as $AI = GI \cup I$. Also, let there be a relation $LR \subseteq GI \times I$ such that $(g, i) \in LR$ iff there is a path from g to i. And, let there be a relation $GR \subseteq AI \times AI$ such that $(d, f) \in GR$ iff there is a path from d to f. An **itemset** X is a set of items such that $X \subseteq AI$ and $\nexists i, j \in X | i \in Desc(j, \tau)$. An itemset X is a **generalized itemset** iff $\exists g \in X$ such that $g \in GI$.*

Example 1. Table 1 depicts a database, which will be used as running example, having seven transactions $(T_1, T_2, \ldots, T_7)$ and five items $(I = \{a, b, c, d, e\})$, where internal utilities (e.g. quantities) are shown as integers beside items. For instance, transaction T_3 indicates that 1, 5, 1, 3 and 1 units of items a, b, c, d, e were bought, respectively. Table 2 indicates that the external utilities (unit profits) of these items are 5, 2, 1, 2, 3. Figure 1 depicts an item taxonomy for this database. For instance, items a and b are aggregated into the generalized item Y.

Definition 3 (Descendant/Specialization/Sibling). *The **leaf items** of a generalized item g in a taxonomy τ are all the leaves that can be reached by following paths starting from g. This set is formally defined as $Leaf(g, \tau) =$*

Table 1. A transaction database

TID	Transaction
T_1	$(a,1),(c,1)$
T_2	$(e,1)$
T_3	$(a,1),(b,5),(c,1),(d,3),(e,1)$
T_4	$(b,4),(c,3),(d,3),(e,1)$
T_5	$(a,1),(c,1),(d,1)$
T_6	$(a,2),(c,6),(e,2)$
T_7	$(b,2),(c,2),(e,1)$

Table 2. External utility values

Item	Unit profit
a	5
b	2
c	1
d	2
e	3

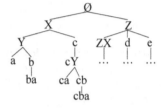

Fig. 1. A taxonomy of items

Fig. 2. A part of the search space

$\{i|(g,i) \in LR\}$. The **descendant items** of a (generalized) item d is the set $Desc(d,\tau) = \{f|(d,f) \in GR\}$. An itemset X is called a **descendant itemset** of an itemset Y if $|X| = |Y|$ and $\forall f \in X, \exists d \in Y|f \in Desc(d,\tau)$. An itemset X is called a **specialization** of an itemset Y if $|X| = |Y|$ and $\forall f \in X, f \in Y \vee \exists d \in Y|f \in Desc(d,\tau)$. Two items a and b are said to be **siblings** if they have the same parent in τ. Furthermore, let $level(d)$ denotes the number of edges to be traversed to reach an item d starting from the root of τ.

Example 2. In the taxonomy of Fig. 1, $Leaf(\{X\},\tau) = \{a,b,c\}$ and $Desc(\{X\}, \tau) = \{Y,a,b,c\}$, The itemset $\{Y,d\}$ is a descendant of $\{X,Z\}$, and a specialization of $\{Y,Z\}$. Moreover, the items Y and c are siblings, and $level(Y) = 2$.

Definition 4 (Utility of an item/itemset). *The utility of an item i in a transaction T_c is denoted as $u(i,T_c)$ and defined as $p(i) \times q(i,T_c)$. The utility of an itemset X (a group of items $X \subseteq I$) in a transaction T_c is denoted as $u(X,T_c)$ and defined as $u(X,T_c) = \sum_{i \in X} u(i,T_c)$. The utility of an itemset X (in a database) is denoted as $u(X)$ and defined as $u(X) = \sum_{T_c \in g(X)} u(X,T_c)$, where $g(X)$ is the set of transactions containing X* [13].

Example 3. The utility of a in T_3 is $u(a,T_3) = 5 \times 1 = 5$. The utility of $\{a,c\}$ in T_3 is $u(\{a,c\},T_3) = u(a,T_3) + u(c,T_3) = 5 \times 1 + 1 \times 1 = 6$. The utility of $\{a,c\}$ in the database is $u(\{a,c\}) = u(a) + u(c) = u(a,T_1) + u(a,T_3) + u(a,T_5) + u(a,T_6) + u(c,T_1) + u(c,T_3) + u(c,T_5) + u(c,T_6) = 5 + 5 + 5 + 10 + 1 + 1 + 1 + 6 = 34$.

Definition 5 (Utility of a generalized item/itemset). *The utility of a generalized item g in a transaction T_c is denoted and defined as $u(g, T_c) = \sum_{i \in Leaf(g,\tau)} p(i) \times q(i, T_c)$. The utility of a generalized itemset GX in a transaction T_c is defined as $u(GX, T_c) = \sum_{d \in GX} u(d, T_c)$. The utility of a generalized itemset GX (in a database) is denoted and defined as $u(GX) = \sum_{T_c \in g(GX)} u(GX, T_c)$, where $g(GX) = \{T_c \in D | \exists X \subseteq T_c \wedge X$ is a descendant of $GX\}$.*

Example 4. The utility of the generalized item Y in T_3 is $u(Y, T_3) = u(a, T_3) + u(b, T_3) = 1 \times 5 + 5 \times 2 = 15$. The utility of the generalized itemset $\{Y, c\}$ in T_3 is $u(\{Y, c\}, T_3) = u(Y, T_3) + u(c, T_3) = 15 + 1 = 16$. The utility of the generalized itemset $\{Y, c\}$ is $u(\{Y, c\}) = u(\{Y, c\}, T_1) + u(\{Y, c\}, T_3) + u(\{Y, c\}, T_4) + u(\{Y, c\}, T_5) + u(\{Y, c\}, T_6) + u(\{Y, c\}, T_7) = 6 + 16 + 11 + 6 + 16 + 6 = 61$.

Definition 6 (Cross-level high utility mining). *The problem of **cross-level high utility mining** is defined as finding all cross-level high-utility itemsets (CLHUIs). A (generalized) itemset X is a CLHUI if $u(X) \geq minutil$ for a user-specified minutil threshold.*

Example 5. If $minutil = 60$, the cross-level high utility itemsets in the database of the running example are $\{X\}, \{Y, c\}, \{Z, X\}, \{Z, Y\}, \{Z, Y, c\}, \{e, X\}, \{e, Y, c\}$ with respectively a utility of $61, 61, 84, 71, 84, 64, 64$.

4   Proposed Algorithm

To efficiently discover all cross-level HUIs, a novel algorithm is proposed, named CLH-Miner. It assumes that a processing order $\succ$ is defined on items of AI. That total order $\succ$ is defined such that $a \prec b$ for two items $a, b \in AI$, if $level(a) < level(b)$ or $level(a) = level(b) \wedge GWU(a) < GWU(b)$, where GWU is a utility measure. This ordering between levels ensures that the algorithm considers items that are higher in the taxonomy before those that are lower, which is important as it enables to prune specializations of itemsets using techniques that will be presented in this section. The algorithm explores the search space of all itemsets by recursively performing two types of extensions on itemsets, defined as follows.

Definition 7 (Extension). *Let X be an itemset. The **join-based extensions** of X are the itemsets obtained by appending an item y to X such that $y \in AI$, $y \succ i, \forall i \in X, y \notin Desc(i, \tau)$. The **tax-based extensions** of X are the itemsets obtained by replacing the last item y of X according to $\succ$ by a descendant of y.*

For example, if $Z \succ X$, the itemset $\{X, d\}$ is a tax-based extension of $\{X, Z\}$, which is a join-extension of $\{X\}$ with item Z. A part of the search space of the running example is shown in Fig. 2.

To quickly calculate the utility of an itemset and those of its extensions, the CLH-Miner algorithm creates a *tax-utility-list* structure for each itemset visited in the search space. This vertical structure extends the utility-list structure used in traditional HUIM [19] with taxonomy information, and is defined as follows.

Definition 8 (Tax-Utility-list). *Let there be an itemset X, a database D, and the total order $\succ$. The **tax-utility-list** $tuList(X)$ of X contains a tuple $(tid, iutil, rutil)$ for each transaction T_{tid} containing X. The iutil element of a tuple is the utility of X in T_{tid}. i.e., $u(X, T_{tid})$. The rutil element of a tuple is defined as $\sum_{i \in T_{tid} \wedge i \succ x \forall x \in X} u(i, T_{tid})$. Also, $tuList(X)$ contains childs, a set of pointers to the tax-utility-list of the child items of X in the taxonomy.*

For example, assume that $X \prec Z \prec Y \prec d \prec b \prec a \prec e \prec c$. The tax-utility-list of itemset $\{X, d\}$ is $tuList(\{X, d\}) = \{(3, 22, 3), (4, 17, 3), (5, 8, 0)\}$.

The proposed algorithm initially constructs the tax-utility-lists of single items by scanning the database, and then applies a modified version of the construct procedure of HUI-Miner [19] to create the tax-utility-lists of their extensions by joining the tax-utility-lists of other itemsets. The difference between the construct procedure of CLH-Miner and that of HUI-Miner is that *childs* is updated. A tax-utility list of an itemset X allows to quickly obtain the utility of X without scanning the database, as $u(X) = \sum_{e \in tuList(X)} e.iutil$ [19]. For example, based on $tuList(\{X, d\})$, the utility of $\{X, d\}$ is $u(\{X, d\}) = 22 + 17 + 8 = 47$.

By doing an exhaustive search of all transitive extensions of single items and calculating their utility using their tax-utility-lists, all cross-level HUIs can be found. But this approach is time-consuming if the number of items is large, which is the case for many real-life databases. To efficiently find cross-level HUIs, two search space pruning techniques are introduced next. The first one is a novel technique generalizing the TWU measure used in traditional HUIM [13].

Definition 9 (The GWU measure). *The **transaction utility** (TU) of a transaction T_c is the sum of the utilities of all the items in T_c. i.e. $TU(T_c) = \sum_{x \in T_c} u(x, T_c)$. The **generalized-weighted utilization** (GWU) of a (generalized) itemset $X \subseteq AI$ is defined as the sum of the transaction utilities of transactions containing X, i.e. $GWU(X) = \sum_{T_c \in g(X)} TU(T_c)$.*

Property 1 (The GWU is a monotone upper bound on the utility measure). Let there be two itemsets X and Y such that Y is a specialization of a superset of X. The GWU of X is no less than its utility $(GWU(X) \geq u(X))$. Moreover, the GWU of X is no less than the utility of its supersets $(GWU(X) \geq u(Y) \forall Y \supset X)$.

Proof. If $Y \subset X$, then $u(Y) \leq GWU(Y) \leq GWU(X) < minutil$. If Y is a specialization of X, then $u(Y) \leq u(X) \leq GWU(X) < minutil$.

To reduce the search space, the next property is derived from Property 1.

Property 2 (Pruning the search space using the GWU). For any itemset X, if $GWU(X) < minutil$, then X is a low utility itemset as well as all its join-based and tax-based extensions.

For example, if $minutil = 50$, $GWU(\{c, d, e\}) = 45 < minutil$, and thus $\{c, d, e\}$ and all its extensions can be ignored, as they are not CLHUIs.

The third pruning technique utilizes information from the tax-utility-list of an itemset to prune its transitive extensions. This technique generalizes the remaining utility pruning technique of HUI-Miner for a taxonomy.

Property 3 (Pruning the search space using the remaining utility). Given the total order $\prec$ on a taxonomy, the *remaining utility upper bound* of an itemset X is the sum of the *iutil* and *rutil* values in its tax-utility-list. Formally, this upper bound is defined as $reu(X) = \sum_{e \in tuList(X)} (e.iutil + e.rutil)$. If $reu(X) <$ *minutil*, then X and all its tax-based extensions are not CLHUIs.

Example 6. If *minutil* = 25, $reu(\{c,d,e\}) = 19 <$ *minutil*, and thus $\{c,d,e\}$ and all its tax-based extensions can be ignored, as they are not CLHUIs.

The proposed CLH-Miner algorithm takes as input a transaction database with utility, a taxonomy τ, and the user-defined *minutil* threshold. CLH-Miner outputs the set of all CLHUIs. The pseudocode is shown in Algorithm 1. The algorithm first reads the database to calculate the GWU of each item and generalized item. Then, the algorithm identifies the set I^* of each item i whose GWU is no less than *minutil* and for which $GWU(g) \geq$ *minutil* if g is a generalized item such that $i \in Desc(g, \tau)$. Next the algorithm identifies the set GI^* of all generalized items having a GWU no less than *minutil*. Thereafter, all items not in I^* and GI^* and their extensions are ignored according to pruning Property 2. Then, item taxonomy levels and GWU values are used to calculate the total order $\prec$ on items (items first sorted by ascending taxonomy levels, and then by ascending GWU values). A second database scan is then performed to reorder items in transactions according to the $\succ$ order, and build the tax-utility-lists of each item $i \in I^*$ and generalized item $g \in GI^*$. Then, the depth-first search of itemsets starts by calling the recursive *Search* procedure with the set $TULs$ of generalized items of taxonomy level 1, and the *minutil* threshold.

Algorithm 1. The CLH-Miner algorithm

---

input : D: a transaction database, τ: a taxonomy, *minutil*: the threshold
output: the CLHUIs

1 Scan D and τ to calculate the GWU of each item $i \in I$ and each generalized item $g \in GI$;
2 $I^* \leftarrow \{i | i \in I \wedge GWU(i) \geq minutil \wedge \forall g \in GI$ such that $Desc(g, \tau) \ni i$,
    $GWU(g) \geq minutil\}$;
3 $GI^* \leftarrow \{g | g \in GI \wedge GWU(g) \geq minutil\}$;
4 Calculate the total order $\prec$ on $I^* \cup GI^*$;
5 Scan D and τ to build the $tuList$ of each item $i \in I^*$ and of generalized item $g \in GI^*$;
6 $TULs \leftarrow \{g | g \in GI^* \wedge level(g) = 1\}$;
7 Search $(TULs, minutil)$;

The *Search* procedure takes as input (1) $PULs$, a set of itemsets of the form Px that extend an itemset P with some item x (initially $P = \emptyset$), and (2) the user-specified *minutil* threshold. The search procedure is applied as follows. For each itemset $Px \in TULs$, if the sum of the *iutil* values of the $tuList(Px)$ is no less than *minutil*, then Px is a CLHUI and it is output. Then, *join-based extensions* of Px are generated by joining Px with all extensions Py of $TULs$ such that $x \succ y$ to generate extensions of the form Pxy containing $|Px| + 1$ items. If the GWU of the *join-based extensions* is larger or equal than *minutil*, they are added to $ExtensionsOfX$. The $tuList(Pxy)$ is then constructed by applying a

---

Algorithm 2. The *Search* procedure

---

input : *TULs*: a set of extensions of an itemset P, *minutil*: the threshold
output: the CLHUIs that are transitive extensions of P

1 **foreach** itemset $X \in TULs$ **do**
2   **if** $SUM(X.tuList.iutils) \geq minutil$ **then** Output X;
3       $ExtensionsOfX \leftarrow \emptyset$;
4   **foreach** itemset $Y \in TULs$ such that $X \prec Y$ **do**
5           $JoinExtension.tuList \leftarrow Construct(X, Y)$;
6   **if** $GWU(JoinExtension) \geq minutil$ **then**
         $ExtensionsOfX \leftarrow ExtensionsOfX \cup \{JoinExtension\}$;
7   **end**
8   **if** $SUM(X.tuList.iutils) + SUM(X.tuList.rutils) \geq minutil$ **then**
9   **foreach** itemset $T \in X.childs$ **do**
10              $TaxExtension.tuList \leftarrow Construct(X, T)$;
11   **if** $GWU(TaxExtension) \geq minutil$ **then**
             $ExtensionsOfX \leftarrow ExtensionsOfX \cup \{TaxExtension\}$;
12   **end**
13   **end**
14   Search $(ExtensionsOfX, minutil)$;
15 **end**

---

modified version of the *Construct* procedure of HUI-Miner to join the $tuList(P)$, $tuList(Px)$ and $tuList(Py)$. The difference with the original *Construct* procedure is that the *tax-utility-list* is added as a child item of y. This procedure is not shown due to the space limitation. Then, if the sum of *iutil* and *rutil* values in the $tuList(Px)$ is no less than *minutil*, *tax-based-extensions* of Px should be explored (by Property 3). This is done by replacing item x with each child item of x. Also, if the *GWU* of *tax-based extensions* is larger or equal to *minutil*, they are added to $ExtensionsOfX$. Then, a recursive call to the *Search* procedure with extensions of Px is done to calculate its utility and explore its extension(s). When the algorithm terminates all the cross-level high utility itemsets have been output. Since the algorithm only eliminate itemsets using Properties 2 and 3, it can be proven that the algorithm outputs all CLHUIs.

5 Experiment

To evaluate CLH-Miner's performance, experiments were done on a computer equipped with an Intel(R) Xeon(R) CPU W-2123 3.60 GHz, 32 GB of RAM, and running Windows 10. We compared the performance of CLH-Miner for CLHUI mining with ML-HUI Miner for multi-level HUIM, and HUI-Miner for HUIM. Algorithms were implemented in Java, and runtime and memory were measured using the standard Java API. Two real-world datasets were used, named *Liquor* ($|D| = 9,284, |I| = 2,626, |GI| = 77, T_{max} = 5, T_{avg} = 2.7, Level = 7$) and *Fruithut* ($|D| = 181,970, |I| = 1,265, |GI| = 43, T_{max} = 36, T_{avg} = 3.58, Level = 4$), where $|D|$ is the transaction count, $|I|$ is the item count, $|GI|$ is the generalized item count, T_{max} is the maximum transaction length, T_{avg} is the average transaction length, and *Level* is the maximum level. *Liquor* and *Fruithut* contain transactions from US liquor stores and grocery stores, respectively.

In a first experiment, *minutil* was varied to evaluate its influence on the performance of CLH-Miner. Three versions of CLH-Miner were compared: (1) CLH-Miner, (2) CLH-Miner without the GWU pruning strategy, and (3) CLH-Miner without the remaining utility pruning strategy. Results for runtime and peak memory usage are shown in Fig. 3 for the two datasets. It is found that as *minutil* increases runtime decreases. This is reasonable since the lower *minutil* is, the more CLHUIs are found, and the larger the search space is. For example, on *Liquor*, for *minutil* values of 40,000 and 50,000, there are 939 and 469 CLHUIs, and the runtimes are 47.76 and 19.83 s, respectively. It is also found that *GWU*-based pruning generally greatly decreases runtime, and memory usage. For instance, on *Fruithut*, when *minutil* = 20,000,000, CLH-Miner with *GWU*-based pruning is up to 3.74 times faster than CLH-Miner without it and uses up to 1.68 times less memory. Also, it is found that the *remaining utility* pruning strategy reduces memory usage, and slightly reduces runtime. For example, on *Fruithut*, when *minutil* = 25,000,000, memory is reduced by up to 47%. The *remaining utility* pruning strategy reduces runtime and memory because when the strategy is applicable, less extensions are stored in memory.

We also evaluated the proposed algorithm's scalability in terms of runtime and pattern count when transaction count is varied. For this experiment, *Fruithut* was divided into five parts and the algorithm's performance was measured after adding each part to the previous ones. Figure 4 shows the results for *minutil* = 10,000,000. It is found that runtime and pattern count increase with database size. This is because the utility of itemsets may be greater in a larger database, which increases the number of tax-utility-list to be created for itemsets, and more time is needed to search a larger search space.

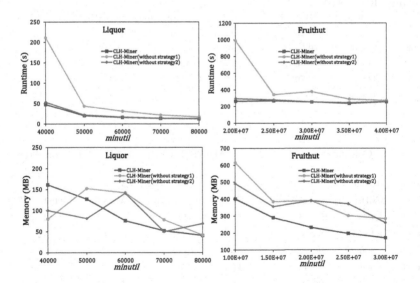

Fig. 3. Runtime and peak memory usage of three versions of CLH-Miner

In a third experiment, CLH-Miner's performance was compared with that of HUI-Miner and ML-HUI Miner. Table 3 and 4 show the runtime and peak memory usage for the two datasets. It is found that CLH-Miner takes more time than HUI-Miner and ML-HUI Miner, and that CLH-Miner generally consumes more memory than HUI-Miner and ML-HUI Miner on *Liquor*, and less memory than ML-HUI Miner on *Fruithut*. This is reasonable since the proposed problem of CLHUI mining has a much larger search space than HUIM and multi-level HUIM. Thus, more patterns are considered and kept in memory.

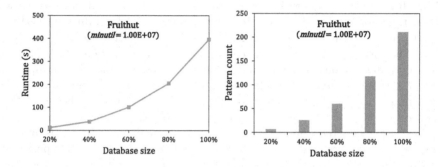

Fig. 4. Scalability of CLH-Miner when variying the database size

Table 3. Comparison of CLH-Miner, ML-HUI Miner and HUI-Miner on *Liquor*

minutil	Runtime (s)			Memory (MB)		
	CLH-Miner	ML-HUI Miner	HUI-Miner	CLH-Miner	ML-HUI Miner	HUI-Miner
9k	877.74	39.23	0.08	222	130	59
10k	738.77	40.65	0.08	146	67	104
11k	515.81	39.00	0.06	183	111	27
12k	448.82	39.76	0.05	229	180	73
13k	406.77	40.74	0.05	150	316	118

Table 4. Comparison of CLH-Miner, ML-HUI Miner and HUI-Miner on *Fruithut*

minutil	Runtime (s)			Memory (MB)		
	CLH-Miner	ML-HUI Miner	HUI-Miner	CLH-Miner	ML-HUI Miner	HUI-Miner
4E+06	1,198.20	76.36	1.43	938	378	140
5E+06	1,166.74	83.26	1.16	778	1,788	129
6E+06	806.20	92.22	1.03	668	1,892	121
7E+06	694.89	97.96	0.69	581	2,162	100
8E+06	603.69	96.70	0.55	520	1,803	92

Table 5. Comparison of pattern count on *Liquor* and *Fruithut*

Liquor				Fruithut			
minutil	CLHUIs	Multi-level HUIs	HUIs	minutil	CLHUIs	Multi-level HUIs	HUIs
10k	5537	234	241	4E+06	1248	55	11
20k	2136	119	86	5E+06	833	45	8
30k	1391	79	43	6E+06	586	36	4
40k	939	56	19	7E+06	435	30	2
50k	469	49	12	8E+06	331	23	1

The number of pattern found by the algorithms was also compared for different *minutil* values. Results are shown in Table 5. It is observed that CLH-Miner finds more patterns than HUI-Miner (HUIs) and ML-HUI Miner (Multi-level HUIs). For example, for *Liquor* and *minutil* = 50,000, the number of CLHUIs is 9.57 times and 39.08 times greater than that of multi-level HUIs and HUIs, respectively. Hence, CLH-Miner finds some patterns that HUI-Miner and ML-HUI Miner cannot find. For example, for *minutil* = 500,000, CLH-Miner can find interesting HUIs containing items of multiple taxonomy levels in *Liquor* such as {*Cordials& Liqueurs, Neutral Grain Spirits*} and {*Liqueurs, Distilled Spirits Specialty*} that HUI-Miner and ML-HUI Miner cannot find.

6 Conclusion

This paper has defined a novel problem of mining cross-level HUIs, studied its properties and presented a novel algorithm named CLH-Miner to efficiently mine all cross-level HUIs. The algorithm integrates novel pruning strategies to reduce the search space by taking advantages of the relationships between abstraction levels. An extensive experimental evaluation has shown that the algorithm has excellent performance, that optimizations improves its performance, and that interesting patterns are found in real-life retail data. Source code and datasets of CLH-Miner are available in the SPMF data mining library [4]. In future work, we will generalize the proposed problem to consider using a hierarchy in other pattern mining problems such as episode mining [8,9], subgraph mining [3], and high-average utility mining [24].

References

1. Cagliero, L., Cerquitelli, T., Garza, P., Grimaudo, L.: Misleading generalized itemset discovery. Expert Syst. Appl. **41**, 1400–1410 (2014)
2. Cagliero, L., Chiusano, S., Garza, P., Ricupero, G.: Discovering high-utility itemsets at multiple abstraction levels. In: Proceedings of 21st European Conference on Advances in Databases and Information Systems, pp. 224–234 (2017)

3. Fournier-Viger, P., Cheng, C., Lin, J.C.-W., Yun, U., Kiran, R.U.: TKG: efficient mining of top-K frequent subgraphs. In: Madria, S., Fournier-Viger, P., Chaudhary, S., Reddy, P.K. (eds.) BDA 2019. LNCS, vol. 11932, pp. 209–226. Springer, Cham (2019). https://doi.org/10.1007/978-3-030-37188-3_13

4. Fournier-Viger, P., et al.: The SPMF open-source data mining library version 2. In: Berendt, B., et al. (eds.) ECML PKDD 2016. LNCS (LNAI), vol. 9853, pp. 36–40. Springer, Cham (2016). https://doi.org/10.1007/978-3-319-46131-1_8

5. Fournier-Viger, P., Chun-Wei Lin, J., Truong-Chi, T., Nkambou, R.: A survey of high utility itemset mining. In: Fournier-Viger, P., Lin, J.C.-W., Nkambou, R., Vo, B., Tseng, V.S. (eds.) High-Utility Pattern Mining. SBD, vol. 51, pp. 1–45. Springer, Cham (2019). https://doi.org/10.1007/978-3-030-04921-8_1

6. Fournier-Viger, P., Lin, J.C.W., Vo, B., Chi, T.T., Zhang, J., Le, B.: A survey of itemset mining. WIREs Data Min. Knowl. Discov. 7(4), e1207 (2017)

7. Fournier-Viger, P., Wu, C.-W., Zida, S., Tseng, V.S.: FHM: faster high-utility itemset mining using estimated utility co-occurrence pruning. In: Andreasen, T., Christiansen, H., Cubero, J.-C., Raś, Z.W. (eds.) ISMIS 2014. LNCS (LNAI), vol. 8502, pp. 83–92. Springer, Cham (2014). https://doi.org/10.1007/978-3-319-08326-1_9

8. Fournier-Viger, P., Yang, P., Lin, J.C.-W., Yun, U.: HUE-span: fast high utility episode mining. In: Li, J., Wang, S., Qin, S., Li, X., Wang, S. (eds.) ADMA 2019. LNCS (LNAI), vol. 11888, pp. 169–184. Springer, Cham (2019). https://doi.org/10.1007/978-3-030-35231-8_12

9. Fournier-Viger, P., Yang, Y., Yang, P., Lin, J.C.W., Yun, U.: Tke: Mining top-k frequent episodes. In: Proceedings of 33rd International Conference on Industrial, Engineering and Other Applications of Applied Intelligent Systems, pp. 832–845. Springer (2020)

10. Han, J., Fu, Y.: Mining multiple-level association rules in large databases. IEEE Trans. Knowl. Data Eng. 11, 798–804 (1999)

11. Hashem, T., Ahmed, C.F., Samiullah, M., Akther, S., Jeong, B.S., Jeon, S.: An efficient approach for mining cross-level closed itemsets and minimal association rules using closed itemset lattices. Expert Syst. Appl. 41, 2914–2938 (2014)

12. Hipp, J., Myka, A., Wirth, R., Güntzer, U.: A new algorithm for faster mining of generalized association rules. In: Proceedings of 2nd Pacific-Asia Conference on Knowledge Discovery and Data Mining, pp. 74–82 (1998)

13. Liu, Y., Keng Liao, W., Choudhary, A.N.: A two-phase algorithm for fast discovery of high utility itemsets. In: Proceedings of 9th Pacific-Asia Conference on Knowledge Discovery and Data Mining, pp. 689–695 (2005)

14. Lui, C.L., Chung, K.F.L.: Discovery of generalized association rules with multiple minimum supports. In: Proceedings of 4th European Conference Principles of Data Mining and Knowledge Discovery, pp. 510–515 (2000)

15. Luna, J.M., Fournier-Viger, P., Ventura, S.: Frequent itemset mining: a 25 years review. WIREs Data Min. Knowl. Discov. 9(6), e1329 (2019)

16. Mao, Y.X., Shi, B.L.: AFOPT-Tax: an efficient method for mining generalized frequent itemsets. In: Proceedings of 2nd International Conference on Intelligent Information and Database Systems, pp. 82–92 (2010)

17. Leong Ong, K., Ng, W.K., Lim, E.P.: Mining multi-level rules with recurrent items using fp'-tree. In: Proceedings of 3rd International Conference Information Communication and Signal Processing (2001)

18. Pramudiono, I.: Fp-tax: tree structure based generalized association rule mining. In: Proceedings of ACM/SIGMOD International Workshop on Research Issues on Data Mining and Knowledge Discovery, pp. 60–63 (2004)

19. Qu, J.-F., Liu, M., Fournier-Viger, P.: Efficient algorithms for high utility itemset mining without candidate generation. In: Fournier-Viger, P., Lin, J.C.-W., Nkambou, R., Vo, B., Tseng, V.S. (eds.) High-Utility Pattern Mining. SBD, vol. 51, pp. 131–160. Springer, Cham (2019). https://doi.org/10.1007/978-3-030-04921-8_5

20. Rajkumar, N.D., Karthik, M.R., Sivanandam, S.N.: Fast algorithm for mining multilevel association rules. In: Proceedings of 2003 TENCON Conference on Convergent Technologies for Asia-Pacific Region, vol. 2, pp. 688–692 (2003)

21. Srikant, R., Agrawal, R.: Mining generalized association rules. In: Proceedings of 21th International Conference on Very Large Data Bases (1995)

22. Sriphaew, K., Theeramunkong, T.: A new method for finding generalized frequent itemsets in generalized association rule mining. In: Proceedings of 7th International Conference Knowledge Based Intelligent Information and Engineering Systems, pp. 476–484 (2002)

23. Xu, Y., Zeng, M., Liu, Q., Wang, X.: A genetic algorithm based multilevel association rules mining for big datasets. Math. Prob. Eng. **2014**, 9 (2014)

24. Yun, U., Kim, D., Yoon, E., Fujita, H.: Damped window based high average utility pattern mining over data streams. Knowl.-Based Syst. **144**, 188–205 (2018)

Efficient Mining of Pareto-Front High Expected Utility Patterns

Usman Ahmed[1], Jerry Chun-Wei Lin[1(✉)], Jimmy Ming-Tai Wu[2],
Youcef Djenouri[3], Gautam Srivastava[4], and Suresh Kumar Mukhiya[1]

[1] Department of Computer Science, Electrical Engineering and Mathematical
Sciences, Western Norway University of Applied Sciences, 5063 Bergen, Norway
{usman.ahmed,skmu}@hvl.no, jerrylin@ieee.org
[2] School of Computer Science and Engineering, Shandong University of Science
and Technology, Qingdao, China
wmt@wmt35.idv.tw
[3] SINTEF Digital, Mathematics and Cybernetics, Oslo, Norway
youcef.djenouri@sintef.no
[4] Department of Mathematics and Computer Science, Brandon University,
Brandon, Canada
srivastavag@brandonu.ca

Abstract. In this paper, we present a model called MHEUPM to efficiently mine the interesting high expected utility patterns (HEUPs) by employing the multi-objective evolutionary framework. The model considers both uncertainty and utility factors to discover meaningful HEUPMs without requiring pre-defined threshold values (such as minimum utility and minimum uncertainty). The effectiveness of the model is validated using two encoding methodologies. The proposed MHEUPM model can discover a set of HEUPs within a limited period. The efficiency of the proposed model is determined through rigorous analysis and compared to the standard pattern-mining methods in terms of hypervolume, convergence, and number of the discovered patterns.

Keywords: Evolutionary computation · Uncertain databases · Data mining · High expected utility pattern mining · Multi-objective optimization

1 Introduction

The pattern-mining algorithms [1] have proven as one of the effective methods to discover knowledge for decision making. Among all the existing algorithms, Apriori [1] is the most widely employed technique which identifies associativity between transactions of databases then discovers the set of association rules (ARs) according to the pre-defined minimum support and confidence thresholds. Association rule mining (ARM) is used to find the ARs in the pattern mining area, which has been employed in varying scenarios of different disciplines and

© Springer Nature Switzerland AG 2020
H. Fujita et al. (Eds.): IEA/AIE 2020, LNAI 12144, pp. 872–883, 2020.
https://doi.org/10.1007/978-3-030-55789-8_74

played a significant role in knowledge discovery and decision making. However, Apriori overlooks potential factors of items such as importantness, quantity, weight, or interestingness.

This limitation of Apriori has been overcome by high-utility itemset mining (HUIM) [2]. HUIMs is an evolving topic that contemplates both quantity and unit profit of items to derive high utility itemsets (HUIs). The prime focus of HUIM is to discover those patterns that hold high utility to the users. HUIM also utilizes threshold value for minimum utility. An item is said to be HUIs if its utility is equal to or maximum than the threshold value of minimum utility. This functionality of HUIM provides more insights into the items through which a profitable and meaningful relationship could be discovered for accurate decision making. Many works of HUIM have been extensively studied and developed.

However, most of the contemporary pattern mining algorithms require a threshold value. Deciding an appropriate value for a threshold is a non-trivial process since it requires the involvement of domain and knowledge experts to look into the issues to be avoided, such as "combinational explosion" and "rare-item". To overcome the above limitations, this paper presents 1). A novel model called **M**ulti-**O**bjective **H**igh **E**xpected **U**tility **P**attern **M**ining (MHEUPM) is presented to focus on considering both uncertainty and utility objects to mine potential high expected utility patterns (HEUPs) in the context of evolutionary computation from uncertain databases; 2). There is no need to have a predefined value for minimum utility threshold or minimum uncertain threshold to mine the patterns; rather, non-dominated patterns can be mined without predefined threshold; 3). Two encoding schemas are used to validate the potential of the proposed model; and 4). The outcomes revealed that proposed MHE-UPM outperforms the conventional pattern-mining algorithms in the context of convergence, hypervolume, and number of the derived patterns.

2   Related Work

Association rule mining (ARM) is one of the most suitable methods to discover knowledge from databases. Apriori [1] is the first ARM algorithm proposed by Agrawal and Srikant to capture relationships among items in a transaction. Various extensions [3] of Apriori have been presented to incorporate other useful aspects. The insights to the discovered patterns cannot be known based on the frequency value, i.e., the profitable of an itemset. High-utility itemset mining (HUIM) [3] is one of such methods identifying profitable items from the databases. The HUIM considers both quantity and unit profit of items to derive high-utility itemsets (HUIs). The downward closure property was not part of the original HUIM. Therefore, a two-phase model called transaction-weighted utilization (TWU) was presented to form the downward closure property by employing high transaction-weighted utilization itemsets (HTWUIs) to ensure the completeness and correctness of the identified HUIs. A vast amount of pattern-mining algorithms such as HUIM or ARM required a pre-defined value of minimum threshold to filter non-worthy or unpromising patterns. Some algorithms are presented to mine HUIs without predefining threshold value such as top-k HUIM [4,5].

Evolutionary computation (EC) is a method that follows a meta-heuristic approach to address optimization and NP-hard problems efficiently. The most widely adopted EC-based approaches include Genetic Algorithm (GA) [6–9], Particle Swarm Optimization (PSO) [10], and Ant-Colony Optimization (ACO) [11]. Subsequently, Lin et al. [10] refined the PSO-based model to mine HUIs. To verify different solutions pertaining to evolutionary progress, OR/NOR tree [12] was presented to produce accurate and complete HUIs. Another algorithm, named HUIM-ACS [13], was designed to identify HUIs efficiently. MOEA/D has convergence capabilities than other MOEA algorithms compared to the NSGA-II [6] An evolutionary progress model was developed by Zhang et al. [6] to mine the utility and frequent patterns. The model does not require prior knowledge and identifies non-dominated patterns. However, the model identifies a very small amount of patterns and misses some important patterns for decision-making.

3   Preliminaries and Problem Statements

Let us consider $I = \{i_1, i_2, \ldots, i_m\}$ as a set of items, and a set of transactions from uncertain databases are $D = \{T_1, T_2, \ldots, T_n\}$. An uncertain value is associated to each item of a transaction such as $uv(i_k, T_c)$. Examples of uncertain and quantitative databases are illustrated in Table 1. The items' unit profit in the database is also shown in the table.

Consider Tables 1 and 2 as an example that will be performed according to the proposed algorithm step-by-step. There are five transactions in the table $(T_1, T_2, T_3, T_4, T_5)$. For instance, transaction T_2 exhibits the items (c) and (d); and quantities of their purchase which are 4 and 2 respectively. In T_2, the uncertain values are 0.75 and 0.9. The unit of profit earned against each sold item of the transaction is shown in Table 2, i.e., \$8 profit is earned by a retailer as a profit of the sold item (a).

Table 1. The quantitative and uncertain database

T_{id}	Item: quantity, probability	Total utility
T_1	$(a{:}5, 0.3); (b{:}3, 0.40); (c{:}6, 0.9)$	\$97
T_2	$(c{:}4, 0.75); (d{:}2, 0.9)$	\$38
T_3	$(a{:}7, 1.0); (b{:}8, 1.0); (e{:}2, 0.75)$	\$90
T_4	$(a{:}3, 0.9); (c{:}1, 0.9)$	\$32
T_5	$(b{:}2, 1.0); (c{:}4, 0.95); (e{:}4, 1.0)$	\$58

Table 2. The unit profit of the items

Item	Profit
a	8
b	3
c	8
d	3
e	5

Definition 1 *(Item Utility in a Transaction).* *The utility of an item i_k is $u(i_k, T_c)$ in T_c transaction is defined as follows:*

$$u(i_k, T_c) = pr(i_k) \times q(i_k, T_c) \tag{1}$$

Definition 2 *(Itemset Utility in a Transaction).* *In a transaction T_c, the itemset X has the utility $u(X, T_c)$, wherein $i_k \in X \subseteq T_c$. It is defined as:*

$$u(X, T_c) = \sum_{ik \in X} u(i_k, T_c) \tag{2}$$

Definition 3 *(Itemset Utility in a Database).* *In the database D, the utility of X is denoted as $u(X)$ and defined as:*

$$u(X) = \sum_{X \subseteq T_c \wedge T_c \in D} u(X, T_c) \tag{3}$$

Definition 4 *(Itemset Uncertainty in a Transaction).* *An itemset X has an uncertain value in transaction which is denoted as $uc(X, T_c)$ and defined as:*

$$p(X, T_c) = \Pi_{x_i \in X} p(x_i, T_c) \tag{4}$$

Most of the contemporary generic pattern-mining algorithms concerned to validate the patterns using a threshold value for whether utility, uncertainty, confidence or support factor. Moreover, for pattern evaluation, it is a non-trivial process to define the appropriate threshold value. The problem statement of this paper is described below.

Problem Statement: This paper focuses on incorporating both utility and uncertainty factors of quantitative uncertain databases to mine the non-dominated high expected utility patterns without a pre-defined threshold value. To effectuate the process for handling big data in terms of minimized computation complexity, and mining the up-to-date information, the multi-objective evolutionary computation (MOEA) model is employed to solve the traditional limitations of pattern mining.

4   Designed MHEUPM Model

This section encompasses the details regarding the MOEA-based model known as MHEUPM to mine non-dominated high expected utility patterns (HEUP) from uncertain and quantitative databases. The proposed algorithm first considers both uncertainty and utility-based factors of uncertain databases to simultaneously derive the set of the non-dominated high expected utility patterns (HEUPs). These two measures, uncertainty and utility, have some conflicts. For instance, in some scenarios, patterns having high utility do not necessarily own high value for uncertainty and vice versa. The consideration of these two factors without requiring prior knowledge could be deemed as a two-objective optimization problem. The proposed model focuses on identifying only highly meaningful patterns for decision making than the existing approaches that produce a vast number of patterns based on a threshold value. The two objectives should be maximized as MHEUPM considers both utility and uncertainty as exhibited in Eq. 5. Details of the designed framework are explained as follow:

$$Max \, F(X) \, = \, \{max \, (utility(X), uncertainty \, (X)^T\} \qquad (5)$$

4.1   Initialization

The first step of proposed MHEUPM is an initialization of problem-specific initialization strategy. Two methods, **meta-itemset-selection** and **transaction-itemset-selection** are employed for population selection. 50% of the individuals from the population are selected based on utility or probability for the transaction-itemset-selection. This process validates that those individuals that have been selected are the ideal solutions that exist in the database. For remaining 50% of the individuals in the population of the meta-itemset-selection, based on the probability of uncertainty, only one item in a transaction is encoded. To rapidly determine the optimal solutions in the evolutionary progress (i.e., used in mutation and crossover operations), transaction-itemset can play a useful role. Therefore, in this step, meta-itemset's uncertain probability is computed first (Algorithm 1, lines 1 to 3), which is illustrated in Table 3. After that, the utility probability of the transaction-itemset is computed (Algorithm 1, lines 4 to 6)

Table 3. Meta-itemset-selection strategy

Itemset	Uncertainty	Uncertain probability
a	1.3	0.16
b	1.4	0.17
c	2.75	0.34
d	0.9	0.11
e	1.75	0.22

and the results are shown in Table 4. Afterwards, all 50% of the individuals from Tables 3 and 4 are produced to form a population (Algorithm 1, line 7). The initialization algorithm is exhibited in Algorithm 1.

Algorithm 1. MHEUPM: initialization

INPUT: The transaction data set D, population pop size, weight vector $\{w_1, w_2, \ldots, w_{pop}\}$, the size of neighbors n_s, crossover probability p_c, mutation probability p_m.

OUTPUT: Initialize individuals in a population.

1: **foreach** $i \in meta_{itemset}$ **do**
2:   | $M \leftarrow (\frac{uncertain(D_i)}{uncertain(D)})$.
3: **end foreach**
4: **foreach** $j \in transaction_{itemset}$ **do**

5:   | $T \leftarrow (\frac{utility(D_j)}{utility(D)})$ ▷ T are transaction itemsets with
                              utility probability

6: **end foreach**
7: $P \leftarrow initial(M, T, pop)$
8: **Return** P

Table 4. Transaction-itemset-selection strategy

T_{ID}	a	b	c	d	e	Utility	Utility probability
T_1	5	3	6	0	0	97	0.31
T_2	0	0	4	2	0	38	0.12
T_3	7	8	0	0	2	90	0.29
T_4	3	0	1	0	0	32	0.10
T_5	0	2	4	0	4	58	0.18

- **Encoding:** We have critically analyzed different encoding methods for evolutionary computation from different domains and applications. It has been reported that binary and value encoding methods have widely been employed as they hold the potential for higher convergence and diversity of the derived solutions [6]. Our model also adopts the binary and value encoding methods, which is performed after population initialization (Algorithm 2, line 1). In binary encoding, a value of 1 is defined for an item in encoding schema if the item exists in a transaction; otherwise, we set it as 0. The quantity of the items is set as the encoding value in the schema for the value encoding. In transactions, the value of a purchased item is considered as quantity. The value of uncertainty can also be considered as an encoding value. However, no direct relationship exists between uncertain values.

- **Reference point:** The Tchebycheff reference point [6] is employed in the proposed MHEUPM to discover non-dominated solutions relying on uncertainty and utility factors (Algorithm 2, line 2). The Tchebycheff value is utilized to validate the significance of the encoding schema [6], which is described in Eq. 6. Tchebycheff value is used to produce useful sets that can assist in identifying non-dominated high expected utility patterns.

$$min \, g^{te}(X|w_i, z^*) \leftarrow max_{j=1}^2 \left\{ w_i^j \cdot (|F_j(X) - z^*|) \right\} \tag{6}$$

To show the progress for identifying the reference point, an example is illustrated below. Let us consider that the *pop* has the number of sub-problem and set W to be a set of even weights such that $(w_1, w_2, \ldots, w_n)$, wherein $w_i^1 + w_i^2 = 1$. Let us suppose encoding individual like $\{1, 0, 1, 0, 0\}$ having utility value as 22 and 03 value for uncertainty value. Also, assume that 107 is the maximum utility of the database and 0.715 is the value of maximum uncertainty. Relying on the Eq. 6, the maximum utility and uncertainty values of the individuals will be considered as reference point z^* as shown in (Algorithm 2, lines 3 to 5). Let us consider that w_i^1 is 0.34 and w_i^2 is 0.66 and these values are considered as even weights for uncertainty and utility respectively. The utility value for an individual is computed using Eq. 6 as $0.34 \times |22 - 107| = 28.9$, and the uncertainty value of an individual is $0.66 \times |0.13 - 0.715| = 0.38$. Thus, the $max(28.9, 0.38)$ produced 28.9, which is the Tchebycheff value.

- **Neighbor exploration:** After performing the steps stated above, the neighbors for each individual in the transaction are identified using (Algorithm 2, lines 3 to 5). For each weight vector w_i ($i < pop$), wherein *pop* is the population size which was initialized at the start. The Euclidean distance for all individuals between weight vector w_i in the population (*pop*) is computed as follows: $\sqrt{(|u(ind) - w_i^1|^2 + |p(ind) - w_i^2|^2)}$. The neighbor performs two-way crossover for the mutated child produced while population evolution. The method of computing calculation of the reference point and the neighbor exploration is explained in Algorithm 2.

Algorithm 2. MHEUPM: reference point calculation and neighbor exploration

---

INPUT: P, the size of neighbors n_s, crossover probability p_c and mutation probability p_m.
OUTPUT: Population with neighbors.
1: $P \leftarrow Encode(P)$. $z^* \leftarrow$ initialize reference point.
2: **for** all $p \in pop$ **do**
3: | $N_i \leftarrow P$.
4: **end for**
5: **Return** N_i.

4.2   Population Evolution

In the evolution progress, the MOEA/D method is utilized in the proposed MHEUPM model (Algorithm 3, lines 1 to 6). One individual $Pop_i^{'}$ is randomly selected from N_j (i.e., neighbors of Pop_i for all the individuals in Pop_i. After that, among Pop_i, Pop'_i, two-way crossover operator is utilized. The mutation operator is also incorporated in the same manner. The *Tchebyshev* value for offsprings is computed and compared to the value N hold by the neighbor. The good value of *Tchebyshev* value of the individual is harnessed to substitute the Pop_i with the offspring, and the reference point is updated z^* as described in Algorithm 3, lines 2 to 5. The same process is performed iteratively until achieving the termination of max_{gen}. The *Tchebyshev* value [6] is one alternative approach that identifies non-dominated solutions. This step is to diminish the need for prior defined thresholds value.

4.3 Population Selection

After performing all the above steps, the final population is formed. For the final results, sorting of all individuals in the population is done of whom non-dominated solutions are discovered. These steps are described in Algorithm 3.

Algorithm 3. MHEUPM: population evolution

INPUT: P, N_i and number of generations max_{gen}
OUTPUT: Non-dominated solutions
1: **while** max_{gen} **do**
2:   | **for** *all* $i \in pop$ **do**
3:   | | $P'_i \leftarrow$ *Randomly select an individual from* N_i.
4:   | | *child* $\leftarrow CrossMutation(P'_i, P_i)$.
5:   | **end for**
6: **end while**
7: $Final_{solution} \leftarrow SelectNonDominatedItemsets(P)$.
8: **Return** $Final_{solution}$.

---

5   Experimental Evaluation

This section presents the comparison results of the MHEUPM model with two baseline algorithms, the U-Apriori algorithm [14], and EFIM [15]. The U-Apriori [14] utilizes the minimum uncertainty threshold, and EFIM [15] utilizes the minimum utility threshold. The implementation of the proposed algorithm was implemented by python, and the baseline methods have been developed by Java language. The platforms for experiments conduction were Windows 10 PC with AMD Ryzen 5 PRO 3500U processor and 16 GB of RAM. The value for population size was 100, the value of the maximal generation was 50, the size of neighbors was 10, the value for crossover probability was 1.0 and 0.01 for mutation probability.

To evaluate the approach, we use two respectively called chess and mushroom databases for evaluation. These databases can be accessed using the SPMF library [16]. For each transaction itemset, the individual probability was randomly assigned between 0 to 1. To evaluate the effectiveness of the discovered patterns, two renowned measurements hypervolume (HV) and Coverage (Cov) were employed [6] . HV ensures the convergence of the derived solutions [6], and the hypervolume analyzes the distribution and convergence of the mined item-sets within the search space of the objective, i.e., *in our case, the values of uncertain and utility.*The convergence equation is defined as $Cov = N_d/N$ where N is the total number of discovered patterns and N_d is the different patterns.

5.1   Encoding Schema Analysis

The comparison of two encoding schemas, i.e., value and binary encoding, is performed in terms of different generations for the Cov and HV metrics. The evaluation of HV and Cov is done to validate the effectiveness of two encoding schemas. The outcomes for the two variants of encoding schemas are exhibited in Fig. 1 and 2. The high value of CoV and HV of an algorithm shows its significance, among others. From Fig. 1, the value encoding is converged with the generation having low numbers. The performance of binary encoding becomes better if multiple generations are attained.

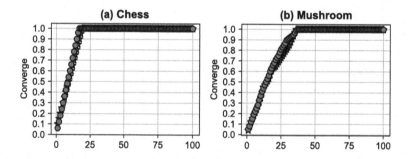

Fig. 1. The Convergence of two encoding schema

Figure 2 represents the HV that determines similar behavior as Cov. From the results it can showed that the performance of two encoding schemas.

5.2   Pattern Analysis

To the best of our knowledge, there does not exist any algorithm on the evolutionary progress which incorporates utility and uncertainty factors together. The performance of the proposed model is compared to standard mining methods like U-Apriori [14] and EFIM [15] in terms of HV and Cov. Since these methods require predefined value of a threshold, therefore, for utility or uncertainty, we have set threshold values as 20% to 80% respectively with 20% increment

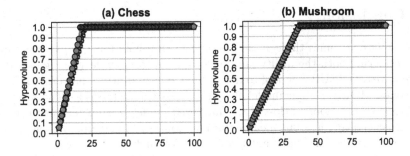

Fig. 2. The HV of two encoding schema

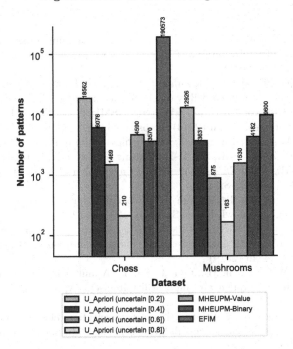

Fig. 3. The number of valid patterns

each time for the U-Apriori, and set 20% for the EFIM since the resting thresholds (i.e., 40%, 60%, and 80%) generate empty patterns (HUIs) in the mining progress. The quantity of identified patterns is exhibited in Fig. 3.

The patterns produced by MHEUPM are 5 to 10 times less than the patterns produced by U-Apriori and EFIM. These patterns are generated without a threshold value. On the other hand, *U-Apriori* produced a minimum of 12,926 and a maximum of 35,035. Similarly, *EFIM* produced a minimum of 67 and a maximum of 190,573. These outcomes indicate that U-Apriori and EFIM will consume more time than the proposed algorithm. The proposed algorithm easily gets converged and produced comparatively less and useful patterns for decision-making.

6   Conclusion

This paper overcomes the generic pre-defined threshold issue in pattern mining by considering both the utility and uncertainty factors by developing evolutionary progress MHEUPM model to derive high expected-utility patterns (HEUPs) relying on MOEA/D. The model also introduces two methods to generate useful individuals in the population of the proposed model. The experimental results yield that creating less but meaningful non-dominated patterns without threshold value can play an important role in decision-making.

References

1. Agrawal, R., Srikant, R., et al.: Fast algorithms for mining association rules. In: Proceedings 20th International Conference Very Large Data Bases, VLDB, vol. 1215, pp. 487–499 (1994)
2. Fournier-Viger, P., Wu, C.W., Zida, S., Tseng, V.S.: FHM: faster high-utility itemset mining using estimated utility co-occurrence pruning. In: Andreasen, T., Christiansen, H., Cubero, J.C., Raś, Z.W. (eds.) ISMIS 2014. LNCS (LNAI), vol. 8502, pp. 83–92. Springer, Cham (2014). https://doi.org/10.1007/978-3-319-08326-1_9
3. Lin, Y.C., Wu, C.W., Tseng, V.S.: Mining high utility itemsets in big data. In: Cao, T., Lim, E.P., Zhou, Z.-H., Ho, T.-B., Cheung, D., Motoda, H. (eds.) PAKDD 2015. LNCS (LNAI), vol. 9078, pp. 649–661. Springer, Cham (2015). https://doi.org/10.1007/978-3-319-18032-8_51
4. Ryang, H., Yun, U.: Top-k high utility pattern mining with effective threshold raising strategies. Knowl.-Based Syst. **76**, 109–126 (2015)
5. Chennupati Saideep, K.R., Toyoda, M., Kitsuregawa, M.: Finding periodic-frequent patterns in temporal databases using periodic summaries. Data Sci. Pattern Recogn. **3**(2), 24–46 (2019)
6. Zhang, L., Fu, G., Cheng, F., Qiu, J., Su, Y.: A multi-objective evolutionary approach for mining frequent and high utility itemsets. Appl. Soft Comput. **62**, 974–986 (2018)
7. Frank, J., Cooper, K.: Multiobjective feature selection: classification using educational datasets in an ensemble validation scheme. Data Sci. Pattern Recogn. **3**(1), 9–34 (2019)
8. Xingsi Xue, H.Y., Zhang, J.: Using population-based incremental learning algorithm for matching class diagrams. Data Sci. Pattern Recogn. **3**(1), 1–8 (2019)
9. Ahmed, U., Aleem, M., Noman Khalid, Y., Arshad Islam, M., Azhar Iqbal, M.: RALB-HC: a resource-aware load balancer for heterogeneous cluster. Concurrency Comput. Pract. Exp., e5606 (2019)
10. Lin, J.C.W., et al.: Mining high-utility itemsets based on particle swarm optimization. Eng. Appl. Artif. Intell. **55**, 320–330 (2016)
11. Dorigo, M., Stützle, T.: Ant colony optimization: overview and recent advances. In: Gendreau, M., Potvin, J.Y. (eds.) Handbook of Metaheuristics, pp. 311–351. Springer, Boston (2019). https://doi.org/10.1007/978-1-4419-1665-5_8
12. Lin, J.C.W., Yang, L., Fournier-Viger, P., Hong, T.P., Voznak, M.: A binary PSO approach to mine high-utility itemsets. Soft Comput. **21**(17), 5103–5121 (2017)
13. Wu, J.M.T., Zhan, J., Lin, J.C.W.: An ACO-based approach to mine high-utility itemsets. Knowl.-Based Syst. **116**, 102–113 (2017)

14. Hooshsadat, M., Bayat, S., Naeimi, P., Mirian, M.S., Zaiane, O.R.: UAPRIORI: an algorithm for finding sequential patterns in probabilistic data. In: Uncertainty Modeling in Knowledge Engineering and Decision Making, pp. 907–912. World Scientific (2012)
15. Zida, S., Fournier-Viger, P., Lin, J.C.W., Wu, C.W., Tseng, V.S.: EFIM: a highly efficient algorithm for high-utility itemset mining. In: Sidorov, G., Galicia-Haro, S.N. (eds.) MICAI 2015. LNCS (LNAI), vol. 9413, pp. 530–546. Springer, Cham (2015). https://doi.org/10.1007/978-3-319-27060-9_44
16. Fournier-Viger, P., et al.: The SPMF open-source data mining library version 2. In: Berendt, B., et al. (eds.) ECML PKDD 2016. LNCS (LNAI), vol. 9853, pp. 36–40. Springer, Cham (2016). https://doi.org/10.1007/978-3-319-46131-1_8

Maintenance of Prelarge High Average-Utility Patterns in Incremental Databases

Jimmy Ming-Tai Wu[1], Qian Teng[1], Jerry Chun-Wei Lin[2(✉)],
Philippe Fournier-Viger[3], and Chien-Fu Cheng[4]

[1] School of Computer Science and Engineering, Shandong University of Science
and Technology, Qingdao, China
wmt@wmt35.idv.tw, 2970618915@qq.com
[2] Department of Computer Science, Electrical Engineering and Mathematical
Sciences, Western Norway University of Applied Sciences, Bergen, Norway
jerrylin@ieee.org
[3] School of Humanities and Social Sciences, Harbin Institute of Technology
(Shenzhen), Shenzhen, China
philfv8@yahoo.com
[4] Department of Computer Science and Information Engineering,
Tamkang University, New Taipei, Taiwan
cfcheng@mail.tku.edu.tw

Abstract. Most existing data mining algorithms focused on mining the
information from the static database. In this paper, the principle of pre-
large is used to update the newly discovered HAUIs and reduce the time
of the rescanning process. To further improve the performance of the suggested algorithm, two new upper-bounds are also proposed to decrease
the number of candidates for HAUIs. The experimental results show the
proposed algorithm has excellent performance and good potential to be
applied in real applications.

Keywords: Pre-large · High average-utility itemset mining · Dynamic
database · Lead partial maximal utility upper bound · Incremental

1 Introduction

Knowledge discovery in database (KDD) [1,4,7] and data analytics [14,25,27]
are the process of extracting practical, potentially and useful knowledge from
the original for analytics, predication or recommendation. The first Apriori [1]
utilizes the generate-and-check approach for mining ARs through a level-wise
process in KDD. Many extensions have provided the information needed in
different knowledge domains, like HUIM [8,29] or HAUIM [13], based on the
Apriori-like or DC property. However, frequent itemset mining (FIM) only considers an item's frequency and assumes that all items occur at most once in a
transaction. High-utility itemset mining (HUIM) has been considered to be an

© Springer Nature Switzerland AG 2020
H. Fujita et al. (Eds.): IEA/AIE 2020, LNAI 12144, pp. 884–895, 2020.
https://doi.org/10.1007/978-3-030-55789-8_75

effective decision-making method for displaying profitable products and itemsets. That is, the quantity and utility of the itemsets are concerned with revealing high-utility itemsets (HUIs). However, HUIMs yield a more difficult problem than FIM because the utility measure is not anti-monotonic. To overcome this problem, Liu et al. subsequently implemented a transaction weighted utilization (TWU) model to retain the transaction-weighted downward closure (TWDC) property. Therefore, it can further reduce the search space for revealing HUIs. A high-utility pattern (HUP)-tree [18] was proposed by Lin et al. to decrease the computational costs considerably. Several previous methods [19,23,26,28,30] have been extensively studied and discussed, and most of them are focused on the TWU model to reveal HUIs.

Even though HUIM can find more information in the decision-making process, the utility of an itemset is increased especially when the length of the pattern is very long. High average-utility itemset mining (HAUIM) [13] was investigated to evaluate high utility patterns by considering the length of an itemset. To improve mining performance, Lin et al. proposed the high average utility pattern (HAUP) tree structure [17] to maintain the 1-HAUUBIs in the compressed tree structure. The current HUIM or HAUIM algorithms were developed on the maintenance of the static database, but size of the database has changed dynamically in realistic situations. In this article, we utilize the pre-large concept [15] for efficiently handling incremental mining while some transactions are inserted. Two new upper-bounds and a linked list structure are proposed for the incremental operation of transaction insertion. Experiments showed that the designed model outperforms the generic HAUIM algorithms running on the static databases.

2   Literature Review

The basic association-rule mining (ARM) algorithm [1] is based on two minimum thresholds to find the relationship between the itemsets, forming as the association rules (ARs), and the original algorithm is called Apriori [1]. However, Apriori only works with binary databases; therefore, other influences, such as weight, importance, and quantity, are ignored in ARM. To find more meaningful information from the discovered knowledge, HUIM (high-utility itemset mining) [18] was designed to evaluate two factors that are unit profit and the quantity of items to reveal useful information in the database. However, traditional HUIM cannot solve the problem of combined explosions. Therefore, the search space is too large to find the required HUIs.

To solve this problem and reduce the size of search space during the mining process, a TWU model [31] that retains high transaction-weighted utilization itemsets (HTWUIs) is proposed to maintain the downward closure property. Moreover, utility pattern (UP)-growth+ [26] was proposed to apply the UP-tree structure and several pruning strategies for mining HUIs to accelerate the mining efficiency. Besides, a utility-list (UL)-structure and the HUI-Miner algorithm [17] were designed to generate the k-candidate HUIs easily through a straightforward join operation. Several HUIM [19,23,28,30] experiments have been extensively discussed, and most of them focused on the TWU model to reduce the number of candidates for mining the HUIs.

During HUIM, the value of the utility increases with the length of the itemset. Therefore, any combination of an HUI, such as caviar or diamonds, is also called a HUI in the databases. High average-utility itemset mining (HAUIM) [13,17] was extended from HUIM regarding the length of the itemset for evaluation. The first algorithm in HAUIM was called TPAU [13], which discoveredthe Apriori approach by presenting the *auub* model to keep the downward closure property. Lin et al. then presented a HAUP-tree [17] to keep 1-HTWUIs in a compressed tree structure, thereby significantly reducing the computational cost. Several extensions of HAUIM [20,24] were respectively studied and most of them relied on the *auub* model to find the set of HAUIs.

For traditional pattern mining, including ARM [1], HUIM [29,31], and sequential pattern mining [7], they mostly reply on mining the meaningful patterns from the static databases. When the database size is changed, such as insertion, deletion, or modification, for the batch and generic algorithms, even if there is a small change (i.e., tiny 1–5 transactions are inserted into the database), they must process the updated database. Cheung et al. developed the Fast UPdate (FUP) concept [3] to handle dynamic data mining for transaction insertion to update the discovered frequent itemsets. Considering the discvoered patterns from both original databases and newly inserted transactions, the found frequent itemsets are divided into four parts, and each part is then maintained and processed using the designed approaches. This concept has been utilized in different domains for knowledge discovery, such as ARM [10], sequential pattern mining [21], HUIM [16], and HAUIM [12].

To better maintain the discovered knowledge and avoid the multiple database scans, Hong et al. developed the pre-large concept [11,15], which is used to set up two thresholds for knowledge maintenance. These two thresholds (upper and lower) are used to keep not only the frequent itemsets but also the pre-large itemsets for pattern maintenance. Thus, the pre-large itemsets will be kept as the buffer to avoid multiple database scans, especially for a pattern which was small in the original database but is large in the newly inserted transactions. An equation is also defined to determine whether the number of the newly inserted transactions is less than the safety bound. Some cases for database rescans can be avoided, but the completeness and correctness of the discovered knowledge can still be maintained. Figure 1 shows the pre-large concept with nine cases.

3   Proposed Incremental Framework

The designed Apriori-based HAUP with pre-large concept (APHAUP) is based on Apriori [1] includes some specific improved designs to increase the performance and to fit in with an incremental environment. First, to reduce the size of component itemsets, a stricter upper-bound than *auub*, called partial maximal utility upper bound (*pmuub*), is designed. It utilizes the similar concept to *mfuub*.

In the traditional Apriori, a component itemset is a large-n itemset and used to produce the candidate itemsets where the length is $n+1$. In APHAUP, a component itemset is called *hpmuubi* (high *pmuub* itemset), which means its utility is larger than the *pmuub*. Second, APHAUP selects a subset from *hpmuubi*, called *lhpmuubi* (lead-*hpmuubi*). In fact, APHAUP produces these two sets at the same time, but it applies a smaller upper-bound than *pmuub*. Therefore, *lhpmuubis* is a subset for *hpmuubis* in the proposed method. This smaller upper-bound is called *lhpmuub* (lead-*hpmuub*). The proposed *lhpmuubi* can further reduce the size of the candidate itemsets effectively. Third, to make sure that each transaction will be scanned at most one time in each round, a linked list structure is proposed. Finally, the incremental updating process for the proposed algorithm is proposed. The detailed definitions and algorithms for each part will be described in the following subsections.

3.1 Designed Partial Maximal Utility Upper-Bound Models

In previous algorithms, *auub* was widely applied to reveal HAUIs. However, *auub* is too large and keeps many unpromising itemsets for later mining progress. The proposed *pmuub* was designed to solve this issue and the detailed definition is given below.

Definition 1. *An active item (ai) and active itemset (ais) in the proposed Apriori-like process. Thus, in HAUIM, APHAUP follows the traditional Apriori-like model, and for each iteration, it generates the candidate hpmuubis where the length is n from the hpmuubis including $n-1$ items. Assuming that APHAUP handles the process to estimate hpmuubis where the length is n, then an active item (ai) indicates that this item exists, at least, in one of the hpmuubi where the length is $n-1$. The definition of ais_n (active itemset of the candidate hpmuubis with the length is n), which can be defined as:*

$$ais_n = \left\{ ai \in x \,\middle|\, x \in hpmuubis^{n-1} \right\}, \tag{1}$$

where $hpmuubis^{n-1}$ is the hpmuubis in which the length is $n-1$.

Definition 2. *Assume an itemset $i = \{i_1, i_2, \ldots, i_n\}$ in a transaction $t = \{i_t, u_t\}$. The $i_t = \{i_{t1}, i_{t2}, \ldots, i_{tm}\}$ is the purchase items in this transaction, $u_t = \{u_t(i_{t1}) = u_{t1}, u_{t2}, \ldots, u_{tm}\}$ is the corresponding utility for each item in this transaction, and the active itemsets is ais. Thus, the remaining maximal utility of the active itemset in a transaction (rmua) of i in t is denoted as $rmua(i,t)$, which is described as:*

$$rmua(i,t) = max \left\{ u_t(i) \,\middle|\, i \in (ais \cap i_t) \setminus i \right\}. \tag{2}$$

Definition 3. *Assume an itemset $i = \{i_1, i_2, \ldots, i_n\}$ in a transaction $t = \{i_t, u_t\}$. $i_t = \{i_{t1}, i_{t2}, \ldots, i_{tm}\}$ is the purchase items in this transaction, $u_t = \{u_t(i_{t1}) = u_{t1}, u_{t2}, \ldots, u_{tm}\}$ is the corresponding utility for each item in this* transaction, and the active itemset is ais. The partial maximal utility upper bound (pmuub) of i in t is denoted as pb_i^t, which is defined as:

$$pb_i^t = \begin{cases} \frac{u(i,t)+m \times rmua(i,t)}{|i|+m} & ,if\ rmua(i,t) > au(i,t) \\ \frac{u(i,t)+rmua(i,t)}{|i|+1} & ,if\ 0 < rmua(i,t) \leq au(i,t) \\ 0 & ,if\ rmua(i,t) = 0, \end{cases} \quad (3)$$

where m is the number of $(ais \cap i_t) \setminus i$.

Definition 4. Assume an itemset i and a transaction dataset D. The partial maximal utility upper bound (pmuub) of i in D is denoted as pb_i, and defined as:

$$pb_i = \sum_{t \in D} pb_i^t \quad (4)$$

Lemma 1 (Anti-monotonicity property of pmuub in a transaction). Assume an itemset $i = \{i_1, i_2, \ldots, i_n\}$ in a transaction $t = \{i_t, u_t\}$, active itemsets $= ais$, and a superset of i, $i' = \{i_1, i_2, \ldots, i_n, i_{n+1}, \ldots, i_o\}$, in which $\forall x \in \{i_{n+1}, \ldots, i_o\} \to x \in ais \setminus i$, such that $tmu(i,t) \geq pb_i^t \geq au(i',t)$.

Lemma 2 (The closure property of pmuub). Assume an itemset $i = \{i_1, i_2, \ldots, i_n\}$ in a transaction dataset $d = \{t_1, t_2, \ldots, t_z\}$, active itemsets $= ais$, and a superset of i, $i' = \{i_1, i_2, \ldots, i_n, i_{n+1}, \ldots, i_o\}$, in which $\forall x \in \{i_{n+1}, \ldots, i_o\} \to x \in ais \setminus i$, such that $auub(i) \geq pb_i \geq au(i')$.

Definition 5. If an itemset i is a high partial maximal utility upper bound itemset (hpmuubi) and a predefined threshold is r, then it indicates that $pb_i \geq r$.

By Lemma 2, the proposed APHAUP would maintain a set of pmuubi to generate the candidate itemsets. This technique has ability to effectively reduce the search space.

3.2 Designed Lead Partial Maximal Utility Upper-Bound Models

In this section, a subset of hpmuubi is introduced to reduce the search space of the candidate itemsets. This subset is called the lead partial maximal utility upper bound (lead-pmuub). A predefined order of items in an itemset needed to be set to apply this technique. Assume that a predefined item order l is given and a candidate itemset $i_c = \{i_1, i_2, \ldots, i_n\}$ follows the order, l. In the Apriori-like process, this candidate itemset must be made up of a set of hpmuubi. This set of hpmuubi is $\{\{i_1, i_2, \ldots, i_{n-1}\}, \{i_1, i_2, \ldots, i_{n-2}, i_n\}, \ldots, \{i_2, i_3, \ldots, i_n\}\}$. The first of this set is one of the lead-hpmuubis. Due to this limitation, a smaller upper bound than pmuub can be defined and described in the following section, which can further reduce the size of the candidate itemsets.

Definition 6. Assume that an itemset $i = \{i_1, i_2, \ldots, i_n\}$, a transaction $t = \{i_t, u_t\}$. $i_t = \{i_{t1}, i_{t2}, \ldots, i_{tm}\}$ is the purchase items in this transaction, $u_t = $

$\{u_t(i_{t1}) = u_{t1}, u_{t2}, \ldots, u_{tm}\}$ *is the corresponding utility for each item in this transaction, and the active itemset is ais and a predefined item order* $l = \{i_1^l, i_2^l, \ldots, i_p^l\}$. *Assume that* $i_n = i_w^l$, *and set an itemset* $s = \{i_{w+1}^l, i_{w+2}^l, \ldots, i_p^l\}$, *the lead remaining maximal utility of active itemset in a transaction (lrmua) of* i *in* t *is denoted as* $lrmua(i,t)$ *and is defined as:*

$$lrmua(i,t) = max\left\{u_t(i) \,\middle|\, i \in (ais \cap i_t \cap s) \setminus i\right\}. \tag{5}$$

Definition 7. *Assume an itemset* $i = \{i_1, i_2, \ldots, i_n\}$ *in a transaction* $t = \{i_t, u_t\}$. $i_t = \{i_{t1}, i_{t2}, \ldots, i_{tm}\}$ *is the purchase items in this transaction,* $u_t = \{u_t(i_{t1}) = u_{t1}, u_{t2}, \ldots, u_{tm}\}$ *is the corresponding utility for each item in this transaction, the active itemsets is ais, and a predefined item order* $l = \{i_1^l, i_2^l, \ldots, i_p^l\}$. *The lead partial maximal utility upper bound (lead-pmuub) of* i *in* t *is denoted as* lpb_i^t, *and defined as:*

$$lpb_i^t = \begin{cases} \dfrac{u(i,t) + m \times rmua(i,t)}{|i| + m} \\ \quad, if\ lrmua(i,t) > au(i,t) \\ \dfrac{u(i,t) + rmua(i,t)}{|i| + 1} \\ \quad, if\ 0 < lrmua(i,t) \leq au(i,t) \\ 0 \\ \quad, if\ lrmua(i,t) = 0, \end{cases} \tag{6}$$

where m *is the number of* $(ais \cap i_t \cap s) \setminus i$ *(the definition of* s *is same as the one in Definition 6).*

Definition 8. *The lead partial maximal utility upper bound (lead-pmuub) in a transaction dataset is denoted as* lpb_i, *and defined as:*

$$lpb_i = \sum_{t \in D} lpb_i^t \tag{7}$$

Definition 9. *For the lead high partial maximal utility upper bound itemset (lead-hpmuubi), if an itemset* i *is a lead high partial maximal utility upper bound itmeset (lead-hpmuubi) and a predefined threshold is* r, *then* $pb_i \geq r$.

The lead high partial maximal utility upper bound itemset *lead-hpmuubi* has a similar definition to *hpmuubi*. Therefore, *head-hpmuubi* also has the closure property like *hpmuubi*. The set of *lead-hpmuubi* is a subset of the set of *hpmuubi* due to the stricter limitation than *hpmuubi*. It can further reduce the size of the candidate itemsets in the proposed APHAUP.

4   Apriori-Based HAUP with Pre-large Concept, APHAUP

Different from traditional HAUI mining, the incremental process needs to set a threshold for the pre-large patterns. In the beginning, the rescan threshold would be set in line 3. APHAUP applies *auub* to initialize *hpmuubi* and *lead-hpmuubi*. At this time, all of the itemsets in *hpmuubi* and *lead-hpmuubi* have one item. APHAUP also calculates the average utility for all the itemsets (with 1 item) at the same time. Then, it is determined whether the average utilities is larger than the high average utility threshold and not the pre-large utility threshold. This description is in lines 4–25. There is a while loop for each iteration to reveal the high average itemsets and pre-large itemsets in lines 26–50. In lines 30–32, each itemset in *lead-hpmuubis* is combined with the itemsets in *hpmuubis* to generate the candidate itemsets. Then, APHAUP scans the dataset again to obtain the utility information for each itemset in the candidate itemsets. Finally, at the end of this while loop, APHAUP updates the set of *hpmuubis*, *lead-hpmuubis*, P, and H. If *hpmuubis* and *lead-hpmuubis* are all empty-set, then the loop is stopped and output the set of HAUIs and pre-large itemsets, and then the threshold is recalculated. The detailed pseudo-codes are respectively given in Algorithms 1 and 2.

Algorithm 1. Apriori-based HAUP with pre-large concept, APHAUP

Input: a transaction dataset, D,
         a high average utility threshold, t
         and a pre-large utility threshold, p.
Output: a set of HAUI,
         a set of Prelarge itemsets
         and a rescan threshold.

1: set $H = \emptyset$ for HAUIs;
2: set $P = \emptyset$ for PreLarge Itemsets;
3: count rescan threshold r for D;
4: set array A for *auub*;
5: **for** each transaction t in D **do**
6:     find maximal utility m in t;
7:     **for** each item i in t **do**
8:         $A[i]+ = m$;
9:         **if** $au(i) \geq p$ **then**
10:             $P \leftarrow i$;
11:         **end if**
12:         **if** $au(i) \geq p$ **then**
13:             $H \leftarrow i$;
14:         **end if**
15:     **end for**
16:     accumulate average utility (utility) for each itemset(item) i in t;
17: **end for**
18: set $hpmuubis \leftarrow \emptyset$;
19: set $lead\text{-}hpmuubis \leftarrow \emptyset$;
20: **for** each item i **do**
21:     **if** $A[i] \geq p$ **then**
22:         $hpmuubis \leftarrow i$;
23:         $lead\text{-}hpmuubis \leftarrow i$;
24:     **end if**
25: **end for**
26: **while** $hpmuubis \neq \emptyset$ and $lead\text{-}hpmuubis \neq \emptyset$ **do**

```
27:     set C = ∅ for the candidate itemsets;
28:     set hpmuubis ← ∅;
29:     set lead-hpmuubis ← ∅;
30:     for each itemset e in lead-hpmuubis do
31:         search hpmuubis by e to generate candidate itemsets → C
32:     end for
33:     for each transaction t in D do
34:         scan utility information for each candidate itemsets in t;
35:     end for
36:     for each candidate itemset c in C do
37:         if pb_c ≥ p then
38:             hpmuubis ← c;
39:         end if
40:         if lpb_c ≥ p then
41:             lead-hpmuubis ← c;
42:         end if
43:         if au(c) ≥ p then
44:             P ← c;
45:         end if
46:         if au(c) ≥ p then
47:             H ← c;
48:         end if
49:     end for
50: end while
51: return H, P, r;
```

5   Experimental Results

This section describes two implementations of the proposed APHAUP. One is with the *lhpmuub* threshold and another is without the *lhpmuub* threshold. The experimental results showed the performance of APHAUP with six different real datasets [6] (retail, mushroom, foodmart, chess, BMS and accidents) using different parameter settings.

5.1   Runtime with Different High Average Utility Threshold

This section compares the runtimes for the proposed APHAPU with different high average utility thresholds in six real datasets. The experimental results are shown in Fig. 1.

The results showed that the Apriori(A, lhpmuub) and Apriori(I, lhpmuub) were both good for the mushroom, chess, and accidents datasets. It should be noticed that these databases were dense datasets. Furthermore, it can be observed that with the increasing of the minimum threshold, the runtime of the algorithm was gradually reduced. This is reasonable since if the threshold is larger, fewer itemsets are then discovered, thus less computational cost is needed. Note that the performance of Apriori(I) in mushroom and chess datasets is very

Algorithm 2. Incremental process for APHAUP
***

Input: a incremental transaction dataset, D',
            a remaining rescan threshold, r,
            a high average utility threshold, t
            and a pre-large utility threshold, p.
Output: a set of HAUI,
            a set of Prelarge
            and updated rescan threshold.

1: $inc =$ the size of the input D';
2: update $r = r - inc$;
3: **if** $r \geq 0$ **then**
4:      apply Algorithm 1 for D', t and p;
5:      obtain $\rightarrow P', H'$;
6:      set $S = \emptyset$ for rescan itemsets;
7:      update utility from $H, P, H', P' \rightarrow H, P$
8:      $S \leftarrow (H \cup P) \cap \sim (H' \cup P')$;
9:      scan utility information for S in $D' \rightarrow H'', P''$;
10:     update utility from $H, P, H'', P'' \rightarrow H, P$
11: **else**
12:     update $D \leftarrow D + D'$;
13:     rescan whole datasert by Algorithm 1 for D, t and p;
14:     update $\rightarrow H, P, r$;
15: **end if**
16: **return** H, P, r;
***

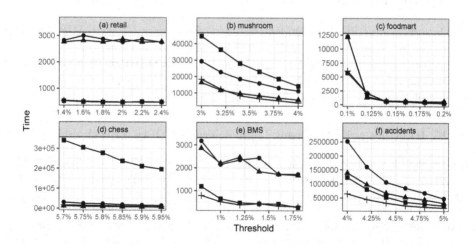

Method ● Apriori(A) ▲ Apriori(A, lhpmuub) ■ Apriori(I) + Apriori(I, lhpmuub)

Fig. 1. The runtimes for APHAUP in six datasets with different thresholds.

poor. That is because Apriori(I) applies a loosen upper-bound and produces too many candidate itemsets in scanning a new dataset (shown in Fig. 2). Each incremental process needs to evaluate a huge number of candidate itemsets. Thus, to rescan the whole dataset might spend less CPU time than the incremental-based algorithms.

5.2 The Number of Candidates for Proposed Method and FUP-Based Model

The numbers of candidates depending on whether using *lhpmuub* in six real datasets are shown in Fig. 2.

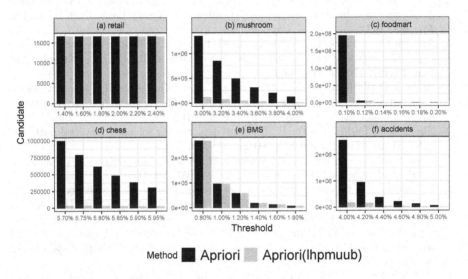

Fig. 2. The numbers of the candidate itemsets for APHAUP in six datasets with different minimal support.

Comparing with Apriori, the cost of rescanning the whole updated dataset does not require a lot in the Apriori(lhpmuub) in terms of some datasets. It is very suitable to be applied in a stream environment. On the other hand, the traditional upper-bound value cannot select potential candidate itemsets precisely, especially for mushroom and chess datasets. For those datasets, there are a vast number of itemsets between two upper-bounds, and it causes that the traditional Apriori-like approach needs to perform the evaluation process for a lot of unpromising itemsets. Thus, Apriori(I) has the worst performance in mushroom and chess datasets.

6 Conclusion

We designed an incremental transaction insertion algorithm based on the prelarge concept for high average-utility itemset mining. Furthermore, we proposed two new upper-bounds to reduce the size of candidates, called *pmuub* and *lead-pmuub*. The results of experiments showed that the developed APHAUP with *lead-pmuub* could significantly reduce the execution times for updating the discovered HAUIs compared to APHAUP with *pmuub* in dense databases. Moreover, the number of determined candidates was much less than APHAUP with *pmuub*. Thus, it showed that the pre-large concept has capacities to improve the mining performance for HAUIM.

References

1. Agrawal, R., Srikant, R.: Fast algorithms for mining association rules in large databases. In: The 20th International Conference on Very Large Data Bases, pp. 487–499 (1994)
2. Ahmed, C.F., Tanbeer, S.K., Jeong, B.S., Lee, Y.K.: Efficient tree structures for high utility pattern mining in incremental databases. IEEE Trans. Knowl. Data Eng. **21**(12), 1708–1721 (2009)
3. Cheung, D.W., Han, J., Ng, V.T., Wang, C.Y.: Maintenance of discovered association rules in large databases: an incremental updating technique. In: The International Conference on Data Engineering, pp. 106–114 (2002)
4. Chen, M.S., Park, J.S., Yu, P.: Efficient data mining for path traversal patterns. IEEE Trans. Knowl. Data Eng. **10**(2), 209–221 (1998)
5. Chen, C.M., Xiang, B., Liu, Y., Wang, K.H.: A secure authentication protocol for intermet of vehicles. IEEE Access **7**, 12047–12057 (2019)
6. Fournier-Viger, P., et al.: The SPMF open-source data mining library version 2. In: Joint European Conference on Machine Learning and Knowledge Discovery in Databases, pp. 36–40 (2016)
7. Gan, W., Lin, J.C.W., Fournier-Viger, P., Chao, H.C., Yu, S.: A survey of parallel sequential pattern mining. ACM Trans. Knowl. Disc. Data **13**(3), 1–34 (2019). Article 25
8. Gan, W., Lin, J.C.W., Fournier-Viger, P., Chao, H.C., Tseng, V., Yu, P.: A survey of utility-oriented pattern mining. IEEE Trans. Knowl. Data Eng. (2019)
9. Han, J., Pei, J., Yin, Y., Mao, R.: Mining frequent patterns without candidate generation: a frequent-pattern tree approach. Data Min. Knowl. Disc. **8**, 53–87 (2004)
10. Hong, T.P., Lin, C.W., Wu, Y.L.: Incrementally fast updated frequent pattern trees. Expert Syst. Appl. **34**, 2424–2435 (2008)
11. Hong, T.P., Wang, C.Y., Tao, Y.H.: A new incremental data mining algorithm using pre-large itemsets. Intell. Data Anal. **5**(2), 111–129 (2001)
12. Hong, T.P., Lee, C.H., Wang, S.L.: An incremental mining algorithm for high average-utility itemsets. In: The International Symposium on Pervasive Systems, Algorithms, and Networks, pp. 421–425 (2009)
13. Hong, T.P., Lee, C.H., Wang, S.L.: Effective utility mining with the measure of average utility. Expert Syst. Appl. **38**(7), 8259–8265 (2011)
14. Kim, N.V.: Some determinants affecting purchase intention of domestic products at local markets in Tien Giang province, Vietnam. Data Sci. Pattern Recogn. **1**(2), 43–52 (2017)
15. Lin, C.W., Hong, T.P., Lu, W.H.: The pre-FUFP algorithm for incremental mining. Expert Syst. Appl. **36**(5), 9498–9505 (2009)

16. Lin, C.W., Lan, G.C., Hong, T.P.: An incremental mining algorithm for high utility itemsets. Expert Syst. Appl. **39**(8), 7173–7180 (2009)
17. Lin, C.W., Hong, T.P., Lu, W.H.: Efficiently mining high average utility itemsets with a tree structure. In: The Asian Conference on Intelligent Information and Database Systems, pp. 131–139 (2010)
18. Lin, C.W., Hong, T.P., Lu, W.H.: An effective tree structure for mining high utility itemsets. Expert Syst. Appl. **38**(6), 7419–7424 (2011)
19. Liu, M., Qu, J.: Mining high utility itemsets without candidate generation. In: ACM International Conference on Information and Knowledge Management, pp. 55–64 (2012)
20. Lan, G.C., Hong, T.P., Tseng, V.: Efficient mining high average-utility itemsets with an improved upper-bound strategy. Int. J. Inform. Technol. Decis. Making **11**, 1009–1030 (2012)
21. Lin, C.W., Hong, T.P., Lin, W.Y., Lan, G.C.: Efficient updating of sequential patterns with transaction insertion. Intell. Data Anal. **18**, 1013–1026 (2014)
22. Lu, T., Vo, B., Nguyen, H.T., Hong, T.P.: A new method for mining high average utility itemsets. In: IFIP International Conference on Computer Information Systems and Industrial Management, pp. 33–42 (2015)
23. Liu, J., Wang, K., Fung, B.C.M.: Mining high utility patterns in one phase without generating candidates. IEEE Trans. Knowl. Data Eng. **28**(5), 1245–1257 (2016)
24. Lin, J.C.W., Ren, S., Fournier-Viger, P., Hong, T.P.: EHAUPM: efficient high average-utility pattern mining with tighter upper-bounds. IEEE Access **5**, 12927–12940 (2017)
25. Su, J.H., Chang, W.Y., Tseng, V.S.: Integrated mining of social and collaborative information for music recommendation. Data Sci. Pattern Recogn. **1**(1), 13–30 (2017)
26. Tseng, V., Shie, B.E., Wu, C.W., Yu, P.: Efficient algorithms for mining high utility itemsets from transactional databases. IEEE Trans. Knowl. Data Eng. **25**, 1772–1786 (2013)
27. Wang, Z., Chen, K., He, L.: AsySIM: modeling asymmetric social influence for rating prediction. Data Sci. Pattern Recogn. **2**(1), 25–40 (2018)
28. Wu, J.M.T., Lin, J.C.W., Tamrakar, A.: High-utility itemset mining with effective pruning strategies. ACM Trans. Knowl. Disc. Data **13**(6), 1–22 (2019)
29. Yao, H., Hamilton, H.J., Butz, C.J.: A foundational approach to mining itemset utilities from databases. In: The SIAM International Conference on Data Mining, pp. 215–221 (2004)
30. Zida, S., Fournier-Viger, P., Lin, J.C.-W., Wu, C.-W., Tseng, V.S.: EFIM: a fast and memory efficient algorithm for high-utility itemset mining. Knowl. Inf. Syst. **51**(2), 595–625 (2016). https://doi.org/10.1007/s10115-016-0986-0
31. Liu, Y., Liao, W.K., Choudhary, A.: A two-phase algorithm for fast discovery of high utility itemsets. In: The Pacific-Asia Conference on Advances in Knowledge Discovery and Data Mining, pp. 689–695 (2005)

System Control, Classification, and Fault Diagnosis

Development and Research of a Terminal Controller for Marine Robots

V. Pshikhopov and Boris Gurenko$^{(\boxtimes)}$

Southern Federal University, Rostov-on-Don 344006, Russian Federation
`borisgurenko@sfedu.ru`

Abstract. Marine robots (unmanned surface vehicles) are actively used for various civil applications. One of the interesting tasks – automatic transportation of goods and people along rivers, seas and oceans – is being solved now. Control of the ship's movement at the modern technological level allows to implement complex modes of operation with good speed and accuracy. However, if unmanned ship is used in ports and canals, it is very important to ensure that the task is performed both automatically and within certain time limits. These two requirements are specific for so-called terminal control that authors try to implement for marine robots taking the problem of surface vehicle docking as an example. The proposed solution consists of two stages and include obstacle avoidance method. Procedure of terminal control synthesis is provided as well as movement algorithm. Experiments were based upon the mathematical model used in real surface mini vessel "Neptune" and include simulation of docking in several different scenes with obstacles. Method of obstacle avoidance, proposed in this paper, is based on unstable mode. The unstable mode can be used directly but can also be used in a hybrid version that includes intelligent analysis of current situation and unstable mode of movement. In general, the unstable mode is based on a bionic approach; in particular, such an approach to avoiding obstacles can be observed in the behavior of fish.

Keywords: Surface vehicles · Terminal control · Bionics · Unstable mode

1 Introduction

Marine robotics is actively used in the civilian fleet for seabed mapping, surveillance and diagnostics of underwater communications (pipelines, data transmission cables), transportation of goods, etc. There is an increasing interest for implementation of robotic complexes for passenger and cargo transportation along rivers, seas and even oceans.

Taking countries with vast river networks, one can see that rivers are used mainly for energy generation but not for transport. For example, in Russian Federation, the length of inland waters is over 100 thousand km, and the length of the sea border is 42 thousand km. At the same time, the volume of sea and river transportations makes up less than

---

The original version of this chapter was revised: the funding statement was corrected. The correction to this chapter is available at https://doi.org/10.1007/978-3-030-55789-8_80

© Springer Nature Switzerland AG 2020, corrected publication 2020
H. Fujita et al. (Eds.): IEA/AIE 2020, LNAI 12144, pp. 899–906, 2020.
https://doi.org/10.1007/978-3-030-55789-8_76

5% of the total volume of cargo transportation within the country It is obvious that this ratio will increase in conjunction with the flow of goods along the rivers. We consider that aquatic autonomous revolution will soon follow the driverless cars.

Automatic control of the ship's movement at the modern technological level allows to implement difficult modes of operation, for example, maneuvering the vessel in conditions of traffic jams and maneuver restrictions, and to increase the speed and accuracy of planning and following the trajectory [1–9].

However, existing systems mostly function remotely, and only limited number of the tasks can be performed autonomously without the intervention of an operator. In particular, if we assume that an unmanned ship will operate in conjunction with the habitable systems, it is very important, especially if operating in ports and canals, to ensure that the task is performed automatically within certain time limits and at a given moment of time. Thus, the task of bringing unmanned ship from its initial position to some desired final state in the space of coordinates at a given moment in time (terminal control task) is very relevant.

To solve the problem of terminal control, an approach that involves two stages is used most often: the choice and calculation of a program trajectory as a certain time function; and solution of the stabilization problem ensuring the motion of a mobile object along a given program trajectory [12–14]. The difficulty of terminal control law synthesis is greatly increased if a nonlinear model of motion of a mobile object is used. Consequently, authors solve the problem of terminal control for an unmanned surface vehicle (USV), the model of which is represented by a system of nonlinear differential equations. If we are talking about the real process of docking USV to the pier, there can be obstacles along the way while performing such operation. There are many methods to circumvent obstacles [10]. The method of obstacle avoidance proposed in this paper is based on unstable mode. The unstable mode can be used directly, but can be used in a hybrid version, including intelligent analysis of the current situation and unstable mode of movement. In general, an unstable mode is based on a bionic approach; in particular, such an approach to avoiding obstacles can be observed in the behavior of fish.

2 Synthesis of Terminal Control

Let the mathematical model of the USV be given, in accordance with the coordinate system shown in Fig. 1, has the following form [12]:

$$\dot{Y} = \begin{bmatrix} \cos\varphi & \sin\varphi & 0 \\ -\sin\varphi & \cos\varphi & 0 \\ 0 & 0 & 1 \end{bmatrix} X, \tag{1}$$

$$\dot{X} = F_u(X, Y, d, 1, t) + F_d(P, V, W) + F_v(G, A, R), \tag{2}$$

where $Y = \begin{bmatrix} x_g & y_g & \varphi \end{bmatrix}^T$ is a vector of the coordinates of the center of gravity of the ship in a fixed coordinate system; $X = \begin{bmatrix} V_x & V_z & \omega_y \end{bmatrix}^T$ - the projection of the velocity vector on the axis of the coordinate system X Y Z associated with the mini ship; φ - current rate, and ω_y- the angular velocity of the ship relative to its vertical axis OY; $F_d(P, V, W)$ - (3×1) vector of nonlinear dynamics elements including Coriolis forces,

$F_v(G, A, R)$- (3×1) - a vector of measured and unmeasurable external disturbances, M - (3×3) -matrix weight and inertia parameters whose elements are the mass, moments of inertia, attached masses. $F_u(X, Y, \delta, 1, t)$- (3×1) - vector control forces and moments are 1 - vector of design parameters, $P = \begin{bmatrix} x_g & y_g \end{bmatrix}^T$. The mathematical model is described in more detail in [11, 12].

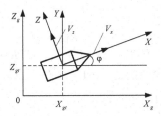

Fig. 1. USV coordinate system

For the terminal control, it is required to synthesize a control algorithm that would ensure the movement of the ship from the initial position $P_0 = (x_0, z_0)$ to the final $P_k = (x_k, z_k)$ at a given time $t = T_k$, and the velocity at time $t = T_k$ must equal zero, i.e.:

$$X_k = \left(V_x, V_z, \omega_y \right)^T = 0_{3x1}. \tag{3}$$

To solve the problem of synthesis, we set the programed trajectory of the motion of the USV in polynomial form and represent it in the matrix form:

$$Y^* = CL, \tag{4}$$

where $C = \begin{bmatrix} x_0 & 0 & \frac{3}{T_k^2}(x_k - x_0) & \frac{3}{T_k^3}(x_k - x_0) \\ z_0 & 0 & \frac{3}{T_k^2}(z_k - z_0) & \frac{3}{T_k^3}(z_k - z_0) \end{bmatrix}$, $L = \begin{bmatrix} 1 \\ t \\ t^2 \\ t^3 \end{bmatrix}$,

C is the matrix of constant coefficients, which depend on the initial and final positions of USV, as well as by T_k- positioning a predetermined time. It can be shown that $\dot{Y}^* = CD_1L$, $\ddot{Y}^* = CD_2L$, where $D1 = \begin{bmatrix} 0 & 0 & 0 & 0 \\ 1 & 0 & 0 & 0 \\ 0 & 2 & 0 & 0 \\ 0 & 0 & 3 & 0 \end{bmatrix}$, $D2 = \begin{bmatrix} 0 & 0 & 0 & 0 \\ 0 & 0 & 0 & 0 \\ 2 & 0 & 0 & 0 \\ 0 & 6 & 0 & 0 \end{bmatrix}$.

Formation of the law of terminal control is implemented on the basis of position-trajectory algorithms [15]. In order to do so, we set the error of the deviation of the real trajectory of motion from the specified program one as: $\psi_{tr} = P - P^* = P - CL$.

We introduce an additional variable $\psi = \dot{\psi}_{tr} + T_1 * \psi_{tr}$, where T_1- constant diagonal matrix of dimension (2×2), $\dot{\psi}_{tr} = \dot{Y} - CD_1L$. And we define the desired behavior of a closed system as follows: $\dot{\psi} + T_2 * \psi = 0$, where T_2- constant diagonal matrix of dimension (2×2). Thus, the desired behavior of the system is expressed by ψ_{tr} the following form:

$$\ddot{\psi}_{tr} + T_1 * \dot{\psi}_{tr} + T_2 * \left(\dot{\psi}_{tr} + T_1 * \psi_{tr} \right) = 0, \tag{5}$$

where $\dot{\psi}_{tr} = \dot{Y} - CD_1L$, $\ddot{\psi}_{tr} = \ddot{Y} - CD_2L$.

Substituting the equation of a mathematical model of USV to Eq. 5, and expressing an appropriate vector control forces and moments F_u, will get the following control law:

$$F_u = -M_1 * R_1^{-1}[\dot{R}_1X_1 - CD_2L + T_1 * \dot{\psi}_{tr} + T_2 * (\dot{\psi}_{tr} + T_1 + \psi_{tr})] - F_d' \qquad (6)$$

For stability analysis, we substitute the resulting Eq. 6 into the equation of the mathematical model.

$$\begin{cases} \dot{P} = R_1X_1 \\ \dot{X}_1 = R_1^{-1}\left[\begin{array}{c} \dot{R}_1X_1 - CD_2L + T_1 * (R_1R_1 - CD_1L) \\ + T_2 * (\dot{R}_1X_1 + R_1\dot{X}_1 - CD_2L + T_1 * (R_1X - CD_1L)) \end{array} \right] \end{cases} \qquad (7)$$

We define the Lyapunov function in the form:

$$V(x) = \psi^T Q\psi, \qquad (8)$$

where Q a positive definite diagonal matrix of size (2×2). Then the derivative of the Lyapunov function is: $\dot{V}(x) = 2[R_1X_1 - CD_1L + T_1 * \psi_{tr}]Q[-T_2(X_1 - CD_1L + T_1 * \psi_{tr})]$.

Let's define matrix Q such that $Q T_2 = C$, where C is a diagonal positive definite matrix of size (2×2), and then we obtain the following expression for the derivative $\dot{V}(x)$:

$$\dot{V}(x) = -2[R_1X_1 - CD_1L + T_1 * \psi_{tr}]G[R_1X_1 - CD_1L + T_1 * \psi_{tr}] \qquad (9)$$

As G – positive definite matrix, then the function $\dot{V}(x)$ is negative definite at all points except $R_1X_1 - CD_1L + T_1 * \psi_{tr} = 0$. It's enough just to show that the point $X_1 = -R_1^{-1}(CD_1L + T_1 * \psi_{tr})$ is not a solution of Eq. 7 and it can be asserted that the system Eq. 7 is asymptotically stable with allowance for the criteria [16].

3   The Obstacle Avoidance Algorithm

In the real situation, there may be obstacles on the USV path. There are quite a lot of intellectual methods for avoiding obstacles [1–10]. In [10] it was shown that from the point of view of reducing the requirements for the sensor subsystem, the most effective method is to use the obstacle avoidance method based on unstable mode. Such a bionic approach to circumventing an obstacle does not require pre-mapping, and thus reduces the requirements for the intellectual provision of the USV. The essence of the method is as follows: introduce the bifurcation parameter β into the controller structure, the value of which depends on the distance to the obstacle. If this distance is more than acceptable R^*, the bifurcation parameter $\beta = 0$ and a closed system is in a stable condition. If the distance is less than the acceptable R^*, then $\beta \neq 0$ and the system goes into an unstable mode. Since this instability depends on the distance to the obstacle, naturally, the unstable mode deflects the USV from the trajectory on which the obstacle is located, in order to increase this distance to the obstacle.

We define the bifurcation parameter by the following method:

$$\beta = |R - R^*| - (R - R^*),$$

where R – the distance to obstacle, measured by sensor, R^* – acceptance distance to the obstacle, i.e., the distance at which you want to start the process of divergence with an obstacle. It is easy to show that if $R \geq R^*$, then $\beta = 0$. If the distance to the obstacle is less than acceptable, $R < R^*$ then $\beta = -2(R - R^*)$. Considering that $(R - R^*) < 0$, then $\beta > 0$.

As noted, behavior of the system will be stable if the matrix is positive definite. It follows that the behavior of the system will be stable if the matrix $G = Q\,T_2$ is positive definite. Consequently, for the derivation of the system to an unstable mode, one of the diagonal elements of the matrix T_2 must be made negative, for example, $T_2(1, 1) = -\beta$.

The proposed controller functions as follows. The algorithm runs if USV reaches the end point P_k. If USV meets the obstacle on its path, and the obstacle is out of the danger zone, i.e., P_k, i.e. $\beta = 0$, the conventional terminal control calculated according to the Eq. 6. If condition $R > R^*$ fails, and $\beta \neq 0$ then there is the calculation of the bifurcation parameter, and USV switches to an unstable mode. In case of return to normal terminal control mode, it is necessary to form a new vector of initial conditions for the terminal control algorithm (Eq. 6). As soon as $P = P_k$ control is reset or must transfer control to stabilize the relative mode P_k.

4   Experiments and Discussion

Mathematical model of the surface mini vessel "Neptune" [12] was used for experiments with the proposed controller. The appearance of the surface mini vessel is shown in Fig. 2. The actuators are two drives based on brushless asynchronous motors and servos.

Fig. 2. The appearance of the surface mini vessel "Neptune" and its propulsion steering complex

The motors and propellers are mounted on the movable frame and can deviate from the longitudinal axis by the same angle. The motors and servo drive are controlled by local controllers whose inputs are fed with a PWM signal. The inertia of the motors and the servo can be neglected in comparison with the inertia of the object.

The sensors installed on the nose of the USV are three range finders. Their rays diverge at an angle of $30°$. The length of each beam is 5 m, i.e. $R^* = 5$ m. In the simulation, the sensor model checks the intersection of each ray with each obstacle and, in the presence of an intersection, produces a minimum distance, as well as a flag indicating the presence of an obstacle.

In the experiment, we added two obstacles in the form of a circle with a radius of 5 m. The center of the first circle is at the point (20, 20), the center of the second circle is (40, 40). The results are shown in Fig. 3.

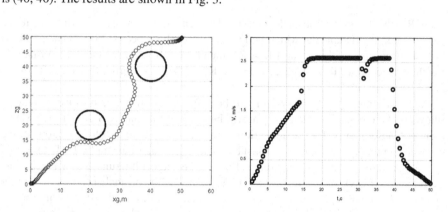

Fig. 3. USV trajectory in an environment with two obstacles and change of USV speed

The results of modeling for other location of static obstacles are shown in Fig. 4.

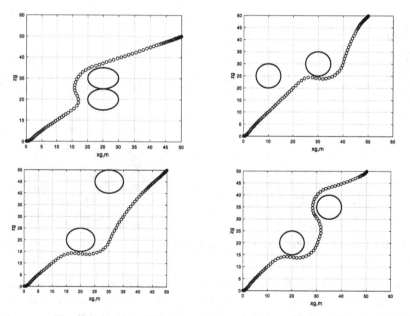

Fig. 4. Examples of obstacle avoidance using an unstable mode

Simulation results allow to conclude the efficiency of the proposed approach. Comparing the charts of velocity changes with and without obstacles, one can see that in the absence of obstacles the maximum speed of the USV does not exceed 2.2 m/s and reaches its maximum in the middle of the path. With obstacles, speed raises to 2.5 m/s, and this value holds quite a long time. This is understandable, because the trajectory itself in the second case is longer, and in order to reach the final point at a given time, the average speed, and therefore the maximum, should be higher. Consequently, the trajectory has some maximum length, increasing that the USV will not have enough energy to reach the goal point at a stated moment of time.

5 Conclusion

The solution of terminal control problem proposed in the paper can be applied to docking of autonomous unmanned boat to the pier. A model of the USV is given as system of nonlinear differential equations. Unstable mode is used to avoid possible obstacles. The proposed algorithm allows to minimize the requirements for the sensor subsystem and eliminates the need for preliminary mapping. However, it is necessary to understand that in order to solve the terminal problem, calculation of time T_k should take into account the energy capabilities of the boat itself, i.e. limitations on the control actions and the power of the actuators. The task of trajectory generation (Eq. 4), that considers constraints on the control actions and the power of the executive mechanisms, is planned for the next paper. It is also necessary to estimate the effect of external disturbances, such as currents, and structural disturbances (power deviation, the change in mass) on the accuracy of the terminal task solution. In the future, we plan to experiment the proposed algorithm with moving obstacles.

Funding Statement. This work was supported by the Southern Federal University of the Russian Federation, internal grant No VnGr-07/2017-19, grants of the President of the Russian Federation for the state support of young Russian scientists MK- 3099.2019.8 "Methods for developing of an intelligent group control system of autonomous marine robotics".

References

1. Klinger, W.B., Bertaska, I.R., von Ellen-rieder, K.D., Dhanak, M.R.: Control of an unmanned surface vehicle with uncertain displacement and drag. IEEE J. Ocean Eng. **42**(2), 458–476 (2017)
2. Villa, J.L., Paez, J., Quintero, C., Yime, E., Cabrera, J.: Design and control of an unmanned surface vehicle for environmental monitoring applications. In: 2016 IEEE Colombian Conference on Robotics and Automation (CCRA) (2016)
3. Feng, K., Wang, N., Lill, D., Er, M.J.: Adaptive fuzzy trajectory tracking control of unmanned surface vehicles with unknown dynamics. In: 2016 3rd International Conference on Informative and Cybernetics for Computational Social Systems (ICCSS) (2016)
4. Meng, W., Sheng, L.H., Qing, M., et al.: Intelligent control algorithm for ship dynamic positioning. Archives Control Sci. **24**(4), 479–497 (2014)
5. Chen, C.H., Chen, G.Y., Chen, J.J.: Design and implementation for USV based on fuzzy control. In: CACS International Automatic Control Conference, pp. 345–349 (2013)
6. Nicolescu, M., et al.: A training simulation system with realistic autonomous ship control. Comput. Intell. **23**(4), 497–516 (2007)

7. Gaiduk, A., Gurenko, B., Plaksienko, E., Shapovalov, I., Beresnev, M.: Development of algorithms for control of motorboat as multidimensional nonlinear object. In: MATEC Web of Conferences, vol. 34 (2015). http://dx.doi.org/10.1051/matecconf/20153404005

8. Li, B., Xu, Y., Liu, C., Fan, S., Xu, W.: Terminal navigation and control for docking an underactuated autonomous underwater vehicle. In: 2015 IEEE International Conference on Cyber Technology in Automation Control and Intelligent Systems (CYBER), pp. 25–30 (2015). https://doi.org/10.1109/cyber.2015.7287904

9. Shikai, W., Hongzhang, J., Lingwei, M.: Trajectory tracking for underactuated UUV using terminal sliding mode control. In: 2016 Chinese Control and Decision Conference (CCDC), pp. 6833–6837 (2016). https://doi.org/10.1109/ccdc.2016.7532229

10. Pshikhopov, V.: Path Planning for Vehicles Operating in Uncertain 2D Environments, 1st February 2017. eBook ISBN 9780128123065, Paperback ISBN 9780128123058, Imprint: Butterworth-Heinemann, 312 p.

11. Gurenko, B.: Mathematical model of autonomous underwater vehicle. In: Proceedings of the Second International Conference on Advances in Mechanical and Robotics Engineering - AMRE 2014, pp. 84–87 (2014). https://doi.org/10.15224/978-1-63248-031-6-156

12. Vladimir, K., Boris, G., Andrey, M.: Mathematical model of the surface mini vessel. In: ICMCE 2016 Proceedings of the 5th International Conference on Mechatronics and Control Engineering, Venice, Italy, 14–17 December 2016, pp. 57–60. ACM, New York (2016). ©2016 table of contents. ISBN 978-1-4503-5215-4. https://doi.org/10.1145/3036932.3036947

13. Pshikhopov, V.K., Fedotov, A.A., Medvedev, M.Y., Medvedeva, T.N., Gurenko, B.V.: Position-trajectory system of direct adaptive control marine autonomous vehicles. In: 4th International Workshop on Computer Science and Engineering - Summer, WCSE (2014)

14. Fedorenko R., Gurenko B., Shevchenko V.: Research of autonomous surface vehicle control system. In: Proceedings of the 4th International Conference on Control, Mechatronics and Automation (ICCMA 2016), pp. 131–135 (2016). https://doi.org/10.1145/3029610.3029642

15. Medvedev, M.Y., Pshikhopov, V.Kh.: Robust control of nonlinear dynamic systems. In: Proceedings of 2010 IEEE Latin-American Conference on Communications (ANDERSON), 14–17 September Bogota, pp. 1–7 (2010). https://doi.org/10.1109/andescon.2010.5633481

16. Kalman, R.E.: Lyapunov functions for the problem of Lur'e in automatic control. Proc. Nat. Acad. Sci. 49(2), 201–205 (1963)

A Machine Learning Approach for Classifying Movement Styles Based on UHF-RFID Detections

Christoph Uran[✉], Markus Prossegger, Sebastian Vock, and Helmut Wöllik

Carinthia University of Applied Sciences, Klagenfurt, Austria
{c.uran,m.prossegger,s.vock,h.woellik}@cuas.at

Abstract. At large scale sport events such as various running competitions and triathlons, the determination of split and finish times of athletes is mostly done with the help of Radio Frequency Identification (RFID). This work focuses on Ultra High Frequency (UHF) RFID systems, which consist of RFID transponders and RFID readers with appropriate antennas along the track. Recent improvements in the UHF-RFID technology open up new possibilities for analyzing additional data on top of the timing data. By modifying the placements of transponders and antennas and analyzing additional signal properties using Machine Learning algorithms, this paper outlines a way to classify the athlete's movements based on RFID detections from an individual sports timing system. Findings derived from experiments with a proposed antenna and transponder setup in a lab environment are presented. A comparison of the k-Nearest Neighbor approach and the Decision Tree approach shows that they both yield a movement style classification accuracy of approximately 70%.

Keywords: Radio frequency identification · Machine learning · Movement classification · Sports timing

1 Introduction

The determination of an athlete's finish time at sport events, such as marathons, triathlons, or bike races, is typically achieved with the help of Radio Frequency Identification (RFID) [10,11]. This process involves a set of complex tasks in the areas of RFID, computer science, communication technology, and logistics [1]. Typical RFID-based measurement stations are configured in a way, that they determine a timestamp at which a transponder was positioned at a certain position along the track. One or more RFID transponder(s) is/are attached to each athlete and the linkage between transponder(s) and athlete is stored in the timer's database.

Special heuristics as part of the timing software deal with multiple detections of a transponder at a certain measurement station within a certain timeframe, thereby assuring comparable results across all athletes. Depending on the timing

© Springer Nature Switzerland AG 2020
H. Fujita et al. (Eds.): IEA/AIE 2020, LNAI 12144, pp. 907–913, 2020.
https://doi.org/10.1007/978-3-030-55789-8_77

software and the firmware of the RFID readers, there might be different heuristics in place [5]. However, especially RFID readers that work on Ultra High Frequencies (UHF) are known to generate many more timestamps than actually needed for a comparable result, thereby making the aforementioned heuristics even more important [8].

At the same time, this opens up new possibilities in the area of movement classification by mining as many RFID detections as possible and deducing the kind of movement from them, which is what this paper is about. We have defined the following nine distinctive kinds of movements: walking, jogging, running with feet high up the buttocks, running with high knees, sprinting, running with raised arms, running while jumping high, limping, and walking sideways. This paper describes the measurement setup and the analysis methods necessary to classify the movement, solely based on UHF-RFID detections. Eventually, this can be used to determine the athlete's degree of exhaustion or the ideal moment to take a finisher picture.

The remainder of the paper is organized as follows. In Sect. 2, the measurement setup, the measurement procedure, and the approaches to analyze the generated data are described. Section 3 covers the achieved results in the movement prediction based on the generated RFID detections. Interpretations of the results as well as our conclusions are shown in Sect. 4.

2   Materials and Methods

This section describes the measurement setup, the measurement procedure, and the approaches that were used to analyze the generated data.

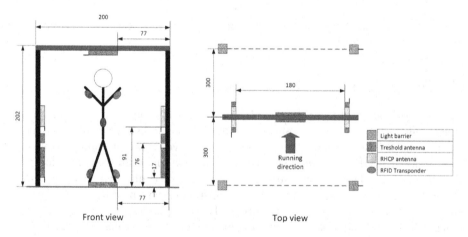

Fig. 1. The full measurement setup, showing the placement of the 5 UHF-RFID transponders, the 8 UHF-RFID antennas (6 with linear polarization and 2 with circular polarization), and the 2 light barriers. All distances are given in centimeters.

As can be seen in Fig. 1, there are several pieces of equipment needed to determine the motion of an athlete with the help of UHF-RFID. In order to identify various body parts, UHF-RFID transponders[1] were attached to the test person's arms, legs, and chest. This idea is in contrast to [6], whose authors focused exclusively on the legs. The RFID measurement stations[2] allow access to all unprocessed RFID detections. The set of eight UHF-RFID antennas consists of six linear polarized antennas, mounted in different orientations, and two right-hand circular polarized antennas. Furthermore, Alge PR1a light barriers are placed before and after the setup to determine the exact timeframe of a test run.

2.1 Data Analysis

All captured data has been stored in a database using a self-developed timing software. The data contains more than 500,000 UHF-RFID detections – including unique transponder number, receiving antenna, signal strength, and timestamp – and 900 light barrier timestamps, representing the exact start and end times of the 450 test runs. Due to the nature of UHF-RFID, many so-called cross reads occurred while the RFID transponders are outside the testing area between the two light barriers and therefore are to be ignored. Hence, not all of the 500,000 RFID detections can be used for predicting the kind of movement.

Each of the nine movement styles described has been executed 50 times. According to a preliminary exploratory data analysis, this generated enough data to be able to differentiate between the defined movement styles.

2.2 Machine Learning

Machine Learning algorithms are commonly used for classification tasks like the one at hand [2,7]. The challenge is to find suitable algorithms and the appropriate parameters. A thorough exploratory data analysis shows that both the k-Nearest Neighbor (kNN) [4] and the Decision Tree (DT) [9] classifiers might be viable options. This is due to the observation that each movement style has a distinctive detection pattern.

Before being able to test the two classifiers with the recorded data, it has to be preprocessed to get a well-defined input. Each test run has to be split up into n timeframes for each receiving antenna a (8 in the given setup) and detected transponder t (5 in the given setup). For 10 timeframes, this results in 400 data points. Each data point contains the average signal strength of transponder t, received by antenna a in timeframe n. If t is never detected by a in n, an extremely low signal strength (-85 dBm) is assumed.

After transforming the data, it is randomly split up into training data and test data using a common splitting ratio of 70% training data and 30% test data.

---

[1] Smartrac DogBone, Impinj Monza R6 chip, 860–960 MHz, linear polarization.

[2] Two time synchronized Impinj Speedway Revolution R420 readers (custom firmware).

The training data is used to train the algorithm on how to allocate a given set of data points to a movement style. The test data is afterwards used to measure the performance of the algorithm, which is the number of correct predictions in relation to the number of tests.

As mentioned earlier, two different classifiers were used and compared to each other. The k-Nearest Neighbor (kNN) approach derives the class of a data point from its nearest neighbors. In the context of movement classification this implies that detection patterns that closely resemble each other are representing the same movement style. The similarity of detection patterns is measured by a comparison of the average signal strengths of each transponder t at each antenna a in a given timeframe n. The nearest neighbor is the one with the lowest Euclidean distance in $(t * a * n)$-dimensional space. This approach intuitively makes sense for allocating detection patterns to movement styles. The most influential parameter is the number of neighbors k, which are taken into account when classifying a given set of data points. This is further discussed in Sect. 3.

The Decision Tree (DT) approach assumes that a set of nested alternatives can be derived from the training data, which can be used to classify sets of data points. The algorithm chooses the conditions of these alternatives based on one of two metrics, that being Gini and Entropy [3]. Both of them choose conditions for the alternatives so that the process of narrowing the prediction down to a single movement style is as accurate as possible. However, they use different computations to do so and therefore produce slightly different results and are differently computationally intensive.

3   Results

As described in Sect. 2, each test run has to be split up into n timeframes in order to make the test runs comparable to each other. Initial tests have been conducted with an arbitrarily chosen number of timeframes between 3 and 100. Furthermore, the number of neighbors k of the kNN approach and the criterion for deciding on the best conditions (Gini or Entropy) of the DT approach has been chosen arbitrarily. In order to find the actually best parameters, 100 tests each with 1–10 timeframes, the kNN approach (2–15 neighbors) and the DT approach (Gini and Entropy criteria) have been conducted.

This results in a total of 16,000 tests with randomly split training and test data sets, a different number of timeframes and different classification approaches with different parameters. Figure 2 shows the average accuracies of the predictions produced by the algorithms. The accuracy of a given approach is measured by the number of correct predictions in relation to the number of test data sets.

The results show that splitting up the test runs into 6 timeframes yields the best results. Furthermore, the kNN approach performs best when taking 5 neighbors into account and the DT approach performs best using the Entropy criterion. Based on that knowledge and in order to better compare the kNN and DT approaches, more tests have been conducted.

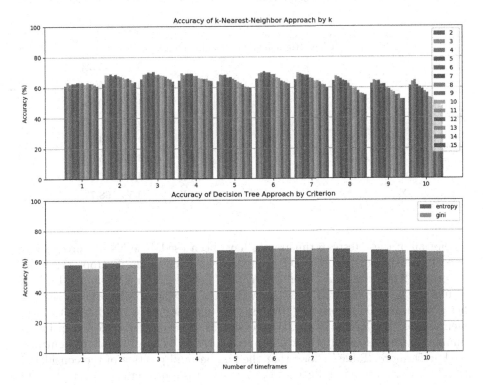

Fig. 2. Respective accuracies with 1–10 timeframes for kNN (2–15 neighbors k) and DT (using Entropy or Gini criterion) approaches, averaged over 100 tests.

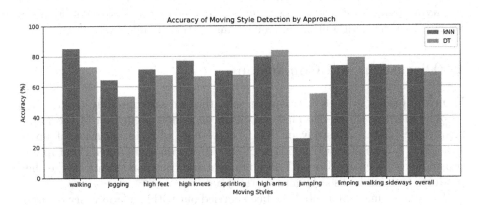

Fig. 3. Accuracy of the prediction of different movement styles and overall for the kNN and the DT approach respectively.

Table 1. Overall accuracies depending on the used algorithm and the considered tag(s).

Position					Algorithm	
Right leg	Left leg	Right arm	Left arm	Chest	kNN	DT
X					50.4%	45.3%
	X				40.5%	35.8%
				X	54.1%	50.2%
X				X	65.2%	51.0%
	X			X	59.7%	49.5%
X			X	X	66.5%	55.0%
	X	X		X	69.5%	67.2%
X	X	X	X	X	70.5%	68.7%

For all of these tests, 6 timeframes have been used. The kNN approach was fixed to 5 neighbors and the DT approach was fixed to the Entropy criterion. A total of 2,000 tests have been conducted (1,000 each for kNN and DT). The results can be seen in Fig. 3. They show that the kNN approach performs slightly better than the DT approach (70% vs. 68% accuracy). For the different movement styles, the results are relatively similar. The only notable exception is jumping, which especially the kNN approach seems to have trouble recognizing.

Furthermore, Table 1 shows that the number and position of the considered transponders massively influences the quality of the prediction. In a real-world scenario with only a single transponder (typically on one leg or on the backside of the bib number), the proposed system would achieve prediction accuracies between 40 and 54% using the kNN approach. A slightly expanded, but still realistic scenario with one transponder on the right leg and one transponder on the chest would improve the prediction accuracy to 65%, which is only 5% below the atypical scenario with five transponders. Therefore, the proposed system can be used by timers without extensively changing their familiar workflows.

4   Discussion and Conclusions

From the results in Sect. 3, it can be seen that an appropriately trained and parameterized Machine Learning algorithm delivers a prediction of the movement style based exclusively on RFID detections with an accuracy of up to 70%. This can be achieved in conjunction with appropriate antenna setup and transponder placement. Both RFID equipment and transponders are off-the-shelf products and can be used by any sports timer.

Up to now, measurements have been carried out within a laboratory environment. Furthermore, the sample size of 450 test runs is quite limited. Our work is therefore considered as a proof of concept. We have shown that this kind of classification is possible with UHF-RFID.

The next steps to be taken are a significant increase in the number of test runs and the number of test subjects. Our goal is that the resulting classifier should work at actual sport events, so the antenna and transponder setups will be optimized to resemble the setups of finish lines at sport events more closely.

Finally, our Machine Learning approach will be optimized and biomechanics and RFID researchers will add existing domain knowledge to it. This might require switching to other Machine Learning approaches.

References

1. Andersons, A., Ritter, S.: Advanced RFID applications for sports events management: the case of sportident in Latvia. Procedia Comput. Sci. **43**, 78–85 (2015)
2. Annasaheb, A.B., Verma, V.K.: Data mining classification techniques: a recent survey. Int. J. Emerg. Technol. Eng. Res. (IJETER) **4**, 51–54 (2016)
3. Breiman, L.: Some properties of splitting criteria. Mach. Learn. **24**(1), 41–47 (1996)
4. Cunningham, P., Delany, S.J.: k-nearest neighbour classifiers. Multiple Classifier Syst. **34**(8), 1–17 (2007)
5. Kolaja, J., Ehlerova, J.K.: Effectivity of sports timing RFID system, field study. In: 2019 IEEE International Conference on RFID Technology and Applications (RFID-TA), pp. 220–223. IEEE (2019)
6. Krigslund, R., Dosen, S., Popovski, P., Dideriksen, J.L., Pedersen, G.F., Farina, D.: A novel technology for motion capture using passive UHF RFID tags. IEEE Trans. Biomed. Eng. **60**(5), 1453–1457 (2012)
7. Nikam, S.S.: A comparative study of classification techniques in data mining algorithms. Orient. J. Comput. Sci. Technol. **8**(1), 13–19 (2015)
8. Polivka, M., Svanda, M., Hudec, P.: The optimization of the RFID system for the identification of sportsmen in mass races. In: 2007 European Microwave Conference, pp. 732–735. IEEE (2007)
9. Safavian, S.R., Landgrebe, D.: A survey of decision tree classifier methodology. IEEE Trans. Syst. Man Cybern. **21**(3), 660–674 (1991)
10. Tuttle, J.R.: Traditional and emerging technologies and applications in the radio frequency identification (RFID) industry. In: 1997 IEEE Radio Frequency Integrated Circuits (RFIC) Symposium. Digest of Technical Papers, pp. 5–8. IEEE (1997)
11. Woellik, H., Mueller, A., Herriger, J.: Permanent RFID timing system in a track and field athletic stadium for training and analysing purposes. Procedia Eng. **72**, 202–207 (2014)

Process Decomposition and Test Selection for Distributed Fault Diagnosis

Elodie Chanthery[1]([⊠]), Anna Sztyber[2], Louise Travé-Massuyès[1], and Carlos Gustavo Pérez-Zuñiga[3]

[1] LAAS-CNRS, University of Toulouse, CNRS, INSA, Toulouse, France
{elodie.chanthery,louise}@laas.fr
[2] Institute of Automatic Control and Robotics, Warsaw University of Technology, Warsaw, Poland
anna.sztyber@pw.edu.pl
[3] Engineering Department, Pontifical Catholic University of Peru, PUCP, Lima, Peru
gustavo.perez@pucp.pe

Abstract. Decomposing is one way to gain efficiency when dealing with large scale systems. In addition, the breakdown into subsystems may be mandatory to reflect some geographic or confidentiality constraints. In this context, the selection of diagnostic tests must comply with decomposition and it is desired to minimize the number of subsystem interconnections while still guaranteeing maximal diagnosability. On the other hand, it should be noticed that there is often some flexibility in the way to decompose a system. By placing itself in the context of structural analysis, this paper provides a solution to the double overlinked problem of choosing the decomposition of the system by leveraging existing flexibility and of selecting the set of diagnostic tests so as to minimize subsystem interconnections while maximizing diagnosability.

Keywords: Structural analysis · Diagnosis test selection · System decomposition

1 Introduction

One way to manage complexity is to use the precept of "divide and conquer". This is all the more true as the current systems are complex and large-scale. However, when constructing health monitoring processes such as diagnosers for such systems, decomposition influences feasible diagnostic tests. Indeed, these must be evaluable by measurements local to each subsystem or, if this is impossible to reach the desired level of diagnosability, minimize the interactions required between subsystems.

In the literature, there are two points of view. Either the breakdown of the system is fixed by strong constraints, whether geographic, functional, or confidential, or it is completely flexible (cf. the related work section). In the first

This project is supported by ANITI through the French "Investing for the Future – PIA3" program under the Grant agreement n°ANR-19-PI3A-0004.

H. Fujita et al. (Eds.): IEA/AIE 2020, LNAI 12144, pp. 914–925, 2020.
https://doi.org/10.1007/978-3-030-55789-8_78

case, diagnostic tests must adapt to the decomposition while in the later, it is the decomposition which adapts to the chosen diagnostic test. However, these two extreme cases do not reflect the reality of most systems. In practice, the constraints only define a partial decomposition and there remains a certain flexibility. By placing itself in the context of structural analysis, this paper provides a solution to the double overlinked problem of choosing the decomposition of the system by leveraging existing flexibility and of selecting the set of diagnostic tests so as to minimize the required subsystem interconnections while maximizing diagnosability. This is performed thanks to an iterative algorithm that guarantees to improve the solution at each iteration, providing a nice and efficient distributed fault diagnosis architecture.

The paper is organized as follow. Section 2 presents the works that are related to the proposal of this paper. Section 3 presents the formulation of the problem. Section 4 recalls the prerequisites concerning structural analysis for fault diagnosis and the graph partitioning problem that is used for system decomposition. Section 5 explains our iterative *DecSel* algorithm that decomposes the system and selects optimal tests, improving the solution at each iterative step. Finally, Sect. 6 concludes the paper.

2   Related Work

Diagnosis and test selection are crucial problems to maintain the health status of a system, and consequently, a large body of research is devoted to these problems. Part of these works position themselves in a distributed framework in which the system is decomposed in subsystems. Another subset formulates the problem in a structural analysis framework like we do. It is by these two aspects that these works are related to this paper. Two main tracks exist: either the breakdown of the system is fixed by strong constraints, whether geographic, functional, or confidential, which limit the feasible diagnostic tests or it is completely flexible and based on the selected diagnostic tests. To the best of our knowledge, no prior work attempted to consider the situation in which constraints only define a partial decomposition and there remains a certain flexibility.

In the first track, we find [3] which presents an approach for decentralized fault-focused residual generation based on structural analysis. The work in [13] is a direct continuation of [3], the diagnostic tests calculated with the decentralized architecture are the same as those computed in a centralized architecture. However, a more efficient algorithm (BILP) is used for test selection.

In [8,11] distributed diagnosis approaches are proposed, where only local models and limited information of neighboring systems are required. [14] shows a design of communication network for distributed diagnosis, where subsystems are given and contain subsets of variables, communication - transferring values of known variables using BILP formulation.

The work of [15] proposes a fuzzy diagnostic inference in the two-level decentralized structure where subsystems contain faults and residuals and information is transferred to second level if needed. Decomposition of the diagnostic system relies on association of diagnostic subsystems with the separate technological nodes. Nevertheless, all these methods consider pre-existing constraints of confidentiality, distance, or information access limitations through inter-level communications as mandatory and therefore considers totally predefined subsystems.

In the second track, we find the possible conflicts approach proposed in [2] where possible conflicts are presented as a way to decompose the system. Possible conflicts can be used to generate diagnostic tests, hence the decomposition is such that no interconnection is needed between subsystems. As a drawback, no practical constraint is taken into account by the decomposition.

Concerning the decomposition process, the authors of [4] present an optimal decomposition for distributed fault detection applied to linear models with known parameters. It is based on a cost function to measure the detection performance of the fault diagnosis decomposition method. The authors argue that minimization of coupling is not necessarily the best (based on numerical results) and minimize detection time.

An approach for diagnostic system decomposition minimizing interconnections between subsystems was proposed in [9]. The method is based on qualitative models in the form of graphs of a process (GP), introduced in [16], and diagnosis system. Variables and faults are divided into disjoint subsystems. Decomposition is based on causal links and residuals are assigned to subsystems after decomposition in a way minimizing interconnections.

As it was shown, the development of fault diagnosis systems applying distributed and decentralized architectures has become an active research area, however, most of these works consider predefined subsystems, with few approaches proposing solutions to the problem of decomposition.

3   Problem Formulation

Let us consider a continuous system described by a set of n_e equations $\Sigma(z, x, \mathbf{f})$ including known (or measured) variables given by a vector z, unknown variables given by a vector x and possible faults that may occur on the system given in the vector $\mathbf{f}$. The possibility to generate diagnosis tests from $\Sigma(z, x, \mathbf{f})$ relies on the analytical redundancy embedded in the model. Diagnosis tests, or residual generators, can be implemented by relations that only involved measured variables and their derivatives of the form $arr(z', \dot{z}', \ddot{z}', ...) = r$, where r is a scalar signal named *residual*.

Definition 1 (Diagnosis test for $\Sigma(z, x, \mathbf{f})$). *A relation of the form $arr(z', \dot{z}', \ddot{z}', ...) = r$, with input z' a subvector of z and output r, a scalar signal named residual, is a residual generator for the model $\Sigma(z, x, f)$ if, for all z consistent with $\Sigma(z, x, f)$, it holds that $\lim_{t \to \infty} r(t) = 0$.*

The relation $arr(z', \dot{z}', \ddot{z}', ...) = r$ is qualified as an Analytical Redundancy Relation (ARR). The resulting diagnosis test can be used to check whether the measured variables z are consistent with the system model.

As detailed later in Sect. 4.1, each diagnosis test arises from a subset of equations of $\Sigma(z, x, \mathbf{f})$, also called its *equation support*, and it can be designed so as to be sensitive to a specific subset of faults qualified as its *fault support*. When the residual is non zero, it means that at least one of the faults of the fault support has occurred.

Without loss of generality, a decomposition $\mathcal{D}$ of the system leads to subsystems denoted $\Sigma_i(z_i, x_i, \mathbf{f}_i)$, with $i = 1, ..., n_s$, where z_i is the vector of known variables, x_i the vector of unknown variables, and $\mathbf{f}_i$ refers to the vector of faults. The corresponding sets of unknown variables, known variables, and faults are defined as $X_i \in X$, $Z_i \in Z$, $F_i \in F$ of Σ_i.

Definition 2 (Shared variables in $\mathcal{D}$). x^s *is said to be a shared variable in $\mathcal{D}$ if there exist two subsystems Σ_i and Σ_j such that $x^s \in X_i \cap X_j$. The set of shared variables of the whole system Σ in a given decomposition $\mathcal{D}$ is denoted by $X_{\mathcal{D}}^s$.*

Note that a variable that is involved in two equations that belong to the same subsystem is not a shared variable and it is qualified as a *joint variable*.

Knowing the decomposition $\mathcal{D}$ and the set of diagnosis tests Φ for Σ, it is possible to evaluate the total number of interconnections between subsystems required to evaluate diagnosis tests.

Definition 3 (Number of interconnections between two equations in $\mathcal{D}$). *Let e_i and e_j be two different equations of Σ. The number of interconnections between e_i and e_j in $\mathcal{D}$, denoted $I_{\mathcal{D}}(e_i, e_j)$ is the number of shared variables between e_i and e_j.*

Definition 4 (Number of interconnections of φ in $\mathcal{D}$). *Let φ be a subset of equations of Σ. The number of interconnections of φ in $\mathcal{D}$ is:*

$$I_{\mathcal{D}}(\varphi) = \sum_{\substack{e_i, e_j \in \varphi \\ i \neq j}} I_{\mathcal{D}}(e_i, e_j) \tag{1}$$

Let $\mathcal{S} \subset \Phi$ be a selection of diagnosis tests $\{\varphi\}$, then the total number of interconnections for $\mathcal{S}$ in $\mathcal{D}$ is defined as:

$$I_{\mathcal{D}}(\mathcal{S}) = \sum_{\varphi \in \mathcal{S}} I_{\mathcal{D}}(\varphi) \tag{2}$$

Knowing $\mathcal{D}$, the goal is to select the subset $\mathcal{S}^*$ of diagnosis tests that minimizes the number of interconnections and maximizes diagnosability, i.e. the number isolable fault pairs (cf. Definition 8), also called the *isolability degree*.

In several works like [13], it is hypothesized that the decomposition in subsystems is guided by functional constraints. This paper aims at removing this hypothesis and searching for a decomposition that improves the efficiency of the architecture by decreasing the number of interconnections while keeping maximal diagnosability.

Furthermore, some additional constraints are added to the problem formulation. The first constraint is that the number of subsystems in the decomposition is fixed at the beginning of the process. This is an important hypothesis that could be studied in future works. The second constraint is that the number of equations in the subsystems defined by the decomposition $\mathcal{D}$ has to be as much balanced as possible. This constraint aims at avoiding the algorithm to put all the equations in the same subsystem, hence trivially minimizing the number of interconnections.

As far as the criterion is concerned, the objective is to minimize the number of interconnections, but we have added a second term to avoid adding "free" structurally independent diagnosis tests, i.e. without any interconnections.

The formal definition of the criterion can be written as follows. Once a decomposition $\mathcal{D}$ of n_s subsystems is chosen, the goal is to select a subset $\mathcal{S}$ of diagnosis tests that minimizes $\mathcal{C}(\mathcal{S})$ defined by:

$$\mathcal{C}(\mathcal{S}) = I_{\mathcal{D}}(\mathcal{S}) + \alpha|\mathcal{S}| \tag{3}$$

where $\alpha \in \mathbb{R}^+$ is such that the term $\alpha|\mathcal{S}|$ is lower than 1. For example, it is always possible to take $\alpha = \frac{1}{|\Phi|}$.

The goal of this paper is to present a method for finding simultaneously the decomposition $\mathcal{D}$ and the subset $\mathcal{S}$ of diagnosis tests to minimize the total number of interconnections while still guaranteeing maximal diagnosability. The selection of $\mathcal{D}$ and $\mathcal{S}$ is a challenging task, because the optimal $\mathcal{D}$ depends on the optimal $\mathcal{S}$ and vice versa. The problem of simultaneously selecting decomposition and diagnosis tests is computationally intensive. Therefore we propose an iterative algorithm, named *DecSel*, which computes $\mathcal{D}$ and $\mathcal{S}$ in alternating steps, improving the solution at each step.

The solution that we propose relies on graph theory to formalize the decomposition search and on structural analysis to formalize the diagnosis tests search. These two frameworks are perfectly consistent given that the structural model of a system Σ can be represented by a graph and that the generation of diagnosis tests has a graphical interpretation.

4   Prerequisites

4.1   Structural Analysis for Fault Diagnosis

Structural analysis consists in abstracting a model by keeping only the links between equations and variables. The main advantages of structural analysis are that the approach can be applied to large scale systems, linear or non linear, even under uncertainty. The structural model of $\Sigma(z, x, \mathbf{f})$ can be represented by a bipartite graph $\mathcal{G}(\Sigma \cup X \cup Z, \mathcal{E})$, where $\mathcal{E}$ is a set of edges linking equations of Σ and variables of X and Z. An edge exists if the variable is involved in the equation.

When used for fault diagnosis purposes, structural analysis can be used to find subsets of equations endowed with redundancy. Structural redundancy is defined as follow.

Definition 5 (Structural redundancy). *The structural redundancy $\rho_{\Sigma'}$ of a set of equations $\Sigma' \subseteq \Sigma$ is defined as the difference between the number of equations and the number of unknown variables.*

Actually, subsets of equations endowed with structural redundancy have been proved to be the equation support of diagnostic tests that may take the form of ARRs. ARRs are designed offline [1] and then the corresponding diagnosis tests are used on-line to check the consistency of the observations with respect to the model.

4.2 Fault-Driven Minimal Structurally Overdetermined (FMSO) Sets

Previous work [10] defined the concept of Fault-Driven Minimal Structurally Overdetermined (FMSO) sets that we recall briefly here.

Let us define F_φ as the set of faults that are involved in a set of equations $\varphi \subseteq \Sigma(z, x, \mathbf{f})$.

Definition 6 (FMSO set). *A subset of equations $\varphi \subseteq \Sigma(z, x, f)$ is an FMSO set of $\Sigma(z, x, f)$ if (1) $F_\varphi \neq \emptyset$ and $\rho_\varphi = 1$ that means $|\varphi| = |X_\varphi| + 1$, (2) no subset of φ is overdetermined, i.e. with more equations than unknown variables. The set of FMSO sets of Σ is denoted Φ.*

FMSO sets are proved to be of special interest for diagnosis purpose because they can be turned into ARR. Indeed, by definition, all the unknown variables of X_φ can be determined using $|\varphi| - 1$ equations and can then be substituted in the $|\varphi|^{th}$ equation to generate an ARR off-line. They are used on-line as diagnosis tests.

The concept of FMSO set is also relevant to define detectable fault, isolable faults and ambiguity set. We recall here these definitions.

Definition 7 (Detectable fault). *A fault $f \in F$ is detectable in the system $\Sigma(z, x, f)$ if there exists an FMSO set $\varphi \in \Phi$ such that $f \in F_\varphi$.*

Definition 8 (Isolable faults). *Given two detectable faults f and f' of F, $f \neq f'$, f is isolable from f' if there exists an FMSO set $\varphi \in \Phi$ such that $f \in F_\varphi$ and $f' \notin F_\varphi$.*

Definition 9 (Ambiguity set). *An ambiguity set is a set of faults not isolable two by two. The set of ambiguity sets is denoted $\mathcal{A}$.*

4.3 Graph Partitioning

Given the structural model of the system $\Sigma(z, x, \mathbf{f})$, one can represent the interconnections existing among equations by a graph $G(\Sigma, A)$, where the set of nodes Σ is the set of equations and A is the set of edges linking equations of Σ. An edge (e_i, e_j) exists if there is a shared variable between equation e_i and e_j. The graph $G(\Sigma, A)$ can be used to support system decomposition thanks to graph partitioning algorithms.

The graph partitioning problem consists in dividing nodes of a graph into n_s parts, as equal as possible, in a way that minimizes interconnections between the parts. If graph partitioning is applied to $G(\Sigma, A)$, it hence breaks down the system into subsystems that are connected as loosely as possible.

The edges of the graph can be weighted according to some property of the corresponding links. Formally, the graph partitioning algorithm partitions Σ into n_s subsets $\Sigma_1, \ldots, \Sigma_{n_s}$, where $\Sigma_i \cap \Sigma_j = \emptyset$ for $i \neq j$, $\bigcup_i \Sigma_i = \Sigma$, minimizing:

$$n_{cuts} = \Sigma w(e_i, e_j) : e_i \in \Sigma_k, e_j \in \Sigma_l, l \neq k \tag{4}$$

where $w(e_i, e_j)$ is the weight of the edge between e_i and e_j.

Graph partitioning is an NP-complete problem, but there exist algorithms for finding approximate solutions [6,7].

5   Algorithms

5.1   Main Loop Algorithm

As said before, the problem of simultaneously selecting decomposition and diagnosis tests via FMSO sets is computationally intensive. In the following we propose the algorithm *DecSel* (Decompose and Select) given by Algorithm 1 that is guaranteed to improve the solution at each iteration k.

Algorithm 1: *DecSel* Algorithm

/* Initialization   */
1 $S_{-1} = \emptyset$, $k = 0$;
   /* Section 5.2   */
2 $\mathcal{D}_0 \leftarrow$ Compute initial decomposition taking into account existing constraints ;
   /* Section 5.3   */
3 $S_0 \leftarrow$ Select FMSO sets minimizing interconnections from the decomposition $\mathcal{D}_0$;
   /* Main loop   */
4 **while** $(S_k \neq S_{k-1})$ **do**
       /* Section 5.2   */
5      $\quad G_{k+1}(V, A_{k+1}) \leftarrow$ Build a new graph from the selected FMSO sets S_k ;
6      $\quad \mathcal{D}_{k+1} \leftarrow$ Compute decomposition on G_{k+1} ;
       /* Section 5.3   */
7      $\quad S_{k+1} \leftarrow$ Select FMSO sets minimizing interconnections from the decomposition $\mathcal{D}_{k+1}$;
8      $\quad k = k + 1$;

Section 5.2 presents the way to obtain the initial decomposition of the system, named $\mathcal{D}_0$. It also presents how to obtain, at each iteration k, a new decomposition of the system $\mathcal{D}_{k+1}$ given the current selection of FMSO sets S_k. Then,

Sect. 5.3 presents how to obtain a new selection $\mathcal{S}_{k+1}$ of FMSO sets minimizing interconnections between the subsystems given the decomposition $\mathcal{D}_{k+1}$.

5.2 System Decomposition

Initial System Decomposition. We build a graph $G_0(\Sigma, A_0)$ such that Σ is the set of equations in the system model and A_0 the set of edges of the graph. An edge (e_i, e_j) exists if there is a joint variable to equation e_i and e_j. The weight of edge (e_i, e_j) is the number of joint variables to e_i and e_j.

Complex process decompositions have to take into account constraints related to physical components, geographical location and confidentiality. In this work we interpret this requirement by assuming that some equations must be assigned to some specific subsystem *a priory*, leaving flexibility on the assignment of the remaining equations. The graph partitioning algorithm accounts for these constraints by adding some artificially big weights to the edges between equations which should be in the same subsystem.

System Decomposition Given Selected FMSO Sets. To account for the selected FMSO sets, we use a graph with edges representing only active connections, i.e. edges between equations which are both in the support of one of the selected FMSO sets.

From the selected FMSO sets $\mathcal{S}_k$, we build a new graph $G_{k+1}(V, A_{k+1})$ such that V is the set of equations in the system model and A_{k+1} the set of edges of the graph. One edge exists if there is a joint variable to equation e_i and e_j and if e_i and e_j are in the support of the same selected FMSO set. The weight of the edge (e_i, e_j) is the number of joint variables to e_i and e_j. The graph partitioning algorithm is applied to $G_{k+1}(V, A_{k+1})$ and a new system decomposition $\mathcal{D}_{k+1}$ is computed.

5.3 FMSO Set Selection

This subsection explains the FMSO sets selection given the decomposition $\mathcal{D}_k$. The selection is performed by an A^* (A-star) algorithm. This idea has been initially proposed in [11]. The A^* algorithm [5] is an algorithm that is widely used in path-finding to find a shortest path in a graph between a start node n_0 and one end node included in a set of goal nodes denoted V_f. We recall here the main concepts used in A^*, but interested readers should refer to [5] for further information.

Considering a search weighted graph $G^*(V, E)$ with nodes V and edges E weighted by a cost, the A^* algorithm is a best-first search that searches a path associating to each node v a value $f(v) = g(v) + h(v)$, where:

- $g(v)$ is the cost of the path from the initial node v_0 to the current node v,
- $h(v)$ is a heuristic estimate of the cost of a path from node v to a node included in V_f. This value is a mere estimation rather than an exact value. The more accurate the heuristic the faster the goal state is reached and the higher the accuracy.

– $f(v) = g(v) + h(v)$ is the current approximation of the shortest path to a goal state. f is called the *evaluation function*. $f(v)$ is calculated for any node v to determine which node should be expanded next.

At each iteration of the algorithm, the node with lowest evaluation function value is chosen for expansion.

In our FMSO set selection problem, the weighted graph $G^* = (V, E)$ is built from the current decomposition $\mathcal{D}_k$ and the set Φ of system's FMSO sets. A path identifies a selection $\mathcal{S} \subseteq \Phi$ of FMSO sets. The originality of the work is that each node $v_i \in V$ has a state that corresponds to a set of ambiguity sets $\mathcal{A}_i$ and edges correspond to the assignment of one FMSO set φ_i to $\mathcal{S}$. The initial node v_0 has state $\mathcal{A}_0 = F = \{\{f_0, f_1, \ldots, f_{n_f}\}\}$, which is reduced to one single ambiguity set representing the whole set of faults. A node v_i of the search graph is identified by the ambiguity state $\mathcal{A}_i$ resulting from the FMSO sets that have been selected on the path from the start node v_0 to node v_i. In the case where all the faults are isolable, a goal node $v_f \in V_f$ has state $\mathcal{A}_f = \{\{f_0\}, \{f_1\}, \ldots, \{f_n\}\}$.

The neighbors of a node v_i are all the nodes that can be reached by selecting one additional FMSO set to be added to $\mathcal{S}$ that increases the cardinal of the ambiguity set $\mathcal{A}_i$.

The weight of the edge representing the selection of the FMSO set φ_i is equal to:

$$c(\varphi_i) = I_{\mathcal{D}_k}(\varphi_i) + \alpha, \tag{5}$$

with α defined as in Eq. 3.

$g(v_i)$ is the sum of weights in the path from v_0 to v_i. For example, let $\mathcal{S} = \{\varphi_{i1}, \varphi_{i2}, \ldots, \varphi_{ik}\}$ be a path from the initial node v_0 to the current node v_i, then:

$$g(v_i) = \sum_{l=1}^{k} c(\varphi_{il}). \tag{6}$$

A heuristic estimate for a node has been proposed in [11] assuming that the cost of each test is 1. It uses the notion of dichotomic cut that has been proved to be the most efficient manner to increase the cardinal of the set of isolability sets. The same heuristic is used in this work, adapted to the cost given Eq. 5. The heuristic uses the minimum number of FMSO sets that are necessary to disambiguate all the sets $\mathcal{A}_i^j$ of the ambiguity set $\mathcal{A}_i$ and is given by:

$$h(v_i) = Max_j \left\lceil \frac{ln(|\mathcal{A}_i^j|)}{ln(2)} \right\rceil (\min_{\varphi \in \phi \setminus \mathcal{S}} I_{\mathcal{D}_k}(\varphi) + \alpha) \tag{7}$$

where the $\mathcal{A}_i^j$ are the different ambiguity sets of the set $\mathcal{A}_i$ at step i.

Running the A^* algorithm on a decomposition $\mathcal{D}_k$, results in a selection of FMSO sets denoted $\mathcal{S}_k$.

5.4   Application

We use a four tank system, illustrated in Fig. 1, to test the proposed algorithm. The model for this system is composed of 20 equations e_1 to e_{20} given in [8].

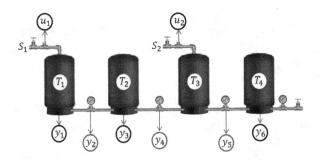

Fig. 1. Four tank system

All decompositions in the paper were calculated using the approximate graph partitioning algorithm proposed in [6]. Computations are done on Ubuntu run in VirtualBox on Windows with Intel i7-5500U CPU 2.40 GHz.

We assume that the number of subsystems is equal to 4. We also assume that measurements and faults are assigned to the same subsystems as in [8] and [12]. The sets of equations specified by *constr* should belong to the same subsystem, the rest of the equations is left free and will be assigned to subsystems by the successive decompositions in a way minimizing the number of interconnections.

Figure 2 illustrates all steps of *DecSel* algorithm. The initial decomposition $\mathcal{D}_0$ with constraints *constr* is shown in Fig. 2(a). The equations in each of the subsets indicated by *constr* must be assigned to the same subsystem by the decomposition whereas the other equations are free. The obtained subsystems are depicted by colors. The cost of decomposition $\mathcal{D}_0$, given by Eq. 4, is $n_{cuts} = 19$, which provides the sum of weights between nodes in different subsystems, i.e. the number of shared variables in total. The FMSO sets selection $\mathcal{S}_0$ for decomposition $\mathcal{D}_0$ and graph G_1 are shown in Fig. 2(b). There are four FMSO sets, each indicated by a different color by bold edges between equations. The number of interconnections for decomposition $\mathcal{D}_0$ given by $I_{\mathcal{D}_0}(\mathcal{S}_0)$ is equal to 10. Decomposition $\mathcal{D}_1$ (Fig. 2(c)) is a graph partitioning based on the FMSO selection $\mathcal{S}_0$. Note that $\mathcal{D}_1$ differs from $\mathcal{D}_0$ in that equation $e_1 9$ has been reassigned. This new decomposition decreases the number of interconnections (i.e. number of shared variables, cf. Definition 3) to $I_{\mathcal{D}_1}(\mathcal{S}_0) = 7$. Then, the FMSO sets of $\mathcal{S}_1$ are selected based on decomposition $\mathcal{D}_1$ (Fig. 2(d)); because the set of selected FMSO sets remains the same as in the previous step, i.e. $\mathcal{S}_0 = \mathcal{S}_1$, the algorithm stops.

The running time of the *DecSel* algorithm applied to the four tanks example is 14.1 s ± 78 ms per loop (mean ± standard deviation of 7 runs, one loop each), which would be 2 or 3 times faster if run directly on Intel i7-5500U CPU 2.40 GHz instead of in VirtualBox.

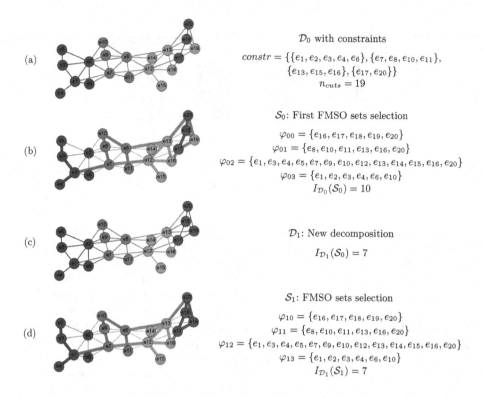

(a)

(b)

(c)

(d)

$\mathcal{D}_0$ with constraints

$$constr = \{\{e_1, e_2, e_3, e_4, e_6\}, \{e_7, e_8, e_{10}, e_{11}\},$$
$$\{e_{13}, e_{15}, e_{16}\}, \{e_{17}, e_{20}\}\}$$
$$n_{cuts} = 19$$

$\mathcal{S}_0$: First FMSO sets selection

$$\varphi_{00} = \{e_{16}, e_{17}, e_{18}, e_{19}, e_{20}\}$$
$$\varphi_{01} = \{e_8, e_{10}, e_{11}, e_{13}, e_{16}, e_{20}\}$$
$$\varphi_{02} = \{e_1, e_3, e_4, e_5, e_7, e_9, e_{10}, e_{12}, e_{13}, e_{14}, e_{15}, e_{16}, e_{20}\}$$
$$\varphi_{03} = \{e_1, e_2, e_3, e_4, e_6, e_{10}\}$$
$$I_{\mathcal{D}_0}(\mathcal{S}_0) = 10$$

$\mathcal{D}_1$: New decomposition

$$I_{\mathcal{D}_1}(\mathcal{S}_0) = 7$$

$\mathcal{S}_1$: FMSO sets selection

$$\varphi_{10} = \{e_{16}, e_{17}, e_{18}, e_{19}, e_{20}\}$$
$$\varphi_{11} = \{e_8, e_{10}, e_{11}, e_{13}, e_{16}, e_{20}\}$$
$$\varphi_{12} = \{e_1, e_3, e_4, e_5, e_7, e_9, e_{10}, e_{12}, e_{13}, e_{14}, e_{15}, e_{16}, e_{20}\}$$
$$\varphi_{13} = \{e_1, e_2, e_3, e_4, e_6, e_{10}\}$$
$$I_{\mathcal{D}_1}(\mathcal{S}_1) = 7$$

Fig. 2. Results for four tanks system example

6   Conclusion

In the paper we consider the double overlinked problem of choosing the decomposition of a system by leveraging existing flexibility and of selecting a set of diagnostic tests so as to minimize their number and the communication between subsystems. The algorithm *DecSel* that we propose articulates diagnosis test selection solved with a specific A* algorithm and system decomposition solved with a graph partitioning algorithm. These two methods are used iteratively to improve the solution at each step. When most of the works proposed in the literature about distributed diagnosis assume that the decomposition is given or that it is totally flexible, in this paper we assume a more realistic situation in which there are some constraints on decomposition, but there is some flexibility. We propose a solution to design subsystems taking into account the existing constraints on decomposition, fault isolability and communication.

However the formulated optimization problem guarantees to improve the solution at each iteration but it does not guarantee to provide the globally optimal solution and the algorithm requires the computation of all FMSO sets. These limitations should be considered in future work.

References

1. Blanke, M., Kinnaert, M., Lunze, J., Staroswiecki, M.: Diagnosis and Fault-Tolerant Control. Springer, Heidelberg (2006). https://doi.org/10.1007/978-3-540-35653-0
2. Bregon, A., Daigle, M., Roychoudhury, I., Biswas, G., Koutsoukos, X., Pulido, B.: An event-based distributed diagnosis framework using structural model decomposition. Artif. Intell. **210**, 1–35 (2014)
3. Chanthery, E., Travé-Massuyès, L., Indra, S.: Fault isolation on request based on decentralized residual generation. IEEE Trans. Syst. Man Cybern. Syst. **46**(5), 598–610 (2016)
4. Gei, C., Boem, F., Parisini, T.: Optimal system decomposition for distributed fault detection: insights and numerical results. IFAC-PapersOnLine **51**(24), 578–585 (2018)
5. Hart, P.E., Nilsson, N.J., Raphael, B.: A formal basis for the heuristic determination of minimum cost paths. IEEE Trans. Syst. Sci. Cybern. **4**(2), 100–107 (1968)
6. Karypis, G., Kumar, V.: A fast and high quality multilevel scheme for partitioning irregular graphs. SIAM J. Sci. Comput. **20**(1), 359–392 (1998)
7. Kernighan, B.W., Lin, S.: An efficient heuristic procedure for partitioning graphs. Bell Syst. Tech. J. **49**(2), 291–307 (1970)
8. Khorasgani, H., Biswas, G.: Jung: structural methodologies for distributed fault detection and isolation. Appl. Sci. **9**(7), 1286 (2019)
9. Kościelny, J., Sztyber, A.: Decomposition of complex diagnostic systems. IFAC-PapersOnLine **51**(24), 755–762 (2018)
10. Pérez, C.G., Travé-Massuyès, L., Chanthery, E., Sotomayor, J.: Decentralized diagnosis in a spacecraft attitude determination and control system. J. Phys. Conf. Ser. **659**(1), 1–12 (2015)
11. Perez-Zuñiga, C.G.: Structural analysis for the diagnosis of distributed systems. PhD thesis of INSA Toulouse (France) (2017). English. NNT : 2017ISAT0024
12. Perez-Zuñiga, C.G., Chanthery, E., Travé-Massuyès, L., Sotomayor, J.: Fault-driven structural diagnosis approach in a distributed context. IFAC-PapersOnLine **50**(1), 14254–14259 (2017)
13. Pérez-Zuñiga, C.G., Chanthery, E., Travé-Massuyès, L., Sotomayor, J., Artigues, C.: Decentralized diagnosis via structural analysis and integer programming. IFAC-PapersOnLine **51**(24), 168–175 (2018)
14. Rosich, A., Voos, H., Pan, L.: Network design for distributed model-based fault detection and isolation. In: Svartholm, N. (ed.) IEEE Multi-Conference on Systems and Control, MSC 2014. IEEE (2014)
15. Syfert, M., Bartyś, M., Kościelny, J.: Refinement of fuzzy diagnosis in decentralized two-level diagnostic structure. IFAC-PapersOnLine **51**(24), 160–167 (2018)
16. Sztyber, A., Ostasz, A., Koscielny, J.M.: Graph of a process - a new tool for finding model structures in a model-based diagnosis. IEEE Trans. Syst. Man Cybern. Syst. **45**, 1004–1017 (2015)

Managing Situations with High Number of Elements in Group Decision Making

J. A. Morente-Molinera[1](✉)⑩, S. Alonso[2]⑩, S. Ríos-Aguilar[2]⑩,
R. González[2]⑩, and E. Herrera-Viedma[1]⑩

[1] Andalusian Research Institute on Data Science and Computational Intelligence,
University of Granada, Granada, Spain
{jamoren,viedma}@decsai.ugr.es
[2] Department of Engineering, School of Engineering and Technology,
Universidad Internacional de la Rioja (UNIR), Logroño, Spain
zerjioi@ugr.es

Abstract. Group Decision Making environments have completely changed. The number of information that the experts have available and that, therefore, they can use to discuss about is constantly increasing. There is a need of new Group Decision Making methods, like the one developed in this paper, that are capable of dealing with environments where the number of alternatives is high. In this paper, clustering methods are used in order to sort alternatives in categories and help experts in the task of making a decision.

Keywords: Clustering methods · Large-scale Group Decision Making · Human-computer interaction

1 Introduction

Internet has gone under a profound change. Now, users have a main role in the information that is present on the Web. Also, Internet is now available to everyone while, in its beginnings, it was only accessible to a small percent of the population.

This new Internet has also provoked a profound change in the way that Group Decision Making (hereafter in referred to as GDM) methods [4,8,9] are performed. Therefore, there is a need for these methods to undergo a profound change in order to meet the experts' new requirements. For instance, it is necessary to find new GDM methods that are capable of managing the large number of elements that the experts have to discuss about. The number of elements that experts can debate about at the same time is reduced. Therefore, computational

This work has been supported by the 'Juan de la Cierva Incorporacion' grant from the Spanish Ministry of Economy and Competitiveness and by the Grant from the FEDER funds provided by the Spanish Ministry of Economy and Competitiveness (No. TIN2016-75850-R).

systems need to provide means to help the experts in analyzing all the available information. It is important to allow the experts to work on a organized way. If not, experts would end up making bad decisions.

One way of allowing the experts to work with a reduced number of elements is by removing unnecessary information. This way, experts would only have to debate about the more interesting alternatives. Nevertheless, there are framework where applying this is not possible due to the fact that there is no information that can be used for detecting which alternatives are the most promising ones. In order to solve this situation, a novel Group Decision approach based on clustering methods [1,3] is proposed. The proposed approach uses the experts' aid in order to reduce the number of information that they have to deal with at the same time. Thanks to this, it is possible to sort the alternatives in different categories and the experts can discuss which categories are the best ones.

This article proposes the conversion of an unique GDM method with a high number of alternatives, into two different methods that have a reduced set of alternatives. In the first one, experts discuss about the alternatives' categories. Finally, in the second one, experts decide the best alternative among the ones belonging to the category that has been chosen. Thanks to this process, no external information is required about the alternatives. On the other way around, the experts are the one making the categories and making the decision.

The paper employs the following organization. In Sect. 2, GDM basis are presented. In Sect. 3, the proposed method is thoroughly described. In Sect. 4, an example in which the method is applied is shown. Finally, the paper ends with a conclusion section.

2   Group Decision Making Basis

GDM is quite popular in the recent literature [6,7,10,13]. A typical GDM process can be defined as follows:

Let have two sets, of experts and another for alternatives. That is, $E = \{e_1, \ldots, e_n\}$ and $X = \{x_1, \ldots, x_m\}$, respectively. A GDM problem consists on ranking alternatives according to a set of preferences provided by the experts, $P = \{p_{ij}^k\}$. p_{ij}^k indicates how much expert e_k prefers x_i over x_j.

Generally, a typical GDM process follows the next steps in order to reach a solution:

- **Sharing preferences to the system**: Experts interact with the system in order to provide their preferences.
- **Calculating the collective preference matrix**: Preferences are aggregated in order to obtain a collective preference matrix that represents the overall opinion of the experts.
- **Obtaining the alternatives ranking**: Using the matrix calculated on the previous step, the ranking of the alternatives is obtained.
- **Measuring consensus among the experts**: It is important that decisions are consensual. Therefore, it is important to measure the consensus before reaching a final solution. If the consensus is low, experts can be encouraged to carry out more debate.

3   Handling a Big Set of Alternatives in Group Decision Making Processes

The main goal of the presented method is to convert a single GDM process that have a high number of alternatives into two GDM processes that have a reduced set of elements to discuss about. The main purpose of achieving this goal is to allow the experts to make decisions in a comfortable environment. The novel presented method is carried out by following the next steps:

- **Providing experts' groups**: Each expert creates the number of groups that he/she prefers and classify the alternatives onto them.
- **Going under a consensual grouping process**: An hierarchical clustering method is used in order to generate groups of alternatives that uses all the grouping options provided by the experts.
- **Deciding the most preferred category**: A GDM process is performed in order to select the most promising group of alternatives.
- **Selecting the most preferred alternative**: A GDM process is performed in order to select the most adequate alternative.

This process is exposed in detail in the following subsections.

3.1   Calculating the Categories

In order to classify the alternatives in different groups, all the grouping options of the experts are taken into account. Using that information, a co-occurrence matrix, CO, is built. Each position, co_{ij}, indicates how many times the experts have classified elements i and j together in the same group.

An hierarchical clustering process [5,12] is applied over the co-occurrence matrix in order to obtain different grouping options for the alternatives. On it, the Lance–Williams dissimilarity update formula is used. By selecting the number of categories that the experts prefer, they obtain the categories and assigned alternatives that they will use in the GDM process.

3.2   Carrying Out the Two Group Decision Making Processes

Once that categories and alternatives are defined, experts go through two GDM processes. In the first one, the choose their most preferred category while in the second one, they select the most preferred alternative within the group. In order to perform this task, the scheme exposed in 2 is employed. The following elements are taken into account:

- **Preference providing means**: Experts use preference relations matrices in order to share information with the system.
- **Aggregation operator**: In order to aggregate the information, the mean operator is used:

$$c_{ij} = \sum_k w_k \cdot p_{ij}^k \tag{1}$$

where w_k is the weight associated to expert e_k. It is important to notice that the weights fulfill that $\sum w_i = 1$.

- **Ranking alternatives**: Guided dominance degree (GDD) and guided non-dominance selection operators (GNDD) [2,11] are used in order to generate the final ranking of alternatives.

4 Example

In this section, an application example is used. For a better understanding, an alternative set of 14 elements is used. Three experts, $E = \{e_1, e_2, e_3\}$ decide where to invest some funds. They have 14 different choices according to their advisers: $X = \{x_1, \cdots, x_{14}\}$. First of all, experts must create their own groups of similar alternatives. After that, the co-occurrence matrix is generated (Table 1).

Table 1. Obtained co-occurrence matrix.

	x_1	x_2	x_3	x_4	x_5	x_6	x_7	x_8	x_9	x_{10}	x_{11}	x_{12}	x_{13}	x_{14}
x_1	–	–	–	–	–	–	–	–	–	–	–	–	–	–
x_2	3	–	–	–	–	–	–	–	–	–	–	–	–	–
x_3	0	0	–	–	–	–	–	–	–	–	–	–	–	–
x_4	0	0	0	–	–	–	–	–	–	–	–	–	–	–
x_5	0	0	3	0	–	–	–	–	–	–	–	–	–	–
x_6	0	0	3	0	3	–	–	–	–	–	–	–	–	–
x_7	0	0	0	0	0	0	–	–	–	–	–	–	–	–
x_8	0	0	0	3	0	0	0	–	–	–	–	–	–	–
x_9	0	0	0	0	0	0	3	0	–	–	–	–	–	–
x_{10}	2	2	0	1	0	0	0	1	0	–	–	–	–	–
x_{11}	0	0	0	1	0	0	0	1	0	0	–	–	–	–
x_{12}	0	0	2	0	2	2	1	0	1	0	0	–	–	–
x_{13}	2	2	0	1	0	0	0	1	0	1	0	0	–	–
x_{14}	0	0	0	2	0	0	0	2	0	0	2	0	1	–

Experts decide to generate 5 different categories. They are shown in Table 2.

Table 2. Categories and assigned alternatives.

Cat. 1	Cat. 2	Cat. 3	Cat. 4	Cat. 5
x_8	x_6	x_4	x_3	x_1
x_{11}	x_{12}	x_7	x_5	x_2
x_{14}		x_9		x_{10}
				x_{13}

Experts start debating about the categories. First of all, they provide preferences by using the linguistic label set $S = \{s_1, s_2, s_3, s_4, s_5\}$. After that, the preferences are aggregated into a single collective value. The resulting matrix is shown below:

$$P_c = \begin{pmatrix} - & 2 & 1 & 1.333 & 2.667 \\ 1.667 & - & 2 & 2.333 & 1.667 \\ 5 & 4.667 & - & 5 & 4.333 \\ 1 & 2.667 & 3 & - & 2.333 \\ 2 & 1.667 & 1.667 & 2 & - \end{pmatrix}$$

By using the chosen selection operators, the ranking of alternatives groups is performed. All the values are expressed on Table 3 by using the interval [0,1].

Table 3. Ranking of categories.

	x_1	x_2	x_3	x_4	x_5
GNDD	0.8	0.833	1	0.8833	0.8166
GDD	0.15	0.2166	0.933	0.366	0.2166
Ranking value	0.475	0.525	0.966	0.625	0.5166
Ranking	x_3	x_4	x_2	x_5	x_1

The chosen group contains the following alternatives: $\{x_7, x_9, x_{12}\}$. Now, a new GDM process is carried out over these alternatives. After debating, x_7 is selected as the best alternative.

5   Conclusions

In this paper, a novel GDM method whose goal is to deal with environments that have a high number of alternatives is presented. This method employs clustering methods in order to create categories of alternatives that the experts can use in order to discuss the information in an organized manner. An initial unmanageable GDM process is converted into two debating processes that the experts can handle. First, they sort the alternatives in categories. Next, they choose a category. Finally, they select the best alternative of the selected category.

References

1. Arabie, P., Hubert, L.J., De Soete, G.: Clustering and Classification. World Scientific, Singapore (1996)
2. Cabrerizo, F.J., Herrera-Viedma, E., Pedrycz, W.: A method based on pso and granular computing of linguistic information to solve group decision making problems defined in heterogeneous contexts. Eur. J. Oper. Res. **230**, 624–633 (2013)
3. de la Fuente-Tomas, L., et al.: Classification of patients with bipolar disorder using k-means clustering. PloS One **14**(1), e0210314 (2019)

4. Herrera, F., Alonso, S., Chiclana, F., Herrera-Viedma, E.: Computing with words in decision making: foundations, trends and prospects. Fuzzy Optim. Decis. Mak. **8**(4), 337–364 (2009)
5. Kamis, N.H., Chiclana, F., Levesley, J.: Geo-uninorm consistency control module for preference similarity network hierarchical clustering based consensus model. Knowl.-Based Syst. **162**, 103–114 (2018)
6. Li, C.C., Dong, Y., Herrera, F.: A consensus model for large-scale linguistic group decision making with a feedback recommendation based on clustered personalized individual semantics and opposing consensus groups. IEEE Trans. Fuzzy Syst. **27**(2), 221–233 (2019)
7. Liu, X., Xu, Y., Montes, R., Ding, R.X., Herrera, F.: Alternative ranking-based clustering and reliability index-based consensus reaching process for hesitant fuzzy large scale group decision making. IEEE Trans. Fuzzy Syst. **27**(1), 159–171 (2019)
8. Liu, Y., Dong, Y., Liang, H., Chiclana, F., Herrera-Viedma, E.: Multiple attribute strategic weight manipulation with minimum cost in a group decision making context with interval attribute weights information. IEEE Trans. Syst. Man Cybern. Syst. **49**(10), 1981–1992 (2018)
9. Morente-Molinera, J.A., Al-Hmouz, R., Morfeq, A., Balamash, A.S., Herrera-Viedma, E.: A decision support system for decision making in changeable and multi-granular fuzzy linguistic contexts. J. Multiple-Valued Logic Soft Comput. **26**, 485–514 (2016)
10. Morente-Molinera, J.A., Kou, G., Samuylov, K., Ureña, R., Herrera-Viedma, E.: Carrying out consensual group decision making processes under social networks using sentiment analysis over comparative expressions. Knowl.-Based Syst. **165**, 335–345 (2019)
11. Pérez, I.J., Cabrerizo, F.J., Herrera-Viedma, E.: A mobile decision support system for dynamic group decision-making problems. IEEE Trans. Syst. Man Cybern. Part A Syst. Hum. **40**(6), 1244–1256 (2010)
12. Siless, V., Chang, K., Fischl, B., Yendiki, A.: Anatomicuts: hierarchical clustering of tractography streamlines based on anatomical similarity. Neuroimage **166**, 32–45 (2018)
13. Zhang, H., Dong, Y., Chiclana, F., Yu, S.: Consensus efficiency in group decision making: a comprehensive comparative study and its optimal design. Eur. J. Oper. Res. **275**(2), 580–598 (2019)

Correction to: Development and Research of a Terminal Controller for Marine Robots

V. Pshikhopov and Boris Gurenko

Correction to:
Chapter "Development and Research of a Terminal Controller for Marine Robots" in: H. Fujita et al. (Eds.):
Trends in Artificial Intelligence Theory and Applications, **LNAI 12144,**
https://doi.org/10.1007/978-3-030-55789-8_76

The original version of this chapter was revised. The number in the funding statement was corrected to MK- 3099.2019.8.

The updated version of this chapter can be found at
https://doi.org/10.1007/978-3-030-55789-8_76

© Springer Nature Switzerland AG 2020
H. Fujita et al. (Eds.): IEA/AIE 2020, LNAI 12144, p. C1, 2020.
https://doi.org/10.1007/978-3-030-55789-8_80

Author Index

Printed in the United States
By Bookmasters